Lecture Notes in Computer Science 12707

More information about this subseries at http://www.springer.com/series/7407

Mohit Singh · David P. Williamson (Eds.)

Integer Programming and Combinatorial Optimization

22nd International Conference, IPCO 2021
Atlanta, GA, USA, May 19–21, 2021
Proceedings

 Springer

Editors
Mohit Singh
Georgia Institute of Technology
Atlanta, GA, USA

David P. Williamson ⓘ
Cornell University
Ithaca, NY, USA

ISSN 0302-9743   ISSN 1611-3349   (electronic)
Lecture Notes in Computer Science
ISBN 978-3-030-73878-5   ISBN 978-3-030-73879-2   (eBook)
https://doi.org/10.1007/978-3-030-73879-2

LNCS Sublibrary: SL1 – Theoretical Computer Science and General Issues

This Springer imprint is published by the registered company Springer Nature Switzerland AG
The registered company address is: Gewerbestrasse 11, 6330 Cham, Switzerland

Preface

This volume collects the 33 extended abstracts presented at IPCO 2021, the 22nd Conference on Integer Programming and Combinatorial Optimization, held May 19–21, 2021, in an online format. IPCO is under the auspices of the Mathematical Optimization Society, and it is an important forum for presenting the latest results on the theory and practice of the various aspects of discrete optimization. The first IPCO conference took place at the University of Waterloo in May 1990, and the Georgia Institute of Technology organized the 22nd such event.

The conference had a Program Committee consisting of 17 members. In response to the Call for Papers, we received more than 90 submissions. Each submission was reviewed by at least three Program Committee members. Because of the limited number of time slots for presentations, many excellent submissions could not be accepted. The page limit for contributions to this proceedings was set to 15. We expect the full versions of the extended abstracts appearing in this *Lecture Notes in Computer Science* volume to be submitted for publication in refereed journals, and a special issue of *Mathematical Programming Series B* containing such versions is in process.

For the second time, IPCO had a Best Paper Award. The IPCO 2021 Best Paper Award was given to Jannis Blauth, Vera Traub, and Jens Vygen for their paper *Improving the Approximation Ratio for Capacitated Vehicle Routing*.

This year, IPCO was preceded by a Summer School held May 17–18, 2021, with lectures by Moon Duchin (Tufts), Simge Küçükyavuz (Northwestern), and László Végh (LSE). We thank them warmly for their contributions. We would also like to thank

- The authors who submitted their research to IPCO;
- The members of the Program Committee, who spent much time and energy reviewing the submissions;
- The expert additional reviewers whose opinions were crucial in the paper selection;
- The members of the Local Organizing Committee, who made this conference possible;
- The Mathematical Optimization Society and in particular the members of its IPCO Steering Committee, Oktay Günlük, Jochen Könemann, and Giacomo Zambelli, for their help and advice;
- EasyChair for making paper management simple and effective; and
- Springer for their efficient cooperation in producing this volume, and for financial support for the Best Paper Award.

We would further like to thank the following sponsors for their financial support: Gurobi, FICO, MOSEK, SAS, and the Georgia Institute of Technology.

March 2021

Mohit Singh
David P. Williamson

Conference Organization

Program Committee

José Correa	University of Chile
Sanjeeb Dash	IBM Research
Jesús A. De Loera	University of California, Davis
Friedrich Eisenbrand	École Polytechnique Fédérale de Lausanne
Oktay Günlük	Cornell University
Satoru Iwata	University of Tokyo
Volker Kaibel	Otto von Guericke University of Magdeburg
Andrea Lodi	Polytechnique Montreal
Jim Luedtke	University of Wisconsin-Madison
Viswanath Nagarajan	University of Michigan
Alantha Newman	Université Grenoble Alpes
Britta Peis	RWTH Aachen University
Mohit Singh	Georgia Institute of Technology
László Végh	London School of Economics and Political Science
Juan Pablo Vielma	Google
David P. Williamson (Chair)	Cornell University
Rico Zenklusen	ETH Zürich

Local Organizing Committee

Santanu S. Dey	Georgia Institute of Technology
Swati Gupta	Georgia Institute of Technology
Mohit Singh	Georgia Institute of Technology
Alejandro Toriello	Georgia Institute of Technology

Conference Sponsors

Contents

Improving the Approximation Ratio for Capacitated Vehicle Routing

Jannis Blauth[1], Vera Traub[2(✉)], and Jens Vygen[1]

[1] Research Institute for Discrete Mathematics, Hausdorff Center for Mathematics, University of Bonn, Bonn, Germany
{blauth,vygen}@or.uni-bonn.de
[2] Department of Mathematics, ETH Zurich, Zurich, Switzerland
vera.traub@ifor.math.ethz.ch

Abstract. We devise a new approximation algorithm for capacitated vehicle routing. Our algorithm yields a better approximation ratio for general capacitated vehicle routing as well as for the unit-demand case and the splittable variant. Our results hold in arbitrary metric spaces. This is the first improvement on the classical tour partitioning algorithm by Haimovich and Rinnooy Kan [16] and Altinkemer and Gavish [2].

1 Introduction

In the CAPACITATED VEHICLE ROUTING problem, we are given a metric space with a depot and customers, each with a positive demand. The goal is to design tours of minimum total length such that each tour contains the depot, each customer is served by some tour, and the total demand of the customers in one tour does not exceed 1 (after scaling, this is the vehicle capacity). CAPACITATED VEHICLE ROUTING generalizes the famous traveling salesman problem and has obvious applications in logistics. There is a huge body of literature studying heuristics, mixed-integer programming models, and application scenarios.

The so far best known approximation algorithm is more than 30 years old and quite simple: it first computes a traveling salesman tour (ignoring the capacity constraint) and then partitions the tour optimally into segments of total demand at most 1, each of which is then served by a separate tour from the depot. The approximation ratio of this algorithm is $\alpha + 2$, where α is the approximation ratio of an algorithm computing the traveling salesman tour. Essentially the same algorithm has been the best known for the unit-demand special case (where all customers have the same demand), and also for the variant where a customer's demand can be split and served by more than one tour. For these special cases, the approximation ratio is $\alpha + 1$.

These algorithms have been proposed and analyzed in the 1980s by Altinkemer and Gavish [2] and Haimovich and Rinnooy Kan [16]. Despite many efforts and progress in special cases (cf. Sect. 1.3), they have not been improved, except

Supported by Swiss National Science Foundation grant 200021_184622.

M. Singh and D. P. Williamson (Eds.): IPCO 2021, LNCS 12707, pp. 1–14, 2021.
https://doi.org/10.1007/978-3-030-73879-2_1

that the traveling salesman tour can now be computed by the Karlin–Klein–Oveis Gharan algorithm [18] instead of the Christofides–Serdjukov algorithm [12,21], which improves α to slightly less than $\frac{3}{2}$ if one allows randomization.

In this paper we improve upon the classical algorithms of [2] and [16]. Our result is a better black-box reduction to the traveling salesman problem. Therefore, our new algorithm has a better approximation ratio than the classical algorithms of [2] and [16], and this will remain true if the approximation ratio for the traveling salesman problem will be improved further. Here is our main result:

Theorem 1. *For every* $\alpha > 1$ *there is an* $\varepsilon > 0$ *such that the following holds. If there is an* α-*approximation algorithm for the traveling salesman problem, then there are an* $(\alpha + 2 \cdot (1 - \varepsilon))$-*approximation algorithm for* CAPACITATED VEHICLE ROUTING *and an* $(\alpha + 1 - \varepsilon)$-*approximation algorithm for* UNIT-DEMAND CAPACITATED VEHICLE ROUTING *and* SPLITTABLE CAPACITATED VEHICLE ROUTING. *For* $\alpha = \frac{3}{2}$ *we have* $\varepsilon > \frac{1}{3000}$.

1.1 Formal Problem Description

Given a depot s and a set V of customers, we want to design tours serving all customers. For now, a *tour* is a cycle that contains s and a subset of customers (later we will also consider tours that begin in s but do not end in s). To measure the cost of a tour, we have a semi-metric $c : (\{s\} \cup V) \times (\{s\} \cup V) \rightarrow \mathbb{R}_{\geq 0}$, i.e., c is symmetric and satisfies the triangle inequality. We interpret a tour Q as an undirected graph with vertex set $V(Q)$ and edge set $E(Q)$. We write $c(Q) = \sum_{\{v,w\} \in E(Q)} c(v, w)$ for the cost of Q. Moreover, each customer has a demand, and the total demand of the customers served by a tour must not exceed the vehicle capacity, which we can assume to be 1 (by scaling). Then the problem can be described as follows.

Definition 1 (CAPACITATED VEHICLE ROUTING). *An instance consists of*

– *a finite set* V *(of customers)*,
– *a depot* s, *not belonging to* V,
– *a semi-metric* c *on* $\{s\} \cup V$, *defining distances (or cost)*,
– *a demand* $d(v) \in [0, 1]$ *for each customer* $v \in V$.

A feasible solution is a set $\mathcal{Q}$ *of tours such that*

– *every tour* $Q \in \mathcal{Q}$ *is a cycle that contains* s,
– *every customer belongs to exactly one tour, and*
– $\sum_{v \in V(Q) \setminus \{s\}} d(v) \leq 1$ *for all* $Q \in \mathcal{Q}$.

The task is to minimize the total cost $c(\mathcal{Q}) := \sum_{Q \in \mathcal{Q}} c(Q)$.

We denote by OPT($\mathcal{I}$) or simply OPT the minimum cost of a feasible solution to a given instance $\mathcal{I}$. We note the following well-known lower bound:

Proposition 1. OPT $\geq \sum_{v \in V} 2d(v)c(s, v)$.

Proof. Let $\mathcal{Q}$ be a feasible solution. For each $v \in V$ we obtain two s-v-paths by splitting the tour $Q \in \mathcal{Q}$ that contains v. By the triangle inequality, each of these paths has length at least $c(s, v)$, and hence $2c(s, v) \leq c(Q)$. Summation yields $\sum_{v \in V} 2d(v)c(s, v) \leq \sum_{Q \in \mathcal{Q}} \sum_{v \in V(Q) \setminus \{s\}} d(v)c(Q) \leq c(\mathcal{Q})$. $\square$

If $d(v) = \frac{1}{k}$ for all $v \in V$, where k is some positive integer, we speak of UNIT-DEMAND CAPACITATED VEHICLE ROUTING (then every tour can serve up to k customers). This is closely related to SPLITTABLE CAPACITATED VEHICLE ROUTING: here the demand of a customer is arbitrary but can be split into several parts, each of which is served by a different tour.

All variants include the traveling salesman problem as special case and are thus APX-hard. CAPACITATED VEHICLE ROUTING also includes bin packing; hence there is no approximation algorithm with ratio less than $\frac{3}{2}$ unless $P = NP$.

1.2   Outline

To obtain our results, we analyze instances for which the approximation guarantees of [2] and [16] are almost tight and exploit their structure to design better solutions. We will call such instances *difficult*.

We view every tour in a solution to a CAPACITATED VEHICLE ROUTING instance as the union of two paths from the depot to the *peak* of the tour: the point farthest away from the depot. Our first observation is that the performance of the classical algorithms can be close to the worst case guarantee only if, for most tours, these two paths have small detour, i.e., they are approximately shortest paths from the depot to the peak.

Definition 2 (detour). *For $v, w \in V \cup \{s\}$ we define*

$$\text{detour}(v, w) := c(v, w) + c(s, v) - c(s, w).$$

For a directed path P that starts at the depot s and ends in t we have $\text{detour}(P) = c(P) - c(s, t)$. The detour was called *excess* by [9] and *regret metric* by [14]. The detour is not symmetric, but it is non-negative and fulfills the triangle inequality.

We will compute an even number of paths that all start at the depot such that all customers are visited by some path. Then we combine pairs of these paths to tours by adding an edge between their endpoints.

If there exists a set of paths with small total detour (like the one induced by an optimum solution to a difficult instance), then we can find a set of paths that is not much longer in polynomial time. In fact, this problem is closely related to regret-bounded vehicle routing, a problem that has been studied by Friggstad and Swamy [14, 15]. Here, one asks for a minimum number of paths serving all customers such that the detour of any path is bounded.

However, combining pairs of paths to tours can be too expensive. We need to ensure that a cheap matching of the endpoints of the paths exists. Ideally, two paths end at the peak of each tour in an optimum solution, then the matching would not cost anything. But of course we do not know these peaks.

Therefore we try to "guess" them, by exploiting another property of difficult instances: in almost all tours of an optimum solution, the total demand of customers near the peak is almost 1 (the vehicle capacity). Consequently, we can assume that "most" customers are clustered, and we can force two paths to end in each cluster.

However, another difficulty arises because the clusters are not necessarily clearly separated from one another. Still we can identify groups of nearby clusters, and estimate the number of tours whose peak is in that group. Instead of prescribing the endpoints of the paths, we only specify the total number of paths that must end in each group. This number will always be even, in order to ensure that we can find a matching within each group. Although customers in the same group can be far away from each other if there is a chain of pairwise overlapping clusters, we will prove that a relatively cheap matching exists.

The key subproblem therefore asks to find an appropriate number of paths that begin at the depot and end in these target groups, such that all customers (including those that do not belong to any group) are served by some tour. We call this problem VEHICLE ROUTING WITH TARGET GROUPS. The instance of VEHICLE ROUTING WITH TARGET GROUPS that we compute has the property that it has a solution that is cheap and has small total detour. This will enable us to find a cheap set of paths in polynomial time: either by a simple and fast combinatorial algorithm, or alternatively by leveraging an LP-based approach suggested for regret-bounded vehicle routing by Friggstad and Swamy [15].

Once we have these paths, we compute a cheapest matching of their endpoints and combine them to tours. These tours will generally still not meet the capacity constraint, but we can simply concatenate all these tours (and shortcut) to obtain a traveling salesman tour. Since this tour will be not much more expensive than an optimum solution to our CAPACITATED VEHICLE ROUTING instance, applying the classical tour partitioning algorithm finishes the job.

A full version of this paper with all proofs can be found in [8].

1.3 Related Work

Despite of a huge amount of research on vehicle routing, the best known approximation ratio for CAPACITATED VEHICLE ROUTING (as well as for the unit-demand and splittable variants) has not been improved in more than 30 years. However, there has been progress on several special cases. First of all, the tour partitioning algorithm (cf. Sect. 1.4) already yields a slightly better approximation guarantee if the least common denominator k of all demands is bounded (this is often called the *bounded capacity* case). Compared to $\alpha + 2$ and $\alpha + 1$, the approximation ratios reduce by $\frac{2\alpha}{k}$ (for general demands) and $\frac{\alpha}{k}$ (for unit demands). Bompadre, Dror and Orlin [11] gain another $\Omega(\frac{1}{k^3})$.

There are also several results for geometric instances for the unit-demand case (in which $d(v) = \frac{1}{k}$ for all $v \in V$). In the Euclidean plane a PTAS is known for constant k (Haimovich and Rinnooy Kan [16]), for $k = O(\log(n)/\log\log(n))$, where $n = |V|$ (Asano, Katoh, Tamaki, and Tokuyama [3]), and for $k \leq 2^{\log^{f(\varepsilon)}(n)}$

(Adamszek, Czumaj, and Lingas [1]). The latter uses a result by Das and Mathieu [13], who provided a quasi-polynomial time approximation scheme for the Euclidean plane and unbounded k. For higher dimensional Euclidean metrics, Khachay and Dubinin [19] found a PTAS for fixed dimension l and $k = O(\log^{\frac{1}{l}}(n))$.

Better approximation ratios have also been found for graph metrics arising from graphs with a special structure. For the unit-demand case with constant k, Becker, Klein and Schild [7] devised a PTAS in planar graphs, and Becker, Klein and Saulpic [6] found a PTAS in graphs with bounded highway dimension. Becker [5] designed a $\frac{4}{3}$-approximation algorithm for SPLITTABLE CAPACITATED VEHICLE ROUTING in tree metrics, improving on results by Hamaguchi and Katoh [17] and Asano, Kawashima and Katoh [4].

For general CAPACITATED VEHICLE ROUTING, no improvement on the classical approximation algorithm [2] has been found except for tree metrics. For tree metrics, Labbé, Laporte and Mercure [20] gave a 2-approximation algorithm. If the tree is a path (i.e., on the line), Wu and Lu [22] described a $\frac{5}{3}$-approximation algorithm for CAPACITATED VEHICLE ROUTING. Note that the unit-demand case is polynomially solvable on the line.

One part of our proof is leveraging an LP relaxation that was proposed by Friggstad and Swamy [15] for regret-bounded vehicle routing. In (additive) regret-bounded vehicle routing, the goal is to find a minimum number of paths starting at the depot and covering all customers such that none of the tours has a detour (or regret) more than a given bound. Friggstad and Swamy [14] provided the first constant-factor approximation algorithm for this problem. They improved the approximation ratio from 31 to 15 in [15].

Regret-bounded vehicle routing is a special case of the school bus problem. In the school bus problem, there is the additional constraint that no tour can serve more customers than a given bound (the vehicle capacity). Bock, Grant, Könemann and Sanità [10] observed that any μ-approximation algorithm for regret-bounded vehicle routing implies a $(\mu+1)$-approximation algorithm for the school bus problem. They gave a 3-approximation algorithm for regret-bounded vehicle routing on trees and thus obtain a 4-approximation algorithm for the school bus problem on trees. The later results by Friggstad and Swamy mentioned above imply a 16-approximation algorithm for the general school bus problem.[1]

1.4   Review of the Classical Algorithms

In this section we review the classical algorithms by Altinkemer and Gavish [2] and Haimovich and Rinnooy Kan [16], and we do this for two reasons. First, we will exploit properties of instances in which their analysis is tight. Second, the final step of our new algorithm will be identical to these classical algorithms.

The classical algorithms [2,16] consist of two steps. The first step simply runs an approximation algorithm for the traveling salesman problem, and we

---

[1] As a by-product, we obtain a 10-approximation algorithm for regret-bounded vehicle routing and thus an 11-approximation algorithm for the school bus problem.

denote by α the approximation ratio of this algorithm. The classical Christofides–Serdjukov algorithm [12,21] obtains $\alpha = \frac{3}{2}$, the new randomized algorithm by Karlin, Klein, and Oveis Gharan [18] improves on this by a tiny constant.

A *traveling salesman tour* (for a given instance (V, s, c, d) of CAPACITATED VEHICLE ROUTING) is a cycle with vertex set $\{s\} \cup V$. The minimum length of a traveling salesman tour is a lower bound on OPT; hence the first step yields a traveling salesman tour of length $\alpha \cdot$ OPT.

In the second step, this traveling salesman tour is partitioned optimally into segments of total demand at most 1, each of which is then served by a separate tour from the depot. This step is summarized by the following theorem:

Theorem 2 ([2,16]). *Given an instance (V, s, c, d) of* CAPACITATED VEHICLE ROUTING *and a traveling salesman tour Q, one can compute a feasible solution of cost at most $c(Q) + \sum_{v \in V} 4d(v)c(s, v)$ in $O(n^2)$ time, where $n = |V|$. For* UNIT-DEMAND CAPACITATED VEHICLE ROUTING *and* SPLITTABLE CAPACITATED VEHICLE ROUTING *the bound improves to $c(Q) + \sum_{v \in V} 2d(v)c(s, v)$.*

Together with Proposition 1, Theorem 2 immediately implies the approximation ratio $\alpha + 2$ for CAPACITATED VEHICLE ROUTING [2], and the approximation ratio $\alpha + 1$ for UNIT-DEMAND CAPACITATED VEHICLE ROUTING and SPLITTABLE CAPACITATED VEHICLE ROUTING [16]. In this paper, we present the first improvement on these more-than-thirty-year-old results.

2 Difficult Instances

In the following we concentrate on instances where Proposition 1 and thus the analysis of the classical algorithms is almost tight. We will call such instances *difficult*. In order to give a formal definition, we fix a constant $0 < \varepsilon < 1$.

Definition 3. *An instance (V, s, c, d) of* CAPACITATED VEHICLE ROUTING *is called* difficult *if*

$$2 \cdot \sum_{v \in V} d(v) \cdot c(s, v) > (1 - \varepsilon) \cdot \text{OPT}.$$

If an instance is not difficult, the classical tour partitioning algorithm yields a cheap solution (of cost at most $(\alpha + 2 \cdot (1 - \varepsilon)) \cdot$ OPT) by Theorem 2. For difficult instances we will compute a cheap traveling salesman tour and then apply Theorem 2 to this. We will prove:

Theorem 3. *There is a function $f : \mathbb{R}_{>0} \to \mathbb{R}_{>0}$ with $\lim_{\varepsilon \to 0} f(\varepsilon) = 0$ and a polynomial-time algorithm that returns a traveling salesman tour of cost at most $(1 + f(\varepsilon)) \cdot \text{OPT}(\mathcal{I})$ for any given difficult instance $\mathcal{I}$ of* CAPACITATED VEHICLE ROUTING.

From this traveling salesman tour, we get a solution of cost at most $(3 + f(\varepsilon)) \cdot$OPT for any given difficult instance by Theorem 2. Choose $\varepsilon > 0$ such that

$3 + f(\varepsilon) \leq \alpha + 2 \cdot (1 - \varepsilon)$. Running both the classical and our new algorithm and returning the better solution hence proves Theorem 1 (except for the constant).

In the following, we prove Theorem 3. Informally, in an optimum solution to a difficult instance almost every tour must have the following two properties. First, the tour Q is not much longer than $2 \cdot c(s, \mathrm{peak}(Q))$, where $\mathrm{peak}(Q)$ is the vertex in Q that is farthest away from the depot s. Second, the total demand served by the tour is almost 1, and almost all of it is close to the peak.

To formalize the second property we introduce the notion of the *peak cluster*. Let $0 < \tau, \rho \leq \frac{1}{6}$ be constants (that depend on ε).

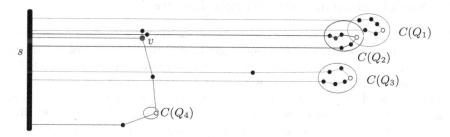

Fig. 1. An instance of CAPACITATED VEHICLE ROUTING with a solution. The customer v (shown in red) has demand $\frac{4}{7}$ and all other customers have demand $\frac{1}{7}$. The peaks of the tours are shown as empty circles. The peak clusters are drawn for $\rho = \frac{1}{10}$ and $\tau = \frac{1}{6}$ and Euclidean distances. The peak clusters $C(Q_1)$, $C(Q_2)$ and $C(Q_3)$ are large. (Color figure online)

Definition 4. *Let (V, s, c, d) be an instance of* CAPACITATED VEHICLE ROUT-ING. *Let Q be a cycle with $s \in V(Q)$. Then we define* $\mathrm{peak}(Q)$ *to be a vertex $v \in V(Q)$ with $c(s, v)$ maximal, and the* peak cluster *to be*

$$C(Q) := \{ u \in V(Q) : c(u, \mathrm{peak}(Q)) + \kappa \cdot \mathrm{detour}(u, \mathrm{peak}(Q)) < \rho \cdot c(s, \mathrm{peak}(Q)) \}.$$

where $\kappa := \frac{1 - 2\tau - \tau \cdot \rho}{2\tau}$.

We call the peak cluster *large if $d(C(Q)) > 1 - \tau$ and small otherwise. Here and in the following we abbreviate $d(C(Q)) = \sum_{v \in C(Q)} d(v)$.*

In an optimum solution to a difficult instance, the total length of tours with small peak cluster is small compared to the total length of all tours. This can be derived from the following lemma, which gives a lower bound on $c(Q) - 2 \cdot \sum_{v \in V(Q)} d(v) \cdot c(s, v)$ for tours Q with a small peak cluster. Because in a difficult instance the sum of these expressions over all tours is small (it is less than $\varepsilon \cdot \mathrm{OPT}$), the lemma implies that in an optimum solution the total length of such tours with small peak clusters must be small. Note that we will choose τ and ρ such that $\tau \cdot \rho$ is much larger than ε.

Lemma 1. *Let Q be a tour with small peak cluster. Then*

$$c(Q) - 2 \cdot \sum_{v \in V(Q) \setminus \{s\}} d(v) \cdot c(s, v) \geq \tau \cdot \rho \cdot c(Q).$$

We remark that the peak cluster $C(Q)$ is the smallest possible set such that Lemma 1 holds. Choosing $\kappa = 0$ would already be sufficient to improve the approximation ratio of CAPACITATED VEHICLE ROUTING. In this case the peak cluster of a tour Q is a ball of radius $\rho \cdot c(s, \mathrm{peak}(Q))$ around $\mathrm{peak}(Q)$. However, choosing $\kappa = 0$ would yield a smaller improvement in the approximation ratio.

3 Vehicle Routing with Target Groups

The first step of our algorithm aims at identifying the peak clusters. Since these clusters can be close to each other and might be difficult to distinguish, we merge such clusters into larger ones. We describe our clustering algorithm in Sect. 4.

Having identified clusters, we want to find paths starting at the depot and ending in the clusters, such that every customer is visited by one such path, regardless of whether the customer is part of a cluster or not. To find such paths we compute a solution to an instance of VEHICLE ROUTING WITH TARGET GROUPS, which is a new problem we introduce. The *targets* will be some customers inside the clusters, where targets in the same cluster belong to the same *target group*.

Definition 5 (VEHICLE ROUTING WITH TARGET GROUPS). *An instance consists of*

- *disjoint finite sets V (of customers) and $\bar{T}$ (of targets),*
- *a depot s, not belonging to $V \cup \bar{T}$,*
- *a semi-metric c on $\{s\} \cup V \cup \bar{T}$,*
- *a partition $\mathcal{T}$ of the target set $\bar{T}$ into target groups, and*
- *numbers $b : \mathcal{T} \to \mathbb{Z}_{>0}$ telling how many tours must end in each target group.*

A feasible solution is a set $\mathcal{P}$ of tours such that

- *every tour $P \in \mathcal{P}$ is either an s-t-path for some target $t \in \bar{T}$ or a cycle containing s, and all other vertices of P belong to V,*
- *every element of V belongs to at least one of these tours, and*
- *for every target group $T \in \mathcal{T}$, exactly $b(T)$ of these tours end in an element of T.*

The task is to minimize the total cost $c(\mathcal{P}) := \sum_{P \in \mathcal{P}} c(P)$.

For a target group T, we will set the number $b(T)$ to roughly twice the demand of the cluster containing T. The number $b(T)$ of paths ending in T will always be even. Therefore, we can turn a solution to our instance of VEHICLE ROUTING WITH TARGET GROUPS into a cheap traveling salesman tour as follows. For each target group T we pair up the paths and complete a pair of paths to a tour by adding an edge between their endpoints. In the end we concatenate all these tours and shortcut to obtain a traveling salesman tour as needed to prove Theorem 3.

4 Clustering Algorithm

In this section we describe the algorithm we use to construct an instance of VEHICLE ROUTING PROBLEM WITH TARGET GROUPS. Before giving a formal description of the algorithm, let us informally explain some important properties. Ideally, our algorithm would choose the targets as the peaks of the tours of an optimum solution; then two tours should end in each peak t, i.e. $b(\{t\}) = 2$. Our algorithm will not always identify the peaks of the optimum tours correctly. However, for every tour Q with a large peak cluster, we will select a target vertex that is not far away from peak(Q).

For every target $t \in \bar{T}$, we then consider an area B_t around t that is large enough to guarantee that each large peak cluster of an optimum solution is fully contained in one of these areas. Our algorithm might also select targets that are not close to any of the peaks of optimum tours, but we can show that for difficult instances this happens rarely.

We want to set the numbers b in the VEHICLE ROUTING WITH TARGET GROUPS instance large enough so that for every tour with a large peak cluster, two paths are allowed to end in a target close to the peak. Thus, the number of paths ending in the targets will depend on the total demand d in the area B_t around t. In order to avoid requiring a too high number of paths, we want

---

Algorithm 1.

Input: Instance $\mathcal{I} = (V, s, c, d)$ of CAPACITATED VEHICLE ROUTING.
Output: Instance $\mathcal{J} = (V, \bar{T}, s, c, \mathcal{T}, b)$ of VEHICLE ROUTING WITH TARGET GROUPS.

1. **Choose the set $\bar{T}$ of targets:**
   Number the customers in V such that $c(s, v_1) \le c(s, v_2) \le \cdots \le c(s, v_n)$.
   Initialize $\bar{T} := \emptyset$ and $Y := \emptyset$.

   for $v = v_n, v_{n-1}, \ldots, v_1$ **do**
     Define $C_v := \{u \in V : c(u, v) + \kappa \cdot \mathrm{detour}(u, v) < \rho \cdot c(s, v)\}$.
     if $v \notin Y$ **and** $d(C_v \setminus Y) > 1 - \tau$ **then**
       Set $\bar{T} := \bar{T} \cup \{v\}$. Set $Y := Y \cup C_v$.

2. **Partition $\bar{T}$ into target groups:**
   For $t \in \bar{T}$ define

   $$B_t := \left\{v \in V : c(v, t) < \frac{3\rho}{1 - \rho} \cdot c(s, v),\ c(v, t) < 6\rho \cdot c(s, t) - \frac{3\rho}{1 - \rho} \cdot c(s, v)\right\}.$$

   Let E_B be the set of all edges $\{t, t'\}$ with $t, t' \in \bar{T}$ and $B_t \cap B_{t'} \neq \emptyset$.
   Let $\mathcal{T}$ be the set of vertex sets of connected components of $(\bar{T}, E_B)$.

3. **Determine the number of tours that should end in each target group:**
   For $T \in \mathcal{T}$ define $B_T := \bigcup_{t \in T} B_t$.
   Define $b(T) := 2 \cdot \left\lfloor \frac{d(B_T)}{1 - \tau} \right\rfloor$ for all $T \in \mathcal{T}$.

4. **Output** $(V, \bar{T}, s, c, \mathcal{T}, b)$.

---

to avoid that the demand $d(v)$ of a customer v is counted twice here when v is contained in two of the areas B_t. Therefore, if the areas B_t for different targets t overlap, e.g. because the peaks of two tours are close to each other, we merge the areas B_t and the corresponding targets will form a group.

Algorithm 1 describes our algorithm to construct an instance of VEHICLE ROUTING WITH TARGET GROUPS. See also Fig. 2. Note that the set $\bar{T}$ of targets is represented by a subset of V, and formally contains a copy of each such customer because by Definition 5 the sets V and $\bar{T}$ should be disjoint.

The precise definition of B_t is chosen in order to optimize the approximation ratio of our algorithm. As mentioned above, a crucial property is that every large peak cluster is contained in one of the sets B_t.

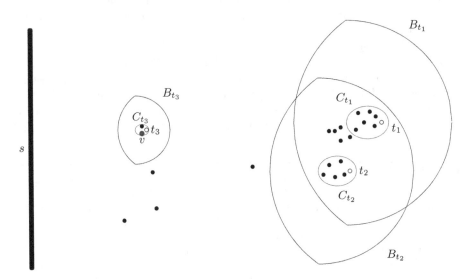

Fig. 2. The output $\mathcal{I} = (V, \bar{T}, s, c, \mathcal{T}, b)$ of Algorithm 1 applied to the instance from Fig. 1. The targets $\bar{T} = \{t_1, t_2, t_3\}$ (shown as empty circles here) are partitioned into target groups $\mathcal{T} = \{\{t_1, t_2\}, \{t_3\}\}$. A solution of $\mathcal{I}$ consists of $b(\{t_1, t_2\}) = 6$ tours ending in $\{t_1, t_2\}$ and $b(\{t_3\}) = 2$ tours ending in t_3. We will see that it does not harm that t_3 was selected as a target although it does not belong to any peak cluster. Note that peak(Q_2) was not identified as a target even though Q_2 has a large peak cluster. However, the peak cluster of Q_2 lies completely in $B_{\{t_1,t_2\}}$, which will guarantee that $b(\{t_1, t_2\})$ is chosen large enough so that $\mathcal{Q}$ can be easily transformed into a weak fractional solution of $\mathcal{I}$ that costs not much more than $\mathcal{Q}$.

Lemma 2. *Let $\mathcal{I}$ be an instance of* CAPACITATED VEHICLE ROUTING, *and let* $(V, \bar{T}, s, c, \mathcal{T}, b)$ *be the output of Algorithm 1 with input $\mathcal{I}$. Then for every tour Q with large peak cluster there exists a target $t \in \bar{T}$ such that $C(Q) \subseteq B_t$.*

Proof (sketch). The definition of C_v in step 1 of Algorithm 1 is such that for every tour Q with a large peak cluster, the following holds. If $Y \cap C(Q) = \emptyset$ when we consider $v = \mathrm{peak}(Q)$, then $\mathrm{peak}(Q)$ is selected as a target. Hence, in any case we have $C_t \cap C(Q) \neq \emptyset$ for some target t with $c(s,t) \geq c(s, \mathrm{peak}(Q))$ (by the order in which we consider the customers). One can show that this implies $C(Q) \subseteq B_t$. □

Another important property is that our choice of the target groups guarantees that the matching that we use to pair up paths from our solution of VEHICLE ROUTING WITH TARGET GROUPS is cheap.

Lemma 3. *Let $\mathfrak{I}$ be an instance of CAPACITATED VEHICLE ROUTING and let $(V, \bar{T}, s, c, \mathcal{T}, b)$ be the instance of VEHICLE ROUTING WITH TARGET GROUPS computed by Algorithm 1 applied to $\mathfrak{I}$. Let $U \subseteq \bar{T}$ such that $|U \cap T|$ is even for every target group $T \in \mathcal{T}$. Then there is a perfect matching on U with cost at most $\frac{3\rho}{(1-\rho)\cdot(1-\tau)} \cdot \mathrm{OPT}(\mathfrak{I})$.*

Proof. (sketch). We consider a fixed target group T. We can bound the cost of a minimum-cost perfect matching on $T \cap U$ by the cost $c(S)$ of any tree (T, S). Because T is the vertex set of a connected component of $(\bar{T}, E_B)$, one can show that there is such a tree with $c(S) \leq 6\rho \cdot \sum_{t \in T} c(s,t)$. We use Y_t to denote the set Y at the beginning of the iteration of Algorithm 1 in which we add t to $\bar{T}$. Then the sets $C_t \setminus Y_t$ for $t \in \bar{T}$ in Algorithm 1 are disjoint. Moreover $c(s,v) > (1 - \rho)c(s,t)$ whenever $v \in C_t$. Hence

$$(1 - \tau) \cdot (1 - \rho) \cdot \sum_{t \in \bar{T}} c(s,t) < (1 - \rho) \cdot \sum_{t \in \bar{T}} c(s,t) \cdot d(C_t \setminus Y_t)$$

$$< \sum_{t \in \bar{T}} \sum_{v \in C_t \setminus Y_t} c(s,v) \cdot d(v) \leq \sum_{v \in V} c(s,v) \cdot d(v) \leq \tfrac{1}{2}\mathrm{OPT}(\mathfrak{I}).$$

□

5   Weak Fractional Solutions

A key part of our proof is to show that we can compute a cheap solution to our instance $\mathfrak{J}$ of VEHICLE ROUTING WITH TARGET GROUPS. First, we want to show that if our CAPACITATED VEHICLE ROUTING instance $\mathfrak{I}$ is difficult, the instance $\mathfrak{J}$ that we construct by Algorithm 1 has a solution that is not much more expensive than $\mathrm{OPT}(\mathfrak{I})$, and has small total detour. It simplifies our proofs and leads to better approximation ratios to consider what we call *weak fractional solutions* of $\mathfrak{J}$ instead of actual solutions.

Weak fractional solutions are composed of walks in the digraph $G(\mathfrak{J})$ with vertex set $V(\mathfrak{J}) = \{s\} \cup V \cup \bar{T}$ and edge set $E(\mathfrak{J}) = E_1(\mathfrak{J}) \cup E_2(\mathfrak{J})$, where

$$E_1(\mathfrak{J}) = \{(v, w) : v \in \{s\} \cup V, \, w \in V, \, v \neq w\},$$
$$E_2(\mathfrak{J}) = \{(v, w) : v \in \{s\} \cup V, \, w \in \{s\} \cup \bar{T}, \, v \neq w\}.$$

Definition 6 (Weak Fractional Solution). *A weak fractional solution to an instance* $\mathfrak{J} = (V, \bar{T}, s, c, \mathcal{T}, b)$ *of* VEHICLE ROUTING WITH TARGET GROUPS *is a vector* $x \in \mathbb{R}_{\geq 0}^{E(\mathfrak{J})}$ *such that*

$$x = \sum_{P \in \mathcal{P}} \lambda_P \cdot \chi^{E(P)},$$

where $\mathcal{P}$ *is a set of walks in* $G(\mathfrak{J})$ *and* $\lambda_P \in \mathbb{R}_{\geq 0}$ *for all* $P \in \mathcal{P}$ *such that*

- *every walk* $P \in \mathcal{P}$ *begins in* s *and ends in* $\{s\} \cup \bar{T}$, *and all inner vertices belong to* V,
- *for every* $v \in V$, *we have* $\sum_{P \in \mathcal{P}: v \in V(P)} \lambda_P \geq 1$, *and*
- *for every target group* $T \in \mathcal{T}$, *the total weight of the walks ending in* T *is*

$$\sum_{P \in \mathcal{P}:\, P \text{ ends in } T} \lambda_P = (1 - \tau) \cdot b(T).$$

Here $\chi^{E(P)}$ *is the incidence vector of* $E(P)$. *For any vector* $x \in \mathbb{R}_{\geq 0}^{E(\mathfrak{J})}$ *we write* $c(x) := \sum_{(v,w) \in E(\mathfrak{J})} c(v,w) x_{(v,w)}$ *and* $\mathrm{detour}(x) := \sum_{(v,w) \in E(\mathfrak{J})} \mathrm{detour}(v,w) x_{(v,w)}$.

We reduce the amount to arrive in each target group slightly because we cannot assume that the total demand of a tour near its peak is 1, but only $1 - \tau$.

We show that the instance $\mathfrak{J}$ that we construct in Algorithm 1 has a weak fractional solution that has small detour and is not much more expensive than OPT($\mathfrak{J}$), assuming $\mathfrak{J}$ is difficult. To this end, we start with an optimum solution to our CAPACITATED VEHICLE ROUTING instance $\mathfrak{J}$. Then for every customer v that is contained in a tour Q of this solution, we partition Q into two paths starting at the depot s and ending at v. We extend these paths by an edge connecting v either to a close-by target or to the depot s. The latter will happen rarely, as we can show using Lemma 2. The resulting walks P will then contribute with weight $\lambda_P = d(v)$ to the weak fractional solution.

6 Solving Vehicle Routing with Target Groups

While VEHICLE ROUTING WITH TARGET GROUPS in general is at least as hard as the traveling salesman problem, we can compute solutions that are not much more expensive than the best weak fractional solution with small detour. This is formally stated in the following theorem. For small detour, we can choose a small value of η. Then the factor on the cost of the given weak fractional solution is close to 1 because we will choose the constant $\tau \in (0, 1)$ to be close to 0.

Theorem 4. *There is a polynomial-time algorithm for* VEHICLE ROUTING WITH TARGET GROUPS *that computes for every instance* $\mathfrak{J}$ *and any given* $\eta \in (0, 1]$ *a feasible solution* $\mathcal{P}$ *of* $\mathfrak{J}$ *such that*

$$c(\mathcal{P}) \;<\; \left(\tfrac{1}{1-\tau} + \eta\right) \cdot c(x) + O(\tfrac{1}{\eta}) \cdot \mathrm{detour}(x|_{E_1(\mathfrak{J})}),$$

for every weak fractional solution x *of* $\mathfrak{J}$.

We propose two approaches for solving VEHICLE ROUTING WITH TARGET GROUPS, both implying Theorem 4. In both approaches we compute a forest and a network flow to obtain tours that visit not necessarily every customer, but every connected component of the forest. In the network flow problem we ensure that the number of tours ending in each target group meets the requirements. Doubling the edges of the forest and shortcutting yields the desired solution.

The first approach is a simple and fast combinatorial algorithm. We can compute a cheapest set of walks from s to the targets with the property that every vertex v has at least one predecessor that is closer to the depot than v. This problem can be reduced to a network flow problem. We compute such a "forward walk solution" for a subset of customers for which we can guarantee that this solution is not too expensive. To find such a subset of customers we use a simple greedy algorithm. Finally, we connect the remaining customers that are not yet visited by a minimum-cost forest.

The second approach leverages a sophisticated LP relaxation for regret-bounded vehicle routing due to Friggstad and Swamy [15]. In contrast to the combinatorial approach, here the network flow only uses edges moving away from the depot, i.e. edges (v, w) with $c(s, w) > c(s, v)$, (and some edges entering s). Both the forest and the network flow are obtained from an optimum LP solution. We combine the rounding approach by [15] with a new construction of a fractional solution, which also enables us to obtain a better approximation ratio for regret-bounded vehicle routing and the school bus problem.

See [8] for complete proofs.

References

1. Adamaszek, A., Czumaj, A., Lingas, A.: PTAS for k-tour cover problem on the plane for moderately large values of k. Int. J. Found. Comput. Sci. **21**, 893–904 (2010)
2. Altinkemer, K., Gavish, B.: Heuristics for unequal weight delivery problems with a fixed error guarantee. Oper. Res. Lett. **6**, 149–158 (1987)
3. Asano, T., Katoh, N., Tamaki, H., Tokuyama, T.: Covering points in the plane by k-tours: towards a polynomial time approximation scheme for general k. In: Proceedings of the Annual ACM Symposium on Theory of Computing (STOC), pp. 275–283 (1997)
4. Asano, T., Katoh, N., Kawashima, K.: A new approximation algorithm for the capacitated vehicle routing problem on a tree. J. Comb. Optim. **5**, 213–231 (2001)
5. Becker, A.: A tight 4/3 approximation for capacitated vehicle routing in trees. In: Approximation, Randomization, and Combinatorial Optimization. Algorithms and Techniques (APPROX/RANDOM), pp. 3:1–3:15 (2018)
6. Becker, A., Klein, P.N., Saulpic, D.: Polynomial-time approximation schemes for k-center, k-median, and capacitated vehicle routing in bounded highway dimension. In: 26th Annual European Symposium on Algorithms (ESA), pp. 8:1–8:15 (2018)
7. Becker, A., Klein, P.N., Schild, A.: A PTAS for bounded-capacity vehicle routing in planar graphs. In: Friggstad, Z., Sack, J.-R., Salavatipour, M.R. (eds.) WADS 2019. LNCS, vol. 11646, pp. 99–111. Springer, Cham (2019). https://doi.org/10.1007/978-3-030-24766-9_8

8. Blauth, J., Traub, V., Vygen, J.: Improving the approximation ratio for capacitated vehicle routing. arXiv:2011.05235 (2020)
9. Blum, A., Chawla, S., Karger, D.R., Lane, T., Meyerson, A., Minkoff, M.: Approximation algorithms for orienteering and discounted-reward TSP. SIAM J. Comput. **37**, 653–670 (2007)
10. Bock, A., Grant, E., Könemann, J., Sanità, L.: The school bus problem on trees. Algorithmica **67**, 10–19 (2011)
11. Bompadre, A., Dror, M., Orlin, J.B.: Improved bounds for vehicle routing solutions. Discrete Optim. **3**, 299–316 (2006)
12. Christofides, N.: Worst-case analysis of a new heuristic for the traveling salesman problem. Technical report, Carnegie-Mellon University (1976)
13. Das, A., Mathieu, C.: A quasi-polynomial time approximation scheme for Euclidean capacitated vehicle routing. Algorithmica **73**, 115–142 (2015)
14. Friggstad, Z., Swamy, C.: Approximation algorithms for regret-bounded vehicle routing and applications to distance-constrained vehicle routing. In: Proceedings of the Annual ACM Symposium on Theory of Computing (STOC), pp. 744–753 (2014)
15. Friggstad, Z., Swamy, C.: Compact, provably-good LPs for orienteering and regret-bounded vehicle routing. In: Eisenbrand, F., Koenemann, J. (eds.) IPCO 2017. LNCS, vol. 10328, pp. 199–211. Springer, Cham (2017). https://doi.org/10.1007/978-3-319-59250-3_17
16. Haimovich, M., Rinnooy Kan, A.H.G.: Bounds and heuristics for capacitated routing problems. Math. Oper. Res. **10**, 527–542 (1985)
17. Hamaguchi, S., Katoh, N.: A capacitated vehicle routing problem on a tree. In: Chwa, K.-Y., Ibarra, O.H. (eds.) ISAAC 1998. LNCS, vol. 1533, pp. 399–407. Springer, Heidelberg (1998). https://doi.org/10.1007/3-540-49381-6_42
18. Karlin, A.R., Klein, N., Gharan, S.O.: A (slightly) improved approximation algorithm for metric TSP. arXiv:2007.01409 (2020)
19. Khachay, M., Dubinin, R.: PTAS for the Euclidean capacitated vehicle routing problem in R^d. In: Kochetov, Y., Khachay, M., Beresnev, V., Nurminski, E., Pardalos, P. (eds.) DOOR 2016. LNCS, vol. 9869, pp. 193–205. Springer, Cham (2016). https://doi.org/10.1007/978-3-319-44914-2_16
20. Labbé, M., Laporte, G., Mercure, H.: Capacitated vehicle routing on trees. Oper. Res. **39**, 616–622 (1991)
21. Serdjukov, A.: Some extremal bypasses in graphs. Upravlyaemye Sistemy **17**, 76–79 (1978). (in Russian)
22. Wu, Y., Lu, X.: Capacitated vehicle routing problem on line with unsplittable demands. J. Comb. Optim. 1–11 (2020). https://doi.org/10.1007/s10878-020-00565-5

Online k-Taxi via Double Coverage and Time-Reverse Primal-Dual

Niv Buchbinder[1], Christian Coester[2(✉)], and Joseph (Seffi) Naor[3]

[1] Tel Aviv University, Tel Aviv, Israel
nivb@tauex.tau.ac.il
[2] CWI, Amsterdam, Netherlands
christian.coester@cwi.nl
[3] Computer Science Department, Technion, Haifa, Israel
naor@cs.technion.ac.il

Abstract. We consider the online k-taxi problem, a generalization of the k-server problem, in which k servers are located in a metric space. A sequence of requests is revealed one by one, where each request is a pair of two points, representing the start and destination of a travel request by a passenger. The goal is to serve all requests while minimizing the distance traveled *without carrying a passenger*.

We show that the classic *Double Coverage* algorithm has competitive ratio $2^k - 1$ on HSTs, matching a recent lower bound for deterministic algorithms. For bounded depth HSTs, the competitive ratio turns out to be much better and we obtain tight bounds. When the depth is $d \ll k$, these bounds are approximately $k^d/d!$. By standard embedding results, we obtain a randomized algorithm for arbitrary n-point metrics with (polynomial) competitive ratio $O(k^c \Delta^{1/c} \log_\Delta n)$, where Δ is the aspect ratio and $c \geq 1$ is an arbitrary positive integer constant. The only previous known bound was $O(2^k \log n)$. For general (weighted) tree metrics, we prove the competitive ratio of Double Coverage to be $\Theta(k^d)$ for any fixed depth d, but unlike on HSTs it is not bounded by $2^k - 1$.

We obtain our results by a dual fitting analysis where the dual solution is constructed step-by-step *backwards* in time. Unlike the forward-time approach typical of online primal-dual analyses, this allows us to combine information from the past and the future when assigning dual variables. We believe this method can be useful also for other problems. Using this technique, we also provide a dual fitting proof of the k-competitiveness of Double Coverage for the k-server problem on trees.

Keywords: Online algorithms · k-taxi · k-server · Dual fitting

1 Introduction

The k-taxi problem, proposed three decades ago as a natural generalization of the k-server problem by Fiat et al. [13], has gained renewed interest recently.

This research was supported in part by US-Israel BSF grant 2018352, by ISF grant 2233/19 (2027511) and by NWO VICI grant 639.023.812.

M. Singh and D. P. Williamson (Eds.): IPCO 2021, LNCS 12707, pp. 15–29, 2021.
https://doi.org/10.1007/978-3-030-73879-2_2

In this problem there are k servers, or taxis, which are located in a metric space containing n points. A sequence of requests is revealed one by one to an online algorithm, where each request is a pair of two points, representing the start and destination of a travel request by a passenger. An online algorithm must serve each request (by selecting a server that travels first to its start and then its destination) without knowledge of future requests. The goal is to minimize the total distance traveled by the servers *without carrying a passenger*. The motivation for not taking into account the distance the servers travel with a passenger is that any algorithm needs to travel from the start to the destination, independently of the algorithm's decisions. Thus, the k-taxi problem seeks to only minimize the overhead travel that depends on the algorithm's decisions. While this does not affect the optimal (offline) assignment, it affects the competitive factor.

Besides scheduling taxi rides, the k-taxi problem also models tasks such as scheduling elevators (the metric space is the line), and other applications where objects need to be transported between locations.

The extensively studied and influential k-server problem is the special case of the k-taxi problem where for each request, the start equals the destination. A classical algorithm for the k-server problem on tree metrics is DOUBLECOV-ERAGE. This algorithm is described as follows. A server s is called *unobstructed* if there is no other server on the unique path from s to the current request. To serve the request, DOUBLECOVERAGE moves *all* unobstructed servers towards the request at equal speed, until one of them reaches the request. If a server becomes obstructed during this process, it stops while the others keep moving.

DOUBLECOVERAGE was originally proposed for the line metric, to which it owes its name, as there are at most two servers moving at once. For a line metric it achieves the optimal competitive ratio of k [7], and this result was later generalized to tree metrics [8].

Given the simplicity and elegance of DOUBLECOVERAGE, it is only natural to analyze its performance for the k-taxi problem. Here, we use it only for bringing a server to the start vertex of a request.

1.1   Related Work and Known Results

For the k-server problem, the best known deterministic competitive factor on general metrics is $2k-1$ [17]; with randomization, on hierarchically well-separated trees (HSTs)[1] the best known bound is $O(\log^2 k)$ [4,5]. By a standard embedding argument, this implies a bound of $O(\log^2 k \log n)$ for n-point metrics, and it was also shown in [4] that a dynamic embedding yields a bound of $O(\log^3 k \log \Delta)$ for metrics with aspect ratio Δ. In [18], a more involved dynamic embedding was proposed to achieve a polylog(k)-competitive algorithm for general metrics.[2] Contrast these upper bounds with the known deterministic lower bound of k [21] and the randomized lower bound of $\Omega(\log k)$ [12]. More information about the k-server problem can be found in [16].

---

[1] See Sect. 2 for an exact definition of HSTs.
[2] There is a gap in the version posted to the arXiv on February 21, 2018 [19,20].

Surprisingly, until recently very little was known about the k-taxi problem, in contrast to the extensive work on the k-server problem. Coester and Koutsoupias [9] provided a $(2^k - 1)$-competitive memoryless randomized algorithm for the k-taxi problem on HSTs against an adaptive online adversary. This result implies: (i) the existence of a 4^k-competitive deterministic algorithm for HSTs via a known reduction [3], although this argument is non-constructive; (ii) an $O(2^k \log n)$-competitive randomized algorithm for general metric spaces (against an oblivious adversary). Both bounds currently constitute the state of the art. Coester and Koutsoupias also provided a lower bound of $2^k - 1$ on the competitive factor of any deterministic algorithm for the k-taxi problem on HSTs, thus proving that the problem is substantially harder than the k-server problem. However, large gaps still remain in our understanding of the k-taxi problem, and many problems remain open in both deterministic and randomized settings. For general metrics, an algorithm with competitive factor depending only on k is known only if $k = 2$, and for the line metric only if $k \leq 3$ [9]. Both of these algorithms can be viewed as variants of DOUBLECOVERAGE.

The version of the problem where the start-to-destination distances also contribute to the objective function was called the "easy" k-taxi problem in [9,15], whereas the version we are considering here is the "hard" k-taxi problem. The easy version has the same competitive factor as the k-server problem [9]. The k-taxi problem was recently reintroduced as the Uber problem in [10], who studied the easy version in a stochastic setting.

1.2   Our Contribution

We provide the following bounds on the competitive ratio of DOUBLECOVERAGE for the k-taxi problem.

Theorem 1. *The competitive ratio of* DOUBLECOVERAGE *for the k-taxi problem is at most*

(a) $\left(c_{kd} = \sum_{h=1}^{\min\{k,d\}} \binom{k}{h} \right)$ *on HSTs of depth d.*

(b) $O(k^d)$-*competitive on general (weighted) tree metrics of depth d.*

We complement these upper bounds by the following lower bounds:

Theorem 2. *The competitive ratio of* DOUBLECOVERAGE *for the k-taxi problem is at least*

(a) $\left(c_{kd} = \sum_{h=1}^{\min\{k,d\}} \binom{k}{h} \right)$ *on HSTs of depth d.*

(b) $\Omega(k^d)$ *on (even unweighted) tree metrics of constant depth d.*

When the depth d of the HST is at least k, the upper bound $c_{kd} = 2^k - 1$ exactly matches the lower bound of [9] that holds even for randomized algorithms against an adaptive online adversary. Note that for fixed d, c_{kd} is roughly $k^d/d!$ up to a multiplicative error that tends to 1 as $k \to \infty$. The $\Omega(k^d)$ lower bound on general trees is hiding a constant factor that depends on d. Since the root in general trees can be chosen arbitrarily, d is essentially half the hop-diameter.

By well-known embedding techniques of general metrics into HSTs [2,11], slightly adapted to HSTs of bounded depth (see Theorem 5 in Sect. 2), we obtain the following result for general metrics.

Corollary 1. *There is a randomized $O(k^c \Delta^{1/c} \log_\Delta n)$-competitive algorithm for the k-taxi problem for every n-point metric, where Δ is the aspect ratio of the metric, and $c \geq 1$ is an arbitrary positive integer. In particular, setting $c = \left\lceil \sqrt{\frac{\log \Delta}{\log k}} \right\rceil$, the competitiveness is $2^{O\left(\sqrt{\log k \log \Delta}\right)} \log_\Delta n$.*

Compared to the $O(2^k \log n)$ upper bound of [9], our bound has only a polynomial dependence on k at the expense of some dependence on the aspect ratio. Since $c_{kd} \leq 2^k - 1$ for all d, we still recover the same $O(2^k \log n)$ competitive factor. The bounds in Corollary 1 actually hide another division by $(c-1)!$ if $c \leq k$. Therefore, whenever Δ is at most $2^{O(k^2)}$, our bound results in an improvement.

Techniques. For the k-server problem, there exists a simple potential function analysis of DOUBLECOVERAGE. The potential value depends on the relative distances of the server locations, which, in the k-taxi problem, can change arbitrarily by relocation requests even though the algorithm does not incur any cost. Therefore, such a potential cannot work for the k-taxi problem. In [9], the $2^k - 1$ upper bound for the randomized HST algorithm is proved via a potential function that is $2^k - 1$ times the minimum matching between the online and offline servers. As is stated there, the same potential can be used to obtain the same bound for DOUBLECOVERAGE when $k = 2$, but it fails already when $k = 3$. Nonetheless, they conjectured that DOUBLECOVERAGE achieves the competitive ratio of $2^k - 1$ on HSTs.

We are able to prove that this is the case (and give the more refined bound of c_{kd}) with a primal-dual approach (which still uses an auxiliary potential function as well). The primal solution is the output of the DOUBLECOVERAGE algorithm. A dual solution is constructed to provide a lower bound on the optimal cost. The typical way a dual solution is constructed in the online primal-dual framework is forward in time, step-by-step, along with the decisions of the online algorithm (see e.g. [1,6,14]). By showing that the objective values of the constructed primal and dual solutions are within a factor c of each other, one gets that the primal solution is c-competitive and the dual solution is $1/c$-competitive. For the LP formulation of the k-taxi problem we consider, we show that a pure forward-time approach (producing a dual solution as well) is doomed to fail:

Theorem 3. *There exists no competitive online algorithm for the dual problem of the k-taxi LP as defined in Sect. 3, even for $k = 1$.*

Our main conceptual contribution is a novel way to overcome this problem by constructing the dual solution backwards in time. Our assignment of dual variables for time t combines knowledge about both the future *and* the past: It incorporates knowledge about the future simply due to the time-reversal; knowledge about the past is also used because the dual assignments are guided by the movement of DOUBLECOVERAGE, which is a forward-time (online) algorithm. Our method can be seen as a restricted form of online dual fitting, which is more "local", and hence easier to analyze step by step, similarly to primal-dual algorithms. We believe that this time-reversed method of constructing a dual solution may be useful for analyzing additional online problems, especially when information about the future helps to construct better dual solutions. Using this technique, we also provide a primal-dual proof of the k-competitiveness of DOUBLECOVERAGE for the k-server problem on trees. To the best of our knowledge, a primal-dual proof of this classical result was not known before.

Theorem 4 ([8]). DOUBLECOVERAGE *is* k-*competitive for the* k-*server problem on trees.*

Due to space constraints, most proofs are omitted from this extended abstract and we focus on showing the upper bound for k-taxi on HSTs (Theorem 1(a)).

2 Preliminaries

The k-Taxi Problem. The k-taxi problem is formally defined as follows. We are given a metric space with point set V, where $|V| = n$. Initially, k taxis, or servers, are located at points of V. At each time t we get a request (s_t, d_t), where $s_t, d_t \in V$. To serve the request, one of the servers must move to s_t and the cost paid by the algorithm is the distance traveled by the server. Then, one of the servers from s_t is relocated to the point d_t. There is no cost for relocating the server from s_t to d_t. The goal is to minimize the cost.

Without loss of generality, we can split each request (s_t, d_t) into two requests: a *simple* request and a *relocation* request. Thus, at each time t, request (s_t, d_t) is either one of the following:

- *Simple request* $(s_t = d_t)$: a server needs to move to s_t, if there is no server there already. The cost is the distance traveled by the server.
- *Relocation request* $(s_t = d_{t-1})$: a server is relocated from s_t to d_t. There is no relocation cost.

We can then partition the time horizon into two sets T_s, T_r (odd and even times). For times in $T_s = \{1, 3, 5, \ldots, 2T - 1\}$ we have simple requests, and for times in $T_r = \{2, 4, 6, \ldots, 2T\}$ we have relocation requests. The k-server problem is the special case of the k-taxi problem without relocation requests.

Trees and HSTs. Consider now a tree $\mathbb{T} = (V, E)$ and let $\mathbb{r}$ denote its root. There is a positive weight function defined over the edge set E, and without loss of generality all edge weights are integral. The *distance* between vertices u and v is the sum of the weights of the edges on the (unique) path between them in $\mathbb{T}$, which induces a metric. The *combinatorial depth* of a vertex $v \in V$ is defined to be the number of edges on the path from $\mathbb{r}$ to v. The combinatorial depth of $\mathbb{T}$ is the maximum combinatorial depth among all vertices. At times it will be convenient to assume that all edges in E have unit length by breaking edges into unit length parts called *short edges*. We then refer to the original edges of $\mathbb{T}$ as *long edges*. However, the combinatorial depth of $\mathbb{T}$ is still defined in terms of the long edges. We define the *weighted depth* of a vertex u as the number of *short* edges on the path from $\mathbb{r}$ to u. For $u \in V$, let $V_u \subseteq V$ be the vertices of the subtree rooted at u. In trees where all leaves are at the same weighted/combinatorial depth (namely, HSTs, see below), we define the *weighted/combinatorial height* of a vertex u as the number of short/long edges on the path from u to any leaf in V_u. We let $v \prec u$ denote that v is a child of u, and let $p(u)$ denote the parent of u.

Hierarchically well-separated trees (HSTs), introduced by Bartal [2], are special trees that can be used to approximate arbitrary finite metrics. For $\alpha \geq 1$, an α-HST is a tree where every leaf is at the same combinatorial depth d and the edge weights along any root-to-leaf path decrease by a factor α in each step. The associated metric space of the HST is only the set of its leaves. Hence, for the k-taxi problem on HSTs, the requested points s_t and d_t are always leaves. Any n-point metric space can be embedded into a random α-HST such that (i) the distance between any two points can only be larger in the HST and (ii) the expected blow-up of each distance is $O(\alpha \log_\alpha n)$ [11]. The latter quantity is also called the *distortion* of the embedding. The depth of the random HST constructed in the embedding is at most $\lceil \log_\alpha \Delta \rceil$, where Δ is the aspect ratio, i.e., the ratio between the longest and shortest non-zero distance. Choosing $\alpha = \Delta^{1/d}$, we obtain an HST of depth d and with distortion $O\left(d\Delta^{1/d} \log_\Delta n\right)$.

Theorem 5 (Corollary to [11]). *Any metric with n points and aspect ratio Δ can be embedded into a random HST of combinatorial depth d with distortion $O\left(d\Delta^{1/d} \log_\Delta n\right)$.*

The Double Coverage Algorithm. We define DOUBLECOVERAGE in a way that will suit our definition of short edges of length 1 later. Consider the arrival of a simple request at location s_t. A server located at vertex v is *unobstructed* if there are no other servers on the path between v and s_t. If other servers on this path exist only at v, we consider only one of them unobstructed (chosen arbitrarily). Serving the request is done in several small steps, as follows:

DOUBLECOVERAGE (upon a simple request at s_t):
While no server is at s_t: all currently unobstructed servers move distance 1 towards s_t.

Upon a relocation request (s_t, d_t), we simply relocate a server from s_t to d_t.

For a given small step (i.e., iteration of the while-loop), we denote by U and B the sets of servers moving towards the root (**upwards** in the tree) and away from the root (towards the **bottom** of the tree), respectively.

Observation 1. In any small step:

- B is either a singleton ($B = \{j\}$ for a server j) or empty ($B = \emptyset$).
- The subtrees rooted at servers of U are disjoint and do not contain s_t. If $B = \{j\}$, then these subtrees and s_t are inside the subtree rooted at j.

3 LP Formulations

We formulate a linear program (LP) for the k-taxi problem along with its dual that we use for the purpose of analysis. We assume for ease of exposition that all edges are short edges. As already mentioned, when considering the k-taxi problem on HSTs, requests appear only at the leaves. It is easy to see that in this case the upward movement cost (i.e., movement towards the root) is the same as the downward movement cost, up to an additive error of k times the distance from the root to any leaf. The same is true of the k-server problem on general trees (but not for the k-taxi problem on general trees). Hence, for the k-taxi problem on HSTs and for the k-server problem, we can use an LP that only takes into account the upward movement cost. (A slightly different LP is needed for the k-taxi problem on general trees.)

The LP below is a relaxation of the problem as it allows for fractional values of the variables. For $u \in V$, let variable x_{ut} denote the number of servers in V_u after the request at time t has been served. Variable $y_{ut} \geq 0$ denotes the number of servers that left subtree V_u (moving upwards) at time t. For $u = \mathfrak{r}$, $x_{\mathfrak{r}t}$ is defined to be the *constant* k. (It is not a variable.) The primal LP is the following:

$$\min \quad \sum_{t \in T_s} \sum_{u \neq \mathfrak{r}} y_{ut}$$

$$x_{ut} \geq \mathbb{1}_{\{u = s_t \text{ and } t \in T_s\}} + \sum_{v \lessdot u} x_{vt} \quad \forall u \in V, t \in T_s \cup T_r$$

$$y_{ut} \geq x_{u,t-1} - x_{ut} \qquad\qquad \forall u \neq \mathfrak{r}, t \in T_s$$

$$x_{ut} = x_{u,t-1} + \xi_{ut} \qquad\qquad \forall u \neq \mathfrak{r}, t \in T_r$$

$$y_{ut} \geq 0 \qquad\qquad\qquad\quad \forall u \neq \mathfrak{r}, t \in T_s,$$

where

$$\xi_{ut} := \begin{cases} -1 & \text{if } s_t \in V_u, d_t \notin V_u \\ 1 & \text{if } s_t \notin V_u, d_t \in V_u \\ 0 & \text{otherwise.} \end{cases}$$

For technical reasons, we will add the additional constraint

$$0 = x_{u,2T} - \bar{x}_{u,2T} \qquad \forall u \neq \mathfrak{r}$$

to the primal LP, where $\bar{x}_{u,2T}$ are *constants* specifying the configuration of DOUBLECOVERAGE at the last time step. Clearly, this affects the optimal value by only an additive constant. We will also view $x_{u0} = \bar{x}_{u0}$ as constants describing the initial configuration of the servers. The corresponding dual LP is the following.

$$\max \quad \sum_{t \in T_s} \lambda_{s_t t} + \sum_{t \in T_r} \sum_{u \neq \mathfrak{r}} \xi_{ut} b_{u,t-1} - \sum_{t \in T_s \cup T_r} k\lambda_{\mathfrak{r}t} + \sum_{u \neq \mathfrak{r}} \left(\bar{x}_{u0} b_{u0} - \bar{x}_{u,2T} b_{u,2T} \right)$$

$$\lambda_{ut} - \lambda_{p(u)t} = b_{ut} - b_{u,t-1} \qquad\qquad \forall u \neq \mathfrak{r}, t \in T_s \cup T_r$$

$$b_{u,t-1} \in [0,1] \qquad\qquad\qquad\qquad \forall u \neq \mathfrak{r}, t \in T_s$$

$$\lambda_{ut} \geq 0 \qquad\qquad\qquad\qquad\quad \forall u \in V, t \in T_s \cup T_r$$

We can use the same primal and dual formulation for the k-server problem, except that the set T_r is then empty.[3]

3.1 Dual Transformation

The dual LP is not very intuitive. By a transformation of variables, we get a simpler equivalent dual LP, which we can interpret as building a mountain structure on the tree over time. This new dual LP has only one variable A_{ut} for each vertex u and time t. We interpret A_{ut} as the *altitude* of u at time t. We denote by $\Delta_t A_u = A_{ut} - A_{u,t-1}$ the change of altitude of vertex u at time t. For a server i of DOUBLECOVERAGE, we denote by v_{it} its location at time t, and define similarly $\Delta_t A_i := A_{v_{it}t} - A_{v_{i,t-1}t-1}$ as the change of altitude of server i at time t. The new dual LP is the following:

$$\max \quad \sum_{t \in T_s}\left[\Delta_t A_{s_t} - \sum_{i=1}^{k}\Delta_t A_i\right] - \sum_{t \in T_r}\sum_{i=1}^{k}\Delta_t A_{v_{it}}$$

$$A_{ut} - A_{p(u)t} \in [0,1] \quad \forall u \neq \mathfrak{r}, t+1 \in T_s \tag{1}$$

$$\Delta_t A_u \geq 0 \quad \forall u \in V, t \in T_s \cup T_r \tag{2}$$

The constraints of the LP stipulate that altitudes are non-decreasing over time and (at time steps before a simple request) along root-to-leaf paths, with the difference in altitude of two adjacent nodes being at most 1. The objective function measures changes in the altitudes of request and server locations.

We define

$$D_t := \begin{cases} \Delta_t A_{s_t} - \sum_{i=1}^{k}\Delta_t A_i & t \in T_s \\ -\sum_{i=1}^{k}\Delta_t A_{v_{it}} & t \in T_r, \end{cases} \tag{3}$$

so that the dual objective function is equal to $D := \sum_{t \in T_s \cup T_r} D_t$.

The following lemma allows us to use this new LP for our analyses.

Lemma 1. *The two dual LPs are equivalent. That is, any feasible solution to one of them can be translated (online) to a feasible solution to the other with the same objective function value.*

A proof of this lemma is given in the full version of our paper. It is based on a transformation of variables satisfying

$$A_{ut} - A_{p(u)t} = b_{ut}$$
$$\Delta_t A_u = \lambda_{ut}.$$

---

[3] We note that our LP for the k-server problem is different from LPs used in the context of polylogarithmically-competitive randomized algorithms for the k-server problem. In our context of deterministic algorithms for k-taxi (and k-server), we show that we can work with this simpler formulation.

Moreover, the comparably simple dual objective function is obtained by expanding the terms $\bar{x}_{u0}b_{u0} - \bar{x}_{u,2T}b_{u,2T}$ in the objective of the original dual to a telescoping sum.

4 The k-Taxi Problem on HSTs

In this section we analyze DOUBLECOVERAGE on HSTs, proving Theorem 1(a).

Besides constructing a dual solution, our analysis will also employ a potential function Ψ (that depends on the state of the system). The choice of Ψ will be the only difference between the analyses of the k-server and k-taxi problems. The dual solution and potential will be such that for all $t \in T_s \cup T_r$,

$$cost_t^\uparrow + \Psi_t - \Psi_{t-1} \leq c \cdot D_t, \tag{4}$$

where $cost_t^\uparrow$ is the cost of movement *towards the root* by DOUBLECOVERAGE's servers while serving the tth request, c is the desired competitive ratio, and D_t is the increase of the dual objective function at time t, as given by (3). As discussed in Sect. 2 we may use in this case the dual of the program that only measures movement cost towards the root. Thus, summing (4) for all times will then imply that DOUBLECOVERAGE is c-competitive.

Recall that for a simple request ($t \in T_s$), DOUBLECOVERAGE breaks the movement of the servers into small steps in which the servers in $U \cup B$ move distance 1 towards the request. We will break the construction of a dual solution into these same small steps. We will denote by $\Delta\Psi$ the change of Ψ during the step and by ΔD the contribution of the step to D_t. The cost paid by the servers (for moving towards the root) in the step is $|U|$. Using this notation, we satisfy (4) for simple requests if we show for each step that:

$$|U| + \Delta\Psi \leq c \cdot \Delta D. \tag{5}$$

In Sect. 4.1 we describe how we construct the dual solution by going backwards in time. We also mention a simple potential function to prove the k-competitiveness of k-server on trees. In Sect. 4.2 we describe a more involved potential function proving the competitiveness for k-taxi on HSTs.

4.1 Constructing the Dual Solution

As already mentioned, we break the construction of a dual solution into the same small steps that already partition the movement of DOUBLECOVERAGE. That is, we will define altitudes also for the times between two successive small steps. We will call a dual solution where altitudes are also defined for times between small steps an *extended dual solution*.

We will construct this dual solution by induction backwards in time. For a given point in time, let A_u be the altitude of a vertex u at this time as determined by the induction hypothesis, and let v_i be the location of server i at this time. We will denote by A'_u and v'_i the new values of these quantities at the next point in reverse-time. We denote by $\Delta A_u := A_u - A'_u$ and $\Delta A_i := A_{v_i} - A'_{v'_i}$ the change of the altitude of vertex u and server i, respectively, in *forward-time* direction. For the update due to a small step when serving a simple request s_t, define

$$\Delta D := \Delta A_{s_t} - \sum_i \Delta A_i.$$

Thus, the sum of the quantities ΔD for all small steps corresponding to a simple request at time t is precisely D_t. In reverse-time, we can think of ΔD as the amount by which the request's altitude *decreases* plus the amount by which the sum of server altitudes *increases*.

We will update altitudes so as to satisfy the following two rules:

(i) $\Delta A_u \geq 0$ for all $u \in V$ (constraint (2) is satisfied): In reverse-time, we only *decrease* the altitude of any vertex (or leave it unchanged).
(ii) $A_u - A_{p(u)} \in \{0, 1\}$ for all $u \neq \mathfrak{r}$ at all times (constraint (1) is satisfied): The altitude of u and $p(u)$ is the same, or the altitude of u is higher by one than the altitude of $p(u)$. Overall, altitudes are non-increasing towards the root.

Lemma 2. *There exists a feasible extended dual solution satisfying:*

- *For a relocation request at time t: $D_t = 0$.*
- *For a small step where $B = \emptyset$: $\Delta D \geq 1$.*
- *For a small step where $B = \{j\}$: $\Delta D \geq 0$.*

Proof. For the base of the reverse time induction, let A be some arbitrary constant and define the altitude of every vertex $u \in V$ at the time after the final request to be A. This trivially satisfies rule (ii).

Relocation Requests $(t \in T_r)$: We guarantee $D_t = 0$ by simply keeping all altitudes unchanged.

Simple Requests $(t \in T_s)$: Consider a small step of the simple request to s_t. In reverse-time, any server $i \in U$ moves from $v_i = p(v'_i)$ to v'_i during the small step. Then, $\Delta A_i = A_{p(v'_i)} - A'_{v'_i}$. By rule (ii) of the induction hypothesis, we have $A_{p(v'_i)} - A_{v'_i} \in \{-1, 0\}$. Similarly, if $B = \{j\}$, then j moves from v_j to $v'_j = p(v_j)$ in reverse-time, and $\Delta A_j = A_{v_j} - A'_{p(v_j)}$.

Case 1: $B = \emptyset$: If for at least one server $i \in U$ we have $A_{v'_i} - A_{p(v'_i)} = 1$, then we set $A'_u := A_u$ for all vertices. In this case, $\Delta_t A_i = -1$ for the aforementioned server i, and for all other servers in $i \in U$, $\Delta_t A_i \leq 0$. Overall, $\Delta D \geq 1$.

Otherwise, for all $i \in U$, $A_{v'_i} - A_{p(v'_i)} = 0$ meaning that every edge along which a server moves during this small step connects two vertices of the same altitude (an example of the update of the dual for this case is shown in Fig. 1). Let $V' = V \setminus \bigcup_{i \in U} V_{v'_i}$ be the connected component containing s_t when cutting all edges traversed by a server in this step. Notice that V' does not contain v'_i even for servers i that are not moving during the step, since those are located in subtrees below the servers of U. For each $u \in V'$, we set $A'_u := A_u - 1$ (or $\Delta A_u = 1$), and otherwise we keep the altitudes unchanged. In particular, rule (i) is satisfied.

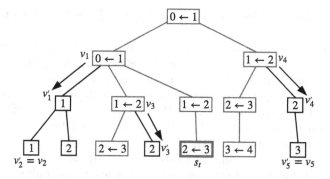

Fig. 1. Example of a dual update for a small step of a simple request when $B = \emptyset$. The current request is s_t, and the vertices colored red are the component V'. The arrows show the movement of servers (in reverse time direction). Numbers in boxes represent altitudes, where $b \leftarrow a$ means that the altitude is updated from a to b (in reverse time). (Color figure online)

Since for all cut edges $A_{v'_i} - A_{p(v'_i)} = 0$, then for these edges $A'_{v'_i} - A'_{p(v'_i)} = 1$, and rule (ii) also remains satisfied. As stated, servers only moved along edges connecting vertices of the same altitude (before the update), and all server positions v'_i are outside the component V', so $\Delta A_i = 0$ for each server. But the component contains the request s_t, so $\Delta A_{s_t} = A_{s_t} - A'_{s_t} = 1$. Overall, we get $\Delta D = 1$.

Case 2: $B = \{j\}$: If some server $i \in U$ moves in reverse-time to a vertex of higher altitude $(A_{p(v'_i)} - A_{v'_i} = -1)$ *or* server j moves to a vertex of the same altitude $(A_{p(v_j)} - A_{v_j} = 0)$, then we set $A'_u := A_u$ for each $u \in V$. In this case, $\Delta A_i \in \{-1, 0\}$ for all $i \in U$, and $\Delta A_j \in \{0, 1\}$, and the aforementioned condition translates to the

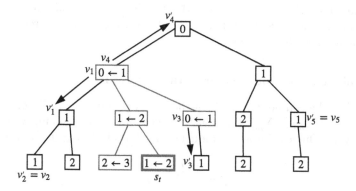

Fig. 2. Example of a dual update for a small step of a simple request when $B = \{j\}$. The current request is s_t, and the vertices colored red are the component V'. The arrows show the movement of servers (in reverse time direction). Numbers in boxes represent altitudes, where $b \leftarrow a$ means that the altitude is updated from a to b (in reverse time). (Color figure online)

condition that $\Delta A_i = -1$ for some $i \in U$ *or* $\Delta A_j = 0$. Either case then guarantees that $\Delta D \geq 0$.

Otherwise, j moves (in reverse-time) to a vertex of lower altitude ($A_{p(v_j)} - A_{v_j} = -1$) *and* all servers in U move along edges of unchanging altitude ($A_{p(v_i')} - A_{v_i'} = 0$) (an example of the update of the dual for this case is shown in Fig. 2). Let $V' = V_{v_j} \setminus \bigcup_{i \in U} V_{v_i'}$ be the connected component containing s_t when cutting all edges traversed by servers in this step. We decrease the altitudes of all vertices in this component by 1 ($A_u' := A_u - 1$ for $u \in V'$) and leave other altitudes unchanged, satisfying rule (i). As $A_{p(v_j)} - A_{v_j} = -1$ and $A_{p(v_i')} - A_{v_i'} = 0$ for $i \in U$, also rule (ii) is satisfied. Again, the locations v_i' of any server i (moving or not) are outside the component, so the update of altitudes does not affect ΔA_i. Thus, $\Delta A_j = A_{v_j} - A_{p(v_j)} = 1$ and, for each server $i \neq j$, $\Delta A_i = 0$. But the altitude of s_t is decreasing by 1 in reverse-time, so $\Delta_t A_{s_t} = 1$. Overall, we get that $\Delta D = 0$. □

Potential Function Requirements. Based on Lemma 2, we conclude that the following requirements of a potential function Ψ are sufficient to conclude inequality (4) (resp. (5)) and therefore c-competitiveness of DOUBLECOVERAGE.

Observation 2. DOUBLECOVERAGE is c-competitive if there is a potential Ψ satisfying:

- For a relocation request at time t: $\Psi_t = \Psi_{t-1}$.
- In a single step of a simple request, where no server is going downwards: $|U| + \Delta\Psi \leq c$.
- In a single step of a simple request, where there is a server moving downwards: $|U| + \Delta\Psi \leq 0$.

In the k-server problem there are no relocation requests, and one can satisfy the two requirements involving simple requests in Observation 2 with the potential function $\Psi = -\sum_{i<j} d_{\text{lca}(i,j)}$, where the sum is taken over all pairs of servers $\{i, j\}$ and $d_{\text{lca}(i,j)}$ denotes the weighted depth (distance from the root $\mathfrak{r}$) of the least common ancestor of i and j.

4.2   Finalizing the Analysis for k-Taxi on HSTs

We show that the competitive ratio of the k-taxi problem on HSTs of combinatorial depth d is $c_{kd} = \sum_{h=1}^{k \wedge d} \binom{k}{h}$, which is the number of non-empty sets of at most d servers. One can prove by induction on k that for $k \geq 0$ and $d \geq 1$,

$$c_{kd} = k + \sum_{i=0}^{k-1} c_{i,d-1}. \tag{6}$$

For a given point in time, fix a naming of the servers by the numbers $0, \ldots, k-1$ such that their heights $h_0 \leq \cdots \leq h_{k-1}$ are non-decreasing. If we consider a (small) step, we choose this numbering such that $h_0 \leq \cdots \leq h_{k-1}$ holds both *before* and *after* the step. Since it is not possible that a server is strictly higher than another server before the step and then strictly lower afterwards, such a numbering exists. Let $U, B \subseteq \{0, \ldots, k-1\}$ be the sets of servers that move upwards (i.e., towards the root) and downwards (away from the root), respectively. Note that by our earlier observation, B is either a singleton $\{j\}$ (one server moves downwards) or the empty set (no server moves downwards).

Let α_ℓ denote the weighted height of the node layer at combinatorial height ℓ in the HST (i.e., the distance of these vertices from the leaf layer). Thus, $0 = \alpha_0 < \alpha_1 < \cdots < \alpha_d$. We use the following potential function at time t:

$$\Psi_t = \sum_{i=0}^{k-1} \sum_{\ell=0}^{d-1} c_{i\ell} \cdot \max \{\alpha_\ell, h_{it} \wedge \alpha_{\ell+1}\},$$

where $h_{0t} \leq \cdots \leq h_{k-1,t}$ are the weighted heights of the servers. We next show that Ψ satisfies the requirements of Observation 2 with $c = c_{kd}$.

A Relocation Request, $t \in T_r$: Since all requests are at the leaves and thus the height of the moving server is 0, we have $\Psi_t = \Psi_{t-1}$.

A Small Step of a Simple Request, $t \in T_s$: Let ℓ_i be such that the edge traversed by server i during the step is located between node layers of combinatorial heights ℓ_i and ℓ_i+1. Thus, the weighted height of i lies in $[\alpha_{\ell_i}, \alpha_{\ell_i+1}]$. Then

$$\Delta\Psi = \sum_{i \in U} c_{i\ell_i} - \sum_{j \in B} c_{j\ell_j}. \tag{7}$$

Case 1: $B = \emptyset$:

$$|U| + \Delta\Psi \leq k + \sum_{i=0}^{k-1} c_{i\ell_i} \leq k + \sum_{i=0}^{k-1} c_{i,d-1} = c_{kd}.$$

The first inequality follows from (7) and since $|U| \leq k$. The second inequality follows from the definition of c_{kd}. The final equation is due to (6).

Case 2: $B = \{j\}$: In this case $U \subseteq \{0, \ldots, j-1\}$ and $\ell_i \leq \ell_j - 1$ for each $i \in U$. Therefore,

$$|U| + \Delta\Psi \leq j - c_{j\ell_j} + \sum_{i \in U} c_{i\ell_i} \leq j - c_{j\ell_j} + \sum_{i=0}^{j-1} c_{i,\ell_j-1} = 0.$$

The first inequality follows from (7) and since $U \subseteq \{0, \ldots, j-1\}$. The second inequality follows from the definition of c_{kd}. The equation is due to (6).

5   The k-Taxi Problem on Weighted Trees

We briefly summarize the main ways in which the proof of part (b) of Theorem 1 differs from that of part (a). There are two reasons why our analysis for HSTs fails on general weighted trees:

1. The costs of movement towards and away from the root no longer need to be within a constant of each other. E.g., if relocation repeatedly brings servers closer to the root, then most cost would be incurred while moving away from the root.
2. The potential is no longer constant under relocation requests, because servers can be relocated to and from internal vertices, affecting their height.

To address the first issue, we use an LP formulation that measures movement cost both towards and away from the root. This is achieved in the primal LP by introducing additional variables $z_{ut} \geq 0$ for downwards movement, replacing y_{ut} by $y_{ut} + z_{ut}$ in objective function, and replacing the constraints $y_{ut} \geq x_{u,t-1} - x_{ut}$ by $y_{ut} - z_{ut} = x_{u,t-1} - x_{ut}$. The only change this causes in the (transformed) dual is that the constraint $A_{ut} - A_{p(u)t} \in [0,1]$ becomes $A_{ut} - A_{p(u)t} \in [-1,1]$.

To address the second issue, we eliminate the potential function from our proof. Instead, we construct a dual solution that bounds the cost of DOUBLE-COVERAGE in each step, but it may violate the constraints $A_{ut} - A_{p(u)t} \in [-1,1]$. However, it will still satisfy $A_{ut} - A_{p(u)t} \in [-c,c]$ for some c. Thus, dividing all dual variables by c yields a feasible dual solution, and c is our competitive ratio. A complete proof is given in the full version of our paper.

References

1. Azar, Y., et al.: Online algorithms for covering and packing problems with convex objectives. In: IEEE 57th Annual Symposium on Foundations of Computer Science, FOCS 2016, pp. 148–157. IEEE Computer Society (2016)
2. Bartal, Y.: Probabilistic approximation of metric spaces and its algorithmic applications. In: 37th Annual Symposium on Foundations of Computer Science, FOCS 1996, pp. 184–193 (1996)
3. Ben-David, S., Borodin, A., Karp, R.M., Tardos, G., Wigderson, A.: On the power of randomization in on-line algorithms. Algorithmica 11(1), 2–14 (1994)
4. Bubeck, S., Cohen, M.B., Lee, Y.T., Lee, J.R., Madry, A.: k-server via multiscale entropic regularization. In: Proceedings of the 50th Annual ACM SIGACT Symposium on Theory of Computing, STOC 2018, pp. 3–16. ACM (2018)
5. Buchbinder, N., Gupta, A., Molinaro, M., Naor, J.S.: k-servers with a smile: online algorithms via projections. In: Proceedings of the Thirtieth Annual ACM-SIAM Symposium on Discrete Algorithms, SODA 2019, pp. 98–116. SIAM (2019)
6. Buchbinder, N., Naor, J.: The design of competitive online algorithms via a primal-dual approach. Found. Trends Theor. Comput. Sci. 3(2–3), 93–263 (2009)
7. Chrobak, M., Karloff, H., Payne, T., Vishwanathan, S.: New results on server problems. SIAM J. Discrete Math. 4(2), 172–181 (1991)
8. Chrobak, M., Larmore, L.L.: An optimal on-line algorithm for k servers on trees. SIAM J. Comput. 20(1), 144–148 (1991)
9. Coester, C., Koutsoupias, E.: The online k-taxi problem. In: Proceedings of the 51st Annual ACM SIGACT Symposium on Theory of Computing, STOC 2019, pp. 1136–1147. ACM (2019)
10. Dehghani, S., Ehsani, S., Hajiaghayi, M., Liaghat, V., Seddighin, S.: Stochastic k-server: how should Uber work? In: 44th International Colloquium on Automata, Languages, and Programming (ICALP 2017), pp. 126:1–126:14 (2017)

11. Fakcharoenphol, J., Rao, S., Talwar, K.: A tight bound on approximating arbitrary metrics by tree metrics. J. Comput. Syst. Sci. **69**(3), 485–497 (2004)
12. Fiat, A., Karp, R.M., Luby, M., McGeoch, L.A., Sleator, D.D., Young, N.E.: Competitive paging algorithms. J. Algorithms **12**(4), 685–699 (1991)
13. Fiat, A., Rabani, Y., Ravid, Y.: Competitive k-server algorithms (extended abstract). In: 31st Annual Symposium on Foundations of Computer Science, FOCS 1990, pp. 454–463 (1990)
14. Gupta, A., Nagarajan, V.: Approximating sparse covering integer programs online. Math. Oper. Res. **39**(4), 998–1011 (2014)
15. Kosoresow, A.P.: Design and analysis of online algorithms for mobile server applications. Ph.D. thesis, Stanford University (1996)
16. Koutsoupias, E.: The k-server problem. Comput. Sci. Rev. **3**(2), 105–118 (2009)
17. Koutsoupias, E., Papadimitriou, C.H.: On the k-server conjecture. J. ACM **42**(5), 971–983 (1995)
18. Lee, J.R.: Fusible HSTs and the randomized k-server conjecture. In: Proceedings of the 59th Annual IEEE Symposium on Foundations of Computer Science, FOCS 2018, pp. 438–449 (2018)
19. Lee, J.R.: Fusible HSTs and the randomized k-server conjecture. arXiv:1711.01789v2, February 2018
20. Lee, J.R.: Personal Communication (2019)
21. Manasse, M., McGeoch, L., Sleator, D.: Competitive algorithms for on-line problems. In: Proceedings of the Twentieth Annual ACM Symposium on Theory of Computing, STOC 1988, pp. 322–333. ACM (1988)

Approximating the Discrete Time-Cost Tradeoff Problem with Bounded Depth

Siad Daboul[(✉)], Stephan Held, and Jens Vygen

Research Institute for Discrete Mathematics and Hausdorff Center for Mathematics,
University of Bonn, Bonn, Germany
{daboul,held,vygen}@dm.uni-bonn.de

Abstract. We revisit the deadline version of the discrete time-cost tradeoff problem for the special case of bounded depth. Such instances occur for example in VLSI design. The depth of an instance is the number of jobs in a longest chain and is denoted by d. We prove new upper and lower bounds on the approximability.

First we observe that the problem can be regarded as a special case of finding a minimum-weight vertex cover in a d-partite hypergraph. Next, we study the natural LP relaxation, which can be solved in polynomial time for fixed d and—for time-cost tradeoff instances—up to an arbitrarily small error in general. Improving on prior work of Lovász and of Aharoni, Holzman and Krivelevich, we describe a deterministic algorithm with approximation ratio slightly less than $\frac{d}{2}$ for minimum-weight vertex cover in d-partite hypergraphs for fixed d and given d-partition. This is tight and yields also a $\frac{d}{2}$-approximation algorithm for general time-cost tradeoff instances.

We also study the inapproximability and show that no better approximation ratio than $\frac{d+2}{4}$ is possible, assuming the Unique Games Conjecture and P $\neq$ NP. This strengthens a result of Svensson [17], who showed that under the same assumptions no constant-factor approximation algorithm exists for general time-cost tradeoff instances (of unbounded depth). Previously, only APX-hardness was known for bounded depth.

1 Introduction

The (deadline version of the discrete) time-cost tradeoff problem was introduced in the context of project planning and scheduling more than 60 years ago [14]. An instance of the *time-cost tradeoff problem* consists of a finite set V of jobs, a partial order $(V, \prec)$, a deadline $T > 0$, and for every job v a finite nonempty set $S_v \subseteq \mathbb{R}^2_{\geq 0}$ of time/cost pairs. An element $(t, c) \in S_v$ corresponds to a possible choice of performing job v with delay t and cost c. The task is to choose a pair $(t_v, c_v) \in S_v$ for each $v \in V$ such that $\sum_{v \in P} t_v \leq T$ for every chain P (equivalently: the jobs can be scheduled within a time interval of length T, respecting the precedence constraints), and the goal is to minimize $\sum_{v \in V} c_v$.

The partial order can be described by an acyclic digraph $G = (V, E)$, where $(v, w) \in E$ if and only if $v \prec w$. Every chain of jobs corresponds to a path in G, and vice versa.

© Springer Nature Switzerland AG 2021
M. Singh and D. P. Williamson (Eds.): IPCO 2021, LNCS 12707, pp. 30–42, 2021.
https://doi.org/10.1007/978-3-030-73879-2_3

De et al. [6] proved that this problem is strongly NP-hard. Indeed, there is an approximation-preserving reduction from vertex cover [10], which implies that, unless P = NP, there is no 1.3606-approximation algorithm [8]. Assuming the Unique Games Conjecture and P $\neq$ NP, Svensson [17] could show that no constant-factor approximation algorithm exists.

Even though the time-cost tradeoff has been extensively studied due to its numerous practical applications, only few positive results about approximation algorithms are known. Skutella [16] described an algorithm that works if all delays are natural numbers in the range $\{0, \ldots, l\}$ and returns an l-approximation. If one is willing to relax the deadline, one can use Skutella's bicriteria approximation algorithm [16]. For a fixed parameter $0 < \mu < 1$, it computes a solution in polynomial time such that the optimum cost is exceeded by a factor of at most $\frac{1}{1-\mu}$ and the deadline T is exceeded by a factor of at most $\frac{1}{\mu}$. Unfortunately, for many applications, including VLSI design, relaxing the deadline is out of the question.

The instances of the time-cost tradeoff problem that arise in the context of VLSI design usually have a constant upper bound d on the number of vertices on any path [5]. This is due to a given target frequency of the chip, which can only be achieved if the logic depth is bounded. For this important special case, we will describe better approximation algorithms.

The special case $d = 2$ reduces to weighted bipartite matching and can thus be solved optimally in polynomial time. However, already the case $d = 3$ is strongly NP-hard [6]. The case $d = 3$ is even APX-hard, because Deĭneko and Woeginger [7] devised an approximation-preserving reduction from vertex cover in cubic graphs (which is known to be APX-hard [2]).

On the other hand, it is easy to obtain a d-approximation algorithm: either by applying the Bar-Yehuda–Even algorithm for set covering [3,5] or (for fixed d) by simple LP rounding; see the end of Sect. 3.

As we will observe in Sect. 3, the time-cost tradeoff problem with depth d can be viewed as a special case of finding a minimum-weight vertex cover in a d-partite hypergraph. Lovász [15] studied the unweighted case and proved that the natural LP has integrality gap $\frac{d}{2}$. Aharoni, Holzman and Krivelevich [1] showed this ratio for more general unweighted hypergraphs by randomly rounding a given LP solution. Guruswami, Sachdeva and Saket [11] proved that approximating the vertex cover problem in d-partite hypergraphs with a better ratio than $\frac{d}{2} - 1 + \frac{1}{2d}$ is NP-hard, and better than $\frac{d}{2}$ is NP-hard if the Unique Games Conjecture holds.

2   Results and Outline

In this paper, we first reduce the time-cost tradeoff problem with depth d to finding a minimum-weight vertex cover in a d-partite hypergraph. Then we simplify and derandomize the LP rounding algorithm of Lovász [15] and Aharoni et al. [1] and show that it works for general nonnegative weights. This yields a simple deterministic $\frac{d}{2}$-approximation algorithm for minimum-weight vertex cover in d-partite hypergraphs for fixed d, given d-partition, and given LP solution. To obtain a $\frac{d}{2}$-approximation algorithm for the time-cost tradeoff problem, we develop a slightly stronger bound for rounding the LP solution, because the vertex cover LP can only be solved approximately (unless d is fixed). This will imply our first main result:

Theorem 1. *There is a polynomial-time $\frac{d}{2}$-approximation algorithm for the time-cost tradeoff problem, where d denotes the depth of the instance.*

The algorithm is based on rounding an approximate solution to the vertex cover LP. The basic idea is quite simple: we partition the jobs into levels and carefully choose an individual threshold for every level, then we accelerate all jobs for which the LP solution is above the threshold of its level. We get a solution that costs slightly less than $\frac{d}{2}$ times the LP value. Since the integrality gap is $\frac{d}{2}$ [1,15] (even for time-cost tradeoff instances; see Sect. 3), this ratio is tight.

The results by [11] suggest that this approximation guarantee is essentially best possible for general instances of the vertex cover problem in d-partite hypergraphs. Still, better algorithms might exist for special cases such as the time-cost tradeoff problem. However, we show that much better approximation algorithms are unlikely to exist even for time-cost tradeoff instances. More precisely:

Theorem 2. *Let $d \in \mathbb{N}$ with $d \geq 2$ and $\rho < \frac{d+2}{4}$ be constants. Assuming the Unique Games Conjecture and $P \neq NP$, there is no polynomial-time ρ-approximation algorithm for time-cost tradeoff instances with depth d.*

This gives strong evidence that our approximation algorithm is best possible up to a factor of 2. To obtain our inapproximability result, we leverage Svensson's theorem on the hardness of vertex deletion to destroy long paths in an acyclic digraph [17] and strengthen it to instances of bounded depth by a novel compression technique.

Section 3 introduces the vertex cover LP and explains why the time-cost tradeoff problem with depth d can be viewed as a special case of finding a minimum-weight vertex cover in a d-partite hypergraph. In Sect. 4 we describe our approximation algorithm, which rounds a solution to this LP. Then, in Sects. 5 and 6 we prove our inapproximability result. We omit some relatively easy proofs due to lack of space; see arXiv:2011.02446 for the full version.

3 The Vertex Cover LP

Let us define the *depth* of an instance of the time-cost tradeoff problem to be the number of jobs in the longest chain in $(V, \prec)$, or equivalently the number of vertices in the longest path in the associated acyclic digraph $G = (V, E)$. We write $n = |V|$, and the depth will be denoted by d throughout this paper.

First, we note that one can restrict attention to instances with a simple structure, where every job has only two alternatives and the task is to decide which jobs to accelerate. This has been observed already by Skutella [16]. The following definition describes the structure that we will work with.

Definition 3. *An instance I of the time-cost tradeoff problem is called normalized if for each job $v \in V$ the set of time/cost pairs is of the form $S_v = \{(0, c), (t, 0)\}$ for some $c, t \in \mathbb{R}_+ \cup \{\infty\}$.*

In a normalized instance, every job has only two possible ways of being executed. The slow execution is free and the fast execution has a delay of zero. Therefore, the time-cost tradeoff problem is equivalent to finding a subset $F \subseteq V$ of jobs that are to be executed fast. The objective is to minimize the total cost of jobs in F. Note that for notational convenience we allow one of the alternatives to have infinite delay or cost, but of course such an alternative can never be chosen in a feasible solution of finite cost, and it could be as well excluded.

We call two instances I and I' of the time-cost tradeoff problem *equivalent* if any feasible solution to I can be transformed in polynomial time to a feasible solution to I' with the same cost and vice-versa.

Proposition 4 (Skutella [16]). *For any instance I of the time-cost tradeoff problem one can construct an equivalent normalized instance I' of the same depth in polynomial time.*

The structure of only allowing two execution times per job gives rise to a useful property, as we will now see. As noted above, for a normalized instance I the solutions correspond to subsets of jobs $F \subseteq V$ to be accelerated. Consider the clutter $\mathcal{C}$ of inclusion-wise minimal feasible solutions to I. Denote by $\mathcal{B} = \mathrm{bl}(\mathcal{C})$ the blocker of $\mathcal{C}$, i.e., the clutter over the same ground set V whose members are minimal subsets of jobs that have nonempty intersection with every element of $\mathcal{C}$.

Let $T > 0$ be the deadline of our normalized time-cost tradeoff instance and t_v denote the slow delay of executing job $v \in V$. By the properties of a normalized instance, the elements of $\mathcal{B}$ are the minimal chains $P \subseteq V$ with $\sum_{v \in P} t_v > T$. The well-known fact that $\mathrm{bl}(\mathrm{bl}(\mathcal{C})) = \mathcal{C}$ [9,12] immediately implies the next proposition, which also follows from an elementary calculation.

Proposition 5. *A set $F \subseteq V$ is a feasible solution to a normalized instance I of the time-cost tradeoff problem if and only if $P \cap F \neq \emptyset$ for all $P \in \mathcal{B}$.*   □

Therefore, our problem is to find a minimum-weight vertex cover in the hypergraph $(V, \mathcal{B})$. If our time-cost tradeoff instance has depth d, this hypergraph is d-partite[1] and a d-partition can be computed easily:

Proposition 6. *Given a time-cost tradeoff instance with depth d, we can partition the set of jobs in polynomial time into sets $V_1, \ldots, V_d$ (called* layers*) such that $v \prec w$ implies that $v \in V_i$ and $w \in V_j$ for some $i < j$. Then, $|P \cap V_i| \leq 1$ for all $P \in \mathcal{B}$ and $i = 1, \ldots, d$.*   □

---

[1] A hypergraph $(V, \mathcal{B})$ is d-partite if there exists a partition $V = V_1 \dot\cup V_2 \ldots \dot\cup V_d$ such that $|P \cap V_i| \leq 1$ for all $P \in \mathcal{B}$ and $i \in \{1, \ldots, d\}$. We call $\{V_1, \ldots, V_d\}$ a d-partition. We do not require the hypergraph to be d-uniform.

This also leads to a simple description as an integer linear program. The feasible solutions correspond to the vectors $x \in \{0,1\}^V$ with $\sum_{v \in P} x_v \geq 1$ for all $P \in \mathcal{B}$. We consider the following linear programming relaxation, which we call the *vertex cover LP*:

$$
\begin{aligned}
\text{minimize:} \quad & \sum_{v \in V} c_v \cdot x_v \\
\text{subject to:} \quad & \sum_{v \in P} x_v \geq 1 \qquad \text{for all } P \in \mathcal{B} \qquad (1) \\
& x_v \geq 0 \qquad\qquad \text{for all } v \in V.
\end{aligned}
$$

Let LP denote the value of this linear program (for a given instance). It is easy to see that the PARTITION problem reduces to the separation problem of this linear program. Hence:

Proposition 7. *If the vertex cover LP* (1) *can be solved in polynomial time for normalized time-cost tradeoff instances, then* P = NP.

However, we can solve the LP up to an arbitrarily small error; in fact, there is a fully polynomial approximation scheme (as essentially shown by [13]):

Proposition 8. *For normalized instances of the time-cost tradeoff problem with bounded depth, the vertex cover LP* (1) *can be solved in polynomial time. For general normalized instances and any given $\epsilon > 0$, a feasible solution of cost at most $(1 + \epsilon)$LP can be found in time bounded by a polynomial in n and $\frac{1}{\epsilon}$.*

We remark that the d-partite hypergraph vertex cover instances given by [1] can be also considered as normalized instances of the time-cost tradeoff problem. This shows that the integrality gap of LP (1) is at least $\frac{d}{2}$.

Since $|P| \leq d$ for all $P \in \mathcal{B}$, the Bar-Yehuda–Even algorithm [3] can be used to find an integral solution to the time-cost tradeoff instance of cost at most $d \cdot$ LP, and can be implemented to run in polynomial time because for integral vectors x there is a linear-time separation oracle [5]. A d-approximation can also be obtained by rounding up all $x_v \geq \frac{1}{d}$. In the following we will improve on this.

4   Rounding Fractional Vertex Covers in d-Partite Hypergraphs

In this section, we show how to round a fractional vertex cover in a d-partite hypergraph with given d-partition. Together with the results of the previous section, this yields an approximation algorithm for time-cost tradeoff instances and will prove Theorem 1.

Our algorithm does not need an explicit list of the edge set of the hypergraph, which is interesting if d is not constant and there can be exponentially many hyperedges. The algorithm only requires the vertex set, a d-partition, and a feasible solution to the LP (a fractional vertex cover). For normalized instances of the time-cost tradeoff problem such a fractional vertex cover can be obtained as in Proposition 8, and a d-partition by Proposition 6.

Our algorithm builds on two previous works for the unweighted d-partite hypergraph vertex cover problem. For rounding a given fractional solution, Lovász [15] obtained a deterministic polynomial-time $(\frac{d}{2} + \epsilon)$-approximation algorithm for any $\epsilon > 0$. Based on this, Aharoni, Holzman and Krivelevich [1] described a randomized recursive algorithm that works in more general unweighted hypergraphs. We simplify their algorithm for d-partite hypergraphs, which will allow us to obtain a deterministic polynomial-time algorithm that also works for the weighted problem and always computes a $\frac{d}{2}$-approximation. At the end of this section, we will slightly improve on this guarantee in order to compensate for an only approximate LP solution.

We will first describe the algorithm in the even simpler randomized form. This algorithm computes a random threshold for each layer to determine whether a variable x_v is rounded up or down. We will use the following probability distribution, which is easily seen to exist:

Lemma 9. *For any $d \geq 2$, there is a probability distribution that selects $a_1, \ldots, a_d$, such that $\sum_{i=1}^{d} a_i = 1$ and a_i is uniformly distributed in $[\frac{2(i-1)}{d^2}, \frac{2i}{d^2}]$. For any i, j such that $|i - j| \geq 3$, the random variables corresponding to a_i and a_j are independent.*

Theorem 10. *Let x be a fractional vertex cover in a d-partite hypergraph with given d-partition. There is a randomized linear-time algorithm that computes an integral solution $\bar{x}$ of expected cost $\mathbb{E}[\sum_{v \in V} c_v \cdot \bar{x}_v] \leq \frac{d}{2} \sum_{v \in V} c_v \cdot x_v$.*

Proof. Let $V_1, \ldots, V_d$ be the given d-partition of our hypergraph $(V, \mathcal{B})$, so $|P \cap V_i| \leq 1$ for all $i = 1, \ldots, d$ and every hyperedge $P \in \mathcal{B}$. We write $l(v) = i$ if $v \in V_i$ and call V_i a *layer* of the given hypergraph.

Now consider the following randomized algorithm, which is also illustrated in Fig. 1: Choose a random permutation $\sigma : \{1, \ldots, d\} \to \{1, \ldots, d\}$ and choose random numbers a_i uniformly distributed in $[\frac{2(\sigma(i)-1)}{d^2}, \frac{2\sigma(i)}{d^2}]$ for $i = 1, \ldots, d$ such that $\sum_{i=1}^{d} a_i = 1$, as stated in Lemma 9. Then, for all $v \in V$, set $\bar{x}_v := 1$ if $x_v \geq a_{l(v)}$ and $\bar{x}_v := 0$ if $x_v < a_{l(v)}$.

To show that $\bar{x}$ is a feasible solution, observe that any hyperedge $P \in \mathcal{B}$ has $\sum_{v \in P} x_v \geq 1 = \sum_{i=1}^{d} a_i \geq \sum_{v \in P} a_{l(v)}$ and hence $x_v \geq a_{l(v)}$ for some $v \in P$.

It is also easy to see that the probability that $\bar{x}_v$ is set to 1 is exactly $\min\{1, \frac{d}{2}x_v\}$. Indeed, if $x_v \geq \frac{2}{d}$, we surely set $\bar{x}_v = 1$. Otherwise, $x_v \in [\frac{2(j-1)}{d^2}, \frac{2j}{d^2}]$ for some $j \in \{1, \ldots, d\}$; then we set $\bar{x}_v = 1$ if and only if $\sigma(l(v)) < j$ or $(\sigma(l(v)) = j$ and $a_{l(v)} \leq x_v)$, which happens with probability $\frac{j-1}{d} + \frac{1}{d}(x_v - \frac{2(j-1)}{d^2})\frac{d^2}{2} = \frac{d}{2}x_v$. Hence the expected cost $\mathbb{E}[\sum_{v \in V} c_v \cdot \bar{x}_v]$ is at most $\frac{d}{2} \sum_{v \in V} c_v \cdot x_v$. □

This algorithm can be derandomized using bipartite matching to compute σ:

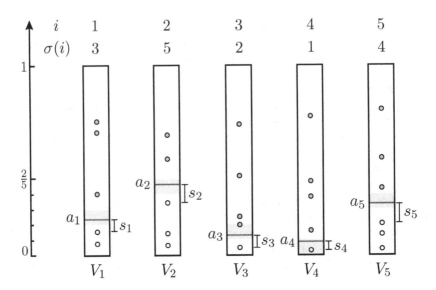

Fig. 1. A sketch of thresholds $a_1, \ldots, a_5$ chosen by our randomized algorithm in Theorem 10 for the case $d = 5$. The circles represent vertices in the hypergraph, drawn by their position in the partition and the value of their corresponding variable in the LP. Suppose the permutation $(\sigma(1), \ldots, \sigma(5)) = (3, 5, 2, 1, 4)$ is chosen. Then the thresholds a_i are randomly chosen in the light blue intervals $\left[\frac{2(\sigma(i)-1)}{d^2}, \frac{2\sigma(i)}{d^2}\right]$; moreover, the thresholds a_1, a_3, a_4 are chosen independently of the thresholds a_2, a_5, as indicated by their color. The points above the thresholds are filled; these variables are rounded up to 1, while the empty circles represent variables that are rounded down to 0. Finally, the figure also shows "slack" values $s_1, \ldots, s_5$, telling how much each threshold could be lowered without changing the solution returned by our algorithm. These will play a key role to improve the approximation guarantee in Theorem 12. (Color figure online)

Theorem 11. *Let x be a fractional vertex cover in a d-partite hypergraph with given d-partition. There is a deterministic algorithm that computes an integral solution $\bar{x}$ of cost $\sum_{v \in V} c_v \cdot \bar{x}_v \leq \frac{d}{2} \sum_{v \in V} c_v \cdot x_v$ in time $\mathcal{O}(n^3)$.*

We omit the proof here. In order to obtain a true $\frac{d}{2}$-approximation algorithm (and thus prove Theorem 1), we need a slightly stronger bound.

Theorem 12. *Let $d \geq 4$. Let x be a fractional vertex cover in a d-partite hypergraph with given d-partition. There is a randomized linear-time algorithm that computes an integral solution $\bar{x}$ of expected cost $\sum_{v \in V} c_v \cdot \bar{x}_v \leq \left(\frac{d}{2} - \frac{d}{64n}\right) \sum_{v \in V} c_v \cdot x_v$.*

Proof. First we choose the permutation σ and thresholds $a_1, \ldots, a_d$ with sum $\sum_{i=1}^{d} a_i = 1$ randomly as above such that the thresholds are independent except within groups of two or three. For $i \in \{1, \ldots, d\}$ denote the *slack* of level i by $s_i := \min\{\frac{1}{d}, a_i, a_i - \max\{x_v : v \in V_i, x_v < a_i\}\}$. The slack is always non-negative. Lowering the threshold a_i by less than s_i would yield the same solution $\bar{x}$. The reason for cutting off the slack at $\frac{1}{d}$ will become clear only below.

Next we randomly select one level $\lambda \in \{1, \ldots, d\}$. Let Λ be the corresponding group (cf. Lemma 9), i.e., $\lambda \in \Lambda \subseteq \{1, \ldots, d\}$, $|\Lambda| \leq 3$, and a_i is independent of a_λ whenever $i \notin \Lambda$. Now raise the threshold a_λ to $a'_\lambda = a_\lambda + \sum_{i \notin \Lambda} s_i$. Set $a'_i = a_i$ for $i \in \{1, \ldots, d\} \setminus \{\lambda\}$.

As before, for all $v \in V$, set $\bar{x}_v := 1$ if $x_v \geq a'_{l(v)}$ and $\bar{x}_v := 0$ if $x_v < a'_{l(v)}$. We first observe that $\bar{x}$ is feasible. Indeed, if there were any hyperedge $P \in \mathcal{B}$ with $x_v < a'_{l(v)}$ for all $v \in P$, we would get $1 \leq \sum_{v \in P} x_v < \sum_{v \in P: l(v) \notin \Lambda} (a_{l(v)} - s_{l(v)}) + \sum_{v \in P: l(v) \in \Lambda} a'_\lambda \leq \sum_{i \notin \Lambda} (a_i - s_i) + \sum_{i \in \Lambda \setminus \{\lambda\}} a_i + a'_\lambda = \sum_{i=1}^{d} a_i = 1$, a contradiction.

We now bound the expected cost of $\bar{x}$. Let $v \in V$. With probability $\frac{d-1}{d}$ we have $l(v) \neq \lambda$ and, conditioned on this, an expectation $\mathbb{E}[\bar{x}_v \mid \lambda \neq l(v)] = \frac{d}{2} \min\{x_v, \frac{2}{d}\} \leq \frac{d}{2} x_v$ as before. Now we condition on $l(v) = \lambda$ and in addition, for any S with $0 \leq S \leq \frac{d-2}{d}$, on $\sum_{i \notin \Lambda} s_i = S$; note that a_λ is independent of S. The probability that $\bar{x}_v$ is set to 1 is $\frac{d}{2} \max\{0, \min\{x_v - S, \frac{2}{d}\}\} \leq \frac{x_v}{\frac{2}{d}+S} \leq \frac{d}{2}(1-S)x_v$ in this case. In the last inequality we used $S \leq \frac{d-2}{d}$, and this was the reason to cut off the slacks. In total we have for all $v \in V$:

$$
\begin{aligned}
\mathbb{E}[\bar{x}_v] &= \frac{d-1}{d} \cdot \mathbb{E}[\bar{x}_v \mid \lambda \neq l(v)] \\
&\quad + \frac{1}{d} \cdot \int_0^{\frac{d-2}{d}} \mathbb{P}\left[\sum_{i \notin \Lambda} s_i = S \mid \lambda = l(v)\right] \cdot \mathbb{E}\left[\bar{x}_v \mid \lambda = l(v), \sum_{i \notin \Lambda} s_i = S\right] dS \\
&\leq \frac{d-1}{d} \cdot \frac{d}{2} x_v + \frac{1}{d} \cdot \int_0^{\frac{d-2}{d}} \mathbb{P}\left[\sum_{i \notin \Lambda} s_i = S \mid \lambda = l(v)\right] \cdot \frac{d}{2}(1-S)x_v \, dS \\
&\leq \frac{d}{2}\left(1 - \frac{1}{d}\int_0^{\frac{d-2}{d}} \mathbb{P}\left[\sum_{i \notin \Lambda} s_i = S \mid \lambda = l(v)\right] \cdot S \, dS\right) \cdot x_v \\
&= \frac{d}{2}\left(1 - \frac{1}{d} \cdot \mathbb{E}[S \mid \lambda = l(v)]\right) x_v.
\end{aligned}
$$

Let $\Lambda[v]$ be the set Λ in the event $\lambda = l(v)$. We estimate

$$
\mathbb{E}[S \mid \lambda = l(v)] = \sum_{i \notin \Lambda(v)} \mathbb{E}[s_i] \geq \sum_{i \notin \Lambda(v)} \frac{1}{d(n_i + 1)} \geq \frac{(d-3)^2}{d(n+d)} \geq \frac{d}{32n}.
$$

Here $n_i = |V_i|$, and the first inequality holds because $\mathbb{E}[s_i]$ is maximal if $\{x_v : v \in V_i\} = \{\frac{2j}{d(n_i+1)} : j = 1, \ldots, n_i\}$. We conclude $\mathbb{E}\left[\sum_{v \in V} c_v \cdot \bar{x}_v\right] \leq (\frac{d}{2} - \frac{d}{64n}) \sum_{v \in V} c_v \cdot x_v$. $\qquad \square$

Let us now derandomize this algorithm. This is easy as we can afford to lose a little again.

Theorem 13. *Let $d \geq 4$. Let x be a fractional vertex cover in a d-partite hyper-graph with given d-partition. There is a deterministic algorithm that computes an integral solution $\bar{x}$ of cost $\sum_{v \in V} c_v \cdot \bar{x}_v \leq (\frac{d}{2} - \frac{d}{128n}) \sum_{v \in V} c_v \cdot x_v$ in time $\mathcal{O}(n^3)$.*

Proof. (Sketch) Round down the costs to integer multiples of $\frac{d \, \text{LP}}{128n^2}$ and compute the best possible choice of threshold values a_i for $i \in \{1, \ldots, d\}$ such that $\sum_{j=1}^{d} a_j \leq 1$ and $\sum_{j=1}^{d} \sum_{v \in V_j, x_v \geq a_j} c_v'$ is minimized; this is a simple dynamic program. By Theorem 12 there is such a solution with cost $\sum_{j=1}^{d} \sum_{v \in V_j, x_v \geq a_j} c_v \leq (\frac{d}{2} - \frac{d}{64n}) \text{LP}$. $\qquad\qquad\qquad\square$

As explained above, together with Propositions 6 and 8 (with $\epsilon = \frac{1}{128n}$), Theorem 13 implies Theorem 1.

5   Inapproximability

Guruswami, Sachdeva and Saket [11] proved that approximating the vertex cover problem in d-partite hypergraphs with a better ratio than $\frac{d}{2}$ is NP-hard under the Unique Games Conjecture. We show that even for the special case of time-cost tradeoff instances[2], the problem is hard to approximate by a factor of $\frac{d+2}{4}$. (Theorem 2). Instead of starting from k-uniform hypergraphs like [11], we devise a reduction from the vertex deletion problem in acyclic digraphs, which Svensson [17] called DVD.[3] Let k be a positive integer; then DVD(k) is defined as follows: given an acyclic digraph, compute a minimum-cardinality set of vertices whose deletion destroys all paths with k vertices. This problem is easily seen to admit a k-approximation algorithm.

Svensson proved that anything better than this simple approximation algorithm would solve the unique games problem:

Theorem 14 ([17]). *Let $k \in \mathbb{N}$ with $k \geq 2$ and $\rho < k$ be constants. Let OPT denote the size of an optimum solution for a given DVD(k) instance. Assuming the Unique Games Conjecture it is NP-hard to compute a number $l \in \mathbb{R}_+$ such that $l \leq OPT \leq \rho l$.*

This is the starting point of our proof. Svensson [17] already observed that DVD(k) can be regarded as a special case of the time-cost tradeoff problem. Note that this does not imply Theorem 2 because the hard instances of DVD(k) constructed in the proof of Theorem 14 have unbounded depth even for fixed k. (Recall that the *depth* of an acyclic digraph is the number of vertices in a longest path.) The following is a variant (and slight strengthening) of Svensson's observation.

---

[2] Note that this is really a special case: for example the 3-partite hypergraph with vertex set $\{1, 2, 3, 4, 5, 6\}$ and hyperedges $\{1, 4, 6\}, \{2, 3, 6\}$, and $\{2, 4, 5\}$ does not result from a time-cost tradeoff instance of depth 3 with our construction.

[3] An undirected version of this problem has been called k-*path vertex cover*[4] or *vertex cover* P_k [18].

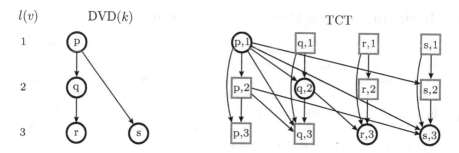

Fig. 2. Lemma 15: an instance of DVD(k) is transformed into an equivalent instance of the time-cost tradeoff problem. Blue squares represent jobs with fixed execution time. (Color figure online)

Lemma 15. *Any instance of* DVD(k) *(for any k) can be transformed in linear time to an equivalent instance of the time-cost tradeoff problem, with the same depth and the same optimum value.*

Proof. Let $G = (V, E)$ be an instance of DVD(k), an acyclic digraph, say of depth d. Let $l(v) \in \{1, \ldots, d\}$ for $v \in V$ such that $l(v) < l(w)$ for all $(v, w) \in E$. Let $J := \{(v, i) : v \in V, i \in \{1, \ldots, d\}\}$ be the set of jobs of our time-cost tradeoff instance. Job (v, i) must precede job (w, j) if ($v = w$ and $i < j$) or (($v, w) \in E$ and $l(v) \le i < j$). Let $\prec$ be the transitive closure of these precedence constraints. For $v \in V$, the job $(v, l(v))$ is called *variable* and has a fast execution time 0 at cost 1 and a slow execution time $d + 1$ at cost 0. All other jobs are *fixed*; they have a fixed execution time d at cost 0. The deadline is $d^2 + k - 1$. See Fig. 2. Then, any set of variable jobs whose acceleration constitutes a feasible solution of this time-cost tradeoff instance corresponds to a set of vertices whose deletion destroys all paths in G with k vertices, and vice versa. □

Therefore a hardness result for DVD(k) for bounded depth instances transfers to a hardness result for the time-cost tradeoff problem with bounded depth. We will show the following strengthening of Theorem 14:

Theorem 16. *Let $k, d \in \mathbb{N}$ with $2 \le k \le d$ and $\rho < \frac{k(d+1-k)}{d}$ be constants. Let OPT denote the size of an optimum solution for a given DVD(k) instance. Assuming the Unique Games Conjecture it is NP-hard to compute a number $l \in \mathbb{R}_+$ such that $l \le OPT \le \rho l$.*

It is easy to see that Theorem 16 and Lemma 15 imply Theorem 2. Indeed, let $d \in \mathbb{N}$ with $d \ge 2$ and $\rho < \frac{d+2}{4}$, and suppose that a ρ-approximation algorithm $\mathcal{A}$ exists for time-cost tradeoff instances of depth d. Let $k := \lceil \frac{d+1}{2} \rceil$ and consider an instance of DVD(k) with depth d. Transform this instance to an equivalent time-cost tradeoff instance by Lemma 15 and apply algorithm $\mathcal{A}$. This constitutes a ρ-approximation algorithm for DVD(k) with depth d. Since $\rho < \frac{d+2}{4} \le \frac{k(d+1-k)}{d}$, Theorem 16 then implies that the Unique Games Conjecture is false or P = NP.

It remains to prove Theorem 16, which will be the subject of the next section.

6 Reducing Vertex Deletion to Constant Depth

In this section we prove Theorem 16. The idea is to reduce the depth of a digraph by transforming it to another digraph with small depth but related vertex deletion number. Let $k, d \in \mathbb{N}$ with $2 \le k \le d$, and let G be a digraph. We construct an acyclic digraph G^d of depth at most d by taking the tensor product with the acyclic tournament on d vertices: $G^d = (V^d, E^d)$, where $V^d = V \times \{1, \ldots, d\}$ and $E^d = \{((v,i),(w,j)) : (v,w) \in E \text{ and } i < j\}$. It is obvious that G^d has depth d. Here is our key lemma:

Lemma 17. *Let G be an acyclic directed graph and $k, d \in \mathbb{N}$ with $2 \le k \le d$. If we denote by $OPT(G, k)$ the minimum number of vertices of G hitting all paths with k vertices, then*

$$(d + 1 - k) \cdot OPT(G, k) \le OPT(G^d, k) \le d \cdot OPT(G, k). \qquad (2)$$

Lemma 17, together with Theorem 14, immediately implies Theorem 16: assuming a ρ-approximation algorithm for $DVD(k)$ instances with depth d, with $\rho < \frac{k(d+1-k)}{d}$, we can compute $OPT(G, k)$ up to a factor less than k for any digraph G. By Theorem 14, this would contradict the Unique Games Conjecture or $P \ne NP$. One can show that the bounds are sharp.

Proof. (Lemma 17) Let G be an acyclic digraph. The upper bound of (2) is trivial: for any set $W \subseteq V$ that hits all k-vertex paths in G we can take $X := W \times \{1, \ldots, d\}$ to obtain a solution to the $DVD(k)$ instance G^d.

To show the lower bound, we fix a minimal solution X to the $DVD(k)$ instance G^d. Let Q be a path in G^d with at most k vertices. We write $\text{start}(Q) = i$ if Q begins in a vertex (v, i). We define $\mathcal{Q}$ as the set of paths in G^d with exactly k vertices. For $Q \in \mathcal{Q}$ let $\text{lasthit}(Q)$ denote the last vertex of Q that belongs to X. For $x \in X$ we define

$$\varphi(x) := \max\{\text{start}(Q) : Q \in \mathcal{Q}, \text{lasthit}(Q) = x\}.$$

Note that this is well-defined due to the minimality of X, and $1 \le \varphi(x) \le d+1-k$ for all $x \in X$. We will show that for $j = 1, \ldots, d+1-k$,

$$S_j := \{v \in V : (v, i) \in X \text{ and } \varphi((v, i)) = j \text{ for some } i \in \{1, \ldots, d\}\}$$

hits all k-vertex paths in G. This shows the lower bound in (2) because then $OPT(G, k) \le \min_{j=1}^{d+1-k} |S_j| \le \frac{|X|}{d+1-k}$.

Let P be a path in G with k vertices $v_1, \ldots, v_k$ in this order. Consider d "diagonal" copies $D_1, \ldots, D_d$ of (suffixes of) P in G^d: the path D_i consists of the vertices $(v_s, s + i - k), \ldots, (v_k, i)$, where $s = \max\{1, k+1-i\}$. Note that the paths $D_1, \ldots, D_{k-1}$ have fewer than k vertices.

We show that for each $j = 1, \ldots, d + 1 - k$, at least one of these diagonal paths contains a vertex $x \in X$ with $\varphi(x) = j$. This implies that $S_j \cap P \ne \emptyset$ and concludes the proof.

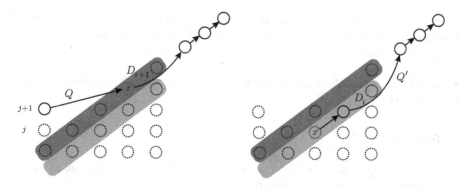

Fig. 3. The Claim in the proof of Lemma 17 asserts that if D_{i+1} contains a vertex $x \in X$ with $\varphi(x) = j + 1$, then D_i contains a vertex $x' \in X$ with $\varphi(x') \geq j$. The upper diagonal D_{i+1} is colored in light green, the lower diagonal D_i is depicted in dark green. We start by selecting a path Q with lasthit$(Q) \in D_{i+1}$ and start$(Q) = j + 1$. This path is depicted on the left; the vertex $x = $ lasthit(Q) is highlighted in red. We construct a path Q' (shown on the right) such that $x' = $ lasthit$(Q') \in D_i$ and start$(Q') = $ start$(Q) - 1$. This path Q' results from appending the end of path Q to an appropriate subpath of the next lower diagonal D_i. (Color figure online)

First, D_d contains a vertex in $x \in X$ with $\varphi(x) = d+1-k$, namely lasthit(D_d). Now we show for $i = 1, \ldots, d - 1$ and $j = 1, \ldots, d - k$:

Claim: If D_{i+1} contains a vertex $x \in X$ with $\varphi(x) = j + 1$, then D_i contains a vertex $x' \in X$ with $\varphi(x') \geq j$.

This Claim implies the theorem because D_1 consists of a single vertex $(v_k, 1)$, and if it belongs to X, then $\varphi((v_k, 1)) = 1$.

To prove the Claim (see Fig. 3 for an illustration), let $x = (v_h, l(x)) \in X \cap D_{i+1}$ and $\varphi(x) \geq j+1$, and let x be the last such vertex on D_{i+1}. We have $\varphi(x) \geq$ start(D_{i+1}) for otherwise we have start$(D_{i+1}) > 1$, so D_{i+1} contains k vertices and we should have chosen $x = $ lasthit(D_{i+1}); note that $\varphi($lasthit$(D_{i+1})) \geq$ start(D_{i+1}).

Let $Q \in \mathcal{Q}$ be a path attaining the maximum in the definition of $\varphi(x)$. So start$(Q) = \varphi(x)$ and lasthit$(Q) = x$. Suppose x is the p-th vertex of Q; then

$$p \leq 1 + l(x) - \varphi(x) \tag{3}$$

because Q starts on level $\varphi(x)$, rises at least one level with every vertex, and reaches level $l(x)$ at its p-th vertex.

Now consider the following path Q'. It begins with part of the diagonal D_i, namely $(v_{h+1-p}, l(x)-p), \ldots, (v_h, l(x)-1)$, and continues with the $k-p$ vertices from the part of Q after x. Note that by (3)

$$l(x) - p \geq \varphi(x) - 1 \geq \max\{j, \text{start}(D_{i+1}) - 1\} \geq \max\{1, \text{start}(D_i)\},$$

so Q' is well-defined. The second part of Q' does not contain any vertex from X because lasthit$(Q) = x$. Hence $x' := \text{lasthit}(Q')$ is in the diagonal part of Q', i.e., in D_i. By definition, $\varphi(x') \geq \text{start}(Q') = l(x) - p \geq j$.   □

Acknowledgement. We thank Nikhil Bansal for fruitful discussions at an early stage of this project.

References

1. Aharoni, R., Holzman, R., Krivelevich, M.: On a theorem of Lovász on covers in r-partite hypergraphs. Combinatorica **16**(2), 149–174 (1996)
2. Alimonti, P., Kann, V.: Some APX-completeness results for cubic graphs. Theoret. Comput. Sci. **237**(1–2), 123–134 (2000)
3. Bar-Yehuda, R., Even, S.: A linear-time approximation algorithm for the weighted vertex cover problem. J. Algorithms **2**(2), 198–203 (1981)
4. Brešar, B., Kardoš, F., Katrenič, J., Semanišin, G.: Minimum k-path vertex cover. Discrete Appl. Math. **159**(12), 1189–1195 (2011)
5. Daboul, S., Held, S., Vygen, J., Wittke, S.: An approximation algorithm for threshold voltage optimization. TODAES **23**(6), 1–16 (2018). Article no. 68
6. De, P., Dunne, E.J., Ghosh, J.B., Wells, C.E.: Complexity of the discrete time-cost tradeoff problem for project networks. Oper. Res. **45**(2), 302–306 (1997)
7. Deĭneko, V.G., Woeginger, G.J.: Hardness of approximation of the discrete time-cost tradeoff problem. OR Lett. **29**(5), 207–210 (2001)
8. Dinur, I., Safra, S.: On the hardness of approximating minimum vertex cover. Ann. Math. **162**(1), 439–485 (2005)
9. Edmonds, J., Fulkerson, D.R.: Bottleneck extrema. J. Comb. Theory **8**(3), 299–306 (1970)
10. Grigoriev, A., Woeginger, G.J.: Project scheduling with irregular costs: complexity, approximability, and algorithms. Acta Inf. **41**(2), 83–97 (2004)
11. Guruswami, V., Sachdeva, S., Saket, R.: Inapproximability of minimum vertex cover on k-uniform k-partite hypergraphs. SIAM J. Discrete Math. **29**(1), 36–58 (2015)
12. Isbell, J.R.: A class of simple games. Duke Math. J. **25**(3), 423–439 (1958)
13. Karmarkar, N., Karp, R.M.: An efficient approximation scheme for the one-dimensional bin-packing problem. In: FOCS 1982, pp. 312–320 (1982)
14. Kelley, J.E., Walker, M.R.: Critical-path planning and scheduling. In: Proceedings of the AIEE-ACM 1959, pp. 160–173 (1959)
15. Lovász, L.: On minmax theorems of combinatorics. Mathematikai Lapok **26**, 209–264 (1975). Doctoral thesis (in Hungarian)
16. Skutella, M.: Approximation algorithms for the discrete time-cost tradeoff problem. Math. Oper. Res. **23**(4), 909–929 (1998)
17. Svensson, O.: Hardness of vertex deletion and project scheduling. Theory Comput. **9**(24), 759–781 (2013)
18. Tu, J., Zhou, W.: A primal-dual approximation algorithm for the vertex cover P3 problem. Theoret. Comput. Sci. **412**(50), 7044–7048 (2011)

Sum-of-Squares Hierarchies for Binary Polynomial Optimization

Lucas Slot[1]([✉]) and Monique Laurent[1,2]

[1] Centrum Wiskunde & Informatica (CWI), Amsterdam, The Netherlands
{lucas,monique}@cwi.nl
[2] Tilburg University, Tilburg, The Netherlands

Abstract. We consider the sum-of-squares hierarchy of approximations for the problem of minimizing a polynomial f over the boolean hypercube $\mathbb{B}^n = \{0,1\}^n$. This hierarchy provides for each integer $r \in \mathbb{N}$ a lower bound $f_{(r)}$ on the minimum $f_{\min}$ of f, given by the largest scalar λ for which the polynomial $f - \lambda$ is a sum-of-squares on $\mathbb{B}^n$ with degree at most $2r$. We analyze the quality of these bounds by estimating the worst-case error $f_{\min} - f_{(r)}$ in terms of the least roots of the Krawtchouk polynomials. As a consequence, for fixed $t \in [0, 1/2]$, we can show that this worst-case error in the regime $r \approx t \cdot n$ is of the order $1/2 - \sqrt{t(1-t)}$ as n tends to ∞. Our proof combines classical Fourier analysis on $\mathbb{B}^n$ with the polynomial kernel technique and existing results on the extremal roots of Krawtchouk polynomials. This link to roots of orthogonal polynomials relies on a connection between the hierarchy of lower bounds $f_{(r)}$ and another hierarchy of upper bounds $f^{(r)}$, for which we are also able to establish the same error analysis. Our analysis extends to the minimization of a polynomial over the q-ary cube $(\mathbb{Z}/q\mathbb{Z})^n$.

Keywords: Binary polynomial optimization · Lasserre hierarchy ·
Sum-of-squares polynomials · Fourier analysis · Krawtchouk
polynomials · Polynomial kernels · Semidefinite programming

1 Introduction

We consider the problem of minimizing a polynomial $f \in \mathbb{R}[x]$ of degree $d \leq n$ over the n-dimensional boolean hypercube $\mathbb{B}^n = \{0,1\}^n$, i.e., of computing

$$f_{\min} := \min_{x \in \mathbb{B}^n} f(x). \tag{1}$$

This optimization problem is NP-hard in general, already for $d = 2$. Indeed, as is well-known, one can model an instance of MAX-CUT on the complete graph K_n with edge weights $w = (w_{ij})$ as a problem of the form (1) by setting:

$$f(x) = -\sum_{1 \leq i < j \leq n} w_{ij}(x_i - x_j)^2,$$

© Springer Nature Switzerland AG 2021
M. Singh and D. P. Williamson (Eds.): IPCO 2021, LNCS 12707, pp. 43–57, 2021.
https://doi.org/10.1007/978-3-030-73879-2_4

As another example one can compute the stability number $\alpha(G)$ of a graph $G = (V, E)$ via the program

$$\alpha(G) = \max_{x \in \mathbb{B}^{|V|}} \sum_{i \in V} x_i - \sum_{\{i,j\} \in E} x_i x_j.$$

One may replace the boolean cube $\mathbb{B}^n = \{0,1\}^n$ by the discrete cube $\{\pm 1\}^n$, in which case maximizing a quadratic polynomial $x^T A x$ has many other applications, e.g., to MAX-CUT [13], to the cut norm [1], or to correlation clustering [4]. Approximation algorithms are known depending on the structure of the matrix A (see [1,6,13]), but the problem is known to be NP-hard to approximate within any factor less than $13/11$ [2].

Problem (1) also permits to capture polynomial optimization over a general region of the form $\mathbb{B}^n \cap P$ where P is a polyhedron [17] and thus a broad range of combinatorial optimization problems. The general intractability of problem (1) motivates the search for tractable bounds on the minimum value in (1). For this, several lift-and-project methods have been proposed, based on lifting the problem to higher dimension by introducing new variables modelling higher degree monomials. Such methods also apply to constrained problems on $\mathbb{B}^n$ where the constraints can be linear or polynomial; see, e.g., [3,18,27,33,36,40]. In [21] it is shown that the sum-of-squares hierarchy of Lasserre [18] in fact refines the other proposed hierarchies. As a consequence the sum-of-squares approach for polynomial optimization over $\mathbb{B}^n$ has received a great deal of attention in the recent years and there is a vast literature on this topic. Among many other results, let us just mention its use to show lower bounds on the size of semidefinite programming relaxations for combinatorial problems such as max-cut, maximum stable sets and TSP in [25], and the links to the Unique Game Conjecture in [5]. For background about the sum-of-squares hierarchy applied to polynomial optimization over general semi-algebraic sets we refer to [16,19,23,30] and further references therein.

This motivates the interest in gaining a better understanding of the quality of the bounds produced by the sum-of-squares hierarchy. Our objective in this paper is to investigate such an error analysis for this hierarchy applied to binary polynomial optimization as in (1).

1.1   The Sum-of-Squares Hierarchy on the Boolean Cube

The *sum-of-squares hierarchy* was introduced by Lasserre [16,18] and Parrilo [30] as a tool to produce tractable lower bounds for polynomial optimization problems. When applied to problem (1) it provides for any integer $r \in \mathbb{N}$ a lower bound $f_{(r)} \leq f_{\min}$ on $f_{\min}$, given by:

$$f_{(r)} := \sup_{\lambda \in \mathbb{R}} \{ f(x) - \lambda \text{ is a sum-of-squares of degree at most } 2r \text{ on } \mathbb{B}^n \}. \quad (2)$$

Throughout, Σ_r denotes the set of sum-of-squares polynomials with degree at most $2r$, i.e., of the form $\sum_i p_i^2$ with $p_i \in \mathbb{R}[x]_r$. In program (2), the condition

'$f(x) - \lambda$ is a sum-of-squares of degree at most $2r$ on $\mathbb{B}^n$' means that there exists a sum-of-squares polynomial $s \in \Sigma_r$ such that $f(x) - \lambda = s(x)$ for all $x \in \mathbb{B}^n$, or, equivalently, that the polynomial $f - \lambda - s$ belongs to the ideal generated by the polynomials $x_1 - x_1^2, \ldots, x_n - x_n^2$.

As sums of squares of polynomials can be modelled using semidefinite programming, problem (2) can be reformulated as a semidefinite program of size polynomial in n for fixed r [16, 30]. In the case of unconstrained boolean optimization, the resulting semidefinite program is known to have an optimum solution with small coefficients (see [29] and [31]). For fixed r, the parameter $f_{(r)}$ may therefore be computed efficiently (up to any precision).

The bounds $f_{(r)}$ have finite convergence: $f_{(r)} = f_{\min}$ for $r \geq n$ [18]. In fact, it has been shown in [34] that the bound $f_{(r)}$ is exact already for $2r \geq n + d - 1$. That is,

$$f_{(r)} = f_{\min} \text{ for } r \geq \frac{n + d - 1}{2}. \tag{3}$$

In addition, it is shown in [34] that the bound $f_{(r)}$ is exact for $2r \geq n + d - 2$ when the polynomial f has only monomials of even degree. This extends an earlier result of [12] shown for quadratic forms ($d = 2$), which applies in particular to the case of MAX-CUT. Furthermore, this result is tight for MAX-CUT, since one needs to go up to order $2r \geq n$ in order to reach finite convergence (in the cardinality case when all edge weights are 1) [22]. Similarly, the result (3) is tight when d is even and n is odd [15].

The main contribution of this work is an analysis of the quality of the bounds $f_{(r)}$ when $2r < n + d - 1$. The following is our main result, which expresses the error of the bound $f_{(r)}$ in terms of the least roots of Krawtchouk polynomials.

Theorem 1. *Fix $d \leq n$ and let $f \in \mathbb{R}[x]$ be a polynomial of degree d. For $r, n \in \mathbb{N}$, let ξ_r^n be the least root of the degree r Krawtchouk polynomial (11) with parameter n. Then, if $(r + 1)/n \leq 1/2$ and $d(d + 1) \cdot (\xi_{r+1}^n/n) \leq 1/2$, we have:*

$$\frac{f_{\min} - f_{(r)}}{\|f\|_\infty} \leq 2C_d \cdot \xi_{r+1}^n/n. \tag{4}$$

Here $C_d > 0$ is an absolute constant depending only on d and we set $\|f\|_\infty := \max_{x \in \mathbb{B}^n} |f(x)|$.

The extremal roots of Krawtchouk polynomials are well-studied in the literature. The following result of Levenshtein [26] shows their asymptotic behaviour.

Theorem 2 ([26], Sect. 5). *For $t \in [0, 1/2]$, define the function*

$$\varphi(t) = 1/2 - \sqrt{t(1 - t)}. \tag{5}$$

Then the least root ξ_r^n of the degree r Krawtchouk polynomial with parameter n satisfies

$$\xi_r^n/n \leq \varphi(r/n) + c \cdot (r/n)^{-1/6} \cdot n^{-2/3} \tag{6}$$

for some universal constant $c > 0$.

Applying (6) to (4), we find that the relative error of the bound $f_{(r)}$ in the regime $r \approx t \cdot n$ behaves as the function $\varphi(t) = 1/2 - \sqrt{t(1-t)}$, up to a noise term in $O(1/n^{2/3})$, which vanishes as n tends to ∞.

1.2   A Second Hierarchy of Bounds

In addition to the *lower* bound $f_{(r)}$, Lasserre [20] also defines an *upper* bound $f^{(r)} \geq f_{\min}$ on $f_{\min}$ as follows:

$$f^{(r)} := \inf_{s \in \Sigma_r} \left\{ \int_{\mathbb{B}^n} f(x) \cdot s(x) d\mu(x) : \int_{\mathbb{B}^n} s(x) d\mu(x) = 1 \right\}, \qquad (7)$$

where μ is the uniform probability measure on $\mathbb{B}^n$. For fixed r, similarly to $f_{(r)}$, one may compute $f^{(r)}$ (up to any precision) efficiently by reformulating (7) as a semidefinite program [20]. Furthermore, as shown in [20] the bound is exact for some order r, and it is not difficult to see that the bound $f^{(r)}$ is exact at order $r = n$ and that this is tight.

Essentially as a side result in the proof of our main Theorem 1, we get the following analog of Theorem 1 for the upper bounds $f^{(r)}$.

Theorem 3. *Fix $d \leq n$ and let $f \in \mathbb{R}[x]$ be a polynomial of degree d. Then, for any $r, n \in \mathbb{N}$ with $(r+1)/n \leq 1/2$, we have:*

$$\frac{f^{(r)} - f_{\min}}{\|f\|_\infty} \leq C_d \cdot \xi_{r+1}^n / n,$$

where $C_d > 0$ is the constant of Theorem 1.

So we have the same estimate of the relative error for the upper bounds $f^{(r)}$ as for the lower bounds $f_{(r)}$ (up to a constant factor 2) and indeed we will see that our proof relies on an intimate connection between both hierarchies. Note that the above analysis of $f^{(r)}$ does not require any condition on the size of ξ_{r+1}^n as was necessary for the analysis of $f_{(r)}$ in Theorem 1. Indeed, this condition on ξ_{r+1}^n follows from a technical argument which is not required in the proof of Theorem 3 (namely, the condition $\Lambda \leq 1/2$ above relation (16) in Sect. 2.3).

1.3   Asymptotic Analysis for Both Hierarchies

The results above imply that the relative error of both hierarchies is bounded asymptotically by the function $\varphi(t)$ from (5) in the regime $r \approx t \cdot n$. This is summarized in the following corollary which can be seen as an asymptotic version of Theorem 1 and Theorem 3.

Corollary 1. *Fix $d \leq n$ and for $n, r \in \mathbb{N}$ write*

$$E_{(r)}(n) := \sup_{f \in \mathbb{R}[x]_d} \{ f_{\min} - f_{(r)} : \|f\|_\infty = 1 \},$$

$$E^{(r)}(n) := \sup_{f \in \mathbb{R}[x]_d} \{ f^{(r)} - f_{\min} : \|f\|_\infty = 1 \}.$$

Let C_d be the constant of Theorem 1 and let $\varphi(t)$ be the function from (5). Then, for any $t \in [0, 1/2]$, we have:

$$\lim_{r/n \to t} E^{(r)}(n) \leq C_d \cdot \varphi(t)$$

and, if $d(d+1) \cdot \varphi(t) \leq 1/2$, we also have:

$$\lim_{r/n \to t} E_{(r)}(n) \leq 2 \cdot C_d \cdot \varphi(t).$$

Here, the limit notation $r/n \to t$ means that the claimed convergence holds for all sequences $(n_j)_j$ and $(r_j)_j$ of integers such that $\lim_{j \to \infty} n_j = \infty$ and $\lim_{j \to \infty} r_j/n_j = t$.

We close with some remarks. First, note that $\varphi(1/2) = 0$. Hence Corollary 1 tells us that the relative error of both hierarchies tends to 0 as $r/n \to 1/2$. We thus 'asymptotically' recover the exactness result (3) of [34].

Our results in Theorems 1 and 3 and Corollary 1 extend directly to the case of polynomial optimization over the discrete cube $\{\pm 1\}^n$ instead of the boolean cube $\mathbb{B}^n = \{0, 1\}^n$, as can easily be seen by applying a change of variables $x \in \{0, 1\} \mapsto 2x - 1 \in \{\pm 1\}$. In addition, our results extend to the case of polynomial optimization over the q-ary cube $\{0, 1, \ldots, q-1\}^n$ for $q > 2$ (see the extended version of this work in [38]).

Clearly, we may also obtain upper (resp., lower) bounds on the *maximum* $f_{\max}$ of f over $\mathbb{B}^n$ by using $f_{(r)}$ (resp., $f^{(r)}$) applied to $-f$. To avoid possible confusion we will also refer to $f_{(r)}$ as the *outer* Lasserre hierarchy, whereas we will refer to $f^{(r)}$ as the *inner* Lasserre hierarchy. This terminology (borrowed from [7]) is motivated by the following observations. One can reformulate $f_{\min}$ via optimization over the set $\mathcal{M}$ of Borel measures on $\mathbb{B}^n$:

$$f_{\min} = \min \left\{ \int_{\mathbb{B}^n} f(x) d\nu(x) : \nu \in \mathcal{M}, \int_{\mathbb{B}^n} d\nu(x) = 1 \right\}.$$

If we replace the set $\mathcal{M}$ by its *inner* approximation consisting of all measures $\nu(x) = s(x) d\mu(x)$ with polynomial density $s \in \Sigma_r$ with respect to a given fixed measure μ, then we obtain the bound $f^{(r)}$. On the other hand, any $\nu \in \mathcal{M}$ gives a linear functional $L_\nu : p \in \mathbb{R}[x]_{2r} \mapsto \int_{\mathbb{B}^n} p(x) d\nu(x)$ which is nonnegative on sum-of-squares on $\mathbb{B}^n$. These linear functionals thus provide an *outer* approximation for $\mathcal{M}$ and maximizing $L_\nu(p)$ over it gives the bound $f_{(r)}$ (in dual formulation).

1.4   Related Work

As mentioned above, the bounds $f_{(r)}$ from (2) are exact when $2r \geq n+d-1$. The case $d = 2$ (which includes MAX-CUT) was treated in [12], positively answering a question posed in [22]. Extending the strategy of [12], the general case was settled in [34]. These exactness results are best possible for d even and n odd [15].

In [14], the sum-of-squares hierarchy is considered for approximating instances of KNAPSACK. This can be seen as a variation on the problem (1), restricting to a linear polynomial objective with positive coefficients, but introducing a single, linear constraint, of the form $a_1 x_1 + \ldots + a_n x_n \leq b$ with $a_i > 0$. There, the authors show that the outer hierarchy has relative error at most $1/(r-1)$ for any $r \geq 2$. To the best of our knowledge this is the only known case where one can analyze the quality of the outer bounds for *all* orders $r \leq n$.

For optimization over sets other than the boolean cube, the following results on the quality of the outer hierarchy $f_{(r)}$ are available. When considering general semi-algebraic sets (satisfying a compactness condition), it has been shown in [28] that there exists a constant $c > 0$ (depending on the semi-algebraic set) such that $f_{(r)}$ converges to $f_{\min}$ at a rate in $O(1/\log(r/c)^{1/c})$ as r tends to ∞. This rate can be improved to $O(1/r^{1/c})$ if one considers a variation of the sum-of-squares hierarchy which is stronger (based on the preordering instead of the quadratic module), but much more computationally intensive [35]. Specializing to the hypersphere S^{n-1}, better rates in $O(1/r)$ were shown in [10,32], and recently improved to $O(1/r^2)$ in [11]. Similar improved results exist also for the case of polynomial optimization on the simplex and the continuous hypercube $[-1,1]^n$; we refer, e.g., to [7] for an overview.

The results for semi-algebraic sets other than $\mathbb{B}^n$ mentioned above all apply in the asymptotic regime where the dimension n is fixed and $r \to \infty$. This makes it difficult to compare them directly to our new results. Indeed, we have to consider a different regime in the case of the boolean cube $\mathbb{B}^n$, as the hierarchy always converges in at most n steps. The regime where we are able to provide an analysis in this paper is when $r \approx t \cdot n$ with $0 < t \leq 1/2$.

Turning now to the *inner* hierarchy (7), as far as we are aware, nothing is known about the behaviour of the bounds $f^{(r)}$ on $\mathbb{B}^n$. For full-dimensional compact sets, however, results are available. It has been shown that, on the hypersphere [8], the unit ball and the simplex [37], and the unit box [9], the bound $f^{(r)}$ converges at a rate in $O(1/r^2)$. A slightly weaker convergence rate in $O(\log^2 r/r^2)$ is known for general (full-dimensional) semi-algebraic sets [24,37]. Again, these results are all asymptotic in r, and thus hard to compare directly to our analysis on $\mathbb{B}^n$.

1.5   Overview of the Proof

We give here a broad overview of the main ideas that we use to show our results. Our broad strategy follows the one employed in [11] to obtain information on the sum-of-squares hierarchy on the hypersphere. The following four ingredients will play a key role in our proof:

1. we use the *polynomial kernel technique* in order to produce low-degree sum-of-squares representations of polynomials that are positive on $\mathbb{B}^n$, thus allowing an analysis of $f_{\min} - f_{(r)}$;
2. using classical *Fourier analysis* on the boolean cube $\mathbb{B}^n$ we are able to exploit symmetry and reduce the search for a *multivariate kernel* to a *univariate sum-of-squares polynomial* on the discrete set $[0:n] := \{0, 1, \ldots, n\}$;

3. we find this univariate sum-of-squares by applying the *inner* Lasserre hierarchy to an appropriate univariate optimization problem on $[0:n]$;
4. finally, we exploit a known connection between the inner hierarchy and the *extremal roots of corresponding orthogonal polynomials* (in our case, the Krawtchouk polynomials).

Following these steps we are able to analyze the sum-of-squares hierarchy $f_{(r)}$ as well as the inner hierarchy $f^{(r)}$. In the next section we will sketch in some more detail how our proof articulates along these four main steps.

2   Sketch of Proof

Here we sketch the main arguments needed to prove Theorem 1. It turns out that the proof for Theorem 3 follows essentially from some of these arguments. For a complete detailed proof we refer to the extended version [38] of this work. This section is organized along the four main steps outlined in Sect. 1.5.

2.1   The Polynomial Kernel Technique

Let $f \in \mathbb{R}[x]_d$ be the polynomial with degree d for which we wish to analyze the bounds $f_{(r)}$ and $f^{(r)}$. After rescaling, and up to a change of coordinates, we may assume w.l.o.g. that f attains its minimum over $\mathbb{B}^n$ at $0 \in \mathbb{B}^n$ and that $f_{\min} = 0$ and $f_{\max} = 1$. So we have $\|f\|_\infty = 1$. To simplify notation, we will make these assumptions throughout.

The first key idea is to consider a *polynomial kernel* K on $\mathbb{B}^n$ of the form:

$$K(x, y) = u^2(d(x, y)) \quad (x, y \in \mathbb{B}^n), \tag{8}$$

where $u \in \mathbb{R}[t]_r$ is a univariate polynomial of degree at most r and $d(x, y)$ is the Hamming distance between x and y. Such a kernel K induces an operator $\mathbf{K}$, which acts linearly on the space of polynomials on $\mathbb{B}^n$ by:

$$p \in \mathbb{R}[x] \mapsto \mathbf{K}p(x) := \int_{\mathbb{B}^n} p(y)K(x, y)d\mu(y) = \frac{1}{2^n} \sum_{y \in \mathbb{B}^n} p(y)K(x, y).$$

Recall that μ is the uniform probability distribution on $\mathbb{B}^n$. An easy but important observation is that, if p is nonnegative on $\mathbb{B}^n$, then $\mathbf{K}p$ is a sum-of-squares (on $\mathbb{B}^n$) of degree at most $2r$. We use this observation as follows.

Given a scalar $\delta \geq 0$, define the polynomial $\tilde{f} := f + \delta$. Assuming that the operator $\mathbf{K}$ is non-singular, we can express $\tilde{f}$ as $\tilde{f} = \mathbf{K}(\mathbf{K}^{-1}\tilde{f})$. Therefore, if $\mathbf{K}^{-1}\tilde{f}$ is nonnegative on $\mathbb{B}^n$, we find that $\tilde{f}$ is a sum-of-squares on $\mathbb{B}^n$ with degree at most $2r$, and thus that $f_{\min} - f_{(r)} \leq \delta$.

One way to guarantee that $\mathbf{K}^{-1}\tilde{f}$ is indeed nonnegative on $\mathbb{B}^n$ is to select the operator $\mathbf{K}$ in such a way that $\mathbf{K}(1) = 1$ and

$$\|\mathbf{K}^{-1} - I\| := \sup_{p \in \mathbb{R}[x]_d} \frac{\|\mathbf{K}^{-1}p - p\|_\infty}{\|p\|_\infty} \leq \delta. \tag{9}$$

We collect this as a lemma for further reference.

Lemma 1. *If the kernel operator* $\mathbf{K}$ *associated to* $u \in \mathbb{R}[t]_r$ *via relation (8) satisfies* $\mathbf{K}(1) = 1$ *and* $\|\mathbf{K}^{-1} - I\| \leq \delta$, *then we have* $f_{\min} - f_{(r)} \leq \delta$.

Proof. With $\tilde{f} = f + \delta$, we have: $\|\mathbf{K}^{-1}\tilde{f} - \tilde{f}\|_\infty = \|\mathbf{K}^{-1}f - f\|_\infty \leq \delta\|f\|_\infty = \delta$. Therefore we obtain that $\mathbf{K}^{-1}\tilde{f}(x) \geq \tilde{f}(x) - \delta = f(x) \geq f_{\min} = 0$ on $\mathbb{B}^n$. $\square$

In light of Lemma 1, we want to choose $u \in \mathbb{R}[t]_r$ in such a way that the operator $\mathbf{K}^{-1}$ (and thus $\mathbf{K}$) is 'close to the identity operator' in a certain sense. In other words, we want the *eigenvalues* of $\mathbf{K}$ to be as close as possible to 1.

2.2   Fourier Analysis on $\mathbb{B}^n$ and the Funk-Hecke Formula

As kernels of the form (8) are invariant under the symmetries of $\mathbb{B}^n$, we are able to use classical Fourier analysis on the boolean cube to express the eigenvalues of $\mathbf{K}$ in terms of the polynomial u. More precisely, it turns out that the eigenvalues of $\mathbf{K}$ are given by the coefficients of the expansion of u^2 in the basis of *Krawtchouk polynomials*. This link is known as the *Funk-Hecke formula* (cf. Theorem 4 below).

The Character Basis. Consider the space $\mathcal{R}[x]$ of polynomials on $\mathbb{B}^n$, defined as the quotient of $\mathbb{R}[x]_r$ under the relation $p \sim q$ if $p(x) = q(x)$ for all $x \in \mathbb{B}^n$. Equip this space $\mathcal{R}[x]$ with the inner product: $\langle p, q \rangle_\mu = \int p(x)q(x)d\mu(x)$, where μ is the uniform probability measure on $\mathbb{B}^n$. W.r.t. this inner product the space $\mathcal{R}[x]$ has an orthonormal basis given by the set of *characters*:

$$\chi_a(x) := (-1)^{a \cdot x} \quad (a \in \mathbb{B}^n).$$

The group $\mathrm{Aut}(\mathbb{B}^n)$ of automorphisms of $\mathbb{B}^n$ is generated by the coordinate permutations, of the form $x \mapsto \sigma(x) := (x_{\sigma(1)}, \ldots, x_{\sigma(n)})$ for $\sigma \in \mathrm{Sym}(n)$, and the permutations corresponding to bit-flips, of the form $x \in \mathbb{B}^n \mapsto x \oplus a \in \mathbb{B}^n$ for any $a \in \mathbb{B}^n$. If we set

$$H_k := \mathrm{span}\{\chi_a : |a| = k\} \quad (0 \leq k \leq n),$$

then each H_k is an irreducible, $\mathrm{Aut}(\mathbb{B}^n)$-invariant subspace of $\mathcal{R}[x]$ of dimension $\binom{n}{k}$. We may then decompose $\mathcal{R}[x]$ as the direct sum

$$\mathcal{R}[x] = H_0 \perp H_1 \perp \cdots \perp H_n,$$

where the subspaces H_k are pairwise orthogonal w.r.t. $\langle \cdot, \cdot \rangle_\mu$. In fact, we have that $\mathcal{R}[x]_d = H_0 \perp H_1 \perp \cdots \perp H_d$ for all $d \leq n$, and we may thus write any $p \in \mathcal{R}[x]_d$ (in a unique way) as

$$p = p_0 + p_1 + \cdots + p_d \quad (p_k \in H_k). \tag{10}$$

The Funk-Hecke Formula. For $k \in \mathbb{N}$, the *Krawtchouk polynomial* of degree k (and with parameter n) is the univariate polynomial in t given by:

$$\mathcal{K}_k^n(t) := \sum_{i=0}^{k} (-1)^i \binom{t}{i} \binom{n-t}{k-i} \tag{11}$$

(see, e.g. [39]). Here $\binom{t}{i} := t(t-1)\dots(t-i+1)/i!$. The Krawtchouk polynomials form an orthogonal basis for $\mathbb{R}[t]$ with respect to the inner product $\langle \cdot, \cdot \rangle_\omega$ given by the following discrete probability measure on the set $[0:n] = \{0, 1, \dots, n\}$:

$$\omega := \frac{1}{2^n} \sum_{t=0}^{n} \binom{n}{t} \delta_t. \tag{12}$$

The following lemma explains the connection between the Krawtchouk polynomials and the character basis on $\mathcal{R}[x]$.

Lemma 2. *Let $t \in [0:n]$ and choose $x, y \in \mathbb{B}^n$ so that $d(x, y) = t$. Then for any $0 \le k \le n$ we have:*

$$\mathcal{K}_k^n(t) = \sum_{|a|=k} \chi_a(x)\chi_a(y). \tag{13}$$

Using Lemma 2, one is then able to show the Funk-Hecke formula.

Theorem 4 (Funk-Hecke). *Given $u \in \mathbb{R}[t]_r$, decompose u^2 in the basis of Krawtchouk polynomials as $u^2 = \sum_{i=0}^{2r} \lambda_i \mathcal{K}_i^n$ and consider the kernel operator $\mathbf{K}$ associated to u via (8). For any $p \in \mathcal{R}[x]_d$ with harmonic decomposition $p = p_0 + p_1 + \dots + p_d$ as in (10), we have:*

$$\mathbf{K}p = \lambda_0 p_0 + \lambda_1 p_1 + \dots + \lambda_d p_d. \tag{14}$$

2.3   Optimizing the Choice of the Univariate Polynomial u

Recall that in light of Lemma 1 we wish to bound the quantity $\|\mathbf{K}^{-1} - I\|$ from (9). To define such $\mathbf{K}$ we need to suitably select the polynomial $u \in \mathbb{R}[t]_r$. Assume we choose $u \in \mathbb{R}[t]_r$ such that $u^2 = \sum_{i=0}^r \lambda_i \mathcal{K}_i^n$ with $\lambda_0 = 1$ and $\lambda_i \neq 0$ for all i. Then, for any $p = p_0 + \dots + p_d$ with $\|p\|_\infty = 1$, we find that

$$\|\mathbf{K}^{-1}p - p\|_\infty \le \sum_{i=1}^{d} |1 - \lambda_i^{-1}| \cdot \|p_i\|_\infty \le \gamma_d \sum_{i=1}^{d} |1 - \lambda_i^{-1}|, \tag{15}$$

where $\gamma_d > 0$ is a constant depending only on d. The left most inequality follows after an application of the Funk-Hecke formula (14). The right most inequality is a result of the following technical lemma (see [38] for the proof).

Lemma 3. *There exists a constant $\gamma_d > 0$, depending only on d, such that for any $p = p_0 + p_1 + \dots + p_d \in \mathcal{R}[x]_d$, we have:*

$$\|p_k\|_\infty \le \gamma_d \|p\|_\infty \text{ for all } 0 \le k \le d.$$

The key fact here is that the constant γ_d in Lemma 3 does not depend on the dimension n.

The quantity $\sum_{i=1}^{d} |1 - \lambda_i^{-1}|$ in (15) is still difficult to analyze. Following [11], we therefore consider the following 'linearized' version instead:

$$\Lambda := \sum_{i=1}^{d} (1 - \lambda_i).$$

It turns out that, as long as $\Lambda \leq 1/2$, we have $\sum_{i=1}^{d} |1 - \lambda_i^{-1}| \leq 2\Lambda$, implying:

$$\|\mathbf{K}^{-1} - I\| \leq 2\gamma_d \cdot \Lambda. \tag{16}$$

Recall that the Krawtchouk polynomials are orthogonal w.r.t. the inner product $\langle \cdot, \cdot \rangle_\omega$, where ω is the discrete probability measure on $[0:n]$ of (12). Therefore, we may express the scalars λ_i as:

$$\lambda_i = \langle \widehat{\mathcal{K}}_i^n, u^2 \rangle_\omega, \text{ with } \widehat{\mathcal{K}}_i^n := \mathcal{K}_i^n / \|\mathcal{K}_i^n\|_\omega^2.$$

We thus wish to find a univariate polynomial $u \in \mathbb{R}[t]_r$ for which:

$$\lambda_0 = \langle 1, u^2 \rangle_\omega = 1, \text{ and}$$

$$\Lambda = d - \sum_{i=1}^{d} \lambda_i = d - \sum_{i=1}^{d} \langle \widehat{\mathcal{K}}_i^n, u^2 \rangle_\omega \text{ is small.}$$

Unpacking the definition of $\langle \cdot, \cdot \rangle_\omega$, we thus need to solve the following optimization problem:

$$\inf_{u \in \mathbb{R}[t]_r} \left\{ \Lambda := \int g \cdot u^2 d\omega : \int u^2 d\omega = 1 \right\}, \text{ where } g(t) := d - \sum_{i=1}^{d} \widehat{\mathcal{K}}_i^n(t). \tag{17}$$

We recognize this program to be the analog of the program (7), where we now consider the inner Lasserre bound of order r for the minimum $g_{\min} = g(0) = 0$ of the polynomial g over the set $[0:n]$, computed with respect to the measure $d\omega(t) = 2^{-n}\binom{n}{t}$ on $[0:n]$. Hence the optimal value of (17) is equal to $g^{(r)}$ and, using (16), we may conclude the following result, which tells us how to select the polynomial u (and thus $\mathbf{K}$).

Theorem 5. *Let g be as in (17). Assume that $g^{(r)} - g_{\min} \leq 1/2$. Then there exists a polynomial $u \in \mathbb{R}[t]_r$ such that $\lambda_0 = 1$ and*

$$\|\mathbf{K}^{-1} - I\| \leq 2\gamma_d \cdot (g^{(r)} - g_{\min}).$$

Here, $g^{(r)}$ is the inner Lasserre bound on $g_{\min}$ of order r, computed on $[0:n]$ w.r.t. ω, via the program (17), and γ_d is the constant of Lemma 3.

2.4   The Inner Lasserre Hierarchy and Orthogonal Polynomials

In order to finish the proof of Theorem 1, it now remains to analyze the range $g^{(r)} - g_{\min}$ for the polynomial g in (17).

We recall a technique that may be used to perform such an analysis, which was developed in [9] and further employed for this purpose, e.g., in [8,37]. We present it here for the special case for optimization over $[0:n]$ w.r.t. the measure ω, but it actually applies to univariate optimization w.r.t. general measures.

First, we observe that we may replace g by a suitable *upper estimator* $\widehat{g}$ which satisfies $\widehat{g}_{\min} = g_{\min}$ and $\widehat{g}(t) \geq g(t)$ for all $t \in [0:n]$. Indeed, then we have:

$$g^{(r)} - g_{\min} \leq \widehat{g}^{(r)} - g_{\min} = \widehat{g}^{(r)} - \widehat{g}_{\min}.$$

Next, we use the following crucial link to the roots of Krawtchouk polynomials. This is a special case of a result by de Klerk and Laurent [9], applied to optimization over the set $[0:n]$ equipped with the measure ω, so that the corresponding orthogonal polynomials are given by the Krawtchouk polynomials $\mathcal{K}_k^n$.

Theorem 6 ([9]). *Suppose $\widehat{g}(t) = ct$ is a linear polynomial with $c > 0$. Then the Lasserre inner bound $\widehat{g}^{(r)}$ of order r for minimization of $\widehat{g}(t)$ on $[0:n]$ w.r.t. the measure ω can be reformulated in terms of the smallest root ξ_{r+1}^n of $\mathcal{K}_{r+1}^n$ as:*

$$\widehat{g}^{(r)} = c \cdot \xi_{r+1}^n.$$

The upshot is that if we can upper bound the function g in (17) by some linear polynomial $\widehat{g}(t) = ct$ with $c > 0$, we then find:

$$g^{(r)} - g_{\min} \leq \widehat{g}^{(r)} - \widehat{g}_{\min} \leq c \cdot \xi_{r+1}^n.$$

Indeed, g can be upper bounded on $[0:n]$ by its linear approximation at $t = 0$:

$$g(t) \leq \widehat{g}(t) := d(d+1) \cdot (t/n) \quad \forall t \in [0:n].$$

This inequality can be obtained by combining basic identities and inequalities concerning Krawtchouk polynomials (see [38]). We have thus shown that:

$$g^{(r)} - g_{\min} \leq d(d+1) \cdot (\xi_{r+1}^n/n). \tag{18}$$

We have now gathered all the tools required to prove Theorem 1.

Theorem 7 (Restatement of Theorem 1). *Fix $d \leq n$ and let $f \in \mathbb{R}[x]$ be a polynomial of degree d. Then we have:*

$$\frac{f_{\min} - f_{(r)}}{\|f\|_\infty} \leq 2\gamma_d \cdot d(d+1) \cdot (\xi_{r+1}^n/n),$$

whenever $d(d+1) \cdot (\xi_{r+1}^n/n) \leq 1/2$. Here, ξ_{r+1}^n is the smallest root of $\mathcal{K}_{r+1}^n$ and γ_d is the constant of Lemma 3.

Proof. Combining Theorem 5 with (18), we find that we may choose $u \in \mathbb{R}[t]_r$ such that $\lambda_0 = 1$ and:

$$\|\mathbf{K}^{-1} - I\| \leq 2\gamma_d \cdot (g^{(r)} - g_{\min}) \leq 2\gamma_d \cdot d(d+1) \cdot (\xi_{r+1}^n/n).$$

Using the Funk-Hecke formula (14), we see that $\lambda_0 = 1$ implies that $\mathbf{K}(1) = 1$. We may thus use Lemma 1 to conclude the proof, obtaining Theorem 1 with $C_d := \gamma_d \cdot d(d+1)$. □

3   Concluding Remarks

Summary. We have shown a theoretical guarantee on the quality of the sum-of-squares hierarchy $f_{(r)} \leq f_{\min}$ for approximating the minimum of a polynomial f of degree d over the boolean cube $\mathbb{B}^n$. As far as we are aware, this is the first such analysis that applies to values of r smaller than $(n+d)/2$, i.e., when the hierarchy is not exact. Additionally, our guarantee applies to a second, measure-based hierarchy of bounds $f^{(r)} \geq f_{\min}$. Our result may therefore also be interpreted as bounding the range $f^{(r)} - f_{(r)}$.

A limitation of the present work is that no information is gained for low levels of the hierarchy, when r is fixed and the dimension n grows. Indeed, our results apply only in the regime $r \approx t \cdot n$, where $t \in [0, 1/2]$ is a fixed fraction. They are therefore of limited practical value, as computation beyond the first few levels of the hierarchy is currently infeasible.

Our analysis also applies to polynomial optimization over the cube $\{\pm 1\}^n$ (by a simple change of variables). Furthermore, the techniques we use on the *binary* cube $\mathbb{B}^n$ generalize naturally to the q-ary cube $(\mathbb{Z}/q\mathbb{Z})^n = \{0, 1, \ldots, q-1\}^n$ for $q > 2$. As a result we are able to show close analogs of our results on $\mathbb{B}^n$ in this more general setting as well. We present this generalization in the expanded version [38] of this work.

The Constant γ_d. The strength of our results depends in large part on the size of the constant γ_d appearing in Theorem 1 and Theorem 3, where we may set $C_d = d(d+1)\gamma_d$. In [38] we show the existence of this constant γ_d, but the resulting dependence on d there is quite bad. This dependence, however, seems to be mostly an artifact of our proof. As we explain in [38], it is possible to compute explicit upper bounds on γ_d for small values of d. Table 1 lists some of these upper bounds, which appear much more reasonable than our theoretical guarantee would suggest.

Table 1. Upper bounds on γ_d. Values rounded to indicated precision.

d	1	2	3	4	5	6	7	8	9	10	11	12
γ_d	1.00	2.00	4.00	8.00	20.0	48.1	112	258	578	1306	2992	6377

Computing Extremal Roots of Krawtchouk Polynomials. Although Theorem 2 provides only an asymptotic bound on the least root ξ_r^n of $\mathcal{K}_r^n$, it should be noted that ξ_r^n can be computed explicitly for small values of r, n, thus allowing for a concrete estimate of the error of both Lasserre hierarchies via Theorem 1 and Theorem 3, respectively. Indeed, as is well-known, the root ξ_{r+1}^n is equal to the smallest eigenvalue of the $(r+1) \times (r+1)$ matrix A (aka Jacobi matrix), whose entries are given by $A_{i,j} = \langle t\widehat{\mathcal{K}}_i^n(t), \widehat{\mathcal{K}}_j^n(t)\rangle_\omega$ for $i, j \in \{0, 1, \ldots, r\}$. See, e.g., [39] for more details.

Connecting the Hierarchies. Our analysis of the *outer* hierarchy $f_{(r)}$ on $\mathbb{B}^n$ relies essentially on knowledge of the *inner* hierarchy $f^{(r)}$. Although not explicitly mentioned there, this is the case for the analysis on S^{n-1} in [11] as well. As the behaviour of $f^{(r)}$ is generally quite well understood, this suggests a potential avenue for proving further results on $f_{(r)}$ in other settings.

For instance, the inner hierarchy $f^{(r)}$ is known to converge at a rate in $O(1/r^2)$ on the unit ball B^n or the unit box $[-1, 1]^n$, but matching results on the outer hierarchy $f_{(r)}$ are not available. The question is thus whether the strategy used for the hypersphere S^{n-1} in [11] and for the boolean cube $\mathbb{B}^n$ here might be extended to these cases as well.

Although B^n and $[-1, 1]^n$ have similar symmetric structure to S^{n-1} and $\mathbb{B}^n$, respectively, the accompanying Fourier analysis is significantly more complicated. In particular, a direct analog of the Funk-Hecke formula (14) is not available. New ideas are therefore needed to define the kernel $K(x, y)$ (cf. (8)) and analyze its eigenvalues.

Acknowledgments. This work is supported by the European Union's Framework Programme for Research and Innovation Horizon 2020 under the Marie Skłodowska-Curie Actions Grant Agreement No. 764759 (MINOA). We wish to thank Sven Polak and Pepijn Roos Hoefgeest for several useful discussions, as well as the anonymous referees for their helpful suggestions.

References

1. Alon, N., Naor, A.: Approximating the cut-norm via Grothendieck's inequality. In: 36th Annual ACM Symposium on Theory of Computing, pp. 72–80 (2004)
2. Arora, S., Berger, E., Hazan, E., Kindler, G., Safra, M.: On non-approximability for quadratic programs. In: Proceedings of the 46th Annual IEEE Symposium on Foundations of Computer Science, pp. 206–215 (2005)
3. Balas, E., Ceria, S., Cornuéjols, G.: A lift-and-project cutting plane algorithm for mixed 0–1 programs. Math. Program. **58**, 295–324 (1993)
4. Bansal, N., Blum, A., Chawla, S.: Correlation clustering. Mach. Learn. **46**(1–3), 89–113 (2004)
5. Barak, B., Steurer, D.: Sum-of-squares proofs and the quest toward optimal algorithms. In: Proceedings of International Congress of Mathematicians (ICM) (2014)
6. Charikar, M., Wirth, A.: Maximizing quadratic programs: extending Grothendieck's inequality. In: Proceedings of the 45th Annual IEEE Symposium on Foundations of Computer Science, pp. 54–60 (2004)

7. Klerk, E., Laurent, M.: A survey of semidefinite programming approaches to the generalized problem of moments and their error analysis. In: Araujo, C., Benkart, G., Praeger, C.E., Tanbay, B. (eds.) World Women in Mathematics 2018. AWMS, vol. 20, pp. 17–56. Springer, Cham (2019). https://doi.org/10.1007/978-3-030-21170-7_1

8. de Klerk, E., Laurent, M.: Convergence analysis of a Lasserre hierarchy of upper bounds for polynomial minimization on the sphere. Math. Program. (2020). https://doi.org/10.1007/s10107-019-01465-1

9. de Klerk, E., Laurent, M.: Worst-case examples for Lasserre's measure-based hierarchy for polynomial optimization on the hypercube. Math. Oper. Res. **45**(1), 86–98 (2020)

10. Doherty, A.C., Wehner, S.: Convergence of SDP hierarchies for polynomial optimization on the hypersphere. arXiv:1210.5048v2 (2013)

11. Fang, K., Fawzi, H.: The sum-of-squares hierarchy on the sphere, and applications in quantum information theory. Math. Program. (2020). https://doi.org/10.1007/s10107-020-01537-7

12. Fawzi, H., Saunderson, J., Parrilo, P.A.: Sparse sums of squares on finite abelian groups and improved semidefinite lifts. Math. Program. **160**(1–2), 149–191 (2016). https://doi.org/10.1007/s10107-015-0977-z

13. Goemans, M., Williamson, D.: Improved approximation algorithms for maximum cut and satisfiability problems using semidefinite programming. J. Assoc. Comput. Mach. **42**(6), 1115–1145 (1995)

14. Karlin, A.R., Mathieu, C., Nguyen, C.T.: Integrality gaps of linear and semidefinite programming relaxations for knapsack. In: Günlük, O., Woeginger, G.J. (eds.) IPCO 2011. LNCS, vol. 6655, pp. 301–314. Springer, Heidelberg (2011). https://doi.org/10.1007/978-3-642-20807-2_24

15. Kurpisz, A., Leppänen, S., Mastrolilli, M.: Tight sum-of-squares lower bounds for binary polynomial optimization problems. In: Chatzigiannakis, I., et al. (eds.) 43rd International Colloquium on Automata, Languages, and Programming (ICALP 2016), vol. 78, pp. 1–14 (2016)

16. Lasserre, J.B.: Global optimization with polynomials and the problem of moments. SIAM J. Optim. **11**(3), 796–817 (2001)

17. Lasserre, J.B.: A max-cut formulation of 0/1 programs. Oper. Res. Lett. **44**, 158–164 (2016)

18. Lasserre, J.B.: An explicit exact SDP relaxation for nonlinear 0-1 programs. In: Aardal, K., Gerards, B. (eds.) IPCO 2001. LNCS, vol. 2081, pp. 293–303. Springer, Heidelberg (2001). https://doi.org/10.1007/3-540-45535-3_23

19. Lasserre, J.B.: Moments, Positive Polynomials and Their Applications. Imperial College Press, London (2009)

20. Lasserre, J.B.: A new look at nonnegativity on closed sets and polynomial optimization. SIAM J. Optim. **21**(3), 864–885 (2010)

21. Laurent, M.: A comparison of the Sherali-Adams, Lovász-Schrijver and Lasserre relaxations for 0–1 programming. Math. Oper. Res. **28**(3), 470–496 (2003)

22. Laurent, M.: Lower bound for the number of iterations in semidefinite hierarchies for the cut polytope. Math. Oper. Res. **28**(4), 871–883 (2003)

23. Laurent, M.: Sums of squares, moment matrices and optimization over polynomials. In: Putinar, M., Sullivant, S. (eds.) Emerging Applications of Algebraic Geometry. The IMA Volumes in Mathematics and Its Applications, vol. 149, pp. 157–270. Springer, New York (2009). https://doi.org/10.1007/978-0-387-09686-5_7

24. Laurent, M., Slot, L.: Near-optimal analysis of of Lasserre's univariate measure-based bounds for multivariate polynomial optimization. Math. Program. (2020). https://doi.org/10.1007/s10107-020-01586-y
25. Lee, J.R., Raghavendra, P., Steurer, D.: Lower bounds on the size of semidefinite programming relaxations. In: STOC 2015: Proceedings of the Forty-Seventh Annual ACM Symposium on Theory of Computing, pp. 567–576 (2015)
26. Levenshtein, V.I.: Universal bounds for codes and designs. In: Handbook of Coding Theory, vol. 9, pp. 499–648. North-Holland, Amsterdam (1998)
27. Lovász, L., Schrijver, A.: Cones of matrices and set-functions and 0–1 optimization. SIAM J. Optim. **1**, 166–190 (1991)
28. Nie, J., Schweighofer, M.: On the complexity of Putinar's positivstellensatz. J. Complex. **23**(1), 135–150 (2007)
29. O'Donnell, R.: SOS is not obviously automatizable, even approximately. In: 8th Innovations in Theoretical Computer Science Conference, vol. 59, pp. 1–10 (2017)
30. Parrilo, P.A.: Structured semidefinite programs and semialgebraic geometry methods in robustness and optimization. Ph.D. thesis, California Institute of Technology (2000)
31. Raghavendra, P., Weitz, B.: On the bit complexity of sum-of-squares proofs. In: 44th International Colloquium on Automata, Languages, and Programming, vol. 80, pp. 1–13 (2017)
32. Reznick, B.: Uniform denominators in Hilbert's seventeenth problem. Mathematische Zeitschrift **220**(1), 75–97 (1995)
33. Rothvoss, T.: The Lasserre hierarchy in approximation algorithms. Lecture Notes for the MAPSP 2013 Tutorial (2013)
34. Sakaue, S., Takeda, A., Kim, S., Ito, N.: Exact semidefinite programming relaxations with truncated moment matrix for binary polynomial optimization problems. SIAM J. Optim. **27**(1), 565–582 (2017)
35. Schweighofer, M.: On the complexity of Schmüdgen's positivstellensatz. J. Complex. **20**(4), 529–543 (2004)
36. Sherali, H.D., Adams, W.P.: A hierarchy of relaxations between the continuous and convex hull representations for zero-one programming problems. SIAM J. Discrete Math. **3**, 411–430 (1990)
37. Slot, L., Laurent, M.: Improved convergence analysis of Lasserre's measure-based upper bounds for polynomial minimization on compact sets. Math. Program. (2020). https://doi.org/10.1007/s10107-020-01468-3
38. Slot, L., Laurent, M.: Sum-of-squares hierarchies for binary polynomial optimization. arXiv:2011.04027 (2020)
39. Szegö, G.: Orthogonal Polynomials. American Mathematical Society Colloquium Publications, vol. 23. American Mathematical Society (1959)
40. Tunçel, L.: Polyhedral and Semidefinite Programming Methods in Combinatorial Optimization, Fields Institute Monograph. American Mathematical Society, Providence (2010)

Complexity, Exactness, and Rationality in Polynomial Optimization

Daniel Bienstock[1], Alberto Del Pia[2], and Robert Hildebrand[3]([✉])

[1] Departments of IEOR, APAM and EE, Columbia University, New York, NY, USA
`dano@columbia.edu`
[2] Department of Industrial and Systems Engineering and Wisconsin Institute for Discovery, University of Wisconsin-Madison, Madison, USA
`delpia@wisc.edu`
[3] Grado Department of Industrial and Systems Engineering, Virginia Tech, Blacksburg, USA
`rhil@vt.edu`

Abstract. We focus on rational solutions or nearly-feasible rational solutions that serve as certificates of feasibility for polynomial optimization problems. We show that, under some separability conditions, certain cubic polynomially constrained sets admit rational solutions. However, we show in other cases that it is NP Hard to detect if rational solutions exist or if they exist of any reasonable size. Lastly, we show that in fixed dimension, the feasibility problem over a set defined by polynomial inequalities is in NP.

Keywords: Polynomial optimization · Rational solutions · NP

1 Introduction

This paper addresses basic questions of precise certification of feasibility and optimality, for optimization problems with polynomial constraints, in polynomial time, under the Turing model of computation. Recent progress in polynomial optimization and mixed-integer nonlinear programming has produced elegant methodologies and effective implementations; however such implementations may produce *imprecise* solutions whose actual quality can be difficult to rigorously certify, even approximately. The work we address is motivated by these issues, and can be summarized as follows:

A. Del Pia is partially funded by ONR grant N00014-19-1-2322. D. Bienstock is partially funded by ONR grant N00014-16-1-2889. R. Hildebrand is partially funded by ONR grant N00014-20-1-2156 and by AFOSR grant FA9550-21-0107. Any opinions, findings, and conclusions or recommendations expressed in this material are those of the authors and do not necessarily reflect the views of the Office of Naval Research or the Air Force Office of Scientific Research.

M. Singh and D. P. Williamson (Eds.): IPCO 2021, LNCS 12707, pp. 58–72, 2021.
https://doi.org/10.1007/978-3-030-73879-2_5

Question: Given a polynomially constrained problem, what can be said about the existence of feasible or approximately feasible rational solutions of polynomial size (bit encoding length)[1], and more generally the existence of rational, feasible or approximately-feasible solutions that are also approximately-optimal for a given polynomial objective?

As is well-known, Linear Programming is polynomially solvable [13,15], and, moreover, every face of a rational polyhedron contains a point of polynomial size. [22]. If we instead optimize a quadratic function over linear constraints, the problem becomes NP-Hard [18], but perhaps surprisingly, Vavasis [24] proved that a feasible system consisting of linear inequalities and just one quadratic inequality, all with rational coefficients, always has a rational feasible solution of polynomial size. This was extended by Del Pia, Dey, and Molinaro [9] to show that the same result holds in the mixed-integer setting. See also [12] discussion of mixed-integer nonlinear optimization problems with linear constraints.

On the negative side, there are classical examples of SOCPs all of whose feasible solutions require exponential size [2,16,20], or all of whose feasible solutions are irrational (likely, a folklore result. See full paper). In the nonconvex setting, there are examples of quadratically constrained, linear objective problems, on n bounded variables, and with coefficients of magnitude $O(1)$, that admit solutions with maximum additive infeasibility $O(2^{-2^{\Theta(n)}})$ but multiplicative (or additive) superoptimality $\Theta(1)$ (see full paper). O'Donnell [17] questions whether SDPs associated with fixed-rank iterates of sums-of-squares hierarchies (which relax nonconvex polynomially constrained problem) can be solved in polynomial time, because optimization certificates might require exponential size. The issue of accuracy in solutions is not just of theoretical interest. As an example, [25] describes instances of SDPs (again, in the sums-of-squares setting) where a solution is very nearly certified as optimal, and yet proves substantially suboptimal.

Vavasis' result suggests looking at systems of two or more quadratic constraints, or (to some extent equivalently) optimization problems where the objective is quadratic, and at least one constraint is quadratic, with all other constraints linear. The problem of optimizing a quadratic subject to one quadratic constraint (and no linear constraints) can be solved in polynomial time using semidefinite-programming techniques [19], to positive tolerance. When the constraint is positive definite (i.e. a ball constraint) the problem can be solved to tolerance ϵ in time $\log \log \epsilon^{-1}$ [26], [14] (in other words $O(k)$ computations guarantee accuracy 2^{-2^k}). Vavasis [23] proved, on the other hand, that *exact* feasibility of a system of two quadratics can be tested in polynomial time.

With regards to systems of more than two quadratic constraints, Barvinok [3] proved a fundamental result: for each fixed integer m there is an algorithm that, given $n \times n$ rational matrices A_i $(1 \leqslant i \leqslant m)$ tests, in polynomial-time, feasibility of the system of equations

---

[1] Throughout we will use the concept of *size* of rational numbers, vectors, linear inequalities, and formulations. For these standard definitions we refer the reader to Section 2.1 in [22].

$$x^T A_i x = 0 \quad \text{for } 1 \leqslant i \leqslant m, \quad x \in \mathbb{R}^n, \ \|x\|_2 = 1.$$

A feature of this algorithm is that certification does not rely on producing a feasible vector; indeed, all feasible solutions may be irrational. As a corollary of this result, [6] proves that, for each fixed integer m there is an algorithm that solves, in polynomial time, an optimization problem of the form

$$\min f_0(x), \quad \text{s.t.} \quad f_i(x) \leqslant 0 \quad \text{for } 1 \leqslant i \leqslant m$$

where for $0 \leqslant i \leqslant m$, $f_i(x)$ is an n-variate quadratic polynomial, and we assume that the quadratic part of $f_1(x)$ positive-definite; moreover a rational vector that is (additively) both ϵ-feasible and -optimal can be computed in time polynomial in the size of the formulation and $\log \epsilon^{-1}$. An important point with regards to [3] and [6] is that the analyses do not apply to systems of arbitrarily many *linear* inequalities and just two quadratic inequalities.

De Loera et al. [8] use the *Nullstellensatz* to provide feasibility and infeasibility certificates to systems of polynomial equations through solving a sequence of large linear equations. Bounds on the size of the certificates are obtained [11]. This technique does not seem amenable to systems with a large number of linear inequalities due to the necessary transformation into equations and then blow up of the number of variables used. Another approach is to use the Positivestellensatz and compute an infeasibility certificate using sums of squares hierarchies. As mentioned above, see [17] for a discussion if exactness and size of these hierarchies needed for a certificate.

Renegar [21] shows that the problem of deciding whether a system of polynomial inequalities is nonempty can be decided in polynomial time provided that the dimension is considered fixed. This is a landmark result, however, the algorithm and techniques are quite complicated. Another technique to obtain a similar result is *Cylindrical Algebraic Decomposition*. See, e.g., [4]. In this work, we aim to avoid these techniques and provide an extremely simple certificate that shows the feasibility question is in NP.

Our Results. The main topic we address in this paper is whether a system of polynomial inequalities admits rational, feasible or near-feasible solutions of polynomial size. First, we show that in dimension 2, with one separable cubic inequality and linear inequalities, there exists a rational solution of polynomial size (Theorem 2). In the next section, we show that this result fails without separability. Using this motivating example, we show that it is strongly NP-hard to test if a system of quadratic inequalities that has feasible rational solutions, admits feasible rational solutions of polynomial size (Theorem 4). And it is also hard to test if a feasible system of quadratic inequalities has a rational solution (Theorem 5).

We next show that, given a system of polynomial inequalities on n variables that is known to have a bounded, nonempty feasible region, we can produce as a certificate of feasibility a rational, near-feasible vector that has polynomial size, for fixed n (Theorem 7). This certificate yields a direct proof that, in fixed dimension, the feasibility problem over a system of polynomial inequalities is in NP.

Omitted proofs can be found in the extended version of the paper [7].

2   Existence of Rational Feasible Solutions

Local Minimizers of Cubic Polynomials. We will prove results about the
rationality of local minimizers of cubic polynomials. This will be used in the next
section to argue that rational solutions exist to certain feasibility problems.

We first show that local minimizers of separable cubic polynomials are rational provided that the function value is rational.

We will need the following 3 lemmas. The first result is known by
Nicolo de Brescia, a.k.a., Tartaglia. It shows that in a univariate cubic polynomial, shifting by a constant allows us to assume that the x^2 term has a zero
coefficient.

Lemma 1 (Rational shift of cubic). *Let $f(x) = ax^3 + bx^2 + cx + d$. Then
$f(y - \frac{b}{3a}) = ay^3 + \tilde{c}y + \tilde{d}$ where $\tilde{c} = \frac{27a^2c - 9ab^2}{27a^2}$ and $\tilde{d} = \frac{27a^2d - 9abc + 2b^3}{27a^2}$.*

The next lemma provides bounds on the roots of a univariate polynomial.
We attribute this result to Cauchy; a proof can be found in Theorem 10.2 of [4].

Lemma 2 (Cauchy - size of roots). *Let $f(x) = a_nx^n + \cdots + a_1x + a_0$, where
$a_n, a_0 \neq 0$. Let $\bar{x} \neq 0$ such that $f(\bar{x}) = 0$. Then $L \leqslant |\bar{x}| \leqslant U$, where*

$$U = 1 + \max\left\{\left|\frac{a_0}{a_n}\right|, \ldots, \left|\frac{a_{n-1}}{a_n}\right|\right\}, \quad \frac{1}{L} = 1 + \max\left\{\left|\frac{a_1}{a_0}\right|, \ldots, \left|\frac{a_n}{a_0}\right|\right\}.$$

The next lemma is a special case of Theorem 2.9 in [1].

Lemma 3. *Let $n \geqslant 1$. Let $r_i \in \mathbb{Q}_+$ for $i = 0, 1, \ldots, n$, $q_i \in \mathbb{Q}_+$ for $i = 1, \ldots, n$.
If $\sum_{i=1}^n r_i\sqrt{q_i} = r_0$, then $\sqrt{q_i} \in \mathbb{Q}$ for all $i = 1, \ldots, n$. Furthermore, the size of
$\sqrt{q_i}$ is polynomial in the size of q_i.*

Theorem 1 (Rational local minimum). *Let $f(x) = \sum_{i=1}^n f_i(x_i)$ where
$f_i(x_i) = a_ix_i^3 + b_ix_i^2 + c_ix_i + d_i \in \mathbb{Z}[x_i]$ and $a_i \neq 0$ for all $i \in [n]$. Assume
that the absolute value of the coefficients of f is at most H. Suppose x^* is the
unique local minimum of f and $\gamma^* := f(x^*)$ is rational. Then x^* is rational and
has size that is polynomial in $\log H$ and in the size of γ^*.*

Proof. For every $i \in [n]$, Let $\tilde{c}_i, \tilde{d}_i$ be defined as in Lemma 1, let $g_i(y_i) :=
a_iy_i^3 + \tilde{c}_iy_i$, and define $g(y) = \sum_{i=1}^n g_i(y_i)$. Then $y^* \in \mathbb{R}^n$ defined by $y_i^* := x_i^* + \frac{b_i}{3a_i}$,
$i \in [n]$, is the unique local minimum of $g(y)$ and $g(y^*) = \gamma^* - \sum_{i=1}^n \tilde{d}_i$.

We now work with the gradient. Since y^* is a local minimum of g, we have
$\nabla g(y^*) = 0$. Since $g(y)$ is separable, we obtain that for every $i \in [n]$,

$$g_i'(y_i^*) = 0 \quad \Rightarrow \quad y_i^* = \pm\sqrt{\frac{-\tilde{c}_i}{3a_i}}. \tag{1}$$

Furthermore, we will need to look at the second derivative. Since y^* is a local minimizer, then $\nabla^2 g(y^*) \geqslant 0$. Again, since $g(y)$ is separable, this implies that $g_i''(y_i) \geqslant 0$ for every $i \in [n]$. Hence we have

$$g_i''(y_i) \geqslant 0 \quad \Rightarrow \quad 6a_i y_i^* \geqslant 0 \quad \Rightarrow \quad a_i \left(\pm \sqrt{\frac{-\tilde{c}_i}{3a_i}} \right) \geqslant 0. \tag{2}$$

Also, notice that we must have $\frac{-\tilde{c}_i}{3a_i} \geqslant 0$ for $\sqrt{\frac{-\tilde{c}_i}{3a_i}}$ to be a real number. Thus,

$$\operatorname{sign}(-\tilde{c}_i) = \operatorname{sign}(a_i) = \operatorname{sign}\left(\pm \sqrt{\frac{-\tilde{c}_i}{3a_i}} \right). \tag{3}$$

Finally, we relate this to $g(y^*)$.

$$\gamma^* - \sum_{i=1}^{n} \tilde{d}_i = g(y^*) = \sum_{i=1}^{n} \left(a_i(\tfrac{-\tilde{c}_i}{3a_i}) \left(\pm \sqrt{\tfrac{-\tilde{c}_i}{3a_i}} \right) + \tilde{c}_i \left(\pm \sqrt{\tfrac{-\tilde{c}_i}{3a_i}} \right) \right) = -\frac{2}{3} \sum_{i=1}^{n} |\tilde{c}_i| \sqrt{\frac{-\tilde{c}_i}{3a_i}}, \tag{4}$$

where the last equality comes from comparing the signs of the data from (3). Hence, we have

$$\sum_{i=1}^{n} |\tilde{c}_i| \sqrt{\frac{-\tilde{c}_i}{3a_i}} = -\frac{3}{2}(\gamma^* - \sum_{i=1}^{n} \tilde{d}_i). \tag{5}$$

By Lemma 3, for every $i \in [n]$, $\sqrt{\frac{-\tilde{c}_i}{3a_i}}$ is rational and has size polynomial in $\log H$ and in the size of γ^*. From (1), so does y^*, and hence x^*. $\quad\square$

Rational Solutions to Nice Cubic Feasibility Problems.

We denote by $\mathbb{Z}[x_1, \ldots, x_n]$ the set of all polynomial functions from $\mathbb{R}^n$ to $\mathbb{R}$ with integer coefficients. We next provide a standard result about Lipschitz continuity of a polynomial on a bounded region.

Lemma 4 (Lipschitz continuity of a polynomial on a box). *Let $g \in \mathbb{Z}[x_1, \ldots, x_n]$ be a polynomial of degree at most d with coefficients of absolute value at most H. Let $y, z \in [-M, M]^n$ for some $M > 0$. Then*

$$|g(y) - g(z)| \leqslant L\|y - z\|_\infty \tag{6}$$

where $L := ndHM^{d-1}(n+d)^{d-1}$.

Next, we present a result that employs Lemma 4 and that will be used in a couple of proofs in the remainder of the paper.

Proposition 1. *Let $f_i \in \mathbb{Z}[x_1, \ldots, x_n]$, for $i \in [m]$, of degree one. Let $g_j \in \mathbb{Z}[x_1, \ldots, x_n]$, for $j \in [\ell]$, of degree bounded by an integer d. Assume that the absolute value of the coefficients of f_i, $i \in [m]$, and of g_j, $j \in [\ell]$, is at most H. Let δ be a positive integer. Let $P := \{x \in \mathbb{R}^n \mid f_i(x) \leqslant 0, i \in [m]\}$ and consider the sets*

$$R := \{x \in P \mid g_j(x) \leqslant 0, j \in [\ell]\}, \quad S := \{x \in P \mid \ell \delta g_j(x) \leqslant 1, j \in [\ell]\}.$$

Assume that P is bounded. If R is nonempty, then there exists a rational vector in S of size bounded by a polynomial in $n, d, \log \ell, \log H, \log \delta$.

Proof. Since P is bounded, it follows from Lemma 8.2 in [5] that $P \subseteq [-M, M]^n$, where $M = (nH)^n$. Let L be defined as in Lemma 4, i.e.,

$$L := ndHM^{d-1}(n+d)^{d-1} = ndH(nH)^{n(d-1)}(n+d)^{d-1}.$$

Note that $\log L$ is bounded by a polynomial in $n, d, \log H$. Let $\varphi := \lceil LM\ell\delta \rceil$. Therefore $\log \varphi$ is bounded by a polynomial in $n, d, \log \ell, \log H, \log \delta$.
    We define the following $(2\varphi)^n$ boxes in $\mathbb{R}^n$ with $j_1, \dots, j_n \in \{-\varphi, \dots, \varphi - 1\}$:

$$C_{j_1,\dots,j_n} := \left\{ x \in \mathbb{R}^n \mid \frac{M}{\varphi} j_i \leq x_i \leq \frac{M}{\varphi}(j_i + 1), i \in [n] \right\}. \qquad (7)$$

Note that the union of these $(2\varphi)^n$ boxes is the polytope $[-M, M]^n$ which contains the polytope P. Furthermore, each of the $2n$ inequalities defining a box (7) has size polynomial in $n, d, \log \ell, \log H, \log \delta$.
    Let $\tilde{x}$ be a vector in R, according to the statement of the theorem. Since $\tilde{x} \in P$, there exists a box among (7), say $\tilde{C}$, that contains $\tilde{x}$. Let $\bar{x}$ be a vertex of the polytope $P \cap \tilde{C}$. Since each inequality defining P or $\tilde{C}$ has size polynomial in $n, d, \log \ell, \log H, \log \delta$, it follows from Theorem 10.2 in [22] that also $\bar{x}$ has size polynomial in $n, d, \log \ell, \log H, \log \delta$.
    To conclude the proof of the theorem we only need to show $\bar{x} \in S$. Since $\bar{x}, \tilde{x} \in \tilde{C}$, we have $\|\bar{x} - \tilde{x}\|_\infty \leq \frac{M}{\varphi}$. Then, from Lemma 4 we obtain that for each $j \in [\ell]$,

$$|g_j(\bar{x}) - g_j(\tilde{x})| \leq L\|\bar{x} - \tilde{x}\|_\infty \leq \frac{LM}{\varphi} \leq \frac{1}{\ell\delta}.$$

If $g_j(\bar{x}) \leq 0$ we directly obtain $g_j(\bar{x}) \leq \frac{1}{\ell\delta}$ since $\frac{1}{\ell\delta} > 0$. Otherwise we have $g_j(\bar{x}) > 0$. Since $g_j(\tilde{x}) \leq 0$, we obtain $g_j(\bar{x}) \leq |g_j(\bar{x}) - g_j(\tilde{x})| \leq \frac{1}{\ell\delta}$. We have shown that $\bar{x} \in S$, and this concludes the proof of the theorem.

$\square$

We are now ready to prove our main result of this section.

Theorem 2. *Let $n \in \{1, 2\}$. Let $f_i \in \mathbb{Z}[x_1, \dots, x_n]$, for $i \in [m]$, of degree one. Let $g(x) = \sum_{i=1}^n (a_i x_i^3 + b_i x_i^2 + c_i x_i + d_i) \in \mathbb{Z}[x_1, \dots, x_n]$ with $a_i \neq 0$ for $i \in [n]$. Assume that the absolute value of the coefficients of $g, f_i, i \in [m]$, is at most H. Consider the set*

$$R := \{x \in \mathbb{R}^n \mid g(x) \leq 0, \ f_i(x) \leq 0, i \in [m]\}.$$

If R is nonempty, then it contains a rational vector of size bounded by a polynomial in $\log H$. This vector provides a certificate of feasibility for R that can be checked in a number of operations that is bounded by a polynomial in $m, \log H$.

Proof. Define $P := \{x \in \mathbb{R}^n \mid f_i(x) \leq 0, i \in [m]\}$, let x^* be a vector in P minimizing $g(x)$, and let $\gamma^* := g(x^*)$. We prove separately the cases $n = 1, 2$.
    First we consider the case $n = 1$. If x^* is in the boundary of P, then, since P is an interval described by rational data, we must have x^* is an endpoint of

this interval and hence is rational and of size bounded by a polynomial in $\log H$. Thus, in the remainder of the proof we suppose that x^* is in the interior of P. In particular, x^* is the unique local minimum of g.

Since R is nonempty, we have $\gamma^* \leqslant 0$. If we have $\gamma^* = 0$, then Theorem 1 implies that the size of x^* is bounded by a polynomial in $\log H$, thus the result holds. Therefore, in the remainder of the proof we assume $\gamma^* < 0$.

Let $\tilde{c}, \tilde{d}$ be defined as in Lemma 1. Following the calculation of Theorem 1,

(5) $\gamma^* - \tilde{d} = -\frac{2}{3}|\tilde{c}|\left(\pm\sqrt{\frac{-\tilde{c}}{3a}}\right)$. Hence, $(\gamma^* - \tilde{d})^2 = -\frac{4\tilde{c}^3}{27a}$, that is, γ^* is a non-zero root of the above quadratic equation. From Lemma 2, we have that $|\gamma^*| \geqslant \frac{1}{\delta}$, where δ is an integer and $\log \delta$ is bounded by a polynomial in $\log H$. Since $\gamma^* < 0$, we have thereby shown that x^* is a vector in P satisfying $g(x^*) \leqslant -\frac{1}{\delta}$.

Clearly x^* is a root of the quadratic equation $g'(x) = 0$, thus using again Lemma 2 we obtain that $-M \leqslant x^* \leqslant M$, where M is an integer and $\log M$ is bounded by a polynomial in $\log H$. We apply Proposition 1 to the polytope $\{x \in \mathbb{R} \mid f_i(x) \leqslant 0, i \in [m], -M \leqslant x \leqslant M\}$, with $j := 1$, and with $g_1(x) := g(x) + \frac{1}{\delta}$. Proposition 1 then implies that there exists a vector $\bar{x} \in P$ with $g_1(\bar{x}) \leqslant \frac{1}{\delta}$, or equivalently $g(\bar{x}) \leqslant 0$, of size bounded by a polynomial in $\log H$. Such a vector is our certificate of feasibility.

Next, we consider the case $n = 2$. If x^* is in the boundary of P, then we can restrict to a face of P and reformulate the problem in dimension one. Then the first part of the proof ($n = 1$) secures the result. Thus, in the remainder of the proof we suppose that x^* is in the interior of P. In particular, x^* is the unique local minimum of g. As in the proof of the case $n = 1$, due to Theorem 1, we can assume $\gamma^* < 0$.

For every $i \in [n]$, let $\tilde{c}_i, \tilde{d}_i$ be defined as in Lemma 1. Following the calculation of Theorem 1, (5) $\sum_{i=1}^{2} |\tilde{c}_i|\sqrt{\frac{-\tilde{c}_i}{3a_i}} = -\frac{3}{2}(\gamma^* - \sum_{i=1}^{2} \tilde{d}_i)$. Squaring both sides we obtain

$$\sum_{i=1}^{2} \frac{-\tilde{c}_i^3}{3a_i} + 2|\tilde{c}_1||\tilde{c}_2|\sqrt{\frac{\tilde{c}_1\tilde{c}_2}{9a_1a_2}} = \frac{9}{4}(\gamma^* - \sum_{i=1}^{2} \tilde{d}_i)^2.$$

If we isolate the square root, and then square again both sides of the equation, we obtain that γ^* is a non-zero root of a quartic equation with rational coefficients. From Lemma 2, we have that $|\gamma^*| \geqslant \frac{1}{\delta}$, where δ is an integer and $\log \delta$ is bounded by a polynomial in $\log H$. Since $\gamma^* < 0$, we have thereby shown that x^* is a vector in P satisfying $g(x^*) \leqslant -\frac{1}{\delta}$.

Clearly, for $i \in [2]$, x_i^* is a root of the quadratic equation

$$(a_i x_i^3 + b_i x_i^2 + c_i x_i + d_i)' = 3a_i x_i^2 + 2b_i x_i + c_i = 0,$$

thus using again Lemma 2 we obtain that $-M \leqslant x_i^* \leqslant M$, where M is an integer and $\log M$ is bounded by a polynomial in $\log H$. We apply Proposition 1 to the polytope $\{x \in \mathbb{R}^2 \mid f_i(x) \leqslant 0, i \in [m], -M \leqslant x_i \leqslant M, i \in [2]\}$, with $j := 1$, and with $g_1(x) := g(x) + \frac{1}{\delta}$. Proposition 1 then implies that there exists a vector $\bar{x} \in P$ with $g_1(\bar{x}) \leqslant \frac{1}{\delta}$, or equivalently $g(\bar{x}) \leqslant 0$, of size bounded by a polynomial in $\log H$. Such a vector is our certificate of feasibility.

To conclude the proof for both cases $n = 1$ and $n = 2$, we bound the number of operations needed to check if $\bar{x}$ is in R by substituting $\bar{x}$ in the $m + 1$ inequalities defining R. It is simple to check that $\bar{x}$ satisfies these $m + 1$ inequalities in a number of operations that is bounded by a polynomial in $m, \log H$.   □

In Example 1 in Sect. 3, we will see that Theorem 2 is best possible. We also remark that Theorem 2 implies that the corresponding feasibility problem is in the complexity class NP.

3   NP-Hardness of Determining Existence of Rational Feasible Solutions

We begin with two, previously unknown, motivating examples.

Example 1 (Feasible system with no rational feasible vector). Define

$$h(y) := 2y_1^3 + y_2^3 - 6y_1y_2 + 4, \quad R_\gamma := [1.259 - \gamma, 1.26] \times [1.587, 1.59]. \quad (8)$$

Then $\{y \in R_0 : h(y) \leqslant 0\} = \{y^*\}$ where $y^* = (2^{\frac{1}{3}}, 2^{\frac{2}{3}}) \approx (1.2599, 1.5874)$.

In particular, for this bi-variate set described by one cubic constraint and linear inequalities there is a unique feasible solution, and it is irrational.   ◇

Observation 3. *(1) The point $y^* \in R_\gamma$ for all $\gamma \geqslant 0$. (2) y^* is the unique minimizer of $h(y)$ for $y \in \mathbb{R}_+^2$. (3) For $y \in R_4$, $h(y) > -12$. (4) The point $\bar{y} = (-2.74, 1.588) \in R_4$ attains $h(\bar{y}) < -7$.*

Example 2 (Exponentially Small Solutions). Next, consider the following system of quadratic inequalities on variables $d_1, \ldots, d_N$ and s such that

$$d_1 \leqslant \frac{1}{2}, s \leqslant d_N^2, \quad s \geqslant 0, \quad d_{k+1} \leqslant d_k^2 \text{ for } k \in [N-1], \ d_k \geqslant 0 \text{ for } k \in [N]. \quad (9)$$

In any feasible solution to system (9) we either have $s = 0$, or $0 < s \leqslant 2^{-2^N}$ and in this case if s is rational then we need more than 2^N bits to represent it. Further, there are rational solutions to system (9) with $s > 0$.

Remark. The literature abounds with examples of SOCPs all of whose solutions are doubly exponentially large, see e.g. [2], [20]. Our example is similar, however it is *non-convex*. Further, unlike in the SOCP examples, the solutions have magnitude that is upper-bounded by a value independent on n and N and exhibit an 'either-or' behavior.   ◇

NP-Hardness Construction. We will show that it is strongly NP-hard to test whether a system of quadratic inequalities which is known to have feasible rational solutions, admits feasible rational solutions of polynomial size (Theorem 4). The same proof technique shows that it is hard to test whether a feasible system of quadratic inequalities has a rational solution (Theorem 5).

The main reduction is from the problem 3SAT. An instance of this problem is defined by n *literals* $w_1, \ldots, w_n$ as well as their negations $\bar{w}_1, \ldots, \bar{w}_n$, and a set of m clauses $C_1, \ldots C_m$ where each clause C_i is of the form $(u_{i1} \vee u_{i2} \vee u_{i3})$. Here, each u_{ij} is a literal or its negation, and $\vee$ means 'or'. The problem is to find 'true' or 'false' values for each literal, and corresponding values for their negations, so that the formula

$$C_1 \wedge C_2 \ldots \wedge C_m \tag{10}$$

is true, where $\wedge$ means 'and'.

Let $N \geqslant 1$ be an integer. For now N is generic; below we will discuss particular choices. Given an instance of 3SAT as above, we construct a system of quadratic inequalities on the following $2n + N + 7$ variables:

- For each literal w_j we have a variable x_j; for $\bar{w}_j$ we use variable x_{n+j}.
- Additional variables $\gamma, \Delta, y_1, y_2, s$, and $d_1, \ldots, d_N$.

We describe the constraints in our quadratically constrained problem[2]. For each clause $C_i = (u_{i_1} \vee u_{i_2} \vee u_{i_3})$ associate the variable x_{i_k} is with u_{i_k} for $1 \leqslant k \leqslant 3$.

$$-1 \leqslant x_j \leqslant 1 \text{ for } j \in [2n], \quad x_j + x_{n+j} = 0, \quad \text{for } j \in [n]. \tag{11a}$$

$$x_{i_1} + x_{i_2} + x_{i_3} \geqslant -1 - \Delta, \text{ for each clause } C_i = (u_{i_1} \vee u_{i_2} \vee u_{i_3}) \tag{11b}$$

$$0 \leqslant \gamma, \ 0 \leqslant \Delta \leqslant 2, \ \Delta + \frac{\gamma}{2} \leqslant 2, \ (y_1, y_2) \in R_\gamma, \tag{11c}$$

$$-n^5 \sum_{j=1}^{n} x_j^2 + h(y) - s \leqslant -n^6. \tag{11d}$$

$$d_1 \leqslant \frac{1}{2}, s \leqslant d_N^2, \quad s \geqslant 0, \quad d_{k+1} \leqslant d_k^2 \text{ for } k \in [N-1], \ d_k \geqslant 0 \ \text{ for } k \in [N], \tag{11e}$$

Theorem 4. *Let $n \geqslant 3$ and $N \geqslant 2$. The formula* (10) *is satisfiable if and only if there is a rational solution to* (11) *of size polynomial in n, m and N. As a result, it is strongly NP-hard to test if a system of quadratic inequalities has a rational feasible solution of polynomial size, even if the system admits rational feasible solutions.*

Comment. The choice $N = n$ in the above construction is natural and yields a result directly interpretable in terms of the formula (10).

Proof. **Claim 1:** *System* (11) *has a rational feasible solution.* To see this, set $x_j = 1 = -x_{n+j}$ for $j \in [n]$, $\Delta = 2$, $\gamma = 0$, $d_k = 2^{-2^{k-1}}$ for $k \in [N]$ and $s = 2^{-2^N}$. By inspection these rational values satisfy (11a), (11b), (11c), and (11e). Since y^* and y^* is in the interior of R_0, there exists a small ball $B \subseteq R_0$ containing y^*. Since h is continuous, y^* is the unique local minimizer of h, $h(y^*) = 0$, and

---

[2] Constraint (11e) as written is cubic, but is equivalent to three quadratic constraints by defining new variables $y_1^2 = y_{12}, y_2^2 = y_{22}$, and rewriting the constraint as $-n^5 \sum_{j=1}^{n} x_j^2 + y_{12} y_1 + y_{22} y_2 - 6 y_1 y_2 + 4 - s \leqslant -n^6$.

the set $\mathbb{Q}^2$ is dense in $\mathbb{R}^2$, there exists a feasible rational choice of y such that $h(y) \leqslant s$. Hence, (11d) is also satisfied.

Claim 2: *Suppose* $(x, \gamma, \Delta, y, d, s)$ *is feasible for* (11). *Then for* $1 \leqslant j \leqslant 2n$, $|x_j| \geqslant 1 - \frac{12}{n^5} - \frac{2^{-2^N}}{n^5}$.

To begin, let $\sigma^2 := \min_j\{x_j^2\}$. Then $-n^5(n-1) - n^5\sigma^2 + h(y) - s \leqslant -n^6$, so $\sigma^2 \geqslant 1 + (h(y) - s)/n^5 \geqslant 1 - \frac{12}{n^5} - \frac{2^{-2^N}}{n^5}$, where the last inequality follows from Observation 3 and the fact that (11d) implies that $s \leqslant 2^{-2^N}$. $\qquad\square$

Claim 3: *Suppose formula* (10) *is satisfiable. Then* (11) *has a rational feasible solution of polynomial size.*

For $1 \leqslant j \leqslant n$ set $x_j = 1 = -x_{n+j}$ if w_j is true, else set $x_j = -1 = -x_{n+j}$. Set $\Delta = 0$, $\gamma = 4$, and $d_1 = \ldots = d_N = s = 0$. Finally (Observation 3) we set $(y_1, y_2) = (-2.74, 1.588)$. $\qquad\square$

The next result concludes the proof of Theorem 4.

Claim 4: *Suppose formula* (10) *is not satisfiable. Then in every feasible rational solution to* (11), *s has size at least* 2^N.

Let $(x, \gamma, \Delta, y, d, s)$ be feasible. For $1 \leqslant j \leqslant n$ set w_j to be true if $x_j > 0$ and false otherwise. It follows that there is at least one clause $C_i = (u_{i1} \vee u_{i2} \vee u_{i3})$ such that every u_{ik} (for $1 \leqslant k \leqslant 3$) is false, i.e. each $x_{ik} < 0$. Using constraint (11b) and Claim 2, we obtain

$$-3 + \frac{36}{n^5} + 3\frac{2^{-2^N}}{n^5} \geqslant -1 - \Delta, \text{ and by (11c) } \gamma \leqslant \frac{72}{n^5} + 6\frac{2^{-2^N}}{n^5} < 1.$$

This fact has two implications. First, since $(y_1, y_2) \in R_\gamma \subset \mathbb{R}^2_+$, Observation 3 implies $h(y) > 0$ (because y is rational). Second, constraint (11d) implies $-n^5 \sum_{j=1}^n x_j^2 + h(y) - s \leqslant -n^5$ and, therefore, $h(y) \leqslant s$. So $s > 0$ and since $s \leqslant 2^{-2^N}$ (by constraint (11c)) the proof is complete. $\qquad\square$

As an (easy) extension of Theorem 4 we have the following theorem.

Theorem 5. *It is strongly NP-hard to test if there exists a rational solution to a system of the form*

$$f(x) \leqslant 0, \quad Ax \leqslant b,$$

where $f \in \mathbb{Z}[x_1, \ldots, x_n]$ *is of degree 3, and* $A \in \mathbb{Q}^{m \times n}$, $b \in \mathbb{Q}^m$.

Proof sketch. We proceed with a transformation from 3SAT just as above, except that we dispense with the variables $d_1, \ldots, d_N$ and s and, rather than constraint (11d) we impose

$$-n^5 \sum_{j=1}^n x_j^2 + h(y) \leqslant -n^6. \qquad (12)$$

In the analog of Claim 4 we conclude that $y \in \mathbb{R}^2_+$ while also $h(y) \leqslant 0$ (no term $-s$) which yields that $y = y^*$. $\qquad\square$

4   Short Certificate of Feasibility: An Almost Feasible Point

In this section we are interested in the existence of *short* certificates of feasibility for systems of polynomial inequalities, i.e., certificates of feasibility of size bounded by a polynomial in the size of the system.

We will be using several times the functions ϵ and δ defined as follows:

$$\epsilon(n, m, d, H) := (2^{4-\frac{n}{2}} \max\{H, 2n + 2m\}d^n)^{-n2^n d^n},$$

$$\delta(n, m, d, H) := \lceil 2\epsilon^{-1}(n, m, d, H)\rceil = \lceil 2(2^{4-\frac{n}{2}} \max\{H, 2n + 2m\}d^n)^{n2^n d^n}\rceil.$$

A fundamental ingredient in our arguments is the following result by Geronimo, Perrucci, and Tsigaridas, which follows from Theorem 1 in [10].

Theorem 6. *Let $n \geqslant 2$. Let $g, f_i \in \mathbb{Z}[x_1, \ldots, x_n]$, for $i \in [m]$, of degree bounded by an even integer d. Assume that the absolute value of the coefficients of g, f_i, $i \in [m]$, is at most H. Let $T := \{x \in \mathbb{R}^n \mid f_i(x) \leqslant 0, i \in [m]\}$, and let C be a compact connected component of T. Then, the minimum value that g takes over C, is either zero, or its absolute value is greater than or equal to $\epsilon(n, m, d, H)$.*

Using Theorem 6 we obtain the following lemma.

Lemma 5. *Let $n \geqslant 2$. Let $g, f_i \in \mathbb{Z}[x_1, \ldots, x_n]$, for $i \in [m]$, of degree bounded by an even integer d. Assume that the absolute value of the coefficients of g, f_i, $i \in [m]$, is at most H. Let $\delta := \delta(n, m, d, H)$. Let $T := \{x \mid f_i(x) \leqslant 0, i \in [m]\}$ and consider the sets*

$$R := \{x \in T \mid g(x) \leqslant 0\}, \quad S := \{x \in T \mid \delta g(x) \leqslant 1\}.$$

Assume that T is bounded. Then R is nonempty if and only if S is nonempty.

Proof. Since $\delta > 0$ we have $R \subseteq S$, therefore if R is nonempty also S is nonempty. Hence we assume that S is nonempty and we show that R is nonempty.

Since S is nonempty, there exists a vector $\bar{x} \in T$ with $g(\bar{x}) \leqslant 1/\delta < \epsilon(n, m, d, H)$. Let C be a connected component of T containing $\bar{x}$. Since T is compact, we have that C is compact as well. In particular, the minimum value that g takes over C is less than $\epsilon(n, m, d, H)$. The contrapositive of Theorem 6 implies that the minimum value that g takes over C is less than or equal to zero. Thus there exists $\tilde{x} \in C$ with $g(\tilde{x}) \leqslant 0$. Hence the set R is nonempty. $\square$

From Lemma 5 we obtain the following result.

Proposition 2. *Let $n \geqslant 2$. Let $f_i, g_j \in \mathbb{Z}[x_1, \ldots, x_n]$, for $i \in [m], j \in [\ell]$, of degree bounded by an even integer d. Assume that the absolute value of the coefficients of f_i, $i \in [m]$, and of g_j, $j \in [\ell]$, is at most H. Let $\delta := \delta(n, m + \ell, 2d, \ell H^2)$. Let $T := \{x \mid f_i(x) \leqslant 0, i \in [m]\}$ and consider the sets*

$$R := \{x \in T \mid g_j(x) \leqslant 0, j \in [\ell]\}, \quad S := \{x \in T \mid \ell \delta g_j(x) \leqslant 1, j \in [\ell]\}.$$

Assume that T is bounded. Then R is nonempty if and only if S is nonempty.

Proof. Since $\ell\delta > 0$ we have $R \subseteq S$, therefore if R is nonempty also S is nonempty. Hence we assume that S is nonempty and we show that R is nonempty.

Let $\bar{x} \in S$, and define the index set $J := \{j \in [\ell] : g_j(\bar{x}) > 0\}$. We introduce the polynomial function $g \in \mathbb{Z}[x_1, \ldots, x_n]$ defined by $g(x) := \sum_{j \in J} g_j^2(x)$. Note that the degree of g is bounded by $2d$. The absolute value of the coefficients of each g_j^2 is at most H^2, hence the absolute value of the coefficients of g is at most ℓH^2. Next, let $T' := \{x \in T \mid g_j(x) \leqslant 0, j \in [\ell] \setminus J\}$ and

$$R' := \{x \in T' \mid g(x) \leqslant 0\}, \quad S' := \{x \in T' \mid \delta g(x) \leqslant 1\}.$$

First, we show that the vector $\bar{x}$ is in the set S', implying that S' is nonempty. Clearly $\bar{x} \in T$, and for every $j \in [\ell] \setminus J$ we have that $g_j(\bar{x}) \leqslant 0$, thus we have $\bar{x} \in T'$. For every $j \in J$, we have $0 < g_j(\bar{x}) \leqslant \frac{1}{\ell\delta}$, and since $\ell\delta \geqslant 1$, we have $0 < g_j^2(\bar{x}) \leqslant \frac{1}{\ell\delta}$. Thus, we obtain $g(\bar{x}) \leqslant \frac{\ell}{\ell\delta} = \frac{1}{\delta}$. We have thus proved $\bar{x} \in S'$, and so S' is nonempty.

Next, we show that the set R' is nonempty. To do so, we apply Lemma 5 to the sets T', R', S'. The number of inequalities that define T' is a number m' with $m \leqslant m' \leqslant m + \ell$. The degree of f_i, g_j, g, for $i \in [m]$, $j \in [\ell] \setminus J$, is bounded by $2d$. The absolute value of the coefficients of f_i, g_j, g, for $i \in [m]$, $j \in [\ell] \setminus J$, is at most ℓH^2. Since the function $\delta(n, m, d, H)$ is increasing in m and $m' \leqslant m + \ell$, we obtain from Lemma 5 that R' is nonempty if and only if S' is nonempty. Since S' is nonempty, we obtain that R' is nonempty.

Finally, we show that the set R is nonempty. Since R' is nonempty, let $\tilde{x} \in R'$. From the definition of R' we then know $\tilde{x} \in T$, $g_j(\tilde{x}) \leqslant 0$, for $j \in [\ell] \setminus J$, and $g(\tilde{x}) \leqslant 0$. Since g is a sum of squares, $g(\tilde{x}) \leqslant 0$ implies $g(\tilde{x}) = 0$, and this in turn implies $g_j(\tilde{x}) = 0$ for every $j \in J$. Hence $\tilde{x} \in R$, and R is nonempty. $\qquad\square$

Proposition 2 and Proposition 1 directly yield our following main result.

Theorem 7 (Certificate of polynomial size). *Let $n \geqslant 2$. Let $f_i \in \mathbb{Z}[x_1, \ldots, x_n]$, for $i \in [m]$, of degree one. Let $g_j \in \mathbb{Z}[x_1, \ldots, x_n]$, for $j \in [\ell]$, of degree bounded by an even integer d. Assume that the absolute value of the coefficients of f_i, $i \in [m]$, and of g_j, $j \in [\ell]$, is at most H. Let $\delta := \delta(n, m+\ell, 2d, \ell H^2)$. Let $P := \{x \in \mathbb{R}^n \mid f_i(x) \leqslant 0, i \in [m]\}$ and consider the sets*

$$R := \{x \in P \mid g_j(x) \leqslant 0, j \in [\ell]\}, \quad S := \{x \in P \mid \ell\delta g_j(x) \leqslant 1, j \in [\ell]\}.$$

Assume that P is bounded. Denote by s the maximum number of terms of g_j, $j \in [\ell]$, with nonzero coefficients. If R is nonempty, then there exists a rational vector in S of size bounded by a polynomial in $d, \log m, \log \ell, \log H$, for n fixed. This vector is a certificate of feasibility for R that can be checked in a number of operations that is bounded by a polynomial in $s, d, m, \ell, \log H$, for n fixed.

Proof. From the definition of δ, we have that $\log \delta$ is bounded by a polynomial in $d, \log m, \log \ell, \log H$, for n fixed. From Proposition 1, there exists a vector $\bar{x} \in S$ of size bounded by a polynomial in $d, \log m, \log \ell, \log H$, for n fixed. Such a vector is our certificate of feasibility. In fact, from Proposition 2 (applied to the sets $T = P, R, S$), we know that S nonempty implies R nonempty.

To conclude the proof we bound the number of operations needed to check if the vector $\bar{x}$ is in S by substituting $\bar{x}$ in the $m + \ell$ inequalities defining S.

The absolute value of the coefficients of $f_i(x) \leqslant 0$, $i \in [m]$, is at most H. Thus, it can be checked that $\bar{x}$ satisfies these m inequalities in a number of operations that is bounded by a polynomial in $d, m, \log \ell, \log H$, for n fixed.

Next, we focus on the inequalities $\ell \delta g_j(x) \leqslant 1$, $j \in [\ell]$. Note that the total number of terms of $\ell \delta g_j$, $j \in [\ell]$, with nonzero coefficients is bounded by $s\ell$. The logarithm of the absolute value of each nonzero coefficient of $\ell \delta g_j(x)$, $j \in [\ell]$, is bounded by $\log(\ell \delta H)$, which in turn is bounded by a polynomial in $d, \log m, \log \ell, \log H$, for n fixed. Therefore, it can be checked that $\bar{x}$ satisfies these ℓ inequalities in a number of operations that is bounded by a polynomial in $s, d, \ell, \log m, \log H$, for n fixed. □

In particular, Theorem 7 implies that polynomial optimization is in NP, provided that we fix the number of variables.

As mentioned in the introduction, this fact is not new. In fact, it follows from Theorem 1.1 in Renegar [21] that the problem of deciding whether the set R, as defined in Theorem 7, is nonempty can be solved in a number of operations that is bounded by a polynomial in $s, d, m, \ell, \log H$, for n fixed. Therefore, Renegar's algorithm, together with its proof, provides a certificate of feasibility of size bounded by a polynomial in the size of the system, which in turns implies that the decision problem is in NP.

The main advantages of Theorem 7 over Renegar's result are that (i) our certificate of feasibility is simply a vector in S of polynomial size, and (ii) the feasibility of the system can be checked by simply plugging the vector into the system of inequalities defining S. The advantages of Renegar's result over our Theorem 7 are: (iii) Renegar does not need to assume that the feasible region is bounded, while we do need that assumption, and (iv) Renegar shows that the decision problem is in P, while we show that it is in the larger class NP.

References

1. Albu, T.: The irrationality of sums of radicals via Cogalois theory. Analele stiintifice ale Universitatii Ovidius Constanta **19**(2), 15–36 (2011)
2. Alizadeh, F.: Interior point methods in semidefinite programming with applications to combinatorial optimization. SIAM J. Optim. **5**, 13–51 (1995)
3. Barvinok, A.I.: Feasibility testing for systems of real quadratic equations. Discrete Comput. Geom. **10**(1), 1–13 (1993). https://doi.org/10.1007/BF02573959
4. Basu, S., Pollack, R., Roy, M.F.: Algorithms in Real Algebraic Geometry. AACIM, vol. 10. Springer, Heidelberg (2006). https://doi.org/10.1007/3-540-33099-2
5. Bertsimas, D., Tsitsiklis, J.: Introduction to Linear Optimization. Athena Scientific, Belmont (1997)
6. Bienstock, D.: A note on polynomial solvability of the CDT problem. SIAM J. Optim. **26**, 486–496 (2016)

7. Bienstock, D., Del Pia, A., Hildebrand, R.: Complexity, exactness, and rationality in polynomial optimization. Optimization (2020). http://www.optimization-online.org/DB_HTML/2020/11/8105.html. 2020/11/8105

8. De Loera, J.A., Lee, J., Malkin, P.N., Margulies, S.: Computing infeasibility certificates for combinatorial problems through Hilbert's Nullstellensatz. J. Symbolic Comput. **46**(11), 1260–1283 (2011). https://doi.org/10.1016/j.jsc.2011.08.007. http://www.sciencedirect.com/science/article/pii/S0747717111001192

9. Del Pia, A., Dey, S.S., Molinaro, M.: Mixed-integer quadratic programming is in NP. Math. Program. **162**(1–2), 225–240 (2016). https://doi.org/10.1007/s10107-016-1036-0

10. Geronimo, G., Perrucci, D., Tsigaridas, E.: On the minimum of a polynomial function on a basic closed semialgebraic set and applications. SIAM J. Optim. **23**(1), 241–255 (2013). https://doi.org/10.1137/110857751

11. Grigoriev, D., Vorobjov, N.: Complexity of Null- and Positivstellensatz proofs. Ann. Pure Appl. Logic **113**(1), 153–160 (2001). https://doi.org/10.1016/S0168-0072(01)00055-0. http://www.sciencedirect.com/science/article/pii/S0168007201000550. First St. Petersburg Conference on Days of Logic and Computability

12. Hochbaum, D.S.: Complexity and algorithms for nonlinear optimization problems. Ann. Oper. Res. **153**(1), 257–296 (2007). https://doi.org/10.1007/s10479-007-0172-6

13. Karmarkar, N.: A new polynomial-time algorithm for linear programming. Combinatorica **4**(4), 373–395 (1984). https://doi.org/10.1007/bf02579150

14. Karmarkar, N.: An interior-point approach to NP-complete problems (1989). manuscript

15. Khachiyan, L.: Polynomial algorithms in linear programming. USSR Comput. Math. Math. Phys. **20**(1), 53–72 (1980). https://doi.org/10.1016/0041-5553(80)90061-0. http://www.sciencedirect.com/science/article/pii/0041555380900610

16. Letchford, A., Parkes, A.J.: A guide to conic optimisation and its applications. RAIRO-Oper. Res. **52**, 1087–1106 (2018)

17. O'Donnell, R.: SOS is not obviously automatizable, even approximately. In: Papadimitriou, C.H. (ed.) 8th Innovations in Theoretical Computer Science Conference (ITCS 2017). Leibniz International Proceedings in Informatics (LIPIcs), vol. 67, pp. 59:1–59:10. Dagstuhl, Germany (2017). https://doi.org/10.4230/LIPIcs.ITCS.2017.59, http://drops.dagstuhl.de/opus/volltexte/2017/8198

18. Pardalos, P.M., Vavasis, S.A.: Quadratic programming with one negative eigenvalue is NP-hard. J. Global Optim. **1**(1), 15–22 (1991). https://doi.org/10.1007/BF00120662

19. Pólik, S., Terlaky, T.: A survey of the S-lemma. SIAM Rev. **49**, 371–418 (2007)

20. Ramana, M.: An exact duality theory for semidefinite programming and its complexity implications. Math. Program. **77**, 129–162 (1997)

21. Renegar, J.: On the computational complexity and geometry of the first-order theory of the reals. Part I: introduction. Preliminaries. The geometry of semi-algebraic sets. The decision problem for the existential theory of the reals. J. Symbolic Comput. **13**, 255–299 (1992)

22. Schrijver, A.: Theory of Linear and Integer Programming. Wiley, New York (1986)

23. Vavasis, S., Zippel, R.: Proving polynomial-time for sphere-constrained quadratic programming. Technical report 90–1182, Department of Computer Science, Cornell University (1990)

24. Vavasis, S.A.: Quadratic programming is in NP. Inf. Process. Lett. **36**(2), 73–77 (1990). https://doi.org/10.1016/0020-0190(90)90100-C. http://www.science direct.com/science/article/pii/002001909090100C
25. Waki, H., Nakata, M., Muramatsu, M.: Strange behaviors of interior-point methods for solving semidefinite programming problems in polynomial optimization. Comput. Optim. Appl. **53**, 823–844 (2012)
26. Ye, Y.: A new complexity result on minimization of a quadratic function with a sphere constraint. In: Floudas, A., Pardalos, P. (eds.) Recent Advances in Global Optimization, pp. 19–31. Princeton University Press, Princeton (1992)

On the Geometry of Symmetry Breaking Inequalities

José Verschae[1]($\boxtimes$) (iD), Matías Villagra[2,3]($\boxtimes$) (iD), and
Léonard von Niederhäusern[4,5]($\boxtimes$) (iD)

[1] Institute for Mathematical and Computational Engineering,
Faculty of Mathematics and School of Engineering,
Pontificia Universidad Católica de Chile, Santiago, Chile
jverschae@uc.cl
[2] Faculty of Mathematics, Pontificia Universidad Católica de Chile, Santiago, Chile
mjvillagra@uc.cl
[3] IEOR, Columbia University, New York City, USA
[4] Institute for Engineering Sciences, Universidad de O'Higgins, O'Higgins, Chile
leonard.vonniederhausern@uoh.cl
[5] Centro de Modelamiento Matemático (AFB170001 - CNRS UMI 2807),
Universidad de Chile, Santiago, Chile

Abstract. Breaking symmetries is a popular way of speeding up the branch-and-bound method for symmetric integer programs. We study polyhedra that break all symmetries, namely, fundamental domains. Our long-term goal is to understand the relationship between the complexity of such polyhedra and their symmetry breaking capability.

Borrowing ideas from geometric group theory, we provide structural properties that relate the action of the group with the geometry of the facets of fundamental domains. Inspired by these insights, we provide a new generalized construction for fundamental domains, which we call generalized Dirichlet domain (GDD). Our construction is recursive and exploits the coset decomposition of the subgroups that fix given vectors in $\mathbb{R}^n$. We use this construction to analyze a recently introduced set of symmetry breaking inequalities by Salvagnin [23] and Liberti and Ostrowski [15], called Schreier-Sims inequalities. In particular, this shows that every permutation group admits a fundamental domain with less than n facets. We also show that this bound is tight.

Finally, we prove that the Schreier-Sims inequalities can contain an exponential number of isomorphic binary vectors for a given permutation group G, which provides evidence of the lack of symmetry breaking effectiveness of this fundamental domain. Conversely, a suitably constructed GDD for G has linearly many inequalities and contains unique representatives for isomorphic binary vectors.

Keywords: Symmetry breaking inequalities · Fundamental domains · Polyhedral theory · Orthogonal groups

© Springer Nature Switzerland AG 2021
M. Singh and D. P. Williamson (Eds.): IPCO 2021, LNCS 12707, pp. 73–88, 2021.
https://doi.org/10.1007/978-3-030-73879-2_6

1 Introduction

Symmetries are mappings from one object into itself that preserve its structure. Their study has proven fruitful across a myriad of fields, including integer programming, where symmetries are commonly present. For instance, almost 30% of mixed-integer linear programs (MILP) in the model library used by the solver CPLEX are considerably affected by symmetry [1]. Moreover, symmetry exploitation techniques are of importance in various situations. In particular, they help to avoid traversing symmetric branches of the tree considered by a branch-and-bound algorithm.

Roughly speaking, the symmetry group G of an optimization problem is the set of functions in $\mathbb{R}^n$ that leave the feasible region and the objective function invariant (see Sect. 2 for a precise definition). The symmetry group G, or any of its subgroups, partitions $\mathbb{R}^n$ into G-orbits, which are sets of isomorphic solutions. A natural technique for handling symmetries is to add a static set of symmetry breaking inequalities. That is, we add extra inequalities that remove isomorphic solutions while leaving at least one representative per G-orbit. This well established approach has been studied extensively, both in general settings and different applications; see e.g. [8–10,12–15,20,23,26]. In most of these works, the symmetry breaking inequalities select the lexicographically maximal vector in each G-orbit of binary vectors. However, this constitutes a major drawback when dealing with general permutation groups: selecting the lexicographically maximal vector in a G-orbit is an NP-hard problem [2]. Hence, the separation problem of the corresponding symmetry breaking inequalities is also NP-hard. On the other hand, there is nothing preventing us to select orbits' representatives with a different criteria.

In this article, we are interested in understanding *fundamental domains* of a given group G, which are sets that break all the symmetries of G. Ideally, a fundamental domain F contains a unique representative per G-orbit. However, such a set does not necessarily exist for every group. Instead, a fundamental domain F only contains a unique representative for G-orbits that intersect F in its interior, while it can contain one or more representatives of a G-orbit intersecting its boundary. Despite this, F breaks all symmetries in a group, as any proper closed subset of F leaves some G-orbit unrepresented. On the other hand, a given symmetry group can admit inherently different fundamental domains. While all fundamental domains for finite orthogonal groups, including permutation groups (the main focus when considering mixed integer linear programs), are polyhedral cones, their polyhedral structure and complexity might differ greatly.

Our long term and ambitious goal is to understand the tension (and potential trade-offs) between the symmetry breaking effectiveness and the complexity of fundamental domains. The complexity can be measured in several ways: from the sizes of the coefficients in its matrix description, the number of facets, or even its extension complexity. On the other hand, the symmetry breaking effectiveness is related to the number of representatives that each orbit contains. Hence, the boundary of a fundamental domain, which can contain overrepresented G-orbits, becomes problematic in particular if our points of interest (e.g., binary points in a binary integer program) can lie within it.

More precisely, we contribute to the following essential questions: (i) Which groups admit fundamental domains in $\mathbb{R}^n$ with poly(n) facets? (ii) What is the structure of these facets? (iii) Which algorithmic methods can we use to construct different fundamental domains? (iv) Which fundamental domains contain unique representatives for every orbit?

Related Work. The concept of fundamental domain traces back to the 19th century; see [21] and the references therein. In particular Dirichlet [7] gives a construction which implies the existence of a fundamental domain in a general context, including all groups of isometries in $\mathbb{R}^n$.

Kaibel and Pfetsch [13] introduce the concept of *orbitopes* as the convex-hull of 0–1 matrices that are lexicographically maximal under column permutations, and give a complete description of the facets for the cyclic group and the symmetric group. Friedman [9] considers general permutation groups. Based on the Dirichlet Domain, he introduces the idea of a universal ordering vector, which yields a fundamental domain with unique representatives of binary points. On the other hand, this fundamental domain has an exponential number of facets, its defining inequalities can contain exponentially large coefficients in n, and the separation problem is NP-hard for general permutation groups [2]. Liberti [14] and later Dias and Liberti [6] also consider general permutation groups G and derive a class of symmetry breaking constraints by studying the orbits of G acting on $[n] = \{1, \ldots, n\}$. Liberti and Ostrowski [15], and independently Salvagnin [23], extend this construction and introduce a set of symmetry breaking inequalities based on a chain of pointwise coordinate stabilizers. We will refer to this set as the Schreier-Sims inequalities, as they are strongly related to the *Schreier-Sims table* from computational group theory [25]. Hojny and Pfetsch [12] study *symretopes*, defined as the convex hulls of lexicographically maximal vectors in binary orbits. They obtain a linear time algorithm for separating the convex hull of polytopes derived by a single lexicographic order enforcing inequality and show how to exploit this construction computationally.

For integer programming techniques, dynamic methods have been used to deal with symmetries within the Branch-and-Bound tree. Some methods are *Orbital Fixing* [17], *Isomorphism Pruning* [16] and *Orbital Branching* [19]. A more geometric approach for solving symmetric integer programs relies on the theory of *core points* [5,11]. Schürmann [24] provides an interesting overview of symmetry exploitation techniques for areas beyond integer programming.

Our Contribution. In this article we focus on finite orthogonal groups in $\mathbb{R}^n$, that is, groups of linear isometries. We start by presenting basic structural results from the theory of fundamental domains for a given orthogonal group G. A basic observation is that each facet is related to a group element g. We also show the following new property of the facets: for an interesting class of fundamental domains, which we call *subgroup consistent*, the vector defining a facet must be orthogonal to the invariant subspace of g. This implies that each inequality is of the form $\alpha^t x \geq \alpha^t(gx)$ for some vector $\alpha \in \mathbb{R}^n$ and some element $g \in G$. In other words, the inequalities of any subgroup consistent fundamental domain have the same structure as inequalities of Dirichlet domains.

Inspired by these new insights, we state our main contribution: a generalized construction of fundamental domains for any finite orthogonal group, including permutation groups. Our method is based on choosing a vector α and finding the coset decomposition using the stabilizer subgroup $G_\alpha = \{g \in G : g\alpha = \alpha\}$. We add inequalities to our fundamental domain, one for each member in the coset decomposition. Therefore, for a well chosen α the number of cosets can be bounded, yielding a polynomial number of inequalities. Then, we proceed recursively on the subgroup G_α. We say that a fundamental domain obtained with this method is a *generalized Dirichlet domain* (GDD), as it generalizes the classical construction by Dirichlet [7]. To the best of our knowledge, this construction generalizes all convex fundamental domains found in the literature. For the special case of permutation groups, our construction can be computed in quasi-polynomial time[1]. Polynomial time can even be guaranteed for appropriate choices of the vectors α in the recursive construction.

A natural way of breaking symmetries is to choose the lexicographically maximal element for *every* G-orbit in $\mathbb{R}^n$ (not only binary vectors, as in the construction by Friedman [9]). However, it is not hard to see that the obtained set is not necessarily closed. On the other hand, the set is convex. We show that the closure of this set coincides with the Schreier-Sims inequalities studied by Salvagnin [23] and Liberti and Ostrowski [15]. Moreover, we show that this set is a GDD, which implies that it is a fundamental domain. Finally, we give a stronger bound on the number of facets for this fundamental domain, implying that all permutation groups admit a fundamental domain, and hence a set that breaks all symmetries, with at most $n - 1$ inequalities. We also show that there are groups for which any fundamental domain has $\Omega(n)$ facets.

Salvagnin [23] recognizes that the symmetry breaking efficiency of the Schreier-Sims inequalities might be limited: the orbit of a binary vector can be overrepresented in the set. We give a specific example of a permutation group in which an orbit of binary vectors can have up to $2^{\Omega(n)}$ many representatives. Using the flexibility given by our GDD construction, we exhibit a fundamental domain for the same group with a unique representative for each binary orbit, while having at most $O(n)$ facets. This exemplifies that exploiting the structure of the given group can yield a relevant improvement in the way symmetries are broken. Moreover, we show that the only groups that admit a fundamental domain with a unique representative for *every* orbit are reflection groups. Finally, we propose a new way of measuring the effectiveness of fundamental domains, which we hope will pave the road for future work in deriving fundamental domains that exploit the structure of the groups involved. Due to space reasons we refer several proofs and details to the full version of this paper [27].

---

[1] Under the common assumption that a group is described by a set of generators [25].

2   Preliminaries

Throughout the whole paper, G denotes a group, and $H \leq G$ means that H is a subgroup of G. The element id $\in G$ denotes the identity. For a subset S of G, $\langle S \rangle$ is the smallest group containing S. $O_n(\mathbb{R})$ denotes the orthogonal group in $\mathbb{R}^n$, that is, the group of all $n \times n$ orthogonal matrices (equivalently, linear isometries). Hence, it holds that if $g \in O_n(\mathbb{R})$ then the inverse g^{-1} equals the transpose g^t. All groups considered in what follows are finite subgroups of $O_n(\mathbb{R})$. Also, $G_{(S)}$ denotes the pointwise stabilizer of the set S, that is $G_{(S)} := \{g \in G : x = gx \ \forall x \in S\}$. If $S := \{x\}$, we write $G_x := G_{(S)}$. The set fix(g) denotes the invariant subspace of $g \in G$, i.e. fix$(g) := \{x \in \mathbb{R}^n : gx = x\}$. For $H \leq G$, a *transversal* for H in G is a set of representatives from the left cosets of H in G, the set of left cosets being $\{gH : g \in G\}$. Given a set of elements $S \subseteq G$, we denote by $S^{-1} := \{g^{-1} : g \in S\}$. Given $x \in \mathbb{R}^n$, the G-orbit of x is the set $\mathrm{Orb}_G(x) := \{gx : g \in G\}$.

For an exhaustive introduction to group theory, see for instance Rotman [22]. For an exposition on computational aspects of group theory, in particular permutation groups, see Seress [25].

We denote $[n] := \{1, \ldots, n\}$ for all $n \in \mathbb{N}$, and S_n denotes the symmetric group, that is the group of all permutations over $[n]$. For $G \leq S_n$, each element $g \in G$ acts on $\mathbb{R}^n$ by the mapping $x \mapsto gx := \left(x_{g^{-1}(i)}\right)_{i=1}^n$. Equivalently, we consider $G \leq S_n$ as a group of isometries where each $g \in G$ is interpreted as the corresponding permutation matrix.

An optimization problem $\min\{f(x) : x \in X\}$ is *G-invariant* if for all feasible x and $g \in G$,

1. $f(x) = f(gx)$,
2. gx is feasible.

For an introduction and overview on techniques regarding symmetry handling in mixed integer linear programming see [18].

Given a G-invariant optimization problem, we can use the group G to restrict the search of solutions to a subset of $\mathbb{R}^n$, namely a *fundamental domain*.

Definition 1. *A subset F of $\mathbb{R}^n$ is a* fundamental domain *for $G \leq O_n(\mathbb{R})$ if*

1. *the set F is closed and convex[2],*
2. *the members of $\{\mathrm{int}(gF) : g \in G\}$ are pairwise disjoint,*
3. *$\mathbb{R}^n = \bigcup_{g \in G} gF$.*

Notice that for any $x \in \mathbb{R}^n$, its G-orbit $\mathrm{Orb}_G(x)$ satisfies that $|\mathrm{Orb}_G(x) \cap F| \geq 1$. Also, $|\mathrm{Orb}_G(x) \cap F| = 1$ if $x \in \mathrm{int}\,(F)$. It is not hard to see that all fundamental domains for a finite subgroup of $O_n(\mathbb{R})$ are full-dimensional sets.

Definition 2. *A subset R of $\mathbb{R}^n$ is a* fundamental set *for a group $G \leq O_n(\mathbb{R})$ if it contains exactly one representative of each G-orbit in $\mathbb{R}^n$.*

---

[2] Notice that in part of the literature, e.g. [21], convexity is not part of the definition.

3   The Geometric Structure of Fundamental Domains

In this section we review some basic geometric properties of fundamental domains and derive new properties. Propositions 1, and 3 are well known; their proof can be found in [21, Ch. 6]. Proposition 2 was known under stronger assumptions (namely for *exact* fundamental domains [21, Ch. 6]). All the proofs in this section, including known ones, can be found in the full version of this manuscript [27].

The following proposition, together with the existence of a vector α whose stabilizer is trivial [21, Thm. 6.6.10.], guarantees the existence of a fundamental domain for any $G \leq O_n(\mathbb{R})$. We will refer to the construction F_α in the proposition as a *Dirichlet domain*.

Proposition 1. *Let $G \leq O_n(\mathbb{R})$ be finite and non-trivial, and let α be a point in $\mathbb{R}^n$ whose stabilizer G_α is trivial. Then the following set is a fundamental domain for G,*

$$F_\alpha = \{x \in \mathbb{R}^n : \alpha^t x \geq \alpha^t g x, \ \forall g \in G\}.$$

A specific kind of Dirichlet domains are *k-fundamental domains*. For any integer $k \geq 2$, we define $\overline{k} := \left(k^{n-1}, k^{n-2}, \ldots, 1\right)$ as the *k-universal ordering vector*. The set $F_{\overline{k}}$ is the k-fundamental domain for the symmetry group G. It has been proven in [9] that $F_{\overline{2}}$ contains a unique representative per G-orbit of binary points in $\mathbb{R}^n$. This fact easily generalizes for points $x \in \{0, \ldots, k-1\}^n$ with the k-ordering vector (see [18]).

Given a fundamental domain F and $g \in G \setminus \{\mathrm{id}\}$, let H_g be any closed half-space that separates F and gF. More precisely, it holds that $F \subseteq H_g$ and $gF \subseteq \overline{H_g^c}$. The existence of this half-space follows from the convex separation theorem. We say that a collection $\{H_g\}_{g \in G}$ *represents* F if for every $g \in G$, the set H_g is a closed half-space that separates F and gF. Notice that representations are non unique.

Let us denote by $H_g^= := \partial(H_g)$ the hyperplane defining H_g. We let γ_g be some defining vector for H_g, i.e., $H_g = \{x \in \mathbb{R}^n : \gamma_g^t x \geq 0\}$, and thus $H_g^= = \{x \in \mathbb{R}^n : \gamma_g^t x = 0\}$.

Proposition 2. *Let $G \leq O_n(\mathbb{R})$ be finite, and let F be a fundamental domain. Then $F = \bigcap_{g \in G} H_g$. In particular F is a polyhedral cone. Moreover, if $A \subseteq G$ is a minimal set such that $F = \bigcap_{g \in A} H_g$, then A is a set of generators of G.*

Note that since A is minimal, the facets of F are $F \cap H_g^=$ for each $g \in A$.

We introduce a new type of fundamental domains and characterize their facial structure.

Definition 3. *A fundamental domain F is said to be subgroup consistent if there exists a collection $\{H_g\}_{g \in G}$ representing F such that for every subgroup $G' \leq G$ the set $F' = \bigcap_{g \in G'} H_g$ is a fundamental domain for G'.*

It is not hard to see that Dirichlet domains are subgroup consistent. More-over, subgroup consistent fundamental domains are amenable to be constructed iteratively, either by starting the construction of a fundamental domain for a sub-group and extending it to larger subgroups (bottom-up), or adding inequalities for G and recurse to smaller subgroups (top-down, as our technique in Sect. 4).

With the help of the following lemmas, we show a close relationship between supporting hyperplanes of a subgroup consistent fundamental domain F: all facet-defining inequalities of F are of the form $\alpha^t x \geq \alpha^t gx$ for some α and $g \in G$. In this case we say that the inequality is of *Dirichlet type*. To the best of our knowledge this property of facets is new.

Lemma 1. *Let $g \in G$. Then $(\mathrm{fix}(g) \cap F) \setminus H_g^= = \emptyset$.*

Proof. Let $x \in \mathrm{fix}(g)$. If $x \in F \setminus H_g^=$, then $\gamma_g^t x > 0$. Moreover, $\gamma_g^t(gx) \leq 0$ since $gx \in gF$. But this is a contradiction as $gx = x$. □

Lemma 2. *If G is Abelian, then for every $g \in G$, the set $\mathrm{fix}(g)$ is G-invariant, i.e., $h \cdot \mathrm{fix}(g) = \mathrm{fix}(g)$ for all $h \in G$.*

Proof. Let $g, h \in G$. We show that $h\,\mathrm{fix}(g) = \mathrm{fix}(g)$. Indeed, if $y \in h\,\mathrm{fix}(g)$, i.e., $y = hx$ for some $x \in \mathrm{fix}(g)$, then $gy = g(hx) = h(gx) = hx = y$. Therefore, $y \in \mathrm{fix}(g)$, and thus $h\,\mathrm{fix}(g) \subseteq \mathrm{fix}(g)$. The inclusion $\mathrm{fix}(g) \subseteq h\,\mathrm{fix}(g)$ follows by applying the previous argument to h^{-1}, implying that $h^{-1}\,\mathrm{fix}(g) \subseteq \mathrm{fix}(g)$. □

Lemma 3. *Let $F \subseteq \mathbb{R}^n$ be a subgroup consistent fundamental domain for G and $\{H_g\}_{g \in G}$ a collection representing F where $H_g = \{x \in \mathbb{R}^n : \gamma_g^t x \geq 0\}$. Then γ_g belongs to the orthogonal complement of the fixed space of g, i.e.,*

$$\gamma_g \in \mathrm{fix}(g)^\perp := \{x \in \mathbb{R}^n \ : \ gx = x\}^\perp.$$

Proof. We start by showing the lemma for the case that G is Abelian. By Lemma 2 we have that $\mathrm{fix}(g)$ is G-invariant for every $g \in G$, and hence $h\,\mathrm{fix}(g) = \mathrm{fix}(g)$ for any $h \in G$. Therefore,

$$\mathrm{fix}(g) = \mathrm{fix}(g) \cap \left(\underbrace{\bigcup_{h \in G} hF} \right) = \bigcup_{h \in G} (\mathrm{fix}(g) \cap hF) = \bigcup_{h \in G} h(\mathrm{fix}(g) \cap F).$$

Let $\langle S \rangle$ denote the linear span of a set S. Notice that $\dim(\langle \mathrm{fix}(g) \cap F \rangle) = \dim(\mathrm{fix}(g))$, otherwise, $\mathrm{fix}(g)$ would be contained in the union of finitely many subspaces of strictly smaller dimension, which is clearly a contradiction. Since $F \cap \mathrm{fix}(g) \subseteq \mathrm{fix}(g)$, we conclude that $\langle F \cap \mathrm{fix}(g) \rangle = \mathrm{fix}(g)$. As by Lemma 1 we have that $F \cap \mathrm{fix}(g) \subseteq H_g^=$, this implies that $\mathrm{fix}(g) = \langle F \cap \mathrm{fix}(g) \rangle \subseteq H_g^=$. Since by definition γ_g is orthogonal to every vector in $H_g^=$, we conclude that $\gamma_g \in \mathrm{fix}(g)^\perp$. The lemma follows if G is Abelian.

For the general case, assume that F is subgroup consistent and $F = \bigcap_{h \in G} H_h$. Therefore, the Abelian subgroup $G' = \langle g \rangle$ has $F' = \bigcap_{h \in G'} H_h$ as a fundamental domain. Then our argument for the Abelian case implies that $\gamma_g \in \mathrm{fix}(g)^\perp$. □

Theorem 1. *Let $F \subseteq \mathbb{R}^n$ be a subgroup consistent fundamental domain for G and let $\{H_g\}_{g \in G}$ be a collection representing F, where $H_g = \{x : \gamma_g^t x \geq 0\}$. Then there exists $\alpha_g \in \mathbb{R}^n$ such that $\gamma_g = (\mathrm{id} - g)\alpha_g$. In particular, any facet-defining inequality for F is of the form $\alpha_g^t x \geq \alpha_g^t g^{-1} x$ for some $g \in G$, and hence of Dirichlet type.*

Proof. Recall that any automorphism f of $\mathbb{R}^n$ satisfies $\mathrm{Im}(f)^\perp = \ker(f^t)$. Since $\mathrm{fix}(g) = \ker(\mathrm{id} - g)$, by Lemma 3 we have that

$$\gamma_g \in \mathrm{fix}(g)^\perp = \mathrm{fix}(g^{-1})^\perp = \ker(\mathrm{id} - g^t)^\perp = \mathrm{Im}\,(\mathrm{id} - g).$$

Hence, there exists $\alpha_g \in \mathbb{R}^n$ such that $\gamma_g = (\mathrm{id} - g)\alpha_g$. □

We say that a fundamental domain F is *exact* if for every facet S of F there exists a group element $g \in G$ such that $S = F \cap gF$. In this case we say that g defines a facet of F. Notice that it also holds that $S = F \cap H_g^=$. Exact fundamental domains are well structured and studied extensively [21]. It is worth noticing that Dirichlet domains are exact.

For exact fundamental domains, facets come in pairs, i.e., if g defines a facet of F, then g^{-1} also does.

Proposition 3. *Let $F \subseteq \mathbb{R}^n$ be an exact fundamental domain for $G \leq O_n(\mathbb{R})$ finite. If S is a facet of F, then there is a unique non-trivial element $g \in G$ such that $S = F \cap gF$, moreover $g^{-1}S$ is a facet of F.*

Proposition 3 and Theorem 1 together imply the following corollary.

Corollary 1. *Let $F \subseteq \mathbb{R}^n$ be an exact and subgroup consistent fundamental domain for $G \leq O_n(\mathbb{R})$ finite. Suppose that $\gamma_g^t x \geq 0$ defines the facet $F \cap gF$, and let α_g be a vector such that $\gamma_g = (\mathrm{id} - g)\alpha_g$. Then $\gamma_{g^{-1}} = (\mathrm{id} - g^t)\alpha_g$ and hence we can take $\alpha_{g^{-1}} = \alpha_g$.*

4   Generalized Dirichlet Domains

In this section we present our main contribution, an algorithm which constructs a fundamental domain for an arbitrary finite orthogonal group. We use the insights gained from the geometric properties of the previous section to guide our search for new constructions. In particular we create fundamental domains based on a sequence of nested stabilizers of the G action on $\mathbb{R}^n$. This construction generalizes Dirichlet domains, and hence k-fundamental domains, as well as the *Schreier-Sims fundamental domain*, presented in Sect. 4.1. Both types of fundamental domains can be easily constructed using our algorithm. Moreover, in Sect. 5 we use the flexibility of our construction to define a new fundamental domain with better properties for a specific group.

Theorem 1 and Corollary 1 suggest that we should consider vectors α_g for some $g \in G$ and consider inequalities of the form $\alpha_g^t x \geq \alpha_g^t g x$ and $\alpha_g^t x \geq \alpha_g^t g^{-1} x$, although it seems hard to decide whether we should pick different vectors α_g

for each pair g, g^{-1}, and if so, how to choose them. For instance, as a special case, if we fix a vector $\gamma = \alpha_g$ for all $g \in G$, we would obtain a Dirichlet domain. However, if γ's stabilizer is not trivial, then all inequalities $\alpha_g^t x \geq \alpha_g^t g^{-1} x$ in a coset of G_γ are equivalent. This hints that we should choose a vector γ, apply a coset decomposition using a stabilizer subgroup, and add the Dirichlet inequalities related to all members of the decomposition. Then, we recurse on $G = G_\gamma$. Formally, this process is described in Algorithm 1.

Algorithm 1. Construction of a generalized Dirichlet domain (GDD)

    Input: A set of generators S_G of a finite orthogonal group G
    Output: A fundamental domain F for G
1: Set $F := \mathbb{R}^n$, $G_0 := G$, and $i := 1$
2: **while** $i \leq n$ and G_{i-1} does not fix $\mathbb{R}^n$ pointwise **do**
3:     Choose $\gamma_i \in \mathbb{R}^n$ such that $g\gamma_i \neq \gamma_i$ for some $g \in G_{i-1}$
4:     Compute $G_i := \{g \in G_{i-1} : g\gamma_i = \gamma_i\}$
5:     Choose a transversal H_i for G_i in G_{i-1} and add the inverses $H_i := H_i \cup H_i^{-1}$
6:     Set $F_i := \{x \in \mathbb{R}^n : \gamma_i^t x \geq \gamma_i^t h x \quad \forall h \in H_i\}$
7:     $F := F \cap F_i$ and $i := i + 1$
8: **end while**
9: **return** F

Theorem 2. *Algorithm 1 outputs a fundamental domain F.*

Proof. Let $x \in \text{int}(F)$ and let g be a non-trivial element of G. We will show that $gx \notin F$. There are two cases. First, suppose that $g \notin G_1$, and hence $g^{-1} \notin G_1$. Let $h \in H_1$ be such that $g^{-1} \in hG_1$, which implies that $g^{-1}\gamma_1 = h\gamma_1$. Recall that isometries correspond to orthogonal matrices, and thus for $r \in G$, it holds that $r^{-1} = r^t$. As $h^{-1} \in H_1$, we have that

$$\gamma_1^t x > \gamma_1^t(h^{-1}x) = \gamma_1^t(h^t x) = (h\gamma_1)^t x = (g^{-1}\gamma_1)^t x = \gamma_1^t gx. \tag{1}$$

Similarly, let $s \in H_1$ such that $g \in sG_1$, and thus $g\gamma_1 = s\gamma_1$. Thus, by Eq. (1),

$$\gamma_1^t(gx) < \gamma_1^t x = \gamma_1^t g^{-1}(gx) = (g\gamma_1)^t(gx) = (s\gamma_1)^t(gx) = \gamma_1^t s^{-1}(gx),$$

where $s^{-1} \in H_1$ by construction. We conclude that $gx \notin F$. If $g \in G_1$, consider recursively $G = G_1$ to show that $x \notin \bigcap_{i>1} F_i$.

Now we show that for all $x \in \mathbb{R}^n$, there exists $g \in G$ such that $gx \in F$. We also have two cases. First assume that $x \notin F_1$ and let us show that there exists $g \in G$ such that $gx \in F_1$. Let $\bar{h} \in \text{argmax}\{\gamma_1^t h x : h \in H_1\}$. We show that $\bar{h}x \in F_1$. Indeed, for all $h \in H_1$,

$$\gamma_1^t \bar{h}x \geq \gamma_1^t h x = \gamma_1^t(h\bar{h}^{-1})\bar{h}x = (\bar{h}h^t\gamma_1)^t \bar{h}x. \tag{2}$$

Consider an arbitrary $s \in H_1$. We will show that there exists $h_s \in H_1$ s.t. $s\gamma_1 = \bar{h}h_s^t \gamma_1$. This together with (2) implies that $\bar{h}x \in F_1$. To show that h_s exists, let $r \in H_1$ be a representative of the coset of $\bar{h}^{-1}s$, that is, $\bar{h}^{-1}s \in rG_1$. Hence, $\bar{h}^{-1}s = rg_1$ for some $g_1 \in G_1$, and therefore $\bar{h}^{-1}s\gamma_1 = r\gamma_1$. Hence, it suffices to take $h_s = r^t = r^{-1} \in H_1$. This concludes that $gx \in F_1$. Now we can replicate the argument with gx as x, G_1 as G and F_2 as F_1. Hence, there exists $g' \in G_1$ such that $g'(gx) \in F_2$. Notice that it also holds that $g'(gx) \in F_1$, as $gx \in F_1$ and $g' \in G_1$. Indeed, simply notice that $\gamma_1^t g'(gx) = \gamma_1^t gx = \max\{\gamma_1^t h(gx) : h \in H_1\}$. Hence we can follow the argument as above. Therefore $g'(gx) \in F_2 \cap F_1$. Iterating the argument shows that $rx \in F$ for some $r \in G$.   □

It is worth noticing that if we take γ_1 such that G_1 is trivial, then the algorithm finishes in one iteration. Indeed, the obtained fundamental domain is the Dirichlet domain F_{γ_1}. This justifies the name generalized Dirichlet domain.

Complexity. First notice that the algorithm terminates after at most n iterations, as the vectors in $\{\gamma_i\}_i$ must be linearly independent. For the rest of the runtime analysis we focus on permutation groups. Lines 4 and 5 are the most challenging with respect to the algorithm's computational complexity. The result of the computation in line 4 is a setwise stabilizer of the coordinates of γ. Computing a set of generators for this subgroup can be performed in quasi-polynomial time with the breakthrough result by Babai [3,4] for String Isomorphism. In general, however, H_1 might be of exponential size, and hence line 5 can take exponential time. Indeed, the number of cosets of G_1 in G equals the size of the orbit $|\mathrm{Orb}_G(\gamma_1)|$, by the Orbit-Stabilizer Theorem. If we choose γ_1 with pair-wise different coordinates, then $|\mathrm{Orb}_G(\gamma_1)| = |G|$. This is exactly the case for the Dirichlet domain.

On the other hand, we can choose the γ_i vectors carefully in order to avoid the exponential time complexity. Consider, for example, a vector γ of the form $(\gamma^{(1)}, \ldots, \gamma^{(k)}, 0, \ldots, 0)$ such that $\gamma^{(i)} \neq \gamma^{(j)}$ for $i \neq j$ in $[k]$, and $\gamma^{(i)} \neq 0$ for $i \in [k]$. Hence, a set of generators for the stabilizer G_γ can be computed in polynomial time. Indeed, it corresponds to the pointwise stabilizers of coordinates 1 to k [25, Section 5.1.1]. Moreover, the number of cosets is $O(n^k)$, as again the number of cosets equals the cardinality of the orbit of γ. Therefore, we get a polynomial running time if k is a constant.

4.1   The Lex-Max Fundamental Domain

In this section, we study a natural idea for breaking symmetries: in any orbit, choose the vector that is lexicographically maximal.

Let $\succ$ denote a lexicographic order on $\mathbb{R}^n$, that is for any pair $x, y \in \mathbb{R}^n$, $y \succ x$ if and only if there exists $j \in [n]$ such that $y_j > x_j$ and $y_i = x_i$ for all $i < j$. Therefore $\succeq$ defines a total order on $\mathbb{R}^n$, where, $y \succeq x$ if and only if $y \succ x$ or $y = x$. Given a group $G \leq S_n$ acting on $\mathbb{R}^n$, we define

$$\mathrm{Lex}_G := \{x \in \mathbb{R}^n : x \succeq gx, \forall g \in G\}.$$

In what follows we show an alternative characterization of the set of lexicographically maximal points using k-fundamental domains. Recall that k-fundamental domains are Dirichlet domains $F_{\bar{k}}$ for some integer $k \geq 2$.

Lemma 4. *Let $n \in \mathbb{N}$ and $x, y \in \mathbb{R}^n$. If $x \succ y$, then there exists $N \in \mathbb{N}$ such that for every $k \in \mathbb{N}$ greater or equal than N, $\overline{k}^t x > \overline{k}^t y$, where $\overline{k} := (k^{n-1}, k^{n-2}, \ldots, k, 1)$ for $k \geq 2$ integer.*

With the help of the previous lemma, we provide an alternative characterization of Lex_G. Recall that a *fundamental set* is a set that contains exactly one representative for each G-orbit.

Lemma 5. *Let $G \leq S_n$. Then Lex_G is a convex fundamental set and*

$$\mathrm{Lex}_G = \liminf_{k \to \infty} F_{\overline{k}} = \bigcup_{i=1}^{\infty} \bigcap_{k=i}^{\infty} F_{\overline{k}}.$$

Since Lex_G is a convex fundamental set for any permutation group G, its closure seems a reasonable candidate for being a fundamental domain. This is actually the case, which implies that $\overline{\mathrm{Lex}_G}$ is a polyhedral cone by Proposition 2.

Theorem 3. *For any $G \leq S_n$, the closure of Lex_G is a fundamental domain.*

A Characterization of $\overline{\mathrm{Lex}_G}$ Using the Schreier-Sims Table. In what follows, we provide a characterization of $\overline{\mathrm{Lex}_G}$, which in particular allows to compute its facets efficiently. Indeed, we show that its description coincides with the Schreier-Sims inequalities for G [23].

The *Schreier-Sims table* is a representation of a permutation group $G \leq S_n$. The construction is as follows. Consider the chain of nested pointwise stabilizers defined as: $G^0 := G$ and $G^i := \{g \in G^{i-1} : g(i) = i\}$ for each $i \in [n]$. Note that the chain is not necessarily strictly decreasing (properly), and we always have that $G^{n-1} = \{\mathrm{id}\}$. For a given $i \in [n]$ and $j \in \mathrm{Orb}_{G^{i-1}}(i)$, let $h_{i,j}$ be any permutation in G^{i-1} which maps i to j. Hence, $U_i := \{h_{i,j} : j \in \mathrm{Orb}_{G^{i-1}}(i)\}$ is a transversal for the cosets of G^i in G^{i-1}.

We arrange the permutations in the sets U_i, for $i \in [n]$, in an $n \times n$ table T where $T_{i,j} = h_{i,j}$ if $j \in \mathrm{Orb}_{G^{i-1}}(i)$ and $T_{i,j} = \emptyset$ otherwise.

The most interesting property of this representation of G is that each $g \in G$ can be uniquely written as $g = g_1 g_2 \cdots g_n$ with $g_i \in U_i$, for $i \in [n]$. Therefore, the permutations in the table form a set of generators of G which is called a *strong generating set (SGS)* for G [25].

The *Schreier-Sims polyhedron*, denoted by SS_G, is the polyhedron given by the inequalities $x_i \geq x_j$ for all $T_{i,j} \neq \emptyset$. The following theorem states that $\overline{\mathrm{Lex}_G} = SS_G$. A crucial observation to prove this is that for any vector $x \in \mathbb{R}^n$, x is in the closure of Lex_G if and only if x can be perturbed into the interior of $\overline{\mathrm{Lex}_G}$, where the perturbed vector is lexicographically maximal in its orbit.

Theorem 4. *Let $G \leq S_n$. Then $\overline{\mathrm{Lex}_G} = SS_G$.*

The next result exhibits the generality of our GDD method for constructing fundamental domains. It shows that by choosing γ_i as the cannonical vectors in our GDD construction the algorithm outputs SS_G. We note that this also gives an alternative proof to Theorem 3.

Proposition 4. *For any group $G \leq S_n$ the set SS_G is a GDD.*

We finish this section by showing that several inequalities in the description of SS_G are redundant, and that at most $n - 1$ of them define facets.

Theorem 5. *Let $G \leq S_n$ and f denote the number of G-orbits in $[n]$. Then SS_G is a polyhedron with at most $n - f$ facets.*

Proof. Let $D = ([n], E)$ be a directed graph defined as follows. For each $i \in [n]$ we have that $(i, j) \in E$ for each $j \in \mathrm{Orb}_{G^{i-1}}(i) \setminus \{i\}$. By construction, D is a topological sort, and hence it is a directed acyclic graph (DAG).

Claim: Let $j \in [n]$. If $(i, j), (k, j) \in E$ then either $(i, k) \in E$ or $(k, i) \in E$.

Indeed, without loss of generality, let us assume that $i < k$. As $(i, j) \in E$ then $j \in \mathrm{Orb}_{G^{i-1}}(i)$. Similarly, it holds that $j \in \mathrm{Orb}_{G^{k-1}}(k) \subseteq \mathrm{Orb}_{G^{i-1}}(k)$. Therefore, by transitivity, $k \in \mathrm{Orb}_{G^{i-1}}(i)$, and hence $(i, k) \in E$. This shows the claim.

Let $\tilde{D} = ([n], \tilde{E})$ be the *minimum equivalent graph* of D, that is, a subgraph with a minimum number of edges that preserves the reachability of D. Hence, there exists a (u, v)-dipath in D if and only if there exist a (u, v)-dipath in $\tilde{D}$. Notice that

$$SS_G = \{x \ : \ x_i \geq x_j \text{ for all } (i, j) \in E\}.$$

Let us define

$$\widetilde{SS}_G = \{x \ : \ x_i \geq x_j \text{ for all } (i, j) \in \tilde{E}\}.$$

Now we show that $SS_G = \widetilde{SS}_G$. Clearly we have that $SS_G \subseteq \widetilde{SS}_G$. On the other hand, if $x_i \geq x_j$ is an inequality of SS_G, then there exists an (i, j)-dipath in $\tilde{D}$ and hence $x_i \geq x_{i_1} \geq x_{i_2} \geq \ldots \geq x_{i_k} \geq x_j$ is a valid set of inequalities for $\widetilde{SS}_G$, for certain nodes $i_1, \ldots, i_k$. We conclude that $SS_G = \widetilde{SS}_G$.

Now we argue that $\tilde{D}$ is a collection of at least f out-trees. Indeed, let us assume by contradiction that for $j \in [n]$ there exists two distinct nodes i, k such that $(i, j), (k, j) \in \tilde{E}$. By our previous claim, k is reachable from i in D (or analogously i is reachable from k), and hence the same is true in $\tilde{D}$. This is a contradiction as the edge (i, j) could be removed from $\tilde{D}$ preserving the reachability. As $\tilde{D}$ is a DAG, then $\tilde{D}$ must be a collection of node-disjoint out-trees. Finally, note that the smallest element in each orbit of G in $[n]$ has in-degree 0 in D, and hence also in $\tilde{D}$. Therefore $\tilde{D}$ has at least f different trees, which implies that $\tilde{D}$ has at most $n - f$ edges. We conclude that $\widetilde{SS}_G$ is defined by at most $n - f$ inequalities. □

This means that every permutation group admits a fundamental domain with at most $n - 1$ facets. We complement this result by showing that this bound is tight.

Proposition 5. *Any fundamental domain for S_n has $n - 1$ facets.*

5   Overrepresentation of Orbit Representatives

A desirable property of symmetry breaking polyhedra is that they select a unique representative per G-orbit. In general, the definition of fundamental domains only guarantees this for vectors in their interior. Recall that a subset R of $\mathbb{R}^n$ which contains exactly one point from each G-orbit is called a *fundamental set*. The following result shows that closed convex fundamental sets are only attained by reflection groups. In other words, the only groups that admit fundamental domains containing unique representatives for every orbit are reflection groups.

Theorem 6. *Let $G \le O_n(\mathbb{R})$ finite. Then G admits a fundamental domain F with $|F \cap O| = 1$ for every G-orbit $O \subseteq \mathbb{R}^n$ if and only if G is a reflection group.*

As a corollary we can characterize when the fundamental set Lex_G is closed. Alternatively, this characterizes when SS_G contains a unique representative for *every* orbit.

Corollary 2. *Let $G \le S_n$ be such that it partitions $[n]$ into the orbits $O_1, \ldots, O_m$. Then Lex_G is closed if and only if $G \simeq S_{|O_1|} \times \cdots \times S_{|O_m|}$ acts on $\mathbb{R}^n = \mathbb{R}^{|O_1|} \times \cdots \times \mathbb{R}^{|O_m|}$, where $S_{|O_i|}$ acts on $\mathbb{R}^{|O_i|}$ by permuting its coordinates.*

In integer programming problems, we are concerned about the number of representatives of binary orbits in a fundamental domain. The Schreier-Sims domain can be weak in this regard, as shown in the following example.

Example 1. Let $n \in \mathbb{N}$ be divisible by 3, and consider the direct product $G := C^1 \times C^2 \times \cdots \times C^{n/3}$ where C^i for $i \in [n/3]$ is the cyclic group on the triplet $(3(i-1)+1, 3(i-1)+2, 3(i-1)+3)$. Consider the binary vector $x := (1,1,0,1,1,0,\ldots,1,1,0)$. For each vector in the G-orbit of x, there are three possible values for each triplet: $(1,1,0)$, $(0,1,1)$ or $(1,0,1)$. Therefore, the orbit of x has cardinality $3^{n/3}$. The fundamental domain SS_G for G can be described by the set of inequalities $x_{3(i-1)+1} \ge x_{3(i-1)+2}$ and $x_{3(i-1)+1} \ge x_{3(i-1)+3}$ for all $i \in [n/3]$. It is clear that each vector in $\text{SS}_G \cap \text{Orb}_G(x)$ admits two options for its index triplets: $1,1,0$ and $1,0,1$. As a result $|\text{SS}_G \cap \text{Orb}_G(x)| = 2^{n/3}$. ∇

Let X be a G-invariant subset of $\mathbb{R}^n$ (e.g. $X = \{0,1\}^n$). Let $\mathcal{O}(G, X)$ be the set of all G-orbits in X. Motivated by our previous discussion, we define, for a fixed G, the *worst-case effectiveness of F on X* as

$$\Lambda_{G,X}(F) := \max_{O \in \mathcal{O}(G,X)} |F \cap O|.$$

Now, we use our GDD algorithm to obtain a suitable fundamental domain in Example 1 with $\Lambda_{G,\{0,1\}^n}(F) = 1$ while $\Lambda_{G,\{0,1\}^n}(\text{SS}_G) = 2^{\Omega(n)}$.

Example 1 (continued). We construct a GDD F with $\Lambda_{G,\{0,1\}^n}(F) = 1$. First, note that G has $n/3$ orbits in $[n]$ given by: $\Delta_i := \{3(i-1)+1, 3(i-1)+2, 3(i-1)+3\}$ for $i \in [n/3]$. In our GDD construction we choose for every $i \in [n/3]$

a vector $\gamma_i = (0^{3(i-1)}, 4, 2, 1, 0^{n-3i})$, where 0^r is an r-dimension 0 vector. We obtain that $\mathrm{Orb}_G(x) \cap F = \{x\}$ for any $x \in \{0,1\}^n$. The number of cosets in each iteration is 3. Omitting the trivial coset, the number of inequalities that defines our GDD is $2 \cdot (n/3)$. $\triangledown$

6   Future Work

Our work leaves several major questions.

Q1: Does our GDD construction exhaust all possible fundamental domains for a group of isometries, or are there other fundamental domains that are not GDDs?

Any light on this question can help creating new fundamental domains with potential practical relevance, or help us show impossibility results. This can also have consequences regarding our long term goal: understanding the tension (potentially trade-off) between the symmetry breaking effectiveness of a polyhedron and its complexity.

Q2: Does every group of isometries admit a fundamental domain with a single representative of each binary orbit, and with a polynomial number of facets?

It is not hard to imagine other interesting variants of this question. For example, we could be interested either in the extension complexity or complexity of the separation problem, instead of the number of facets. At the moment, the only information we have is that blindly choosing lexicographically maximal binary vectors as representatives should not help, as finding them is NP-hard [2]. It is worth noticing that an answer to Q1 might help answering Q2, either positively or negatively. Alternatively, the relation between $\Lambda_{G,X}(F)$ and the number of facets of a fundamental domain F is of interest, for example for $X = \{0,1\}^n$. On the other hand, we know that only reflection groups admit fundamental domains with $\Lambda_{G,\mathbb{R}^n}(F) = 1$. Characterizing, for example, the class of groups that allows for $\Lambda_{G,\mathbb{R}^n}(F) = O(1)$ might also give us a better understanding on the limitations of symmetry breaking polyhedra.

Acknowledgements. This work was partially funded by Fondecyt Proyect Nr. 1181527 and ANID – Millennium Science Initiative Program – NCN17_059. Part of this work was done while the first author was affiliated to the University of O'Higgins, Chile. Léonard von Niederhäusern was supported by CMM ANID PIA AFB170001 from ANID (Chile). We are greatly indebted to C. Hojny, M. Pfetsch, A. Behn, and V. Verdugo for fruitful discussions on the topic of this paper. We are also thankful for the insightful comments of anonymous reviewers that helped improving the quality of this manuscript.

References

1. Achterberg, T., Wunderling, R.: Mixed integer programming: analyzing 12 years of progress. In: Jünger, M., Reinelt, G. (eds.) Facets of Combinatorial Optimization, pp. 449–481. Springer, Heidelberg (2013). https://doi.org/10.1007/978-3-642-38189-8_18

2. Babai, L., Luks, E.M.: Canonical labeling of graphs. In: Proceedings of the Fifteenth Annual ACM Symposium on Theory of Computing (STOC 1983), pp. 171–183 (1983)
3. Babai, L.: Graph isomorphism in quasipolynomial time. arXiv preprint arXiv:1512.03547v2 (2016)
4. Babai, L.: Graph isomorphism in quasipolynomial time. In: Proceedings of the Forty-Eighth Annual ACM Symposium on Theory of Computing, pp. 684–697 (2016)
5. Bödi, R., Herr, K., Joswig, M.: Algorithms for highly symmetric linear and integer programs. Math. Program. Ser. A **137**, 65–90 (2013)
6. Dias, G., Liberti, L.: Exploiting symmetries in mathematical programming via orbital independence. Ann. Oper. Res. **298**, 149–182 (2019)
7. Dirichlet, G.L.: Über die reduction der positiven quadratischen formen mit drei unbestimmten ganzen zahlen. J. für die reine und angewandte Mathematik **1850**(40), 209–227 (1850)
8. Faenza, Y., Kaibel, V.: Extended formulations for packing and partitioning orbitopes. Math. Oper. Res. **34**(3), 686–697 (2009)
9. Friedman, E.J.: Fundamental domains for integer programs with symmetries. In: Dress, A., Xu, Y., Zhu, B. (eds.) COCOA 2007. LNCS, vol. 4616, pp. 146–153. Springer, Heidelberg (2007). https://doi.org/10.1007/978-3-540-73556-4_17
10. Ghoniem, A., Sherali, H.D.: Defeating symmetry in combinatorial optimization via objective perturbations and hierarchical constraints. IIE Trans. **43**, 575–588 (2011)
11. Herr, K., Rehn, T., Schürmann, A.: Exploiting symmetry in integer convex optimization using core points. Oper. Res. Lett. **41**, 298–304 (2013)
12. Hojny, C., Pfetsch, M.: Polytopes associated with Symmetry handling. Math. Program. Ser. A **175**, 197–240 (2018)
13. Kaibel, V., Pfetsch, M.: Packing and partitioning orbitopes. Math. Program. **114**, 1–36 (2008)
14. Liberti, L.: Reformulations in mathematical programming: automatic symmetry detection and exploitation. Math. Program. Ser. A **131**, 273–304 (2012)
15. Liberti, L., Ostrowski, J.: Stabilizer-based Symmetry breaking constraints for mathematical programs. J. Glob. Optim. **60**, 183–194 (2014)
16. Margot, F.: Pruning by isomorphism in branch-and-cut. Math. Program. Ser. A **94**, 71–90 (2002)
17. Margot, F.: Exploiting orbits in symmetric integer linear program. Math. Program. Ser. B **98**, 3–21 (2003)
18. Margot, F.: Symmetry in integer linear programming. In: Jünger, M., et al. (eds.) 50 Years of Integer Programming 1958-2008, pp. 647–686. Springer, Heidelberg (2010). https://doi.org/10.1007/978-3-540-68279-0_17
19. Ostrowski, J., Linderoth, J., Rossi, F., Smriglio, S.: Orbital branching. Math. Program. Ser. A **126**, 147–178 (2011)
20. Ostrowski, J., Anjos, M.F., Vannelli, A.: Symmetry in scheduling problems (2010). cahier du GERAD G-2010-69
21. Ratcliffe, J.G.: Foundation of Hyperbolic Manifolds, 3rd edn. Springer, Heidelberg (2019). https://doi.org/10.1007/978-3-030-31597-9
22. Rotman, J.J.: An Introduction to the Theory of Groups, 4th edn. Springer, Heidelberg (1995). https://doi.org/10.1007/978-1-4612-4176-8
23. Salvagnin, D.: Symmetry breaking inequalities from the schreier-sims table. In: International Conference on the Integration of Constraint Programming, Artificial Intelligence, and Operations Research, pp. 521–529 (2018)

24. Schürmann, A.: Exploiting symmetries in polyhedral computations. In: Bezdek, K., Deza, A., Ye, Y. (eds.) Discrete Geometry and Optimization, vol. 69, pp. 265–278. Springer, Heidelberg (2013). https://doi.org/10.1007/978-3-319-00200-2_15
25. Seress, A.: Permutation Group Algorithms. Cambridge University Press, Cambridge (2003)
26. Sherali, H.D., Smith, J.C.: Improving discrete model representations via symmetry considerations. Manag. Sci. **47**, 1396–1407 (2001)
27. Verschae, J., Villagra, M., von Niederhäusern, L.: On the Geometry of Symmetry Breaking Inequalities. arXiv:2011.09641 (2020)

Affinely Representable Lattices, Stable Matchings, and Choice Functions

Yuri Faenza and Xuan Zhang$^{(\boxtimes)}$

Columbia University, New York, NY 10027, USA
{yf2414,xz2569}@columbia.edu

Abstract. Birkhoff's representation theorem [11] defines a bijection between elements of a distributive lattice $\mathcal{L}$ and the family of upper sets of an associated poset $\mathcal{B}$. When elements of $\mathcal{L}$ are the stable matchings in an instance of Gale and Shapley's marriage model, Irving et al. [22] showed how to use $\mathcal{B}$ to devise a combinatorial algorithm for maximizing a linear function over the set of stable matchings. In this paper, we introduce a general property of distributive lattices, which we term as affine representability, and show its role in efficiently solving linear optimization problems over the elements of a distributive lattice, as well as describing the convex hull of the characteristic vectors of lattice elements. We apply this concept to the stable matching model with path-independent quota-filling choice functions, thus giving efficient algorithms and a compact polyhedral description for this model. To the best of our knowledge, this model generalizes all models from the literature for which similar results were known, and our paper is the first that proposes efficient algorithms for stable matchings with choice functions, beyond extension of the Deferred Acceptance algorithm [31].

Keywords: Stable matching · Choice function · Distributive lattice · Birkhoff's representation theorem

1 Introduction

Since Gale and Shapley's seminal publication [17], the concept of stability in matching markets has been widely studied by the optimization community. With minor modifications, the one-to-many version of Gale and Shapley's original stable *marriage* model is currently employed in the National Resident Matching Program [30], which assigns medical residents to hospitals in the US, and for matching eighth-graders to public high schools in many major cities in the US [1].

In this paper, matching markets have two sides, which we call firms F and workers W. In the marriage model, every agent from $F \cup W$ has a *strict preference list* that ranks agents from the opposite side of the market. The problem asks for a *stable matching*, which is a matching where no pair of agents prefer each other to their assigned partner. A stable matching can be found efficiently via the Deferred Acceptance algorithm [17].

© Springer Nature Switzerland AG 2021
M. Singh and D. P. Williamson (Eds.): IPCO 2021, LNCS 12707, pp. 89–103, 2021.
https://doi.org/10.1007/978-3-030-73879-2_7

Although successful, the marriage model does not capture features that have become of crucial importance both inside and outside academia. For instance, there is growing attention to models that can increase diversity in school cohorts [28,37]. Such constraints cannot be represented in the original model, or even its one-to-many or many-to-many generalizations, since admission decisions with diversity concerns cannot be captured by a strict preference list.

To model these and other markets, every agent $a \in F \cup W$ is endowed with a *choice function* C_a that picks a team she prefers the best from a given set of potential partners. See, e.g., [7,14,23] for more applications of models with choice functions, and the literature review section for more references. *Mutatis mutandis*, one can define a concept of stability in this model as well (for this and the other technical definitions mentioned below, see Sect. 2). Two classical assumptions on choices functions are *substitutability* and *consistency*, under which the existence of stable matchings is guaranteed [6,20]. Clearly, existence results are not enough for applications (and for optimizers). Interestingly, little is known about efficient algorithms in models with choice functions. Only extensions of the classical Deferred Acceptance algorithm for finding the one-side optimal matching have been studied for this model [13,31].

The goal of this paper is to study algorithms for optimizing a linear function w over the set of stable matchings in models with choice functions, where w is defined over firm-worker pairs. Such algorithms can be used to obtain a stable matching that is e.g., *egalitarian*, *profit-optimal*, and *minimum regret* [25]. We focus in particular on the model where choice functions are assumed to be substitutable, consistent, and *quota-filling*. This model (QF-MODEL) generalizes all classical models where agents have strict preference lists, on which results for the question above were known. For this model, Alkan [3] has shown that stable matchings form a distributive lattice. As we argue next, this is a fundamental property that allows us to solve our optimization problem efficiently. For missing proofs, extended discussions, and examples, see the full version of the paper [15].

Our contributions and techniques. We give a high-level description of our approach and results. For the standard notions of posets, distributive lattices, and related definitions, see [19]. All sets considered in this paper are finite.

Let $\mathcal{L} = (\mathcal{X}, \succeq)$ be a distributive lattice, where all elements of $\mathcal{X}$ are distinct subsets of a base set E and $\succeq$ is a partial order on $\mathcal{X}$. We refer to $S \in \mathcal{X}$ as an *element* (of the lattice). Birkhoff's theorem [11] implies that we can associate[1] to every distributive lattice $\mathcal{L}$ a poset $\mathcal{B} = (Y, \succeq^\star)$ such that there is a bijection $\psi : \mathcal{X} \to \mathcal{U}(\mathcal{B})$, where $\mathcal{U}(\mathcal{B})$ is the family of *upper sets* of $\mathcal{B}$. $U \subseteq Y$ is an upper set of $\mathcal{B}$ if $y \in U$ and $y' \succeq^\star y$ for some $y' \in Y$ implies $y' \in U$. We say therefore that $\mathcal{B}$ is a *representation poset* for $\mathcal{L}$ with *representation function* ψ. See Example 1 for a demonstration. $\mathcal{B}$ may contain much fewer elements than the lattice $\mathcal{L}$ it represents, thus giving a possibly "compact" description of $\mathcal{L}$.

The representation function ψ satisfies that for $S, S' \in \mathcal{X}$, $S \succeq S'$ if and only if $\psi(S) \subseteq \psi(S')$. Albeit $\mathcal{B}$ and ψ explain how elements of $\mathcal{X}$ are related to each

---

[1] The result proved by Birkhoff is actually a bijection between the families of lattices and posets, but in this paper we shall not need it in full generality.

other with respect to $\succeq$, they do not contain any information on which items from E are contained in each lattice element. We introduce therefore Definition 1. For $S \in \mathcal{X}$ and $U \in \mathcal{U}(\mathcal{B})$, we write $\chi^S \in \{0,1\}^E$ and $\chi^U \in \{0,1\}^Y$ to denote their characteristic vectors, respectively.

Definition 1. *Let $\mathcal{L} = (\mathcal{X}, \succeq)$ be a distributive lattice on a base set E and $\mathcal{B} = (Y, \succeq^*)$ be a representation poset for $\mathcal{L}$ with representation function ψ. $\mathcal{B}$ is an* affine *representation of $\mathcal{L}$ if there exists an affine function $g : \mathbb{R}^Y \to \mathbb{R}^E$ such that $g(\chi^U) = \chi^{\psi^{-1}(U)}$, for all $U \in \mathcal{U}(B)$. In this case, we also say that $\mathcal{B}$* affinely represents $\mathcal{L}$ via affine function g and that $\mathcal{L}$ is *affinely representable.*

Note that in Definition 1, we can always assume $g(u) = Au + x_0$, where x_0 is the characteristic vector of the maximal element of $\mathcal{L}$ and $A \in \{0, \pm 1\}^{E \times Y}$.

Example 1. Consider the distributive lattice $\mathcal{L} = (\mathcal{X}, \succeq)$ with base set $E = \{1, 2, 3, 4\}$ whose Hasse diagram is given below.

$$A = \begin{pmatrix} 0 & 0 \\ -1 & 0 \\ 1 & 0 \\ 0 & 1 \end{pmatrix}$$

The representation poset $\mathcal{B} = (Y, \succeq^*)$ of $\mathcal{L}$ is composed of two non-comparable elements, y_1 and y_2. The representation function ψ is defined as

$$\psi(S_1) = \emptyset =: U_1; \quad \psi(S_2) = \{y_1\} =: U_2; \quad \psi(S_3) = \{y_2\} =: U_3; \quad \psi(S_4) = Y =: U_4.$$

That is, $\mathcal{U}(\mathcal{B}) = \{U_1, U_2, U_3, U_4\}$. One can think of y_1 as the operation of adding $\{3\}$ and removing $\{2\}$, and y_2 as the operation of adding $\{4\}$. $\mathcal{B}$ affinely represents $\mathcal{L}$ via the function $g(\chi^U) = A\chi^U + \chi^{S_1}$, with matrix A given above.

Now consider the distributive lattice $\mathcal{L}'$ obtained from $\mathcal{L}$ by switching S_3 and S_4. One can check that $\mathcal{L}'$ is not affinely representable [15]. $\triangle$

As we show next, affine representability allows one to efficiently solve linear optimization problems over elements of a distributive lattice. In particular, it generalizes a property that is at the backbone of combinatorial algorithms for optimizing a linear function over the set of stable matchings in the marriage model and its one-to-many and many-to-many generalizations (see, e.g., [10,22]). In the marriage model, the base set E is the set of pairs of agents from two sides of the market, $\mathcal{X}$ is the set of stable matchings, and for $S, S' \in \mathcal{X}$, $S \succeq S'$ if every firm prefers its partner in S to its partner in S' or is indifferent between the two. Elements of its representation poset are certain (trading) cycles, called *rotations*.

Lemma 1. *Assume poset $\mathcal{B} = (Y, \succeq^*)$ affinely represents lattice $\mathcal{L} = (\mathcal{X}, \succeq)$. Let $w : E \to \mathbb{R}$ be a linear function over the base set E of $\mathcal{L}$. Then the problem $\max\{w^\intercal \chi^S : S \in \mathcal{X}\}$ can be solved in time* min-cut$(|Y| + 2)$, *where* min-cut(k) *is the time complexity required to solve a minimum cut problem with nonnegative weights in a digraph with k nodes.*

Proof. Let $g(u) = Au + x_0$ be the affine function for the representation. Then,

$$\max_{S \in \mathcal{X}} w^{\intercal} \chi^S = \max_{U \in \mathcal{U}(\mathcal{B})} w^{\intercal} g(\chi^U) = \max_{U \in \mathcal{U}(\mathcal{B})} w^{\intercal} (A\chi^U + x_0) = w^{\intercal} x_0 + \max_{U \in \mathcal{U}(\mathcal{B})} (w^{\intercal} A)\chi^U.$$

Thus, our problem boils down to the optimization of a linear function over the upper sets of $\mathcal{B}$. It is well-known that the latter problem is equivalent to computing a minimum cut in a digraph with $|Y| + 2$ nodes [29].

We want to apply Lemma 1 to the QF-MODEL model. Observe that a choice function may be defined on all the (exponentially many) subsets of agents from the opposite side of the market. We avoid this computational concern by modeling choice functions via an oracle model. That is, choice functions can be thought of as agents' private information. The complexity of our algorithms will therefore be expressed in terms of $|F|$, $|W|$, and the time required to compute the choice function $\mathcal{C}_a(X)$ of an agent $a \in F \cup W$, where the set X is in the domain of $\mathcal{C}_a$. The latter running time is denoted by `oracle-call` and we assume it to be independent of a and X. Our first result is the following.

Theorem 1. *The distributive lattice* $(\mathcal{S}, \succeq)$ *of stable matchings in the* QF-MODEL *is affinely representable. Its representation poset* $(\Pi, \succeq^\star)$ *has* $O(|F||W|)$ *elements.* $(\Pi, \succeq^\star)$, *as well as its representation function* ψ *and affine function* $g(u) = Au + x_0$, *can be computed in time* $O(|F|^3 |W|^3 \texttt{oracle-call})$. *Moreover, matrix* A *has full column rank.*

In Theorem 1, we assumed that operations, such as checking if two sets coincide and obtaining an entry from the set difference of two sets, take constant time. If this is not the case, the running time needs to be scaled by a factor mildly polynomial in $|F| \cdot |W|$. Observe that Theorem 1 is the union of two statements. First, the distributive lattice of stable matchings in the QF-MODEL is affinely representable. Second, this representation and the corresponding functions ψ and g can be found efficiently. Those two results are proved in Sect. 3 and Sect. 4, respectively. Combining Theorem 1, Lemma 1 and algorithms for finding a minimum cut (see, e.g., [34]), we obtain the following.

Corollary 1. *The problem of optimizing a linear function over the set of stable matchings in the* QF-MODEL *can be solved in time* $O(|F|^3 |W|^3 \texttt{oracle-call})$.

As an interesting consequence of studying a distributive lattice via the poset that affinely represents it, one immediately obtains a linear description of the convex hull of the characteristic vectors of elements of the lattice (see Sect. 5). In contrast, most stable matching literature (see the literature review section) has focused on deducing linear descriptions for special cases of our model via ad-hoc proofs, independently of the lattice structure.

Theorem 2. *Let* $\mathcal{L} = (\mathcal{X}, \succeq)$ *be a distributive lattice and* $\mathcal{B} = (Y, \succeq^\star)$ *be a poset that affinely represents it via the affine function* $g(u) = Au + x_0$. *Then the extension complexity of* $\text{conv}(\mathcal{X}) := \text{conv}\{\chi^S : S \in \mathcal{X}\}$ *is* $O(|Y|^2)$. *If moreover* A *has full column rank, then* $\text{conv}(\mathcal{X})$ *has* $O(|Y|^2)$ *facets.*

Theorem 1 and Theorem 2 imply the following description for the stable matching polytope $\operatorname{conv}(\mathcal{S})$, i.e., the convex hull of the characteristic vectors of stable matchings in the QF-MODEL.

Corollary 2. $\operatorname{conv}(\mathcal{S})$ has $O(|F|^2|W|^2)$ facets.

We conclude with an example of a lattice represented via a non-full-column rank matrix A.

Example 2. Consider the distributive lattice given below.

$$
\begin{array}{|c|}
\hline
S_1 = \{1,2\} \\
\hline
\end{array}
\quad
\begin{array}{|c|}
\hline
S_2 = \{1,3\} \\
\hline
\end{array}
\quad
\begin{array}{|c|}
\hline
S_3 = \{1,2,4\} \\
\hline
\end{array}
\quad
\begin{array}{|c|}
\hline
S_4 = \{1,3,4\} \\
\hline
\end{array}
\qquad
A = \begin{pmatrix} 0 & 0 & 0 \\ -1 & 1 & -1 \\ 1 & -1 & 1 \\ 0 & 1 & 0 \end{pmatrix}
$$

It can be represented via the poset $\mathcal{B} = (Y, \succeq^*)$, that contains three elements y_1, y_2, and y_3 where $y_1 \succeq^* y_2 \succeq^* y_3$. Thus, $\mathcal{U}(\mathcal{B}) = \{\emptyset, \{y_1\}, \{y_1, y_2\}, \{y_1, y_2, y_3\}\}$. In addition, $\mathcal{B}$ affinely represents $\mathcal{L}$ via the function $g(\chi^U) = A\chi^U + \chi^{S_1}$, where A is given below. It is clear that matrix A does not have full column rank. $\triangle$

Relationship with the literature. Gale and Shapley [17] introduced the one-to-one stable marriage (SM-MODEL) and the one-to-many stable admission model (SA-MODEL), and presented an algorithm which finds a stable matching. McVitie and Wilson [27] proposed the break-marriage procedure that finds the full set of stable matchings. Irving et al. [22] presented an efficient algorithm for the maximum-weighted stable matching problem with weights over pairs of agents, using the fact that the set of stable matchings forms a distributive lattice [24] and that its representation poset – an affine representation following our terminology – can be constructed efficiently via the concept of rotations [21]. The above-mentioned structural and algorithm results have been also shown for its many-to-many generalization (MM-MODEL) in [8,10]. A complete survey of results on these models can be found, e.g., in [19,25].

For models with substitutable and consistent choice functions, Roth [31] proved that stable matchings always exist by generalizing the algorithm presented in [17]. Blair [12] proved that stable matchings form a lattice, although not necessarily distributive. Alkan [3] showed that if choice functions are further assumed to be quota-filling, the lattice is distributive. Results on (non-efficient) enumeration algorithms in certain models with choice functions appeared in [26].

It is then natural to investigate whether algorithms from [10,21] can be directly extended to construct the representation poset in the QF-MODEL. However, definition of rotations and techniques in [10,21] rely on the fact that there is a strict ordering of partners, which is not available with choice functions. This, for instance, leads to the fact that the symmetric difference of two stable match-ings that are adjacent in the lattice is always a simple cycle, which is not always true in the QF-MODEL. We take then a more fundamental approach by show-ing a carefully defined ring of sets is isomorphic to the set of stable matchings, and thus we can construct the rotation poset following a maximal chain of the stable matching lattice. This approach conceptually follows the one in [19] for the SM-MODEL and leads to a generalization of the break-marriage procedure from [27]. Again, proofs in [19,27] heavily rely on having a strict ordering of partners, while we need to tackle the challenge of not having one.

Besides the combinatorial perspective, another line of research focuses on the polyhedral aspects. Linear descriptions of the convex hull of the characteris-tic vectors of stable matchings are provided for the SM-MODEL [32,33,38], the SA-MODEL [9], and the MM-MODEL [16]. In this paper, we provide a polyhe-dral description for the QF-MODEL, by drawing connection between the order polytope (i.e., the convex hull of the characteristic vectors of the upper sets of a poset) and Birkhoff's representation theorem of distributive lattices. A similar approach has been proposed in [5]: their result can be seen as a specialization of Theorem 2 to the SM-MODEL.

Aside from the stable matching problem, the feasible spaces of many other combinatorial optimization problems form a distributive lattice. Examples, as pointed out in [18], include feasible rooted trees for the shortest path problem, and market clearing prices for the assignment game [35].

2   The QF-MODEL

Let F and W denote two disjoint finite sets of agents, say firms and workers, respectively. Associated with each firm $f \in F$ is a *choice function* $\mathcal{C}_f : 2^{W(f)} \to 2^{W(f)}$ where $W(f) \subseteq W$ is the set of *acceptable* partners of f and $\mathcal{C}_f$ satisfies the property that for every $S \subseteq W(f)$, $\mathcal{C}_f(S) \subseteq S$. Similarly, a choice function $\mathcal{C}_w : 2^{F(w)} \to 2^{F(w)}$ is associated to each worker w. We assume that for every firm-worker pair (f, w), $f \in F(w)$ if and only if $w \in W(f)$. We let $\mathcal{C}_W$ and $\mathcal{C}_F$ denote the collection of firms' and workers' choice functions respectively. A *matching market* is a tuple $(F, W, \mathcal{C}_F, \mathcal{C}_W)$.

Following Alkan [3], we define the QF-MODEL by assuming that the choice function $\mathcal{C}_a$ of every agent $a \in F \cup W$ satisfies the three properties below.

Definition 2 (Substitutability). *$\mathcal{C}_a$ is substitutable if for any set of partners S, $b \in \mathcal{C}_a(S)$ implies that for all $T \subseteq S$, $b \in \mathcal{C}_a(T \cup \{b\})$.*

Definition 3 (Consistency). *$\mathcal{C}_a$ is consistent if for any sets of partners S and T, $\mathcal{C}_a(S) \subseteq T \subseteq S$ implies $\mathcal{C}_a(S) = \mathcal{C}_a(T)$.*

Definition 4 (Quota-filling). *$\mathcal{C}_a$ is quota-filling if there exists $q_a \in \mathbb{N}$ such that for any set of partners S, $|\mathcal{C}_a(S)| = \min(q_a, |S|)$. We call q_a the quota of a.*

Intuitively, substitutability implies that an agent selected from a set of candidates will also be selected from a smaller subset; consistency is also called "irrelevance of rejected contracts"; and quota-filling means that an agent has a number of positions and she tries to fill those as many as possible. A choice function is substitutable and consistent if and only if it is *path-independent* [2].

Definition 5 (Path-independence). C_a *is path-independent if for any sets of partners* S *and* T, $C_a(S \cup T) = C_a\big(C_a(S) \cup T\big)$.

A *matching* μ is a mapping from $F \cup W$ to $2^{F \cup W}$ such that for all $w \in W$ and all $f \in F$, (1) $\mu(w) \subseteq F(w)$, (2) $\mu(f) \subseteq W(f)$, and (3) $w \in \mu(f)$ if and only if $f \in \mu(w)$. A matching can also be viewed as a collection of firm-worker pairs. That is, $\mu \equiv \{(f, w) : f \in F, w \in \mu(f)\}$. We say a matching μ is *individually rational* if for every agent a, $C_a(\mu(a)) = \mu(a)$. An acceptable firm-worker pair $(f, w) \notin \mu$ is called a *blocking pair* if $w \in C_f(\mu(f) \cup \{w\})$ and $f \in C_w(\mu(w) \cup \{f\})$, and when such pair exists, we say μ is *blocked by* the pair or the pair *blocks* μ. A matching μ is *stable* if it is individually rational and it admits no blocking pairs. If f is matched to w in some stable matching, we say that f (resp. w) is a *stable partner* of w (resp. f). We denote by $\mathcal{S}(C_F, C_W)$ the set of stable matchings in the market (F, W, C_F, C_W). Alkan [3] showed the following.

Theorem 3 ([3]). *Consider a matching market* (F, W, C_F, C_W) *in the* QF-MODEL. *Then,* $\mathcal{S}(C_F, C_W)$ *is a distributive lattice under the partial order* $\succeq$ *where* $\mu_1 \succeq \mu_2$ *if for all* $f \in F$, $C_f(\mu_1(f)) \cup \mu_2(f)) = \mu_1(f)$.

We denote by μ_F and μ_W the firm- and worker-optimal stable matchings, respectively. For every $a \in F \cup W$, let $\Phi_a = \{\mu(a) : \mu \in \mathcal{S}(C_F, C_W)\}$. Alkan [3] showed that for all $S, T \in \Phi_a$ the following holds: $|S| = |T| =: \overline{q}_a$ (equal-quota); and $\overline{q}_a < q_a \implies |\Phi_a| = 1$ (full-quota).

3   Affine Representability of the Stable Matching Lattice

For the rest of the paper, we fix a matching market (F, W, C_F, C_W) and often abbreviate $\mathcal{S} := \mathcal{S}(C_F, C_W)$. In this section, we show that the distributive lattice of stable matchings $(\mathcal{S}, \succeq)$ in the QF-MODEL is affinely representable. Our approach is as follows. First, we show that $(\mathcal{S}, \succeq)$ is isomorphic to a lattice $(\mathcal{P}, \subseteq)$ belonging to a special class, that is called *ring of sets*. We then show that rings of sets are always affinely representable. Next, we show a poset $(\Pi, \succeq^*)$ representing $(\mathcal{S}, \succeq)$. We last show how to combine all those results and "translate" the affine representability of $(\mathcal{P}, \subseteq)$ to the affine representability of $(\mathcal{S}, \succeq)$. We note in passing that in this section we actually rely on weaker assumptions than those from the QF-MODEL, essentially matching those from [4]. That is, instead of quota-filling, we can assume a weaker condition called *cardinal monotonicity*: C_a is cardinal monotone if for all sets of partners $S \subseteq T$, $|C_a(S)| \leq |C_a(T)|$.

Isomorphism between the stable matching lattice and a ring of sets. A family $\mathcal{H}$ of subsets of a *base set* B is a *ring of sets* over B if $\mathcal{H}$ is closed under set union and set intersection [11]. A ring of sets is a distributive lattice with the partial order relation $\subseteq$, and the join ($\vee$) and meet ($\wedge$) operations corresponding to set intersection and set union, respectively.

Let $\phi(a) := \{b : b \in \mu(a)$ for some $\mu \in \mathcal{S}\}$ denote the set of stable partners of agent a. For a stable matching μ, let $P_f(\mu) := \{w \in \phi(f) : w \in \mathcal{C}_f(\mu(f) \cup \{w\})\}$, and define the P-set of μ as $P(\mu) := \{(f,w) : f \in F, \ w \in P_f(\mu)\}$.

The following theorem gives a "description" of the stable matching lattice as a ring of sets. Let $\mathcal{P}(\mathcal{C}_F, \mathcal{C}_W)$ denote the set $\{P(\mu) : \mu \in \mathcal{S}(\mathcal{C}_F, \mathcal{C}_W)\}$, and we often abbreviate $\mathcal{P} := \mathcal{P}(\mathcal{C}_F, \mathcal{C}_W)$.

Theorem 4. *(1) the mapping $P : \mathcal{S} \to \mathcal{P}$ is a bijection;*
(2) $(\mathcal{P}, \subseteq)$ is isomorphic to $(\mathcal{S}, \succeq)$. Moreover, $P(\mu_1 \vee \mu_2) = P(\mu_1) \cap P(\mu_2)$ and $P(\mu_1 \wedge \mu_2) = P(\mu_1) \cup P(\mu_2)$. In particular, $(\mathcal{P}, \subseteq)$ is a ring of sets over the base set $E = \{(f, w) \in F \times W : w \in \phi(f)\}$.

Remark 1. An isomorphism between the lattice of stable matchings and a ring of set is proved in the SM-MODEL as well [19], where the authors define $P_f(\mu) := \{w : f$ prefers w to $\mu(f)\}$, hence including firm-worker pairs that are not stable. Interestingly, there are examples showing that in the QF-MODEL, if we were to use the natural extension of the definition in [19], i.e., $P_f(\mu) := \{w \in W(f) : w \in \mathcal{C}_f(\mu(f) \cup \{w\})\}$, then $\mathcal{P}$ is not a ring of set, see [15].

Affine representability of rings of sets. Consider a poset $(X, \geq)$. Let $a, a' \in X$. If $a' > a$ and there is no $b \in X$ such that $a' > b > a$, we say that a' is an *immediate predecessor* of a in $(X, \geq)$ and that a is an *immediate descendant* of a' in $(X, \geq)$. Fix a ring of set $(\mathcal{H}, \subseteq)$ over a base set B and define set $\mathcal{D}(\mathcal{H}) := \{H \backslash H' : H'$ is an immediate predecessor of H in $(\mathcal{H}, \subseteq)\}$ of *minimal differences* among elements of $\mathcal{H}$. We note that minimal differences are disjoint [19]. We elucidate in Example 3 these definitions and the facts below.

Theorem 5 ([11])**.** *There is a partial order $\sqsupseteq$ over $\mathcal{D}(\mathcal{H})$ such that $(\mathcal{D}(\mathcal{H}), \sqsupseteq)$ is a representation poset for $(\mathcal{H}, \subseteq)$ where the representation function ψ is defined as follows: for any upper set $\overline{\mathcal{D}}$ of $(\mathcal{D}(\mathcal{H}), \sqsupseteq)$, $\psi^{-1}(\overline{\mathcal{D}}) = \bigcup\{K : K \in \overline{\mathcal{D}}\} \cup H_0$ where H_0 is the minimal element of $\mathcal{H}$. Moreover, $|\mathcal{D}(\mathcal{H})| = O(|B|)$.*

From Theorem 5, it is not hard to prove the following.

Theorem 6. *$(\mathcal{D}(\mathcal{H}), \sqsupseteq)$ affinely represents $(\mathcal{H}, \subseteq)$ via the affine function $g(u) = Au + x_0$, where $x_0 = \chi^{H_0}$, and $A \in \{0,1\}^{B \times \mathcal{D}(\mathcal{H})}$ has columns χ^K for each $K \in \mathcal{D}(\mathcal{H})$. Moreover, A has full column rank.*

Example 3. Consider the ring of sets and its representation poset given below.

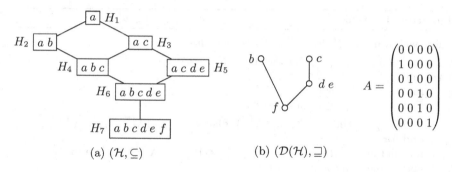

(a) $(\mathcal{H}, \subseteq)$ (b) $(\mathcal{D}(\mathcal{H}), \sqsupseteq)$

Representation function ψ maps $H_1, \cdots, H_7$ to $\emptyset$, $\{\{b\}\}$, $\{\{c\}\}$, $\{\{b\}, \{c\}\}$, $\{\{c\}, \{d, e\}\}$, $\{\{b\}, \{c\}, \{d, e\}\}$, and $\{\{b\}, \{c\}, \{d, e\}, \{f\}\}$, respectively. The affine function is $g(u) = Au + x_0$ with $x_0^\mathsf{T} = (1, 0, 0, 0, 0, 0)$ and matrix A given above. Note that columns of A correspond to $\{b\}, \{c\}, \{d, e\}, \{f\}$ in this order. $\triangle$

Representation of $(\mathcal{S}, \succeq)$ via the poset of rotations. For $\mu' \succeq \mu \in \mathcal{S}$ with μ' being an immediate predecessor of μ in the stable matching lattice, let $\rho^+(\mu', \mu) = \{(f, w) : (f, w) \in \mu \setminus \mu'\}$ and $\rho^-(\mu', \mu) = \{(f, w) : (f, w) \in \mu' \setminus \mu\}$. We call $\rho(\mu', \mu) := (\rho^+(\mu', \mu), \rho^-(\mu', \mu))$ a *rotation* of $(\mathcal{S}, \succeq)$. Let $\Pi(\mathcal{S})$ denote the set of rotations of $(\mathcal{S}, \succeq)$. We abbreviate $\mathcal{D} := \mathcal{D}(\mathcal{P})$ and $\Pi := \Pi(\mathcal{S})$.

Theorem 7. *(1) the mapping $Q : \Pi \to \mathcal{D}$, with $Q(\rho) := \rho^+$, is a bijection;*
(2) $(\mathcal{D}, \sqsupseteq)$ is isomorphic to $(\Pi, \succeq^\star)$ where for two rotations $\rho_1, \rho_2 \in \Pi$, $\rho_1 \succeq^\star \rho_2$ if $Q(\rho_1) \sqsupseteq Q(\rho_2)$;
(3) $(\Pi, \succeq^\star)$ is a representation poset for $(\mathcal{S}, \succeq)$, where the representation function $\psi_\mathcal{S}$ is defined as follows: $\psi_\mathcal{S}^{-1}(\overline{\Pi}) = \mu_F \cup (\bigcup_{\rho \in \overline{\Pi}} \rho^+) \setminus (\bigcup_{\rho \in \overline{\Pi}} \rho^-)$, for any upper set $\overline{\Pi}$ of $(\Pi, \succeq^\star)$.

$(\Pi, \succeq^\star)$ is called the *rotation poset*. By Theorem 6 and Theorem 7, we deduce $|\Pi| = O(|F||W|)$ and the following, proving the structural statement from Theorem 1. The base set E of $(\mathcal{S}, \succeq)$ is the set of acceptable firm-worker pairs.

Theorem 8. *The rotation poset $(\Pi, \succeq^\star)$ affinely represents the stable matching lattice $(\mathcal{S}, \succeq)$ with affine function $g(u) = Au + \mu_F$, where $A \in \{0, \pm 1\}^{E \times \Pi}$ has columns $\chi^{\rho^+} - \chi^{\rho^-}$ for each $\rho \in \Pi$. Moreover, matrix A has full column rank.*

4 Algorithms

To conclude the proof of Theorem 1, we show how to efficiently find the elements of Π, and how they relate to each other via $\succeq^\star$. First, we employ Roth's extension of the Deferred Acceptance algorithm to find a firm- or worker-optimal stable matching. Second, we feed its output to an algorithm that produces a maximal chain $C_0, C_1, \ldots, C_k$ of $(\mathcal{S}, \succeq)$ and the set Π. We then provide an algorithm that, given a maximal chain of a ring of sets, constructs the partial order for the poset

---

Algorithm 1. break-marriage(μ', f', w') with $(f', w') \in \mu' \in \mathcal{S}$

1: **for each** firm $f \neq f'$ **do** $X_f^{(0)} = \overline{X}_f(\mu')$ **end for**
2: let $X_{f'}^{(0)} = \overline{X}_{f'}(\mu') \setminus \{w'\}$; set the step count $s = 0$
3: **repeat**
4:    **for each** worker w **do**
5:       let $X_w^{(s)} = \{f \in F : w \in \mathcal{C}_f(X_f^{(s)})\}$
6:       **if** $w \neq w'$ **then** $Y_w^{(s)} = \mathcal{C}_w(X_w^{(s)})$ **else** $Y_w^{(s)} = \mathcal{C}_w(X_w^{(s)} \cup \{f'\}) \setminus \{f'\}$
7:    **end for**
8:    **for each** firm f **do** $X_f^{(s+1)} = X_f^{(s)} \setminus \{w \in W : f \in X_w^{(s)} \setminus Y_w^{(s)}\}$ **end for**
9:    update the step count $s = s + 1$
10: **until** $X_f^{(s-1)} = X_f^{(s)}$ for every firm f
Output: matching $\overline{\mu}$ with $\overline{\mu}(w) = Y_w^{(s-1)}$ for every worker w

---

of minimal differences. This and previous facts are then exploited to obtain the partial order $\succeq^*$ on rotations of Π. Lastly, we argue on the overall running time.

For a matching μ and $f \in F$, let $\overline{X}_f(\mu) := \{w \in W(f) : \mathcal{C}_f(\mu(f) \cup \{w\}) = \mu(f)\}$. Define the *closure* of μ, denoted by $\overline{X}(\mu)$, as the collection of sets $\{\overline{X}_f(\mu) : f \in F\}$. If μ is individually rational, then $\mu(f) \subseteq \overline{X}_f(\mu)$ for every $f \in F$.

Lemma 2. *Let $\mu_1, \mu_2 \in \mathcal{S}$ such that $\mu_1 \succeq \mu_2$. Then, $\forall f \in F$, $\mu_2(f) \subseteq \overline{X}_f(\mu_1)$.*

Deferred Acceptance Algorithm. Roth [31] generalized to choice function models the algorithm proposed in [17]. There is one side that is proposing – for the following, we let it be F. Initially, for each $f \in F$, let $X_f := W(f)$, i.e., the set of acceptable workers of f. At every step, every $f \in F$ proposes to workers in $\mathcal{C}_f(X_f)$. Then, every $w \in W$ considers the set of firms X_w who made a proposal to w, *temporarily accepts* $Y_w := \mathcal{C}_w(X_w)$, and *rejects* the rest. Afterwards, each firm f removes from X_f all workers that rejected f. Hence, throughout the algorithm, X_f denotes the set of acceptable workers of f that have not rejected f yet. The *firm-proposing* algorithm iterates until there is no rejection.

Theorem 9. *Roth's algorithms outputs μ_F in time $O(|F||W|\mathtt{oracle\text{-}call})$.*

By symmetry, swapping the role of firms and workers, we have the *worker-proposing* deferred acceptance algorithm, which outputs μ_W.

Constructing Π via a maximal chain of $(\mathcal{S}, \succeq)$. A *maximal chain* $C_0, \cdots, C_k$ in $(\mathcal{S}, \succeq)$ is an ordered subset of $\mathcal{S}$ such that C_{i-1} is an immediate predecessor of C_i in $(\mathcal{S}, \succeq)$ for all $i \in [k]$, $C_0 = \mu_F$, and $C_k = \mu_W$.

We now extend to our setting the *break-marriage* idea proposed by McVitie and Wilson [27]. This algorithm produces a matching μ starting from $\mu' \in \mathcal{S}$. A formal description is given in Algorithm 1. Roughly speaking, the algorithm re-initiates the deferred acceptance algorithm from μ' after suitably breaking a matched pair. Repeated applications of Algorithm 1 allow us to obtain an immediate descendant of μ' in $(\mathcal{S}, \succeq)$.

Algorithm 2. Immediate descendant of $\mu' \in \mathcal{S}$

1: set $\mathcal{T} = \emptyset$
2: **for each** $(f', w') \in \mu' \setminus \mu_W$ **do**
3:     run the **break-marriage**(μ', f', w') procedure
4:     **if** the procedure is successful **then** add the output matching $\overline{\mu}$ to $\mathcal{T}$
5: **end for**
Output: a maximal matching μ^* from $\mathcal{T}$ wrt $\succeq$, i.e. $\nexists \mu \in \mathcal{T}$ such that $\mu \succ \mu^*$

It is easy to see that **break-marriage**(μ', f', w') always terminates. We let s^* be the value of step count s at the end of the algorithm. Note that by the termination condition, $\overline{\mu}(f) = C_f(X_f^{(s^*)})$ for every firm f. Let $(f, w) \in F \times W$. We say f is *rejected by w at step s* if $f \in X_w^{(s)} \setminus Y_w^{(s)}$, and we say f is *rejected by w* if f is rejected by w at some step during the break-marriage procedure. Note that a firm f is rejected by all and only the workers in $\overline{X}_f(\mu') \setminus X_f^{(s^*)}$.

Theorem 10. *The running time of Algorithm 1 is $O(|F||W|\text{oracle-call})$.*

Lemma 3. *The matching $\overline{\mu}$ output by Algorithm 1 is individually rational, and for every firm $f \in F$, we have $C_f(\overline{\mu}(f) \cup \mu'(f)) = \mu'(f)$.*

We say **break-marriage**(μ', f', w') is *successful* if $f' \notin C_{w'}(X_{w'}^{(s^*-1)} \cup \{f'\})$.

Lemma 4. *If* **break-marriage**(μ', f', w') *is successful, then $\overline{\mu} \in \mathcal{S}$ and $\mu' \succ \overline{\mu}$.*

Next theorem shows a sufficient condition for the break-marriage procedure to output an immediate descendant in the stable matching lattice.

Theorem 11. *Let $\mu' \succeq \mu \in \mathcal{S}$ and assume μ' is an immediate predecessor of μ in the stable matching lattice. Pick $(f', w') \in \mu' \setminus \mu$ and let $\overline{\mu}$ be the output matching of* **break-marriage**(μ', f', w')*. Then, $\overline{\mu} = \mu$.*

Proof. Note that by Lemma 2, $\mu(f) \subseteq \overline{X}_f(\mu')$ for every $f \in F$. We start by showing that during the algorithm, for every firm f, no worker in $\mu(f)$ rejects f. Assume by contradiction that this is not true. Let s' be the first step where such a rejection happens, with firm f_1 being rejected by worker $w_1 \in \mu(f_1)$.

Claim 1. There exists a firm $f_2 \in Y_{w_1}^{(s')} \setminus \mu(w_1)$ such that $f_2 \in C_{w_1}(\mu(w_1) \cup \{f_2\})$.

Let f_2 be the firm whose existence is guaranteed by Claim 1. In particular, $f_2 \in Y_{w_1}^{(s')}$ implies $w_1 \in C_{f_2}(X_{f_2}^{(s')})$. Note that by our choice of f_1, $\mu(f_2) \subseteq X_{f_2}^{(s')}$. Therefore, using substitutability and $w_1 \in X_{f_2}^{(s')}$, we have $w_1 \in C_{f_2}(\mu(f_2) \cup \{w_1\})$. Thus, (f_2, w_1) is a blocking pair of μ, which contradicts the stability of μ.

Hence, for every firm f, no worker in $\mu(f)$ rejects f and thus, $\mu(f) \subseteq X_f^{(s^*)}$. Because of path-independence and $\overline{\mu}(f) = C_f(X_f^{(s^*)})$, we have $C_f(\overline{\mu}(f) \cup \mu(f)) = C_f(C_f(X_f^{(s^*)}) \cup \mu(f)) = C_f(C_f(X_f^{(s^*)} \cup \mu(f))) = C_f(X_f^{(s^*)}) = \overline{\mu}(f)$ (§). This in

---

Algorithm 3. A maximal chain of $(\mathcal{S}, \succeq)$ and the set of rotations Π

1: set counter $k = 0$; let $C_k = \mu_F$
2: **while** $C_k \neq \mu_W$ **do**
3:    run Algorithm 2 with $\mu' = C_k$, and let μ^* be its output
4:    update counter $k = k + 1$; let $C_k = \mu^*$
5: **end while**
Output: maximal chain $C_0, C_1, \cdots, C_k$; and $\Pi = \{\rho_i := \rho(C_{i-1}, C_i)\}_{i \in [k]}$.

---

particular implies $|\overline{\mu}(f)| \geq |\mu(f)|$ due to individual rationality of μ and quota-filling. Note also that $|\mu(f)| = |\mu'(f)| = |\mathcal{C}_f(\overline{\mu}(f) \cup \mu'(f))| \geq |\mathcal{C}_f(\overline{\mu}(f))| = |\overline{\mu}(f)|$, where the first equality is due to the equal-quota property, the second and the last by Lemma 3, and the inequality by quota-filling. We deduce $|\mu(f)| = |\mu'(f)| = |\overline{\mu}(f)|$ ($\natural$). We now show that break-marriage(μ', f', w') is successful.

Claim 2. $|\overline{\mu}(w)| = |\mu'(w)|$ for every worker $w \neq w'$.

Hence, $|\overline{\mu}(w')| = |\mu'(w')| = \overline{q}_{w'} = q_{w'}$, where the first equality holds from Claim 2 and ($\natural$), the second by the equal-quota property, and the last because $\mu(w') \neq \mu'(w')$ by choice of w' and the full-quota property. Therefore, $f' \notin \mathcal{C}_{w'}(X_{w'}^{(s^*-1)} \cup \{f'\})$ because otherwise $|\overline{\mu}(w')| = |\mathcal{C}_{w'}(X_{w'}^{(s^*-1)} \cup \{f'\}) \setminus \{f'\}| < |\mathcal{C}_{w'}(X_{w'}^{(s^*-1)} \cup \{f'\})| \leq q_{w'}$, where the last inequality holds by quota-filling, a contradiction. Thus, break-marriage(μ', f', w') is successful.

Finally, by Lemma 4, we have $\overline{\mu} \in \mathcal{S}$ and $\mu' \succ \overline{\mu}$. Because of ($\S$), we also have $\overline{\mu} \succeq \mu$. Therefore, it must be that $\overline{\mu} = \mu$ by the choice of μ.

We now present in Algorithm 2 a procedure that finds an immediate descendant for any given stable matching, using the break-marriage procedure.

Lemma 5. Let $\mu_1 \succ \mu_2 \succ \mu_3 \in \mathcal{S}$. If $(f, w) \in \mu_1 \setminus \mu_2$, then $(f, w) \notin \mu_3$.

Theorem 12. The output μ^* of Algorithm 2 is an immediate descendant of μ' in $(\mathcal{S}, \succeq)$. Its running time is $O(|F|^2|W|^2\text{oracle-call})$.

Proof. First note that due to Lemma 4, all matchings in the set $\mathcal{T}$ are stable matchings. Assume by contradiction that the output matching μ^* is not an immediate descendant of μ' in $(\mathcal{S}, \succeq)$. Then, there exists a stable matching μ such that $\mu' \succ \mu \succ \mu^*$. By Lemma 5, for every firm-worker pair $(f', w') \in \mu' \setminus \mu$, we also have $(f', w') \notin \mu_W$. Thus, $\mu \in \mathcal{T}$ due to Theorem 11. However, this means that μ^* is not a maximal matching from $\mathcal{T}$, which is a contradiction. The runtime follows from Theorem 10 and the fact that $|\mu'| = O(|F||W|)$.

Algorithm 3 employs Algorithm 2 to find a maximal chain of the stable matching lattice, as well as the set of rotations.

Theorem 13. Algorithm 3 is correct and runs in time $O(|F|^3|W|^3$ oracle-call$)$.

Algorithm 4. Construction of the rotation poset $(\Pi, \succeq^\star)$

1: Run Roth's algorithm [30], to obtain μ_F and μ_W.
2: Run Algorithm 3 to obtain a maximal chain $C_0, C_1, \cdots, C_k$ of the stable matching lattice $(\mathcal{S}, \succeq)$, and the set of rotations $\Pi \equiv \{\rho_1, \rho_2, \cdots, \rho_k\}$.
3: Run the algorithm from Theorem 14 to obtain the partial order $\succeq^\star$.

Proof. By Theorem 7 and Theorem 12, the maximal chain output is correct. Additionally, by Theorem 5, $|\Pi| = O(|F||W|)$ and the running time follows. By [19, Section 2.4.3], all elements of $\mathcal{D}(\mathcal{P})$ are found on a maximal chain of $\mathcal{P}$. Then, by Theorem 7, Π can also be found on a maximal chain of $\mathcal{S}$. Thus, output Π is correct.

Partial order $\succeq^\star$ over Π. Consider a ring of sets $(\mathcal{H}, \subseteq)$. We can produce an efficient algorithm that obtains the partial order $\sqsupseteq$ of $(\mathcal{D}(\mathcal{H}), \sqsupseteq)$ from a maximal chain of $(\mathcal{H}, \subseteq)$, based on the classical concept of *irreducible* elements: see [15] for details. Together with Theorem 4 and Theorem 7, we have the following.

Theorem 14. *There is an algorithm with runtime $O(|F|^3|W|^3 \texttt{oracle-call})$ that constructs the partial order $\succeq^\star$ given as input the output of Algorithm 3.*

Summary and time complexity analysis. The complete procedure to build the rotation poset is summarized in Algorithm 4. Its correctness and runtime of $O(|F|^3|W|^3 \texttt{oracle-call})$ follow from Theorem 9, Theorem 13, and Theorem 14. This concludes the proof of Theorem 1.

5   The Convex Hull of Lattice Elements: Proof of Theorem 2

The *order polytope* associated with poset $(Y, \succeq^\star)$ is defined as

$$\mathcal{O}(Y, \succeq^\star) := \{y \in [0, 1]^Y : y_i \geq y_j, \; \forall i, j \in Y \text{ s.t. } i \succeq^\star j\}.$$

Stanley [36] showed that the vertices of $\mathcal{O}(Y, \succeq^\star)$ are the characteristic vectors of upper sets of $(Y, \succeq^\star)$, and gave a complete characterization of the $O(|Y|^2)$ facets of $\mathcal{O}(Y, \succeq^\star)$. We claim that

$$\text{conv}(\mathcal{X}) = \{x_0\} \oplus A \cdot \mathcal{O}(Y, \succeq^\star) = \{x \in \mathbb{R}^E : x = x_0 + Ay, \; y \in \mathcal{O}(Y, \succeq^\star)\},$$

where $\oplus$ denotes the Minkowski sum operator. Indeed, g defines a bijection between vertices of $\mathcal{O}(Y, \succeq^\star)$ and vertices of $\text{conv}(\mathcal{X})$. The claim them follows by convexity. As $\mathcal{O}(Y, \succeq^\star)$ has $O(|Y|^2)$ facets, we conclude the first statement from Theorem 2.

Now suppose that A has full column rank. Then, since $\mathcal{O}(Y, \succeq^\star)$ is full-dimensional, $\text{conv}(\mathcal{X})$ is affinely isomorphic to $\mathcal{O}(Y, \succeq^\star)$. Hence, there is a one-to-one correspondence between facets of $\mathcal{O}(Y, \succeq^\star)$ and facets of $\text{conv}(\mathcal{X})$. The second statement then follows from the characterization given in [36]. Example 4 shows that statements above do not hold when A does not have full column-rank.

Example 4. Consider the lattice $(\mathcal{X}, \succeq)$ and its representation poset $(Y, \succeq^{\star})$ from Example 2. Note that

$$\mathrm{conv}(\mathcal{X}) = \{x \in [0,1]^4 : x_1 = 1, \ x_2 + x_3 = 1\}.$$

Thus, $\mathrm{conv}(\mathcal{X})$ has dimension 2. On the other hand, $\mathcal{O}(Y, \succeq^{\star})$ has dimension 3. So the two polytopes are not affinely isomorphic.

More generally, one can easily construct a distributive lattice $(\mathcal{X}, \succeq)$ such that the number of facets of $\mathcal{O}(Y, \succeq^{\star})$ gives no useful information on the number of facets of $\mathrm{conv}(\mathcal{X})$, where $(Y, \succeq^{\star})$ is a poset that affinely represents $(\mathcal{X}, \succeq)$. In fact, the vertices of any 0/1 polytope can be arbitrarily arranged in a chain to form a distributive lattice $(\mathcal{X}, \succeq)$. A poset $\mathcal{O}(Y, \succeq^{\star})$ that affinely represents $(\mathcal{X}, \succeq)$ is given by a chain with $|Y| = |\mathcal{X}| - 1$. It is easy to see that $\mathcal{O}(Y, \succeq^{\star})$ is a simplex and has therefore $|Y| + 1 = |\mathcal{X}|$ facets. However, $\mathrm{conv}(\mathcal{X})$ could have much more (or much less) facets than the number of its vertices. △

References

1. Abdulkadiroğlu, A., Sönmez, T.: School choice: a mechanism design approach. Am. Econ. Rev. **93**(3), 729–747 (2003)
2. Aizerman, M., Malishevski, A.: General theory of best variants choice: some aspects. IEEE Trans. Autom. Control **26**(5), 1030–1040 (1981)
3. Alkan, A.: On preferences over subsets and the lattice structure of stable matchings. Rev. Econ. Design **6**(1), 99–111 (2001)
4. Alkan, A.: A class of multipartner matching markets with a strong lattice structure. Econ. Theor. **19**(4), 737–746 (2002)
5. Aprile, M., Cevallos, A., Faenza, Y.: On 2-level polytopes arising in combinatorial settings. SIAM J. Discret. Mathe. **32**(3), 1857–1886 (2018)
6. Aygün, O., Sönmez, T.: Matching with contracts: comment. Am. Econ. Rev. **103**(5), 2050–51 (2013)
7. Aygün, o., Turhan, B.: Dynamic reserves in matching markets: Theory and applications. Available at SSRN 2743000 (2016)
8. Baïou, M., Balinski, M.: Many-to-many matching: stable polyandrous polygamy (or polygamous polyandry). Discret. Appl. Math. **101**(1–3), 1–12 (2000)
9. Baïou, M., Balinski, M.: The stable admissions polytope. Math. Program. **87**(3), 427–439 (2000). https://doi.org/10.1007/s101070050004
10. Bansal, v., Agrawal, A., Malhotra, V.S.: Polynomial time algorithm for an optimal stable assignment with multiple partners. Theoret. Comput. Sci. **379**(3), 317–328 (2007)
11. Birkhoff, G.: Rings of sets. Duke Math. J. **3**(3), 443–454 (1937)
12. Blair, C.: The lattice structure of the set of stable matchings with multiple partners. Math. Oper. Res. **13**(4), 619–628 (1988)
13. Chambers, C.P., Yenmez, M.B.: Choice and matching. Am. Econ. J. Microecon. **9**(3), 126–47 (2017)
14. Echenique, F., Yenmez, M.B.: How to control controlled school choice. Am. Econ. Rev. **105**(8), 2679–2694 (2015)
15. Faenza, Y., Zhang, X.: Affinely representable lattices, stable matchings, and choice functions (2020). Available on arXiv

16. Fleiner, T.: On the stable b-matching polytope. Math. Soc. Sci. **46**(2), 149–158 (2003)
17. Gale, D., Shapley, L.S.: College admissions and the stability of marriage. Am. Math. Month. **69**(1), 9–15 (1962)
18. Garg, V.K.: Predicate detection to solve combinatorial optimization problems. In: Proceedings of the 32nd ACM Symposium on Parallelism in Algorithms and Architectures, pp. 235–245 (2020)
19. Gusfield, D., Irving, R.W.: The stable marriage problem: structure and algorithms. MIT press (1989)
20. Hatfield, J.W., Milgrom, P.R.: Matching with contracts. Am. Econ. Rev. **95**(4), 913–935 (2005)
21. Irving, R.W., Leather, P.: The complexity of counting stable marriages. SIAM J. Comput. **15**(3), 655–667 (1986)
22. Irving, R.W., Leather, P., Gusfield, D.: An efficient algorithm for the "optimal" stable marriage. J. ACM (JACM) **34**(3), 532–543 (1987)
23. Kamada, Y., Kojima, F.: Efficient matching under distributional constraints: theory and applications. Am. Econ. Rev. **105**(1), 67–99 (2015)
24. Knuth, D.E.: Marriages stables. Technical report (1976)
25. Manlove, D.: Algorithmics of matching under preferences, vol. 2. World Scientific (2013)
26. Martínez, R., Massó, J., Neme, A., Oviedo, J.: An algorithm to compute the full set of many-to-many stable matchings. Math. Soc. Sci. **47**(2), 187–210 (2004)
27. McVitie, D.G., Wilson, L.B.: The stable marriage problem. Commun. ACM **14**(7), 486–490 (1971)
28. Nguyen, T., Vohra, R.: Stable matching with proportionality constraints. Oper. Res. **67**(6), 1503–1519 (2019)
29. Picard, J.-C.: Maximal closure of a graph and applications to combinatorial problems. Manage. Sci. **22**(11), 1268–1272 (1976)
30. Roth, A.E.: The evolution of the labor market for medical interns and residents: a case study in game theory. J. Polit. Econo. **92**(6), 991–1016 (1984)
31. Roth, A.E.: Stability and polarization of interests in job matching. Econom. J. Econ. Soc. **52**, 47–57 (1984)
32. Roth, A.E., Rothblum, U.G., Vate, J.H.V.: Stable matchings, optimal assignments, and linear programming. Math. Oper. Res. **18**(4), 803–828 (1993)
33. Rothblum, U.G.: Characterization of stable matchings as extreme points of a polytope. Math. Program. **54**(1–3), 57–67 (1992)
34. Schrijver, A.: Combinatorial Optimization: Polyhedra and Efficiency, vol. 24. Springer Science & Business Media, Heidelberg (2003)
35. Shapley, L.S., Shubik, M.: The assignment game I: the core. Int. J. Game Theory, **1**(1), 111–130 (1971). https://doi.org/10.1007/BF01753437
36. Stanley, R.P.: Two poset polytopes. Discret. Comput. Geom. **1**(1), 9–23 (1986). https://doi.org/10.1007/BF02187680
37. Tomoeda, K.: Finding a stable matching under type-specific minimum quotas. J. Econ. Theory **176**, 81–117 (2018)
38. Vate, J.H.V.: Linear programming brings marital bliss. Oper. Res. Lett. **8**(3), 147–153 (1989)

A Finite Time Combinatorial Algorithm for Instantaneous Dynamic Equilibrium Flows

Lukas Graf$^{(\boxtimes)}$ (ID) and Tobias Harks

Augsburg University, Institute of Mathematics, Augsburg 86135, Germany
{lukas.graf,tobias.harks}@math.uni-augsburg.de

Abstract. Instantaneous dynamic equilibrium (IDE) is a standard game-theoretic concept in dynamic traffic assignment in which individual flow particles myopically select en route currently shortest paths towards their destination. We analyze IDE within the Vickrey bottleneck model, where current travel times along a path consist of the physical travel times plus the sum of waiting times in all the queues along a path. Although IDE have been studied for decades, no exact finite time algorithm for equilibrium computation is known to date. As our main result we show that a natural extension algorithm needs only finitely many phases to converge leading to the first finite time combinatorial algorithm computing an IDE. We complement this result by several hardness results showing that computing IDE with natural properties is NP-hard.

1 Introduction

Flows over time or dynamic flows are an important mathematical concept in network flow problems with many real world applications such as dynamic traffic assignment, production systems and communication networks (e.g., the Internet). In such applications, flow particles that are sent over an edge require a certain amount of time to travel through each edge and when routing decisions are being made, the dynamic flow propagation leads to later effects in other parts of the network. A key characteristic of such applications, especially in traffic assignment, is that the network edges have a limited flow capacity which, when exceeded, leads to congestion. This phenomenon can be captured by the *fluid queueing model* due to Vickrey [23]. The model is based on a directed graph $G = (V, E)$, where every edge e has an associated physical transit time $\tau_e \in \mathbb{Q}_{>0}$ and a maximal rate capacity $\nu_e \in \mathbb{Q}_{>0}$. If flow enters an edge with higher rate than its capacity, the excess particles start to form a queue at the edge's tail, where they wait until they can be forwarded onto the edge. Thus, the total travel time experienced by a single particle traversing an edge e is the sum of the time spent waiting in the queue of e and the physical transit time τ_e.

This physical flow model then needs to be enhanced with a *behavioral model* prescribing the actions of flow particles. There are two main standard behavioral models in the traffic assignment literature known as *dynamic equilibrium (DE)*

© Springer Nature Switzerland AG 2021
M. Singh and D. P. Williamson (Eds.): IPCO 2021, LNCS 12707, pp. 104–118, 2021.
https://doi.org/10.1007/978-3-030-73879-2_8

(cf. Ran and Boyce [19, Sect. V–VI]) and *instantaneous dynamic equilibrium (IDE)* ([19, Sect. VII–IX]). Under DE, flow particles have complete information on the state of the network for all points in time (including the future evolution of all flow particles) and based on this information travel along a shortest path. The full information assumption is usually justified by assuming that the game is played repeatedly and a DE is then an attractor of a learning process. In an IDE, at every point in time and at every node of the graph, flow particles only enter those edges that lie on a currently shortest path towards their respective sink. The behavioral model of IDE is based on the concept that drivers are informed in real-time about the current traffic situation and, if beneficial, reroute instantaneously no matter how good or bad that route will be in hindsight. IDE has been proposed already in the late 80's (cf. Boyce, Ran and LeBlanc [1,20] and Friesz et al. [7]).

A line of recent works starting with Koch and Skutella [17] and Cominetti, Correa and Larré [3] derived a complementarity description of DE flows via so-called *thin flows with resetting* which leads to an α-extension property stating that for any equilibrium up to time θ, there exists $\alpha > 0$ so that the equilibrium can be extended to time $\theta + \alpha$. An extension that is maximal with respect to α is called a *phase* in the construction of an equilibrium and the existence of equilibria on the whole $\mathbb{R}_{\geq 0}$ then follows by a limit argument over phases using Zorn's lemma. In the same spirit, Graf, Harks and Sering [10] established a similar characterization for IDE flows and also derived an α-extension property.

For both models (DE or IDE), it is an open question whether for constant inflow rates and a finite time horizon, a *finite number of phases* suffices to construct an equilibrium, see [3,10,17]. Proving finiteness of the number of phases would imply an exact finite time combinatorial algorithm. Such an algorithm is not known to date neither for DE nor for IDE.[1] More generally, the computational complexity of equilibrium computation is widely open.

1.1   Our Contribution and Proof Techniques

In this paper, we study IDE flows and derive algorithmic and computational complexity results. As our main result we settle the key question regarding finiteness of the α-extension algorithm.

> Theorem 1: For single-sink networks with piecewise constant inflow rates with bounded support, there is an α-extension algorithm computing an IDE

---

[1] Algorithms for DE or IDE computation used in the transportation science literature are *numerical*, that is, only approximate equilibrium flows are computed given a certain numerical precision using a discretized model, see for example [1,6,11]. While a recent computational study [24] showed some positive results in regards to convergence for DE, Otsubo and Rapoport [18] also reported "significant discrepancies" between the continuous and a discretized solution for the Vickrey model.

> after finitely many extension phases. This implies the first finite time combinatorial exact algorithm computing IDE within the Vickrey model.

The proof of our result is based on the following ideas. We first consider the case of *acyclic* networks and use a topological order of nodes in order to schedule the extension phases in the algorithm. The key argument for the finiteness of the number of extension phases is that for a single node v and any interval with linearly changing distance labels of nodes closer to the sink and constant inflow rate into v, this flow can be redistributed to the outgoing edges in a finite number of phases of constant outflow rates from v. We show this using the properties (derivatives) of suitable edge label functions for the outgoing edges. The overall finiteness of the algorithm follows by induction over the nodes and time. We then generalize to arbitrary single-sink networks by considering *dynamically* changing topological orders depending on the current set of active edges.

We then turn to the computational complexity of IDE flows and show that several natural decision problems about the existence of IDE with certain properties are NP-hard.

> Theorem 2: The following decision problems are all NP-hard:
>
> – Given a specific edge: Is there an IDE using/not using this edge?
> – Given some time horizon T: Is there an IDE that terminates before T?
> – Given some $k \in \mathbb{N}$: Is there an IDE with at most k phases?

1.2   Related Work

The concept of flows over time was studied by Ford and Fulkerson [5]. Shortly after, Vickrey [23] introduced a game-theoretic variant using a deterministic queueing model. Since then, dynamic equilibria have been studied extensively in the transportation science literature, see Friesz et al. [7]. New interest in this model was raised after Koch and Skutella [17] gave a novel characterization of dynamic equilibria in terms of a family of static flows (thin flows). Cominetti et al. [3] refined this characterization and Sering and Vargas-Koch [22] incorporated spillbacks in the fluid queuing model. In a very recent work, Kaiser [16] showed that the thin flows needed for the extension step in computing dynamic equilibria can be determined in polynomial time for series-parallel networks. The papers [3,16] explicitly mention the problem of possible non-finiteness of the extension steps. For further results regarding a discrete packet routing model, we refer to Cao et al. [2], Ismaili [14,15], Scarsini et al. [21], Harks et al. [12] and Hoefer et al. [13].

2   Model and the Extension-Algorithm

In this paper we consider networks $\mathcal{N} = (G, (\nu_e)_{e \in E}, (\tau_e)_{e \in E}, (u_v)_{v \in V \setminus \{t\}}, t)$ given by a directed graph $G = (V, E)$, edge capacities $\nu_e \in \mathbb{Q}_{>0}$, edge travel times $\tau_e \in \mathbb{Q}_{>0}$, and a single sink node $t \in V$ which is reachable from anywhere in the graph. Each node $v \in V \setminus \{t\}$ has a corresponding (network) inflow rate $u_v : \mathbb{R}_{\geq 0} \to \mathbb{Q}_{\geq 0}$ indicating for every time $\theta \in \mathbb{R}_{\geq 0}$ the rate $u_v(\theta)$ at which the infinitesimal small agents enter the network at node v and start traveling through the graph until they leave the network at the common sink node t. We will assume that these network inflow rates are right-constant step functions with bounded support and finitely many, rational jump points.

A *flow over time* in $\mathcal{N}$ is a tuple $f = (f^+, f^-)$ where $f^+, f^- : E \times \mathbb{R}_{\geq 0} \to \mathbb{R}_{\geq 0}$ are integrable functions. For any edge $e \in E$ and time $\theta \in \mathbb{R}_{\geq 0}$ the value $f_e^+(\theta)$ describes the *(edge) inflow rate* into e at time θ and $f_e^-(\theta)$ is the *(edge) outflow rate* from e at time θ. For any such flow over time f we define the *cumulative (edge) in- and outflow rates* F^+ and F^- by $F_e^+(\theta) := \int_0^\theta f_e^+(\zeta)d\zeta$ and $F_e^-(\theta) := \int_0^\theta f_e^-(\zeta)d\zeta$, respectively. The queue length of edge e at time θ is then defined as $q_e(\theta) := F_e^+(\theta) - F_e^-(\theta + \tau_e)$.

Such a flow f is called a *feasible flow* in $\mathcal{N}$, if it satisfies the following constraints (1) to (4). The *flow conservation constraints* are modeled for all nodes $v \neq t$ as

$$\sum_{e \in \delta_v^+} f_e^+(\theta) - \sum_{e \in \delta_v^-} f_e^-(\theta) = u_v(\theta) \quad \text{for all } \theta \in \mathbb{R}_{\geq 0}, \tag{1}$$

where $\delta_v^+ := \{ vu \in E \}$ and $\delta_v^- := \{ uv \in E \}$ are the sets of outgoing edges from v and incoming edges into v, respectively. For the sink node t we require

$$\sum_{e \in \delta_t^+} f_e^+(\theta) - \sum_{e \in \delta_t^-} f_e^-(\theta) \leq 0 \tag{2}$$

and for all edges $e \in E$ we always assume

$$f_e^-(\theta) = 0 \text{ for all } \theta < \tau_e. \tag{3}$$

Finally we assume that the queues operate at capacity which can be modeled by

$$f_e^-(\theta + \tau_e) = \begin{cases} \nu_e, & \text{if } q_e(\theta) > 0 \\ \min\{ f_e^+(\theta), \nu_e \}, & \text{if } q_e(\theta) \leq 0 \end{cases} \quad \text{for all } e \in E, \theta \in \mathbb{R}_{\geq 0}. \tag{4}$$

Following the definition in [10] we call a feasible flow an IDE (flow) if whenever a particle arrives at a node $v \neq t$, it can only ever enter an edge that is the first edge on a currently shortest v-t path. In order to formally describe this property we first define the *current* or *instantaneous travel time* of an edge e at θ by

$$c_e(\theta) := \tau_e + \frac{q_e(\theta)}{\nu_e}. \tag{5}$$

We then define time dependent node labels $\ell_v(\theta)$ corresponding to current shortest path distances from v to the sink t. For $v \in V$ and $\theta \in \mathbb{R}_{\geq 0}$, define

$$\ell_v(\theta) := \begin{cases} 0, & \text{for } v = t \\ \min_{e=vw \in E}\{\ell_w(\theta) + c_e(\theta)\}, & \text{else.} \end{cases} \tag{6}$$

We say that an edge $e = vw$ is *active* at time θ, if $\ell_v(\theta) = \ell_w(\theta) + c_e(\theta)$, denote the set of active edges by $E_\theta \subseteq E$ and call the subgraph induced by E_θ the *active subgraph at time* θ.

Definition 1. A feasible flow over time f is an *instantaneous dynamic equilibrium (IDE)*, if for all $\theta \in \mathbb{R}_{\geq 0}$ and $e \in E$ it satisfies

$$f_e^+(\theta) > 0 \Rightarrow e \in E_\theta. \tag{7}$$

During the computation of an IDE we also temporarily need the concept of a *partial IDE up to some time* $\hat{\theta}$. This is a flow f such that constraints (1) to (4) as well as constraint (7) only hold for all $\theta \in [0, \hat{\theta})$, while $f_e^+(\theta) = f_e^-(\theta + \tau_e) = 0$ for all $\theta \geq \hat{\theta}$. For any such flow, we then define the *gross node inflow rates* b_v^- by setting $b_v^-(\theta) := \sum_{e \in \delta_v^-} f_e^-(\theta) + u_v(\theta)$ for all $v \in V \setminus \{t\}$ and $\theta \in [\hat{\theta}, \hat{\theta} + \tau_{\min})$, where $\tau_{\min} := \min\{\tau_e \mid e \in E\} > 0$.

As shown in [10, Sect. 3] such a partial IDE can always be extended for some additional proper[2] time interval on a node by node basis using constant edge inflow rates. The existence of IDE for the whole $\mathbb{R}_{\geq 0}$ then follows by applying Zorn's lemma. This also leads to the following natural algorithm for computing IDE in single-sink networks:

1. Start with the zero-flow f – a partial IDE up to time 0.
2. While f is not an IDE for all times, extend f for some additional interval.

In the extension step, we first determine a topological order of the nodes in the active subgraph (e.g. sort the nodes w.r.t. to their current node labels ℓ_v). Then we go through the nodes in this order (beginning with the sink node t) and at each node determine a constant distribution of the current gross node inflow rate to the outgoing active edges in such a way that the used edges remain active for some proper time interval into the future. Finally, we take the smallest of these intervals and extend the whole partial IDE over it. For the extension at a single node v at some time θ, we can use a solution to the following convex optimization problem, which can be determined in polynomial time using a simple water filling procedure (see [10, Algorithm 1 (electronic supplementary material)]):

$$\min \quad \sum_{e=vw \in \delta_v^+ \cap E_\theta} \int_0^{x_e} \frac{g_e(z)}{\nu_e} + \partial_+\ell_w(\theta)dz \qquad (\text{OPT-}b_v^-(\theta))$$

$$\text{s.t.} \quad \sum_{e \in \delta_v^+ \cap E_\theta} x_e = b_v^-(\theta), \quad x_e \geq 0 \text{ for all } e \in \delta_v^+ \cap E_\theta,$$

---

[2] We call an interval $[a, b)$ *proper* if $a < b$.

where ∂_+ denotes the right side derivative and $g_e(z) := z - \nu_e$, if $q_e(\theta) > 0$, and $g_e(z) := \max\{z - \nu_e, 0\}$, otherwise. Any solution to $(\text{OPT-}b_v^-(\theta))$ corresponds to a flow distribution to active edges so that for every edge $e = vw \in \delta_v^+ \cap E_\theta$ the following condition is satisfied (see [10, Lemma 3.1])

$$
\begin{aligned}
f_e^+(\theta) > 0 &\implies \partial_+\ell_v(\theta) = \partial_+c_e(\theta) + \partial_+\ell_w(\theta) \\
f_e^+(\theta) = 0 &\implies \partial_+\ell_v(\theta) \leq \partial_+c_e(\theta) + \partial_+\ell_w(\theta).
\end{aligned}
\tag{8}
$$

Because the network inflow rates as well as all already constructed edge inflow rates are piecewise constant and the node label functions as well as the queue length functions are continuous, the used edges will remain active for some proper time interval. As IDE flows in single-sink networks always have a finite termination time ([10, Theorem 4.6]) it suffices to extend the flow for some finite time horizon (in [9] we even provide a way to explicitly compute such a time horizon). Thus, the only possible obstruction for the extension algorithm to terminate within finite time is some Zeno-type behavior of the lengths of the extension phases, e.g. some sequence of extension phases of lengths $\alpha_1, \alpha_2, \ldots$ such that $\sum_{i=1}^{\infty} \alpha_i$ converges to some point strictly before the IDE's termination time. In fact, in the full version of this paper [8], we provide an example of a rather simple network wherein extension phases may indeed become arbitrarily small, provided a long enough lasting network inflow rate.[3] However, this is not a counter example to the finiteness of the extension algorithm, as the shrinking of the extension phases is slow enough to still allow for a finite number of phases to span any fixed time horizon. In the following section we will show that this is in fact true for *all* single-sink networks, i.e. that we can reach any given time horizon within a finite number of phases.

3 Finite IDE-Construction Algorithm

For the proof of our main theorem we will employ two reductions: First, we observe that for acyclic networks, it suffices to consider a single node with constant gross node inflow rate for a given interval and linear node label functions at all the nodes reachable via a single edge from this node. For this situation, we show that the incoming flow can be distributed over active edges using a finite number of phases. Second, we argue that for general networks, we can group the extension phases into finitely many larger intervals such that during each such interval the extension algorithm only has to consider a certain fixed acyclic subgraph (reducing to the first case).

---

[3] This example also shows that the number of (distinct!) extension phases can be exponential in the encoding size of the given instance and that networks with forever lasting network inflow rates may require IDE flows which never reach a stable state. An exponential number of extension phases has also been observed in DE flows while stable states are always reached there (see [4]).

Acyclic Networks. Due to our first reduction, which we will justify afterwards, the proof for acyclic networks essentially rests on the following key lemma.

Lemma 1. *Let $\mathcal{N}$ be a single-sink network on an acyclic graph with some fixed topological order on the nodes, v some node in $\mathcal{N}$ and $\theta_1 < \theta_2 \le \theta_1 + \tau_{\min}$ two times. If f is a flow over time in $\mathcal{N}$ such that*

- *f is a partial IDE up to time θ_1 for nodes at least as far away from t as v,*
- *f is a partial IDE up to time θ_2 for nodes closer to t than v,*
- *during $[\theta_1, \theta_2)$ the gross node inflow rate into v is constant and*
- *the label functions at the nodes reachable via direct edges from v are linear on this interval,*

then we can extend f to a partial IDE up to θ_2 at v in a finite number of phases.

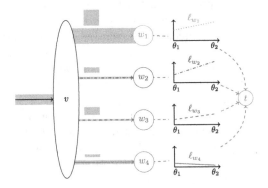

Fig. 1. The situation in Lemma 1: We have an acyclic graph and a partial IDE up to some time θ_2 for all nodes closer to the sink t than v and up to some earlier time θ_1 for v and all nodes further away than v from t. Additionally, over the interval $[\theta_1, \theta_2)$ the edges leading into v have a constant outflow rate (and a physical travel time of at least $\theta_2 - \theta_1$) and the nodes w_i all have affine label functions ℓ_{w_i}. The edges vw_i start with some current queue lengths $q_{vw_i}(\theta_1) \ge 0$.

Proof. Let f be the flow after an, a priori, infinite number of maximal extension steps getting us to a partial IDE up to some $\hat{\theta} \in (\theta_1, \theta_2]$ at node v. Furthermore, let $\delta_v^+ = \{vw_1, \ldots, vw_p\}$ be the set of outgoing edges from v. Then for every such edge vw_i we can define a function $h_i : [\theta_1, \hat{\theta}) \to \mathbb{R}_{\ge 0}, \theta \mapsto c_{vw_i}(\theta) + \ell_{w_i}(\theta)$, denoting for every time $\theta \in [\theta_1, \hat{\theta})$ the shortest current travel time to the sink t for a particle entering edge vw_i at that time. Consequently, during this interval we have $vw_i \in E_\theta$ if and only if $h_i(\theta) = \min\{h_j(\theta) \mid j \in [p]\} = \ell_v(\theta)$. We start by stating several important observations and then proceed by showing two key-properties of the functions h_i and ℓ_v, which are also visualized in Figs. 2 and 3:

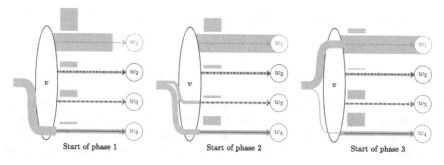

Fig. 2. The first three phases of a possible flow distribution from the node v for the situation depicted in Fig. 1. The corresponding functions h_i are depicted in Fig. 3

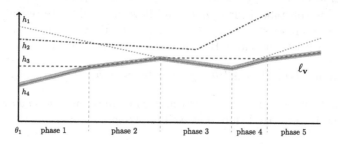

Fig. 3. The functions h_i corresponding to the flow distribution for the situation depicted in Fig. 1 and depicted in Fig. 2 for the first three phases. The second, third and fifth phase start because an edge becomes newly active (edges vw_3, vw_1 and vw_3 again, respectively). The fourth phase starts because the queue on the active edge vw_1 runs empty. These are the only two possible events which can trigger the beginning of a new phase. Edge vw_2 is inactive for the whole time interval and, thus, has a convex graph. The bold gray line marks the graph of the function ℓ_v.

(i) The functions h_i are continuous and piece-wise linear. In particular they are differentiable almost everywhere and their left and right side derivatives $\partial_- h_i$ and $\partial_+ h_i$, respectively, exist everywhere. The same holds for the function ℓ_v.

(ii) A new phase begins at a time $\theta \in [\theta_1, \hat{\theta})$ if and only if at least one of the following two events occurs at time θ: An edge vw_i becomes newly active or the queue of an active edge vw_i runs empty.

(iii) There are uniquely defined numbers $\ell_{I,J}$ for all subsets $J \subseteq I \subseteq [p]$ such that $\ell'_v(\theta) = \ell_{I,J}$ within all phases, where $\{\, vw_i \mid i \in I \,\}$ is the set of active edges in δ_v^+ and $\{\, vw_i \mid i \in J \,\}$ is the subset of such active edges that also have a non-zero queue during this phase.

Claim 1. *If an edge vw_i is inactive during some interval $(a, b) \subseteq [\theta_1, \hat{\theta})$, the graph of h_i is convex on this interval.*

Claim 2. *For any time θ define $I(\theta) := \{\, i \in [p] \mid h_i(\theta) = \ell_v(\theta) \,\}$. Then, we have*

$$\min \{\, \partial_- h_i(\theta) \mid i \in I(\theta) \,\} \leq \partial_+ \ell_v(\theta). \tag{9}$$

If no edge becomes newly active at time θ, we also have $\partial_- \ell_v(\theta) \leq \partial_+ \ell_v(\theta)$.

Proof of Claim 1. By the lemma's assumption ℓ_{w_i} is linear on the whole interval. For an inactive edge vw_i its queue length function consists of at most two linear sections: One where the queue depletes at a constant rate of $-\nu_e$ and one where it remains constant 0. Thus, h_i is convex as sum of two convex functions. ∎

Proof of Claim 2. To show (9), let I' be the set of indices of edges active immediately after θ, i.e. $I' := \{\, i \in I(\theta) \mid \partial_+ h_i(\theta) = \partial_+ \ell_v(\theta) \,\}$. Since the total outflow from node v is constant during $[\theta_1, \hat{\theta})$ and flow may only enter edges vw_i with $i \in I'$ after θ, there exists some $j \in I'$, where the inflow rate into vw_j after θ is the same or larger than before. But then we have $\partial_+ h_j(\theta) \geq \partial_- h_j(\theta)$ and, thus,

$$\min \{\, \partial_- h_i(\theta) \mid i \in I(\theta) \,\} \leq \partial_- h_j(\theta) \leq \partial_+ h_j(\theta) = \partial_+ \ell_v(\theta).$$

If, additionally, no edge becomes newly active at time θ, all edges vw_i with $i \in I'$ have been active directly before θ as well implying

$$\partial_- \ell_v(\theta) = \min \{\, \partial_- h_i(\theta) \mid i \in I(\theta) \,\} \overset{(9)}{\leq} \partial_+ \ell_v(\theta).$$

∎

Using these properties we can now first show a claim which implies that the derivative of ℓ_v can attain the smallest $\ell_{I,J}$ only for a finite number of intervals. Inductively the same then holds for all of the finitely many $\ell_{I,J}$, which by observation (iii) are the only values ℓ_v' can attain. The proof of the lemma finally concludes by observing that an interval with constant derivative of ℓ_v can contain only finitely many phases.

Claim 3. *Let $(a_1, b_1), (a_2, b_2) \subseteq [\theta_1, \hat{\theta})$ be two disjoint maximal non-empty intervals with constant $\ell_v'(\theta) =: c$. If $b_1 < a_2$ and $\ell_v'(\theta) \geq c$ for all $\theta \in (b_1, a_2)$ where the derivative exists, then there exists an edge vw_i such that*

1. the first phase of (a_2, b_2) begins because vw_i becomes newly active and
2. this edge is not active between a_1 and a_2.

In particular, the first phase of (a_1, b_2) is not triggered by vw_i becoming active.

Claim 4. *Let $(a, b) \subseteq [\theta_1, \hat{\theta})$ be an interval during which ℓ_v' is constant. Then (a, b) contains at most $2p$ phases.*

Proof of Claim 3. Since we have $\partial_+ \ell_v(a_2) = c$, Claim 2 implies that there exists some edge vw_i with $h_i(a_2) = \ell_v(a_2)$ and $\partial_- h_i(a_2) \leq c$. As (a_2, b_2) was chosen to be maximal and $\ell_v'(\theta) \geq c$ holds almost everywhere between b_1 and a_2, we have $\partial_- \ell_v(a_2) > c$. Thus, vw_i was inactive before a_2.

Now let $\tilde{\theta} < a_2$ be the last time before a_2, where vw_i was active. By Claim 1 we know then that $h'_i(\theta) \leq c$ holds almost everywhere on $[\tilde{\theta}, a_2]$. At the same time we have $\ell'_v(\theta) \geq c$ almost everywhere on $[a_1, a_2]$ and $\ell'_v(\theta) > c$ for at least some proper subinterval of $[b_1, a_2]$, since the intervals (a_1, b_1) and (a_2, b_2) were chosen to be maximal. Combining these two facts with $\ell_v(\theta) = h_i(\theta)$ implies $\ell_v(\theta) < h_i(\theta)$ for all $\theta \in [\tilde{\theta}, a_2) \cap [a_1, a_2]$. As both functions are continuous we must have $\tilde{\theta} < a_1$. Thus, vw_i is inactive for all of $[a_1, a_2)$.   ∎

Proof of Claim 4. By Claim 1 an edge that changes from active to inactive during the interval (a, b) will remain inactive for the rest of this interval. Thus, at most p phases can start because an edge becomes newly active. By Claim 3 if a phase begins because the queue on an active edge vw_i runs empty at time θ, we have $\partial_+ h_i(\theta) > \partial_- h_i(\theta) = \partial_- \ell_v(\theta) = \partial_+ \ell_v(\theta)$ meaning that this edge will become inactive. Thus, at most p phases start because the queue of an active edge runs empty. Since by observation (ii) these are the only ways to start a new phase, we conclude that there can be no more than $2p$ phases during (a, b).   ∎

Combining Claims 3 and 4 we see that $[\theta_1, \hat{\theta})$ only contains a finite number of phases and, thus, we achieve $\hat{\theta} = \theta_2$ with finitely many extensions.   □

For acyclic networks we can now fix some topological order of the nodes w.r.t. the whole graph at the beginning of the algorithm and then always do the node-wise extensions in this order. Since in a partial IDE up to time $\hat{\theta}$ the gross node inflow rates are already completely determined for the interval $[\hat{\theta}, \hat{\theta} + \tau_{\min})$ we can – for the purpose of the following analysis – slightly rearrange the extension steps, without changing the outcome of the algorithm, by directly extending the partial IDE at each node for this whole interval (using multiple phases). It then suffices to show that these extensions (of constant length $\tau_{\min} > 0$) at a single node only need a finite amount of phases, which follows by repeatedly applying Lemma 1 and using the fact that, by induction, the gross node inflow rate at the current node as well as the label functions ℓ_{w_i} at all nodes closer to the sink are piece-wise constant and piece-wise linear with finitely many breakpoints, respectively.

General Networks. In order to extend this result to general networks, we introduce the concept of a *lazy set of active edges*, which is a time dependent subset of edges $\tilde{E}(\theta)$ satisfying the following properties:

- At every time θ the set $\tilde{E}(\theta)$ contains all currently active edges but no cycle.
- There are (flow independent) constants $C, D > 0$ such that during any time interval of length at most C the set $\tilde{E}(\theta)$ changes at most D times.

This allows us to subdivide the whole time into a finite number of intervals during which $\tilde{E}(\theta)$ does not change and, during those, we can restrict ourselves to considering only edges in the fixed acyclic subgraph induced by the edges in $\tilde{E}(\theta)$. To obtain such a lazy set of active edges we add edges, whenever they become active, but only remove edges if, otherwise, $\tilde{E}(\theta)$ would contain a cycle. Additionally, in this case we always only remove the "most inactive" edge of the cycle. This leads to the final variant of the extension algorithm which is formalized in Algorithm 1 and for which we will now show our main theorem.

Algorithm 1: IDE-Construction Algorithm for single-sink networks

---
Input: A network $\mathcal{N}$ with piecewise constant network inflow rates
Output: An IDE flow f in $\mathcal{N}$
1 Choose T large enough such that all IDE flows in $\mathcal{N}$ terminate before T
2 Let f be the zero flow, set $\theta \leftarrow 0$ and $\tilde{E} \leftarrow E_0$
3 Determine a top. order $t = v_1 < v_2 < \cdots < v_n$ w.r.t. the edges in $\tilde{E}$
4 **while** $\theta < T$ **do**
5   Choose the largest $\alpha_0 > 0$ s.th. all b_v^- are constant over $(\theta, \theta + \alpha_0)$
6   **for** $i = 1, \ldots, n$ **do**
7   Compute a constant distribution to the outgoing active edges from v_i satisfying (8)
8   Determine the largest $\alpha_i \leq \alpha_{i-1}$ such that the set of active edges does not change during $(\theta, \theta + \alpha_i)$
9   **end for**
10   Extend the flow f up to time $\theta + \alpha_n$ and set $\theta \leftarrow \theta + \alpha_n$
11   **if** $E_\theta \setminus \tilde{E} \neq \emptyset$ **then**
12   Define $\tilde{E} \leftarrow \tilde{E} \cup E_\theta$.
13   **while** *there exists a cycle C in $\tilde{E}$* **do**
14   Remove an edge $e = xy \in C$ with maximal $\ell_y(\theta) - \ell_x(\theta)$
15   **end while**
16   Find a top. order $t = v_1 < v_2 < \cdots < v_n$ w.r.t. the edges in $\tilde{E}$
17   **end if**
18 **end while**

Theorem 1. *For any single-sink network with piecewise constant network-inflow rates an IDE flow can be constructed in finite time using the natural extension algorithm (Algorithm 1).*

Proof. As we already have the above theorem for acyclic networks and Algorithm 1 only uses a fixed acyclic subgraph of the whole network as long as the set $\tilde{E}$ used in Algorithm 1 remains unchanged, it suffices to show that this set is indeed a lazy set of active edges. The first property is obvious from the way $\tilde{E}$ is obtained in the algorithm, for the second one we need the following two claims:

Claim 5. *Any edge xy removed from $\tilde{E}$ in line 14 satisfies $\ell_x(\theta) < \ell_y(\theta)$.*

Claim 6. *For any given network there exists some constant $L > 0$ such that for all flows, all nodes v and all times θ we have $|\ell_v'(\theta)| \leq L$.*

Proof of Claim 5. Let $C \subseteq \tilde{E}$ be a cycle containing the removed edge xy. Since $\tilde{E}$ was acyclic before we added the newly active edges in line 12, this cycle also has to contain some edge vw which is currently active, therefore satisfying $\ell_v(\theta) > \ell_w(\theta)$. Thus, summing the differences of the label functions at the two ends of every edge over all edges in C yields the existence of at least one edge $uz \in C$ with $\ell_z(\theta) - \ell_u(\theta) > 0$. This then, in particular, also holds for edge xy. ∎

Proof of Claim 6. For any node v we can bound the maximal inflow rate into this node by the constant $L_v := \sum_{e \in \delta_v^-} \nu_e + \max \{ u_v(\theta) \mid \theta \in \mathbb{R}_{\geq 0} \}$ using constraint (4). Together with flow conservation (1) this, in turn, allows us to upper bound the inflow rates into all edges $e \in \delta_v^+$ and, thus, the rate at which the queue length and the current travel time on these edges can change by $L_e := \frac{L_v}{\nu_e}$. Since this rate of change is also lower bounded by -1 setting $L := \sum_{e \in E} \max \{ 1, L_e \}$ proves the claim, as for all nodes v and times θ, we then have $|\ell_v'(\theta)| \leq \sum_{e \in E} |c_e'(\theta)| \leq \sum_{e \in E} \max \{ 1, L_e \} = L$. ∎

Combining these two claims with the fact that an edge xy is only added to $\tilde{E}$ if it becomes active, i.e. $\ell_x(\theta) = \ell_y(\theta) + c_{xy}(\theta) \geq \ell_y(\theta) + \tau_{xy}$, shows that no edge can enter $\tilde{E}$ twice during any sufficiently small time interval, which implies the second property concluding the proof of the theorem. □

Remark 1. A closer inspection of the proofs allows us to also derive the following rough but explicit bound on the number of phases needed assuming that all edge travel times and capacities are integers (which can always be achieved by rescaling the network):

$$\mathcal{O}\left(P\left(2(\Delta + 1)^{4^{\Delta}+1} \right)^{D/C \cdot T \cdot |V|} \right).$$

Hereby, $\Delta := \max \{ |\delta_v^+| \mid v \in V \}$ is the maximum out-degree of any edge, T the termination time of the IDE and P is the number of intervals with constant network inflow rates. A formal deduction of this bound can be found in the full version of this paper [8].

4   Computational Complexity of IDE

While Theorem 1 shows that IDE can be constructed in finite time, the derived explicit bound is clearly superpolynomial. In this section we complement this result by showing that many natural decision problems about IDE are NP-hard.

Theorem 2. *The following decision problems are NP-hard:*

(i) Given a network and a specific edge: Is there an IDE not using this edge?
(ii) Given a network and a specific edge: Is there an IDE using this edge?
(iii) Given a network and a time T: Is there an IDE that terminates before T?
(iv) Given a network and some $k \in \mathbb{N}$: Is there an IDE with at most k phases?

This theorem can be shown by a reduction from the NP-complete problem 3SAT to the above problems. The main idea of the reduction is as follows: For any given instance of 3SAT we construct a network which contains a source node for each clause with three outgoing edges corresponding to the three literals of the clause. Any satisfying interpretation of the 3SAT-formula translates to a distribution of the network inflow to the literal edges (where sending at least $1/3$ of the flow along an edge corresponds to the respective literal being true), which leads to an IDE flow that passes through the whole network in a straightforward manner. If, on the other hand, the formula is unsatisfiable, every IDE flow will cause a specific type of congestion which will divert a certain amount of flow into a different part of the graph. This part of the graph may contain an otherwise unused edge (for (i)), a gadget blocking access to an otherwise used edge (for (ii)), a gadget which results in a long travel time (for (iii)) or one which produces many phases (for (iv)). The congestion occurs because the flow corresponding to a variable being false forms a queue while the flow corresponding to the same variable being true is delayed. Thus, when the latter finally arrives, the former has already blocked the direct path and diverts the latter away from it. On the other hand, in a flow corresponding to a satisfying interpretation of the formula this does not happen, as only one of the two types of flow is present for every variable. The detailed construction of the described gadgets as well as the formal proof of the reduction's correctness can be found in the full version of this paper [8]. For an illustration of the reduction see Fig. 4.

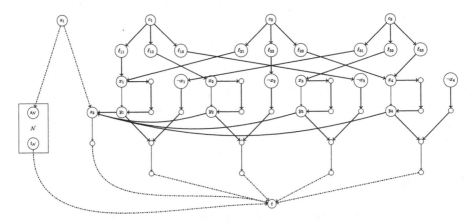

Fig. 4. The whole network for the 3SAT-formula $(x_1 \lor x_2 \lor \neg x_3) \land (x_1 \lor \neg x_2 \lor x_4) \land (\neg x_1 \lor x_3 \lor x_4)$. The bold edges have infinite capacity, while all other edges have capacity 1. The solid edges have a travel time of 1, the dashdotted edges may have variable travel time (depending on the subnetwork $\mathcal{N}$). The network inflow rates are 12 over the interval $[0, 1]$ at all nodes c_i and 0 everywhere else.

5   Conclusion

We showed that Instantaneous Dynamic Equilibria can be computed in finite time for single-sink networks by applying the natural extension algorithm. We complemented this result by showing that several natural decision problems involving IDE flows are NP-hard by describing a reduction from 3SAT. One common observation that can be drawn from many proofs involving IDE flows (in this paper as well as in [9,10]) is that they often allow for some kind of *local analysis* of their structure – something which seems out of reach for DE flows. This was a key aspect of the positive result about the finiteness of the extension algorithm where it allowed us to use inductive reasoning over the single nodes of the given network. At the same time, such local argumentation allows us to analyse the behavior of IDE flows in the rather complex instance resulting from the reduction in the hardness-proof by looking at the local behavior inside the much simpler gadgets from which the larger instance is constructed. We think that this local approach to the analysis of IDE flows might also help to answer further open questions about IDE flows in the future. One such topic might be a further investigation of the computational complexity of IDE flows. While both our upper bound on the number of extension steps as well as our lower bound for the worst case computational complexity are superpolynomial bounds, the latter is at least still polynomial in the termination time of the constructed flow, which is not the case for the former. Thus, there might still be room for improvement on either bound.

Acknowledgments. We are grateful to the anonymous reviewers for their valuable feedback on this paper. Additionally, we thank the Deutsche Forschungsgemeinschaft (DFG) for their financial support. Finally, we want to thank the organizers and participants of the 2020 Dagstuhl seminar on "Mathematical Foundations of Dynamic Nash Flows", where we had many helpful and inspiring discussions on the topic of this paper.

Funding. The research of the authors was funded by the Deutsche Forschungsgemeinschaft (DFG, German Research Foundation) - HA 8041/1-2

References

1. Boyce, D.E., Ran, B., LeBlanc, L.J.: Solving an instantaneous dynamic user-optimal route choice model. Transp. Sci. **29**(2), 128–142 (1995)
2. Cao, Z., Chen, B., Chen, X., Wang, C.: A network game of dynamic traffic. In: Daskalakis, C., Babaioff, M., Moulin, H. (eds.) Proceedings of the 2017 ACM Conference on Economics and Computation, EC 2017, Cambridge, MA, USA, 26–30 June 2017, pp. 695–696. ACM (2017)
3. Cominetti, R., Correa, J., Larré, O.: Dynamic equilibria in fluid queueing networks. Oper. Res. **63**(1), 21–34 (2015)
4. Cominetti, R., Correa, J., Olver, N.: Long term behavior of dynamic equilibria in fluid queuing networks. Oper. Res. (2020, to appear). https://doi.org/10.1287/opre.2020.2081
5. Ford, L.R., Fulkerson, D.R.: Flows in Networks. Princeton University Press, Princeton (1962)
6. Friesz, T.L., Han, K.: The mathematical foundations of dynamic user equilibrium. Transp. Res. Part B Methodol. **126**, 309–328 (2019)

7. Friesz, T.L., Luque, J., Tobin, R.L., Wie, B.-W.: Dynamic network traffic assignment considered as a continuous time optimal control problem. Oper. Res. **37**(6), 893–901 (1989)
8. Graf, L., Harks, T.: A finite time combinatorial algorithm for instantaneous dynamic equilibrium flows. https://arxiv.org/abs/2007.07808 (2020)
9. Graf, L., Harks, T.: The price of anarchy for instantaneous dynamic equilibria. In: Chen, X., Gravin, N., Hoefer, M., Mehta, R. (eds.) WINE 2020. LNCS, vol. 12495, pp. 237–251. Springer, Cham (2020). https://doi.org/10.1007/978-3-030-64946-3_17
10. Graf, L., Harks, T., Sering, L.: Dynamic flows with adaptive route choice. Math. Program. **183**(1), 309–335 (2020). https://doi.org/10.1007/s10107-020-01504-2
11. Han, K., Friesz, T.L., Yao, T.: A partial differential equation formulation of Vickrey's bottleneck model, part ii: numerical analysis and computation. Transp. Res. Part B Methodol. **49**, 75–93 (2013)
12. Harks, T., Peis, B., Schmand, D., Tauer, B., Vargas-Koch, L.: Competitive packet routing with priority lists. ACM Trans. Econ. Comput. **6**(1), 4:1–4:26 (2018)
13. Hoefer, M., Mirrokni, V.S., Röglin, H., Teng, S.-H.: Competitive routing over time. Theor. Comput. Sci. **412**(39), 5420–5432 (2011)
14. Ismaili, A.: Routing games over time with FIFO policy. In: Devanur, N.R., Lu, P. (eds.) WINE 2017. LNCS, vol. 10660, pp. 266–280. Springer, Cham (2017). https://doi.org/10.1007/978-3-319-71924-5_19
15. Ismaili, A.: The complexity of sequential routing games. CoRR, abs/1808.01080 (2018)
16. Kaiser, M.: Computation of dynamic equilibria in series-parallel networks. Math. Oper. Res. (2020, forthcoming)
17. Koch, R., Skutella, M.: Nash equilibria and the price of anarchy for flows over time. Theory Comput. Syst. **49**(1), 71–97 (2011)
18. Otsubo, H., Rapoport, A.: Vickrey's model of traffic congestion discretized. Transp. Res. Part B Methodol. **42**(10), 873–889 (2008)
19. Ran, B., Boyce, D.E.: Dynamic Urban Transportation Network Models: Theory and Implications for Intelligent Vehicle-Highway Systems. Lecture Notes in Economics and Mathematical Systems. Springer, New York (1996). https://doi.org/10.1007/978-3-662-00773-0
20. Ran, B., Boyce, D.E., LeBlanc, L.J.: A new class of instantaneous dynamic user-optimal traffic assignment models. Oper. Res. **41**(1), 192–202 (1993)
21. Scarsini, M., Schröder, M., Tomala, T.: Dynamic atomic congestion games with seasonal flows. Oper. Res. **66**(2), 327–339 (2018)
22. Sering, L., Vargas-Koch, L.: Nash flows over time with spillback. In: Proceedings of the 30th Annual ACM-SIAM Symposium on Discrete Algorithms. ACM (2019)
23. Vickrey, W.S.: Congestion theory and transport investment. Am. Econ. Rev. **59**(2), 251–60 (1969)
24. Ziemke, T., Sering, L., Vargas-Koch, L., Zimmer, M., Nagel, K., Skutella, M.: Flows over time as continuous limits of packet-based network simulations. In: Transportation Research Procedia, vol. 52, pp. 123–130 (2021). https://doi.org/10.1016/j.trpro.2021.01.014

A Combinatorial Algorithm for Computing the Degree of the Determinant of a Generic Partitioned Polynomial Matrix with 2 × 2 Submatrices

Yuni Iwamasa[✉][iD]

Kyoto University, Kyoto 606-8501, Japan
iwamasa@i.kyoto-u.ac.jp

Abstract. In this paper, we consider the problem of computing the degree of the determinant of a block-structured symbolic matrix (a generic partitioned polynomial matrix) $A = (A_{\alpha\beta} x_{\alpha\beta} t^{d_{\alpha\beta}})$, where $A_{\alpha\beta}$ is a 2×2 matrix over a field $\mathbf{F}$, $x_{\alpha\beta}$ is an indeterminate, and $d_{\alpha\beta}$ is an integer for $\alpha, \beta = 1, 2, \ldots, n$, and t is an additional indeterminate. This problem can be viewed as an algebraic generalization of the maximum perfect bipartite matching problem.

The main result of this paper is a combinatorial $O(n^5)$-time algorithm for the deg-det computation of a (2×2)-type generic partitioned polynomial matrix of size $2n \times 2n$. We also present a min-max theorem between the degree of the determinant and a potential defined on vector spaces. Our results generalize the classical primal-dual algorithm (Hungarian method) and min-max formula (Egerváry's theorem) for maximum weight perfect bipartite matching.

Keywords: Generic partitioned polynomial matrix · Weighted Edmonds' problem · Weighted noncommutative Edmonds' problem

1 Introduction

Many of matching-type combinatorial optimization problems have an algebraic formulation as the rank computation of a symbolic matrix. One of the most typical examples is the maximum bipartite matching problem; the maximum cardinality of a matching in a bipartite graph $G = (\{1, 2, \ldots, n\}, \{1, 2, \ldots, n\}; E)$ is equal to the rank of the $n \times n$ symbolic matrix A defined by $(A)_{ij} := x_{ij}$ if $ij \in E$ and zero otherwise, where x_{ij} is a variable for each edge ij. Such an algebraic interpretation is also known for generalizations of maximum bipartite matching, including nonbipartite maximum matching, linear matroid intersection, and

The author was supported by JSPS KAKENHI Grant Number JP17K00029, 20K23323, 20H05795, Japan.

M. Singh and D. P. Williamson (Eds.): IPCO 2021, LNCS 12707, pp. 119–133, 2021.
https://doi.org/10.1007/978-3-030-73879-2_9

linear matroid parity; see [20, 23]. All of the above algebraic formulations are generalized to *Edmonds' problem* [3], which asks to compute the rank of a symbolic matrix A represented by

$$A = A_1 x_1 + A_2 x_2 + \cdots + A_m x_m. \tag{1.1}$$

Here A_i is a matrix over a field $\mathbf{F}$ and x_i is a variable for $i = 1, 2, \ldots, m$. Although a randomized polynomial-time algorithm for Edmonds' problem is known (if $|\mathbf{F}|$ is large) [19, 22], a deterministic polynomial-time algorithm is not known, which is one of the prominent open problems in theoretical computer science (see e.g., [17]).

The computation of the degree of the determinant (deg-det computation) of a symbolic matrix with an additional indeterminate is the weighted analog of the rank computation of a symbolic matrix. Indeed, for a bipartite graph G endowed with edge weights d_{ij} for $ij \in E$, we define the matrix $A(t)$ by $(A(t))_{ij} := x_{ij} t^{d_{ij}}$ if $ij \in E$ and zero otherwise, where t is a new variable. Then the maximum weight of a perfect matching in G is equal to the degree of the determinant of $A(t)$, where we regard the determinant $\det A(t)$ as a polynomial in t. *Weighted Edmonds' problem*, which is a unified algebraic generalization of maximum weight perfect matching problems, asks to compute the degree of the determinant of

$$A(t) = A_1(t) x_1 + A_2(t) x_2 + \cdots + A_m(t) x_m, \tag{1.2}$$

in which $A_k(t)$ is a polynomial matrix with an indeterminate t for each $k = 1, 2, \ldots, m$. As well as Edmonds' problem, the computational complexity of weighted Edmonds' problem is open.

In this paper, we address the deg-det computation (weighted Edmonds' problem) of the following block-structured matrix:

$$A(t) = \begin{pmatrix} A_{11} x_{11} t^{d_{11}} & A_{12} x_{12} t^{d_{12}} & \cdots & A_{1\nu} x_{1n} t^{d_{1n}} \\ A_{21} x_{21} t^{d_{21}} & A_{22} x_{22} t^{d_{22}} & \cdots & A_{2\nu} x_{2n} t^{d_{2n}} \\ \vdots & \vdots & \ddots & \vdots \\ A_{n1} x_{n1} t^{d_{n1}} & A_{n2} x_{n2} t^{d_{n2}} & \cdots & A_{nn} x_{nn} t^{d_{nn}} \end{pmatrix}, \tag{1.3}$$

where $A_{\alpha\beta}$ is a 2×2 matrix over a field $\mathbf{F}$, $x_{\alpha\beta}$ is a variable, and $d_{\alpha\beta}$ is an integer for $\alpha, \beta = 1, 2, \ldots, n$, and t is another variable. The degree of the determinant of $A(t)$, denoted by $\deg \det A(t)$, is with respect to t. A matrix $A(t)$ of the form (1.3) is called a (2×2)-*type generic partitioned polynomial matrix*. We note that the maximum weight perfect bipartite matching problem coincides with the case where each $A_{\alpha\beta}$ in (1.3) is a 1×1 matrix. Our main result is to devise the first combinatorial and strongly polynomial-time algorithm for the deg-det computation of a (2×2)-type generic partitioned polynomial matrix.

Theorem 1. *There exists a combinatorial $O(n^5)$-time algorithm for computing $\deg \det A(t)$ for a (2×2)-type generic partitioned polynomial matrix $A(t)$ of the form (1.3).*

Our problem and result are related to the noncommutative analog of (weighted) Edmonds' problem. *Noncommutative Edmonds' problem* [13] asks to compute the rank of a matrix of the form (1.1), where x_i and x_j are supposed to be noncommutative, i.e., $x_i x_j \neq x_j x_i$ for $i \neq j$. Here the "rank" is defined via the inner rank of a matrix over a free skew field, and is called *noncommutative rank* or *nc-rank*. Surprisingly, the noncommutative setting makes the rank computation easier; Garg, Gurvits, Oliveira, and Wigderson [6], Ivanyos, Qiao, and Subrahmanyam [14], and Hamada and Hirai [7,8] independently developed deterministic polynomial-time algorithms for noncommutative Edmonds' problem. The algorithm of Garg, Gurvits, Oliveira, and Wigderson works for the case of $\mathbf{F} = \mathbf{Q}$, that of Ivanyos, Qiao, and Subrahmanyam for an arbitrary field, and that of Hamada and Hirai in [7] for a field $\mathbf{F}$ provided arithmetic operations on $\mathbf{F}$ can be done in constant time while the bit-length may be unbounded if $\mathbf{F} = \mathbf{Q}$; in [8] they resolve the above bit-length issue. Their algorithms were conceptually different.

Weighted noncommutative Edmonds' problem [9] is the noncommutative analog of weighted Edmonds' problem, or the weighted analog of noncommutative Edmonds' problem. In this problem, given a matrix $A(t)$ of the form (1.2), in which x_i and x_j are supposed to be noncommutative but t is commutative any variable x_i, we are asked to compute the degree of the Dieudonné determinant Det $A(t)$ of $A(t)$. Here the *Dieudonné determinant* [2] (see also [1]) is a determinant concept of a matrix over a skew field. By utilizing the algorithm of Hamada and Hirai [7] for noncommutative Edmonds' problem, Hirai [9] developed a pseudopolynomial-time algorithm for weighted noncommutative Edmonds' problem, provided arithmetic operations on the base field can be done in constant time. Oki [21] devised another pseudopolynomial-time algorithm, via the reduction to the nc-rank computation, that works for an arbitrary field.

Very recently, Hirai and Ikeda [10] have presented a strongly polynomial-time algorithm for computing deg Det $A(t)$ of $A(t)$ having the following special form

$$A(t) = A_1 x_1 t^{d_1} + A_2 x_2 t^{d_2} + \cdots + A_m x_m t^{d_m}, \tag{1.4}$$

where A_i is a matrix over $\mathbf{F}$ and d_i is an integer for $i = 1, 2, \ldots, m$. Note that a (2×2)-type generic partitioned polynomial matrix (with noncommutative variables $x_{\alpha\beta}$) can be represented as (1.4). They also showed that, for a (2×2)-type generic partitioned polynomial matrix $A(t)$, it holds deg det $A(t) = $ deg Det $A(t)$. That is, the strongly polynomial-time solvability of the deg-det computation of a (2×2)-type generic partitioned polynomial matrix follows from their result. Their algorithm is conceptually simple, but is slow and not combinatorial. They first present an $O(n^7 \log D)$-time algorithm via a cost scaling technique, where $D := \log \max_{\alpha,\beta=1,2,\ldots,n} |d_{\alpha\beta}|$. Then, by utilizing the perturbation technique in [5] for $d_{\alpha\beta}$ so that $\log D$ is bounded by $O(n^6)$ in polynomial time, they devise a strongly polynomial-time algorithm. Moreover, in case of $\mathbf{F} = \mathbf{Q}$, their algorithm requires an additional procedure (used in [15]) for bounding the bit-complexity.

Our algorithm is a combinatorial primal-dual augmenting path algorithm, which is the weighted analog of our previous work [11] (see also its full version [12]) on the rank computation of a (2×2)-type generic partitioned matrix.

We introduce, in this article, the concept of *det-matching*, which plays a role as a perfect matching in the maximum weight perfect matching problem. The deg-det computation is reduced to the problem of finding a maximum weight det-matching. We also introduce another matching concept named *pseudo-matching*, which corresponds to a (special) bipartite matching, and define a potential on vector spaces. They enable us to define the auxiliary graph and an augmenting path on it. By repeating the augmenting procedures via an augmenting path, we can finally obtain a maximum weight det-matching, as required. The validity of the algorithm provides a constructive proof of a min-max formula between the degree of the determinant and a potential. The proposed algorithm and min-max formula are algebraically generalizations of the classical primal-dual algorithm, so-called the Hungarian method [18], and minimax theorem, so-called Egerváry's theorem [4], for the maximum weight perfect bipartite matching problem, respectively. In particular, the latter result also generalizes Iwata–Murota's duality theorem [16] on the rank of a (2×2)-type generic partitioned matrix. Our algorithm is simpler and faster than Hirai–Ikeda's algorithm; ours requires no perturbation of the weight and no additional care for bounding the bit size.

All proofs are omitted and will be given in the full version of this paper.

Notations. Let $A(t)$ be a (2×2)-type generic partitioned polynomial matrix of the form (1.3). The matrix $A(t)$ is regarded as a matrix over the field $\mathbf{F}(x, t)$ of rational functions with variables t and $x_{\alpha\beta}$ for $\alpha, \beta = 1, 2, \ldots, n$. If $\mathrm{rank} A(t) < 2n$, then we let $\deg \det A(t) := -\infty$. Symbols α, β, and γ are used to represent a row-block index, column-block index, and row- or column-block index of $A(t)$, respectively. We often drop "$= 1, 2, \ldots, n$" from the notation of "$\alpha, \beta = 1, 2, \ldots, n$" if it is clear from the context. Each α and β is endowed with the 2-dimensional $\mathbf{F}$-vector space $\mathbf{F}^2$, denoted by U_α and V_β, respectively. Each submatrix $A_{\alpha\beta}$ is considered as the bilinear map $U_\alpha \times V_\beta \to \mathbf{F}$ defined by $A_{\alpha\beta}(u, v) := u^\top A_{\alpha\beta} v$ for $u \in U_\alpha$ and $v \in V_\beta$. We denote by $\ker_{\mathrm{L}}(A_{\alpha\beta})$ and $\ker_{\mathrm{R}}(A_{\alpha\beta})$ the left and right kernels of $A_{\alpha\beta}$, respectively. Let us denote by $\mathcal{M}_\alpha$ and $\mathcal{M}_\beta$ the sets of 1-dimensional vector subspaces of U_α and of V_β, respectively.

We define the (undirected) bipartite graph $G := (\{1, 2, \ldots, n\}, \{1, 2, \ldots, n\}; E)$ by $E := \{\alpha\beta \mid A_{\alpha\beta} \neq O\}$. For $I \subseteq E$, let $A_I(t)$ denote the matrix obtained from $A(t)$ by replacing each submatrix $A_{\alpha\beta}$ with $\alpha\beta \notin I$ by the 2×2 zero matrix. An edge $\alpha\beta \in E$ is said to be *rank-k* ($k = 1, 2$) if $\mathrm{rank} A_{\alpha\beta} = k$. For notational simplicity, the subgraph $([n], [n]; I)$ for $I \subseteq E$ is also denoted by I. For a node γ, let $\deg_I(\gamma)$ denote the degree of γ in I, i.e., the number of edges in I incident to γ. An edge $\alpha\beta \in I$ is said to be *isolated* if $\deg_I(\alpha) = \deg_I(\beta) = 1$. An edge subset I is said to be *spanning* if $\deg_I(\gamma) \geq 1$ for any γ.

2   Matchings, Potentials, and Min-Max Formulas

In this section, we introduce the concepts of *det-matching*, *pseudo-matching*, and *potential* for a (2×2)-type generic partitioned polynomial matrix $A(t)$ of the form (1.3). They play a central role in devising our algorithm. We also present a min-max theorem between the degree of the determinant and a potential.

2.1 Matching Concepts

An edge subset $I \subseteq E$ is called a *det-matching* of $A(t)$ if the polynomial $\det A(t)$ has a term $ct^{w(I)} \cdot \prod_{\alpha\beta \in I} x_{\alpha\beta}^{k_{\alpha\beta}}$ for some nonzero $c \in \mathbf{F}$, an integer $w(I)$, and positive integers $k_{\alpha\beta}$, where it holds $w(I) = \sum_{\alpha\beta \in I} k_{\alpha\beta} d_{\alpha\beta}$. We call $w(I)$ the *weight* of I. A det-matching I is said to be *maximum* if $w(I) \geq w(I')$ for every det-matching I', or equivalently, if $w(I) = \deg \det A(t)$. This matching concept is an algebraic generalization of perfect bipartite matching; for a matrix $A(t)$ of the form (1.3) with 1×1 blocks $A_{\alpha\beta}$ and the corresponding bipartite graph G to A, an edge subset I admits a term $ct^{w(I)} \cdot \prod_{\alpha\beta \in I} x_{\alpha\beta}^{k_{\alpha\beta}}$ in $\det A(t)$ if and only if I is a perfect bipartite matching of G (note that $k_{\alpha\beta} = 1$ for each $\alpha\beta \in I$).

Let us return to a (2×2)-type generic partitioned polynomial matrix. For a det-matching I, we have

$$w(I) = \sum \{d_{\alpha\beta} \mid \alpha\beta \in I\} + \sum \{d_{\alpha\beta} \mid \alpha\beta \in I : \alpha\beta \text{ is an isolated rank-2 edge}\}.$$
(2.1)

Indeed, since $ct^{w(I)} \cdot \prod_{\alpha\beta \in I} x_{\alpha\beta}^{k_{\alpha\beta}}$ is a term of $\det A(t)$, I is spanning and consists of isolated rank-2 edges and cycle components. For such I, it clearly holds that $k_{\alpha\beta} = 1$ if $\alpha\beta$ belongs to a cycle component and $k_{\alpha\beta} = 2$ if $\alpha\beta$ forms an isolated rank-2 edge. This implies the identity (2.1).

We mainly consider a special subclass of det-matchings. An edge subset $I \subseteq E$ is called a *pseudo-matching* or *p-matching* if it satisfies the following combinatorial and algebraic conditions (Deg), (Cycle), and (VL):

(Deg) $\deg_I(\gamma) \leq 2$ for each node γ of G.

Suppose that I satisfies (Deg). Then each connected component of I forms a path or a cycle. Thus I is 2-edge-colorable, namely, there are two edge classes such that any two incident edges are in different classes. An edge in one color class is called a $+$-*edge*, and an edge in the other color class is called a $--$-*edge*.

(Cycle) Each cycle component of I has at least one rank-1 edge.

A *labeling* $\mathcal{V} = (\{U_\alpha^+, U_\alpha^-\}, \{V_\beta^+, V_\beta^-\})_{\alpha,\beta}$ for I is a node-labeling that assigns two 1-dimensional subspaces to each node, $U_\alpha^+, U_\alpha^- \in \mathcal{M}_\alpha$ for α and $V_\beta^+, V_\beta^- \in \mathcal{M}_\beta$ for β, such that for each edge $\alpha\beta \in I$, it holds that

$$A_{\alpha\beta}(U_\alpha^+, V_\beta^-) = A_{\alpha\beta}(U_\alpha^-, V_\beta^+) = \{0\},$$
(2.2)

$$(\ker_{\mathrm{L}}(A_{\alpha\beta}), \ker_{\mathrm{R}}(A_{\alpha\beta})) = \begin{cases} (U_\alpha^+, V_\beta^+) & \text{if } \alpha\beta \text{ is a rank-1 } +\text{-edge,} \\ (U_\alpha^-, V_\beta^-) & \text{if } \alpha\beta \text{ is a rank-1 } -\text{-edge.} \end{cases}$$
(2.3)

For α, we refer to U_α^+ and U_α^- as the $+$-*space* and $--$-*space* of α with respect to $\mathcal{V}$, respectively. The same terminology is also used for β. A labeling $\mathcal{V}$ is said to be *valid* if, for each vertex, its $+$-space and $--$-space with respect to $\mathcal{V}$ are different.

(VL) I admits a valid labeling.

In the following, we use the symbol σ as one of the signs $+$ and $-$. The opposite sign of σ is denoted by $\overline{\sigma}$, i.e., $\overline{\sigma} = -$ if $\sigma = +$, and $\overline{\sigma} = +$ if $\sigma = -$.

Remark 1. Let $\alpha\beta$ be a rank-1 σ-edge in I. The condition (2.3) determines U_α^σ and V_β^σ, and the condition (2.2) determines $V_{\beta'}^{\overline{\sigma}}$ and $U_{\alpha'}^\sigma$ (resp. $U_{\alpha'}^{\overline{\sigma}}$ and $V_{\beta'}^\sigma$) for α' and β' belonging to the path in I that starts with α (resp. β) and consists of rank-2 edges.

Suppose that I satisfies (Deg) and (Cycle). For each node in some cycle component of I, its $+$-space and $-$-space are uniquely determined by the above argument, since every cycle component has a rank-1 edge by (Cycle). Let C be a path component of I, which has the end nodes γ and γ' incident to a σ-edge and a σ'-edge, respectively. When we set the $\overline{\sigma}$-space of γ and $\overline{\sigma'}$-space of γ', the $+$-space and $-$-space of every node belonging to C are uniquely determined. ∎

A p-matching I is said to be *perfect* if I is spanning and each connected component of I forms a cycle or an isolated rank-2 edge. The class of perfect p-matchings forms a subclass of det-matchings.

Lemma 1. *A perfect p-matching is a det-matching.*

The following lemma says that a maximum det-matching is attained by a perfect p-matching.

Lemma 2. $\deg \det A(t) = \max\{w(I) \mid I: \text{ perfect p-matching}\}.$

Hence our problem can be reduced to the problem of finding a maximum weight perfect p-matching.

2.2 Minimax Theorems

In this subsection, we provide a min-max formula between the maximum weight of perfect p-matchings (or equivalently $\deg \det A(t)$ by Lemma 2) and the minimum value corresponding to a potential, defined below. This formula is an algebraic generalization of Egerváry's theorem [4] that is the minimax theorem for the maximum perfect bipartite matching problem.

A function $p : \bigcup_\gamma \mathcal{M}_\gamma \to \mathbf{R}$ is called a *potential* if it satisfies $p(X_\alpha) + p(Y_\beta) \geq d_{\alpha\beta}$ for all $\alpha\beta \in E$, $X_\alpha \in \mathcal{M}_\alpha$, and $Y_\beta \in \mathcal{M}_\beta$ such that $A_{\alpha\beta}(X_\alpha, Y_\beta) \neq \{0\}$. The minimax theorem is the following:

Theorem 2.

$$\max\{w(I) \mid I : \text{ perfect p-matching}\}$$
$$= \min\left\{\sum_\alpha \left(p(X_\alpha) + p(\overline{X}_\alpha)\right) + \sum_\beta \left(p(Y_\beta) + p(\overline{Y}_\beta)\right)\right\},$$

where the minimum is taken over all potentials p and all distinct $X_\alpha, \overline{X}_\alpha \in \mathcal{M}_\alpha$ and distinct $Y_\beta, \overline{Y}_\beta \in \mathcal{M}_\beta$ for α and β. In particular, there exists an integer-valued potential p that attains the minimum.

We can refine the above min-max formula by introducing the concept of *compatibility* for potentials. Let I be a p-matching, $\mathcal{V} = \{(U_\alpha^+, U_\alpha^-), (V_\beta^+, V_\beta^-)\}_{\alpha,\beta}$ a valid labeling of I, and p a potential. An edge $\alpha\beta$ is said to be *double-tight* with respect to $(\mathcal{V}, p)$ if there are signs $\sigma, \sigma' \in \{+, -\}$ satisfying $A_{\alpha\beta}(U_\alpha^\sigma, V_\beta^{\sigma'}) \neq \{0\} \neq A_{\alpha\beta}(U_\alpha^{\overline{\sigma}}, V_\beta^{\overline{\sigma'}})$ and $p(U_\alpha^\sigma) + p(V_\beta^{\sigma'}) = d_{\alpha\beta} = p(U_\alpha^{\overline{\sigma}}) + p(V_\beta^{\overline{\sigma'}})$. An edge $\alpha\beta$ is said to be *single-tight* with respect to $(\mathcal{V}, p)$ if there are signs $\sigma, \sigma' \in \{+, -\}$ satisfying $A_{\alpha\beta}(U_\alpha^\sigma, V_\beta^{\sigma'}) \neq \{0\}$ and $p(U_\alpha^\sigma) + p(V_\beta^{\sigma'}) = d_{\alpha\beta}$ but not double-tight. For a path component C of I, an *end edge* is an edge $\alpha\beta \in C$ with $\deg_I(\alpha) = 1$ or $\deg_I(\beta) = 1$.

A potential p is said to be *compatible* with $(I, \mathcal{V})$ if p satisfies the following conditions (Tight), (Reg), and (Path):

(Tight) For $\alpha\beta \in I$,

$$d_{\alpha\beta} = \begin{cases} p(U_\alpha^-) + p(V_\beta^-) & \text{if } \alpha\beta \text{ is a } +\text{-edge,} \\ p(U_\alpha^+) + p(V_\beta^+) & \text{if } \alpha\beta \text{ is a } -\text{-edge.} \end{cases}$$

(Reg) For each α and β,

$$p(X_\alpha) = \max\{p(U_\alpha^+), p(U_\alpha^-)\} \qquad (X_\alpha \in \mathcal{M}_\alpha \setminus \{U_\alpha^+, U_\alpha^-\}),$$
$$p(Y_\beta) = \max\{p(V_\beta^+), p(V_\beta^-)\} \qquad (Y_\beta \in \mathcal{M}_\beta \setminus \{V_\beta^+, V_\beta^-\}).$$

(Path) For each non-isolated path component C of I such that $|C|$ is odd and both the end edges of C are σ-edges, C has a single-tight σ-edge.

The following theorem says that the minimum in the min-max formula in Theorem 2 is attained by a valid labeling and a compatible potential.

Theorem 3.

$$\max\{w(I) \mid I : perfect\ p\text{-}matching\}$$

$$= \min\left\{ \sum_\alpha \left(p(U_\alpha^+) + p(U_\alpha^-)\right) + \sum_\beta \left(p(V_\beta^+) + p(V_\beta^-)\right) \right\},$$

where the minimum is taken over all valid labelings $\mathcal{V} = \{(U_\alpha^+, U_\alpha^-), (V_\beta^+, V_\beta^-)\}_{\alpha,\beta}$ *for I and compatible potentials p with $(I, \mathcal{V})$. In particular, there exists an integer-valued compatible potential p that attains the minimum.*

For a potential p, we define

$$r_p(I) := |I| + \text{the number of isolated double-tight edges in } I$$

The function r_p can be used for an optimality witness of a p-matching:

Lemma 3. *Let p be a potential compatible with $(I, \mathcal{V})$. If $r_p(I) = 2n$, then I is perfect and*

$$w(I) = \sum_\alpha \left(p(U_\alpha^+) + p(U_\alpha^-)\right) + \sum_\beta \left(p(V_\beta^+) + p(V_\beta^-)\right).$$

2.3   Elimination

Let I be a p-matching and p a potential satisfying (Tight) and (Reg) but not (Path). The *elimination* is an operation of modifying I to a p-matching I' with $r_p(I') > r_p(I)$: For each connected component C violating (Path), i.e., each non-isolated path component C of I such that $|C|$ is odd, both the end edges of C are σ-edges, and all σ-edges in C are double-tight, we delete all $\overline{\sigma}$-edges in C from I.

By executing the above deletion to C, the number of edges decreases by $(|C| - 1)/2$ but that of isolated double-tight edges increases by $(|C| + 1)/2$. Namely, r_p increases by one. Clearly $\mathcal{V}$ is still a valid labeling for the resulting I, p satisfies (Tight) and (Reg) for $(I, \mathcal{V})$, and the number of connected components of I violating (Path) strictly decreases. Thus the following holds.

Lemma 4. *Let I be a p-matching, $\mathcal{V}$ a valid labeling for I, p a potential satisfying (Tight) and (Reg) but not (Path), and I^* the p-matching obtained from I by the elimination. Then p is compatible with $(I^*, \mathcal{V})$ and $r_p(I^*) > r_p(I)$.*

3   Augmenting Path

Our proposed algorithm is a primal-dual algorithm, in which we utilize a potential and an *augmenting path* introduced in Sect. 3.1. An outline of our algorithm is as follows; the formal description is given in Sect. 3.2. Let I be a p-matching, $\mathcal{V}$ a valid labeling for I, and p a compatible potential with $(I, \mathcal{V})$. If $r_p(I) = 2n$, then output $\deg \det A(t) = w(I)$ by Lemma 3 and stop the algorithm. Otherwise, we

- verify $\deg \det A(t) = -\infty$ (or equivalently $\operatorname{rank} A(t) < 2n$),
- find a potential satisfying (Tight) and (Reg) but not (Path) for $(I, \mathcal{V})$, or
- find an augmenting path for $(I, \mathcal{V}, p)$.

In the first case, we output $\deg \det A(t) = -\infty$ and stop this procedure. The others are the cases where we augment a p-matching; in the second case, we execute the elimination to I and obtain I^* such that $r_p(I^*) > r_p(I)$ by Lemma 4; in the last case, via the augmenting path, we update I and $\mathcal{V}$ to a p-matching I^* and a valid labeling $\mathcal{V}^*$ for I^* so that p is a compatible potential with $(I^*, \mathcal{V}^*)$ and $r_p(I^*) > r_p(I)$.

As the initialization, we set I as the emptyset, and $\mathcal{V} = \{(U_\alpha^+, U_\alpha^-), (V_\beta^+, V_\beta^-)\}_{\alpha, \beta}$ for I as distinct $U_\alpha^+, U_\alpha^- \in \mathcal{M}_\alpha$ and distinct $V_\beta^+, V_\beta^+ \in \mathcal{M}_\beta$. We define p by $p(X_\alpha) := \max\{d_{\alpha\beta} \mid \alpha\beta \in E\}$ for each α and $X_\alpha \in \mathcal{M}_\alpha$, and $p(Y_\beta) := 0$ for each β and $Y_\beta \in \mathcal{M}_\beta$. It is clear that I is a p-matching, $\mathcal{V}$ is a valid labeling for I, and p is compatible with $(I, \mathcal{V})$.

A non-isolated connected component C of I is said to be *single-counted* if the number of edges in C contributes to $r_p(I)$ only once, or equivalently, C does not form a isolated double-tight edge in I. For a vector space $X \subseteq U_\alpha$, let $X^{\perp_{\alpha\beta}}$ (or $X^{\perp_{\beta\alpha}}$) denote the orthogonal vector space with respect to $A_{\alpha\beta}$:

$$X^{\perp_{\alpha\beta}}(= X^{\perp_{\beta\alpha}}) := \{y \in V_\beta \mid A_{\alpha\beta}(x, y) = 0 \text{ for all } x \in X\}.$$

For a vector space $Y \subseteq V_\beta$, $Y^{\perp_{\alpha\beta}}$ (or $Y^{\perp_{\beta\alpha}}$) is defined analogously.

3.1   Definition

Our augmenting path is defined on the *auxiliary graph* $\mathcal{G}(\mathcal{V}, p)$: The vertex set is $\{\alpha^+, \alpha^- \mid \alpha\} \cup \{\beta^+, \beta^- \mid \beta\}$, and the edge set, denoted by $\mathcal{E}(\mathcal{V}, p)$, is $\{\alpha^\sigma \beta^{\sigma'} \mid A_{\alpha\beta}(U_\alpha^\sigma, V_\beta^{\sigma'}) \neq \{0\}, \; p(U_\alpha^\sigma) + p(V_\beta^{\sigma'}) = d_{\alpha\beta}\}$. We denote by $\mathcal{G}(\mathcal{V}, p)|_I$ the subgraph of $\mathcal{G}(\mathcal{V}, p)$ such that its edge set $\mathcal{E}(\mathcal{V}, p)|_I$ is $\{\alpha^\sigma \beta^{\sigma'} \in \mathcal{E}(\mathcal{V}, p) \mid \alpha\beta \in I\}$. By (2.2), for each $\alpha\beta \in I$, neither $\alpha^+ \beta^-$ nor $\alpha^- \beta^+$ belongs to $\mathcal{E}(\mathcal{V}, p)$. In addition, if $\alpha\beta \in I$ is a σ-edge, then $\alpha^{\overline{\sigma}} \beta^{\overline{\sigma}} \in \mathcal{E}(\mathcal{V}, p)|_I$, since $A_{\alpha\beta}(U_\alpha^{\overline{\sigma}}, V_\beta^{\overline{\sigma}}) \neq \{0\}$ and p satisfies (Tight). The edge $\alpha\beta$ in I is double-tight if and only if both $\alpha^+ \beta^+$ and $\alpha^- \beta^-$ exists in $\mathcal{E}(\mathcal{V}, p)$. A σ-*path* is a path in $\mathcal{G}(\mathcal{V}, p)$ consisting of edges $\alpha^\sigma \beta^\sigma$. For paths $\mathcal{P}$ and $\mathcal{Q}$ in $\mathcal{G}(\mathcal{V}, p)$ such that the last node of $\mathcal{P}$ coincides with the first node of $\mathcal{Q}$, we denote by $\mathcal{P} \circ \mathcal{Q}$ the concatenation of $\mathcal{P}$ and $\mathcal{Q}$.

We next define the *source set* and the *target set* as follows, in which nodes β^σ in the former and $\alpha^{\sigma'}$ in the latter can be the initial and the last nodes of an augmenting path, respectively. Let Γ be the set of the end nodes of some single-counted path component of I. For each $\gamma \in \Gamma$, which is incident to a σ-edge in I, let $\mathcal{C}_\gamma$ be the connected component of $\mathcal{G}(\mathcal{V}, p)|_I$ containing γ^σ. The *source set* $\mathcal{S}(I, \mathcal{V}, p)$ and the *target set* $\mathcal{T}(I, \mathcal{V}, p)$ for $(I, \mathcal{V}, p)$ are defined by

$$\mathcal{S}(I, \mathcal{V}, p) := \{\beta^+, \beta^- \mid \deg_I(\beta) = 0\} \cup \bigcup \{\text{the nodes belonging to } \mathcal{C}_\beta \mid \beta \in \Gamma\},$$

$$\mathcal{T}(I, \mathcal{V}, p) := \{\alpha^+, \alpha^- \mid \deg_I(\alpha) = 0\} \cup \bigcup \{\text{the nodes belonging to } \mathcal{C}_\alpha \mid \alpha \in \Gamma\}.$$

Then the following holds.

Lemma 5. *(1)* $\mathcal{S}(I, \mathcal{V}, p) \cap \mathcal{T}(I, \mathcal{V}, p) = \emptyset$.
(2) For each $\gamma \in \Gamma$ incident to a σ-edge, $\mathcal{C}_\gamma$ forms either an isolated connected component $\{\gamma^\sigma\}$ or an even length σ-path in $\mathcal{G}(\mathcal{V}, p)|_I$ that starts with γ^σ.
(3) Both $|\mathcal{S}(I, \mathcal{V}, p) \cap \{\beta^+, \beta^- \mid \beta\}| - |\mathcal{S}(I, \mathcal{V}, p) \cap \{\alpha^+, \alpha^- \mid \alpha\}|$ and $|\mathcal{T}(I, \mathcal{V}, p) \cap \{\alpha^+, \alpha^- \mid \alpha\}| - |\mathcal{T}(I, \mathcal{V}, p) \cap \{\beta^+, \beta^- \mid \beta\}|$ are equal to $2n - r_p(I)$.

We then define the components of augmenting path. An *outer path* $\mathcal{P}$ for $(I, \mathcal{V}, p)$ is a path in $\mathcal{G}(\mathcal{V}, p)$ of the form

$$(\beta_0^{\sigma_0} \alpha_1^{\sigma_1}, \alpha_1^{\sigma_1} \beta_1^{\sigma_1}, \ldots, \beta_k^{\sigma_k} \alpha_{k+1}^{\sigma_{k+1}})$$

such that

(O1) $\beta_i \alpha_{i+1}$ belongs to $E \setminus I$ for each $i = 0, 1, \ldots, k$ and $\alpha_{i+1} \beta_{i+1}$ is an isolated double-tight edge in I for each $i = 0, 1, \ldots, k - 1$, and

(O2) $A_{\alpha_{i+1} \beta_i}(U_{\alpha_{i+1}}^{\sigma_{i+1}}, V_{\beta_i}^{\sigma_i}) = \{0\}$ for each $i = 0, 1, \ldots, k - 1$.

Note that (O2) does not require $A_{\alpha_{k+1} \beta_k}(U_{\alpha_{k+1}}^{\overline{\sigma_{k+1}}}, V_{\beta_k}^{\sigma_k}) = \{0\}$ on the last edge $\beta_k^{\sigma_k} \alpha_{k+1}^{\sigma_{k+1}}$. The initial vertex $\beta_0^{\sigma_0}$ and last vertex $\alpha_{k+1}^{\sigma_{k+1}}$ are denoted by $\beta(\mathcal{P})$ and $\alpha(\mathcal{P})$, respectively.

An *inner path* $\mathcal{Q}$ for $(I, \mathcal{V}, p)$ is a path in $\mathcal{G}(\mathcal{V}, p)$ of the form

$$(\alpha_0^\sigma \beta_1^\sigma, \beta_1^\sigma \alpha_1^\sigma, \ldots, \alpha_k^\sigma \beta_{k+1}^\sigma),$$

such that

(I1) the underlying path $(\alpha_0\beta_1, \beta_1\alpha_1, \ldots, \alpha_k\beta_{k+1})$ of $\mathcal{Q}$ in G belongs to a single-counted connected component of I, and

(I2) $\alpha_0\beta_1, \alpha_1\beta_2, \ldots, \alpha_k\beta_{k+1}$ are $\overline{\sigma}$-edges and $\beta_1\alpha_1, \beta_2\alpha_2, \ldots, \beta_k\alpha_k$ are σ-edges.

The former condition implies that $\mathcal{Q}$ can also be viewed as a σ-path in $\mathcal{G}(\mathcal{V}, p)|_I$, and the latter implies that the σ-edges $\beta_1\alpha_1, \beta_2\alpha_2, \ldots, \beta_k\alpha_k$ are double-tight. The initial vertex α_0^σ and last vertex β_{k+1}^σ are denoted by $\beta(\mathcal{Q})$ and $\alpha(\mathcal{Q})$, respectively.

We are now ready to define an augmenting path. An *augmenting path* $\mathcal{R}$ for $(I, \mathcal{V}, p)$ is a path in $\mathcal{G}(\mathcal{V}, p)$ such that

(A1) $\mathcal{R}$ is the concatenation $\mathcal{P}_0 \circ \mathcal{Q}_1 \circ \mathcal{P}_1 \circ \cdots \circ \mathcal{Q}_m \circ \mathcal{P}_m$ of outer paths $\mathcal{P}_0, \mathcal{P}_1, \ldots, \mathcal{P}_m$ and inner paths $\mathcal{Q}_1, \ldots, \mathcal{Q}_m$ for $(I, \mathcal{V}, p)$ in which $\alpha(\mathcal{P}_i) = \alpha(\mathcal{Q}_{i+1})$ and $\beta(\mathcal{Q}_{i+1}) = \beta(\mathcal{P}_{i+1})$ for each i, and

(A2) $\beta(\mathcal{P}_0) \in \mathcal{S}(I, \mathcal{V}, p)$, $\alpha(\mathcal{P}_m) \in \mathcal{T}(I, \mathcal{V}, p)$, and all intermediate vertices exit $\mathcal{S}(I, \mathcal{V}, p) \cup \mathcal{T}(I, \mathcal{V}, p)$.

Theorem 4. *For a matching I, a valid labeling $\mathcal{V}$ for I, a potential p compatible with $(I, \mathcal{V})$, and an augmenting path for $(I, \mathcal{V}, p)$, we can obtain a p-matching I^* and a valid labeling $\mathcal{V}^*$ for I^* such that p is compatible with $(I^*, \mathcal{V}^*)$ and $r_p(I^*) > r_p(I)$ in $O(n^4)$ time.*

3.2   Finding an Augmenting Path

In this subsection, we present an algorithm for verifying $\deg \det A(t) = -\infty$, finding a potential satisfying (Tight) and (Reg) but not (Path) for $(I, \mathcal{V})$, or finding an augmenting path for $(I, \mathcal{V}, p)$.

Suppose that we are given as the input a p-matching I_0, a valid labeling $\mathcal{V}_0$ for I_0, and a compatible potential p_0 with $(I_0, \mathcal{V}_0)$ such that $r_{p_0}(I_0) < 2n$. We initialize $I \leftarrow I_0$, $\mathcal{V} \leftarrow \mathcal{V}_0$, and $p \leftarrow p_0$. In addition, during the algorithm, we maintain a forest $\mathcal{F}$ in $\mathcal{G}(\mathcal{V}, p)$ such that each connected component of $\mathcal{F}$ has exactly one node in $\mathcal{S}(I, \mathcal{V}, p)$; we initialize $\mathcal{F} \leftarrow \mathcal{S}(I, \mathcal{V}, p)$, which is nonempty by $r_p(I) < 2n$ and Lemma 5 (3).

The algorithm consists of the primal update and the dual update. While there is an edge $\beta^\sigma \alpha^{\sigma'} \in \mathcal{E}(\mathcal{V}, p)$ such that $\beta^\sigma \in \mathcal{F}$ and $\alpha^{\sigma'} \notin \mathcal{F}$, we execute the primal update. If there is no such edge, then we execute the dual update:

Primal update: We first add an edge $\beta^\sigma \alpha^{\sigma'} \in \mathcal{E}(\mathcal{V}, p)$ such that $\beta^\sigma \in \mathcal{F}$ and $\alpha^{\sigma'} \notin \mathcal{F}$ to $\mathcal{F}$.

(P1) If $\alpha^{\sigma'} \in \mathcal{T}(I, \mathcal{V}, p)$, then output $(I, \mathcal{V}, p)$ and the unique path $\mathcal{R}$ in $\mathcal{F}$ from a vertex in $\mathcal{S}(I, \mathcal{V}, p)$ to $\alpha^{\sigma'}$ as an augmenting path. Stop this procedure.

(P2) Suppose that $\alpha^{\sigma'} \notin \mathcal{T}(I, \mathcal{V}, p)$ and α is incident to an isolated double-tight edge $\alpha\beta'$ in I. Then update the valid labeling $\mathcal{V}$ for I as

$$U_\alpha^{\overline{\sigma'}} \leftarrow (V_\beta^\sigma)^{\perp_{\alpha\beta}}, \qquad V_{\beta'}^{\sigma'} \leftarrow (U_\alpha^{\overline{\sigma'}})^{\perp_{\alpha\beta'}},$$

and add $\alpha^{\sigma'}\beta'^{\sigma'}$ to $\mathcal{F}$. Also update $\mathcal{G}(\mathcal{V}, p)$ for the resulting $\mathcal{V}$. (This case will be an expansion of an outer path.)

(P3) Suppose that $\alpha^{\sigma'} \notin \mathcal{T}(I, \mathcal{V}, p)$ and α belongs to a single-counted connected component of I. Let $\mathcal{Q}$ be the longest inner path in $\mathcal{G}(\mathcal{V}, p)|_I$ starting with $\alpha^{\sigma'}$ such that $\mathcal{Q}$ does not meet $\mathcal{F}$. Then add $\mathcal{Q}$ to $\mathcal{F}$. (This case will be an addition of an inner path.)

Dual update: For each $\beta^{\sigma} \in \mathcal{F}$, define

$$\varepsilon_{\beta^{\sigma}} := \min\{p(U_{\alpha}^{\sigma'}) + p(V_{\beta}^{\sigma}) - d_{\alpha\beta} \mid \alpha^{\sigma'} \notin \mathcal{F}, \ A_{\alpha\beta}(U_{\alpha}^{\sigma'}, V_{\beta}^{\sigma}) \neq \{0\}\},$$

where $\varepsilon_{\beta^{\sigma}} := +\infty$ if there is no $\alpha^{\sigma'} \notin \mathcal{F}$ such that $A_{\alpha\beta}(U_{\alpha}^{\sigma'}, V_{\beta}^{\sigma}) \neq \{0\}$. Let ε be the minimum value of $\varepsilon_{\beta^{\sigma}}$ over $\beta^{\sigma} \in \mathcal{F}$.

(D1) If $\varepsilon = +\infty$, then output "deg det $A(t) = -\infty$" and stop this procedure.

(D2) If $\varepsilon < +\infty$, then update p as

$$p(V_{\beta}^{\sigma}) \leftarrow p(V_{\beta}^{\sigma}) - \varepsilon \quad \text{if } \beta^{\sigma} \in \mathcal{F},$$
$$p(U_{\alpha}^{\sigma}) \leftarrow p(U_{\alpha}^{\sigma}) + \varepsilon \quad \text{if } \alpha^{\sigma} \in \mathcal{F},$$

and adjust p so that p satisfies (Reg), that is, for each α and β,

$$p(X_{\alpha}) \leftarrow \max\{p(U_{\alpha}^{+}), p(U_{\alpha}^{-})\} \quad (X_{\alpha} \in \mathcal{M}_{\alpha} \setminus \{U_{\alpha}^{+}, U_{\alpha}^{-}\}),$$
$$p(Y_{\beta}) \leftarrow \max\{p(V_{\beta}^{+}), p(V_{\beta}^{-})\} \quad (Y_{\beta} \in \mathcal{M}_{\beta} \setminus \{V_{\beta}^{+}, V_{\beta}^{-}\}).$$

(D2-1) Suppose that the resulting p does not satisfy (Path) or that the number of isolated double-tight edges in I with respect to the resulting p increases. If p does not satisfy (Path), then we apply the elimination to I. Output the resulting $(I, \mathcal{V}, p)$; stop this procedure.

(D2-2) Otherwise, suppose that the resulting target set $\mathcal{T}(I, \mathcal{V}, p)$ enlarges. In this case, it holds $\mathcal{F} \cap \mathcal{T}(I, \mathcal{V}, p) \neq \emptyset$. Output $(I, \mathcal{V}, p)$ and a minimal path $\mathcal{R}$ in $\mathcal{F}$ with respect to inclusion from a vertex in $\mathcal{S}(I, \mathcal{V}, p)$ to a vertex in $\mathcal{F} \cap \mathcal{T}(I, \mathcal{V}, p)$ as an augmenting path. Stop this procedure.

(D2-3) Otherwise, update

$$\mathcal{F} \leftarrow \mathcal{F} \cup \mathcal{S}(I, \mathcal{V}, p)$$

if the resulting $\mathcal{S}(I, \mathcal{V}, p)$ enlarges.   ∎

The following theorem says that the above algorithm correctly works.

Theorem 5. (1) *Suppose that the algorithm reaches* (P1) *or* (D2-2). *Then the output I is a p-matching, $\mathcal{V}$ is a valid labeling for I, p is a compatible potential with $(I, \mathcal{V})$, $r_p(I) = r_{p_0}(I_0)$, and $\mathcal{R}$ is an augmenting path for $(I, \mathcal{V}, p)$.*
(2) *If the algorithm reaches* (D1), *then* deg det $A(t) = -\infty$.
(3) *Suppose that the algorithm reaches* (D2-1). *Then the output I is a p-matching, $\mathcal{V}$ is a valid labeling for I, p is a compatible potential with $(I, \mathcal{V})$, and $r_p(I) > r_{p_0}(I_0)$.*
(4) *The running-time of the algorithm is $O(n^3)$.*

Theorems 4 and 5 imply Theorems 1, 2, and 3. Indeed, Theorem 1 follows from that at most $2n$ augmentations occur in the algorithm. Theorems 2 and 3 follow from that the above algorithm detects $\deg \det A(t) = -\infty$ or outputs a p-matching I, a valid labeling $\mathcal{V}$, and a compatible potential p satisfying $r_p(I) = 2n$. In the latter case, the minimum in the min-max formulas in Theorems 2 and 3 is attained by such $\mathcal{V}$ and p by Lemma 3. In particular, since every $d_{\alpha\beta}$ is integer, so is ε in the dual update (D2). Hence p is integer-valued.

4   Augmentation

In this section, we present an overview of the augmentation procedure for a given p-matching I, a valid labeling $\mathcal{V} = \{(U_\alpha^+, U_\alpha^-), (V_\beta^+, V_\beta^-)\}_{\alpha,\beta}$ of I, a potential p compatible with $(I, \mathcal{V})$, and an augmenting path $\mathcal{R} = \mathcal{P}_0 \circ \mathcal{Q}_1 \circ \mathcal{P}_1 \circ \cdots \circ \mathcal{Q}_m \circ \mathcal{P}_m$ for $(I, \mathcal{V}, p)$.

For an outer path $\mathcal{P} = (\beta_0^{\sigma_0} \alpha_1^{\sigma_1}, \alpha_1^{\sigma_1} \beta_1^{\sigma_1}, \dots, \beta_k^{\sigma_k} \alpha_{k+1}^{\sigma_{k+1}})$, we denote by its italic font P the underlying walk $(\beta_0 \alpha_1, \alpha_1 \beta_1, \dots, \beta_k \alpha_{k+1})$ in G. The initial vertex β_0 and last vertex α_{k+1} of P are denoted by $\beta(P)$ and $\alpha(P)$, respectively. Similarly, the underlying path of an inner path $\mathcal{Q}$ is denoted by its italic font Q.

4.1   Base Case: $\mathcal{R} = \mathcal{P}_0$ and P_0 Is a Path

We consider the case where $\mathcal{R}$ consists only of a single outer path $\mathcal{P}_0$ of the form $(\beta_0^{\sigma_0} \alpha_1^{\sigma_1}, \alpha_1^{\sigma_1} \beta_1^{\sigma_1}, \dots, \beta_k^{\sigma_k} \alpha_{k+1}^{\sigma_{k+1}})$ such that its underlying walk $(\beta_0 \alpha_1, \alpha_1 \beta_1, \dots, \beta_k \alpha_{k+1})$ is actually a path in G.

We first modify I so that the degrees of β_0 and of α_{k+1} in I are at most one, β_0 is incident to a σ_0-edge in I if $\deg_I(\beta_0) = 1$, and α_{k+1} is incident to a σ_{k+1} in I if $\deg_I(\alpha_{k+1}) = 1$, as follows. Since $\beta_0^{\sigma_0} \in \mathcal{S}(I, \mathcal{V}, p)$, there is a σ_0-path $\mathcal{P}$ in $\mathcal{G}(\mathcal{V}, p)|_I$ from $\beta_0^{\sigma_0}$ to $\beta_*^{\sigma_0}$, where β_* is an end node of a single-counted path component of I incident to a σ_0-edge in I. Then we delete all $\overline{\sigma_0}$-edges in the underlying path P of $\mathcal{P}$. It is clear that the resulting I is a p-matching, $\mathcal{V}$ is a valid labeling for I, p satisfies (Tight) and (Reg) for $(I, \mathcal{V})$, and $\deg_I(\beta_0) \leq 1$. Furthermore, if $\deg_I(\beta_0) = 1$, then β_0 is incident to a σ_0-edge in I. Since all σ_0-edges in P are double-tight, p satisfies (Path), and hence is compatible with $(I, \mathcal{V})$. Furthermore, $|P|/2$ edges (the $\overline{\sigma_0}$-edges in P) are deleted from I, and $|P|/2$ edges (the σ_0-edges in P) become isolated double-tight edges by this deletion. Hence $r_p(I)$ does not change. Similarly, we can modify I so that $\deg_I(\alpha_{k+1}) \leq 1$ and α_{k+1} is incident to a σ_{k+1}-edge in I if $\deg_I(\alpha_{k+1}) = 1$ without changing r_p. Since $\mathcal{P}_0$ does not meet any newly appeared isolated double-tight edges, $\mathcal{R} = \mathcal{P}_0$ is still an augmenting path for $(I, \mathcal{V}, p)$.

We then redefine +- and −-edges of I and +- and −-spaces of $\mathcal{V}$ so that all $\alpha_1 \beta_1, \alpha_2 \beta_2, \dots, \alpha_k \beta_k$ are +-edges and $\mathcal{P}_0 = (\beta_0^+ \alpha_1^+, \alpha_1^+ \beta_1^+, \dots, \beta_k^+ \alpha_{k+1}^+)$.

We are ready to augment I. Let us define $\hat{I}$ by

$$\hat{I} := I \cup P_0,$$

where P_0 is regarded as an edge set. Since $\deg_I(\beta_i) \leq 1$ and $\deg_I(\alpha_i) \leq 1$ for each i, $\hat{I}$ satisfies (Deg). However $\mathcal{V}$ may be no longer a labeling for $\hat{I}$, particularly, on the connected component $\hat{C}$ of $\hat{I}$ containing P_0. We define a labeling $\hat{\mathcal{V}}$ for $\hat{I}$ as follows: If $\hat{C}$ is a cycle component of $\hat{I}$, then we redefine the $+$- and $-$-spaces of each node in $\hat{C}$ so that (2.2) and (2.3) hold. If $\hat{C}$ is a path component of $\hat{I}$ that has the end nodes are γ and γ' incident to a σ-edge and to a σ'-edge, respectively, then we set the $\overline{\sigma}$-space of γ and the $\overline{\sigma'}$-space of γ' with respect to $\hat{\mathcal{V}}$ as those with respect to $\mathcal{V}$, and define the other spaces of each node in $\hat{C}$ so that (2.2) and (2.3) hold. Recall that such a labeling is uniquely determined (see Remark 1).

The following holds.

Lemma 6. *The edge set $\hat{I}$ is a p-matching, $\hat{\mathcal{V}}$ is a valid labeling for $\hat{I}$, p is a potential satisfying* (Tight) *and* (Reg) *for $(\hat{I}, \hat{\mathcal{V}})$, and $r_p(\hat{I}) \geq r_p(I) + 1$.*

If p also satisfies (Path), then p is compatible with $(\hat{\mathcal{V}}, p)$. If p fails (Path), then we execute the elimination to $\hat{I}$, which makes p compatible and increases r_p by Lemma 4. In the both cases, we can obtain a larger p-matching (in the sense of r_p) with a valid labeling and a compatible potential.

4.2 General Case

Suppose that $\mathcal{R} = \mathcal{P}_0 \circ \mathcal{Q}_1 \circ \mathcal{P}_1 \circ \cdots \circ \mathcal{Q}_m \circ \mathcal{P}_m$ with $m \geq 1$. We can assume that $\mathcal{Q}_m$ forms a $+$-path in $\mathcal{G}(\mathcal{V}, p)|_I$. Let C be the single-counted connected component of I containing $\mathcal{Q}_m$. For the space-limitation, we only explain the procedure for the case where $\mathcal{P}_m$ is a path in G, C is a cycle component having a single-tight $+$-edge, and $\mathcal{P}_k$ and $\mathcal{Q}_k$ for $k < m$ do not meet nodes in C and in $\mathcal{P}_m$.

We modify I so that $\deg_I(\alpha(P_m)) \leq 1$ as in Sect. 4.1. Then $\mathcal{R}$ may no longer be an augmenting path for the resulting I. Indeed, there can exist an outer path $\mathcal{P}_k = (\beta_0^{\sigma_0}\alpha_1^{\sigma_1}, \alpha_1^{\sigma_1}\beta_1^{\sigma_1}, \ldots, \beta_k^{\sigma_k}\alpha_{k+1}^{\sigma_{k+1}})$ in $\mathcal{R}$ such that α_{k+1} is incident to a newly appeared isolated double-tight edge and $A_{\alpha_{k+1}\beta_k}(U_{\alpha_{k+1}}^{\overline{\sigma_{k+1}}}, V_{\beta_k}^{\sigma_k}) \neq \{0\}$. Then the concatenation $\mathcal{P}_k \circ \mathcal{Q}_{k+1} \circ \mathcal{P}_{k+1}$ satisfies the first condition (O1) of an outer path but fails the second condition (O2). Hence, for each such $\mathcal{P}_k$, we update $\mathcal{V}$ as $U_{\alpha_{k+1}}^{\overline{\sigma_{k+1}}} \leftarrow (V_{\beta_k}^{\sigma_k})^{\perp_{\beta_k}\alpha_{k+1}}$ and propagate this update according to $\mathcal{P}_k \circ \mathcal{Q}_{k+1} \circ \mathcal{P}_{k+1}$ so that $\mathcal{P}_k \circ \mathcal{Q}_{k+1} \circ \mathcal{P}_{k+1}$ also satisfies (O2). One can show that $\mathcal{R}$ is an augmenting path for I, p, and the resulting $\mathcal{V}$.

We then define $\hat{I}$ by

$$\hat{I} := (I \setminus \{\text{the last } --\text{edge of } \mathcal{Q}_m\}) \cup P_m,$$

and define a labeling $\hat{\mathcal{V}}$ for $\hat{I}$ as in Sect. 4.1. Similarly to Lemma 6, the following holds.

Lemma 7. *The edge set $\hat{I}$ is a p-matching, $\hat{\mathcal{V}}$ is valid for $\hat{I}$, p is a potential satisfying* (Tight) *and* (Reg) *for $(\hat{I}, \hat{\mathcal{V}})$, and $r_p(\hat{I}) = r_p(I)$.*

If p fails (Path), then we obtain a larger p-matching (in the sense of r_p) with a valid labeling and a compatible potential by executing the elimination to $\hat{I}$ (Lemma 4). If p satisfies (Path), then we define $\hat{\mathcal{R}}$ by

$$\hat{\mathcal{R}} := \mathcal{P}_0 \circ \mathcal{Q}_1 \circ \mathcal{P}_1 \circ \cdots \circ \mathcal{P}_{m-1},$$

which is still a path in $\mathcal{G}(\hat{\mathcal{V}}, p)$.

Lemma 8. *The path $\hat{\mathcal{R}}$ forms an augmenting path for $(\hat{I}, \hat{\mathcal{V}}, p)$.*

In this case, we obtain a new augmenting path for $(\hat{I}, \hat{\mathcal{V}}, p)$ of a shorter length.

In this way, with modifying a p-matching and its valid labeling, we can shorten an augmenting path to reach the base case. There remain several other cases to be dealt with, e.g., P_m has repeated edges, or meets a previous outer alternating path $P_{m'}$, and so on. In these cases, we need further considerations including a "no short-cut property"; similar arguments for the rank computation setting are given in [12]. The remaining arguments for the deg-det computation setting are not explained here by space-limitation and will be given in the full version of this paper.

References

1. Cohn, P.M.: Skew Fields: Theory of General Division Rings. Cambridge University Press, Cambridge (1995)
2. Dieudonné, J.: Les déterminants sur un corps non commutatif. Bull. de la Société Mathématique de France **71**, 27–45 (1943)
3. Edmonds, J.: Systems of distinct representatives and linear algebra. J. Res. Natl. Bureau Stand. **71B**(4), 241–245 (1967)
4. Egerváry, J.: Matrixok kombinatorius tulajdonságairól. In: Matematikai és Fizikai Lapok, pp. 16–28 (1931)
5. Frank, A., Tardos, E.: An application of simultaneous Diophantine approximation in combinatorial optimization. Combinatorica **7**, 49–65 (1987)
6. Garg, A., Gurvits, L., Oliveira, R., Wigderson, A.: Operator scaling: theory and applications. Found. Comput. Math. **20**, 223–290 (2019)
7. Hamada, M., Hirai, H.: Maximum vanishing subspace problem, CAT(0)-space relaxation, and block-triangularization of partitioned matrix (2017). arXiv:1705.02060
8. Hamada, M., Hirai, H.: Computing the nc-rank via discrete convex optimization on CAT(0) spaces (2020). arXiv:2012.13651v1
9. Hirai, H.: Computing the degree of determinants via discrete convex optimization on Euclidean buildings. SIAM J. Appl. Geom. Algebra **3**(3), 523–557 (2019)
10. Hirai, H., Ikeda, M.: A cost-scaling algorithm for computing the degree of determinants (2020). arXiv:2008.11388v2
11. Hirai, H., Iwamasa, Y.: A combinatorial algorithm for computing the rank of a generic partitioned matrix with 2×2 submatrices. In: Proceedings of the 21st Conference on Integer Programming and Combinatorial Optimization (IPCO 2020). LNCS, vol. 12125, pp. 196–208 (2020)
12. Hirai, H., Iwamasa, Y.: A combinatorial algorithm for computing the rank of a generic partitioned matrix with 2 × 2 submatrices (2020). arXiv:2004.10443

13. Ivanyos, G., Qiao, Y., Subrahmanyam, K.V.: Non-commutative Edmonds' problem and matrix semi-invariants. Comput. Complexity **26**, 717–763 (2017)
14. Ivanyos, G., Qiao, Y., Subrahmanyam, K.V.: Constructive non-commutative rank computation is in deterministic polynomial time. Comput. Complexity **27**, 561–593 (2018). https://doi.org/10.1007/s00037-018-0165-7
15. Iwata, S., Kobayashi, Y.: A weighted linear matroid parity algorithm. SIAM J. Comput. (2021)
16. Iwata, S., Murota, K.: A minimax theorem and a Dulmage-Mendelsohn type decomposition for a class of generic partitioned matrices. SIAM J. Matrix Anal. Appl. **16**(3), 719–734 (1995)
17. Kabanets, V., Impagliazzo, R.: Derandomizing polynomial identity tests means proving circuit lower bounds. Comput. Complexity **13**, 1–46 (2004)
18. Kuhn, H.W.: The Hungarian method for assignment problems. Naval Res. Logist. Q. **2**, 83–97 (1955)
19. Lovász, L.: On determinants, matchings, and random algorithms. In: International Symposium on Fundamentals of Computation Theory (FCT 1979) (1979)
20. Lovász, L.: Singular spaces of matrices and their application in combinatorics. Boletim da Sociedade Brasileira de Matemática **20**(1), 87–99 (1989)
21. Oki, T.: On solving (non)commutative weighted Edmonds' problem. In: Proceedings of the 47th International Colloquium on Automata, Languages and Programming (ICALP'20), Leibniz International Proceedings in Informatics (LIPIcs), vol. 168, pp. 89:1–89:14 (2020)
22. Schwartz, J.T.: Fast probabilistic algorithms for verification of polynomial identities. J. ACM **27**(4), 701–717 (1980)
23. Tutte, W.T.: The factorization of linear graphs. J. Lond. Math. Soc. **22**(2), 107–111 (1947)

On the Implementation and Strengthening of Intersection Cuts for QCQPs

Antonia Chmiela[1] , Gonzalo Muñoz[2]([✉]) , and Felipe Serrano[1]

[1] Zuse Institute Berlin, Berlin, Germany
{chmiela,serrano}@zib.de
[2] Universidad de O'Higgins, Rancagua, Chile
gonzalo.munoz@uoh.cl

Abstract. The generation of strong linear inequalities for QCQPs has been recently tackled by a number of authors using the intersection cut paradigm—a highly studied tool in integer programming whose flexibility has triggered these renewed efforts in non-linear settings. In this work, we consider intersection cuts using the recently proposed construction of *maximal quadratic-free sets*. Using these sets, we derive closed-form formulas to compute intersection cuts which allow for quick cut-computations by simply plugging-in parameters associated to an arbitrary quadratic inequality being violated by a vertex of an LP relaxation. Additionally, we implement a cut-strengthening procedure that dates back to Glover and evaluate these techniques with extensive computational experiments.

Keywords: Intersection cuts · QCQP · Quadratic-free sets

1 Introduction

Nowadays, the reach of state-of-the-art optimization solvers is vast and certain classes of non-convex optimization problems that years ago seemed impenetrable can now be solved in moderate running times. However, we still encounter computational challenges preventing us to solve many non-convex optimization instances to provable optimality. In this work, we focus on the generation of general purpose cutting planes for quadratically constrained quadratic programs (QCQPs). These problems can be assumed, without loss of generality, to have the following form

$$\min \quad \bar{c}^\mathsf{T} s \tag{1a}$$

$$\text{s.t.} \quad s \in S \subseteq \mathbb{R}^p, \tag{1b}$$

where $S = \{s \in \mathbb{R}^p : s^\mathsf{T} Q_i s + b_i^\mathsf{T} s + c_i \leq 0, \, i = 1, \ldots, m\}$.

In order to find cutting planes for (1) we follow the *intersection cut* paradigm [4,19,33] which requires the following. We assume we have $\bar{s} \notin S$, a basic optimal

© Springer Nature Switzerland AG 2021
M. Singh and D. P. Williamson (Eds.): IPCO 2021, LNCS 12707, pp. 134–147, 2021.
https://doi.org/10.1007/978-3-030-73879-2_10

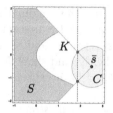 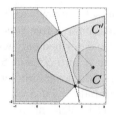

Fig. 1. On the left, an intersection cut (red) separating $\bar{s}$ from S (blue). On the right, the effect of using another S-free set $C' \supsetneq C$. (Color figure online)

solution of a *linear programming* (LP) relaxation of (1)[1]. Such a relaxation can be obtained, for example, using linear over- and under-estimators of the terms $s_i s_j$ [22]. Additionally, we require a simplicial conic relaxation $K \supseteq S$ with apex $\bar{s}$, and an S-free set C—a convex set satisfying $\text{int}(C) \cap S = \emptyset$—such that $\bar{s} \in \text{int}(C)$. With these ingredients, we can find a cutting plane guaranteed to separate $\bar{s}$ from S. In Fig. 1(left) we show a simple intersection cut in the case when all p rays of K intersect the boundary of the S-free set C. In this case, the cut is defined by the hyperplane containing all such intersection points.

When the intersection cuts are computed using the intersection points between the S-free set and the extreme rays of K, the larger the S-free set the better: if two S-free sets C, C' are such that $C \subsetneq C'$, the intersection cut derived from C' is at least as strong as the one derived from C [13]. In Fig. 1(right) we show this phenomenon. This makes inclusion-wise *maximality* of an S-free set a desirable goal. Note that if $\bar{s} \notin S$, there exists $i \in \{1, \dots, m\}$ such that

$$\bar{s} \notin S_i := \{s \in \mathbb{R}^p : s^\top Q_i s + b_i^\top s + c_i \leq 0\},$$

and constructing an S_i-free set containing $\bar{s}$ suffices to ensure separation. We refer to these sets as *quadratic-free sets*. Recently, Muñoz and Serrano [26] provided a method for constructing *maximal* quadratic-free sets for any arbitrary quadratic inequality. Their focus is the construction and maximality proofs of these sets, leaving aside the actual calculation of the cuts and their computational impact.

Contribution. The first contribution of this work is an implementation and extensive testing of intersection cuts based on the maximal quadratic-free sets proposed by Muñoz and Serrano. In [26], they showed how to construct maximal S^h-free and S^g-free sets for the quadratic-representable sets

$$S^h := \{(x, y) \in \mathbb{R}^{n+m} : \|x\| \leq \|y\|\} \tag{2}$$
$$S^g := \{(x, y) \in \mathbb{R}^{n+m} : \|x\| \leq \|y\|, a^\top x + d^\top y = -1\} \tag{3}$$

where $\max\{\|a\|, \|d\|\} = 1$. They also argued that one can always transform

$$S := \{s \in \mathbb{R}^p : s^\top Q s + b^\top s + c \leq 0\}, \tag{4}$$

---

[1] If $\bar{s} \in S$ the problem would be solved.

a generic quadratic set, into S^h or S^g and use the maximal S^h-free and S^g-free to construct maximal S-free sets. The maximal S-free sets of Muñoz and Serrano, however, are described only with respect to S^h and S^g. Here, we provide transformations from S into S^h and S^g explicitly, and show descriptions of the resulting S-free sets. Moreover, we derive closed-form expressions to compute a valid inequality violated by $\bar{s} \notin S$.

Additionally, we implement a well known *cut strengthening* procedure designed to improve the cut coefficient of a ray of K that never intersects the boundary of the S-free set. We show that computing the strengthened cuts involves solving a single-variable convex optimization problems, which we provide explicitly. We also implemented a family of cuts obtained from *implied quadratic constraints* in an extended space. These were used in the maximal *outer-product-free* sets by Bienstock et al. [7,8], which here we reinterpret as maximal quadratic-free sets. All our ideas are tested using the general-purpose solver SCIP [28].

Henceforth, we assume Q in (4) is a symmetric indefinite matrix. Otherwise, S is convex or reverse-convex; in the former case, cutting planes can be computed using supporting hyperplanes, and in the latter case the unique maximal S-free set is the complement of S, which can be used directly in the intersection cuts framework.

1.1 Literature Review

The basic idea behind intersection cuts can be traced back to Tuy [33]. Later on, intersection cuts were introduced in integer programming by Balas [4] and have been largely studied since. See e.g. [6,13,14] for in-depth analyses of the relation of intersection cuts using maximal $\mathbb{Z}^n$-free sets and the generation of facets of conv(S), when S is a mixed-integer set. We also refer the reader to [2,3,5,10,15,20] and references therein. Intersection cuts have also been extended to the mixed-integer conic case: see e.g. [1,21,24,25]. A different, but related, method was proposed by Towle and Luedtke [32].

Lately, there has been a number of methods proposed for the use of the intersection cut framework in non-convex non-linear settings. Fischetti et al. [16] applied intersection cuts to bilevel optimization. Bienstock et al. [7,8] studied outer-product-free sets, which can be used for generating intersection cuts for polynomial optimization. Serrano [31] showed how to construct a concave underestimator of any factorable function and from them one can build intersection cuts. Fischetti and Monaci [17] constructed bilinear-free sets through a bound disjunction and underestimating the bilinear term with McCormick inequalities [22]. Muñoz and Serrano [26] constructed multiple families of maximal quadratic-free sets that can be used to compute intersection cuts for QCQPs.

Alternative cutting-plane-generation approaches to intersection cuts can be obtained from the work of [11,21], and from [27]. We refer to the survey [9] for other efforts of extending cutting planes to the non-linear setting.

2 Maximal Quadratic-Free Sets and Cut Computations

In this section, we show how to use the maximal S^h- and S^g-free sets from [26] in order to construct a maximal quadratic-free for an arbitrary quadratic. Consider S defined in (4), with Q a symmetric matrix, and let $Q = V\Theta V^\mathsf{T}$ be its eigenvalue decomposition. Then, $S = \{s \in \mathbb{R}^p : s^\mathsf{T} V\Theta V^\mathsf{T} s + b^\mathsf{T} s + c \leq 0\}$. Let θ_i, $i = 1,\dots,p$, be the eigenvalues of Q, and define $I_+ = \{i : \theta_i > 0\}$, $I_- = \{i : \theta_i < 0\}$, and $I_0 = \{i : \theta_i = 0\}$. After algebraic manipulations of this last description of S, we obtain the following equivalent description

$$S = \{s \in \mathbb{R}^p : \|x(s)\|^2 - \|y(s)\|^2 + (\bar{b}_{I_0})^\mathsf{T} z(s) + \kappa \leq 0\}.$$

where $\bar{b}_{I_0}$ is the sub-vector of $\bar{b} := V^\mathsf{T} b$ with entries in I_0 and

$$x_i(s) = \sqrt{\theta_i}\, v_i^\mathsf{T}\left(s + \frac{b}{2\theta_i}\right), \quad \forall i \in I_+, \quad y_i(s) = \sqrt{-\theta_i}\, v_i^\mathsf{T}\left(s + \frac{b}{2\theta_i}\right), \qquad \forall i \in I_-,$$

$$z_i(s) = v_i^\mathsf{T} s, \qquad \forall i \in I_0, \qquad \kappa = c - \frac{1}{4}\sum_{i \in I_+ \cup I_-}\frac{(v_i^\mathsf{T} b)^2}{\theta_i},$$

where v_i is the i-th eigenvector, that is, the i-th column of V.

Remark 1. Using an eigenvalue decomposition is not crucial. Other factorizations of the Q matrix can have the same effect, and can lead to other maximal quadratic-free sets. We chose the eigenvalue decomposition since it can be computed efficiently, and it is available within SCIP without extra computations.

Recall that we assume we have a basic solution $\bar{s} \notin S$. In the following, we construct a maximal S-free containing $\bar{s}$ by distinguishing four different scenarios of S. All our maximal S-free sets are described as $\mathcal{C} = \{s : g(s) \leq 0\}$. For a ray r of the simplicial conic relaxation obtained from a basis associated to $\bar{s}$, the cut coefficient of the non-basic variable associated to r is found through the smallest $t > 0$ such that $g(\bar{s} + tr) = 0$ (recall that $g(\bar{s}) < 0$). If t^* is such a root, which is typically called *step-length*, the cut coefficient is $1/t^*$. If no such t exists, then $t^* = \infty$ and the cut coefficient is 0. For convenience, we uniformly use the term "smallest positive root", defining it as ∞ if none exists and consider $1/\infty := 0$.

2.1 Case 1: $\bar{b}_{I_0} = 0$ and $\kappa = 0$

Maximal S-free set. In this case, S simplifies to $S = \{s \in \mathbb{R}^p : \|x(s)\|^2 - \|y(s)\|^2 \leq 0\}$. Using that the map $s \to (x(s), y(s), z(s))$ is affine and invertible, and the maximal S^h-free sets of [26, Theorem 2], we obtain that

$$\mathcal{C} = \left\{s \in \mathbb{R}^p : x(\bar{s})^\mathsf{T} x(s)/\|x(\bar{s})\| \geq \|y(s)\|\right\}$$

is a maximal S-free set and contains $\bar{s}$ in its interior.

Computation of Cut Coefficients. As described above, in order to find the cut-coefficients we need to find the smallest positive solution to

$$\|y(\bar{s} + tr)\| - \frac{x(\bar{s})^{\mathsf{T}}}{\|x(\bar{s})\|} x(\bar{s} + tr) = 0. \tag{5}$$

Below, we show a unified result (Lemma 1) that indicates explicitly how to compute such a root in this and the next two cases.

2.2   Case 2: $\bar{b}_{I_0} = 0$ and $\kappa > 0$

Maximal S-free set. In this case we need to homogenize the quadratic expression using a new variable ζ: $S = \left\{ s \in \mathbb{R}^p : \|x(s)\|^2 - \|y(s)\|^2 + \zeta^2 \leq 0, \frac{\zeta}{\sqrt{\kappa}} = 1 \right\}$. Let $\hat{x}(s) = \frac{x(s)}{\sqrt{\kappa}}$, $\hat{y}(s) = \frac{y(s)}{\sqrt{\kappa}}$, and $\hat{\zeta} = \frac{\zeta}{\sqrt{\kappa}}$. The following reformulation of S shows how it can be mapped to a set of type S^g (see (3)):

$$S = \{ s \in \mathbb{R}^p : \|(\hat{x}(s), \hat{\zeta})\|^2 - \|\hat{y}(s)\|^2 \leq 0, \ a^{\mathsf{T}}(\hat{x}(s), \hat{\zeta}) + d^{\mathsf{T}}\hat{y}(s) = -1 \},$$

where $a = -e_{p_+ + 1}$, $d = 0$, and $p_+ = |I_+|$. Using that $s \to (\hat{x}(z), \hat{y}(z), z(s), 1)$ is affine and one-to-one along with the constructions in [26, Theorem 3], we prove that

$$C = \{ s \in \mathbb{R}^p : \|y(s)\| \leq \lambda^{\mathsf{T}}(x(s), \sqrt{\kappa}) \}$$

with $\lambda = \frac{(x(\bar{s}), \sqrt{\kappa})}{\|(x(\bar{s}), \sqrt{\kappa})\|}$, is a maximal S-free set and contains $\bar{s}$ in its interior.

Computation of Cut Coefficients. Given a ray r, the cut coefficient of the non-basic variable associated to r is given by the smallest $t > 0$ such that

$$\|y(\bar{s} + tr)\| - \frac{(x(\bar{s}), \sqrt{\kappa})}{\|(x(\bar{s}), \sqrt{\kappa})\|}^{\mathsf{T}} (x(\bar{s} + tr), \sqrt{\kappa}) = 0. \tag{6}$$

As for Case 1, in Lemma 1 we show how to compute such a root explicitly.

2.3   Case 3: $\bar{b}_{I_0} = 0$ and $\kappa < 0$

Maximal S-free set. Using a similar homogenization to the last case, and the constructions in [26, Theorem 3], we can show that

$$C = \left\{ s \in \mathbb{R}^p : \|(y(s), \sqrt{-\kappa})\| \leq \lambda^{\mathsf{T}} x(s) \right\},$$

with $\lambda = \frac{x(\bar{s})}{\|x(\bar{s})\|}$, is a maximal S-free set and contains $\bar{s}$ in its interior.

Computation of Cut Coefficients. Similarly to the previous case, we need to find the smallest positive solution of

$$\|(y(\bar{s} + tr), \sqrt{-\kappa})\| - \frac{x(\bar{s})^\mathsf{T}}{\|x(\bar{s})\|} x(\bar{s} + tr) = 0. \tag{7}$$

The next lemma shows how to compute this root, as well as the ones corresponding to cases 1 and 2.

Lemma 1. *Consider S and $\bar{s}$ as defined above and r an arbitrary ray. The step-length associated to r for Cases 1, 2 and 3 is obtained as the smallest positive root of a single-variable quadratic equation of the form*

$$A_r t^2 + B_r t + C_r - (D_r t + E_r)^2 = 0, \tag{8}$$

where the coefficients A_r, B_r, C_r, D_r, E_r for each case are displayed in Table 2. In all cases, Eq. (8) has at most one positive root, and it has no such root if and only if $\sqrt{A_r} \leq D_r$.

Note that computing these roots can be done efficiently. From the coefficients displayed in Table 2, we can see that we just need to compute and store $v_i^\mathsf{T} r$, $v_i^\mathsf{T} \bar{s}$, and $v_i^\mathsf{T} b$ for all $i \in I_+ \cup I_-$. After computing these coefficients, we just need to compute the roots of a single-variable quadratic.

2.4   Case 4: $\bar{b}_{I_0} \neq 0$

Maximal S-free set. For this case, we define $w(s) := (\bar{b}_{I_0})^\mathsf{T} z(s)$. A homogenization and further diagonalization yields the following representation of S

$$S = \{s \in \mathbb{R}^p \ : \ \|\hat{x}(s)\|^2 - \|\hat{y}(s)\|^2 \leq 0, \ a^\mathsf{T}\hat{x}(s) + d^\mathsf{T}\hat{y}(s) = -1\},$$

where

$$\hat{x}(s) = \frac{1}{\sqrt[4]{1+\kappa^2}} \left(x(s), \frac{1}{2\sqrt[4]{1+\kappa^2}} (w(s) + (\kappa + \sqrt{1+\kappa^2})\zeta) \right)$$

$$\hat{y}(s) = \frac{1}{\sqrt[4]{1+\kappa^2}} \left(y(s), \frac{1}{2\sqrt[4]{1+\kappa^2}} (w(s) + (\kappa - \sqrt{1+\kappa^2})\zeta) \right)$$

and $a = -e_{p_+ + 1}$, $d = e_{p_- + 1}$, $p_+ = |I_+|$, and $p_- = |I_-|$. Using [26, Theorem 4], the following maximal S-free set can be obtained

$$\mathcal{C} = \left\{ s : \begin{array}{l} \|\hat{y}(s)\| \leq \lambda^\mathsf{T}\hat{x}(s), \text{ if } -\lambda_{p_+ + 1}\|\hat{y}(s)\| + \hat{y}_{p_- + 1}(s) \leq 0 \\ \frac{\|x(\bar{s})\|}{\sqrt{1+\kappa^2}} \|y(s)\| + \hat{x}_{p_+ + 1}(\bar{s})\hat{y}_{p_- + 1}(s) \leq \hat{x}(\bar{s})^\mathsf{T}\hat{x}(s), \text{ otherwise} \end{array} \right\}.$$

where $\lambda = \frac{\hat{x}(\bar{s})}{\|\hat{x}(\bar{s})\|}$. For brevity, we omit some transformation steps, but we refer the interested reader to the full-length version [12] for these details.

Computation of Cut Coefficients. Since the maximal S-free set of this case is piecewise-defined, the computation of the cut coefficient associated to a ray r involves computing (potentially) two roots:

$$\|\hat{y}(\bar{s}+tr)\| - \lambda^\mathsf{T}\hat{x}(\bar{s}+tr) = 0 \qquad (9)$$

$$\frac{\|x(\bar{s})\|}{\sqrt{1+\kappa^2}}\|y(\bar{s}+tr)\| + \hat{x}_{p_++1}(\bar{s})\hat{y}_{p_-+1}(\bar{s}+tr) - \hat{x}(\bar{s})^\mathsf{T}\hat{x}(\bar{s}+tr) = 0. \qquad (10)$$

Lemma 2. *Both (9) and (10) have at most one positive solution. If (9) does not have a positive solution, neither does (10). If (9) has a positive solution $\bar{t}_1$, then it is the desired step-length if and only if*

$$-\lambda_{p_++1}\|\hat{y}(\bar{s}+\bar{t}_1 r)\| + \hat{y}_{p_-+1}(\bar{s}+\bar{t}_1 r) \le 0. \qquad (11)$$

If (11) does not hold, the smallest positive root $\bar{t}_2$ of (10) is the step-length.

This last lemma indicates how to orderly use both parts in the definition of $\mathcal{C}$ to compute the desired cut coefficient based on the solutions to (9) and (10). The next lemma shows exactly how to compute such solutions.

Lemma 3. *Consider S and $\bar{s}$ as defined above and r an arbitrary ray. The smallest positive root $\bar{t}_1$ of (9) can be found using the quadratic Eq. (8) with the coefficients displayed in Table 3, column "Case 4–1". We have $\bar{t}_1 = \infty$ if and only if $\sqrt{A_r} \le D_r$.*

Similarly, the smallest positive root $\bar{t}_2$ of (10) can be found using the quadratic Eq. (8) with the coefficients displayed in Table 3, column "Case 4–2". We have $\bar{t}_2 = \infty$ if and only if $\sqrt{A_r} \le D_r$.

2.5   Implied Quadratics in an Extended Space

We also incorporated cutting planes using *implied quadratic constraints* of an extended formulation. Most LP relaxations for QCQPs linearize bilinear terms $x_i x_j$ using a new variable $X_{i,j} = x_i x_j$, therefore these variables must satisfy

$$X_{i_1,j_1} X_{i_2,j_2} = X_{i_1,j_2} X_{i_2,j_1}. \qquad (12)$$

We interpret (12) as two inequalities which fall into our Case 1 above. Whenever $i_1 \ne j_1$, $i_1 \ne j_2$, $i_2 \ne j_1$ and $i_2 \ne j_2$, the maximal quadratic-free set we construct is exactly one of the maximal outer-product-free sets constructed by Bienstock et al. [7,8]. Additionally if, for instance, $i_1 = j_1$, we can use the extra valid inequality $X_{i_1,j_1} \ge 0$ to enlarge the (12)-free set. Consider

$$S_M := \{X \,:\, X_{i_1,j_1} X_{i_2,j_2} - X_{i_1,j_2} X_{i_2,j_1} \le 0,\, X_{i_1,j_1} \ge 0\}.$$

This is defined using a homogenous quadratic and a homogenous linear inequality, which is also handled in [26]. Roughly speaking, they show that with the same set used in Case 4 above, one can construct a maximal S_M-free set. For details see the full-length version [12].

3   Strengthening Procedure

When a ray r of the simplicial conic relaxation lies in $\mathrm{rec}(\mathcal{C})$, the corresponding cut-coefficient is 0 and may be strengthened using a *negative edge extension*. This technique was proposed by Glover [18], and many authors have shown its theoretical strength [8,29,30]. Let r^i, $i = 1, \ldots, p$, be the extreme rays of the simplicial conic relaxation, and $\alpha_i^* \in (0, \infty]$ the step-length computed for r^i. Let $F = \{i : \alpha_i^* < \infty\}$. The negative edge extension computes, for each $j \notin F$,

$$\rho_j = \max_{\rho < 0}\{\rho : \alpha_i^* r^i - \rho r^j \in \mathrm{rec}(\mathcal{C}) \quad \forall i \in F\}. \tag{13}$$

and uses the cut-coefficient $1/\rho_j < 0$ instead of 0. It is not hard to see that $\rho_j = \min\{\rho_j^i : i \in F\}$, where

$$\rho_j^i = \max_{\rho < 0}\{\rho : \alpha_i^* r^i - \rho r^j \in \mathrm{rec}(\mathcal{C})\}. \tag{14}$$

We could solve (14) directly through a single-variable convex optimization problem. However, we can reformulate the problem so as to not consider all $\rho < 0$.

Lemma 4. *If r^i and r^j are linearly dependent[2], $\rho_j^i = -\alpha_i^* \|r^i\|/\|r^j\|$. Otherwise, $\rho_j^i = (\bar{\mu} - 1)\alpha_i^*/\bar{\mu}$, where*

$$\bar{\mu} := \max\{\mu \in [0,1] : \mu r^i + (1 - \mu)r^j \in \mathrm{rec}(\mathcal{C})\}. \tag{15}$$

While (14) and (15) look similar, in our experiments the latter proved to be computationally better: the domain of the variable to optimize is bounded, which resulted in a faster and numerically more stable strengthening routine.

Since (15) is a single-variable problem over a bounded domain, we use a binary-search approach to solve it, and thus having an efficient membership oracle of $\mathrm{rec}(\mathcal{C})$ suffices. Using that our S-free sets have the form $\mathcal{C} = \{s : g(s) \le 0\}$, from our previous discussion we further note that $r \in \mathrm{rec}(\mathcal{C})$ if and only if $g(\bar{s} + tr) = 0$ has no positive solution. What follows is based on this fact.

Cases 1, 2 and 3. Since $r \in \mathrm{rec}(\mathcal{C})$ is equivalent to determining if $g(\bar{s}+tr) = 0$ has no positive solution, we see that $r \in \mathrm{rec}(\mathcal{C})$ if and only if $\sqrt{A_r} \le D_r$ (see Lemma 1). Let us denote A_i the coefficient A_{r^i} (according to Table 2), and similarly for the other coefficients. In order to solve (15) we show

Lemma 5. $\mu r^i + (1 - \mu)r^j \in \mathrm{rec}(\mathcal{C})$ *if and only if*

$$\sqrt{\mu^2 A_i + (1 - \mu)^2 A_j + 2\mu(1 - \mu)\sum_{k \in I_-} \theta_k (v_k^\mathsf{T} r^i)(v_k^\mathsf{T} r^j)} - \mu D_i - (1 - \mu)D_j \le 0.$$

---

[2] Since we are considering rays of a simplicial cone of dimension p, they are all linearly independent. However, in practice, the set S is usually of dimension $\ll p$. In these cases, one can either extend the S-free set to dimension p, or restrict the rays to the support of S for computational purposes. The latter might create linear dependence.

This casts (15) as a single-variable single-constraint convex problem.

Case 4. As each part in the definition of $\mathcal{C}$ is a convex function we show:

Lemma 6. Let $\tau(t) = -\lambda_{p_++1}\|\hat{y}(\bar{s} + tr)\| + \hat{y}_{p_-+1}(\bar{s} + tr)$. If $\lim_{t\to\infty} \tau(t) > 0$, then $r \in \text{rec}(\mathcal{C})$ if and only if (9) has a positive root. Otherwise, $r \in \text{rec}(\mathcal{C})$ if and only if (10) has a positive root.

This lemma frames how we can check for a ray r to be in $\text{rec}(\mathcal{C})$. The two following results show precisely how to verify each condition.

Lemma 7. Let A_r, B_r, C_r, D_r, E_r be defined as in Table 3, column "Case 4–1", and $\tau(t)$ as in the previous lemma. Then

$$\lim_{t\to\infty} \tau(t) = \begin{cases} sgn(-\lambda_{p_++1})\infty, & \text{if } \sqrt{\bar{A}_r} > \bar{D}_r \\ sgn(\lambda_{p_++1})\infty, & \text{if } \sqrt{\bar{A}_r} < \bar{D}_r \\ sgn(-\lambda_{p_++1})(\frac{\bar{B}_r}{2\sqrt{\bar{A}_r}} - \bar{E}_r), & \text{if } \sqrt{\bar{A}_r} = \bar{D}_r. \end{cases}$$

where

$$\bar{A}_r := \lambda_{p_++1}^2 A_r, \ \bar{B}_r := \lambda_{p_++1}^2 B_r, \ \bar{C}_r := \lambda_{p_++1}^2 C_r, \ \bar{D}_r := -\frac{w(r)}{2\sqrt[4]{1-\kappa^2}}, \ \bar{E}_r := \hat{y}_{p_-+1}(\bar{s}).$$

Finally, the only missing ingredient is the verification of when (9) or (10) have a positive root for the ray $r = \mu r^i + (1 - \mu)r^j$. We show

Lemma 8. Let $r = \mu r^i + (1 - \lambda)r^j$. Then, (9) has a positive root if and only if

$$\sqrt{\mu^2 A_i + (1-\mu)^2 A_j + 2\mu(1-\mu)\left(\frac{\|x(\bar{s})\|^2}{4(1+\kappa)^2}\sum_{k\in I_-} \theta_k(v_k^\mathsf{T} r^i)(v_k^\mathsf{T} r^j)\right)} - \mu D_i + (1-\mu)D_j \leq 0.$$

where the coefficients are obtained using Table 3, column "Case 4–1". Similarly, (10) *has a positive root if and only if*

$$\sqrt{\mu^2 A_i + (1-\mu)^2 A_j - 2\mu(1-\mu)\left(\frac{w(r^i)w(r^j)}{4(1+\kappa)^2} - \frac{1}{\sqrt{1+\kappa^2}}\sum_{k\in I_-} \theta_k(v_k^\mathsf{T} r^i)(v_k^\mathsf{T} r^j)\right)} - \mu D_i + (1-\mu)D_j \leq 0.$$

where the coefficients are obtained using Table 3, column "Case 4–2".

4   Computational Experiments

To test our approach, we use what is commonly known as *root node experiments*: we start from an LP relaxation of a QCQP (providing a dual bound d_1) and incorporate our cutting planes to SCIP via a *separator*. After SCIP stops, we compute the *gap closed*: if d_2 is the dual bound obtained when the algorithm finishes, and p a reference primal bound, the function $GC(p, d_1, d_2) = \frac{d_2 - d_1}{p - d_1}$ is the *gap closed* improvement of d_2 with respect to d_1.

For our experiments, we used a Linux cluster of Intel Xeon CPU E5-2660 v3 2.60 GHz with 25 MB cache and 128 GB main memory. The time limit in all

Table 1. Summary of gap closed in between intersection cuts and default SCIP. The columns *rel* denote the corresponding relative improvement with respect to default SCIP. The *#solved* row shows the number of instances solved in the root node.

Subset	DEFAULT	ICUTS		ICUTS-S		MINOR		MINOR-S		ICUTS+MINOR		ICUTS+MINOR-B	
	Mean	Mean	Rel	Mean	Rel	Mean	Rel	Mean	Rel	Mean	Rel	Mean	Rel
Clean	0.56	0.61	1.08	0.60	1.07	0.59	1.04	0.58	1.04	0.61	1.09	0.62	1.09
Affected	0.52	0.59	1.12	0.58	1.11	0.55	1.06	0.55	1.06	0.60	1.14	0.60	1.15
#solved	90	116		114		92		91		116		117	

experiments was set to one hour. The test set we used consists of the publicly available instances of the MINLPLib [23]. We selected all non-convex instances with at least one quadratic constraint, and discarded instances for which no primal solution was available, no dual solution was found, or SCIP failed. This resulted in a test set of 587 instances. To compute the gap closed improvement, we used the MINLPLib's best primal bounds as a reference.

Below, we refer to the following settings: DEFAULT refers to SCIP's default settings, ICUTS refers to including our intersections cuts, MINOR refers to including the cuts in extended space obtained from (12), ICUTS-S and MINOR-S refers to their strengthened versions. Finally, MINOR-B refers to cuts obtained from (12) including non-negativity bounds, as discussed in Sect. 2.5. Combinations of these settings are displayed with a '+' sign. Since the root node experiments aim at showing how much gap can be closed by a family of cuts, we did not put any restrictions on the number of cuts added by SCIP. On average, the non-default settings added between 1155–73049 intersection cuts.

The overall best performing setting was ICUTS+MINOR-B, thus we mainly report comparisons of this setting with respect to variations of it. In Table 1 we show summarized results for various settings. On average, we see an improvement of 8% in the gap closed of ICUTS+MINOR-B with respect to DEFAULT. This improvement becomes 12% if we restrict to the 395 *affected* instances, i.e., instances for which at least one of the non-default settings added cutting planes. Considering the heterogeneity of these instances, these improvements are significant. Additionally, using our cutting planes SCIP was able to solve 27 more instances in the root node.

Surprisingly, we can see a (slightly) negative effect of the strengthening. This was an unexpected phenomenon, which we examined in detail on a number of instances. We observe that the cut coefficients are not improved significantly in many cases. If we also consider that the strengthening increases the density of a cut, we conclude that, overall, the coefficient's modest improvement does not compensate the extra difficulties associated to a dense LP. Preliminary experiments we conducted with spatial branch-and-bound support this claim: when using the strengthening routine, around 10% less LP iterations per second were executed compared to default. In contrast, the cuts without the strengthening reduced the LP iterations per second by only 4%.

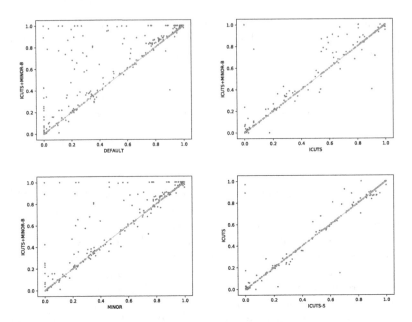

Fig. 2. Scatter plots showing comparisons of gap closed in root node experiments between various pairs of settings.

When comparing ICUTS+MINOR-B with ICUTS and with MINOR, we see that both cutting plane families are complementing each other well. Although the contribution of MINOR on top of ICUTS is modest, including both does make a difference in some cases. Since these two families are using quadratic inequalities that lie in different spaces, combining them is providing significantly different violated constraints. In addition, from ICUTS+MINOR-B and ICUTS+MINOR, we see that also considering the non-negativity bounds has an important positive impact.

In Fig. 2 we show scatter plots comparing different settings. These plots support our previous analysis, and also show that the results are stable: if a setting improves the performance on average, the improvement is relatively consistent among the whole test set. While there are instances where the cutting planes negatively impact SCIP's performance, these were only a small fraction of the total set.

Overall, we believe these results are encouraging for our families of cutting planes. Given the heterogeneity of the instances, and how generic the cutting planes are, the results we are obtaining advocate for our approach as a promising tool for solving QCQPs.

5   Final Remarks

In this work, we have shown an implementation of intersection cuts for QCQPs using the newly developed maximal quadratic-free sets. We show a detailed framework on how to construct cutting planes using any violated quadratic, and the necessary results showing the correctness of our computations. Our results allow for efficient cut computations that any researcher can embed in their optimization routines by simply plugging into our formulas the necessary parameters of a generic quadratic inequality.

Our careful implementation resulted in encouraging results: we were able to close more gap in a significant number of instances and solve more instances in the root node. We also showed that these and the cuts proposed by Bienstock et al. are complementing each other well. While, unfortunately, the strengthening procedure did not yield good results, we believe it still provides valuable insights for the optimization community.

Our current and future work involves a full incorporation of these cutting planes in spatial branch-and-bound. This will require a much more careful handling of the density of the cuts we create, as well as special cut selection rules.

Table 2. Coefficients of equation $A_r t^2 + B_r t + C_r - (D_r t + E_r)^2 = 0$ in the cut-coefficient computations for a ray r in Cases 1, 2 and 3.

Coefficient	Case 1	Case 2	Case 3
A_r	$-\sum_{i\in I_-}\theta_i(v_i^\top r)^2$		
B_r	$-2\sum_{i\in I_-}\theta_i\left(v_i^\top(\bar s + \frac{b}{2\theta_i})\right)(v_i^\top r)$		
C_r	$-\sum_{i\in I_-}\theta_i\left(v_i^\top(\bar s+\frac{b}{2\theta_i})\right)^2$	$-\sum_{i\in I_-}\theta_i\left(v_i^\top(\bar s+\frac{b}{2\theta_i})\right)^2$	$-\kappa-\sum_{i\in I_-}\theta_i\left(v_i^\top(\bar s+\frac{b}{2\theta_i})\right)^2$
D_r	$\frac{1}{E_r}\sum_{i\in I_+}\theta_i\left(v_i^\top(\bar s+\frac{b}{2\theta_i})\right)(v_i^\top r)$		
E_r	$\sqrt{\sum_{i\in I_+}\theta_i\left(v_i^\top(\bar s+\frac{b}{2\theta_i})\right)^2}$	$\sqrt{\kappa+\sum_{i\in I_+}\theta_i\left(v_i^\top(\bar s+\frac{b}{2\theta_i})\right)^2}$	$\sqrt{\sum_{i\in I_+}\theta_i\left(v_i^\top(\bar s+\frac{b}{2\theta_i})\right)^2}$

Table 3. Coefficients of equations $A_r t^2 + B_r t + C_r - (D_r t + E_r)^2 = 0$ in the cut-coefficient computations for a ray r in Case 4.

Coefficient	Case 4–1	Coefficient	Case 4–2
A_r	$\frac{w(r)^2}{4(1+\kappa^2)} - \frac{1}{\sqrt{1+\kappa^2}}\sum_{i\in I_-}\theta_i(v_i^\top r)^2$	A_r	$-\frac{\|x(\bar s)\|^2}{1+\kappa^2}\sum_{i\in I_-}\theta_i(v_i^\top r)^2$
B_r	$2\left(\frac{w(r)}{2\sqrt{1+\kappa^2}}\hat y_{p_-+1}(\bar s) - \frac{2}{\sqrt{1+\kappa^2}}\sum_{i\in I_-}\theta_i\left(v_i^\top(\bar s+\frac{b}{2\theta_i})\right)(v_i^\top r)\right)$	B_r	$-2\frac{\|x(\bar s)\|^2}{1+\kappa^2}\sum_{i\in I_-}\theta_i\left(v_i^\top(\bar s+\frac{b}{2\theta_i})\right)(v_i^\top r)$
C_r	$\hat y_{p_-+1}(\bar s)^2 - \frac{1}{\sqrt{1+\kappa^2}}\sum_{i\in I_-}\theta_i\left(v_i^\top(\bar s+\frac{b}{2\theta_i})\right)^2$	C_r	$-\frac{\|x(\bar s)\|^2}{1+\kappa^2}\sum_{i\in I_-}\theta_i\left(v_i^\top(\bar s+\frac{b}{2\theta_i})\right)^2$
D_r	$\frac{1}{E_r\sqrt{1+\kappa^2}}\left(\sum_{i\in I_+}\theta_i\left(v_i^\top(\bar s+\frac{b}{2\theta_i})\right)(v_i^\top r) + \frac{(w(\bar s)+\kappa+\sqrt{1+\kappa^2})w(r)}{4\sqrt{1+\kappa^2}}\right)$	D_r	$\frac{1}{\sqrt{1+\kappa^2}}\sum_{i\in I_+}\theta_i\left(v_i^\top(\bar s+\frac{b}{2\theta_i})\right)(v_i^\top r)$
E_r	$\frac{1}{\sqrt[4]{1+\kappa^2}}\sqrt{\frac{(w(\bar s)+\kappa+\sqrt{1+\kappa^2})^2}{4\sqrt{1+\kappa^2}} + \sum_{i\in I_+}\theta_i\left(v_i^\top(\bar s+\frac{b}{2\theta_i})\right)^2}$	E_r	$\frac{1}{\sqrt{1+\kappa^2}}\left(\|x(\bar s)\|^2 + \frac{w(\bar s)+\kappa+\sqrt{1+\kappa^2}}{2}\right)$

References

1. Andersen, K., Jensen, A.N.: Intersection cuts for mixed integer conic quadratic sets. In: Goemans, M., Correa, J. (eds.) IPCO 2013. LNCS, vol. 7801, pp. 37–48. Springer, Heidelberg (2013). https://doi.org/10.1007/978-3-642-36694-9_4
2. Andersen, K., Louveaux, Q., Weismantel, R.: An analysis of mixed integer linear sets based on lattice point free convex sets. Math. Oper. Res. **35**(1), 233–256 (2010)

3. Andersen, K., Louveaux, Q., Weismantel, R., Wolsey, L.A.: Inequalities from two rows of a simplex tableau. In: Fischetti, M., Williamson, D.P. (eds.) IPCO 2007. LNCS, vol. 4513, pp. 1–15. Springer, Heidelberg (2007). https://doi.org/10.1007/978-3-540-72792-7_1

4. Balas, E.: Intersection cuts–a new type of cutting planes for integer programming. Oper. Res. **19**(1), 19–39 (1971). https://doi.org/10.1287/opre.19.1.19

5. Basu, A., Conforti, M., Cornuéjols, G., Zambelli, G.: Maximal lattice-free convex sets in linear subspaces. Math. Oper. Res. **35**(3), 704–720 (2010). https://doi.org/10.1287/moor.1100.0461

6. Basu, A., Conforti, M., Cornuéjols, G., Zambelli, G.: Minimal inequalities for an infinite relaxation of integer programs. SIAM J. Discrete Math. **24**(1), 158–168 (2010)

7. Bienstock, D., Chen, C., Muñoz, G.: Intersection cuts for polynomial optimization. In: Lodi, A., Nagarajan, V. (eds.) IPCO 2019. LNCS, vol. 11480, pp. 72–87. Springer, Cham (2019). https://doi.org/10.1007/978-3-030-17953-3_6

8. Bienstock, D., Chen, C., Gonzalo, M., et al.: Outer-product-free sets for polynomial optimization and oracle-based cuts. Math. Program. **183**, 105–148 (2020). https://doi.org/10.1007/s10107-020-01484-3

9. Bonami, P., Linderoth, J., Lodi, A.: Disjunctive cuts for mixed integer nonlinear programming problems. Prog. Comb. Optim., 521–541 (2011)

10. Borozan, V., Cornuéjols, G.: Minimal valid inequalities for integer constraints. Math. Oper. Res. **34**(3), 538–546 (2009). https://doi.org/10.1287/moor.1080.0370

11. Burer, S., Fatma, K.K.: How to convexify the intersection of a second order cone and a nonconvex quadratic. Math. Program. **162**, 393–429 (2016). https://doi.org/10.1007/s10107-016-1045-z

12. Chmiela, A., Muñoz, G., Serrano, F.: On the implementation and strengthening of intersection cuts for QCQPs. Online preprint (2020). https://opus4.kobv.de/opus4-zib/frontdoor/index/index/docId/7999

13. Conforti, M., Cornuéjols, G., Daniilidis, A., Lemaréchal, C., Malick, J.: Cut-generating functions and S-free sets. Math. Oper. Res. **40**(2), 276–391 (2015). https://doi.org/10.1287/moor.2014.0670

14. Cornuéjols, G., Wolsey, L., Yıldız, S.: Sufficiency of cut-generating functions. Math. Program. **152**(1), 643–651 (2014). https://doi.org/10.1007/s10107-014-0780-2

15. Dey, S.S., Wolsey, L.A.: Lifting integer variables in minimal inequalities corresponding to lattice-free triangles. In: Lodi, A., Panconesi, A., Rinaldi, G. (eds.) IPCO 2008. LNCS, vol. 5035, pp. 463–475. Springer, Heidelberg (2008). https://doi.org/10.1007/978-3-540-68891-4_32

16. Fischetti, M., Ljubić, I., Monaci, M., Sinnl, M.: Intersection cuts for bilevel optimization. In: Louveaux, Q., Skutella, M. (eds.) IPCO 2016. LNCS, vol. 9682, pp. 77–88. Springer, Cham (2016). https://doi.org/10.1007/978-3-319-33461-5_7

17. Fischetti, M., Monaci, M.: A branch-and-cut algorithm for mixed-integer bilinear programming. Eur. J. Oper. Res. **282**, 506–514 (2019). https://doi.org/10.1016/j.ejor.2019.09.043

18. Glover, F.: Polyhedral convexity cuts and negative edge extensions. Zeitschrift für Oper. Res. **18**(5), 181–186 (1974)

19. Glover, F.: Convexity cuts and cut search. Oper. Res. **21**(1), 123–134 (1973). https://doi.org/10.1287/opre.21.1.123

20. Gomory, R.E., Johnson, E.L.: Some continuous functions related to corner polyhedra. Math. Program. **3**(1), 23–85 (1972). https://doi.org/10.1007/bf01584976

21. Kılınç-Karzan, F.: On minimal valid inequalities for mixed integer conic programs. Math. Oper. Res. **41**(2), 477–510 (2015)

22. McCormick, G.P.: Computability of global solutions to factorable nonconvex programs: Part i – convex underestimating problems. Math. Program. **10**(1), 147–175 (1976). https://doi.org/10.1007/bf01580665
23. MINLP library. http://www.minlplib.org/
24. Modaresi, S., Kılınç, M.R., Vielma, J.P.: Split cuts and extended formulations for mixed integer conic quadratic programming. Oper. Res. Lett. **43**(1), 10–15 (2015)
25. Modaresi, S., Kılınç, M.R., Vielma, J.P.: Intersection cuts for nonlinear integer programming: convexification techniques for structured sets. Math. Program. **155**(1–2), 575–611 (2016)
26. Muñoz, G., Serrano, F.: Maximal quadratic-free sets. In: Bienstock, D., Zambelli, G. (eds.) IPCO 2020. LNCS, vol. 12125, pp. 307–321. Springer, Cham (2020). https://doi.org/10.1007/978-3-030-45771-6_24
27. Santana, A., Dey, S.S.: The convex hull of a quadratic constraint over a polytope. arXiv preprint arXiv:1812.10160 (2018)
28. SCIP - Solving Constraint Integer Programs. http://scip.zib.de
29. Sen, S., Sherali, H.D.: Facet inequalities from simple disjunctions in cutting plane theory. Math. Program. **34**(1), 72–83 (1986). https://doi.org/10.1007/bf01582164
30. Sen, S., Sherali, H.D.: Nondifferentiable reverse convex programs and facetial convexity cuts via a disjunctive characterization. Math. Program. **37**(2), 169–183 (1987)
31. Serrano, F.: Intersection cuts for factorable MINLP. In: Lodi, A., Nagarajan, V. (eds.) IPCO 2019. LNCS, vol. 11480, pp. 385–398. Springer, Cham (2019). https://doi.org/10.1007/978-3-030-17953-3_29
32. Towle, E., Luedtke, J.: Intersection disjunctions for reverse convex sets. arXiv preprint arXiv:1901.02112 (2019)
33. Tuy, H.: Concave programming with linear constraints. In: Doklady Akademii Nauk, vol. 159, pp. 32–35. Russian Academy of Sciences (1964)

Lifting Convex Inequalities for Bipartite Bilinear Programs

Xiaoyi Gu[1]([✉]), Santanu S. Dey[1], and Jean-Philippe P. Richard[2]

[1] Georgia Institute of Technology, Atlanta, GA 30332, USA
xiaoyigu@gatech.edu, santanu.dey@isye.gatech.edu
[2] University of Minnesota, Minneapolis, MN 55455, USA
jrichar@umn.edu

Abstract. The goal of this paper is to derive new classes of valid convex inequalities for quadratically constrained quadratic programs (QCQPs) through the technique of lifting. Our first main result shows that, for sets described by one bipartite bilinear constraint together with bounds, it is always possible to lift a seed inequality that is valid for a restriction obtained by fixing variables to their bounds, when the lifting is accomplished using affine functions of the fixed variables. In this setting, sequential lifting involves solving a non-convex nonlinear optimization problem each time a variable is lifted, just as in Mixed Integer Linear Programming. To reduce the computational burden associated with this procedure, we develop a framework based on subadditive approximations of lifting functions that permits sequence independent lifting of seed inequalities for separable bipartite bilinear sets. In particular, this framework permits the derivation of closed-form valid inequalities. We then study a separable bipartite bilinear set where the coefficients form a minimal cover with respect to right-hand-side. For this set, we derive a "bilinear cover inequality", which is second-order cone representable. We argue that this bilinear covering inequality is strong by showing that it yields a constant-factor approximation of the convex hull of the original set. We study its lifting function and construct a two-slope subadditive upper bound. Using this subadditive approximation, we lift fixed variable pairs in closed-form, thus deriving a "lifted bilinear cover inequality" that is valid for general separable bipartite bilinear sets with box constraints.

Keywords: Lifting · Bipartite bilinear sets · Subadditivity

1 Introduction

1.1 Generating Strong Cutting Planes Through Lifting

Lifting is a technique that is used to derive or strengthen classes of cutting planes. It was first introduced to optimization in the context of mixed integer linear programming (MILP); see [44] for a review. The lifting process has two steps:

© Springer Nature Switzerland AG 2021
M. Singh and D. P. Williamson (Eds.): IPCO 2021, LNCS 12707, pp. 148–162, 2021.
https://doi.org/10.1007/978-3-030-73879-2_11

– *Fixing* and generation of a *seed inequality*: In the first step, the set S of interest is restricted by fixing a subset of variables, say x^F, to different values (typically to one of their bounds), say $\tilde{x}^F$. A valid inequality $h(x) \geq h_0$, which we call *seed inequality*, is then generated for the restriction $S|_{x^F = \tilde{x}^F}$.

– *Lifting* the seed inequality: The seed inequality $h(x) \geq h_0$, when viewed with "zero coefficients" for the fixed variables $h(x) + 0 \cdot x^F \geq h_0$ is typically not valid for the original set S. The task in the "lifting step" is to generate an inequality $h(x) + g(x^F) \geq h_0 + g_0$, which (i) is valid for S and (ii) satisfies $g(\tilde{x}^F) = g_0$. Under condition (ii), inequality $h(x) + g(x^F) \geq h_0 + g_0$ reduces to inequality $h(x) \geq h_0$ when x^F is set to $\tilde{x}^F$. The process of lifting is often accomplished by rotating or titling the seed inequality [26].

Even though condition (ii) is not strictly necessary to impose, we require it in the remainder of the paper as otherwise $h(x) + g(x^F) \geq h_0 + g_0$ is weak on the face $x^F = \tilde{x}^F$.

Lifting, as a technique for generating cutting-planes in mixed integer programming, has been extensively researched. Originally devised for node packing and knapsack sets [7,10,33,41,42], lifting was extended to general MIP settings [4,29,30,46,47,52,53] and used to derive families of valid inequalities for many sets including [1,3,18,21,27,34,36,54–56] among many other examples. Many of the classes of cutting planes that have yielded significant computational gains can be obtained through lifting. This includes *lifted cover inequalities* [29], lifted tableaux cuts [21,39], and even the *Gomory mixed integer cut* [25]; see [6,9,11–13,13,14,20,23,24,48] for papers related to lifting in the infinite group problem model. Similarly, *mixing inequalities* [31] can be viewed as an outcome of lifting [23].

Significantly fewer articles have focused on studying how lifting can be applied to nonlinear programs and mixed integer nonlinear programs. Exceptions include [49], which develops a general theory for lifting linear inequalities in nonlinear programming, [40] which applies lifting to derive the convex hull of a nonlinear set, [32] which studies lifting for the pooling problem, [5] which uses lifting for conic integer programs, and [19] which develops strong inequalities for mixed integer bilinear programs.

1.2 Goal of This Paper

The goal of this paper is to derive new classes of valid convex inequalities for *quadratically constrained quadratic programs* (QCQPs) through the technique of lifting.

Generating valid inequalities for single row relaxations (together with bounds and integrality restrictions), *i.e.*, for knapsack constraints, was the first, and arguably the most important step in the development of computationally useful cutting-planes in MILP. Motivated by this observation, various cutting-planes and convexification techniques for sets defined by a single non-convex quadratic constraint together with bounds have recently been investigated; see [2,19,51] for classes of valid inequalities for single constraint QCQPs and [22,50] for convex hull results for such sets. The paper [43] studies a set similar to the one we study, albeit with integer variables. Further, [22] demonstrates that cuts obtained from one-row relaxations of QCQPs can be useful computationally. The paradigm of intersection cuts has also been explored to generate cuts for single-constraint QCQPs [16,38]. Due to lack of space, we refrain from describing here the vast literature on convexification techniques for QCQPs and instead refer interested readers to [17,50] and the references therein.

In this paper, we investigate the lifting of a convex seed inequality for a feasible region defined by a single (non-convex) quadratic constraint together with bound constraints. Apart from [5], we are not aware of any paper that attempts to study or employ lifting of convex non-linear inequalities. To the best of our knowledge, this is the first study that derives lifted valid inequalities for general non-convex quadratic constraints with arbitrary number of variables.

1.3 Main Contributions

- *Can we always lift?* We first present a simple example in two variables that illustrates that, even when a set is defined by a convex quadratic constraint, it might not always be possible to lift a linear seed inequality, valid for the restriction obtained by fixing a variable at lower bound, when we assume $g(\cdot) - g_0$ is an affine function of the fixed variable. Our first main result, by contrast, establishes that there exists a large class of sets, described through a single *bipartite bilinear* constraint [22] together with bounds, for which it is always possible to lift when variables were fixed at their bounds. We note that any quadratic constraint can be relaxed to produce a bipartite bilinear constraint.
- *Sequence-independent lifting.* The lifting of a fixed variable requires the solution of a non-convex nonlinear optimization problem. When multiple variables must be lifted one at a time, this process (sometimes referred to as *sequential lifting*) can be computationally prohibitive. In this setting, the form of the lifted inequality obtained will differ depending on the order in which variables are lifted. For MILPs, it was shown in [53] that when the so-called *lifting function* (or a suitable approximation) is subadditive, lifting is far more computationally tractable in part because the form of the lifted inequality is independent of the order in which variables are lifted. We develop a similar general result for sequence-independent lifting of seed inequalities for *separable bipartite bilinear* constraints.
- *Bilinear covering set and bilinear cover inequality.* We next study a *separable bipartite bilinear* set whose coefficients form a minimal cover with respect to right-hand-side. For this set, we derive a *"bilinear cover inequality."* This *second order cone representable inequality* produces a constant-factor approximation of the convex hull of the original set.
- *Sequence-independent lifting for the bilinear cover inequality.* We finally study the lifting function corresponding to the bilinear cover inequality and construct a *two-slope* subadditive upper bound. This function is reminiscent of the two-slope subadditive functions studied in the context of cutting-planes

for the infinite group relaxation [28,35,45], although there is no apparent connection between these sets and the infinite group model. Using this subadditive function, we lift fixed variable pairs in closed-form, thus describing a family of *"lifted bilinear cover inequalities,"* which are valid for general separable bipartite bilinear constraints.

1.4   Notation and Organization of the Paper

Given a positive integer n, we denote the set $\{1, \ldots, n\}$ by $[n]$. Given a set $S \subseteq \mathbb{R}^n$ and $\theta > 0$, we use $\theta \cdot S$ to denote the set $\{\theta x \mid x \in S\}$. We also use $\text{conv}(S)$ to denote the convex hull of set S.

The rest of the paper is organized as follows: In Sect. 2 we present our main results. In Sect. 3 we discuss some key directions for future research. Due to lack of space we do not present proofs.

2   Main Results

2.1   Sufficient Conditions Under Which Seed Inequalities Can Be Lifted

Before we introduce our first main result in Theorem 1, we first give an example of how lifting can be performed for a set defined through a quadratic constraint.

Example 1. Consider the set

$$S := \left\{ (x_1, x_2, x_3) \in [0,1]^3 \mid x_1 x_2 + 2 x_1 x_3 \geq 1 \right\}.$$

– Fixing and seed inequality: Fix $x_3 = 0$ to obtain the restriction $S|_{x_3=0} := \{(x_1, x_2) \mid x_1 x_2 \geq 1\}$. The seed inequality

$$\sqrt{x_1 x_2} \geq 1,$$

is a valid convex inequality for $S|_{x_3=0}$.
– Lifting the seed inequality: Although valid for $S|_{x_3=0}$, the seed inequality is not valid for S, since $(x_1, x_2, x_3) = (1, 0, 1/2)$ violates it while belonging to S. We therefore must introduce variable x_3 into the seed inequality so as to make it valid. In particular we seek to determine whether there exist $\alpha \in \mathbb{R}$ such that

$$\sqrt{x_1 x_2} + \alpha x_3 \geq 1, \tag{1}$$

is valid for S. This question can be answered by solving the problem

$$\alpha^* := \sup \frac{1 - \sqrt{x_1 x_2}}{x_3}$$

$$\text{s.t.} \, x_1 x_2 + 2 x_1 x_3 \geq 1 \tag{2}$$

$$x_3 \in (0,1], \; x_1, x_2 \in [0,1],$$

where the key challenge is to first ascertain that the supremum is finite. When α^* is finite, it is clear that choosing any $\alpha \geq \alpha^*$ in (1) will yield a valid inequality for S. Problem (2) can be analyzed using the following facts: (1) for any fixed value of x_3, we can always assume that an extreme point is the optimal solution, as the objective is to maximize a convex function, and (2) the extreme points of the set where x_3 is fixed to a value within its bounds are well-understood [50]. This suggests that one can inspect all different values of x_3 to establish that the supremum is finite. We illustrate these calculations next.

We obtain α^* by computing the supremum α_1^* of (2) for $x_3 \in [1/2, 1]$ and then computing the supremum α_2^* of (2) for $x_3 \in (0, 1/2]$. When $x_3 \in [1/2, 1]$, one optimal solution is $x_1 = \frac{1}{2x_3}$ and $x_2 = 0$, thus $\alpha_1^* = \sup_{x_3 \in [1/2,1]} \frac{1}{x_3} = 2$. When $x_3 \in (0, 1/2]$, one optimal solution is $x_1 = 1$ and $x_2 = 1 - 2x_3$, thus

$$\alpha_2^* = \sup_{x_3 \in (0, 1/2]} \frac{1 - \sqrt{1 - 2x_3}}{x_3} = \sup_{x_3 \in (0, 1/2]} \frac{2}{1 + \sqrt{1 - 2x_3}} = 2.$$

Choosing any $\alpha \geq \alpha^* = \max\{\alpha_1^*, \alpha_2^*\} = 2$ yields a valid inequality for S. The strongest such valid inequality is

$$\sqrt{x_1 x_2} + 2x_3 \geq 1.$$

The above example might suggest that lifting can always be performed when seeking to derive a linear valid inequality. The following example shows that it is not so, even when fixing inequalities at bounds.

Example 2. Consider the set

$$S = \left\{ (x_1, x_2) \in [0, 1]^2 \mid -x_1^2 - (x_2 - 0.5)^2 \geq -0.5^2 \right\}.$$

The inequality $-x_1 \geq 0$ is valid for the set obtained by fixing $x_2 = 0$. By setting up an optimization problem similar to (2), it is easy to verify that there is no $\alpha \in \mathbb{R}$ for which $-x_1 + \alpha x_2 \geq 0$ is valid for S.

In Example 2, (i) set S is convex, (ii) we are trying to lift a linear inequality, and (iii) x_2 is fixed to a bound. Even then, it is not possible to lift the seed inequality when we insist that the lifting should be accomplished using an affine function of the fixed variable; see Example 4 in Sect. 3 for further discussion.

   Our first result identifies a large class of single row QCQPs where lifting can be accomplished using affine functions of the fixed variables.

Theorem 1. *Consider a set described by one bipartite[1] bilinear constraint and bounds on variables:*

$$S = \{(x, y) \in [0, 1]^m \times [0, 1]^n \mid x^{\mathsf{T}} Q y + a^{\mathsf{T}} x + b^{\mathsf{T}} y \geq c\},$$

---

[1] We use the term bipartite, perhaps redundantly, to highlight that variables can be divided into two groups, such that any degree two term comes from product of variables one each from these two groups [22].

where $Q \in \mathbb{R}^{m \times n}$, $a \in \mathbb{R}^m$, $b \in \mathbb{R}^n$, and $c \in \mathbb{R}$. Given $C \times D \subset [m] \times [n]$ and $\tilde{x}_i, \tilde{y}_j \in \{0,1\}$ for $i \in [m] \backslash C$, $j \in [n] \backslash D$, assume that inequality

$$h(x_C, y_D) \geq r$$

is valid for $\{(x,y) \in S \mid x_{[m] \backslash C} = \tilde{x}_{[m] \backslash C}, y_{[n] \backslash D} = \tilde{y}_{[n] \backslash D}\} \neq \emptyset$, where h is a concave function defined on $[0,1]^{|C|+|D|}$. Then, for any $k \in [m] \backslash C$, there exists a finite $f_k \in (-\infty, \infty)$ for which

$$h(x_C, y_D) + f_k x_k \geq r + f_k \tilde{x}_k$$

is valid for $\{(x,y) \in S \mid x_{([m] \backslash C) \backslash \{k\}} = \tilde{x}_{([m] \backslash C) \backslash \{k\}}, y_{[n] \backslash D} = \tilde{y}_{[n] \backslash D}\}$.

Remark 1. The result of Theorem 1 can be applied iteratively to all the fixed variables one at a time to obtain a valid inequality for S. Also note that Theorem 1 holds even when the bounds on variables are not $[0,1]$, since we can always rescale and translate variables.

The proof of Theorem 1 uses calculations similar to those presented in Example 1. In particular, using a characterization of extreme points of the bipartite bilinear set S [22], the proof reduces to establishing the result for three-variable problems where one of the variables is fixed. For a three-variable problem, a number of cases have to be analyzed to verify that the optimal objective function value of the optimization problem similar to (2) is finite. We mention that this proof can be turned into an algorithm to compute the lifting coefficients, although not necessarily an efficient or practical one.

Theorem 1 assumes that, when variables x and y are fixed, they are fixed at their bounds (either 0 or 1.) When this assumption is not imposed, we show next through an example that lifting may not be possible.

Example 3. Inequality $x \geq 3/4$ is valid for the bipartite bilinear problem obtained by fixing $\hat{x} = 1/2$ in

$$S = \left\{ (x,y,\hat{x}) \in [0,1]^3 \ \middle| \ (x - 1/4)(y - 1/2) \geq \hat{x}/4 + 1/8 \right\}.$$

Further, there is no $\alpha \in \mathbb{R}$ such that $x + \alpha(\hat{x} - 1/2) \geq 3/4$ is valid for S.

2.2   A Framework for Sequence-Independent Lifting

Given a set of variables fixed at their bounds and a seed inequality for the corresponding restriction, a valid inequality for the original problem can be obtained by lifting each fixed variable one at the time. This computationally demanding process requires the solution of a non-convex nonlinear optimization problem, similar to (2), to lift each variable. It results in a lifted inequality whose form depends on the order in which variables are lifted. Next, we study situations where the lifting inequality obtained does not depend on the order in which variables are lifted. In particular, we develop a subadditive theory for lifting in QCQPs that is inspired by that originally developed in MILP in [53]. We consider the special case of the *separable bipartite bilinear constraint* defined below.

Definition 1. *A set Q is said to be a separable bipartite bilinear set if it is of the form*

$$Q := \left\{ (x, y) \in [0, 1]^n \times [0, 1]^n \;\middle|\; \sum_{i=1}^{n} a_i x_i y_i \geq d \right\},$$

where d and $a_i \in \mathbb{R}$ for $i \in [n]$, i.e., each variable x_i or y_i, for $i \in [n]$, appears in only one term.

In the separable case, it is natural to lift each pair of variables x_i and y_i together. Next, we derive conditions on the lifting function of a seed inequality that guarantees that the form of the lifted inequality obtained is independent of the order in which these pairs are lifted. This result is obtained, as is common in MILP, by deriving a subadditive upper bound on the lifting function of the seed inequality, from which all lifting coefficients can be derived.

Proposition 1. *Let Q be a separable bipartite bilinear set. Assume that $\Lambda = \{I, J_0, J_1\}$ is a partition of $[n]$, i.e., $I \cup J_0 \cup J_1 = [n]$ with $I \cap J_0 = I \cap J_1 = J_0 \cap J_1 = \emptyset$, and that*

$$h(x_I, y_I) \geq r$$

is a valid inequality for $\{(x, y) \in Q \mid x_{J_0} = y_{J_0} = 0, x_{J_1} = y_{J_1} = 1\}$ where h is a concave function. For $\delta \in \mathbb{R}$, compute the lifting function

$$\phi(\delta) := \max \left\{ r - h(x_I, y_I) \;\middle|\; \sum_{i \in I} a_i x_i y_i \geq \left(d - \sum_{i \in J_1} a_i \right) - \delta, x, y \in [0, 1]^n \right\}.$$

Let $\psi : \mathbb{R} \mapsto \mathbb{R}$ be such that (i) $\psi(\delta) \geq \phi(\delta)$, $\forall \delta \in \mathbb{R}$; (ii) ψ subadditive, (i.e., $\psi(\delta_1) + \psi(\delta_2) \geq \psi(\delta_1 + \delta_2)$, $\forall \delta_1, \delta_2 \in \mathbb{R}$) with $\psi(0) = 0$; (iii) for any $i \in J_0$, there exists a concave function $\gamma_i(x, y)$ such that $\gamma_i(x, y) \geq \psi(a_i x y)$, $\forall x, y \in [0, 1]$, and for any $i \in J_1$, there exists a concave function $\gamma_i(x, y)$ such that $\gamma_i(x, y) \geq \psi(a_i x y - a_i)$, $\forall x, y \in [0, 1]$. Then, the lifted inequality

$$h(x_I, y_I) + \sum_{i \in J_0 \cup J_1} \gamma_i(x_i, y_i) \geq r$$

is a valid convex inequality for Q.

The statement of Proposition 1 does not specify the type of functional forms $\gamma_i(x_i, y_i)$ to use in ensuring that condition (iii) is satisfied. It is however clear from the definition that choosing $\gamma_i(x, y)$ to be the concave envelope of $\psi(a_i x y)$ over $[0, 1]^2$ when $i \in J_0$, and the concave envelope of $\psi(a_i x y - a_i)$ over $[0, 1]^2$ when $i \in J_1$ is the preferred choice for γ_i.

Remark 2. While we state the result of Proposition 1 for a set Q defined by a single separable bipartite bilinear constraint, a similar result would also hold for sets defined by multiple separable bipartite bilinear constraints.

2.3   A Seed Inequality from a "Minimal Covering Set"

To generate lifted inequalities for separable bipartite bilinear sets, we focus next on a family of restrictions we refer to as *minimal covering sets*. For such minimal covering sets, we introduce a provably strong convex, second-order cone representable valid inequality. We will use this inequality as the seed in our lifting procedures.

Definition 2. *Let $k \in \mathbb{Z}_+$ be a positive integer. We say that $a_i \in \mathbb{R}$, $i \in [k]$ form a minimal cover of $d \in \mathbb{R}$, if*

1. *$a_i > 0$ for all $i \in [k]$, $d > 0$,*
2. *$\sum_{i=1}^{k} a_i > d$*
3. *$\sum_{i \in K \subsetneq [k]} a_i \leq d$.*

For a separable bipartite bilinear set Q, we say that a partition $\Lambda = \{I, J_0, J_1\}$ of $[n]$, where $I \neq \emptyset$, is a minimal cover yielding partition if: a_i, $i \in I$ form a minimal cover of $d^\Lambda := d - \sum_{i \in J_1} a_i$. For a minimal cover yielding partition, we let $J_0^+ := \{i \in J_0 | a_i > 0\}$, $J_0^- := \{i \in J_0 | a_i < 0\}$; we define J_1^+ and J_1^- similarly.

Remark 3. When $k \geq 2$, conditions (2.) and (3.) in the definition of minimal cover imply condition (1.). For example, if $a_i \leq 0$ for some $i \in [k]$, then (2.) implies $\sum_{j \in [k] \setminus \{i\}} a_j > d$, contradicting (3.) Now (3.) together with $a_i > 0$ for $i \in [k]$ implies $d > 0$.

Our overall plan is the following. We will fix $x_i y_i = 0$ for $i \in J_0$ and $x_i y_i = 1$ for $i \in J_1$. Then, we will find a valid seed inequality for the set where the coefficients form a minimal cover (I). Finally, we will lift this seed inequality. One key reason to generate cuts from a seed inequality corresponding to a minimal cover is the following result.

Theorem 2. *For a separable bilinear set Q, either there exists at least one minimal cover yielding partition or we have that $Q = \emptyset$ or that $\mathrm{conv}(Q)$ is polyhedral.*

Loosely speaking, the proof of the above theorem is based on showing that if there is no minimal cover yielding partition, then Q is "almost" a packing-type set, *i.e.*, a set of the form $\{x, y \in [0,1]^n \mid \sum_{i=1}^{n} a_i x_i y_i \leq d\}$ where a_is are non-negative.[2] For packing sets Q, it is shown in [49] that $\mathrm{conv}(Q) = \mathrm{proj}_{x,y}(G)$ where

$$G = \left\{ (x, y, w) \in [0,1]^{3n} \ \middle| \ \sum_{i=1}^{n} a_i w_i \leq d, \ x_i + y_i - 1 \leq w_i, \ \forall i \in [n] \right\}.$$

---

[2] We say "almost", since there are non-packing examples, such as $S := \{x, y \in [0,1]^2 \mid x_1 y_1 - 100 x_2 y_2 \geq -98\}$, where there is no partition that yields a minimal cover. Such sets are "overwhelmingly" like a packing set; in the case of the example, it is a perturbation of the packing set $\{x_2, y_2 \in [0,1] \mid 100 x_2 y_2 \leq 98\}$. For such sets it is not difficult to show that $\mathrm{conv}(S)$ is polyhedral.

Since the main focus of this paper is the study of lifting convex (nonlinear) inequalities and since in the packing case the convex hull is trivially obtained using McCormick inequalities [37], the remainder of the paper will concentrate on the case where there exists a minimal cover yielding partition.

Associated with a minimal cover is a specific convex valid inequality that we present next.

Theorem 3. *Consider the bipartite bilinear minimal covering set*

$$Q := \left\{ (x, y) \in [0, 1]^n \times [0, 1]^n \mid \sum_{i=1}^{n} a_i x_i y_i \geq d \right\}, \tag{3}$$

where the a_i, $i \in [n]$ form a minimal cover of d. Then the inequality

$$\sum_{i=1}^{n} \frac{\sqrt{a_i}}{\sqrt{a_i} - \sqrt{d_i}} \left(\sqrt{x_i y_i} - 1 \right) \geq -1, \tag{4}$$

which we refer to as bilinear cover inequality is valid for Q, where $d_i = d - \sum_{j \in [n] \setminus \{i\}} a_j$.

Our proof of Theorem 2 uses techniques from disjunctive programming [8] and an "approximate version" of Fourier-Motzkin projection. In particular, using the minimal covering property of the coefficients in (3) and a characterization of the extreme points of bipartite bilinear sets [22], we obtain n second order cone representable sets whose union contains all the extreme points of (3). Next we set up an extended formulation [8,15] of the convex hull of the union of these sets. Finally, we use the Fourier-Motzkin procedure to project out the auxiliary variables of the extended formulation one at a time. This procedure works to project out most of the variables. The last steps however require a relaxation to be constructed so that projection can be carried in closed-form. We obtain (4).

It can be easily verified that (4) does not produce the convex hull of Q. However there are a number of reasons to use this inequality as a seed for lifting. The first reason is that, not only is inequality (4) second-order cone representable, we only need to introduce one extra variable representing $\sqrt{x_i y_i}$ for each $i \in [n]$, to write it as a second order cone representable set. Apart from the convenience of using this inequality within modern conic solvers, the main reason for considering it as a seed inequality is its strength. In particular, we prove next that (4) provides a constant factor approximation of the convex hull of the original set.

Theorem 4. *Let Q be a bipartite bilinear minimal covering set as described in (3). Let $R := \{(x, y) \in \mathbb{R}_+^n \times \mathbb{R}_+^n \mid (4)\}$. Then*

$$(4 \cdot R) \cap [0, \ 1]^n \subseteq \text{conv}(Q) \subseteq R \cap [0, \ 1]^n.$$

Since R is a covering-type set (that is, the recession cone is the non-negative orthant), we have that $4 \cdot R \subseteq R$. To give intuition as to why Theorem 4 holds, we note that in any feasible solution of (3), $x_i y_i \geq d_i / a_i$ for $i \in [n]$. This condition is also enforced by the bilinear cover inequality (4). Therefore, there is no point in the set R that is very close to the origin, so that it must be substantially rescaled to be in convex hull of (3). The proof of Theorem 4 is based on optimizing linear functions with non-negative coefficients on R and Q and proving a bound of 4 on the ratio of their optimal objective function values.

A set similar to (3) is studied in [51], except that there are no upper bounds on the variables. In this case, the convex hull can be described in closed form using a single nonlinear inequality that is similar in structure to (4). We can formally verify, however, that inequality (4), which also uses the information of upper bounds, dominates the inequality presented in [51]. Moreover, if $n \geq 2$ and their exists $i \in [n]$ such that $d_i > 0$, then (4) strictly dominates the inequality presented in [51].

2.4 Lifting the Bilinear Cover Inequality (4)

We now follow the steps of Proposition 1 to perform sequence-independent lifting of the bilinear cover inequality. The first step is to study the lifting function associated with (4).

Theorem 5. *Consider the lifting function for valid inequality (4):*

$$\phi(\delta) := \max \sum_{i=1}^{n} \frac{\sqrt{a_i}}{\sqrt{a_i} - \sqrt{d_i}} \left(1 - \sqrt{x_i y_i}\right) - 1$$

$$\text{s.t.} \sum_{i=1}^{n} a_i x_i y_i \geq d - \delta, \ x, y \in [0,1]^n.$$

Let $\Delta := \sum_{i=1}^{n} a_i - d$ and let $a_{i_0} = \min\{a_i | a_i > \Delta\}$ if it exists. Define

$$\psi(\delta) := \begin{cases} l_+ \delta & 0 \leq \delta \\ l_- \delta & -\Delta \leq \delta \leq 0 \\ l_+(\delta + \Delta) - 1 & \delta \leq -\Delta, \end{cases} \qquad (5)$$

where $l_+ = \frac{\sqrt{a_{i_0}} + \sqrt{d_{i_0}}}{\Delta \sqrt{d_{i_0}}}$ if a_{i_0} exists and $l_+ = \frac{1}{\Delta}$ otherwise, and where $l_- = \frac{1}{\Delta}$. Then (i) $l_+ \geq l_- > 0$, (ii) $\psi(\delta)$ is subadditive over $\mathbb{R}$ with $\psi(0) = 0$, and (iii) $\phi(\delta) \leq \psi(\delta)$ for $\delta \in \mathbb{R}$.

Although computing the lifting function for an arbitrary valid inequality, in general, appears to be a difficult task, the bilinear cover inequality (4) has sufficient structure that we can derive a strong subadditive upper bound in Theorem 5. The key to proving Theorem 5 is to first obtain the lifting function exactly in a region around the origin, and to argue that the linear upper bound of the lifting function for this region upper bounds the lifting function globally. Figure 1 presents examples of the lifting function ϕ, and the upper bound ψ we derived in Theorem 5 for the two cases when a_{i_0} exists and when it does not.

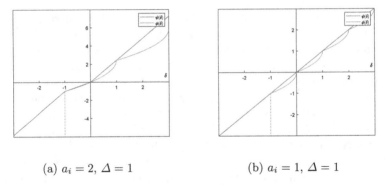

(a) $a_i = 2$, $\Delta = 1$ (b) $a_i = 1$, $\Delta = 1$

Fig. 1. Lifting function $\phi(\delta)$ in red and subadditive upper bound $\psi(\delta)$ in blue (Color figure online)

We observe in Fig. 1 that the lifting function is not subadditive since it is convex in a neighborhood of the origin. Therefore, building a subadditive approximation is required to achieve sequence-independent lifting.

Building on the subadditive upper bound derived in Theorem 5, we are now able to lift the bilinear cover inequality in a sequence-independent manner.

Theorem 6. *Consider the separable bipartite bilinear set presented in Definition 1. Let $\Lambda = \{I, J_0, J_1\}$ be a minimal cover yielding partition and let $\Delta, a_{i_0}, d_i, l_+, l_-$ be defined as in Theorem 3 and Theorem 5 (We clarify that they are calculated using d^Λ instead of d). Let J_0^+, J_0^-, J_1^+, and, J_1^- be as in Definition 2. Then inequality*

$$\sum_{i \in I} \frac{\sqrt{a_i}}{\sqrt{a_i} - \sqrt{d_i}} \left(\sqrt{x_i y_i} - 1 \right) + \sum_{i \notin I} \gamma_i(x_i, y_i) \geq -1, \tag{6}$$

is valid for Q where $\gamma_i : \mathbb{R}^2 \to \mathbb{R}$ for $i \in [n] \setminus I$ are the concave functions:

1. $\gamma_i(x, y) = l_+ a_i \min\{x, y\}$ *for $i \in J_0^+$;*
2. $\gamma_i(x, y) = -l_+ a_i \min\{2 - x - y, 1\}$ *for $i \in J_1^-$;*
3. $\gamma_i(x, y) = \min\{-l_- a_i(1 - x - y), -l_+ a_i(1 - x - y) + l_+ \Delta - 1, 0\}$ *for $i \in J_0^-$;*
4. $\gamma_i(x, y) = \min\{\tilde{g}_i(x, y), \tilde{h}_i(x, y), g_i(x, y), h_i(x, y)\}$, *for $i \in J_1^+$ with $a_i > a_{i_0}$ when a_{i_0} exists, and $\gamma_i(x, y) = \min\{\tilde{g}_i(x, y), \tilde{h}_i(x, y)\}$ in all other cases where $i \in J_1^+$, with*

$$\tilde{g}_i(x, y) = l_+ a_i(\min\{x, y\} - 1) + l_+ \Delta - 1$$
$$\tilde{h}_i(x, y) = l_- a_i(\min\{x, y\} - 1)$$
$$g_i(x, y) = \sqrt{a_i - \Delta}\sqrt{a_i} l_+ + \sqrt{xy} - l_+(a_i - \Delta) - 1$$
$$h_i(x, y) = \frac{\sqrt{a_i}}{\sqrt{a_i} - \sqrt{d_i}}(\sqrt{xy} - 1).$$

We refer to inequality (6) as *lifted bilinear cover inequality*. We note here that the lifted inequality (6) is second order cone representable. As discussed before, the key to proving Theorem 6 is to obtain good approximations of concave envelope of the functions $\psi(\cdot)$. In particular, in the case of J_1^+, is is possible to obtain a $\sqrt{xy}$ type-term which appears to be incomparable with affine terms.

3   Future Directions

The results presented in this paper open up new avenues for generating cutting-planes for QCQPs. They also raise new theoretical and computational questions that can be investigated.

To illustrate this assertion, we revisit next Example 2.

Example 4. Consider $S := \{(x_1, x_2) \in [0, 1]^2 \mid -x_1^2 - (x_2 - 0.5)^2 \geq -0.5^2\}$ with the same fixing as in Example 2, *i.e.*, $x_2 = 0$. For the associated restriction, we consider the seed inequality $-x_1 \geq 0$.

In contrast to our earlier discussion, consider now the problem of lifting this seed inequality into an inequality of the form $-x_1 + \alpha\sqrt{x_2} \geq 0$. Finding the values of α that generate a valid inequality is equivalent to solving the problem

$$\alpha^* := \sup \frac{x_1}{\sqrt{x_2}}$$
$$\text{s.t.} -x_1^2 - (x_2 - 0.5)^2 \geq -0.5^2$$
$$x_1 \in [0, 1], x_2 \in (0, 1].$$

Using constraint $-x_1^2 - (x_2 - 0.5)^2 \geq -0.5^2$ we can bound the objective function as follows:

$$\frac{x_1}{\sqrt{x_2}} \leq \frac{\sqrt{0.5^2 - (x_2 - 0.5)^2}}{\sqrt{x_2}} = \frac{\sqrt{(1 - x_2)(x_2)}}{\sqrt{x_2}} = \sqrt{1 - x_2}.$$

It follows that selecting $\alpha \geq \alpha^* = 1$ yields a valid inequality for S. Note first that $\alpha < 0$ leads to an invalid inequality since $x_1 = 0, x_2 = 0.5$ is a feasible point. Moreover, any $\alpha \in [0, 1)$ yields an invalid inequality, since the point $(x_1 = \sqrt{x_2(1 - x_2)}, x_2 = 1 - ((1 + \alpha)/2)^2)$ is feasible. Therefore, the inequality

$$-x_1 + \sqrt{x_2} \geq 0,$$

is the strongest such lifted inequality.

The above example raises the question of obtaining a complete characterization of when one can accomplish lifting, *i.e.*, of generalizing Theorem 1 to situations where the functional form of the lifted variable is not necessarily linear. It would also be valuable to develop a theory to accomplish sequence-independent lifting in the more general case of bipartite bilinear programs, instead of just the separable case. Even for the separable case, one could also explore the possibility of constructing other subadditive upper bounds to the lifting function. On the computational side, one key question is to understand the complexity of separating the lifted bilinear cover inequality presented in Theorem 6 and to design efficient computational schemes to perform separation. Finally, extensive numerical experiments should be conducted to understand the strength of these inequalities and to determine how useful they can be in the solution of QCQPs. Given the strength of the seed inequality, we are hopeful that these lifted inequalities could yield non-trivial dual bound improvements.

References

1. Agra, A., Constantino, M.F.: Lifting two-integer knapsack inequalities. Math. Program. **109**(1), 115–154 (2007). https://doi.org/10.1007/s10107-006-0705-9
2. Anstreicher, K.M., Burer, S., Park, K.: Convex hull representations for bounded products of variables. arXiv preprint arXiv:2004.07233 (2020)
3. Atamtürk, A.: On the facets of the mixed-integer knapsack polyhedron. Math. Program. **98**(1), 145–175 (2003). https://doi.org/10.1007/s10107-003-0400-z
4. Atamtürk, A.: Sequence independent lifting for mixed-integer programming. Oper. Res. **52**(3), 487–490 (2004)
5. Atamtürk, A., Narayanan, V.: Lifting for conic mixed-integer programming. Math. Program. **126**(2), 351–363 (2011). https://doi.org/10.1007/s10107-009-0282-9
6. Averkov, G., Basu, A.: Lifting properties of maximal lattice-free polyhedra. Math. Program. **154**(1–2), 81–111 (2015). https://doi.org/10.1007/s10107-015-0865-6
7. Balas, E.: Facets of the knapsack polytope. Math. Program. **8**(1), 146–164 (1975). https://doi.org/10.1007/BF01580440
8. Balas, E.: Disjunctive programming: properties of the convex hull of feasible points. Discrete Appl. Math. **89**(1–3), 3–44 (1998)
9. Balas, E., Jeroslow, R.G.: Strengthening cuts for mixed integer programs. Eur. J. Oper. Res. **4**(4), 224–234 (1980)
10. Balas, E., Zemel, E.: Facets of the knapsack polytope from minimal covers. SIAM J. Appl. Math. **34**(1), 119–148 (1978)
11. Basu, A., Campêlo, M., Conforti, M., Cornuéjols, G., Zambelli, G.: Unique lifting of integer variables in minimal inequalities. Math. Program. **141**(1–2), 561–576 (2013). https://doi.org/10.1007/s10107-012-0560-9
12. Basu, A., Cornuéjols, G., Köppe, M.: Unique minimal liftings for simplicial polytopes. Math. Oper. Res. **37**(2), 346–355 (2012)
13. Basu, A., Dey, S.S., Paat, J.: Nonunique lifting of integer variables in minimal inequalities. SIAM J. Discrete Math. **33**(2), 755–783 (2019)
14. Basu, A., Paat, J.: Operations that preserve the covering property of the lifting region. SIAM J. Optim. **25**(4), 2313–2333 (2015)
15. Ben-Tal, A., Nemirovski, A.: Lectures on modern convex optimization: analysis, algorithms, and engineering applications. SIAM (2001)
16. Bienstock, D., Chen, C., Munoz, G.: Outer-product-free sets for polynomial optimization and oracle-based cuts. Math. Program. **183**, 1–44 (2020). https://doi.org/10.1007/s10107-020-01484-3
17. Burer, S.: A gentle, geometric introduction to copositive optimization. Math. Program. **151**(1), 89–116 (2015). https://doi.org/10.1007/s10107-015-0888-z
18. Ceria, S., Cordier, C., Marchand, H., Wolsey, L.A.: Cutting planes for integer programs with general integer variables. Math. Program. **81**(2), 201–214 (1998). https://doi.org/10.1007/BF01581105

19. Chung, K., Richard, J.P.P., Tawarmalani, M.: Lifted inequalities for 0–1 mixed-integer bilinear covering sets. Math. Program. **145**(1–2), 403–450 (2014). https://doi.org/10.1007/s10107-013-0652-1

20. Conforti, M., Cornuéjols, G., Zambelli, G.: A geometric perspective on lifting. Oper. Res. **59**(3), 569–577 (2011)

21. Dey, S.S., Richard, J.P.P.: Linear-programming-based lifting and its application to primal cutting-plane algorithms. INFORMS J. Comput. **21**(1), 137–150 (2009)

22. Dey, S.S., Santana, A., Wang, Y.: New SOCP relaxation and branching rule for bipartite bilinear programs. Optim. Eng. **20**(2), 307–336 (2019)

23. Dey, S.S., Wolsey, L.A.: Composite lifting of group inequalities and an application to two-row mixing inequalities. Discrete Optim. **7**(4), 256–268 (2010)

24. Dey, S.S., Wolsey, L.A.: Constrained infinite group relaxations of MIPs. SIAM J. Optim. **20**(6), 2890–2912 (2010)

25. Dey, S.S., Wolsey, L.A.: Two row mixed-integer cuts via lifting. Math. Program. **124**(1–2), 143–174 (2010). https://doi.org/10.1007/s10107-010-0362-x

26. Espinoza, D., Fukasawa, R., Goycoolea, M.: Lifting, tilting and fractional programming revisited. Oper. Res. Lett. **38**(6), 559–563 (2010)

27. Gómez, A.: Submodularity and valid inequalities in nonlinear optimization with indicator variables (2018)

28. Gomory, R.E., Johnson, E.L.: Some continuous functions related to corner polyhedra. Math. Program. **3**(1), 23–85 (1972). https://doi.org/10.1007/BF01584976

29. Gu, Z., Nemhauser, G.L., Savelsbergh, M.W.P.: Lifted flow cover inequalities for mixed 0–1 integer programs. Math. Program. **85**(3), 439–467 (1999). https://doi.org/10.1007/s101070050067

30. Gu, Z., Nemhauser, G.L., Savelsbergh, M.W.P.: Sequence independent lifting in mixed integer programming. J. Comb. Optim. **4**(1), 109–129 (2000)

31. Günlük, O., Pochet, Y.: Mixing mixed-integer inequalities. Math. Program. **90**(3), 429–457 (2001). https://doi.org/10.1007/PL00011430

32. Gupte, A.: Mixed integer bilinear programming with applications to the pooling problem. Ph.D. thesis, Georgia Institute of Technology (2012)

33. Hammer, P.L., Johnson, E.L., Peled, U.N.: Facet of regular 0–1 polytopes. Math. Program. **8**(1), 179–206 (1975). https://doi.org/10.1007/BF01580442

34. Kaparis, K., Letchford, A.N.: Local and global lifted cover inequalities for the 0–1 multidimensional knapsack problem. Eur. J. Oper. Res. **186**(1), 91–103 (2008)

35. Köppe, M., Zhou, Y.: An electronic compendium of extreme functions for the Gomory-Johnson infinite group problem. Oper. Res. Lett. **43**(4), 438–444 (2015)

36. Martin, A., Weismantel, R.: The intersection of knapsack polyhedra and extensions. In: Bixby, R.E., Boyd, E.A., Ríos-Mercado, R.Z. (eds.) IPCO 1998. LNCS, vol. 1412, pp. 243–256. Springer, Heidelberg (1998). https://doi.org/10.1007/3-540-69346-7_19

37. McCormick, G.P.: Computability of global solutions to factorable nonconvex programs: part I - convex underestimating problems. Math. Program. **10**(1), 147–175 (1976). https://doi.org/10.1007/BF01580665

38. Muñoz, G., Serrano, F.: Maximal quadratic-free sets. In: Bienstock, D., Zambelli, G. (eds.) IPCO 2020. LNCS, vol. 12125, pp. 307–321. Springer, Cham (2020). https://doi.org/10.1007/978-3-030-45771-6_24

39. Narisetty, A.K., Richard, J.P.P., Nemhauser, G.L.: Lifted tableaux inequalities for 0–1 mixed-integer programs: a computational study. INFORMS J. Comput. **23**(3), 416–424 (2011)

40. Nguyen, T.T., Richard, J.P.P., Tawarmalani, M.: Deriving convex hulls through lifting and projection. Math. Program. **169**(2), 377–415 (2018). https://doi.org/10.1007/s10107-017-1138-3

41. Padberg, M.W.: On the facial structure of set packing polyhedra. Math. Program. **5**(1), 199–215 (1973). https://doi.org/10.1007/BF01580121

42. Padberg, M.W.: A note on zero-one programming. Oper. Res. **23**(4), 833–837 (1975)

43. Rahman, H., Mahajan, A.: Facets of a mixed-integer bilinear covering set with bounds on variables. J. Global Optim. **74**(3), 417–442 (2019)

44. Richard, J.P.P.: Lifting techniques for mixed integer programming. In: Wiley Encyclopedia of Operations Research and Management Science (2010)

45. Richard, J.-P.P., Dey, S.S.: The group-theoretic approach in mixed integer programming. In: Jünger, M., et al. (eds.) 50 Years of Integer Programming 1958-2008, pp. 727–801. Springer, Heidelberg (2010). https://doi.org/10.1007/978-3-540-68279-0_19

46. Richard, J.P.P., de Farias Jr, I.R., Nemhauser, G.L.: Lifted inequalities for 0–1 mixed integer programming: basic theory and algorithms. Math. Program. **98**(1–3), 89–113 (2003). https://doi.org/10.1007/s10107-003-0398-2

47. Richard, J.P.P., de Farias Jr, I.R., Nemhauser, G.L.: Lifted inequalities for 0–1 mixed integer programming: superlinear lifting. Math. Program. **98**(1–3), 115–143 (2003). https://doi.org/10.1007/s10107-003-0399-1

48. Richard, J.P.P., Li, Y., Miller, L.A.: Valid inequalities for MIPs and group polyhedra from approximate liftings. Math. Program. **118**(2), 253–277 (2009). https://doi.org/10.1007/s10107-007-0190-9

49. Richard, J.P.P., Tawarmalani, M.: Lifting inequalities: a framework for generating strong cuts for nonlinear programs. Math. Program. **121**(1), 61–104 (2010). https://doi.org/10.1007/s10107-008-0226-9

50. Santana, A., Dey, S.S.: The convex hull of a quadratic constraint over a polytope. SIAM J. Optim. **30**(4), 2983–2997 (2020)

51. Tawarmalani, M., Richard, J.P.P., Chung, K.: Strong valid inequalities for orthogonal disjunctions and bilinear covering sets. Math. Program. **124**(1–2), 481–512 (2010). https://doi.org/10.1007/s10107-010-0374-6

52. Wolsey, L.A.: Facets and strong valid inequalities for integer programs. Oper. Res. **24**(2), 367–372 (1976)

53. Wolsey, L.A.: Valid inequalities and superadditivity for 0–1 integer programs. Math. Oper. Res. **2**(1), 66–77 (1977)

54. Zeng, B., Richard, J.-P.P.: A framework to derive multidimensional superadditive lifting functions and its applications. In: Fischetti, M., Williamson, D.P. (eds.) IPCO 2007. LNCS, vol. 4513, pp. 210–224. Springer, Heidelberg (2007). https://doi.org/10.1007/978-3-540-72792-7_17

55. Zeng, B., Richard, J.P.P.: A polyhedral study on 0–1 knapsack problems with disjoint cardinality constraints: facet-defining inequalities by sequential lifting. Discrete Optim. **8**(2), 277–301 (2011)

56. Zeng, B., Richard, J.P.P.: A polyhedral study on 0–1 knapsack problems with disjoint cardinality constraints: strong valid inequalities by sequence-independent lifting. Discrete Optim. **8**(2), 259–276 (2011)

A Computational Status Update for Exact Rational Mixed Integer Programming

Leon Eifler[1]([✉])(iD) and Ambros Gleixner[1,2](iD)

[1] Zuse Institute Berlin, Takustr. 7, 14195 Berlin, Germany
{eifler,gleixner}@zib.de
[2] HTW Berlin, Treskowallee 8, 10313 Berlin, Germany

Abstract. The last milestone achievement for the roundoff-error-free solution of general mixed integer programs over the rational numbers was a hybrid-precision branch-and-bound algorithm published by Cook, Koch, Steffy, and Wolter in 2013. We describe a substantial revision and extension of this framework that integrates symbolic presolving, features an exact repair step for solutions from primal floating-point heuristics, employs a faster rational LP solver based on LP iterative refinement, and is able to produce independently verifiable certificates of optimality. We study the significantly improved performance and give insights into the computational behavior of the new algorithmic components. On the MIPLIB 2017 benchmark set, we observe an average speedup of 6.6x over the original framework and 2.8 times as many instances solved within a time limit of two hours.

1 Introduction

It is widely accepted that mixed integer programming (MIP) is a powerful tool for solving a broad variety of challenging optimization problems and that state-of-the-art MIP solvers are sophisticated and complex computer programs. However, virtually all established solvers today rely on fast floating-point arithmetic. Hence, their theoretical promise of global optimality is compromised by roundoff errors inherent in this incomplete number system. Though tiny for each single arithmetic operation, these errors can accumulate and result in incorrect claims of optimality for suboptimal integer assignments, or even incorrect claims of infeasibility. Due to the nonconvexity of MIP, even performing an a posteriori analysis of such errors or postprocessing them becomes difficult.

In several applications, these numerical caveats can become actual limitations. This holds in particular when the solution of mixed integer programs is used as a tool in mathematics itself. Examples of recent work that employs MIP

The work for this article has been conducted within the Research Campus Modal funded by the German Federal Ministry of Education and Research (BMBF grant numbers 05M14ZAM, 05M20ZBM).

M. Singh and D. P. Williamson (Eds.): IPCO 2021, LNCS 12707, pp. 163–177, 2021.
https://doi.org/10.1007/978-3-030-73879-2_12

to investigate open mathematical questions include [11,12,18,28,29,32]. Some of these approaches are forced to rely on floating-point solvers because the availability, the flexibility, and most importantly the computational performance of MIP solvers with numerically rigorous guarantees is currently limited. This makes the results of these research efforts not as strong as they could be. Examples for industrial applications where the correctness of results is paramount include hardware verification [1] or compiler optimization [35].

The milestone paper by Cook, Koch, Steffy, and Wolter [16] presents a hybrid-precision branch-and-bound implementation that can still be considered the state of the art for solving general mixed integer programs exactly over the rational numbers. It combines symbolic and numeric computation and applies different dual bounding methods [19,31,33] based on linear programming (LP) in order to dynamically trade off their speed against robustness and quality.

However, beyond advanced strategies for branching and bounding, [16] does not include any of the supplementary techniques that are responsible for the strong performance of floating-point MIP solvers today. In this paper, we make a first step to address this research gap in two main directions.

First, we incorporate a *symbolic presolving* phase, which safely reduces the size and tightens the formulation of the instance to be passed to the branch-and-bound process. This is motivated by the fact that presolving has been identified by several authors as one of the components—if not *the* component—with the largest impact on the performance of floating-point MIP solvers [2,4]. To the best of our knowledge, this is the first time that the impact of symbolic preprocessing routines for general MIP is analyzed in the literature.

Second, we complement the existing dual bounding methods by enabling the use of *primal heuristics*. The motivation for this choice is less to reduce the total solving time, but rather to improve the usability of the exact MIP code in practical settings where finding good solutions earlier may be more relevant than proving optimality eventually. Similar to the dual bounding methods, we follow a hybrid-precision scheme. Primal heuristics are exclusively executed on the floating-point approximation of the rational input data. Whenever they produce a potentially improving solution, this solution is checked for approximate feasibility in floating-point arithmetic. If successful, the solution is postprocessed with an exact repair step that involves an exact LP solve.

Moreover, we integrate the exact LP solver SoPlex, which follows the recently developed scheme of *LP iterative refinement* [23], we extend the logging of *certificates* in the recently developed VIPR format to all available dual bounding methods [13], and produce a thoroughly *revised implementation* of the original framework [16], which improves multiple technical details. Our computational study evaluates the performance of the new algorithmic aspects in detail and indicates a significant overall speedup compared to the original framework.

The overarching goal and contribution of this research is to extend the computational practice of MIP to the level of rigor that has been achieved in recent years, for example, by the field of satisfiability solving [34], while at the same time retaining most of the computational power embedded in floating-point solvers.

In MIP, a similar level of performance and rigor is certainly much more difficult to reach in practice, due to the numerical operations that are inherently involved in solving general mixed integer programs. However, we believe that there is no reason why this vision should be *fundamentally* out of reach for the rich machinery of MIP techniques developed over the last decades. The goal of this paper is to demonstrate the viability of this agenda within a first, small selection of methods. The resulting code is freely available for research purposes as an extension of SCIP 7.0 [17].

2 Numerically Exact Mixed Integer Programming

In the following, we describe related work in numerically exact optimization, including the main ideas and features of the framework that we build upon. Before turning to the most general case, we would like to mention that roundoff-error-free methods are available for several specific classes of pure integer problems. One example for such a combinatorial optimization problem is the traveling salesman problem, for which the branch-and-cut solver Concorde applies safe interval-arithmetic to postprocess LP relaxation solutions and ensures the validity of domain-specific cutting planes by their combinatorial structure [5].

A broader class of such problems, on binary decision variables, is addressed in *satisfiability solving* (SAT) and *pseudo-Boolean optimization* (PBO) [10]. Solvers for these problem classes usually do not suffer from numerical errors and often support solver-independent verification of results [34]. While optimization variants exist, the development of these methods is to a large extent driven by feasibility problems. The broader class of solvers for *satisfiability modulo theories* (SMT), e.g., [30], may also include real-valued variables, in particular for satisfiability modulo the theory of linear arithmetic. However, as pointed out also in [20], the target applications of SMT solvers differ significantly from the motivating use cases in LP and MIP.

Exact optimization over convex polytopes intersected with lattices is also supported by some software libraries for polyhedral analysis [7,8]. These tools are not particularly targeted towards solving LPs or MIPs of larger scale and usually follow the naive approach of simply executing all operations symbolically, in exact rational arithmetic. This yields numerically exact results and can even be highly efficient as long as the size of problems or the encoding length of intermediate numbers is limited. However, as pointed out by [19] and [16], this *purely symbolic approach* quickly becomes prohibitively slow in general.

By contrast, the most effective methods in the literature rely on a *hybrid approach* and combine exact and numeric computation. For solving pure LPs exactly, the most recent methods that follow this paradigm are *incremental precision boosting* [6] and *LP iterative refinement* [23]. In an exact MIP solver, however, it is not always necessary to solve LP relaxations completely, but it often suffices to provide dual bounds that underestimate the optimal relaxation value safely. This can be achieved by postprocessing approximate LP solutions. *Bound-shift* [31] is such a method that only relies on directed rounding and interval arithmetic and is therefore very fast. However, as the name suggests it requires upper and lower bounds on all variables in order to be applicable. A more widely applicable bounding method is *project-and-shift* [33], which uses an interior point or ray of the dual LP. These need to be computed by solving an auxiliary LP exactly in advance, though only once per MIP solve. Subsequently, approximate dual LP solutions can be corrected by projecting them to the feasible region defined by the dual constraints and shifting the result to satisfy sign constraints on the dual multipliers.

The hybrid branch-and-bound method of [16] combines such safe dual bounding methods with a state-of-the-art branching heuristic, reliability branching [3]. It maintains both the exact problem formulation

$$\min\{c^T x \mid Ax \geq b,\ x \in \mathbb{Q}^n,\ x_i \in \mathbb{Z}\ \forall i \in \mathcal{I}\}$$

with rational input data $A \in \mathbb{Q}^{m \times n}, c \in Q^n, b \in \mathbb{Q}^m$, as well as a floating-point approximation with data $\bar{A}, \bar{b}, \bar{c}$, which are defined as the componentwise closest numbers representable in floating-point arithmetic. The set $\mathcal{I} \subseteq \{1, \ldots, n\}$ contains the indices of integer variables.

During the solve, for all LP relaxations, the floating-point approximation is first solved in floating-point arithmetic as an approximation and then postprocessed to generate a valid dual bound. The methods available for this safe bounding step are the previously described *bound-shift* [31], *project-and-shift* [33], and an *exact LP solve* with the exact LP solver QSopt_EX based on incremental precision boosting [6]. (Further dual bounding methods were tested, but reported as less important in [16].) On the primal side, all solutions are checked for feasibility in exact arithmetic before being accepted.

Finally, this exact MIP framework was recently extended by the possibility to generate a *certificate* of correctness [13]. This certificate is a tree-less encoding of the branch-and-bound search, with a set of dual multipliers to prove the dual bound at each node or its infeasibility. Its correctness can be verified independently of the solving process using the checker software VIPR [14].

3 Extending and Improving an Exact MIP Framework

The exact MIP solver presented here extends [16] in four ways: the addition of a symbolic presolving phase, the execution of primal floating-point heuristics coupled with an exact repair step, the use of a recently developed exact LP solver based on LP iterative refinement, and a generally improved integration of the exact solving routines into the core branch-and-bound algorithm.

Symbolic Presolving. The first major extension is the addition of symbolic presolving. To this end, we integrate the newly available presolving library PAPILO [25] for integer and linear programming. PAPILO has several benefits for our purposes.

First, its code base is by design fully templatized with respect to the arithmetic type. This enables us to integrate it with rational numbers as data type for storing the MIP data and all its computations. Second, it provides a large range of presolving techniques already implemented. The ones used in our exact framework are coefficient strengthening, constraint propagation, implicit integer detection, singleton column detection, substitution of variables, simplification of inequalities, parallel row detection, sparsification, probing, dual fixing, dual inference, singleton stuffing, and dominated column detection. For a detailed explanation of these methods, we refer to [2]. Third, PAPILO comes with a sophisticated parallelization scheme that helps to compensate for the increased overhead introduced by the use of rational arithmetic. For details see [21].

When SCIP enters the presolving stage, we pass a rational copy of the problem to PAPILO, which executes its presolving routines iteratively until no sufficiently large reductions are found. Subsequently, we extract the postsolving information provided by PAPILO to transfer the model reductions to SCIP. These include fixings, aggregations, and bound changes of variables and strengthening or deletion of constraints, all of which are performed in rational arithmetic.

Primal Heuristics. The second extension is the safe activation of SCIP's floating-point heuristics and the addition of an exact repair heuristic for their approximate solutions. Heuristics are not known to reduce the overall solving time drastically, but they can be particularly useful on hard instances that cannot be solved at all, and in order to avoid terminating without a feasible solution.

In general, activating SCIP's floating-point heuristics does not interfere with the exactness of the solving process, although care has to be taken that no changes to the model are performed, e.g., the creation of a no-good constraint. However, the chance that these heuristics find a solution that is feasible in the exact sense can be low, especially if equality constraints are present in the model. Thus, we postprocess solutions found by floating-point heuristics in the following way. First, we fix all integer variables to the values found by the floating-point heuristic, rounding slightly fractional values to their nearest integer. Then an exact LP is solved for the remaining continuous subproblem. If that LP is feasible, this produces an exactly feasible solution to the mixed integer program.

Certainly, frequently solving this subproblem exactly can create a significant overhead compared to executing a floating-point heuristic alone, especially when a large percentage of the variables is continuous and thus cannot be fixed. Therefore, we impose working limits on the frequency of running the exact repair heuristic, which are explained in more detail in Sect. 4.

LP Iterative Refinement. Exact linear programming is a crucial part of the exact MIP solving process. Instead of QSOPT_EX, we use SOPLEX as the exact linear programming solver. The reason for this change is that SOPLEX uses LP iterative refinement [24] as the strategy to solve LPs exactly, which compares favorably against incremental precision boosting [23].

Further Enhancements. We improved several algorithmic details in the implementation of the hybrid branch-and-bound method. We would like to highlight two examples for these changes. First, we enable the use of an *objective limit* in the floating-point LP solver, which was not possible in the original framework. Passing the primal bound as an objective limit to the floating-point LP solver allows the LP solver to stop early just after its dual bound exceeds the global primal bound. However, if the overlap is too small, postprocessing this LP solution with safe bounding methods can easily lead to a dual bound that no longer exceeds the objective limit. For this reason, before installing the primal bound as an objective limit in the LP solver, we increase it by a small amount computed from the statistically observed bounding error so far. Only when safe dual bounding fails, the objective limit is solved again without objective limit.

Second, we reduce the time needed for checking exact feasibility of primal solutions by prepending a safe floating-point check. Although checking a single solution for feasibility is fast, this happens often throughout the solve and doing so repeatedly in exact arithmetic can become computationally expensive. To implement such a safe floating-point check, we employ *running error analysis* [27]. Let $x^* \in \mathbb{Q}^n$ be a potential solution and let $\bar{x}^*$ be the floating-point approximation of x^*. Let $a \in \mathbb{Q}^n$ be a row of A with floating-point approximation $\bar{a}$, and right hand side $b_j \in \mathbb{Q}$. Instead of computing $\sum_{i=1}^n a_i x_i^*$ symbolically, we instead compute $\sum_{i=1}^n \bar{a}_i \bar{x}_i^*$ in floating-point arithmetic, and alongside compute a bound on the maximal rounding error that may occur. We adjust the running error analysis described in [27, Alg. 3.2] to also account for roundoff errors $|\bar{x}^* - x^*|$ and $|\bar{a} - a|$. After doing this computation, we can check if either $s - \mu \geq b_j$ or $s + \mu \leq b_j$. In the former, the solution x^* is guaranteed to fulfill $\sum_{i=1}^n a_i x_i^* \geq b_j$; in the latter, we can safely determine that the inequality is violated; only if neither case occurs, we recompute the activity in exact arithmetic.

We note that this could alternatively be achieved by directed rounding, which would give tighter error bounds at a slightly increased computational effort. However, empirically we have observed that most equality or inequality constraints are either satisfied at equality, where an exact arithmetic check cannot be avoided, or they are violated or satisfied by a slack larger than the error bound μ, hence the running error analysis is sufficient to determine feasibility.

4   Computational Study

We conduct a computational analysis to answer three main questions. *First, how does the revised branch-and-bound framework compare to the previous implementation, and to which components can the changes be attributed?* To answer this question, we compare the original framework [16] against our improved implementation, including the exact LP solver SoPlex, but with primal heuristics and exact presolving still disabled. In particular, we analyze the importance and performance of the different dual bounding methods.

Second, what is the impact of the new algorithmic components symbolic presolving and primal heuristics? To answer this question, we compare their impact on the solving time and the number of solved instances, as well as present more in-depth statistics, such as e.g., the primal integral [9] for heuristics or the number of fixings for presolving. In addition, we compare the effectiveness of performing presolving in rational and in floating-point arithmetic.

Finally, what is the overhead for producing and verifying certificates? Here, we consider running times for both the solver and the certificate checker, as well as the overhead in the safe dual bounding methods introduced through enabling certificates. This provides an update for the analysis in [13], which was limited to the two bounding methods project-and-shift and exact LP.

The experiments were performed on a cluster of Intel Xeon CPUs E5-2660 with 2.6 GHz and 128 GB main memory. As in [16], we use CPLEX as floating-point LP solver. Due to compatibility issues, we needed to use CPLEX 12.3.0 for the original and CPLEX 12.8.0 for the new framework. Although these versions are different, they are only used to solve floating-point LPs and have limited impact on the reported results: The vast majority of solving time is spent in the safe dual bounding methods. For exact LP solving, we use the same QSOPT_EX version as in [16] and SOPLEX 5.0.2. For all symbolic computations, we use the GNU Multiple Precision Library (GMP) 6.1.4 [26]. For symbolic presolving, we use PAPILO 1.0.1 [21,25]; all other SCIP presolvers are disabled.

As main test sets, we use the two test sets specifically curated in [16]: one set with 57 instances that were found to be easy for an inexact floating-point branch-and-bound solver (FPEASY), and one set of 50 instances that were found to be numerically challenging, e.g., due to poor conditioning or large coefficient ranges (NUMDIFF). For a detailed description of the selection criteria, we refer to [16]. To complement these test sets with a set of more ambitious and recent instances, we conduct a final comparison on the MIPLIB 2017 [22] benchmark set. All experiments to evaluate the new code are run with three different random seeds, where we treat each instance-seed combination as a single observation. As this feature is not available in the original framework, all comparisons with the original framework were performed with one seed. The time limit was set to 7200 s for all experiments. If not stated otherwise all aggregated numbers are shifted geometric means with a shift of 0.001 s or 100 branch-and-bound nodes, respectively.

The Branch-and-Bound Framework. As a first step, we compare the behavior of the safe branch-and-bound implementation from [16] with QSOPT_EX as the exact LP solver, against its revised implementation with SOPLEX 5.0.2 as exact LP solver. The original framework uses the "Auto-Ileaved" bounding strategy as recommended in [16]. It dynamically chooses the dual bounding method, attempting to employ bound-shift as often as possible. An exact LP is solved whenever a node would be cut off within tolerances, but not with the exact the safe dual bound computed. In the new implementation we use a similar strategy, however we solve the rational LP relaxation every 5 depth levels of the tree, due to improved performance in the exact LP solver.

Table 1 reports the results for solving time, number of nodes, and total time spent in safe dual bounding ("dbtime"), for all instances that could be solved by at least one solver. The new framework could solve 10 instances more on FPEASY and 7 more on NUMDIFF. On FPEASY, we observe a reduction of 69.8% in solving time and of 87.3% in safe dual bounding time. On NUMDIFF, we observe a reduction of 80.3% in solving time, and of 88.3% in the time spent overall in the safe dual bounding methods. We also see this significant performance improvement reflected in the two performance profiles in Fig. 1.

Table 1. Comparison of original and new framework with presolving and primal heuristics disabled

Test set	Size	Original framework				New framework			
		Solved	Time	Nodes	dbtime	Solved	Time	Nodes	dbtime
FPEASY	55	45	128.4	8920.1	86.8	55	38.8	5940.8	11.0
NUMDIFF	21	13	237.0	8882.7	114.6	20	46.6	6219.3	13.4

Table 2. Comparison of safe dual bounding techniques

Test set	Stats	Original framework			New framework		
		bshift	pshift	exlp	bshift	pshift	exlp
FPEASY.	Calls/node	0.92	0.44	0.28	0.53	0.39	0.06
	Time/call [s]	0.0026	0.0022	0.050	0.0072	0.0066	0.010
	Time/solving time	2.9%	40.3%	32.1%	10.8%	27.8%	4.4%
NUMDIFF	Calls/node	0.78	0.36	0.52	0.36	0.39	0.28
	Time/call [s]	0.0055	0.0036	0.4197	0.0247	0.0356	0.1556
	Time/solving time	1.4%	22.7%	62.2%	5.2%	24.8%	40.1%

We identify a more aggressive use of project-and-shift and faster exact LP solves as the two key factors for this improvement. In the original framework, project-and-shift is restricted to instances that had less than 10000 nonzeros. One reason for this limit is that a large auxiliary LP has to be solved by the exact LP solver to compute the relative interior point in project-and-shift. With the improvements in exact LP performance, it proved beneficial to remove this working limit in the new framework.

The effect of this change can also be seen in the detailed analysis of bounding times given in Table 2. For calls per node and the fraction of bounding time per total solving time, which are normalized well, we report the arithmetic means; for time per call, we report geometric means over all instances where the respective bounding method was called at least once.

The fact that time per call for project-and-shift ("pshift") in the new framework increased by a factor of 3 (FPEASY) and 9.9 (NUMDIFF) is for the reason discussed above—it is now also called on larger instances. This is beneficial overall since it replaces many slower exact LP calls. The decrease in exact LP solving time per call ("exlp") by a factor of 2.7 (NUMDIFF) and 5 (FPEASY) can also partly be explained by this change, and partly by an overall performance improvement in exact LP solving due to the use of LP iterative refinement [24]. The increase in bound-shift time ("bshift") is due to implementation details, that will be addressed in future versions of the code, but its fraction of the total solving time is still relatively low. Finally, we observe a decrease in the total number of safe bounding calls per node. One reason is that we now disable bound-shift dynamically if its success rate drops below 20%.

Overall, we see a notable speedup and more solved instances, mainly due to the better management of dual bounding methods and faster exact LP solving.

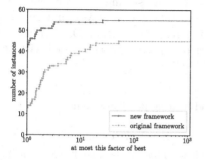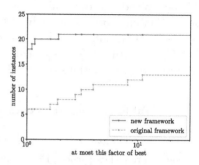

Fig. 1. Performance profiles comparing solving time of original and new framework without presolving and heuristics for FPEASY (left) and NUMDIFF (right)

Symbolic Presolving. Before measuring the overall performance impact of exact presolving, we address the question how effective and how expensive presolving in rational arithmetic is compared to standard floating-point presolving. For both variants, we configured PAPILO to use the same tolerances for determining whether a reduction found is strong enough to be accepted. The only difference in the rational version is that all computations are done in exact arithmetic and the tolerance to compare numbers and the feasibility tolerance are zero. Note that a priori it is unclear whether rational presolving yields more or less reductions. Dominated column detection may be less successful due to the stricter comparison of coefficients; the dual inference presolver might be more successful if it detects earlier that a dual multiplier is strictly bounded away from zero.

Table 3 presents aggregated results for presolving time, the number of presolving rounds, and the number of found fixings, aggregations, and bound changes. We use a shift of 1 for the geometric means of rounds, aggregations, fixings, and bound changes to account for instances where presolving found no

such reductions. Remarkably, both variants yield virtually the same results on
FPEASY. On NUMDIFF, there are small differences, with a slight decrease in the
number of fixings and aggregations and a slight increase in the number of bound
changes for the exact variant. The time spent for exact presolving increases by
more than an order of magnitude but symbolic presolving is still not a performance bottleneck. It consumed only 0.86% (FPEASY) and 2.1% (NUMDIFF)
of the total solving time, as seen in Table 4. Exploiting parallelism in presolving provided no measureable benefit for floating-point presolving, but reduced
symbolic presolving time by 44% (FPEASY) to 43.8% (NUMDIFF). However, this
benefit can be more pronounced on individual instances, e.g., on nw04, where
parallelization reduces the time for rational presolving by a factor of 6.4 from
1770 to 277 s.

To evaluate the impact of exact presolving, we compare the performance of the basic branch-and-bound algorithm established above against
the performance with presolving enabled. The results for all instances that
could be solved to optimality by at least one setting are presented in
Table 4. Enabling presolving solves 3 more instances on FPEASY and 20 more
instances on NUMDIFF. We observe a reduction in solving time of 39.4% (FPEASY)
and 72.9% (NUMDIFF). The stronger impact on NUMDIFF is correlated with the
larger number of reductions observed in Table 3.

Table 3. Comparison of exact and floating-point presolving

Test set	thrds	Floating-point presolving					Exact presolving				
		Time	rnds	Fixed	agg	bdchg	Time	rnds	Fixed	agg	bdchg
FPEASY	1	0.01	3.2	8.5	3.5	10.4	0.25	3.2	8.5	3.5	10.4
	20	0.01	3.2	8.5	3.5	10.4	0.14	3.2	8.5	3.5	10.4
NUMDIFF	1	0.04	8.3	53.8	55.7	51.4	0.89	7.2	41.4	42.9	55.8
	20	0.04	8.3	53.8	55.7	51.4	0.50	7.2	41.4	42.9	55.8

Table 4. Comparison of new framework with and without presolving (3 seeds)

Test set	Size	Presolving disabled			Presolving enabled		
		Solved	Time	Nodes	Solved	Time (presolving)	Nodes
FPEASY	168	165	42.1	6145.3	168	25.5 (0.22)	4724.1
NUMDIFF	91	66	216.6	7237.2	86	58.7 (1.23)	2867.2

Primal Heuristics. To improve primal performance, we enabled all SCIP
heuristics that the floating-point version executes by default. To limit the fraction
of solving time for the repair heuristic described in Sect. 3, the repair heuristic is
only allowed to run at any point in the solve, if it was called at most half as often
as the exact LP calls for safe dual bounding. Furthermore, the repair heuristic is

disabled on instances with more than 80% continuous variables, since the overhead of the exact LP solves can drastically worsen the performance on those instances. Whenever the repair step is not executed, the floating-point solutions are checked directly for exact feasibility.

First, we evaluate the cost and success of the exact repair heuristic over all instances where it was called at least once. The results are presented in Table 5. The repair heuristic is effective at finding feasible solutions with a success rate of 46.9% (FPEASY) and 25.6% (NUMDIFF). The fraction of the solving time spent in the repair heuristic is well below 1%. Nevertheless, the strict working limits we imposed are necessary since there exist outliers for which the repair heuristic takes more than 5% of the total solving time, and performance on these instances would quickly deteriorate if the working limits were relaxed.

Table 5. Statistics of repair heuristic for instances where repair step was called

Test set	Size	Time Total solving	Repair	Fail	Success	Success rate
FPEASY	82	39.8	0.0020	0.0017	0.0003	46.9%
NUMDIFF	42	383.6	0.0187	0.0077	0.0062	25.6%

Table 6. Comparison of new framework with and without primal heuristics (3 seeds, presolving enabled, instances where repair step was called)

Test set	Size	Heuristics disabled			Heuristics enabled		
		Solv. time	Time-to-first	Primal int.	Solv. time	Time-to-first	Primal int.
FPEASY	82	32.5	0.75	2351.8	32.6	0.10	2037.2
NUMDIFF	41	101.7	4.77	8670.7	103.1	1.30	9093.6

Table 6 shows the overall performance impact of enabling heuristics over all instances that could be solved by at least one setting. On both sets, we see almost no change in total solving time. On FPEASY, the time to find the first solution decreases by 86.7% and the primal integral decreases by 13.4%. The picture is slightly different on the numerically difficult test set. Here, the time to find the first solution decreases by 72.7%, while the primal integral increases by 4.9%.

The worse performance and success rate on NUMDIFF is expected, considering that this test set was curated to contain instances with numerical challenges. On those instances floating-point heuristics find solutions that might either not be feasible in exact arithmetic or are not possible to fix for the repair heuristic. In both test sets, the repair heuristic was able to find solutions, while not imposing any significant overhead in solving time.

Producing and Verifying Certificates. The possibility to log certificates as presented in [13] is available in the new framework and is extended to also work when the dual bounding method bound-shift is active. Presolving must currently be disabled, since PAPILO does not yet support generation of certificates.

Besides ensuring correctness of results, certificate generation is valuable to ensure correctness of the solver. Although it does not check the implementation itself, it can help identify and eliminate incorrect results that do not directly lead to fails. For example, on instance x_4 from NUMDIFF, the original framework claimed infeasibility at the root node, and while the instance is indeed infeasible, we found the reasoning for this to be incorrect due to the use of a certificate.

Table 7 reports the performance overhead when enabling certificates. Here we only consider instances that were solved to optimality by both versions since timeouts would bias the results in favor of the certificate. We see an increase in solving time of 101.2% on FPEASY and of 51.4% on NUMDIFF. This confirms the measurements presented in [13]. The increase is explained in part by the effort to keep track of the tree structure and print the exact dual multipliers, and in part by an increase in dual bounding time. The reason for the latter is that bound-shift by default only provides a safe objective value. The dual multipliers needed for the certificate must be computed in a postprocessing step, which introduces the overhead in safe bounding time. This overhead is larger on FPEASY, since bound-shift is called more often. The time spent in the verification of the certificate is on average significantly lower than the time spent in the solving process. Overall, the overhead from printing and checking certificates is significant, but it does not drastically change the solvability of instances.

Table 7. Overhead for producing and verifying certificates on instances solved by both variants

Test set	Size	Certificate disabled		Certificate enabled			
		Solving time	dbtime	Solving time	dbtime	Check time	Overhead
FPEASY	53	32.6	9.1	65.6	16.5	0.9	103.9%
NUMDIFF	21	41.6	11.9	63.0	18.0	0.5	52.6%

Table 8. Comparison on MIPLIB 2017 benchmark set

Test set	Size	Original framework				New framework			
		Solved	Found	Time	Gap	Solved	Found	Time	Gap
All	240	17	74	6003.6	∞	47	167	3928.1	∞
Both	66	16	66	4180.0	67.9%	29	66	1896.2	33.8%
Onesolved	49	17	31	3317.5	∞	47	47	505.1	∞

Performance Comparison on MIBLIB 2017. As a final experiment, we wanted to evaluate the performance on a more ambitious and diverse test set. To that end, we ran both the original framework and the revised framework with presolving and heuristics enabled on the recent MIPLIB 2017 benchmark set. The results in Table 8 show that the new framework solved 30 instances more and the mean solving time decreased by 84.8% on the subset "onesolved" of instances that could be solved to optimality by at least one solver. On more than twice as many instances at least one primal solution was found (167 vs. 74). On the subset of 66 instances that had a finite gap for both versions, the new algorithm achieved a gap of 33.8% in arithmetic mean compared to 67.9% in the original framework.

To conclude, we presented a substantially revised and extended solver for numerically exact mixed integer optimization that significantly improves upon the existing state of the art. We also observe, however, that the performance gap to floating-point solvers is still large. This is not surprising, given that crucial techniques such as numerically safe cutting plane separation, see, e.g., [15], are not yet included. This must be addressed in future research.

Acknowledgements. We wish to thank Dan Steffy for valuable discussions on the revision of the original branch-and-bound framework, Leona Gottwald for creating PaPILO, and Antonia Chmiela for help with implementing the primal repair heuristic.

References

1. Achterberg, T.: Constraint integer programming. Ph.D. thesis, Technische Universität Berlin (2007)
2. Achterberg, T., Bixby, R.E., Gu, Z., Rothberg, E., Weninger, D.: Presolve reductions in mixed integer programming. INFORMS J. Comput. **32**(2), 473–506 (2020). https://doi.org/10.1287/ijoc.2018.0857
3. Achterberg, T., Koch, T., Martin, A.: Branching rules revisited. Oper. Res. Lett. **33**(1), 42–54 (2005). https://doi.org/10.1016/j.orl.2004.04.002
4. Achterberg, T., Wunderling, R.: Mixed integer programming: analyzing 12 years of progress. In: Jünger, M., Reinelt, G. (eds.) Facets of Combinatorial Optimization, pp. 449–481. Springer, Heidelberg (2013). https://doi.org/10.1007/978-3-642-38189-8_18
5. Applegate, D., Bixby, R., Chvatal, V., Cook, W.: Concorde TSP Solver (2006)
6. Applegate, D., Cook, W., Dash, S., Espinoza, D.G.: Exact solutions to linear programming problems. Oper. Res. Lett. **35**(6), 693–699 (2007). https://doi.org/10.1016/j.orl.2006.12.010
7. Assarf, B., et al.: Computing convex hulls and counting integer points with polymake. Math. Program. Comput. **9**(1), 1–38 (2017). https://doi.org/10.1007/s12532-016-0104-z
8. Bagnara, R., Hill, P.M., Zaffanella, E.: The Parma Polyhedra Library: toward a complete set of numerical abstractions for the analysis and verification of hardware and software systems. Sci. Comput. Program. **72**(1–2), 3–21 (2008)
9. Berthold, T.: Measuring the impact of primal heuristics. Oper. Res. Lett. **41**(6), 611–614 (2013). https://doi.org/10.1016/j.orl.2013.08.007

10. Biere, A., Heule, M., van Maaren, H., Walsh, T.: Handbook of Satisfiability: Volume 185 Frontiers in Artificial Intelligence and Applications. IOS Press, Amsterdam (2009)
11. Bofill, M., Manyà, F., Vidal, A., Villaret, M.: New complexity results for Łukasiewicz logic. Soft. Comput. **23**, 2187–2197 (2019). https://doi.org/10.1007/s00500-018-3365-9
12. Burton, B.A., Ozlen, M.: Computing the crosscap number of a knot using integer programming and normal surfaces. ACM Trans. Math. Softw. **39**(1) (2012). https://doi.org/10.1145/2382585.2382589
13. Cheung, K.K.H., Gleixner, A., Steffy, D.E.: Verifying integer programming results. In: Eisenbrand, F., Koenemann, J. (eds.) IPCO 2017. LNCS, vol. 10328, pp. 148–160. Springer, Cham (2017). https://doi.org/10.1007/978-3-319-59250-3_13
14. Cheung, K., Gleixner, A., Steffy, D.: VIPR. Verifying Integer Programming Results. https://github.com/ambros-gleixner/VIPR. Accessed 11 Nov 2020
15. Cook, W., Dash, S., Fukasawa, R., Goycoolea, M.: Numerically safe gomory mixed-integer cuts. INFORMS J. Comput. **21**, 641–649 (2009). https://doi.org/10.1287/ijoc.1090.0324
16. Cook, W., Koch, T., Steffy, D.E., Wolter, K.: A hybrid branch-and-bound approach for exact rational mixed-integer programming. Math. Program. Comput. **5**(3), 305–344 (2013). https://doi.org/10.1007/s12532-013-0055-6
17. Eifler, L., Gleixner, A.: Exact SCIP - a development version. https://github.com/leoneifler/exact-SCIP. Accessed 11 Nov 2020
18. Eifler, L., Gleixner, A., Pulaj, J.: A safe computational framework for integer programming applied to Chvátal's conjecture (2020)
19. Espinoza, D.G.: On linear programming, integer programming and cutting planes. Ph.D. thesis, Georgia Institute of Technology (2006)
20. Faure, G., Nieuwenhuis, R., Oliveras, A., Rodríguez-Carbonell, E.: SAT modulo the theory of linear arithmetic: exact, inexact and commercial solvers. In: Kleine Büning, H., Zhao, X. (eds.) SAT 2008. LNCS, vol. 4996, pp. 77–90. Springer, Heidelberg (2008). https://doi.org/10.1007/978-3-540-79719-7_8
21. Gamrath, G., et al.: The SCIP Optimization Suite 7.0. ZIB-Report 20–10, Zuse Institute Berlin (2020)
22. Gleixner, A., et al.: MIPLIB 2017: data-driven compilation of the 6th mixed-integer programming library. Math. Program. Comput. 1–48 (2021). https://doi.org/10.1007/s12532-020-00194-3
23. Gleixner, A., Steffy, D.E.: Linear programming using limited-precision oracles. Math. Program. **183**, 525–554 (2020). https://doi.org/10.1007/s10107-019-01444-6
24. Gleixner, A., Steffy, D.E., Wolter, K.: Iterative refinement for linear programming. INFORMS J. Comput. **28**(3), 449–464 (2016). https://doi.org/10.1287/ijoc.2016.0692
25. Gottwald, L.: PaPILO – Parallel Presolve for Integer and Linear Optimization. https://github.com/lgottwald/PaPILO. Accessed 9 Sep 2020
26. Granlund, T., Team, G.D.: GNU MP 6.0 Multiple Precision Arithmetic Library. Samurai Media Limited, London, GBR (2015)
27. Higham, N.J.: Accuracy and Stability of Numerical Algorithms, 2nd edn. Society for Industrial and Applied Mathematics, Philadelphia (2002). https://doi.org/10.1137/1.9780898718027

28. Kenter, F., Skipper, D.: Integer-programming bounds on pebbling numbers of Cartesian-product graphs. In: Kim, D., Uma, R.N., Zelikovsky, A. (eds.) COCOA 2018. LNCS, vol. 11346, pp. 681–695. Springer, Cham (2018). https://doi.org/10.1007/978-3-030-04651-4_46

29. Lancia, G., Pippia, E., Rinaldi, F.: Using integer programming to search for counterexamples: a case study. In: Kononov, A., Khachay, M., Kalyagin, V.A., Pardalos, P. (eds.) MOTOR 2020. LNCS, vol. 12095, pp. 69–84. Springer, Cham (2020). https://doi.org/10.1007/978-3-030-49988-4_5

30. de Moura, L., Bjørner, N.: Z3: an efficient SMT solver. In: Ramakrishnan, C.R., Rehof, J. (eds.) TACAS 2008. LNCS, vol. 4963, pp. 337–340. Springer, Heidelberg (2008). https://doi.org/10.1007/978-3-540-78800-3_24

31. Neumaier, A., Shcherbina, O.: Safe bounds in linear and mixed-integer programming. Math. Program. **99**, 283–296 (2002). https://doi.org/10.1007/s10107-003-0433-3

32. Pulaj, J.: Cutting planes for families implying Frankl's conjecture. Math. Comput. **89**(322), 829–857 (2020). https://doi.org/10.1090/mcom/3461

33. Steffy, D.E., Wolter, K.: Valid linear programming bounds for exact mixed-integer programming. INFORMS J. Comput. **25**(2), 271–284 (2013). https://doi.org/10.1287/ijoc.1120.0501

34. Wetzler, N., Heule, M.J.H., Hunt, W.A.: DRAT-trim: efficient checking and trimming using expressive clausal proofs. In: Sinz, C., Egly, U. (eds.) SAT 2014. LNCS, vol. 8561, pp. 422–429. Springer, Cham (2014). https://doi.org/10.1007/978-3-319-09284-3_31

35. Wilken, K., Liu, J., Heffernan, M.: Optimal instruction scheduling using integer programming. SIGPLAN Not. **35**(5), 121–133 (2000). https://doi.org/10.1145/358438.349318

New Exact Techniques Applied to a Class of Network Flow Formulations

Vinícius L. de Lima[1](✉), Manuel Iori[2], and Flávio K. Miyazawa[1]

[1] Institute of Computing, University of Campinas, Campinas, Brazil
{v.loti,fkm}@ic.unicamp.br
[2] DISMI, University of Modena and Reggio Emilia, Reggio Emilia, Italy
manuel.iori@unimore.it

Abstract. We propose a number of solution techniques for general network flow formulations derived from Dantzig-Wolfe decompositions. We present an arc selection method to derive reduced network flow models that may potentially provide good feasible solutions. This method is explored as a variable selection rule for branching. With the aim of improving reduced-cost variable-fixing, we also propose a method to produce different dual solutions of network flow models and provide conditions that guarantee the correctness of the method. We embed the proposed techniques in an innovative branch-and-price method for network flow formulations, and test it on the cutting stock problem. In our computational experiments, 162 out of 237 open benchmark instances are solved to proven optimality within a reasonable computational time, consistently improving previous results in the literature.

Keywords: Network flow models · Variable selection · Variable fixing

1 Introduction

Mixed Integer Linear Programming (MILP) is one of the most popular mathematical programming tools to solve optimization problems. The strength of MILP models has been enhanced in many applications by using the *Dantzig-Wolfe* (DW) decomposition [7], which relies on the fact that every point of a non-empty polyhedron can be represented as a convex combination of its extreme points. The model resulting from a DW decomposition is called *DW model*. A relation between integer solutions of polyhedra with paths in acyclic networks allows us to represent DW models as *network flow formulations* (see, e.g., page 322 of [21]), i.e., formulations that require to determine an optimal flow in a network. According to Ahuja et al. [1], the two main classes of network flow formulations are *path flow formulations*, in which variables correspond to the flow on each path (and cycle) of the network, and *arc flow formulations*, in which variables correspond to flow on individual arcs.

The first and third authors acknowledge the support by CNPq (Proc. 314366/2018-0, 425340/2016-3) and by FAPESP (Proc. 2015/11937-9, 2016/01860-1, 2017/11831-1).

© Springer Nature Switzerland AG 2021
M. Singh and D. P. Williamson (Eds.): IPCO 2021, LNCS 12707, pp. 178–192, 2021.
https://doi.org/10.1007/978-3-030-73879-2_13

In practice, path flow formulations for $\mathcal{NP}$-hard problems have a huge number of variables and are typically solved by sophisticated branch(-and-cut)-and-price (B&P) algorithms (see, e.g., [22]). In contrast, arc flow formulations have a linear number of variables with respect to the number of arcs of the network and have been used to solve medium-sized instances directly by general-purpose MILP solvers (for a recent survey, see [8]). Few authors explore the correspondence between path flow and arc flow models by combining them into a unique solution method. For instance, Pessoa et al. [23] proposed a B&P algorithm based on a path flow model that solves the problem at the root node as an arc flow model by a MILP solver if the network is sufficiently small.

In B&P algorithms, additional constraints derived from branching can impact the pricing problem, making it substantially harder. A number of researches propose general branching schemes that minimize the impact on the pricing problem and at the same time help convergence to optimality (see, e.g., [27–29]). However, not many general branching schemes exploit the network flow representation of DW models. Although branching based on arc flow variables is robust, i.e., it does not change the structure of the pricing, Vanderbeck [27] alerted that the selection of which arcs to branch is a difficult task.

In this paper, we present techniques for network flow formulations derived from DW decomposition. First, we present a non-trivial column generation algorithm that generates multiple columns and can be used to solve the relaxation of both path flow and arc flow models. In order to improve solution methods for network flow formulations, one can consider the reduced-cost variable-fixing method in [16], which uses a dual solution of a path flow model to remove non-optimal arcs, i.e., arcs that do not improve the incumbent integer solution. Notably, different dual solutions usually lead to the removal of different sets of arcs. Here, we propose a method to obtain dual solutions that improves the practical effectiveness of the method in [16].

We also present a method to select arcs that, when removed from the model, lead to a significantly smaller problem that may potentially provide good-quality feasible solutions. The resulting problem may represent an arc flow model of practical size, which could be solved reasonably fast by a MILP solver. This arc selection method leads to a binary branching scheme: the left branch solves the problem resulted from the arc removal as an arc flow model, whereas the right branch solves the problem with an additional linear constraint enforcing that at least one of the arcs removed in the left branch should be in the solution.

All the proposed techniques are embedded into an innovative B&P algorithm that exploits the equivalence between path flow and arc flow formulations and solves a network flow model received in input. The algorithm is applied to the cutting stock problem (CSP). For this problem, we propose a method that may consistently reduce the size of networks by removing arcs without loss of optimality. The effectiveness of the B&P algorithm and its main components is proved by means of extensive computational tests. We also provide hints for applying the developed techniques to other classes of difficult combinatorial optimization problems.

2   Network Flow Formulations

A network $\mathcal{N}$ is a directed graph with a set of nodes $\mathcal{V}$ and a set of arcs $\mathcal{A} \subseteq \mathcal{V} \times \mathcal{V}$. Two special nodes in $\mathcal{V}$ are the source v^+ and the sink v^-, in which no arcs in $\mathcal{A}$ enters v^+ or leaves v^-. The set of all paths from v^+ to v^- is given by $\mathcal{P}$, and the set of all arcs of a path $p \in \mathcal{P}$ is given by $\mathcal{A}_p$. The set of all paths in $\mathcal{P}$ that contain an arc (u,v) is given by $\mathcal{P}_{(u,v)}$. Let variable $\lambda_p \in \mathbb{Z}^+$ represent the flow on path $p \in \mathcal{P}$. The following is a general path flow model:

$$\min \sum_{p \in \mathcal{P}} c_p \lambda_p, \tag{1}$$

$$\text{s.t.:} \sum_{p \in \mathcal{P}} a_p \lambda_p \geq b, \tag{2}$$

$$\lambda_p \in \mathbb{Z}_+, \qquad\qquad \forall p \in \mathcal{P}, \tag{3}$$

where $b \in \mathbb{R}^m$ and, for each $p \in \mathcal{P}$, $a_p \in \mathbb{R}^m$ and $c_p \in \mathbb{R}$. We are interested in path flow formulations derived from DW models, where $\mathcal{N}$ is acyclic and each arc $(u,v) \in \mathcal{A}$ admits unique $c_{(u,v)} \in \mathbb{R}$ and $a_{(u,v)} \in \mathbb{R}^m$, such that $c_p = \sum_{(u,v) \in \mathcal{A}_p} c_{(u,v)}$ and $a_p = \sum_{(u,v) \in \mathcal{A}_p} a_{(u,v)}$, for every $p \in \mathcal{P}$ (see, e.g., [8]). Then, path flow model (1)–(3) admits an equivalent arc flow model:

$$\min \sum_{(u,v) \in \mathcal{A}} c_{(u,v)} \varphi_{(u,v)}, \tag{4}$$

$$\text{s.t.:} \ F_{\mathcal{N},\varphi}(v) = \begin{cases} -z, & \text{if } v = v^+, \\ z, & \text{if } v = v^-, \\ 0, & \text{otherwise,} \end{cases} \qquad \forall v \in \mathcal{V}, \tag{5}$$

$$\sum_{(u,v) \in \mathcal{A}} a_{(u,v)} \varphi_{(u,v)} \geq b, \tag{6}$$

$$\varphi_{(u,v)} \in \mathbb{Z}_+, \qquad\qquad \forall (u,v) \in \mathcal{A}, \tag{7}$$

$$z \in \mathbb{Z}, \tag{8}$$

where $F_{\mathcal{N},\varphi}(v) = \sum_{(u,v) \in \mathcal{A}} \varphi_{(u,v)} - \sum_{(v,w) \in \mathcal{A}} \varphi_{(v,w)}$. Variable $\varphi_{(u,v)}$ corresponds to the flow on arc $(u,v) \in \mathcal{A}$ and variable z corresponds to the total flow in the network. The objective function (4) minimizes the total cost from individual arcs, (5) are the flow conservation constraints, and (6) are general linear constraints equivalent to (2). The flow conservation theorem by Ahuja et al. [1] guarantees a correspondence between the linear relaxation of path flow models and arc flow models that are based on the same network. In this way, formulations (1)–(3) and (4)–(7) model the same problem and have the same primal strength. We exploit this equivalence to derive a number of exact techniques (Sects. 3, 4 and 5) and to produce a method for general network flow formulations based on DW decomposition (Sect. 6).

3   Column Generation

DW models for NP-hard problems are often associated with networks that have pseudo-polynomial or exponential size with respect to the size of the input. Consequently, large instances of both path flow and arc flow models derived from DW decompositions often have a huge number of variables, and their linear relaxation is usually solved by column generation algorithms. In such algorithms, the linear relaxation with a restricted set of columns, called *restricted master problem* (RMP), is iteratively solved. At each iteration, an oracle solves the pricing problem to generate non-basic columns with negative reduced cost. The algorithm halts when none of such columns can be found. For a general discussion on column generation, we refer the interested reader to [11] and [19].

In column generation for the path flow model (1)–(3), given a dual solution $\overline{\beta}$ associated to the linear relaxation of (2), the corresponding pricing problem

$$\min\{c_p - a_p^\top \overline{\beta} : p \in \mathcal{P}\} \tag{9}$$

can be solved as a shortest path problem on $\mathcal{N}$, where the cost of arc $(u, v) \in \mathcal{A}$ is set to $c_{(u,v)} - a_{(u,v)}^\top \overline{\beta}$. Since networks from DW models are acyclic, a shortest path can be efficiently found in $O(|\mathcal{A}|)$ by a topological ordering of the nodes (see, e.g., [1]).

In arc flow models, the oracle may generate a single arc, but in practice generating complete paths (to be decomposed in a set of arcs) can make a significant difference in convergence time (see [26]). Since the dual solutions associated to the linear relaxation of (2) and (6) are equivalent, the pricing problem that generates a complete path for an arc flow model can still be described as (9).

In some problems, generating a single column per iteration may lead to a slow convergence. In network flow formulations, it may be preferable to generate multiple paths at each iteration. For this purpose, we implemented an oracle that generates multiple paths with negative reduced cost (if any exists), and can be used to solve the pricing of both path flow and arc flow models.

The oracle that we developed works as follows. First, we solve a bidirectional *dynamic programming* (DP) model in $O(|\mathcal{A}|)$ to compute, for each arc $(u, v) \in \mathcal{A}$, the minimum reduced cost $\overline{c}_{(u,v)} = \min\{c_p - a_p^\top \overline{\beta} : p \in \mathcal{P}_{(u,v)}\}$ associated with paths in $\mathcal{P}_{(u,v)}$ (for details, we refer to [16]). Then, for each row $k = 1, ..., m$ in (2) or in (6), the algorithm selects an arc (u, v) with minimum $\overline{c}_{(u,v)}$ among the arcs that cover (i.e., has non-zero row coefficient) k, and, by using the DP structure, it generates a path in $\mathcal{P}_{(u,v)}$ of minimum reduced cost. At the end, the oracle has generated for each row k, a path of minimum reduced cost that covers k. For instance, the algorithm generates the columns of minimum reduced cost, such that each item (in packing problems) or each client (in routing problems) is covered by at least one of such columns. Notice that different rows may lead to the same column, so repeated columns are discarded. The overall algorithm has $O(m\zeta + |\mathcal{A}|)$ time complexity, where ζ is the maximum length of a path, and it is preferable to use it when the matrix of (2) is sparse, so that more different columns are generated.

4 Variable Fixing Based on Reduced Costs

Reduced-cost variable-fixing (RCVF) is a domain propagation technique that, for MILP (minimization) models, computes a primal bound z_{ub} on the optimal integer solution value and a dual-feasible solution of objective value z_{lb}. Then, it eliminates from the model any integer variable having a reduced cost greater than or equal to $z_{ub} - z_{lb}$ (see, e.g., [15] and page 389 of [21]). The way in which the dual solution of a path flow model can be used in an RCVF algorithm to remove non-optimal arcs has been discussed in [16]. The authors proposed a bidirectional search to compute, for each arc (u, v), the minimum reduced cost associated to paths in $\mathcal{P}_{(u,v)}$, which is then used to determine whether (u, v) can be removed from the network. An equivalent approach has been implemented by [3] in the general context of Lagrangian bounds for multivalued decision diagrams.

Different dual solutions may correspond to different reduced costs, and this directly affects the effectiveness of the RCVF. Moreover, although counter-intuitive, dual solutions that maximize the effectiveness of RCVF are often suboptimal (for a theoretical discussion, see [24]). Next, we present a method to obtain dual solutions with the aim of maximizing the RCVF effectiveness. Such method improves [16] in terms of reduction effectiveness, but at the cost of additional computational effort (see Sect. 8). The dual of the linear relaxation of the arc flow model (4)–(7) is given by:

$$\max \ b^\top \beta \tag{10}$$

$$\text{s.t.:} \ -\alpha_u + \alpha_v + a_{(u,v)}{}^\top \beta \le c_{(u,v)}, \qquad \forall (u, v) \in \mathcal{A}, \tag{11}$$

$$\alpha_{v^+} - \alpha_{v^-} \le 0, \tag{12}$$

$$\alpha_u \in \mathbb{R}, \qquad \forall u \in \mathcal{V}, \tag{13}$$

$$\beta_k \ge 0, \qquad \forall k = 1, ..., m, \tag{14}$$

where α and β are the dual variables associated with, respectively, (5) and (6). Let $\theta_{(u,v)} = c_{(u,v)} - (-\alpha_u + \alpha_v + a_{(u,v)}{}^\top \beta)$ denote the reduced cost of $(u, v) \in \mathcal{A}$. A dual solution $\overline{\beta}$ which allows the removal of (u, v) by RCVF is one which satisfies $b^\top \overline{\beta} + \theta_{(u,v)} \ge z_{ub}$. If one such solution exists, it can be obtained by solving (10)–(14) with a modified objective function given by $\max b^\top \beta + \theta_{(u,v)}$. However, optimizing a model tailored for each arc can be very time consuming in practice. For this reason, [2] proposed a method to maximize the impact of RCVF on general MILP models, by solving a single MILP model derived from an extension of the dual linear relaxation of the original model. This MILP model has an additional binary variable for each original variable, indicating whether the resulting dual solution is able to eliminate the associated original variable by RCVF. In our case, this method is generally non-practical, since the models we are concerned with usually have a huge number of variables and must rely on methods based on column generation. We thus propose a heuristic approach that solves a single LP model that considers an objective function modified by a subset $\mathcal{A}'$ of arcs, instead of a single one. This model is obtained by replacing (10) with $\max b^\top \beta + \sum_{(u,v)\in\mathcal{A}'} \theta_{(u,v)}$. The resulting objective function has additional terms related to α which lead to a (primal) problem with additional source and sink nodes that are forced to be in the solution. Consequently, from a primal point of view, a very large number of flow conservation constraints may be in the solution base, and this may substantially affect the computational performance. We consider a further simplification obtained by eliminating all additional α variables from the modified objective function. The resulting problem is thus:

$$\max\{b^\top \beta - \sum_{(u,v)\in\mathcal{A}'} a_{(u,v)}^\top \beta : (11),(12),(13),(14)\}. \tag{15}$$

Despite the additional terms in the objective function, (15) is still the dual of a general arc flow model, and can be solved by the column generation algorithm of Sect. 3. Following the correspondence of arc flow and path flow models, we solve the primal of (15) as a path flow model, and then use its dual solution $\bar{\beta}$ as the input of the RCVF method in [16]. In general, depending on the selection of $\mathcal{A}'$, the additional terms in the objective function may lead to an unbounded model (15), but there are special cases that guarantee a bounded model.

Proposition 1. *If the model* (10)–(14) *is bounded, and:* (i) $a_p \geq 0$ *for every* $p \in \mathcal{P}$; *or* (ii) $\mathcal{A}'$ *can be partitioned into a set of complete paths, then the corresponding model* (15) *is also bounded.*

Proof. Suppose that the dual model (10)–(14) is bounded and, consequently, its primal (4)–(7) admits a feasible solution φ'. We show that the corresponding model (15) is bounded under assumptions (i) or (ii) by showing that its primal is feasible. The primal of (15) is equivalent to (4)–(7), where b is incremented by $\sum_{(u,v)\in\mathcal{A}'} a_{(u,v)}$. In this way, to prove that (15) is bounded under any assumption (i) or (ii), it suffices to find a set $\mathcal{P}' \subseteq \mathcal{P}$ whose corresponding columns cover $\sum_{(u,v)\in\mathcal{A}'} a_{(u,v)}$ and the inclusion of one copy of each path $p \in \mathcal{P}'$ in φ' does not lead to a loss of feasibility in (4)–(7).

Suppose that (i) is true. Then, model (4)–(7) can be seen as a covering formulation, which guarantees that feasibility holds under the addition of any set of paths to the solution. In this case, $\mathcal{P}'$ can be given as any set of paths whose corresponding columns cover $\sum_{(u,v)\in\mathcal{A}'} a_{(u,v)}$. A trivial choice is $\mathcal{P}' = \mathcal{P}$.

Now, suppose that (ii) is true. In this case, $\mathcal{P}'$ can be given as the partition of $\mathcal{A}'$ into a set of paths. Then, adding one copy of each path $p \in \mathcal{P}'$ to φ' cancels the additional increment in b given by $\sum_{(u,v)\in\mathcal{A}'} a_{(u,v)}$, and the only significant changes appear in the objective function. As a result, $\sum_{(u,v)\in\mathcal{A}'} a_{(u,v)}$ is covered without losing feasibility in (4)–(7). ∎

5   A Variable-Selection Method Based on Arcs

Exact solutions for DW models represented as path flow models are typically based on B&P, which embeds column generation in a branch-and-bound scheme. On the other hand, exact solutions for arc flow models are typically based on the use of a general-purpose MILP solver. Although the practical efficiency of this method is usually limited to medium-sized instances, the consistent improvement of MILP solvers in the last decades allowed many hard problems to be solved by arc flow models (see, e.g., [8]).

This section presents a variable-selection method that can be useful in the solution of both path flow and arc flow models. It groups arcs that share mutual characteristics into subsets. It then considers arcs in subsets having null flow in the optimal linear solution of a model, and either eliminates all of them or forces at least one of them to be in the solution. This variable selection is then used as a base of a B&P scheme: in the left branch we solve the problem resulted from the arc-elimination as an arc flow model by a MILP solver (the resulting model is indeed expected to be substantially smaller and possibly hold good feasible solutions); whereas in the right branch we add a linear constraint imposing that at least one of the removed arcs is in the solution (hopefully improving the lower bound value). In the following, Sect. 5.1 presents the advantages of branching based on arc flow variables, Sect. 5.2 formally presents our variable-selection method, and Sect. 5.3 gives examples of possible applications.

5.1 Advantages of Branching Based on Arc Flow Variables

Any constraint $\sum_{(u,v)\in\mathcal{A}} a'_{(u,v)} \varphi_{(u,v)} \geq b'$ based on a linear combination of arc flow variables impacts on a pricing solved as a shortest path problem by simply incrementing ${a'_{(u,v)}}^{\top} \beta'$ to the cost of each arc $(u,v) \in \mathcal{A}$, where β' is the dual solution related to the constraint. A consequent result is the following:

Remark 1. Additional constraints based on a linear combination of arc flow variables do not increase the complexity of the pricing problem (9).

Remark 1 guarantees that branching rules based solely on arc flow variables are robust. Furthermore, any arc flow variable $\varphi_{(u,v)}$ can be represented as a sum $\sum_{p\in\mathcal{P}_{(u,v)}} \lambda_p$ of variables from the equivalent path flow model. Consequently, any linear constraint based on arc flow variables can be directly represented as a linear constraint in the equivalent path flow model. On the other hand, it is not always possible to rewrite a linear constraint based on path flow variables as a linear constraint based on arc flow variables. This motivates branching rules based on arc flow variables, as the resulting branching constraints can be easily handled by both path flow and arc flow models.

5.2 Variable-Selection Based on Arcs

We define an *arc family* $\mathcal{F} \subset 2^{\mathcal{A}}$ as a set of mutually disjoint subsets of $\mathcal{A}$. For each $F \in \mathcal{F}$, let variable $\Phi_F = \sum_{(u,v)\in F} \varphi_{(u,v)}$ represent the aggregated sum of arc flow variables associated to arcs in F. Given a linear solution $\overline{\varphi}$ of the arc flow model, we represent as $\overline{\Phi}_F = \sum_{(u,v)\in F} \overline{\varphi}_{(u,v)}$ the cumulated sum of the solution values of arcs in F. We consider three variable-selection methods.

The first variable-selection method (VS1) considers all variables related to the subfamily $\mathcal{B}_1 = \{F \in \mathcal{F} : \overline{\Phi}_F = 0\}$, which are used to create two branches.

In the *left branch*, we implicitly consider the branching constraint $\sum_{F \in \mathcal{B}_1} \Phi_F = 0$ by removing all arcs in each $F \in \mathcal{B}_1$ from the network, and in the *right branch* we add the constraint $\sum_{F \in \mathcal{B}_1} \Phi_F \geq 1$ to the model, implying that at least one arc in $\mathcal{B}$ must be in the solution. Depending on the definition of $\mathcal{F}$, the left branch is expected to lead to a great reduction on the size of the network, while keeping variables that should provide good feasible solutions. The reduced problem may be solved by an alternative method. In particular, we conclude this branch by solving the residual arc flow model by a MILP solver. The domain reduction in the right branch may be weaker, but the branching constraint may behave as a strong cutting plane. In fact, although no arcs are explicitly removed from the network in the right branch, the branching constraint provides a stronger relaxation that may improve the effectiveness of RCVF.

Sometimes, the reduction on the left branch can be too large, so that no good feasible solutions are found. A more balanced reduction of the problem in this branch should increase the chance of finding better feasible solutions. For that, we propose a filtering method to obtain a smaller subfamily $\mathcal{B}_2 \subseteq \mathcal{B}_1$ of branching variables. The aim of the second variable-selection method (VS2) is to remove from $\mathcal{B}_1$ sets of arcs that assume aggregated values greater than or equal to 1 in an alternative optimal linear solution. This is done by solving the path flow model related to the current node with a different objective function, given by $\max \sum_{F \in \mathcal{B}_1} \Phi_F$ and an additional linear constraint $\sum_{(u,v) \in \mathcal{A}} c_{(u,v)} \varphi_{(u,v)} \leq z_{lb}$, where z_{lb} is the optimal linear solution value. The model is optimized by column generation, with an additional criterion to stop whenever a solution with $\overline{\Phi}_F \geq 1$, for some $F \in \mathcal{B}_1$, is found. Then, from the solution $\overline{\Phi}$ obtained, we derive $\mathcal{B}_2 = \mathcal{B}_1 \setminus \{F \in \mathcal{B}_1 : \overline{\Phi}_F \geq 1\}$.

When $\mathcal{B}_1 = \emptyset$, we proceed with the third variable-selection method (VS3), which randomly chooses a set $F' \in \{F \in \mathcal{F} : \overline{\Phi}_F$ is fractional and minimum$\}$. Then, two branches are created, one with the addition of constraint $\Phi_{F'} \leq \lfloor \overline{\Phi}_{F'} \rfloor$, and the other with the addition of constraint $\Phi_{F'} \geq \lceil \overline{\Phi}_{F'} \rceil$. Depending on the definition of $\mathcal{F}$, branching on variables Φ may not be sufficient to achieve integrality. Then, whenever a solution that is integer with respect to the variables Φ_F is found, an alternative method may be used to conclude the enumeration.

5.3   Examples of Arc Families

One may use intuition based on the problem being solved to determine a good choice of $\mathcal{F}$. Next, we give some examples to be explored on some applications:

(i) In many scheduling problems, network flow formulations have arcs relating a job to its start time (see, e.g., [18]). The optimal linear solution could provide a good time interval where a job should be processed in an optimal solution. In such cases, a definition of $\mathcal{F}$ could be based on splitting the time horizon in many parts, and each set in $\mathcal{F}$ is associated with starting the processing of a job in a given split of the time horizon.

(ii) In many vehicle routing problems, network flow formulations have arcs representing the departure of a vehicle with a given capacity from a client to another. Usually, optimal solutions do not consider very long arcs between clients. In such cases, the arc family can be defined by having two sets per client: one having arcs related to the visit of the closest next clients; and one having arcs related to the visit of the farthest next clients. An appropriate ordering of the clients should be considered to guarantee that each arc is in a single set. Then, in the left branch, we expect to eliminate most of the longest arcs. In the right branch, it is enforced that one of the longest arcs should be used. This idea is related to the sparsification heuristic by Fukasawa et al. [12], which eliminates long arcs that are unlikely to be in an optimal solution.

(iii) A general arc family that can be used in any network flow formulation is given by $\mathcal{F} = \{F_u = \{(u, v) \in \mathcal{A}\} : u \in \mathcal{V}\}$. The sets are defined by the nodes of the network, each having all arcs that have a tail on the associated node. Then, the branching over this arc family corresponds to disabling all nodes of a set in the left branch, and enforcing that one node of the set must be in the solution in the right branch.

A number of arc families can be derived for the CSP. However, for the sake of conciseness, our computational experiments below only consider the arc family in (iii), which provided the best results.

6   A Solution Method for Network Flow Formulations

We combine the techniques presented in the previous sections into a solver for network flow formulations. The solver is based on the iterative solution of either path flow or arc flow models by column generation with the oracle discussed in Sect. 3, and of arc flow models by a MILP solver. The input read by the solver is a network $\mathcal{N}$, an arc family $\mathcal{F}$, an upper bound on the optimal solution value, and the parameters of the arc flow model (4)–(7), i.e., the objective coefficients and the linear constraints.

At the root node, the linear relaxation is solved by the column generation algorithm of Sect. 3. Then, RCVF is applied for π_1 iterations. In the first iteration, the dual solution considered is the one obtained at the end of the column generation. In the next $\pi_1 - 1$ iterations, the dual solution is obtained by the method of Sect. 4. The solver uses the branching scheme of Sect. 5 based on the input $\mathcal{F}$, and the branching tree is limited to π_2 levels. At each level of the tree, the left branch, i.e., the branch related to the elimination of arcs from the model, is directly solved as an arc flow model by a MILP solver. Although the intention of the algorithm is to provide relatively easier problems in the left branch, in some cases those problems can be small but still hard enough and consume most of the overall time limit. However, it is important to mention that no additional stopping criterion is used to deal with such cases. In the right branch, the linear relaxation with the additional branching constraint is solved, and it is followed by two iterations of RCVF, the first using the dual solution from the column generation and the second using a dual solution obtained by the method in Sect. 4.

The right branch is to be branched again in the first $\pi_2 - 1$ levels of the tree or to be solved as an arc flow model by a MILP solver in the last level of the tree. The branching at the first $\pi_2 - 1$ levels of the tree are based on VS1, whereas the last level of branching is based on VS2. The tree is explored by breadth-first search by prioritizing left branches. No parallelism is implemented in the exploration of the tree.

7   An Application to the Cutting Stock Problem

We apply the proposed methods to the CSP. In the CSP, we are given an unlimited number of stock rolls of length $W \in \mathbb{Z}_+$ and a set I of items, each $i \in I$ associated to a width $w_i \in \mathbb{Z}_+$ and a demand $d_i \in \mathbb{Z}_+$. The objective is to cut the minimum number of stock rolls in order to obtain the demands of all items. An equivalent problem is the bin packing problem (BPP), where all demands are unitary. We refer the interested reader to [10] for a recent survey and to [9], [22], and [30] for the most recent exact methods that solve the CSP/BPP.

The classical pattern-based model by Gilmore and Gomory [13,14] can be derived from a DW decomposition of the textbook model by Martello and Toth [20]. The DW model in [13,14] can be seen as a path flow model based on the DP network of an unbounded knapsack problem (see, e.g., [8]). The equivalent arc flow model was first solved by Valério de Carvalho [26]. The author proposed reduction criteria to remove arcs from the network, without loss of optimality, by considering that items can always be cut from a stock roll following an ordering of non-increasing widths. Later, Côtê and Iori [6] proposed the *meet-in-the-middle patterns*, which allowed to produce significantly smaller networks.

Based on the following property, we developed a technique that further reduces the network from [26] and may lead to networks smaller than the ones resulting from the meet-in-the-middle patterns.

Property 1. Given a CSP instance, let $\overline{W} \in \mathbb{Z}_+$ be a value ensuring that there exists an optimal solution where the maximum waste of a single stock roll is at most $\overline{W}$. Then, all arcs contained only in paths associated to cutting patterns with a waste larger than $\overline{W}$ can be removed from the network.

A straightforward way to compute $\overline{W}$ considers that in any solution of the CSP with K stock rolls, the maximum waste on each roll is at most $KW - \sum_{i \in I} w_i d_i$. All arcs that only lead to cutting patterns with a waste larger than $\overline{W}$ are computed by a back propagation in the network, similar to the method by Trick [25] to propagate knapsack graphs in a constraint programming context. The reduction effectiveness depends on the maximum waste computed. In particular, instances with a weak continuous lower bound may lead even to no reduction at all. The maximum waste computation that we presented is very basic, but it obtained good results on the instances that we attempted and could even be replaced by more sophisticated methods, as, e.g., LP formulations.

We solve the CSP as a network flow formulation based on the network in [26] with the reductions based on Property 1. We use the method of Sect. 6, with $\pi_1 = 4$ and $\pi_2 = 10$, and solve its linear relaxation as a path flow model. The RMP initial base consists of all cutting patterns with three items and no waste. In our experiments, a simple best-fit heuristic always produced an initial upper bound at most equal to the dual bound plus 1 unit. Indeed, there is no known CSP instance with an optimal solution greater than this (see, e.g., [4] and [17]). We highlight that the matrix of the pattern-based model for the CSP follows the case (i) of Proposition 1. Then, when computing the alternative dual solutions with the method of Sect. 4, we consider the subset $\mathcal{A}'$ equal to $\mathcal{A}$. The branch is based on arc family (iii) of Sect. 5.3, where each $u \in \mathcal{V}$ corresponds to the partial length of a stock roll. The overall algorithm is referred to as CSP-BAP.

8   Computational Experiments

The algorithms were coded in C++ and Gurobi 9.0.3 was used to solve the LP and the MILP models. The experiments were run on a computer with an Intel Xeon X3470 at 2.93 Ghz and 8 GB of RAM. We did not impose an upper limit on the number of threads used by Gurobi. The experiments focus on the most difficult classes of benchmark instances for the CSP, namely the AI and ANI classes proposed in [10]. Each of these classes has 250 instances divided into 5 groups, each composed of 50 instances having the same number of items. Instances in AI have optimal solution value equal to the optimal dual bound z_{lb} of the pattern-based model, whereas instances in ANI have optimal solution value equal to $z_{lb} + 1$. Thus, class AI tests the ability to quickly find better feasible solutions, whereas class ANI tests the ability to increase the lower bound. Overall, the reduction criterion based on Property 1 allowed to obtain networks with an average of 63% less arcs when compared to the meet-in-the-middle networks. This reduction is explained by the fact that the considered instances do not allow waste in any solution of value z_{lb}. At the root node, the basic RCVF presented in [16] eliminated on average 61.4% of arcs. A single additional iteration of RCVF based on a dual solution obtained by the method in Sect. 4 further removed on average 37.2% of the arcs from the residual network, within just an average additional time of 19.6 s.

In Table 1, we compare four variations of CSP-BAP to analyze the different components of the algorithm. Algorithm *Arc Flow* solves the problem directly as an arc flow model by Gurobi, i.e., no custom branching or RCVF is applied. Algorithm *no-branch* solves the root node of CSP-BAP, and after the five iterations of RCVF, solves the resulting model as an arc-flow with Gurobi. Algorithm *no-opt-rcvf* corresponds to CSP-BAP without the last four iterations of RCVF at the root node, and without the second iteration of RCVF at the remaining nodes. Algorithm *no-VS2* corresponds to CSP-BAP where the variable selection at the last level of the tree is based on VS1 instead of VS2. Algorithm *no-property1* corresponds to CSB-BAP without the reduction on the network given by

Property 1. For each algorithm, columns time and $|opt|$ present, for each group of 50 instances, the average solution time and the total number of instances solved to optimality within the time limit of 600 s. Overall, CSP-BAP solved the largest number of instances. Comparing the results in columns no-opt-rcvf and CSP-BAP, we notice that the method in Sect. 4 produced a negative impact on CSP-BAP for the AI class, but it allowed to solve 12 more ANI instances. Comparing columns no-VS2 and CSP-BAP, we notice that the use of VS2 in the last iteration of CSP-BAP allowed to solve 10 additional AI instances. From column no-property1 we notice that the reduction given by Property 1 produced a negative impact on overall results of AI. This may be due to the fact that a larger number of arcs allows a larger number of distinct optimal solutions (as already observed in [9]), and hence, it may be easier to find an optimal solution. However, we can also observe that the reduction is crucial (although not sufficient) to solve a substantially larger number of ANI instances. This happens because for those instances the focus is to prove optimality, i.e., to show that no solution of value z_{lb} exists. In this way, differently from what happens for the AI instances, a more reduced domain is always preferable.

Table 1. Comparison on the efficiency of individual components (time limit: 600s)

Class	$|I|$	Arc flow time	Arc flow $|opt|$	no-branch time	no-branch $|opt|$	no-opt-rcvf time	no-opt-rcvf $|opt|$	no-VS2 time	no-VS2 $|opt|$	no-property1 time	no-property1 $|opt|$	CSP-BAP time	CSP-BAP $|opt|$
AI	202	25.7	**50**	17.8	**50**	0.5	**50**	1.0	**50**	1.8	**50**	0.8	**50**
	403	495.5	16	177.9	37	8.4	**50**	23.3	49	32.2	49	10.2	**50**
	601	600.0	0	313.7	26	46.5	**49**	89.9	46	126.3	45	71.2	**49**
	802	600.0	0	334.1	26	220.5	37	260.6	35	221.3	**39**	260.3	37
	1003	600.0	0	453.4	17	391.2	26	487.8	20	415.5	**30**	458.6	24
Avg./total		464.4	66	259.4	156	133.4	212	172.5	200	159.4	**213**	160.2	210
ANI	201	24.0	**50**	6.5	**50**	15.3	**50**	9.6	**50**	145.3	44	9.7	**50**
	402	586.5	6	61.5	46	75.1	46	65.4	46	395.3	19	60.0	**47**
	600	600.0	0	151.6	39	173.4	39	162.1	41	529.6	7	161.8	40
	801	600.0	0	271.0	**31**	329.5	28	287.0	**31**	600	0	298.2	**31**
	1002	600.0	0	405.3	22	460.3	18	397.8	**25**	600	0	427.8	**25**
Avg./total		482.1	56	179.2	188	210.7	181	184.4	**193**	454.04	70	191.5	**193**
Overall		473.3	122	219.3	344	172.1	393	178.4	393	306.7	283	175.8	**403**

Table 2 presents a comparison of CSP-BAP with the most recent exact results for the BPP/CSP. The method in [9] is based on an enhanced solution of the reflect formulation (a pseudo-polynomial arc flow model for the CSP) and it was solved with an Intel Xeon at 3.10 GHz and 8 GB RAM. The method in [30] is a branch-and-cut-and-price algorithm for the BPP solved with a Intel Xeon E5-1603 at 2.80-GHz and 8 GB RAM. The method in [22] is a branch-and-cut-and-price algorithm for a general class of optimization problems that was solved with a Intel Xeon E5-2680 at 2.50 GHz with 128 GB RAM shared between 8 copies of the algorithm running in parallel. All experiments considered a time limit of 3600 s per instance. In order to compare the performance of each CPU, we provide their passmark indicators (available at www.passmark.com), where higher values are associated with better performance. The computer used in the experiments of [9,22,30], and the present work have single-thread passmark indicators 2132, 1763, 1635, and 1414, respectively. When compared to the best previous results (presented in [22]), CSP-BAP could solve 64 (i.e., 40%) more AI instances and 98 (i.e., 95%) more ANI instances.

Table 2. Comparison with the state-of-the-art for the CSP (time limit:3600 s)

Class	$|I|$	Delorme and Iori [9] Time	$|opt|$	Wei et al. [30] Time	$|opt|$	Pessoa et al. [22] Time	$|opt|$	CSP-BAP Time	$|opt|$
AI	202	8.5	**50**	4.2	**50**	52.3	**50**	0.8	**50**
	403	1205	40	398.1	46	491.4	47	10.2	**50**
	601	–	–	1759.6	27	1454.1	35	72.8	**50**
	802	–	–	2766.3	15	2804.7	28	667.8	**46**
	1003	–	–	3546.1	2	–	–	1825.2	**28**
Avg./Total		606.8	90	1694.9	140	1200.6	160	515.3	**224**
ANI	201	49.3	**50**	13.9	**50**	16.7	**50**	10.1	**50**
	402	2703.9	17	436.2	47	96.0	**50**	229.4	48
	600	–	–	3602.7	0	3512.5	3	659.5	**42**
	801	–	–	3605.9	0	3600.0	0	1318.8	**33**
	1002	–	–	3637.7	0	–	–	1734.2	**28**
Avg./Total		1376.6	67	2259.3	97	1806.3	103	790.4	**201**
Overall		991.7	157	1977.1	237	1503.5	263	652.9	**425**

9   Conclusions

In this paper, we proposed techniques for general network flow formulations derived from DW decompositions. The proposed techniques were combined into a method based on column generation and on the iterative solutions of arc flow models by a MILP solver. The method was then applied to solve the CSP. In our computational experiments, 162 out of 237 open instances for the CSP could be solved to proven optimality for the first time.

We performed preliminary tests on different arc families for the CSP. Some of them obtained promising results, but we presented here only the one that provided the best results. As future research, we will explore additional arc families for the CSP and the combination of branching based on different arc families. We expect that an appropriate combination of arc families may increase the efficiency of the solver in finding good feasible solutions. We also intend to explore the application of the techniques to different problems. From preliminary tests, we verified that the proposed method may improve the solution of a class of scheduling problems. We also intend to study the possibility of embedding our method within an iterative aggregation/disaggregation framework (see, e.g., [5]) to improve the solution of instances with very large networks.

References

1. Ahuja, R., Magnanti, T., Orlin, J.: Network Flows: Theory, Algorithms, and Applications. Prentice-Hall, New Jersey (1993)
2. Bajgiran, O.S., Cire, A.A., Rousseau, L.-M.: A first look at picking dual variables for maximizing reduced cost fixing. In: Salvagnin, D., Lombardi, M. (eds.) CPAIOR 2017. LNCS, vol. 10335, pp. 221–228. Springer, Cham (2017). https://doi.org/10.1007/978-3-319-59776-8_18
3. Bergman, D., Cire, A.A., van Hoeve, W.-J.: Lagrangian bounds from decision diagrams. Constraints **20**(3), 346–361 (2015). https://doi.org/10.1007/s10601-015-9193-y
4. Caprara, A., Dell'Amico, M., Díaz-Díaz, J., Iori, M., Rizzi, R.: Friendly bin packing instances without integer round-up property. Math. Program. **150**, 5–17 (2015)
5. Clautiaux, F., Hanafi, S., Macedo, R., Voge, M.E., Alves, C.: Iterative aggregation and disaggregation algorithm for pseudo-polynomial network flow models with side constraints. Eur. J. Oper. Res. **258**(2), 467–477 (2017)
6. Côté, J.F., Iori, M.: The meet-in-the-middle principle for cutting and packing problems. INFORMS J. Comput. **30**(4), 646–661 (2018)
7. Dantzig, G., Wolfe, P.: The decomposition algorithm for linear programs. Econometrica **29**(4), 767–778 (1961)
8. de Lima, V., Alves, C., Clautiaux, F., Iori, M., Valério de Carvalho, J.: Arc flow formulations based on dynamic programming: Theoretical foundations and applications (2020). https://arxiv.org/abs/2010.00558
9. Delorme, M., Iori, M.: Enhanced pseudo-polynomial formulations for bin packing and cutting stock problems. INFORMS J. Comput. **32**(1), 101–119 (2020)
10. Delorme, M., Iori, M., Martello, S.: Bin packing and cutting stock problems: mathematical models and exact algorithms. Eur. J. Oper. Res. **255**(1), 1–20 (2016)
11. Desaulniers, G., Desrosiers, J., Solomon, M.: Column Generation. Springer, New York (2006)
12. Fukasawa, R., et al.: Robust branch-and-cut-and-price for the capacitated vehicle routing problem. Math. Program. **106**(3), 491–511 (2006). https://doi.org/10.1007/s10107-005-0644-x
13. Gilmore, P., Gomory, R.: A linear programming approach to the cutting-stock problem. Oper. Res. **9**(6), 849–859 (1961)
14. Gilmore, P., Gomory, R.: A linear programming approach to the cutting stock problem - part II. Oper. Res. **11**(6), 863–888 (1963)
15. Hadjar, A., Marcotte, O., Soumis, F.: A branch-and-cut algorithm for the multiple depot vehicle scheduling problem. Oper. Res. **54**(1), 130–149 (2006)
16. Irnich, S., Desaulniers, G., Desrosiers, J., Hadjar, A.: Path-reduced costs for eliminating arcs in routing and scheduling. INFORMS J. Comput. **22**(2), 297–313 (2010)
17. Kartak, V., Ripatti, A., Scheithauer, G., Kurz, S.: Minimal proper non-IRUP instances of the one-dimensional cutting stock problem. Discrete Appl. Math. **187**, 120–129 (2015)
18. Kramer, A., Dell'Amico, M., Iori, M.: Enhanced arc-flow formulations to minimize weighted completion time on identical parallel machines. Eur. J. Oper. Res. **275**(1), 67–79 (2019)
19. Lübbecke, M., Desrosiers, J.: Selected topics in column generation. Oper. Res. **53**(6), 1007–1023 (2005)

192 V. L. de Lima et al.

20. Martello, S., Toth, P.: Knapsack Problems: Algorithms and Computer Implementations. Wiley, New York (1990)
21. Nemhauser, G., Wolsey, L.: Integer and Combinatorial Optimization. Wiley, New York (1988)
22. Pessoa, A., Sadykov, R., Uchoa, E., Vanderbeck, F.: A generic exact solver forvehicle routing and related problems. Math. Program. **183**, 483–523 (2020)
23. Pessoa, A., Uchoa, E., de Aragão, M., Rodrigues, R.: Exact algorithm over an arc-time-indexed formulation for parallel machine scheduling problems. Math. Program. Comput. **2**, 259–290 (2010)
24. Sellmann, M.: Theoretical foundations of CP-based Lagrangian relaxation. In: Wallace, M. (ed.) CP 2004. LNCS, vol. 3258, pp. 634–647. Springer, Heidelberg (2004). https://doi.org/10.1007/978-3-540-30201-8_46
25. Trick, M.: A dynamic programming approach for consistency and propagation for knapsack constraints. Ann. Oper. Res. **118**(1–4), 73–84 (2003)
26. Valério de Carvalho, J.: Exact solution of bin-packing problems using column generation and branch-and-bound. Ann. Oper. Res. **86**, 629–659 (1999)
27. Vanderbeck, F.: On Dantzig-Wolfe decomposition in integer programming and ways to perform branching in a branch-and-price algorithm. Oper. Res. **48**(1), 111–128 (2000)
28. Vanderbeck, F.: Branching in branch-and-price: a generic scheme. Math. Program. **130**, 249–294 (2011)
29. Villeneuve, D., Desrosiers, J., Lübbecke, M., Soumis, F.: On compact formulations for integer programs solved by column generation. Ann. Oper. Res. **139**, 375–388 (2005)
30. Wei, L., Luo, Z., Baldacci, R., Lim, A.: A new branch-and-price-and-cut algorithm for one-dimensional bin-packing problems. INFORMS J. Comput. **32**(2), 428–443 (2020)

Multi-cover Inequalities for Totally-Ordered Multiple Knapsack Sets

Alberto Del Pia[1,2], Jeff Linderoth[1,2], and Haoran Zhu[1(✉)]

[1] Department of Industrial and Systems Engineering,
University of Wisconsin-Madison, Madison, USA
{delpia,linderoth,hzhu94}@wisc.edu
[2] Wisconsin Institute for Discovery, Madison, USA

Abstract. We propose a method to generate cutting-planes from multiple covers of knapsack constraints. The covers may come from different knapsack inequalities if the weights in the inequalities form a totally-ordered set. Thus, we introduce and study the structure of a totally-ordered multiple knapsack set. The valid *multi-cover inequalities* we derive for its convex hull have a number of interesting properties. First, they generalize the well-known $(1, k)$-configuration inequalities. Second, they are not aggregation cuts. Third, they cannot be generated as a rank-1 Chvátal-Gomory cut from the inequality system consisting of the knapsack constraints and all their minimal covers. Finally, we provide conditions under which the inequalities are facet-defining for the convex hull of the totally-ordered knapsack set.

Keywords: Multiple knapsack set · Cutting-planes · Cover inequalities

1 Introduction

We study cutting-planes related to *covers* of 0–1 knapsack sets. For a 0–1 knapsack set

$$K_{\text{KNAP}} := \{x \in \{0,1\}^n \mid a^T x \le b\},$$

with $(a, b) \in \mathbb{Z}_+^{n+1}$, a *cover* is any subset of elements $C \subseteq [n]$ such that $\sum_{j \in C} a_j > b$. The *cover inequality (CI)*

$$\sum_{j \in C} x_j \le |C| - 1$$

is a valid inequality for the *knapsack polytope* $\text{conv}(K_{\text{KNAP}})$ that separates the invalid characteristic vector of C. There is a long and rich literature on (lifted) cover inequalities for the knapsack polytope [1,7,8,10,15], and the readers are directed to the recent survey of [9] for a more complete background.

For the binary-valued set

$$X = \{x \in \{0,1\}^n \mid Ax \le b\},$$

© Springer Nature Switzerland AG 2021
M. Singh and D. P. Williamson (Eds.): IPCO 2021, LNCS 12707, pp. 193–207, 2021.
https://doi.org/10.1007/978-3-030-73879-2_14

where $[A, b] \in \mathbb{Z}_+^{m \times (n+1)}$, a standard and computationally-useful way for generating valid inequalities to improve the linear programming relaxation of X is to generate *lifted cover inequalities* for the knapsack sets defined by the individual constraints of X [4]. In this way, the extensive literature regarding valid inequalities for conv(K_{KNAP}) can be leveraged to solve integer programs whose feasible region is X. In contrast to K_{KNAP}, very little is known about polyhedra that arise as the convex hull of the intersection of *multiple* knapsack sets. In this paper, we introduce a family of cutting-planes, called *(antichain) multi-cover inequalities ((A)MCIs)*, that are derived by simultaneously considering multiple covers which satisfy some particular condition. The covers may come from *any* inequality in the formulation, so long as the weights appearing in the knapsack inequalities are totally-ordered.

More formally, we give a new approach to generate valid inequalities for a special multiple knapsack set, called the *totally-ordered multiple knapsack set (TOMKS)*. Given a constraint matrix $A \in \mathbb{Z}_+^{m \times n}$ whose columns $\{A_1, A_2, \ldots, A_n\}$ form a chain ordered by component-wise order, i.e., $A_1 \geq A_2 \ldots \geq A_n$, and a right-hand-side vector $b \in \mathbb{Z}_+^m$, the TOMKS is the set

$$K = \{x \in \{0,1\}^n \mid Ax \leq b\}. \tag{1}$$

TOMKS can arise in the context of chance-constrained programming. Specifically, consider a knapsack constraint where the weights of the items (a) depend on a random variable (ξ), and we wish to satisfy the chance constraint

$$\mathbb{P}\{a(\xi)^T x \leq \beta\} \geq 1 - \epsilon, \tag{2}$$

selecting a subset of items $(x \in \{0,1\}^n)$ so that the likelihood that these items fit into the knapsack is sufficiently high. In the scenario approximation approach proposed in [3,12], an independent Monte Carlo sample of N realizations of the weights $(a(\xi^1), \ldots a(\xi^N))$ is drawn and the deterministic constraints

$$a(\xi^i)^T x \leq \beta \quad \forall i = 1 \ldots, N \tag{3}$$

are enforced. In [11] it is shown that if the sample size N is sufficiently large:

$$N \geq \frac{1}{2\epsilon^2} \left(\log \left(\frac{1}{\delta} \right) + n \log(2) \right),$$

then any feasible solution to (3) satisfies the constraint (2) with probability at least $1 - \delta$. If the random weights of the items $a_1(\xi), a_2(\xi), \ldots a_n(\xi)$ are independently distributed with means $\mu_1 \geq \mu_2 \ldots \geq \mu_n$, then the feasible region in (3) may either be a TOMKS, or the constraints can be (slightly) relaxed to obey the ordering property.

But the TOMKS may arise in more general situations as well. For a general binary set X, if two knapsack inequalities $a_1^T x \leq b_1$ and $a_2^T x \leq b_2$ have non-zero coefficients in very few of the same variables, their intersection may be totally-ordered, and the (A)MCI would be applicable in this case. In the special case where the multiple covers come from the same knapsack set, the (A)MCI can also produce interesting inequalities. For example, the well-known $(1,k)$-*configuration* inequalities for $\mathrm{conv}(K_{\mathrm{KNAP}})$ [13] are a special case of (A)MCI where all covers come from the same inequality (see Proposition 4). We also give an example where a facet of $\mathrm{conv}(K_{\mathrm{KNAP}})$ found by a new lifting procedure described in [10] is a MCI.

A MCI is generated by a simple algorithm (given in Algorithm 1) that takes as input a special family of covers $\mathscr{C} = \{C_1, C_2, \ldots C_k\}$ that obeys a certain maximality criterion (defined in Definition 3). For many types of cover families $\mathscr{C}$, the MCI may be given in closed-form. In the case that the family of covers $\mathscr{C}$ is an antichain in a certain partial order, the resulting MCI has the interesting property that it simultaneously cuts off at least two of the characteristic vectors of the covers in $\mathscr{C}$. We also give conditions under which the MCI yields a facet for $\mathrm{conv}(K)$ in Sect. 5.

The MCI may be generated by simultaneously considering multiple knapsack inequalities defining K. Another mechanism to generate inequalities taking into account information from multiple constraints of the formulation is to aggregate inequalities together, forming the set

$$\mathcal{A}(A,b) := \bigcap_{\lambda \in \mathbb{R}^m_+} \mathrm{conv}(\{x \in \{0,1\}^n \mid \lambda^T A x \leq \lambda^T b\}).$$

Inequalities valid for $\mathcal{A}(A,b)$ are known as *aggregation cuts*, and have been shown to be quite powerful from both an empirical [6] and theoretical [2] viewpoint. The well-known Chvátal-Gomory (CG) cuts, lifted knapsack cover inequalities, and weight inequalities [14] are all aggregation cuts. In Example 3, we show that MCI are not aggregation cuts. Further, in Example 4, we show that MCI cannot be obtained as a (rank-1) Chvátal-Gomory cut from the linear system consisting of all minimal cover inequalities from K.

The paper is structured as follows: In Sect. 2, we define a certain type of dominance relationship between covers that is necessary for MCI. The MCI is defined in Sect. 3, where we also give many examples to demonstrate that MCIs are not dominated by other well-known families of cutting-planes. In Sect. 4, we propose a strengthening of MCI in the case that the cover-family forms an antichain in a certain partially ordered set. Section 5 provides sufficient condition for the MCI to be a facet-defining inequality for $\mathrm{conv}(K)$.

Notation. For a positive integer n, we denote by $[n] := \{1, \ldots, n\}$. For $x \in \mathbb{R}^n$, $\mathrm{supp}(x) := \{i \in [n] \mid x_i \neq 0\}$ denotes the support of x. The *characteristic vector* of a set $S \subseteq [n]$ is denoted by χ^S. Therefore, given a TOMKS K, we say that a set $S \subseteq [n]$ is a *cover* for K if $\chi^S \notin K$. For a vector $x \in \mathbb{R}^n$ and a set $S \subseteq [n]$, we define $x(S) := \sum_{i \in S} x_i$. This in particular means $x(\emptyset) = 0$. We denote the *power set* of a set S by 2^S, which is the set of all subsets of S.

2   A Dominance Relation

In this section we define and provide some properties of a type of dominance relationship between covers.

Definition 1 (Domination). *For $S_1, S_2 \subseteq [n]$, we say that S_1 dominates S_2 and write $S_1 \rhd S_2$, if there exists an injective function $f : S_2 \to S_1$ with $f(i) \leq i \; \forall i \in S_2$.*

The dominance relation in Definition 1 is reflexive, antisymmetric, and transitive, so $(2^{[n]}, \rhd)$ forms a partially ordered set (poset). For two sets $S_1, S_2 \subseteq [n]$, we say S_1 and S_2 are *comparable* if $S_1 \rhd S_2$ or $S_2 \rhd S_1$.

The dominance relation has a natural use in the context of covers. In fact, if C_2 is a cover for a TOMKS K and C_1 dominates C_2, then C_1 is also a cover for K. Next, we present two technical lemmas. The proofs are technical and can be found in the full version of the paper [5].

Lemma 1. *Let $S_1, S_2 \subseteq [n]$ with $S_1 \neq S_2$. Then for any $S' \subseteq S_1 \cap S_2$, $S_1 \rhd S_2$ if and only if $S_1 \setminus S' \rhd S_2 \setminus S'$.*

Lemma 2. *Let $S \subseteq [n]$ and let $S_1, S_2 \subseteq S$. Then, $S_1 \rhd S_2$ if and only if $S \setminus S_2 \rhd S \setminus S_1$.*

3   Multi-cover Inequalities

Throughout this section, we consider a TOMKS $K := \{x \in \{0,1\}^n \mid Ax \leq b\}$, and we introduce the *multi-cover inequalities (MCIs)*, which form a novel family of valid inequalities for K. Each MCI can be obtained from a special family of covers $\{C_1, \ldots, C_k\}$ for K that we call a multi-cover. In order to define a multi-cover, we first introduce the discrepancy family.

Definition 2 (Discrepancy family). *For a family of sets $\mathscr{C} = \{C_1, \ldots, C_k\}$, we say that $\{C_1 \setminus \cap_{h=1}^k C_h, \ldots, C_k \setminus \cap_{h=1}^k C_h\}$ is the discrepancy family of $\mathscr{C}$, and we denote it by $\mathcal{D}(\mathscr{C})$.*

Now we can define the concept of a multi-cover.

Definition 3 (Multi-cover). *Let $\mathscr{C}$ be a family of covers for K. We then say that $\mathscr{C}$ is a multi-cover for K if for any set $T \subseteq \cup_{D \in \mathcal{D}(\mathscr{C})} D$ with $T \notin \mathcal{D}(\mathscr{C})$, there exists some $D' \in \mathcal{D}(\mathscr{C})$ such that T is comparable with D'.*

For a given family of covers $\{C_1, \ldots, C_k\}$ for K, throughout this paper, for ease of notation we define $C_0 := \cap_{h=1}^k C_h$, $C := \cup_{h=1}^k C_h$, $\bar{C}_h := C \setminus C_h$ for $h \in [k]$, and similarly $\bar{T} := C \setminus T$ for any $T \subseteq C$.

We are now ready to introduce our multi-cover inequalities for K. These inequalities are defined by the following algorithm.

Algorithm 1. Multi-cover inequality (MCI)

Input: A multi-cover $\{C_1, \ldots, C_k\}$ for K.
Output: A multi-cover inequality.
1: Let $C \setminus C_0 = \{i_1, \ldots, i_m\}$, with $i_1 < \ldots < i_m$.
2: Set $\alpha_i := 1$ if $i \in \{i_1, \ldots, i_m\}$, and $\alpha_i := 0$ otherwise.
3: **for** $t = m - 1, \ldots, 1$ **do**
4:     $\alpha_{i_t} := \max_{h \in [k]: i_t \in C_h} \max_{\ell \in \bar{C}_h : \ell > i_t} \alpha_\ell + 1$.
5: **for** $j \in C_0$ **do**
6:     $\alpha_j := \min_{h \in [k]} \max \left\{ \max_{\ell < j, l \in \bar{C}_h} \alpha_\ell, \sum_{t > j, t \in \bar{C}_h} \alpha_t + 1 \right\}$.
7: $\beta := \max_{h=1}^{k} \alpha(C_h) - 1$.
8: **return** the inequality $\alpha^T x \leq \beta$.

We remark that in Algorithm 1, in the case where we take the minimum or maximum over an empty set (see Step 4 and 6), the corresponding minimum or maximum is defaulted to take the value zero.

For the above algorithm, we have the following easy observations.

Observation 1. *Given a multi-cover $\{C_h\}_{h=1}^{k}$, Algorithm 1 performs a number of operations that is polynomial in $|C|$ and k. Furthermore, $\mathrm{supp}(\alpha) = C$.*

The main result of this section is that, given a multi-cover for K, the corresponding MCI is valid for $\mathrm{conv}(K)$. Before presenting the theorem, we will need the following auxiliary result.

Proposition 1. *Let $\{C_h\}_{h=1}^{k}$ be a multi-cover and let $\alpha^T x \leq \beta$ be the associated MCI. If there exists $T \subseteq C \setminus C_0, T \notin \{\bar{C}_h\}_{h=1}^{k}$, with $T \triangleright \bar{C}_{h'}$ for some $h' \in [k]$, then $\alpha(T) > \alpha(\bar{C}_{h'})$.*

Proof. Let T and $\bar{C}_{h'}$ be the sets as assumed in the statement of this proposition, with $T \triangleright \bar{C}_{h'}$. Denote $T_0 := T \cap \bar{C}_{h'}, T_1 := T \setminus T_0$, and $T_2 := \bar{C}_{h'} \setminus T_0$. Then $T = T_0 \cup T_1, \bar{C}_{h'} = T_0 \cup T_2$. Since $T \neq \bar{C}_{h'}$ and $T \triangleright \bar{C}_{h'}$, then we know $T_1 \neq \emptyset$. By Lemma 1, we know that $T_1 \triangleright T_2$. If $T_2 = \emptyset$, then $\alpha(T) = \alpha(T_0) + \alpha(T_1) > \alpha(T_0) = \alpha(\bar{C}_{h'})$. Hence we assume $T_2 \neq \emptyset$. Denote $T_2 := \{j_1, \ldots, j_t\}$. Since $T_1 \triangleright T_2$ and $T_1 \cap T_2 = \emptyset$, we know there exists $\{k_1, \ldots, k_t\} \subseteq T_1$ such that $k_1 < j_1, \ldots, k_t < j_t$.

W.l.o.g., consider k_1 and j_1. By definition, there is $k_1 < j_1, k_1 \notin \bar{C}_{h'}, j_1 \in \bar{C}_{h'}$, which is just saying: $k_1 < j_1, k_1 \in C_{h'}, j_1 \in \bar{C}_{h'}$. Therefore, $j_1 \in \{\ell \mid \ell > k_1, \ell \in \bar{C}_{h'}, k_1 \in C_{h'}\}$. By Step 4 of Algorithm 1, we know that $\alpha_{k_1} > \alpha_{j_1}$. For the remaining j_2 and $j_2, \ldots, k_t$ and j_t we can do the exact same argument and obtain $\alpha_{k_2} > \alpha_{j_2}, \ldots, \alpha_{k_t} > \alpha_{j_t}$.

Therefore, $\alpha(T) = \alpha(T_1) + \alpha(T_0) \geq \alpha_{k_1} + \ldots + \alpha_{k_t} + \alpha(T_0) > \alpha_{j_1} + \ldots + \alpha_{j_t} + \alpha(T_0) = \alpha(\bar{C}_{h'})$, which concludes the proof. □

Now we are ready to present the first main result of this paper.

Theorem 1. *Given a multi-cover $\{C_h\}_{h=1}^{k}$ for a TOMKS K, the MCI produced by Algorithm 1 is valid for $\mathrm{conv}(K)$.*

Proof. Since $\text{supp}(\alpha) = C$, in order to show that $\alpha^T x \leq \beta$ is valid to $\text{conv}(K)$, it suffices to show that, for any $T \subseteq C$ with $\alpha(T) \geq \beta + 1$, T must be a cover to K. Note that from Step 7 there is $\beta + 1 = \max_{h=1}^k \alpha(C_h)$, and for any $T_1, T_2 \subseteq C$, $\alpha(T_1) \geq \alpha(T_2)$ is equivalent to $\alpha(\bar{T}_1) \leq \alpha(\bar{T}_2)$, furthermore from Lemma 2, therefore it suffices for us to show that: for any $T \subseteq C$ with $\alpha(\bar{T}) \leq \min_{h=1}^k \alpha(\bar{C}_h)$, there must exist some $h^* \in [k]$ such that $\bar{C}_{h^*} \triangleright \bar{T}$. We will assume that $T \notin \{C_h\}_{h=1}^k$ since otherwise $\bar{C}_{h^*} \triangleright \bar{T}$ trivially holds. In the following, the proof is subdivided into two cases, depending on whether $\bar{T} \cap C_0 = \emptyset$ or not.

First, we consider the case $\bar{T} \cap C_0 = \emptyset$. In this case, there is $C_0 \subseteq T$. by Definition 3 of multi-cover, we know there must exist $h^* \in [k]$ such that either $C_{h^*} \setminus C_0 \triangleright T \setminus C_0$, or $T \setminus C_0 \triangleright C_{h^*} \setminus C_0$. By the above assumption $C_0 \subseteq T$ and Lemma 1, we know that either $C_{h^*} \triangleright T$ or $T \triangleright C_{h^*}$. If $T \triangleright C_{h^*}$, then Lemma 2 implies $\bar{C}_{h^*} \triangleright \bar{T}$, which completes the proof. So we assume $C_{h^*} \triangleright T$, or equivalently, $\bar{T} \triangleright \bar{C}_{h^*}$. Since $\bar{T} \subseteq C \setminus C_0$ and $\bar{T} \notin \{\bar{C}_h\}_{h=1}^k$, By Proposition 1 we obtain that $\alpha(\bar{T}) > \alpha(\bar{C}_{h^*})$, and this contradicts to the assumption of $\alpha(\bar{T}) \leq \min_{h=1}^k \alpha(\bar{C}_h)$.

Next, we consider the case $\bar{T} \cap C_0 \neq \emptyset$. In this case, we want to construct a $\bar{D} \subseteq C$ with $\bar{D} \cap C_0 = \emptyset, \alpha(\bar{D}) \leq \alpha(\bar{T})$, and $\bar{D} \triangleright \bar{T}$. Then since $\alpha(\bar{T}) \leq \min_{h=1}^k \alpha(\bar{C}_h)$, we have $\alpha(\bar{D}) \leq \min_{h=1}^k \alpha(\bar{C}_h)$ where $\bar{D} \cap C_0 = \emptyset$. According to our discussion in the previous case, we know there exists some $h^* \in [k]$ such that $\bar{C}_{h^*} \triangleright \bar{D}$, which implies $\bar{C}_{h^*} \triangleright \bar{T}$ since $\triangleright$ forms a partial order, and the proof is completed.

Arbitrarily pick $t^* \in \bar{T} \cap C_0$. Then by Step 6, we know there exists $h^* \in [k]$ such that $\alpha_{t^*} = \max\{\min_{\ell < t^*, \ell \in \bar{C}_{h^*}} \alpha_\ell, \sum_{t > t^*, t \in \bar{C}_{h^*}} \alpha_t + 1\}$. If $\{\ell \in \bar{C}_{h^*} \mid \ell < t^*\} \subseteq \bar{T}$, then we have $\alpha(\bar{T}) \geq \sum_{\ell < t^*, \ell \in \bar{C}_{h^*}} \alpha_\ell + \alpha_{t^*}$, which is at least $\sum_{\ell < t^*, \ell \in \bar{C}_{h^*}} \alpha_\ell + \sum_{t > t^*, t \in \bar{C}_{h^*}} \alpha_t + 1$. Since $t^* \notin \bar{C}_{h^*}$, we know that $\sum_{\ell < t^*, \ell \in \bar{C}_{h^*}} \alpha_\ell + \sum_{t > t^*, t \in \bar{C}_{h^*}} \alpha_t + 1 = \alpha(\bar{C}_{h^*}) + 1$. Hence $\alpha(\bar{T}) > \alpha(\bar{C}_{h^*})$, and this contradicts to the initial assumption of $\alpha(\bar{T}) \leq \min_{h=1}^k \alpha(\bar{C}_h)$. Therefore we can find some $\ell^* \in \bar{C}_{h^*}, \ell^* < t^*$ such that $\ell^* \notin \bar{T}$. Now define $\bar{D} := \bar{T} \cup \{\ell^*\} \setminus \{t^*\}$. Since $\ell^* < t^*$, clearly $\bar{D} \triangleright \bar{T}$. Also $\alpha(\bar{T}) - \alpha(\bar{D}) = \alpha_{t^*} - \alpha_{\ell^*}$, since $\alpha_{t^*} \geq \max_{\ell < t^*, \ell \in \bar{C}_{h^*}} \alpha_\ell$, we know that $\alpha(\bar{T}) - \alpha(\bar{D}) \geq 0$. If $\bar{D} \cap C_0 = \emptyset$, then we are done. Otherwise, we can replace $\bar{T}$ by $\bar{D}$, consider any index in $\bar{D} \cap C_0$ and do the above discussion one more time. Every time we are able to obtain a set $\bar{D}$ with $|\bar{D} \cap C_0|$ decreasing by 1. In the end we will obtain a set $\bar{D}$ with the desired property: $\bar{D} \cap C_0 = \emptyset, \alpha(\bar{D}) \leq \alpha(\bar{T})$, and $\bar{D} \triangleright \bar{T}$. This completes the proof for the case $\bar{T} \cap C_0 \neq \emptyset$.

Therefore, from the discussion for the above two cases, we have concluded the proof of MCI $\alpha^T x \leq \beta$ being a valid inequality for $\text{conv}(K)$. $\square$

For some multi-covers with a special discrepancy family, we are able to write the associated MCI in closed form. We provide two examples.

Example 1. Consider $\{C_1, C_2\}$ with discrepancy family $\{\{i_1, i_{t+1}\}, \{i_2, \ldots, i_t\}\}$ for some $t \geq 3$, with $i_1 < \ldots < i_{t+1}$. Easy to verify that such $\{C_1, C_2\}$ is a multi-cover, and the obtained MCI is:

$$\sum_{i<i_1,i\in C}(2t-1)x_i + \sum_{i_1\leq i<i_2,i\in C}(2t-3)x_i + \sum_{\ell=3}^{t}\sum_{i_{\ell-1}<i<i_\ell,i\in C}^{*}(2t-2\ell+3)x_i$$

$$+\sum_{\ell=2}^{t}2x_{i_\ell} + \sum_{i_t<i<i_{t+1},i\in C}2x_i + x_{i_{t+1}} + \sum_{i>i_{t+1},i\in C}2x_i \leq \alpha(C_1)-1, \tag{4}$$

where α is the vector associated with the left-hand-side term. $\diamond$

Example 2. Consider $\{C_1, C_2, C_3\}$ with discrepancy family $\{\{i_1, i_3\}, \{i_1, i_4, i_5\}, \{i_2, i_3, i_5\}\}$, with $i_1 < \ldots < i_5$. Here the family of covers $\{C_1, C_2, C_3\}$ is a multi-cover, and the obtained MCI is:

$$\sum_{i<i_1,i\in C}5x_i + \sum_{i_1\leq i<i_2,i\in C}3x_i + 2x_{i_2} + \sum_{i_2<i<i_3,i\in C}3x_i + 2x_{i_3}$$

$$+\sum_{i_3<i<i_4,i\in C}2x_i + x_{i_4} + \sum_{i_4<i<i_5,i\in C}2x_i + x_{i_5} + \sum_{i>i_5,i\in C}2x_i \leq \alpha(C_1)-1, \tag{5}$$

where α is the vector associated with the left-hand-side term. $\diamond$

Next, we present some illustrative examples to showcase the novelty of MCIs. The first example shows that, unlike lifted cover inequalities or CG cuts, MCIs are not aggregation cuts of the original linear system.

Example 3. Consider the TOMKS:

$$K := \{x \in \{0,1\}^5 \mid 19x_1 + 11x_2 + 5x_3 + 4x_4 + 2x_5 \leq 31,$$
$$16x_1 + 10x_2 + 7x_3 + 5x_4 + 3x_5 \leq 30\}.$$

Then $\{C_1, C_2\} := \{\{1, 2, 5\}, \{1, 3, 4, 5\}\}$ is a multi-cover for K, point χ^{C_1} only violates the first knapsack constraint, and point χ^{C_2} only violates the second knapsack constraint. The associated MCI is

$$3x_1 + 2x_2 + x_3 + x_4 + x_5 \leq 5, \tag{6}$$

and (6) is violated by both χ^{C_1} and χ^{C_2}.

Now consider an aggregation of the knapsack inequalities of K given by $\lambda_1(19, 11, 5, 4, 2)^T x + \lambda_2(16, 10, 7, 5, 3)^T x \leq 31\lambda_1 + 30\lambda_2$, where $\lambda_1, \lambda_2 \geq 0$. For any choice of $\lambda_1 \geq 0, \lambda_2 \geq 0$, it can be verified that C_1 and C_2 cannot both be covers to the knapsack set given by this single inequality, so any aggregation cut for K can cut off at most one of χ^{C_1} and χ^{C_2}. Therefore, the inequality (6) is not an aggregation cut. In some cases, it may be possible to obtain an MCI as a CG cut of the original linear system *augmented* with its minimal cover inequalities. In this example, consider the set

$$K_{CI} := \{x \in \{0,1\}^5 \mid 19x_1 + 11x_2 + 5x_3 + 4x_4 + 2x_5 \leq 31,$$
$$16x_1 + 10x_2 + 7x_3 + 5x_4 + 3x_5 \leq 30,$$
$$x_1 + x_2 + x_3 \leq 2, x_1 + x_2 + x_4 \leq 2,$$
$$x_1 + x_2 + x_5 \leq 2, x_1 + x_3 + x_4 + x_5 \leq 3\}.$$

The inequality (6) is indeed a CG cut with respect to K_{CI}, as shown by multipliers $\frac{1}{12} \cdot (19, 11, 5, 4, 2) + \frac{1}{4} \cdot (1, 1, 1, 0, 0) + \frac{1}{3} \cdot (1, 1, 0, 1, 0) + \frac{1}{2} \cdot (1, 1, 0, 0, 1) + \frac{1}{3} \cdot (1, 0, 1, 1, 1) = (3, 2, 1, 1, 1), \frac{1}{12} \cdot 31 + \frac{1}{4} \cdot 2 + \frac{1}{3} \cdot 2 + \frac{1}{2} \cdot 2 + \frac{1}{3} \cdot 3 = 5.75$. Hence $(3, 2, 1, 1, 1)^T x \leq \lfloor 5.75 \rfloor = 5$ is a CG cut for K_{CI}.   ◇

Example 3 demonstrates that MCI can be obtained from multiple knapsack sets simultaneously. Specifically, the inequality (6) is facet-defining for $\text{conv}(K)$, but it is neither valid for $\{x \in \{0, 1\}^5 \mid 19x_1 + 11x_2 + 5x_3 + 4x_4 + 2x_5 \leq 31\}$ nor $\{x \in \{0, 1\}^5 \mid 16x_1 + 10x_2 + 7x_3 + 5x_4 + 3x_5 \leq 30\}$. Example 3 also shows that an MCI can be a CG cut for the linear system given by the original knapsack constraints along with all their minimal cover inequalities. In the next example, we will see that this is not always the case.

Example 4. Consider the following TOMKS:

$$K := \{x \in \{0, 1\}^8 \mid 28x_1 + 24x_2 + 20x_3 + 19x_4 + 15x_5 + 10x_6 + 7x_7 + 6x_8 \leq 96,$$
$$27x_1 + 24x_2 + 21x_3 + 19x_4 + 13x_5 + 12x_6 + 7x_7 + 4x_8 \leq 96\}.$$

Consider the family of covers $C_1 = \{2, 3, 4, 5, 6, 7, 8\}$, $C_2 = \{1, 3, 4, 5, 6, 8\}$, $C_3 = \{1, 2, 3, 5, 6\}$, $C_4 = \{1, 2, 3, 5, 7, 8\}$. We have $C = [8]$, $C_0 = \{3, 5\}$, and the discrepancy family is $\mathcal{D}(\mathscr{C}) = \{\{2, 4, 6, 7, 8\}, \{1, 4, 6, 8\}, \{1, 2, 6\}, \{1, 2, 7, 8\}\} =: \{D_1, D_2, D_3, D_4\}$.

First, we verify that $\mathscr{C}$ is a multi-cover. For any set $S \subseteq C \setminus C_0$ and $S \notin \mathcal{D}(\mathscr{C})$, if $1 \in S, |S| = 2$, then it is clearly dominated by either D_2, D_3 or D_4. If $1 \in S, |S| = 3$, then either $S \rhd D_3$ or $D_3 \rhd S$. If $1 \in S, |S| = 4$, then S must be comparable with D_2 or D_3. If $1 \in S, |S| = 5$, then $S \rhd D_1$. If $1 \notin S$, then clearly $D_1 \rhd S$ since $S \subseteq D_1$. Hence for any $S \subseteq C \setminus C_0$ and $S \notin \mathcal{D}(\mathscr{C})$, S must be comparable with some set in $\mathcal{D}(\mathscr{C})$. Therefore, $\mathscr{C}$ is a multi-cover.

When Algorithm 1 is applied to $\mathscr{C}$, we obtain the inequality

$$\alpha^T x \leq \beta := 4x_1 + 3x_2 + 3x_3 + 2x_4 + 3x_5 + 2x_6 + x_7 + x_8 \leq 14, \qquad (7)$$

and it can be shown that (7) is a facet-defining inequality for $\text{conv}(K)$.

Consider the linear system given by all the minimal cover inequalities for K, as well as the original two linear constraints. We refer to this linear system as K_{CI}, which consists of 30 inequalities. Solving $\max\{\alpha^T x \mid x \in K_{CI}\}$ gives optimal value 15.307, so any corresponding CG cut with respect to K_{CI} is $\alpha^T x \leq 15$, and it is weaker than inequality (7).   ◇

Even when the cover-family consists of covers all coming from the same knapsack inequality, the MCI can produce interesting inequalities. In the next example, we show a MCI that cannot be obtained as a standard lifted cover inequality, regardless of the lifting order.

Example 5 (Example 3 in [10]). Let $K := \{x \in \{0, 1\}^5 \mid 10x_1 + 7x_2 + 7x_3 + 4x_4 + 4x_5 \leq 16\}$, and consider the multi-cover $\mathscr{C} := \{\{1, 3\}, \{1, 4, 5\}, \{2, 3, 5\}\}$. From inequality (5) of Example 2, we know that the corresponding MCI is

$$3x_1 + 2x_2 + 2x_3 + x_4 + x_5 \leq 4. \qquad (8)$$

The inequality (8) is the same inequality produced by the new lifting procedure described in [10], and the authors of [10] state that (8) is both a facet of conv(K) and cannot be obtained from any cover inequality by standard sequential lifting methods, regardless of the lifting order.   ◇

4   Antichain Multi-cover Inequalities

In this section we propose a way to strengthen MCI when the associated multi-cover forms an antichain in a certain poset. Recall that in order theory, an *antichain* is a subset of a poset such that any two distinct elements in the subset are incomparable, and a *maximal antichain* is an antichain that is not a proper subset of any other antichain.

Definition 4 (Antichain multi-cover). *Let $\mathscr{C}$ be a family of covers for K. Then we say $\mathscr{C}$ is an* antichain multi-cover *for K, if $\mathcal{D}(\mathscr{C})$ is a maximal antichain of the poset $(2^{\cup_{D \in \mathcal{D}(\mathscr{C})} D}, \rhd)$.*

We are now ready to define our *antichain multi-cover inequalities (AMCIs)* by Algorithm 2. AMCIs have the interesting property (proved in Theorem 3) that they cut off at least two characteristic vectors of covers in the antichain multi-cover $\mathscr{C}$.

Algorithm 2. Antichain multi-cover inequality (AMCI)

Input: An antichain multi-cover $\mathscr{C} := \{C_1, \ldots, C_k\}$ for K, and its MCI $\alpha^T x \leq \beta$.
Output: An antichain multi-cover inequality.

1: Let $C \setminus C_0 = \{i_1, \ldots, i_m\}$, with $i_1 < \ldots < i_m$.
2: **if** $\exists h^* \in [k]$ such that $\alpha(C_{h^*})$ is the unique maximum of $\{\alpha(C_h) \mid h \in [k]\}$ **then**
3:     Let i_{t^*} be the minimum of $\{i_1, \ldots, i_m\}$ such that

$$\{h \in [k] : |C_h \cap \{i_1, \ldots, i_{t^*}\}| - |C_{h^*} \cap \{i_1, \ldots, i_{t^*}\}| = 1\} \neq \emptyset.$$

4:     $\delta := \min\{\alpha(C_{h^*}) - \alpha(C_h) : |C_h \cap \{i_1, \ldots, i_{t^*}\}| - |C_{h^*} \cap \{i_1, \ldots, i_{t^*}\}| = 1\}$.
5:     **for** $t = 1, \ldots, t^*$ **do**
6:         $\alpha_{i_t} := \alpha_{i_t} + \delta$.
7:     **for** $j \in C_0$ **do**
8:         $\alpha_j := \min_{h \in [k]} \max \left\{ \max_{\ell < j, l \in \bar{C}_h} \alpha_\ell, \sum_{t > j, t \in \bar{C}_h} \alpha_t + 1 \right\}$.
9:     $\beta := \max_{h=1}^k \alpha(C_h) - 1$.
    return the inequality $\alpha^T x \leq \beta$.

Note that an AMCI is not necessarily different from its corresponding MCI, it depends on if condition 2 is satisfied or not.

First, we show that Algorithm 2 can perform all required steps. The only nontrivial step is Step 3. Thus, we only need to prove the following proposition.

Proposition 2. *In Step 3, for any $h \in [k]$, there exists an index i_{t^*} in $\{i_1, \ldots, i_m\}$ such that $|C_h \cap \{i_1, \ldots, i_{t^*}\}| - |C_{h^*} \cap \{i_1, \ldots, i_{t^*}\}| = 1$.*

Proof. First, we claim that there exists $t^\circ \in [m]$, such that

$$|C_h \cap \{i_1, \ldots, i_{t^\circ}\}| - |C_{h^*} \cap \{i_1, \ldots, i_{t^\circ}\}| \geq 1. \tag{9}$$

We prove this claim by contradiction. Thus we assume that for every $t \in [m]$ we have $|C_h \cap \{i_1, \ldots, i_t\}| - |C_{h^*} \cap \{i_1, \ldots, i_t\}| \leq 0$. Let $C_h \setminus C_0 = \{j_1, \ldots, j_\ell\} \subseteq \{i_1, \ldots, i_m\}$ with $j_1 < \cdots < j_\ell$. To prove our claim it suffices to construct an injective function $f : C_h \setminus C_0 \to C_{h^*} \setminus C_0$ such that $f(j_1) \leq j_1, \ldots, f(j_\ell) \leq j_\ell$. In fact, Definition 1 then implies $C_{h^*} \setminus C_0 \triangleright C_h \setminus C_0$, which gives us a contradiction since by Definition 4 of antichain multi-cover, the discrepancy family $\{C_h \setminus C_0\}_{h \in [k]}$ forms an antichain. Let $s_1 \in [m]$ be an index such that $i_{s_1} = j_1$. From our assumption, we know that $|C_{h^*} \cap \{i_1, \ldots, i_{s_1}\}| \geq |C_h \cap \{i_1, \ldots, i_{s_1}\}| = 1$. So we can find $i_{r_1} \in C_{h^*}$ with $i_{r_1} \leq i_{s_1} = j_1$, and let $f(j_1) := i_{r_1}$. Now let s_2 such that $i_{s_2} = j_2$. From our assumption, we know that $|C_{h^*} \cap \{i_1, \ldots, i_{s_2}\}| \geq |C_h \cap \{i_1, \ldots, i_{s_2}\}| = 2$, thus $|(C_{h^*} \setminus \{i_{r_1}\}) \cap \{i_1, \ldots, i_{s_2}\}| \geq 1$. So we can find $i_{r_2} \in C_{h^*}$ with $i_{r_2} \neq i_{r_1}$ and $i_{r_2} \leq i_{s_2} = j_2$. We then set $f(j_2) := i_{r_2}$. Recursively, we can then construct an injective function $f : C_h \setminus C_0 \to C_{h^*} \setminus C_0$ such that $f(j_1) \leq j_1, \ldots, f(j_\ell) \leq j_\ell$. This concludes the proof of (9).

For every $t \in [m-1]$, we clearly have $0 \leq |C_h \cap \{i_1, \ldots, i_{t+1}\}| - |C_h \cap \{i_1, \ldots, i_t\}| \leq 1$, and the same observation holds if we replace h with h^*. Thus

$$-1 \leq \quad (|C_h \cap \{i_1, \ldots, i_{t+1}\}| - |C_{h^*} \cap \{i_1, \ldots, i_{t+1}\}|) \\ - (|C_h \cap \{i_1, \ldots, i_t\}| - |C_{h^*} \cap \{i_1, \ldots, i_t\}|) \quad \leq 1.$$

From $|C_h \cap \{i_1\}| - |C_{h^*} \cap \{i_1\}| \leq 1$ and (9), we then obtain that there must exist some $t^* \in [t^\circ]$, such that $|C_h \cap \{i_1, \ldots, i_{t^*}\}| - |C_{h^*} \cap \{i_1, \ldots, i_{t^*}\}| = 1$. □

Next, we show that AMCIs are valid for K. The validity proof is analogous to that of Theorem 1, which also requires the following auxiliary result.

Proposition 3. *Let $\{C_h\}_{h=1}^k$ be an antichain multi-cover and let $\alpha^T x \leq \beta$ be the associated AMCI. If there exists $T \subseteq C \setminus C_0, T \notin \{\bar{C}_h\}_{h=1}^k$, with $T \triangleright \bar{C}_{h'}$ for some $h' \in [k]$, then $\alpha(T) > \alpha(\bar{C}_{h'})$.*

Proof. We will assume that the condition at Step 2 in Algorithm 2 is satisfied, since if not, then the AMCI coincides with its MCI, and the statement of this proposition coincides with Proposition 1. Let T and $\bar{C}_{h'}$ be the sets as assumed in the statement of this proposition, with $T \triangleright \bar{C}_{h'}$. Denote $T_0 := T \cap \bar{C}_{h'}, T_1 := T \setminus T_0$, and $T_2 := \bar{C}_{h'} \setminus T_0$. Then $T = T_0 \cup T_1, \bar{C}_{h'} = T_0 \cup T_2$. Since $T \neq \bar{C}_{h'}$ and $T \triangleright \bar{C}_{h'}$, then we know $T_1 \neq \emptyset$. By Lemma 1, we know that $T_1 \triangleright T_2$. If $T_2 = \emptyset$, then $\alpha(T) = \alpha(T_0) + \alpha(T_1) > \alpha(T_0) = \alpha(\bar{C}_{h'})$. Hence we assume $T_2 \neq \emptyset$. Denote $T_2 := \{j_1, \ldots, j_t\}$. Since $T_1 \triangleright T_2$ and $T_1 \cap T_2 = \emptyset$, we know there exists $\{k_1, \ldots, k_t\} \subseteq T_1$ such that $k_1 < j_1, \ldots, k_t < j_t$.

Let $\gamma^T x \leq \theta$ be the MCI of antichain multi-cover $\{C_h\}_{h=1}^k$. From the proof of Proposition 1, we know that $\gamma_{k_1} > \gamma_{j_1}, \ldots, \gamma_{k_t} > \gamma_{j_t}$. By Step 5 and 6, we know for any $i \in C \setminus C_0, \alpha_i = \gamma_i + \delta \cdot \mathbb{1}\{i \le i_{t^*}\}$. Since $k_1 < j_1, \ldots, k_t < j_t$, therefore we have $\mathbb{1}\{k_1 \le i_{t^*}\} \ge \mathbb{1}\{j_1 \le i_{t^*}\}, \ldots, \mathbb{1}\{k_t \le i_{t^*}\} \ge \mathbb{1}\{j_t \le i_{t^*}\}$. Hence $\alpha_{k_1} = \gamma_{k_1} + \delta \cdot \mathbb{1}\{k_1 \le i_{t^*}\} > \gamma_{j_1} + \delta \cdot \mathbb{1}\{j_1 \le i_{t^*}\} = \alpha_{j_1}$, and similarly there is also $\alpha_{k_2} > \alpha_{j_2}, \ldots, \alpha_{k_t} > \alpha_{j_t}$.

Therefore, $\alpha(T) = \alpha(T_1) + \alpha(T_0) \ge \alpha_{k_1} + \ldots + \alpha_{k_t} + \alpha(T_0) > \alpha_{j_1} + \ldots + \alpha_{j_t} + \alpha(T_0) = \alpha(T_2) + \alpha(T_0) = \alpha(\bar{C}_{h'})$. $\qquad\square$

Given the above proposition, the proof of the following theorem is exactly the same as that of Theorem 1. For the completeness of this paper, we present its proof in the following.

Theorem 2. *Given an antichain multi-cover $\mathscr{C}$ for a TOMKS K, the AMCI produced by Algorithm 2 is valid for $\mathrm{conv}(K)$.*

Proof. Since $\mathrm{supp}(\alpha) = C$, in order to show that AMCI $\alpha^T x \le \beta$ is valid to $\mathrm{conv}(K)$, it suffices to show that, for any $T \subseteq C$ with $\alpha(T) \ge \beta + 1$, T must be a cover to K. Note that from Step 7 there is $\beta + 1 = \max_{h=1}^{k} \alpha(C_h)$, and for any $T_1, T_2 \subseteq C$, $\alpha(T_1) \ge \alpha(T_2)$ is equivalent to $\alpha(\bar{T}_1) \le \alpha(\bar{T}_2)$, furthermore from Lemma 2, therefore it suffices for us to show that: for any $T \subseteq C$ with $\alpha(\bar{T}) \le \min_{h=1}^{k} \alpha(\bar{C}_h)$, there must exist some $h^* \in [k]$ such that $\bar{C}_{h^*} \rhd \bar{T}$. We will assume that $T \notin \{C_h\}_{h=1}^{k}$ since otherwise $\bar{C}_{h^*} \rhd \bar{T}$ trivially holds. In the following, the proof is subdivided into two cases, depending on whether $\bar{T} \cap C_0 = \emptyset$ or not.

First, we consider the case $\bar{T} \cap C_0 = \emptyset$. In this case, there is $C_0 \subseteq T$. by Definition 3 of multi-cover, we know there must exist $h^* \in [k]$ such that either $C_{h^*} \setminus C_0 \rhd T \setminus C_0$, or $T \setminus C_0 \rhd C_{h^*} \setminus C_0$. By the above assumption $C_0 \subseteq T$ and Lemma 1, we know that either $C_{h^*} \rhd T$ or $T \rhd C_{h^*}$. If $T \rhd C_{h^*}$, then Lemma 2 implies $\bar{C}_{h^*} \rhd \bar{T}$, which completes the proof. So we assume $C_{h^*} \rhd T$, or equivalently, $\bar{T} \rhd \bar{C}_{h^*}$. Since $\bar{T} \subseteq C \setminus C_0$ and $\bar{T} \notin \{\bar{C}_h\}_{h=1}^{k}$, By Proposition 3 we obtain that $\alpha(\bar{T}) > \alpha(\bar{C}_{h^*})$, and this contradicts to the assumption of $\alpha(\bar{T}) \le \min_{h=1}^{k} \alpha(\bar{C}_h)$.

Next, we consider the case $\bar{T} \cap C_0 \ne \emptyset$. In this case, we want to construct a $\bar{D} \subseteq C$ with $\bar{D} \cap C_0 = \emptyset, \alpha(\bar{D}) \le \alpha(\bar{T})$, and $\bar{D} \rhd \bar{T}$. Then since $\alpha(\bar{T}) \le \min_{h=1}^{k} \alpha(\bar{C}_h)$, we have $\alpha(\bar{D}) \le \min_{h=1}^{k} \alpha(\bar{C}_h)$ where $\bar{D} \cap C_0 = \emptyset$. According to our discussion in the previous case, we know there exists some $h^* \in [k]$ such that $\bar{C}_{h^*} \rhd \bar{D}$, which implies $\bar{C}_{h^*} \rhd \bar{T}$ since $\rhd$ forms a partial order, and the proof is completed.

Arbitrarily pick $t^* \in \bar{T} \cap C_0$. Then by Step 6, we know there exists $h^* \in [k]$ such that $\alpha_{t^*} = \max\left\{\min_{\ell < t^*, \ell \in \bar{C}_{h^*}} \alpha_\ell, \sum_{t > t^*, t \in \bar{C}_{h^*}} \alpha_t + 1\right\}$. If $\{\ell \in \bar{C}_{h^*} \mid \ell < t^*\} \subseteq \bar{T}$, then we have $\alpha(\bar{T}) \ge \sum_{\ell < t^*, \ell \in \bar{C}_{h^*}} \alpha_\ell + \alpha_{t^*}$, which is at least $\sum_{\ell < t^*, \ell \in \bar{C}_{h^*}} \alpha_\ell + \sum_{t > t^*, t \in \bar{C}_{h^*}} \alpha_t + 1$. Since $t^* \notin \bar{C}_{h^*}$, we know that $\sum_{\ell < t^*, \ell \in \bar{C}_{h^*}} \alpha_\ell + \sum_{t > t^*, t \in \bar{C}_{h^*}} \alpha_t + 1 = \alpha(\bar{C}_{h^*}) + 1$. Hence $\alpha(\bar{T}) > \alpha(\bar{C}_{h^*})$, and this contradicts to the initial assumption of $\alpha(\bar{T}) \le \min_{h=1}^{k} \alpha(\bar{C}_h)$. Therefore we can find some $\ell^* \in \bar{C}_{h^*}, \ell^* < t^*$ such that $\ell^* \notin \bar{T}$. Now define $\bar{D} := \bar{T} \cup \{\ell^*\} \setminus \{t^*\}$. Since $\ell^* < t^*$, clearly $\bar{D} \rhd \bar{T}$. Also $\alpha(\bar{T}) - \alpha(\bar{D}) = \alpha_{t^*} - \alpha_{\ell^*}$, since $\alpha_{t^*} \ge \max_{\ell < t^*, \ell \in \bar{C}_{h^*}} \alpha_\ell$, we know that $\alpha(\bar{T}) - \alpha(\bar{D}) \ge 0$. If $\bar{D} \cap C_0 = \emptyset$, then we are done. Otherwise, we can replace $\bar{T}$ by $\bar{D}$, consider any index in $\bar{D} \cap C_0$ and do the above discussion one more time. Every time we are able to obtain a set $\bar{D}$ with $|\bar{D} \cap C_0|$ decreasing by 1. In the end we will obtain a set $\bar{D}$ with the desired property: $\bar{D} \cap C_0 = \emptyset, \alpha(\bar{D}) \leq \alpha(\bar{T})$, and $\bar{D} \triangleright \bar{T}$. This completes the proof of the case $\bar{T} \cap C_0 \neq \emptyset$. □

The next theorem shows that each AMCI cuts off at least two characteristic vectors of covers from the associated antichain multi-cover.

Theorem 3. *Given an antichain multi-cover $\mathscr{C}$, the AMCI produced by Algorithm 2 is violated by at least two characteristic vectors of covers in $\mathscr{C}$.*

Proof. Let $\mathscr{C} := \{C_h\}_{h \in [k]}$. When the "if" condition 2 does not hold, meaning there already exist at least two covers C_{h_1} and C_{h_2} from $\mathscr{C}$, such that $\alpha(C_{h_1}) = \alpha(C_{h_2}) = \max_{h=1}^{k} \alpha(C_h)$. Then according to Step 9, we know that $\alpha^T x \leq \beta$ cuts off $\chi^{C_{h_1}}$ and $\chi^{C_{h_2}}$.

Now assuming the condition 2 is satisfied. For any $i \in C \setminus C_0$, denote the intermediate coefficient of α_i at Step 2 before the updating operation 5 and 6 to be γ_i. Then according to the algorithm, there is $\gamma(C_{h^*}) = \max_{h=1}^{k} \gamma(C_h), \delta = \min\{\gamma(C_{h^*}) - \gamma(C_h) : |C_h \cap \{i_1, \ldots, i_{t^*}\}| - |C_{h^*} \cap \{i_1, \ldots, i_{t^*}\}| = 1\}$ where i_{t^*} is defined by Step 3, and $\alpha_i = \gamma_i + \delta$ for any $i = i_1, \ldots, i_{t^*}$. Let $C_{h^{**}}$ be the cover which satisfies $|C_{h^{**}} \cap \{i_1, \ldots, i_{t^*}\}| - |C_{h^*} \cap \{i_1, \ldots, i_{t^*}\}| = 1$ and $\gamma(C_{h^*}) - \gamma(C_{h^{**}}) = \delta$. Next we are going to show that: $\alpha(C_{h^*}) = \alpha(C_{h^{**}}) = \max_{h=1}^{k} \alpha(C_h)$. Since $\alpha(C_{h^*}) - \alpha(C_{h^{**}}) = \gamma(C_{h^*}) - \gamma(C_{h^{**}}) + \delta \cdot |C_{h^*} \cap \{i_1, \ldots, i_{t^*}\}| - \delta \cdot |C_{h^{**}} \cap \{i_1, \ldots, i_{t^*}\}|$, then we obtain that $\alpha(C_{h^*}) = \alpha(C_{h^{**}})$.

Claim. $\alpha(C_{h^*}) = \max_{h=1}^{k} \alpha(C_h)$.

Proof of claim. Arbitrarily pick $h \in [k], h \neq h^*, h \neq h^{**}$, we want to show that $\alpha(C_{h^*}) \geq \alpha(C_h)$.

If $|C_h \cap \{i_1, \ldots, i_{t^*}\}| - |C_{h^*} \cap \{i_1, \ldots, i_{t^*}\}| = 1$, then by definition of δ, we have $\gamma(C_{h^*}) - \gamma(C_h) \geq \delta$. Therefore $\alpha(C_{h^*}) - \alpha(C_h) = \gamma(C_{h^*}) + \delta \cdot |C_{h^*} \cap \{i_1, \ldots, i_{t^*}\}| - (\gamma(C_h) + \delta \cdot |C_h \cap \{i_1, \ldots, i_{t^*}\}|) = \gamma(C_{h^*}) - \gamma(C_h) - \delta \cdot (|C_h \cap \{i_1, \ldots, i_{t^*}\}| - |C_{h^*} \cap \{i_1, \ldots, i_{t^*}\}|) = \gamma(C_{h^*}) - \gamma(C_h) - \delta \geq 0$.

If $|C_h \cap \{i_1, \ldots, i_{t^*}\}| - |C_{h^*} \cap \{i_1, \ldots, i_{t^*}\}| \leq 0$, then $\alpha(C_{h^*}) - \alpha(C_h) = \gamma(C_{h^*}) - \gamma(C_h) - \delta \cdot (|C_h \cap \{i_1, \ldots, i_{t^*}\}| - |C_{h^*} \cap \{i_1, \ldots, i_{t^*}\}|) \geq 0$. Here the last inequality is because $\gamma(C_{h^*}) = \max_{h=1}^{k} \gamma(C_h)$.

If $|C_h \cap \{i_1, \ldots, i_{t^*}\}| - |C_{h^*} \cap \{i_1, \ldots, i_{t^*}\}| > 1$, then because $|C_h \cap \{i_1\}| - |C_{h^*} \cap \{i_1\}| \leq 1$ and $(|C_h \cap \{i_1, \ldots, i_{t+1}\}| - |C_{h^*} \cap \{i_1, \ldots, i_{t+1}\}|) - (|C_h \cap \{i_1, \ldots, i_t\}| - |C_{h^*} \cap \{i_1, \ldots, i_t\}|) \leq 1$, we know there must exist some $i_{\ell^*} < i_{t^*}$, such that $|C_h \cap \{i_1, \ldots, i_{\ell^*}\}| - |C_{h^*} \cap \{i_1, \ldots, i_{\ell^*}\}| = 1$, which contradicts to the minimum choice of i_{t^*} at Step 3.   ◇

Hence we have shown that $\alpha(C_{h^*}) = \alpha(C_{h^{**}}) = \max_{h=1}^{k} \alpha(C_h)$. According to the definition of β at Step 9, we know that $\alpha^T x \leq \beta$ cuts off $\chi^{C_{h^*}}$ and $\chi^{C_{h^{**}}}$. □

Note that the multi-covers in Example 1 and Example 2 are both antichain multi-covers, and the corresponding MCIs are violated by the characteristic vectors of all covers. Therefore the AMCIs of those examples coincide with their MCIs. Next, we give an example where an AMCI is different from the corresponding MCI.

Example 6. Consider $\{C_1, C_2\}$ with discrepancy family $\{\{i_1\}, \{i_2, \ldots, i_t\}\}$ for some $t \geq 3$, with $i_1 < \ldots < i_t$. $\{C_1, C_2\}$ is obviously an antichain multi-cover, and the obtained MCI is:

$$\gamma^T x := \sum_{i < i_1, i \in C} 3x_i + \sum_{i_1 \leq i < i_2, i \in C} 2x_i + \sum_{\ell=3}^{t} \sum_{i_{\ell-1} < i < i_\ell, i \in C} 2x_i$$

$$+ \sum_{\ell=2}^{t} x_{i_\ell} + \sum_{i > i_t, i \in C} x_i \leq \gamma(C_2) - 1. \tag{10}$$

This MCI is different from the corresponding AMCI obtained by Algorithm 2:

$$\alpha^T x := \sum_{i < i_1, i \in C} tx_i + \sum_{i_1 \leq i < i_2, i \in C} (t-1)x_i + \sum_{\ell=3}^{t} \sum_{i_{\ell-1} < i < i_\ell, i \in C} (t-\ell+2)x_i$$

$$+ \sum_{\ell=2}^{t} x_{i_\ell} + \sum_{i > i_t, i \in C} x_i \leq \alpha(C_1) - 1. \tag{11}$$

◇

The next result states that the well-known $(1, k)$-configuration inequality can be obtained from the AMCI (11) in Example 6.

Proposition 4. *Consider a knapsack set $K = \{x \in \{0,1\}^n \mid a^T x \leq b\}$, a nonempty subset $N \subseteq [n]$, and $t \in [n] \setminus N$. Assume that $\sum_{i \in N} a_i \leq b$ and that $H \cup \{t\}$ is a minimal cover for all $H \subset N$ with $|H| = k$. Then for any $T(r) \subseteq N$ with $|T(r)| = r$, $k \leq r \leq |N|$, the $(1, k)$-configuration inequality $(r - k + 1)x_t + \sum_{j \in T(r)} x_j \leq r$ can be obtained from an AMCI (11) associated with an antichain multi-cover of a knapsack set.*

Proof. When $r = k$, then the above inequality reduces to a cover inequality. Hence we assume $r > k$. W.l.o.g. assuming $a_1 \geq \ldots \geq a_n$. Consider a new knapsack set $K' := \{x \in \{0,1\}^{n+1} \mid a'^T x \leq b\}$, with $a'_i = a_i \; \forall i \leq t, a'_{t+1} = a_t, a'_j = a_{j-1} \; \forall j > t+1$. Then clearly there is also $a'_1 \geq \ldots \geq a'_{n+1}$.

Since for any $H \subset N$ with $|H| = k$, $H \cup \{t\}$ is a cover to K, we know for any $j \in N, N \cup \{t\} \setminus \{j\}$ is also a cover to K, which means $\sum_{i \in N} a_i - a_j + a_t > b$. From the assumption that $\sum_{i \in N} a_i \leq b$, we have $a_t > a_j$, or equivalently, $t < j$ for any $j \in N$. Now for any $T(r) \subseteq N$ with $|T(r)| = r$, $k \leq r \leq |N|$, denote $T(r) := \{j_1, \ldots, j_r\}$ with $j_1 < \ldots < j_r$, so we have $t < j_1$ from above. Then consider $C_1 := \{t\} \cup \{j_{r-k+1}, \ldots, j_r\}, C'_1 := \{t\} \cup \{j_{r-k+1} + 1, \ldots, j_r + 1\}, C'_2 := \{t+1\} \cup \{j_1 + 1, \ldots, j_r + 1\}$. Since $\{j_{r-k+1}, \ldots, j_r\} \subset N$ with $|\{j_{r-k+1}, \ldots, j_r\}| = k$, we know that C_1 is a cover to K, so C'_1 is a cover to K' from the construction of K'. Furthermore it is obvious that C'_2 is also a cover to K' since $a'_{t+1} = a'_t$. Note that the discrepancy family of $\{C'_1, C'_2\}$ is $\{\{t\}, \{t+1, j_1 + 1, \ldots, j_{r-k} + 1\}\}$, then from the AMCI (11) of Example 6, we obtain the AMCI associated with $\{C'_1, C'_2\}$ for K':

$$(r - k + 1)x_t + x_{t+1} + \sum_{\ell=1}^{r} x_{j_\ell + 1} \leq r.$$

Since K can be obtained by simply projecting out of the x_{t+1} variable of K', therefore we obtain that the following inequality is valid for K:

$$(r - k + 1)x_t + \sum_{\ell=1}^{r} x_{j_\ell} = (r - k + 1)x_t + \sum_{j \in T(r)} x_j \leq r.$$

$\square$

5   Facet-Inducing MCI

In this section we provide a sufficient condition for the the MCI to define a facet of conv(K). The proof can be found in the full version of the paper [5].

Given a multi-cover $\{C_1, \ldots, C_k\}$ with its corresponding MCI $\alpha^T x \leq \beta$, we denote by $\{i_{t,1}, \ldots, i_{t,n_t}\} := \{i \in C \setminus C_0 \mid \alpha_i = t\}$, with $i_{t,1} < \ldots < i_{t,n_t}$.

Theorem 4. *Let $\{C_1, \ldots, C_k\}$ be a multi-cover for a TOMKS K, and let $\alpha^T x \leq \beta$ be the associated MCI. Assume that the following conditions hold:*

1. *$C_0 = \emptyset$;*
2. *For each $h \in [k]$, cover C_h is a minimal cover;*
3. *For any $t = 2, \ldots, \max_{i=1}^{n} \alpha_i$, there exist some $i_{t-1,\ell_t} \notin C_{h_t} \in \{C_h\}_{h=1}^{k}$ with $i_{t,1} \in C_{h_t}$ and $i_{1,n_1} \in C_{h_t}$, such that $C_{h_t} \cup \{i_{t-1,\ell_t}\} \setminus \{i_{t,n_t}\}$ is not a cover;*
4. *There exists some $C_{h_1} \in \{C_h\}_{h=1}^{k}$, such that $i_{1,1} \in C_{h_1}$ and for any $i' \notin C$, $C_{h_1} \cup \{i'\} \setminus \{i_{1,1}\}$ is not a cover.*
5. *For any $t = 1, \ldots, \max_{i=1}^{n} \alpha_i$, $\alpha(C_{h_t}) = \beta + 1$.*

Then $\alpha^T x \leq \beta$ is a facet-defining inequality for conv(K).

Example 7. Consider the TOMKS and the multi-cover in Example 5. We have $C_1 = \{1, 3\}$, $C_2 = \{1, 4, 5\}$, $C_3 = \{2, 3, 5\}$, and the associated MCI $\alpha^T x \leq \beta$ is $3x_1 + 2x_2 + 2x_3 + x_4 + x_5 \leq 4$, here $i_{1,1} = 4, i_{1,2} = 5, i_{2,1} = 2, i_{2,2} = 3, i_{3,1} = 1$. Clearly condition 1 in Theorem 4 holds. Since $\alpha(C_1) - \alpha_3 = 10 \leq 16$, $\alpha(C_2) - \alpha_5 = 14 \leq 16$, $\alpha(C_3) - \alpha_5 = 14 \leq 16$, condition 2 holds as well. For $t = 2$, let $C_{h_2} = C_3$, then $i_{1,1} \notin C_{h_2}, i_{1,2} \in C_{h_2}, i_{2,1} \in C_{h_2}$, and $C_{h_2} \cup \{i_{1,1}\} \setminus \{i_{2,2}\} = \{2, 4, 5\}$ is not a cover. For $t = 3$, let $C_{h_3} = C_2$, then $i_{2,1} \notin C_{h_3}, i_{1,2} \in C_{h_3}, i_{3,1} \in C_{h_3}$, and $C_{h_3} \cup \{i_{2,1}\} \setminus \{i_{3,1}\} = \{2, 4, 5\}$ is not a cover. Therefore condition 3 holds. Let $C_{h_1} = C_2$, then $i_{1,1} \in C_{h_1}$, since here $C = [5]$, condition 4 holds. Lastly, $\alpha(C_{h_1}) = \alpha(C_{h_2}) = \alpha(C_{h_3}) = 5$, so condition 5 also holds. Hence Theorem 4 yields that this MCI is a facet-defining inequality for conv(K).

6   Conclusion

In this work, we give a new family of valid inequalities for the intersection of knapsack sets and demonstrate several ways in which the inequalities are not implied by other known cutting-plane methods. We are aware of very little work that explicitly studies the polyhedral structure of the intersection of multiple knapsack sets, and we hope the ideas presented here will give rise to new methods for generating strong valid inequalities for complex binary sets that arise in practical settings.

Acknowledgements. A. Del Pia is partially funded by ONR grant N00014-19-1-2322. Any opinions, findings, and conclusions or recommendations expressed in this material are those of the authors and do not necessarily reflect the views of the Office of Naval Research. J. Linderoth and H. Zhu are in part supported by the U.S. Department of Energy, Office of Science, Office of Advanced Scientific Computing Research (ASCR) under Contract DEAC02-06CH11347.

References

1. Balas, E.: Facets of the knapsack polytope. Math. Program. **8**(1), 146–164 (1975)
2. Bodur, M., Del Pia, A., Dey, S., Molinaro, M., Pokutta, S.: Aggregation-based cutting-planes for packing and covering integer programs. Math. Program. **171**, 331–359 (2018)
3. Calafiore, G., Campi, M.: The scenario approach to robust control design. IEEE Trans. Autom. Control **51**, 742–753 (2006)
4. Crowder, H., Johnson, E.L., Padberg, M.: Solving large-scale zero-one linear programming problems. Oper. Res. **31**(5), 803–834 (1983)
5. Del Pia, A., Linderoth, J., Zhu, H.: Multi-cover inequalities for totally-ordered multiple knapsack sets. Optimization Online (2020). http://www.optimization-online. org/DB_HTML/2020/11/8107.html
6. Fukasawa, R., Goycoolea, M.: On the exact separation of mixed integer knapsack cuts. Math. Program. **128**, 19–41 (2011)
7. Gu, Z., Nemhauser, G.L., Savelsbergh, M.W.: Lifted cover inequalities for 0–1 integer programs: computation. INFORMS J. Comput. **10**(4), 427–437 (1998)
8. Hammer, P.L., Johnson, E.L., Peled, U.N.: Facet of regular 0–1 polytopes. Math. Program. **8**(1), 179–206 (1975)
9. Hojny, C., et al.: Knapsack polytopes: a survey. Ann. Oper. Res. **292**, 469–517 (2020)
10. Letchford, A.N., Souli, G.: On lifted cover inequalities: a new lifting procedure with unusual properties. Oper. Res. Lett. **47**(2), 83–87 (2019)
11. Luedtke, J., Ahmed, S.: A sample approximation approach for optimization with probabilistic constraints. SIAM J. Optim. **19**, 674–699 (2008)
12. Nemirovski, A., Shapiro, A.: Scenario approximation of chance constraints. In: Calafiore, G., Dabbene, F. (eds.) Probabilistic and Randomized Methods for Design Under Uncertainty. pp. 3–48. Springer, London (2005) https://doi.org/10.1007/1-84628-095-8_1
13. Padberg, M.W.: (1, k)-configurations and facets for packing problems. Math. Program. **18**(1), 94–99 (1980)
14. Weismantel, R.: On the 0/1 knapsack polytope. Math.l Program. **77**(3), 49–68 (1997)
15. Wolsey, L.A.: Facets and strong valid inequalities for integer programs. Oper. Res. **24**(2), 367–372 (1976)

Semi-streaming Algorithms
for Submodular Matroid Intersection

Paritosh Garg[(✉)], Linus Jordan, and Ola Svensson

EPFL, Lausanne, Switzerland
{paritosh.garg,ola.svensson}@epfl.ch, linus.jordan@bluewin.ch

Abstract. While the basic greedy algorithm gives a semi-streaming algorithm with an approximation guarantee of 2 for the *unweighted* matching problem, it was only recently that Paz and Schwartzman obtained an analogous result for weighted instances. Their approach is based on the versatile local ratio technique and also applies to generalizations such as weighted hypergraph matchings. However, the framework for the analysis fails for the related problem of weighted matroid intersection and as a result, the approximation guarantee for weighted instances did not match the factor 2 achieved by the greedy algorithm for unweighted instances. Our main result closes this gap by developing a semi-streaming algorithm with an approximation guarantee of $2+\varepsilon$ for *weighted* matroid intersection, improving upon the previous best guarantee of $4 + \varepsilon$. Our techniques also allow us to generalize recent results by Levin and Wajc on submodular maximization subject to matching constraints to that of matroid-intersection constraints.

While our algorithm is an adaptation of the local ratio technique used in previous works, the analysis deviates significantly and relies on structural properties of matroid intersection, called kernels. Finally, we also conjecture that our algorithm gives a $(k+\varepsilon)$ approximation for the intersection of k matroids but prove that new tools are needed in the analysis as the used structural properties fail for $k \geq 3$.

1 Introduction

For large problems, it is often not realistic that the entire input can be stored in random access memory so more memory efficient algorithms are preferable. A popular model for such algorithms is the (semi-)streaming model (see e.g. [13]): the elements of the input are fed to the algorithm in a stream and the algorithm is required to have a small memory footprint.

This research was supported by the Swiss National Science Foundation project 200021-184656 "Randomness in Problem Instances and Randomized Algorithms."

© Springer Nature Switzerland AG 2021
M. Singh and D. P. Williamson (Eds.): IPCO 2021, LNCS 12707, pp. 208–222, 2021.
https://doi.org/10.1007/978-3-030-73879-2_15

Consider the classic maximum matching problem in an undirected graph $G = (V, E)$. An algorithm in the semi-streaming model[1] is fed the edges one-by-one in a stream $e_1, e_2, \ldots, e_{|E|}$ and at any point of time the algorithm is only allowed $O(|V| \operatorname{polylog}(|V|))$ bits of storage. The goal is to output a large matching $M \subseteq E$ at the end of the stream. Note that the allowed memory usage is sufficient for the algorithm to store a solution M but in general it is much smaller than the size of the input since the number of edges may be as many as $|V|^2/2$. Indeed, the intuitive difficulty in designing a semi-streaming algorithm is that the algorithm needs to discard many of the seen edges (due to the memory restriction) without knowing the future edges and still return a good solution at the end of the stream.

For the unweighted matching problem, the best known semi-streaming algorithm is the basic greedy approach:

> Initially, let $M = \emptyset$. Then for each edge e in the stream, add it to M if $M \cup \{e\}$ is a feasible solution, i.e., a matching; otherwise the edge e is discarded.

The algorithm uses space $O(|V| \log |V|)$ and a simple proof shows that it returns a 2-approximate solution in the *unweighted* case, i.e., a matching of size at least half the size of an maximum matching. However, this basic approach fails to achieve any approximation guarantee for *weighted graphs*.

Indeed, for weighted matchings, it is non-trivial to even get a small constant-factor approximation. One way to do so is to replace edges if we have a much heavier edge. This is formalized in [6] who get a 6-approximation. Later, [12] improved this algorithm to find a 5.828-approximation; and, with a more involved technique, [4] provided a $(4 + \varepsilon)$-approximation.

It was only in recent breakthrough work [14] that the gap in the approximation guarantee between unweighted and weighted matchings was closed. Specifically, [14] gave a semi-streaming algorithm for weighted matchings with an approximation guarantee of $2 + \varepsilon$ for every $\varepsilon > 0$. Shortly after, [9] came up with a simplified analysis of their algorithm, reducing the memory requirement from $O_\varepsilon(|V| \log^2 |V|)$ to $O_\varepsilon(|V| \log |V|)$. These results for weighted matchings are tight (up to the ε) in the sense that any improvement would also improve the state-of-the-art in the unweighted case, which is a long-standing open problem.

The algorithm of [14] is an elegant use of the local ratio technique ([1]) ([2]) in the semi-streaming setting. While this technique is very versatile and it readily generalizes to weighted hypergraph matchings, it is much harder to use it for the related problem of weighted matroid intersection. This is perhaps surprising as many of the prior results for the matching problem also applies to the matroid intersection problem in the semi-streaming model (see Sect. 2 for definitions). Indeed, the greedy algorithm still returns a 2-approximate solution

---

[1] This model can also be considered in the multi-pass setting when the algorithm is allowed to take several passes over the stream. However, in this work we focus on the most basic and widely studied setting in which the algorithm takes a single pass over the stream.

in the unweighted case and the algorithm in [4] returns a $(4 + \varepsilon)$-approximate solution for weighted instances. So, prior to our work, the status of the matroid intersection problem was that of the matching problem *before* [14].

We now describe on a high-level the reason that the techniques from [14] are not easily applicable to matroid intersection and our approach for dealing with this difficulty. The approach in [14] works in two parts, first certain elements of the stream are selected and added to a set S, and then at the end of the stream a matching M is computed by the greedy algorithm that inspects the edges of S in the reverse order in which they were added. This way of constructing the solution M greedily by going backwards in time is a standard framework for analyzing algorithms based on the local ratio technique. Now in order to adapt their algorithm to matroid intersection, recall that the bipartite matching problem can be formulated as the intersection of two partition matroids. We can thus reinterpret their algorithm and analysis in this setting. Furthermore, after this reinterpretation, it is not too hard to define an algorithm that works for the intersection of any two matroids. However, bipartite matching is a *special* case of matroid intersection which captures a rich set of seemingly more complex problems. This added expressiveness causes the analysis and the standard framework for analyzing local ratio algorithms to fail. Specifically, we prove that a solution formed by running the greedy algorithm on S in the reverse order (as done for the matching problem) fails to give any constant-factor approximation guarantee for the matroid intersection problem. To overcome this and to obtain our main result, we make a connection to a concept called matroid kernels (see [7] for more details about kernels), which allows us to, in a more complex way, identify a subset of S with an approximation guarantee of $2 + \varepsilon$.

Finally, for the intersection of more than two matroids, the same approach in the analysis does not work, because the notion of matroid kernel does not generalize to more than two matroids. However, we conjecture that the subset S generated for the intersection of k matroids still contains a $(k+\varepsilon)$-approximation. Currently, the best approximation results are a $(k^2 + \varepsilon)$-approximation from [4] and a $(2(k + \sqrt{k(k-1)}) - 1)$-approximation from [3]. For $k = 3$, the former is better, giving a $(9 + \varepsilon)$-approximation. For $k > 3$, the latter is better, giving an $O(k)$-approximation.

Generalization to Submodular Functions. Very recently, Levin and Wajc [11] obtained improved approximation ratios for matching and b-matching problems in the semi-streaming model with respect to submodular functions. Specifically, they get a $(3 + 2\sqrt{2})$-approximation for monotone submodular b-matching, $(4 + 3\sqrt{2})$-approximation for non-monotone submodular matching, and a $(3 + \varepsilon)$-approximation for maximum weight (linear) b-matching. In our paper, we are able to extend our algorithm for weighted matroid intersection to work with submodular functions by combining our and their ideas. In fact, we are able to generalize all their results to the case of matroid intersection with better or equal[2] approximation ratios: we get $(3 + 2\sqrt{2} + \delta)$-approximation for monotone

---

[2] One can get rid of the δ factor if we assume that the function value is polynomially bounded by $|E|$, an assumption made by [11].

submodular matroid intersection, $(4+3\sqrt{2}+\delta)$-approximation for non-monotone submodular matroid intersection and $(2+\varepsilon)$-approximation for maximum weight (linear) matroid intersection. Due to space limitations, we refer the reader to the full version [8] of our paper for this generalization.

Outline. In Sect. 2, we introduce basic matroid concepts and we formally define the weighted matroid intersection problem in the semi-streaming model. Section 3 is devoted to our main result. Here, we adapt the algorithm of [14] without worrying about the memory requirements, show why the standard analysis fails, and then give our new analysis to get a 2-approximation. We then make the obtained algorithm memory efficient in Sect. 4. Finally, in Sect. 5, we discuss the case of more than two matroids.

2 Preliminaries

Matroids. We define and give a brief overview of the basic concepts related to matroids that we use in this paper. For a more comprehensive treatment, we refer the reader to [15]. A *matroid* is a tuple $M = (E, I)$ consisting of a finite ground set E and a family $I \subseteq 2^E$ of subsets of E satisfying:

- if $X \subseteq Y, Y \in I$, then $X \in I$; and
- if $X \in I, Y \in I$ and $|Y| > |X|$, then $\exists\, e \in Y \setminus X$ such that $X \cup \{e\} \in I$.

The elements in I (that are subsets of E) are referred to as the *independent sets* of the matroid and the set E is referred to as the *ground set*. With a matroid $M = (E, I)$, we associate the *rank function* $\mathrm{rank}_M : 2^E \to \mathbb{N}$ and the *span function* $\mathrm{span}_M : 2^E \to 2^E$ defined as follows for every $E' \subseteq E$,

$$\mathrm{rank}_M(E') = \max\{|X| \mid X \subseteq E' \text{ and } X \in I\},$$
$$\mathrm{span}_M(E') = \{e \in E \mid \mathrm{rank}_M(E' \cup \{e\}) = \mathrm{rank}_M(E')\}.$$

We simply write $\mathrm{rank}(\cdot)$ and $\mathrm{span}(\cdot)$ when the matroid M is clear from the context. In words, the rank function equals the size of the largest independent set when restricted to E' and the span function equals the elements in E' and all elements that cannot be added to a maximum cardinality independent set of E' while maintaining independence. The *rank of the matroid* equals $\mathrm{rank}(E)$, i.e., the size of the largest independent set.

The Weighted Matroid Intersection Problem in the Semi-Streaming Model. In the *weighted matroid intersection problem*, we are given two matroids $M_1 = (E, I_1), M_2 = (E, I_2)$ on a common ground set E and a non-negative weight function $w : E \to \mathbb{R}_{\geq 0}$ on the elements of the ground set. The goal is to find a subset $X \subseteq E$ that is independent in both matroids, i.e., $X \in I_1$ and $X \in I_2$, and whose weight $w(X) = \sum_{e \in X} w(e)$ is maximized.

In seminal work [5], Edmonds gave a polynomial-time algorithm for solving the weighted matroid intersection problem to optimality in the classic model of computation when the whole input is available to the algorithm throughout the computation. In contrast, the problem becomes significantly harder and tight results are still eluding us in the semi-streaming model where the memory footprint of the algorithm and its access pattern to the input are restricted. Specifically, in the *semi-streaming* model the ground set E is revealed in a stream $e_1, e_2, \ldots, e_{|E|}$ and at time i the algorithm gets access to e_i and can perform computation based on e_i and its current memory but without knowledge of future elements $e_{i+1}, \ldots, e_{|E|}$. The algorithm has independence-oracle access to the matroids M_1 and M_2 restricted to the elements stored in the memory, i.e., for a set of such elements, the algorithm can query whether the set is independent in each matroid. The goal is to design an algorithm such that (i) the memory usage is near-linear $O((r_1 + r_2)\operatorname{polylog}(r_1 + r_2))$ at any time, where r_1 and r_2 denote the ranks of the input matroids M_1 and M_2, respectively, and (ii) at the end of the stream the algorithm should output a feasible solution $X \subseteq E$, i.e., a subset X that satisfies $X \in I_1$ and $X \in I_2$, of large weight $w(X)$. We remark that the memory requirement $O((r_1 + r_2)\operatorname{polylog}(r_1 + r_2))$ is natural as $r_1 + r_2 = |V|$ when formulating a bipartite matching problem as the intersection of two matroids[3].

The difficulty in designing a good semi-streaming algorithm is that the memory requirement is much smaller than the size of the ground set E and thus the algorithm must intuitively discard many of the elements without knowledge of the future and without significantly deteriorating the weight of the final solution X. The quality of the algorithm is measured in terms of its approximation guarantee: an algorithm is said to have an *approximation guarantee* of α if it is guaranteed to output a solution X, no matter the input and the order of the stream, such that $w(X) \geq \mathrm{OPT}/\alpha$ where OPT denotes the weight of an optimal solution to the instance. As aforementioned, our main result in this paper is a semi-streaming algorithm with an approximation guarantee of $2 + \varepsilon$, for every $\varepsilon > 0$, improving upon the previous best guarantee of $4 + \varepsilon$ [4].

3 The Local Ratio Technique for Weighted Matroid Intersection

In this section, we first present the local ratio algorithm for the weighted matching problem that forms the basis of the semi-streaming algorithm in [14]. We then adapt it to the weighted matroid intersection problem. While the algorithm is fairly natural to adapt to this setting, we give an example in Sect. 3.2 that shows that the same techniques as used for analyzing the algorithm for

---

[3] The considered problem can also be formulated as the problem of finding an independent set in one matroid, say M_1, and maximizing a submodular function which would be the (weighted) rank function of M_2. For that problem, [10] recently gave a streaming algorithm with an approximation guarantee of $(2+\varepsilon)$. However, the space requirement of their algorithm is exponential in the rank of M_1 (which would correspond to be exponential in $|V|$ in the matching case) and thus it does not provide a meaningful guarantee for our setting.

matchings does not work for matroid intersection. Instead, our analysis, which is presented in Sect. 3.3, deviates from the standard framework for analyzing local ratio algorithms and it heavily relies on a structural property of matroid intersection known as kernels. We remark that the algorithms considered in this section do not have a small memory footprint.

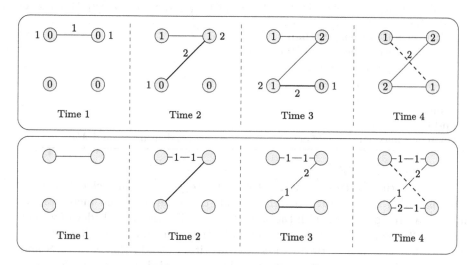

Fig. 1. The top part shows an example execution of the local ratio technique for weighted matchings The bottom part shows how to adapt this (bipartite) example to the language of weighted matroid intersection (Algorithm 1).

3.1 Local-Ratio Technique for Weighted Matching

The local ratio algorithm for the weighted matching problem is as follows. The algorithm maintains vertex potentials $w(u)$ for every vertex u, a set S of selected edges, and an auxiliary weight function $g : S \to \mathbb{R}_{\geq 0}$ of the selected edges. Initially the vertex potentials are set to 0 and the set S is empty. When an edge $e = \{u, v\}$ arrives, the algorithm computes how much it gains compared to the previous edges, by taking its weight minus the weight/potential of its endpoints $(g(e) = w(e) - w(u) - w(v))$. If the gain is positive, then we add the edge to S, and add the gain to the weight of the endpoints, that is, we set $w(u) = w(u) + g(e)$ and $w(v) = w(v) + g(e)$. At the end, we return a maximum weight matching M among the edges stored on the stack S.

For a better intuition of the algorithm, consider the example depicted on the top of Fig. 1. The stream consists of four edges e_1, e_2, e_3, e_4 with weights $w(e_1) = 1$ and $w(e_2) = w(e_3) = w(e_4) = 2$. At each time step i, we depict the arriving edge e_i in thick along with its weight; the vertex potentials before the algorithm considers this edge is written on the vertices, and the updated vertex potentials (if any) after considering e_i are depicted next to the incident vertices. The edges that are added to S are solid and those that are not added to S are dashed.

Algorithm 1. Local ratio for matroid intersection

Input: A stream of the elements of the common ground set of matroids $M_1 = (E, I_1), M_2 = (E, I_2)$.
Output: A set $X \subseteq E$ that is independent in both matroids.

$\quad S \leftarrow \emptyset$
$\quad$**for** element e in the stream **do**
$\qquad$ calculate $w_i^*(e) = \max\left(\{0\} \cup \{\theta : e \in \text{span}_{M_i}(\{f \in S \mid w_i(f) \geq \theta\})\}\right)$ for $i \in \{1, 2\}$.
$\qquad$**if** $w(e) > w_1^*(e) + w_2^*(e)$ **then**
$\qquad\quad g(e) \leftarrow w(e) - w_1^*(e) - w_2^*(e)$
$\qquad\quad w_1(e) \leftarrow w_1^*(e) + g(e)$
$\qquad\quad w_2(e) \leftarrow w_2^*(e) + g(e)$
$\qquad\quad S \leftarrow S \cup \{e\}$
$\qquad$**end if**
$\quad$**end for**
$\quad$**return** a maximum weight set $T \subseteq S$ that is independent in M_1 and M_2

At the arrival of the first edge of weight $w(e_1) = 1$, both incident vertices have potential 0 and so the algorithm adds this edge to S and increases the incident vertex potentials with the gain $g(e_1) = 1$. For the second edge of weight $w(e_2) = 2$, the sum of incident vertex potentials is 1 and so the gain of e_2 is $g(e_2) = 2 - 1$, which in turn causes the algorithm to add this edge to S and to increase the incident vertex potentials by 1. The third time step is similar to the second. At the last time step, edge e_4 arrives of weight $w(e_4) = 2$. As the incident vertex potentials sum up to 2 the gain of e_4 is not strictly positive and so this edge is *not* added to S and no vertex potentials are updated. Finally, the algorithm returns the maximum weight matching in S which in this case consists of edges $\{e_1, e_3\}$ and has weight 3. Note that the optimal matching of this instance had weight 4 and we thus found a 4/3-approximate solution.

In general, the algorithm has an approximation guarantee of 2. This is proved using a common framework to analyze algorithms based on the local ratio technique: We ignore the weights and greedily construct a matching M by inspecting the edges in S in reverse order, i.e., we first consider the edges that were added last. An easy proof (see e.g. [9]) then shows that the matching M constructed in this way has weight at least half the optimum weight.

In the next section, we adapt the above described algorithm to the context of matroid intersections. We also give an example that the above framework for the analysis fails to give any constant-factor approximation guarantee. Our alternative (tight) analysis of this algorithm is then given in Sect. 3.3.

3.2 Adaptation to Weighted Matroid Intersection

When adapting the local ratio algorithm for weighted matching to matroid intersection to obtain Algorithm 1, the first problem we encounter is the fact that matroids do not have a notion of vertices, so we cannot keep a weight/potential for each vertex. To describe how we overcome this issue, it is helpful to consider the case of bipartite matching and in particular the example depicted in Fig. 1. It is well known that the weighted matching problem on a bipartite graph with edge set E and bipartition V_1, V_2 can be modelled as a weighted matroid intersection problem on matroids $M_1 = (E, I_1)$ and $M_2 = (E, I_2)$ where for $i \in \{1, 2\}$

$$I_i = \{E' \subseteq E \mid \text{each vertex } v \in V_i \text{ is incident to at most one vertex in } E'\} .$$

Instead of keeping a weight for each vertex, we will maintain two weight functions w_1 and w_2, one for each matroid. These weight functions will be set so that the following holds in the special case of bipartite matching: on the arrival of a new edge e, let $T_i \subseteq S$ be an independent set in I_i of selected edges that maximizes the weight function w_i. Then we have that

$$\min_{f \in T_i : T_i \setminus \{f\} \cup \{e\} \in I_i} w_i(f) \qquad \text{if } T_i \cup \{e\} \notin I_i \text{ and } 0 \text{ otherwise} \tag{1}$$

equals the vertex potential of the incident vertex V_i when running the local ratio algorithm for weighted matching. It is well-known (e.g. by the optimality of the greedy algorithm for matroids) that the cheapest element f to remove from T_i to make $T_i \setminus \{f\} \cup \{e\}$ an independent set equals the largest weight θ so that the elements of weight at least θ spans e. We thus have that (1) equals

$$\max \left(\{0\} \cup \{\theta : e \in \text{span}_{M_i} (\{f \in S \mid w_i(f) \geq \theta\})\} \right)$$

and it follows that the quantities $w_1^*(e)$ and $w_2^*(e)$ in Algorithm 1 equal the incident vertex potentials in V_1 and V_2 of the local ratio algorithm in the special case of bipartite matching. To see this, let us return to our example in Fig. 1 and let V_1 be the two vertices on the left and V_2 be the two vertices on the right. In the bottom part of the figure, the weight functions w_1 and w_2 are depicted (at the corresponding side of the edge) after the arrival of each edge. At time step 1, e_1 does not need to replace any elements in any of the matroids and so $w_1^*(e_1) = w_2^*(e_1) = 0$. We therefore have that its gain is $g(e_1) = 1$ and the algorithm sets $w_1(e_1) = w_2(e_1) = 1$. At time 2, edge e_2 of weight 2 arrives. It is not spanned in the first matroid whereas it is spanned by edge e_1 of weight 1 in the second matroid. It follows that $w_1^*(e_2) = 0$ and $w_2^*(e_2) = w_2(e_1) = 1$ and so e_2 has positive gain $g(e_2) = 1$ and it sets $w_1(e_2) = 1$ and $w_2(e_2) = w_2(e_1) + 1 = 2$. The third time step is similar to the second. At the last time step, e_4 of weight 2 arrives. However, since it is spanned by e_1 with $w_1(e_1) = 1$ in the first matroid and by e_3 with $w_2(e_3) = 1$ in the second matroid, its gain is 0 and it is thus not added to the set S. Note that throughout this example, and in general for bipartite graphs, Algorithm 1 is identical to algorithm for weighted matching. One may therefore expect that the analysis of the latter also generalizes to Algorithm 1. We explain next that this is not the case for general matroids.

Counter Example to Same Approach in Analysis. We give a simple example showing that the greedy selection (as done in the analysis for local ratio algorithm for weighted matching) does not work for matroid intersection. Still, it turns out that the set S generated by Algorithm 1 always contains a 2-approximation but the selection process is more involved.

Lemma 1. *There exist two matroids $M_1 = (E, I_1)$ and $M_2 = (E, I_2)$ on a common ground set E and a weight function $w : E \to \mathbb{R}_{\geq 0}$ such that a greedy algorithm that considers the elements in the set S in the reverse order of when they were added by Algorithm 1 does not provide any constant-factor approximation.*

Proof. The example consists of the ground set $E = \{a, b, c, d\}$ with weights $w(a) = 1, w(b) = 1 + \varepsilon, w(c) = 2\varepsilon, w(d) = 3\varepsilon$ for a small $\varepsilon > 0$ (the approximation guarantee will be at least $\Omega(1/\varepsilon)$). The matroids $M_1 = (E, I_1)$ and $M_2 = (E, I_2)$ are defined by

- a subset of E is in I_1 if and only if it does not contain $\{a, b\}$; and
- a subset of E is in I_2 if and only if it contains at most two elements.

To see that M_1 and M_2 are matroids, note that M_1 is a partition matroid with partitions $\{a, b\}, \{c\}, \{d\}$, and M_2 is the 2-uniform matroid (alternatively, one can easily check that M_1 and M_2 satisfy the definition of a matroid). Now consider the execution of Algorithm 1 when given the elements of E in the order a, b, c, d:

- Element a has weight 1, and $\{a\}$ is independent both in M_1 and M_2, so we set $w_1(a) = w_2(a) = g(a) = 1$ and a is added to S.
- Element b is spanned by a in M_1 and not spanned by any element in M_2. So we get $g(b) = w(b) - w_1^*(b) - w_2^*(b) = 1 + \varepsilon - 1 - 0 = \varepsilon$. As $\varepsilon > 0$, we add b to S, and set $w_1(b) = w_1(a) + \varepsilon = 1 + \varepsilon$ and $w_2(b) = \varepsilon$.
- Element c is not spanned by any element in M_1 but is spanned by $\{a, b\}$ in M_2. As b has the smallest w_2 weight, $w_2^*(c) = w_2(b) = \varepsilon$. So we have $g(c) = 2\varepsilon - w_1^*(c) - w_2^*(c) = 2\varepsilon - 0 - \varepsilon = \varepsilon > 0$, and we set $w_1(c) = \varepsilon$ and $w_2(c) = 2\varepsilon$ and add c to S.
- Element d is similar to c. We have $g(d) = 3\varepsilon - 0 - 2\varepsilon = \varepsilon > 0$ and so we set $w_1(d) = \varepsilon$ and $w_2(d) = 3\varepsilon$ and add d to S.

As the algorithm selected all the elements, we have $S = E$. It follows that the greedy algorithm on S (in the reverse order of when elements were added) will select d and c, after which the set is a maximal independent set in M_2. This gives a weight of 5ε, even though a and b both have weight at least 1, which shows that this algorithm does not guarantee any constant factor approximation.   □

3.3   Analysis of Algorithm 1

We prove that Algorithm 1 has an approximation guarantee of 2.

Theorem 1. *Let S be the subset generated by Algorithm 1 on a stream E of elements, matroids $M_1 = (E, I_1), M_2 = (E, I_2)$ and weight function $w : E \to \mathbb{R}_{\geq 0}$. Then there exists a subset $T \subseteq S$ independent in M_1 and in M_2 whose weight $w(T)$ is at least $w(S^*)/2$, where S^* denotes an optimal solution to the weighted matroid intersection problem.*

Throughout the analysis we fix the input matroids $M_1 = (E, I_1), M_2 = (E, I_2)$, the weight function $w : R \rightarrow \mathbb{R}_{\geq 0}$, and the order of the elements in the stream. While Algorithm 1 only defines the weight functions w_1 and w_2 for the elements added to the set S, we extend them in the analysis by, for $i \in \{1, 2\}$, letting $w_i(e) = w_i^*(e)$ for the elements e not added to S.

We now prove Theorem 1 by showing that $g(S) \geq w(S^*)/2$ (Lemma 3) and that there is a solution $T \subseteq S$ such that $w(T) \geq g(S)$ (Lemma 4). In the proof of both these lemmas, we use the following properties of the computed set S.

Lemma 2. *Let S be the set generated by Algorithm 1 and $S' \subseteq S$ any subset. Consider one of the matroids M_i with $i \in \{1, 2\}$. There exists a subset $T' \subseteq S'$ that is independent in M_i, i.e., $T' \in I_i$, and $w_i(T') \geq g(S')$. Furthermore, the maximum weight independent set in M_i over the whole ground set E can be selected to be a subset of S, i.e. $T_i \subseteq S$, and it satisfies $w_i(T_i) = g(S)$.*

Proof. Consider matroid M_1 (the proof is identical for M_2) and fix $S' \subseteq S$. The set $T_1' \subseteq S'$ that is independent in M_1 and that maximizes $w_1(T_1')$ satisfies

$$w_1(T_1') = \int_0^\infty \text{rank}(\{e \in T_1' \mid w_1(e) \geq \theta\}) \, d\theta = \int_0^\infty \text{rank}(\{e \in S' \mid w_1(e) \geq \theta\}) \, d\theta \, .$$

The second equality follows from the fact that the greedy algorithm that considers the elements in decreasing order of weight is optimal for matroids and thus we have $\text{rank}(\{e \in T_1' \mid w_1(e) \geq \theta\}) = \text{rank}(\{e \in S' \mid w_1(e) \geq \theta\})$ for any $\theta \in \mathbb{R}$.

Now index the elements of $S' = \{e_1, e_2, \ldots, e_\ell\}$ in the order they were added to S by Algorithm 1 and let $S_j' = \{e_1, \ldots, e_j\}$ for $j = 0, 1, \ldots, \ell$ (where $S_0' = \emptyset$). By the above equalities and by telescoping,

$$w_1(T_1') = \sum_{i=1}^\ell \int_0^\infty \left(\text{rank}(\{e \in S_i' \mid w_1(e) \geq \theta\}) - \text{rank}(\{e \in S_{i-1}' \mid w_1(e) \geq \theta\}) \right) d\theta \, .$$

We have that $\text{rank}(\{e \in S_i' \mid w_1(e) \geq \theta\}) - \text{rank}(\{e \in S_{i-1}' \mid w_1(e) \geq \theta\})$ equals 1 if $w(e_i) \geq \theta$ and $e_i \notin \text{span}(\{e \in S_{i-1}' \mid w_1(e) \geq \theta\})$ and it equals 0 otherwise. Therefore, by the definition of $w_1^*(\cdot)$, the gain $g(\cdot)$ and $w_1(e_i) = w_1^*(e_i) + g(e_i)$ in Algorithm 1 we have

$$w_1(T_1') = \sum_{i=1}^\ell \left[w_1(e_i) - \max \left(\{0\} \cup \{\theta : e_i \in \text{span} \left(\{f \in S_{i-1}' \mid w_i(f) \geq \theta\} \right) \} \right) \right]$$

$$\geq \sum_{i=1}^\ell g(e_i) = g(S') \, .$$

The inequality holds because S_{i-1}' is a subset of the set S at the time when Algorithm 1 considers element e_i. Moreover, if $S' = S$, then S_{i-1}' equals the set S at that point and so we then have

$$w_1^*(e_i) = \max \left(\{0\} \cup \{\theta : e_i \in \text{span} \left(\{f \in S_{i-1}' \mid w_i(f) \geq \theta\} \right) \} \right) .$$

This implies that the above inequality holds with equality in that case. We can thus also conclude that a maximum weight independent set $T_1 \subseteq S$ satisfies

$w_1(T_1) = g(S)$. Finally, we can observe that T_1 is also a maximum weight independent set over the whole ground set since we have rank($\{e \in S \mid w_1(e) \geq \theta\}$) = rank($\{e \in E \mid w_1(e) \geq \theta\}$) for every $\theta > 0$, which holds because, by the extension of w_1, an element $e \notin S$ satisfies $e \in \text{span}(\{f \in S : w_1(f) \geq w_1(e)\})$.
□

We can now relate the gain of the elements in S with the weight of an optimal solution.

Lemma 3. *Let S be the subset generated by Algorithm 1. Then $g(S) \geq w(S^*)/2$.*

Proof. We first observe that $w_1(e) + w_2(e) \geq w(e)$ for every element $e \in E$. Indeed, for an element $e \in S$, we have by definition $w(e) = g(e) + w_1^*(e) + w_2^*(e)$, and $w_i(e) = g(e) + w_i^*(e)$, so $w_1(e) + w_2(e) = 2g(e) + w_1^*(e) + w_2^*(e) = w(e) + g(e) > w(e)$. In the other case, when $e \notin S$ then $w_1^*(e) + w_2^*(e) \geq w(e)$, and $w_i(e) = w_i^*(e)$, so automatically, $w_1(e) + w_2(e) \geq w(e)$.

The above implies that $w_1(S^*) + w_2(S^*) \geq w(S^*)$. On the other hand, by Lemma 2, we have $w_i(T_i) \geq w_i(S^*)$ (since T_i is a max weight independent set in M_i with respect to w_i) and $w_i(T_i) = g(S)$, thus $g(S) \geq w_i(S^*)$ for $i = 1, 2$. □

We finish the proof of Theorem 1 by proving that there is a $T \subseteq S$ independent in both M_1 and M_2 such that $w(T) \geq g(S)$. As described in Sect. 3.2, we cannot select T using the greedy method. Instead, we select T using the concept of kernels studied in [7].

Lemma 4. *Let S be the subset generated by Algorithm 1. Then there exists a subset $T \subseteq S$ independent in M_1 and in M_2 such that $w(T) \geq g(S)$.*

Proof. Consider one of the matroids M_i with $i \in \{1, 2\}$ and define a total order $<_i$ on E such that $e <_i f$ if $w_i(e) > w_i(f)$ or if $w_i(e) = w_i(f)$ and e appeared later in the stream than f. The pair $(M_i, <_i)$ is known as an ordered matroid. We further say that a subset E' of E dominates element e of E if $e \in E'$ or there is a subset $C_e \subseteq E'$ such that $e \in \text{span}(C_e)$ and $c < e$ for all elements c of C_e. The set of elements dominated by E' is denoted by $D_{M_i}(E')$. Note that if E' is an independent set, then the greedy algorithm that considers the elements of $D_{M_i}(E')$ in the order $<_i$ selects exactly the elements E'.

Theorem 2 in [7] says that for two ordered matroids $(M_1, <_1), (M_2, <_2)$ there always is a set $K \subseteq E$, which is referred to as a $M_1 M_2$-kernel, such that

- K is independent in both M_1 and in M_2; and
- $D_{M_1}(K) \cup D_{M_2}(K) = E$.

We use the above result on M_1 and M_2 restricted to the elements in S. Specifically we select $T \subseteq S$ to be the kernel such that $D_{M_1}(T) \cup D_{M_2}(T) = S$. Let $S_1 = D_{M_1}(T)$ and $S_2 = D_{M_2}(T)$. By Lemma 2, there exists a set $T' \subseteq S_1$ independent in M_1 such that $w_1(T') \geq g(S_1)$. As noted above, the greedy algorithm that considers the element of S_1 in the order $<_i$ (decreasing weights) selects exactly the elements in T. It follows by the optimality of the greedy algorithm for matroids that T is optimal for S_1 in M_1 with weight function w_1, which in turn implies $w_1(T) \geq g(S_1)$. In the same way, we also have $w_2(T) \geq g(S_2)$. By definition, for any $e \in S$, we have $w(e) = w_1(e) + w_2(e) - g(e)$. Together, we have $w(T) = w_1(T) + w_2(T) - g(T) \geq g(S_1) + g(S_2) - g(T)$. As elements from T are in both S_1 and S_2, and all other elements are in at least one of both sets, we have $g(S_1) + g(S_2) \geq g(S) + g(T)$, and thus $w(T) \geq g(S)$. $\square$

4   Making the Algorithm Memory Efficient

We now modify Algorithm 1 to only select elements with a significant gain, parametrized by $\alpha > 1$, and delete elements if we have too many in memory, parametrized by a real number y. Let us call this algorithm **EMI**. If α is close enough to 1 and y is large enough, then EMI is very close to Algorithm 1, and allows for a similar analysis. This method is very similar to the one used in [14] and [9], but our analysis is quite different.

More precisely, we take an element e only if $w(e) > \alpha(w_1^*(e) + w_2^*(e))$ instead of $w(e) > w_1^*(e) + w_2^*(e)$, and we delete all elements e' in the current stack S for which the ratio between the g weight and the maximum g weight i.e., $g_{max} = \max_{e \in S} g(e)$ exceeds y (i.e., $\frac{g_{max}}{g(e')} > y$). For technical purposes, we also need to keep independent sets T_1 and T_2 which maximize the weight functions w_1 and w_2 respectively. If an element with small g weight is in T_1 or T_2, we do not delete it, as this would modify the w_i-weights and selection of coming elements. We show that this algorithm is a semi-streaming algorithm with an approximation guarantee of $(2+\varepsilon)$ for an appropriate selection of the parameters (see Lemma 6 for the space requirement and theorem2 for the approximation guarantee).

Lemma 5. *Let S be the subset generated by EMI with $\alpha \geq 1$ and $y = \infty$. Then $w(S^*) \leq 2\alpha g(S)$.*

Proof. We define $w_\alpha : E \to \mathbb{R}$ by $w_\alpha(e) = w(e)$ if $e \in S$ and $w_\alpha(e) = \frac{w(e)}{\alpha}$ otherwise. By construction, EMI and Algorithm 1 give the same set S, and the same weight function g for this modified weight function. By Lemma 3, $w_\alpha(S^*) \leq 2g(S)$. On the other hand, $w(S^*) \leq \alpha w_\alpha(S^*)$. $\square$

Lemma 6. *Let S be the subset generated generated by EMI with $\alpha = 1 + \varepsilon$ and $y = \frac{\min(r_1, r_2)}{\varepsilon^2}$ and S^* be a maximum weight independent set, where r_1 and r_2 are the ranks of M_1 and M_2 respectively. Then $w(S^*) \leq 2(1 + 2\varepsilon + o(\varepsilon))g(S)$. Furthermore, at any point of time, the size of S is at most $r_1 + r_2 + \min(r_1, r_2) \log_\alpha(\frac{y}{\varepsilon})$.*

Proof. We first prove that the generated set S satisfies $w(S^*) \leq 2(1 + 2\varepsilon + o(\varepsilon))g(S)$ and we then verify the space requirement of the algorithm, i.e., that it is a semi-streaming algorithm.

Let us call S' the set of elements selected by EMI, including the elements deleted later. By Lemma 5, we have $2\alpha g(S') \geq w(S^*)$, so all we have to prove is that $g(S') - g(S) \leq \varepsilon g(S)$. We set $i \in \{1, 2\}$ to be the index of the matroid with smaller rank.

In our analysis, it will be convenient to think that the algorithm maintains the maximum weight independent set T_i of M_i throughout the stream. We have, at the arrival of an element e that is added to S, that the set T_i is updated as follows. If $T_i \cup \{e\} \in I_i$ then e is simply added to T_i. Otherwise, before updating T_i, there is an element $e^* \in T_i$ such that $w_i(e^*) = w_i^*(e)$ and $T_i \setminus \{e^*\} \cup \{e\}$ is maximum weight independent set in M_i with respect to w_i. Thus we can speak of elements which are *replaced* be another element in T_i. By construction, if e replaces f in T_i, then $w_i(e) > \alpha w_i(f)$.

We can now divide the elements of S' into stacks in the following way: If e replaces an element f in T_i, then we add e on top of the stack containing f, otherwise we create a new stack containing only e. At the end of the stream, each element $e \in T_i$ is in a different stack, and each stack contains exactly one element of T_i, so let us call S'_e the stack containing e whenever $e \in T_i$. We define S_e to be the restriction of S'_e to S. In particular, each element from S' is in exactly one S'_e stack, and each element from S is in exactly one S_e stack. For each stack S'_e, we set $e_{del}(S'_e)$ to by the highest element of S'_e which was removed from S. By construction, $g(S'_e) - g(S_e) \leq w_i(e_{del}(S'_e))$. On the other hand, $w_i(f) < \frac{1}{\varepsilon} g(f)$ for any element $f \in S'$ (otherwise we would not have selected it), so $g(S'_e) - g(S_e) < \frac{1}{\varepsilon} g(e_{del}(S'_e))$. As $e_{del}(S'_e)$ was removed from S, we have $g(e_{del}(S'_e)) < \frac{g_{max}}{y}$. As there are exactly r_i stacks, we get

$$g(S') - g(S) < r_i \frac{g_{max}\varepsilon^2}{r_i\varepsilon} = \varepsilon g_{max} \leq \varepsilon g(S).$$

We now have to prove that the algorithm fits the semi-streaming criteria. In fact, the size of S never exceeds $r_1 + r_2 + r_i \log_\alpha(\frac{y}{\varepsilon})$. By the pigeonhole principle, if S has at least $r_i \log_\alpha(\frac{y}{\varepsilon})$ elements, then there is at least one stack S_e which has at least $\log_\alpha(\frac{y}{\varepsilon})$ elements. By construction, the w_i weight increases by a factor of at least α each time we add an element on the same stack, so the w_i weight difference between the lowest and highest element on the biggest stack would be at least $\frac{y}{\varepsilon}$. As $w_i(f) < \frac{1}{\varepsilon} g(f)$, the g weight difference would be at least y, and we would remove the lowest element, unless it was in T_1 or T_2. □

Theorem 2. *Let S be the subset generated by running EMI with $\alpha = 1 + \varepsilon$ and $y = \frac{\min(r_1, r_2)}{\varepsilon^2}$. Then there exists a subset $T \subseteq S$ independent in M_1 and in M_2 such that $w(T) \geq g(S)$. Furthermore, T is a $2(1 + 2\varepsilon + o(\varepsilon))$-approximation for the intersection of two matroids.*

Proof. Let S^* be a maximum weight independent set. By Lemma 6, we have $2(1 + 2\varepsilon + o(\varepsilon))g(S) \geq w(S^*)$. By Lemma 4 we can select an independent set T with $w(T) \geq g(S)$ if the algorithm does not delete elements. Let S' be the set of elements selected by EMI, including the elements deleted later. As long as we do not delete elements from T_1 or T_2, Algorithm 1 restricted to S' will select the same elements, with the same weights, so we can consider S' to be generated by Algorithm 1. We now observe that all the arguments used in Lemma 4 also work for a subset of S', in particular, it is also true for S that we can find an independent set $T \subseteq S$ such that $w(T) \geq g(S)$. □

Remark 1. EMI is not the most efficient possible in terms of memory, but is aimed to be simpler instead. Using the notion of stacks introduced in the proof of Lemma 6, it is possible to modify the algorithm and reduce the memory requirement by a factor $\log(\min(\text{rank}(M_1), \text{rank}(M_2)))$.

Remark 2. The techniques of this section can also be used in the case when the ranks of the matroids are unknown. Specifically, the algorithm can maintain the stacks created in the proof of Lemma 6 and allow for an error ε in the first two stacks created, an error of $\varepsilon/2$ in the next 4 stacks, and in general an error of $\varepsilon/2^i$ in the next 2^i stacks.

Remark 3. It is easy to construct examples where the set S only contains a 2α-approximation (for example with bipartite graphs), so up to a factor ε our analysis is tight.

5   More Than Two Matroids

We can easily extend the algorithm EMI in the previous section to the intersection of k matroids. Now for any element e, if $w(e) > \alpha \sum_{i=1}^{k} w_i^*(e)$, we add it to our stack and update the weight functions $w_1, \ldots, w_k$ similarly as EMI. The only part which does not work is the selection of the independent set from S. Indeed, matroid kernels are very specific to two matroids. We refer the reader to the full version [8] to see why a similar approach fails and that a counter-example can arise. Thus, any attempt to find a k approximation using our techniques must bring some fundamentally new idea. Still, we conjecture that the generated set S contains such an approximation.

Acknowledgements. The authors thank Moran Feldman for pointing us to the recent paper [11].

References

1. Bar-Yehuda, R., Even, S.: A local-ratio theorem for approximating the weighted vertex cover problem. In: Ausiello, G., Lucertini, M. (eds.) Analysis and Design of Algorithms for Combinatorial Problems, North-Holland Mathematics Studies, vol. 109, pp. 27–45. North-Holland (1985). https://doi.org/10.1016/S0304-0208(08)73101-3, http://www.sciencedirect.com/science/article/pii/S0304020808731013

2. Bar-Yehuda, R., Bendel, K., Freund, A., Rawitz, D.: Local ratio: A unified framework for approximation algorithms. in memoriam: Shimon even 1935–2004. ACM Comput. Surv. (CSUR) **36**, 422–463 (2004). https://doi.org/10.1145/1041680.1041683

3. Chakrabarti, A., Kale, S.: Submodular maximization meets streaming: Matchings, matroids, and more. CoRR abs/1309.2038 (2013). http://arxiv.org/abs/1309.2038

4. Crouch, M., Stubbs, D.: Improved streaming algorithms for weighted matching, via unweighted matching. In: Leibniz International Proceedings in Informatics, LIPIcs, vol. 28, pp. 96–104 (2014). https://doi.org/10.4230/LIPIcs.APPROX-RANDOM.2014.96

5. Edmonds, J.: Matroid intersection. In: Discrete Optimization I, Annals of Discrete Mathematics, vol. 4, pp. 39–49. Elsevier (1979)

6. Feigenbaum, J., Kannan, S., McGregor, A., Suri, S., Zhang, J.: On graph problems in a semi-streaming model. Theor. Comput. Sci. **348**(2–3), 207–216 (2005)

7. Fleiner, T.: A matroid generalization of the stable matching polytope. In: Aardal, K., Gerards, B. (eds.) IPCO 2001. LNCS, vol. 2081, pp. 105–114. Springer, Heidelberg (2001). https://doi.org/10.1007/3-540-45535-3_9

8. Garg, P., Jordan, L., Svensson, O.: Semi-streaming algorithms for submodular matroid intersection. arXiv preprint arXiv:2102.04348 (2021)

9. Ghaffari, M., Wajc, D.: Simplified and Space-Optimal Semi-Streaming (2+epsilon)-Approximate Matching. In: Fineman, J.T., Mitzenmacher, M. (eds.) 2nd Symposium on Simplicity in Algorithms (SOSA 2019). OpenAccess Series in Informatics (OASIcs), vol. 69, pp. 13:1–13:8. Schloss Dagstuhl-Leibniz-Zentrum fuer Informatik, Dagstuhl, Germany (2018). https://doi.org/10.4230/OASIcs.SOSA.2019.13, http://drops.dagstuhl.de/opus/volltexte/2018/10039

10. Huang, C., Kakimura, N., Mauras, S., Yoshida, Y.: Approximability of monotone submodular function maximization under cardinality and matroid constraints in the streaming model. CoRR abs/2002.05477 (2020)

11. Levin, R., Wajc, D.: Streaming submodular matching meets the primal-dual method. arXiv preprint arXiv:2008.10062 (2020)

12. McGregor, A.: Finding graph matchings in data streams. In: Chekuri, C., Jansen, K., Rolim, J.D.P., Trevisan, L. (eds.) APPROX/RANDOM -2005. LNCS, vol. 3624, pp. 170–181. Springer, Heidelberg (2005). https://doi.org/10.1007/11538462_15

13. Muthukrishnan, S.: Data streams: algorithms and applications. Found. Trends Theor. Comput. Sci. **1**(2), 117–236 (2005). https://doi.org/10.1561/0400000002

14. Paz, A., Schwartzman, G.: A (2+ε)-approximation for maximum weight matching in the semi-streaming model. CoRR abs/1702.04536 (2017), http://arxiv.org/abs/1702.04536

15. Schrijver, A.: Combinatorial Optimization: Polyhedra and Efficiency. Springer, Algorithms and Combinatorics (2003)

Pfaffian Pairs and Parities: Counting on Linear Matroid Intersection and Parity Problems

Kazuki Matoya and Taihei Oki[✉]

Department of Mathematical Informatics, Graduate School of Information Science and Technology, University of Tokyo, Tokyo 113-8656, Japan
{kazuki_matoya,taihei_oki}@mist.i.u-tokyo.ac.jp

Abstract. Spanning trees are a representative example of linear matroid bases that are efficiently countable. Perfect matchings of Pfaffian bipartite graphs are a countable example of common bases of two matrices. Generalizing these two, Webb (2004) introduced the notion of Pfaffian pairs as a pair of matrices for which counting of their common bases is tractable via the Cauchy–Binet formula.

This paper studies counting on linear matroid problems extending Webb's work. We first introduce "Pfaffian parities" as an extension of Pfaffian pairs to the linear matroid parity problem, which is a common generalization of the linear matroid intersection problem and the matching problem. We show that a large number of efficiently countable discrete structures are interpretable as special cases of Pfaffian pairs and parities.

We also observe that the fastest randomized algorithms for the linear matroid intersection and parity problems by Harvey (2009) and Cheung–Lau–Leung (2014) can be derandomized for Pfaffian pairs and parities. We further present polynomial-time algorithms to count the number of minimum-weight solutions on weighted Pfaffian pairs and parities.

Keywords: Linear matroid intersection · Linear matroid parity · Pfaffian · Matrix-tree theorem · Matching · Pfaffian orientation · $\mathcal{S}$-path · Counting algorithm

1 Introduction

Let A be a totally unimodular matrix of row-full rank; that is, any minor of A is 0 or ± 1. The (generalized) *matrix-tree theorem* [24] claims that the number of column bases of A is equal to $\det AA^\top$. This can be observed by setting $A_1 = A_2 = A$ in the *Cauchy–Binet formula*

$$\det A_1 A_2^\top = \sum_{J \subseteq E: |J| = r} \det A_1[J] \det A_2[J], \qquad (1)$$

The full version of this paper is available at https://arxiv.org/abs/1912.00620.

© Springer Nature Switzerland AG 2021
M. Singh and D. P. Williamson (Eds.): IPCO 2021, LNCS 12707, pp. 223–237, 2021.
https://doi.org/10.1007/978-3-030-73879-2_16

where A_1, A_2 are matrices of size $r \times n$ with common column set E and $A_k[J]$ denotes the submatrix of A_k indexed by columns $J \subseteq E$ for $k = 1, 2$. If A comes from the incidence matrix of an undirected graph, the formula (1) provides the celebrated matrix-tree theorem due to Kirchhoff [18] for counting spanning trees.

From a matroidal point of view, the matrix-tree theorem is regarded as a theorem for counting bases of regular matroids, which are linear matroids represented by totally unimodular matrices. Regular matroids are recognized as the largest class of matroids for which base counting is exactly tractable. For general matroids (even for binary or transversal matroids), base counting is #P-complete [4, 29] and hence approximation algorithms have been well-studied [1, 2].

Another example of a polynomial-time countable object is perfect matchings of graphs with *Pfaffian orientation* [16]. The *Pfaffian* is a polynomial of matrix entries defined for a skew-symmetric matrix S of even order. If S is the Tutte matrix of a graph G, its Pfaffian is the sum over all perfect matchings of G except that each matching has an associated sign as well. Suppose that edges of G are oriented so that all terms in the Pfaffian become $+1$ by assigning $+1$ or -1 to each variable in the Tutte matrix according to the edge direction. This means that there are no cancellations in the Pfaffian, and thus it coincides with the number of perfect matchings of G. Such an orientation is called *Pfaffian*[1] and a graph that admits a Pfaffian orientation is also called *Pfaffian*. Whereas counting of perfect matchings is #P-complete even for bipartite graphs [32], characterizations of Pfaffian graphs and polynomial-time algorithms to give a Pfaffian orientation have been intensively studied [16, 21, 27, 30, 33].

From the viewpoint of matroids again, the bipartite matching problem is generalized to the *linear matroid intersection problem* [5, 6]. This is the problem to find a common column base of two matrices A_1, A_2 of the same size. The *weighted linear matroid intersection problem* is to find a common base of A_1, A_2 that minimizes a given column weight $w : E \rightarrow \mathbb{R}$. Various polynomial-time algorithms have been proposed for both the unweighted and weighted linear matroid intersection problems [5–7, 13]. However, the counting of common bases is intractable even for a pair of totally unimodular matrices, as it includes counting of perfect bipartite matchings.

Commonly generalizing Pfaffian bipartite graphs and regular matroids, Webb [34] introduced the notion of a *Pfaffian (matrix) pair* as a pair of totally unimodular matrices A_1, A_2 such that $\det A_1[B] \det A_2[B]$ is constant for any common base B of A_1 and A_2. This condition means due to the Cauchy–Binet formula (1) that the number of common bases of (A_1, A_2) can be retrieved from $\det A_1 A_2^\top$. For example, bases of a totally unimodular matrix A are clearly common bases of a Pfaffian pair (A, A). Webb [34] indicated that the set of perfect matchings of a Pfaffian bipartite graph can also be represented as common bases of a Pfaffian pair. Although the Pfaffian pairs concept nicely integrates these two

---

[1] An equivalent definition of Pfaffian orientation is as follows: an orientation of G is *Pfaffian* if every even-length cycle C such that $G - V(C)$ has a perfect matching has an odd number of edges directed in either direction.

Table 1. List of discrete structures representable as Pfaffian pairs or parities. The second column indicates those who first showed the polynomial-time countability.

Discrete structures	Countability	Pairs/Parities
Spanning trees	Kirchhoff [18]	Pairs (folklore)
Regular matroid bases	Maurer [24]	Pairs (folklore)
Feasible sets of regular delta-matroids	Webb [34]	Pairs (Webb [34])
Arborescences	Tutte [31]	Pairs (folklore)
Perf. matchings of Pfaffian bip. graphs	Kasteleyn [16,17]	Pairs (Webb [34])
Perf. matchings of Pfaffian graphs	Kasteleyn [16,17]	Parities (**this work**)
Spanning hypertrees of 3-Pfaffian 3-Uniform Hypergraphs	Goodall [12]	Parities (**this work**)
Disjoint S–T paths of DAGs in the LGV position	Lindström [20], Gessel–Viennot [11]	Pairs (**this work**)
Shortest disjoint S–T paths of undirected graphs in the LGV position	**This work**	Weighted pairs (**this work**)
Shortest disjoint S–T–U paths of undirected graphs in the LGV position	**This work**	Weighted parities (**this work**)

celebrated countable objects, its existence and importance do not seem to have been recognized besides the original thesis [34] of Webb.

The linear matroid intersection and the (nonbipartite) matching problems are commonly generalized to the *linear matroid parity problem* [19], which is explained as follows. Let A be a $2r \times 2n$ matrix whose column set is partitioned into pairs, called *lines*. Let L be the set of lines. In this paper, we call (A, L) a *(linear) matroid parity*. The linear matroid parity problem on (A, L) is to find a *parity base* of (A, L), which is a column base of A consisting of lines. The linear matroid parity problem is known to be solvable in polynomial time since the pioneering work of Lovász [22]. Recently, Iwata–Kobayashi [15] presented the first polynomial-time algorithm for the *weighted linear matroid parity problem*, which is to find a parity base of (A, L) that minimizes a given line weight $w : L \to \mathbb{R}$.

In this paper, we explore Pfaffian pairs and their generalization to the linear matroid parity problem, which we call *Pfaffian (linear matroid) parities*. The contributions of this paper are twofold: structural and algorithmic results.

Structural Results. Generalizing Pfaffian pairs, we introduce a new concept "Pfaffian parity" as a matroid parity (A, L) such that $\det A[B]$ is constant for all parity base B of (A, L). As in the case of Pfaffian pairs, this condition ensures that the number of parity bases can be retrieved from the Pfaffian of a skew-symmetric matrix associated with (A, L). This itself is a straightforward consequence of a generalization of the Cauchy–Binet formula (1) to the Pfaffian given by Ishikawa–Wakayama [14] (see Proposition 2.1).

We then present a collection of discrete structures that can be interpreted as common/parity bases of Pfaffian pairs/parities (or minimum-weight ones of weighted Pfaffian pairs/parities), which are summarized in Table 1. The variety of the list illustrates that Pfaffian pairs/parities nicely serve as a unified framework of discrete structures for which counting is tractable. Due to page limitations, this extended abstract describes only prefect matchings and shortest disjoint S–T–U paths. Readers interested in other structures are referred to the full version of this paper.

Algorithmic Results. Let (A_1, A_2) be an $r \times n$ Pfaffian pair and (A, L) a $2r \times 2n$ Pfaffian parity over a field $\mathbb{K}$ with characteristic $\mathrm{ch}(\mathbb{K})$. The definitions of Pfaffian pairs and parities guarantee that one can count the number (modulo $\mathrm{ch}(\mathbb{K})$) of common bases in (A_1, A_2) and of parity bases in (A, L) just by matrix computations. Here, we regard N modulo 0 as N for an integer N. This, however, is valid only when we know the value of $\det A_1[B] \det A_2[B]$ with an arbitrary common base B of (A_1, A_2) and $\det A[B]$ with an arbitrary parity base B of (A, L). These are called *constants*. If we do not know constants, we need to obtain one common/parity base B beforehand by executing linear matroid intersection/parity algorithms. The current best time complexity for the linear matroid intersection is deterministic $\mathrm{O}\bigl(nr^{\frac{5-\omega}{4-\omega}} \log r\bigr)$-time due to Gabow–Xu [9] and randomized $\mathrm{O}\bigl(nr^{\omega-1}\bigr)$-time due to Harvey [13], where ω is the exponent in the complexity of matrix multiplication. Also, the current best time complexity for the linear matroid parity problem is deterministic $\mathrm{O}(nr^\omega)$-time due to Gabow–Stallmann [8] and Orlin [26], and randomized $\mathrm{O}\bigl(nr^{\omega-1}\bigr)$-time due to Cheung–Lau–Leung [3]. We describe that one can derandomize the algorithms of [3, 13] for Pfaffian pairs and parities over $\mathbb{K}$ with $\mathrm{ch}(\mathbb{K}) = 0$ due to "no cancellation" property.

Theorem 1.1. *When* $\mathrm{ch}(\mathbb{K}) = 0$, *we can count the number of common bases of an* $r \times n$ *Pfaffian pair and the number of parity bases of a* $2r \times 2n$ *Pfaffian parity over* $\mathbb{K}$ *in deterministic* $\mathrm{O}\bigl(nr^{\omega-1}\bigr)$-*time.*

We next present a polynomial-time counting algorithm for minimum-weight parity bases of a line-weighted Pfaffian parity. Iwata–Kobayashi [15] gave not only a polynomial-time algorithm but also an algebraic optimality criterion for the weighted linear matroid parity problem. Based on this criterion, we show that the number of minimum-weight parity bases coincides with the leading coefficient of the Pfaffian of a skew-symmetric polynomial matrix. We then apply Murota's upper-tightness testing algorithm [25] to compute the leading coefficient.

Theorem 1.2. *Let* (A, L) *be a* $2r \times 2n$ *Pfaffian parity with line weight* $w : L \to \mathbb{R}$. *We can compute the number of minimum-weight parity bases of* (A, L) *modulo* $\mathrm{ch}(\mathbb{K})$ *in deterministic* $\mathrm{O}\bigl(n^3 r\bigr)$-*time.*

We can also design a more efficient algorithm tailored for weighted Pfaffian pairs, which is based on the weight splitting; see the full version of this paper.

Organization. The rest of this paper is organized as follows. After introducing some preliminaries, Sect. 2 gives formal definitions of Pfaffian pairs and parities. Section 3 present two combinatorial examples of Pfaffian pairs and parities. Finally, Sect. 4 presents our counting algorithms for unweighted and weighted Pfaffian pairs and parities.

2   Pfaffian Pairs and Pfaffian Parities

2.1   Preliminaries

Let $\mathbb{R}$ denote the set of all reals. For a nonnegative integer n, we denote $\{1, 2, \ldots, n\}$ by $[n]$. For a matrix A, denote by $A[I, J]$ the submatrix of A with row subset I and column subset J. If I is all the rows, we denote $A[I, J]$ by $A[J]$. The identity matrix of order n is denoted by I_n. The vector of ones is represented as $\mathbb{1}$.

Let $A \in \mathbb{K}^{r \times n}$ be a matrix with column set E over a field $\mathbb{K}$. The *linear matroid* represented by A is a set family

$$\mathcal{B}(A) := \{B \subseteq E \mid |B| = r, A[B] \text{ is nonsingular}\}.$$

We refer to each element of $\mathcal{B}(A)$ as a *base* of A.

Let $\mathfrak{S}_n$ be the set of all permutations on $[n]$ and $\mathrm{sgn}\,\sigma$ the sign of a permutation $\sigma \in \mathfrak{S}_n$. A square matrix S is said to be *skew-symmetric* if $S^\top = -S$ and all diagonal entries are 0. For a skew-symmetric matrix $S = (S_{i,j})_{i,j \in [2n]} \in \mathbb{K}^{2n \times 2n}$ of even order, the *Pfaffian* of S is defined as

$$\mathrm{Pf}\,S := \sum_{\sigma \in F_{2n}} \mathrm{sgn}\,\sigma \prod_{i=1}^{n} S_{\sigma(2i-1), \sigma(2i)}, \qquad (2)$$

where

$$F_{2n} := \{\sigma \in \mathfrak{S}_{2n} \mid \sigma(1) < \cdots < \sigma(3) < \sigma(2n-1) \text{ and } \sigma(2i-1) < \sigma(2i) \text{ for } i \in [n]\}.$$

It is well-known that $(\mathrm{Pf}\,S)^2 = \det S$ and $\mathrm{Pf}\,ASA^\top = \det A \,\mathrm{Pf}\,S$ hold, where $A \in \mathbb{K}^{2n \times 2n}$ is any square matrix. The following formula is a generalization of the Cauchy–Binet formula (1) to Pfaffian given by Ishikawa–Wakayama [14].

Proposition 2.1 ([14, Theorem 1]). *Let $S \in \mathbb{K}^{2n \times 2n}$ be skew-symmetric with rows and columns E and $A \in \mathbb{K}^{2r \times 2n}$ with columns E. Then it holds*

$$\mathrm{Pf}\,ASA^\top = \sum_{J \subseteq E, |J| = 2r} \det A[J] \mathrm{Pf}\,S[J, J].$$

2.2 Pfaffian Pairs and Parities

Now we define Pfaffian (matrix) pairs slightly generalizing that of Webb [34].

Definition 2.2 (Pfaffian matrix pair; see [34]). *We say that a matrix pair* (A_1, A_2) *of the same size over* $\mathbb{K}$ *is* Pfaffian *if there is* $c \in \mathbb{K} \backslash \{0\}$, *called* constant, *such that* $\det A_1[B] \det A_2[B] = c$ *for all* $B \in \mathcal{B}(A_1, A_2) := \mathcal{B}(A_1) \cap \mathcal{B}(A_2)$.

For a matrix pair (A_1, A_2) with common column set E and a vector $z = (z_j)_{j \in E}$ indexed by E, we define $\Xi(z) := \begin{pmatrix} O & A_1 \\ A_2^\top & D(z) \end{pmatrix}$, where $D(z) := \mathrm{diag}(z_j)_{j \in E}$. Taking the Schur complement, we have $\det \Xi(z) = \det D(z) \det(-A_1 D(z)^{-1} A_2^\top)$. Hence the following holds from the Cauchy–Binet formula (1) and Definition 2.2.

Proposition 2.3. *Let* (A_1, A_2) *be an* $r \times n$ *Pfaffian pair of constant* c *and* $z = (z_j)_{j \in E}$ *a vector indexed by the common column set* E *of* (A_1, A_2). *Then it holds*

$$\det A_1 D(z) A_2^\top = c \sum_{B \in \mathcal{B}(A_1, A_2)} \prod_{j \in B} z_j, \quad \det \Xi(z) = (-1)^r c \sum_{B \in \mathcal{B}(A_1, A_2)} \prod_{j \in E \backslash B} z_j.$$

In particular, the number of common bases of (A_1, A_2) *modulo* $\mathrm{ch}(\mathbb{K})$ *is equal to* $c^{-1} \det A_1 A_2^\top = c^{-1} \det \Xi(\mathbb{1})$.

We next introduce Pfaffian parities. Let (A, L) be a matroid parity over $\mathbb{K}$. We regard parity bases of (A, L) as a subset of L and denote by $\mathcal{B}(A, L)$ the set of all parity bases of A with respect to L. For $J \subseteq L$, we denote by $A[J]$ the submatrix of A consisting of columns corresponding to lines in J.

Definition 2.4 (Pfaffian matroid parity). *We say that a matroid parity* (A, L) *over* $\mathbb{K}$ *is* Pfaffian *if there exists* $c \in \mathbb{K} \backslash \{0\}$ *such that* $\det A[B] = c$ *for all* $B \in \mathcal{B}(A, L)$. *The value* c *is called the* constant.

We abbreviate Pfaffian matroid parity as Pfaffian parity for short. For a vector $z = (z_l)_{l \in L}$ indexed by L, we denote by $\Delta_L(z)$ the $2n \times 2n$ skew-symmetric block-diagonal matrix defined as follows: the row and column sets are indexed by the columns E of A, and each block corresponding to a line $l \in L$ is a 2×2 skew-symmetric matrix $\begin{pmatrix} 0 & +z_l \\ -z_l & 0 \end{pmatrix}$. Then $\Delta(z)[J, J]$ is nonsingular if and only if J consists of lines and $z_l \neq 0$ for $l \in J$. In addition, for $\Phi(z) := \begin{pmatrix} O & A \\ -A^\top & \Delta(z) \end{pmatrix}$, we have $\mathrm{Pf}\,\Phi(z) = \mathrm{Pf}\,\Delta(z) \mathrm{Pf}\, A\Delta(z)^{-1} A^\top$ by the Schur complement again. Then the following is obtained from Proposition 2.1 and Definition 2.4.

Proposition 2.5. *Let* (A, L) *be a Pfaffian parity of constant* c *and* $z = (z_l)_{l \in L}$ *a vector indexed by* L. *Then it holds*

$$\mathrm{Pf}\, A\Delta(z) A^\top = c \sum_{B \in \mathcal{B}(A, L)} \prod_{l \in B} z_l, \quad \mathrm{Pf}\,\Phi(z) = c \sum_{B \in \mathcal{B}(A, L)} \prod_{l \in L \backslash B} z_l.$$

In particular, the number of parity bases of (A, L) *modulo* $\mathrm{ch}(\mathbb{K})$ *is equal to* $c^{-1} \mathrm{Pf}\, A\Delta(\mathbb{1}) A^\top = c^{-1} \mathrm{Pf}\,\Phi(\mathbb{1})$.

The standard reduction of the linear matroid intersection problem to the parity problem [19, Chapter 9.2] retains the property of Pfaffian (see the full paper for proof). Hence Pfaffian parities generalize Pfaffian pairs.

3 Combinatorial Examples

3.1 Perfect Matchings of Pfaffian Graphs

A *matching* of an undirected graph G is an edge subset M such that no two disjoint edges in M share the same end. We also define a matching for a directed graph by ignoring its orientation. A matching M is said to be *perfect* if every vertex of G is covered by some edge in M. Webb [34] observed that perfect bipartite matchings can be represented as a common bases of a Pfaffian pair if the bipartite graph is Pfaffian-oriented. Here we show a generalized relation between general matchings and parity bases of a Pfaffian parity.

Let $G = (V, E)$ be a simple undirected graph that is not necessarily bipartite. Suppose that $|V|$ is even and vertices are ordered as $v_1, \ldots, v_{2n}$. We define $A \in \mathbb{R}^{|V| \times 2|E|}$ as follows: each row is indexed by $v \in V$ and each two columns are associated with an edge $e \in E$. For $v \in V$ and $e = \{v_i, v_j\} \in E$ with $i < j$, the corresponding 1×2 submatrix of A to v and e is defined to be $\begin{pmatrix} +1 & 0 \end{pmatrix}$ if $v = v_i$, $\begin{pmatrix} 0 & +1 \end{pmatrix}$ if $v = v_j$ and O otherwise. We regard each $e \in E$ as a line of A. Then $M \subseteq E$ is a perfect matching of G if and only if $M \in \mathcal{B}(A, E)$ [19, Chapter 9.2].

A perfect matching M of G uniquely corresponds to a permutation $\sigma_M \in F_{2n}$ such that $\{v_{\sigma_M(2i-1)}, v_{\sigma_M(2i)}\} \in M$ for $i \in [n]$. We define the *sign* of M as $\operatorname{sgn} M := \operatorname{sgn} \sigma_M$. Note that $\operatorname{sgn} M$ depends on the ordering of V.

Let $z = (z_e)_{e \in E}$ be a vector of distinct indeterminates indexed by E. The skew-symmetric matrix $A\Delta(z)A^\top$ is called the *Tutte matrix* of G. Its (i, j)th entry is equal to z_e if $e = \{u_i, v_j\} \in E$ and $i < j$, to $-z_e$ if $e = \{u_i, v_j\} \in E$ and $i > j$ and to 0 otherwise for $i, j \in [n]$. By the definition (2) of Pfaffian, it holds

$$\operatorname{Pf} A\Delta(z)A^\top = \sum_{M \in \mathcal{B}(A, E)} \operatorname{sgn} M \prod_{e \in M} z_e.$$

We also have

$$\operatorname{Pf} A\Delta(z)A^\top = \sum_{M \in \mathcal{B}(A, E)} \det A[M] \prod_{e \in M} z_e.$$

by Proposition 2.1. Hence the following holds.

Lemma 3.1. *The sign of a perfect matching M of G is equal to $\det A[M]$.*

We next consider an orientation $\vec{G} = (V, \vec{E})$ of G. Define a vector $s = (s_e)_{e \in E}$ indexed by E as follows: for each $\vec{e} = (v_i, v_j) \in \vec{E}$, we set $s_e := +1$ if $i < j$ and $s_e := -1$ if $i > j$. We also construct a symmetric block-diagonal matrix $X = \operatorname{diag}(X_e)_{e \in E}$, where X_e is a 2×2 matrix defined by $X_e := I_2$ if $s_e = +1$

and $X_e := \begin{pmatrix} 0 & +1 \\ +1 & 0 \end{pmatrix}$ if $s_e = -1$ for $e \in E$. Put $\vec{A} := AX$, i.e., $\vec{A}$ is obtained from A by interchanging two columns associated with each $(v_i, v_j) \in \vec{E}$ with $i > j$. Note that $X\Delta(\mathbb{1})X = \Delta(s)$ and $\mathcal{B}(A, E) = \mathcal{B}(\vec{A}, E)$ hold. Put $S = (S_{i,j})_{i,j \in [2n]} := \vec{A}\Delta(\mathbb{1})\vec{A}^\top = A\Delta(s)A^\top$. It can be confirmed that $S_{i,j} = +1$ if $(v_i, v_j) \in \vec{E}$, -1 if $(v_j, v_i) \in \vec{E}$ and 0 otherwise. For a perfect matching M of G, it holds $s_e = S_{\sigma_M(2i-1), \sigma_M(2i)}$ for every $e = \{v_{\sigma_M(2i-1)}, v_{\sigma_M(2i)}\} \in M$ since $\sigma_M(2i-1) < \sigma_M(2i)$. We define the *sign* of a perfect matching $\vec{M}$ of $\vec{G}$ as

$$\operatorname{sgn} \vec{M} := \operatorname{sgn} M \prod_{e \in M} s_e = \operatorname{sgn} M \prod_{i=1}^{n} S_{\sigma_M(2i-1), \sigma_M(2i)} \in \{+1, -1\}.$$

By Lemma 3.1 and $\det \vec{A}[M] = \det A[M] \prod_{e \in M} s_e$, we have the following.

Lemma 3.2. *The sign of a perfect matching $\vec{M}$ of $\vec{G}$ is equal to $\det \vec{A}[M]$.*

Recall that an orientation $\vec{G}$ of G is called *Pfaffian* if the signs of all perfect matchings of $\vec{G}$ are constant. The following holds from Lemma 3.2.

Theorem 3.3. *Let $G = (V, E)$ be a graph, $\vec{G}$ an orientation of G and $\vec{A}$ the matrix defined above from $\vec{G}$. Then $\mathcal{B}(\vec{A}, E)$ coincides with the set of perfect matchings of G. In addition, if $\vec{G}$ is Pfaffian, so is $(\vec{A}, E)$.*

3.2 Shortest Disjoint S–T–U Paths on Undirected Graphs

We first introduce Mader's disjoint $\mathcal{S}$-path problem [10,23]. Let $G = (V, E)$ be an undirected graph and $\mathcal{S} = \{S_1, \ldots, S_s\}$ a family of disjoint subsets of V. Suppose that $\Sigma := S_1 \cup \cdots \cup S_s$ is of cardinality $2k$ and ordered as $u_1, \ldots, u_{2k}$. Vertices in Σ are called *terminals*. An $\mathcal{S}$-path P of G is the union of k paths $P_1, \ldots, P_k \subseteq E$ of G satisfying the following: there exists $\sigma \in F_{2k}$ such that P_i is a path between $u_{\sigma(2i-1)} \in S_\alpha$ and $u_{\sigma(2i)} \in S_\beta$ with $\alpha \neq \beta$ for each $i \in [k]$ Namely, P is an $\mathcal{S}$-path if the ends of each P_i belong to distinct parts in $\mathcal{S}$ and the ends of P_i and P_j are disjoint for all distinct $i, j \in [k]$. We call an $\mathcal{S}$-path P *disjoint* if P_i and P_j have no common vertices for all distinct $i, j \in [k]$. For a disjoint $\mathcal{S}$-path P, a permutation satisfying the above condition uniquely exists in F_{2k} and we denote it by σ_P. The *sign* of a disjoint $\mathcal{S}$-path P is defined as $\operatorname{sgn} P := \operatorname{sgn} \sigma_P$. The *disjoint $\mathcal{S}$-path problem* on G is to find a disjoint $\mathcal{S}$-path of G. We also consider the situation when G has a positive edge length $l : E \to \mathbb{R}_{>0}$. The *length* of an $\mathcal{S}$-path P is defined as $l(P) := \sum_{e \in P} l(e)$. The *shortest disjoint $\mathcal{S}$-path problem* is to find a disjoint $\mathcal{S}$-path of G with minimum length.

We next describe a reduction of the disjoint $\mathcal{S}$-path problem to the linear matroid parity problem, based on Schrijver's linear representation [28] of Lovász' reduction [22]. We assume that there are no edges connecting terminals. Put $\tilde{V} := V \setminus \Sigma$, $\tilde{E} := \{\{u, v\} \in E \mid u, v \in \tilde{V}\}$, $m := |E|$, and $\tilde{m} := |\tilde{E}|$. Fix two-dimensional row vectors $b_1, \ldots, b_s$ which are pairwise linearly independent. We construct a matrix $X = \begin{pmatrix} X_1 & O \\ X_2 & X_3 \end{pmatrix}$ from G as follows. The size of each block is $2k \times 2\tilde{m}$ for X_1, $(2n - 4k) \times 2\tilde{m}$ for X_2, and $(2n - 4k) \times 2(m - \tilde{m})$ for X_3. Each edge $e \in \tilde{E}$ is associated with two columns of $\begin{pmatrix} X_1 \\ X_2 \end{pmatrix}$ and each $e \in E \setminus \tilde{E}$ is associated with two columns of $\begin{pmatrix} O \\ X_3 \end{pmatrix}$. Each terminal $u_i \in \Sigma$ corresponds to the ith row of $(X_1\ O)$ for $i \in [2k]$ and each $v \in \tilde{V}$ is associated with two rows of $(X_2\ X_3)$. Entries of each block are determined as follows:

- The 1×2 submatrix of X_1 associated with $u_i \in U$ and $e \in E \setminus \tilde{E}$ is b_α if $e \cap S_\alpha = \{u_i\}$ and O otherwise.
- The 2×2 submatrix of X_2 associated with $v \in \tilde{V}$ and $e \in E \setminus \tilde{E}$ is the identify matrix I_2 of order two if $v \in e$ and O otherwise.
- The matrix X_3 is defined to be the Kronecker product $H[\tilde{V}, \tilde{E}] \otimes I_2$, where H is the vertex-edge incidence matrix of any orientation of G. Namely, X_3 is obtained from $H[\tilde{V}, \tilde{E}]$ by replacing ± 1 with $\pm I_2$ and 0 with O.

We regard each edge $e \in E$ as a line of X, which consists of the two columns associated with e.

Lemma 3.4 ([35, Lemma 4]). *An edge subset $B \subseteq E$ is a parity base of (X, E) if and only if B is a spanning forest of G such that every connected component covers exactly two terminals belonging to distinct parts of S.*

Note that if B is a parity base of (X, E), then B has k connected components since B covers all vertices of G by Lemma 3.4. Hence (X, E) has a parity base if and only if G has a disjoint S-path. Unfortunately, this reduction does not yield a one-to-one correspondence between $\mathcal{B}(X, E)$ and disjoint S-paths of G.

Yamaguchi [35] showed that the shortest disjoint S-path problem can be reduced to the weighted linear matroid parity problem. Here we present a simplified reduction for our setting (where an S-path covers all terminals), together with a one-to-one correspondence of optimal solutions. Let G^* be the graph obtained from G by adding a new vertex v^* and an edge set $E' := \{\{v, v^*\} \mid v \in \tilde{V}\}$. We construct a matrix X^* from G^* in the same way as the construction of X from G. Let A be the matrix obtained from X^* by removing the two rows corresponding to v^*. Namely, by an appropriate column permutation and an edge orientation on E', the matrix A is written as $\begin{pmatrix} X_1 & O & O \\ X_2 & X_3 & I_{2n-4k} \end{pmatrix}$, where each two columns of the left, middle and right blocks correspond to an edge in $\tilde{E}$, $E \setminus \tilde{E}$ and E', respectively. Regarding $E^* := E \cup E'$ as the set of lines on A, we set a line weight $w : E^* \to \mathbb{R}$ as $w(e) := l(e)$ for $e \in E$ and as $w(e) = 0$ for $e \in E'$.

Lemma 3.5 (see [35]). *The minimum length of a disjoint S-path of G with respect to l is equal to the minimum weight of a parity base of (A, E^*) with respect to w. In addition, there is a one-to-one correspondence between shortest disjoint S-paths of G and minimum-weight parity bases of (A, E^*).*

We next connect sgn P and det $A[B]$. For a disjoint $\mathcal{S}$-path P, define

$$c_P := (-1)^k \prod_{i=1}^{k} \det \begin{pmatrix} b_{\alpha_{2i-1}} \\ b_{\alpha_{2i}} \end{pmatrix} \in \mathbb{R} \setminus \{0\}, \tag{3}$$

where α_i is the element in $[s]$ such that $u_{\sigma_P(i)} \in S_{\alpha_i}$ for $i \in [2k]$. Then the following holds; refer to the full paper for proof.

Lemma 3.6. *Let P be a disjoint $\mathcal{S}$-path of G and B a parity base of (A, E^*) containing P. Then we have $c_P \operatorname{sgn} P = \det A[B]$.*

We say that $\mathcal{S}$ is in the *LGV position* on G if sgn P is constant for all disjoint $\mathcal{S}$-path P of G. Since det $A[B]$ depends not only on sgn P but on c_P by (3), the matroid parity (A, E^*) might not be Pfaffian even if $\mathcal{S}$ is in the LGV position. Nevertheless, c_P is constant when $|\mathcal{S}| = 3$; see the full paper for proof again. We refer to the (shortest) disjoint $\mathcal{S}$-path problem with $\mathcal{S} = \{S = S_1, T = S_2, U = S_3\}$ as the *(shortest) disjoint S–T–U path problem*. An S–T–U path means an $\{S, T, U\}$-path. We have the following conclusion.

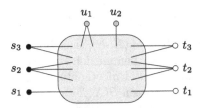

Fig. 1. Example of S, T, U that are in the LGV position, where $S = \{s_1, s_2, s_3\}$, $T = \{t_1, t_2, t_3\}$ and $U = \{u_1, u_2\}$. The gray area represents a planar graph.

Theorem 3.7. *Let G be an undirected graph and take disjoint vertex subsets S, T, U. Let (A, E^*) be the matroid parity defined above. When $\{S, T, U\}$ is in the LGV position, then (A, E^*) is Pfaffian. In addition, when G is equipped with a positive edge length l, there is a one-to-one correspondence between shortest disjoint S–T–U paths of G and minimum-weight parity bases of (A, E^*) with respect to the line weight w defined above.*

We show one example of the LGV position for the S–T–U case (see Fig. 1). Let G be a planar graph and suppose that terminals are aligned on the boundary of one face of G in the order of S, U, T clockwise. Then the connecting pattern of terminals are uniquely determined from $|S|, |T|, |U|$. Hence σ_P is constant for all disjoint S–T–U path P, which means that $\{S, T, U\}$ is in the LGV position.

4   Algorithms

4.1   Counting on Unweighted Pfaffian Pairs and Parities

We describe how we can derandomize the linear matroid intersection algorithm
of Harvey [13] and the linear matroid parity algorithm of Cheung–Lau–Leung [3]
for Pfaffian pairs and parities over $\mathbb{K}$ with $\mathrm{ch}(\mathbb{K}) = 0$. In these algorithms, randomness is used only to find a vector over $\mathbb{K}$ satisfying some genericity conditions,
summarized below. Recall the matrices $\Xi(z)$ and $\Phi(z)$ defined in Sect. 2.2.

Let (A_1, A_2) be a matrix pair with common column set E. A column subset
$J \subseteq E$ is said to be *extensible* if there exists a common base of (A_1, A_2) containing
J. For a vector $z = (z_j)_{j \in E}$ and $J \subseteq E$, let $\phi_J(z)$ denote a vector whose each
component is defined as $\phi_J(z)_j := 0$ if $j \in J$ and z_j otherwise. Then Harvey's
algorithm [13] constructs a common base if a vector $z = (z_j)_{j \in E} \in \mathbb{K}^n$ such that
every $J \subseteq E$ is extensible if and only if $\Xi(\phi_J(z))$ is nonsingular is provided.
Similarly, for a matroid parity (A, L), we call $J \subseteq L$ *extensible* if there exists
a parity base of (A, L) containing J. We also define $\phi_J(z)$ for $z = (z_l)_{l \in L}$ and
$J \subseteq L$ in the same way. Then the algorithm of Cheung–Lau–Leung [3] outputs
a parity base if a vector $z = (z_j)_{j \in L} \in \mathbb{K}^n$ such that every $J \subseteq L$ is extensible if
and only if $\Phi(\phi_J(z))$ is nonsingular is provided.

It is shown in [13, Theorem 4.4], [3, Theorem 6.4] that a vector of distinct
indeterminates satisfies the requirements of z. The algorithms of [3,13] use a
random vector over $\mathbb{K}$ instead of indeterminates to avoid symbolic computations.
For Pfaffian pairs and parities over a field $\mathbb{K}$ with $\mathrm{ch}(\mathbb{K}) = 0$, we can use $\mathbb{1}$ for
z, as the definition of Pfaffian ensures numerical cancellations do not occur; see
the full paper for proof. Theorem 1.1 is obtained as a consequence of this fact.

4.2   Counting on Weighted Pfaffian Parities

Let (A, L) be a $2r \times 2n$ Pfaffian parity of constant c with line weight $w : L \to \mathbb{R}$.
We describe an algorithm to count the number of minimum-weight parity bases
of (A, L) modulo $\mathrm{ch}(\mathbb{K})$. Suppose that (A, L) has at least one parity base. Let
ζ denote the minimum weight of a parity base of (A, L) and N the number of
minimum-weight parity bases modulo $\mathrm{ch}(\mathbb{K})$. Note that N is nonzero if $\mathrm{ch}(\mathbb{K}) =
0$. We put $\delta := w(L) - \zeta$ and $\theta^w := (\theta^{w(l)})_{l \in L}$ for an indeterminate θ. Then the
following holds from Proposition 2.5.

Lemma 4.1. *The coefficient of θ^δ in $\mathrm{Pf}\,\Phi(\theta^w)$ is equal to cN. In addition, it
holds $\delta \geq \deg \mathrm{Pf}\,\Phi(\theta^w)$ and the equality is attained if and only if $N \neq 0$.*

We first obtain a minimum-weight parity base $B \in \mathcal{B}(A, L)$ applying the
algorithm of Iwata–Kobayashi [15]. Then we perform a row transformation and
a line (column) permutation on A so that the left $2r$ columns of A correspond
to B and $A[B] = I_{2r}$. Namely, A is in the form of $A = (I_{2r}\ C)$ for some matrix
$C \in \mathbb{K}^{2r \times (2n-2r)}$. Note that these transformations retain (A, L) Pfaffian but

change the constant to 1. We perform the same transformations on $\Phi(\theta^w)$ (and on $\Delta(\theta^w)$) accordingly. Now the polynomial matrix $\Phi(\theta^w)$ is in the form of

$$
\Phi(\theta^w) = \left(\begin{array}{c|c} O & I_{2r}\ \ C \\ \hline \begin{array}{c} -I_{2r} \\ -C^\mathsf{T} \end{array} & \Delta(\theta^w) \end{array}\right) \begin{array}{l} \leftarrow U \\ \leftarrow B \\ \leftarrow E \setminus B, \end{array}
$$

where U is the row set of A identified with B.

Besides the minimum-weight parity base B, the algorithm of Iwata–Kobayashi [15] output an extra matrix C^*. Its row set U^* and column set E^* contains U and $E \setminus B$, respectively, and elements in $U^* \setminus U$ and $E^* \setminus E = E^* \setminus (E \setminus B)$ are newly introduced ones. The Schur complement of C^* with respect to $Y := C^*[U^* \setminus U, E^* \setminus E]$ coincides with C, i.e., it holds

$$
C = C^*[U, E \setminus B] - C^*[U, E^* \setminus E]Y^{-1}C^*[U^* \setminus U, E \setminus B].
$$

In addition, the cardinalities of U^* and E^* are guaranteed to be $O(n)$. We put $W := U^* \cup B \cup E^*$ and $c^* := \det Y$. Consider the skew-symmetric polynomial matrix $\Phi^*(\theta) = \left(\Phi^*_{u,v}(\theta)\right)_{u,v \in W}$ defined by

$$
\Phi^*(\theta) := \left(\begin{array}{c|c|c} O & \begin{array}{c} O \\ \hline I_{2r} \end{array} & C^* \\ \hline \begin{array}{c} O \\ -C^{*\mathsf{T}} \end{array} \begin{array}{c} -I_{2r} \\ \ \end{array} & \Delta(\theta^w) & \begin{array}{c} O \\ \ \end{array} \\ \hline & O & O \end{array}\right) \begin{array}{l} \leftarrow U^* \setminus U \\ \leftarrow U \\ \leftarrow B \\ \leftarrow E \setminus B \\ \leftarrow E^* \setminus E. \end{array}
$$

Then we have $\mathrm{Pf}\,\Phi^*(\theta) = c^*\mathrm{Pf}\,\Phi(\theta^w)$. Using this equality, Lemma 4.1 can be rephrased in terms of $\Phi^*(\theta)$ as follows.

Lemma 4.2. *The coefficient of θ^δ in $\mathrm{Pf}\,\Phi^*(\theta)$ is equal to $c^* N$. In addition, it holds $\delta \geq \deg \mathrm{Pf}\,\Phi^*(\theta^w)$ and the equality is attained if and only if $N \neq 0$.*

We next define an undirected graph $G = G(\Phi^*)$ associated with $\Phi^*(\theta)$. The vertex set of G is W and the edge set is given by

$$
F := \left\{ \{u, v\} \mid u, v \in W, \ \Phi^*_{u,v}(\theta) \neq 0 \right\}.
$$

We set the weight of every edge $\{u, v\} \in F$ to $\deg \Phi^*_{u,v}(\theta)$. Let $\hat{\delta}(\Phi^*)$ denote the maximum weight of a perfect matching of G. We set $\hat{\delta}(\Phi^*) := -\infty$ if G has no perfect matching. Here we put $\hat{\delta} := \hat{\delta}(\Phi^*)$. From the definition (2) of Pfaffian, $\hat{\delta}$ serves as a combinatorial upper bound on $\deg \mathrm{Pf}\,\Phi^*(\theta)$.

Define $\Omega := \{Z \subseteq W \mid |Z| \text{ is odd and } |Z| \geq 3\}$ and $\Omega_{u,v} := \{Z \in \Omega \mid |Z \cap \{u, v\}| = 1\}$ for $u, v \in W$. The dual problem of the maximum-weight perfect matching problem on G is as follows (see [15] and [28, Theorem 25.1]):

$$
\text{(D)} \quad
\begin{aligned}
&\underset{\pi,\xi}{\text{minimize}} && \sum_{u \in W} \pi(u) - \sum_{Z \in \Omega} \xi(Z) \\
&\text{subject to} && \pi(u) + \pi(v) - \sum_{Z \in \Omega_{u,v}} \xi(Z) \geq \deg \Phi^*_{u,v}(\theta) \; (\{u,v\} \in F), \\
& && \xi(Z) \geq 0 && (Z \in \Omega).
\end{aligned}
$$

The following claim is proved in [15] as a key ingredient of the optimality certi-fication on the weighted linear matroid parity problem.

Proposition 4.3 ([15, Claim 6.3]). *There exists a feasible solution of (D) hav-ing the objective value δ.*

We make use of Proposition 4.3 for the purpose of counting.

Lemma 4.4. $\delta \geq \hat{\delta} \geq \deg \operatorname{Pf} \Phi^*(\theta)$ *holds. The equalities are attained if $N \neq 0$.*

Proof. We have $\delta \geq \hat{\delta}$ by Proposition 4.3 and the weak duality of (D). The latter inequality is due to (2). The equality condition is obtained from Lemma 4.2. $\square$

By Lemma 4.4, we have $N = 0$ if $\delta > \hat{\delta}$. Otherwise, our goal is to compute the coefficient of $\theta^\delta = \theta^{\hat{\delta}}$ in $\operatorname{Pf} \Phi^*(\theta)$ by Lemma 4.2. This is obtained by executing Murota's upper-tightness testing algorithm on combinatorial relaxation for skew-symmetric polynomial matrices [25, Section 4.4] (with det replaced with Pf).

Proposition 4.5 ([25]). *Let $S(\theta)$ be a $2n \times 2n$ skew-symmetric polynomial matrix. We can compute the coefficient of $\theta^{\hat{\delta}(S)}$ in $\operatorname{Pf} S(\theta)$ in $O(n^3)$-time.*

Algorithm 1 shows the entire procedure of our algorithm. The time complex-ity can be estimated as in Theorem 1.2.

Algorithm 1. Counting minimum-weight parity bases of a Pfaffian parity.

Input : An $2r \times 2n$ Pfaffian parity (A, L) and a line weight $w : L \to \mathbb{Z}$
Output: The number of minimum-weight common bases of (A_1, A_2) modulo $\operatorname{ch}(\mathbb{K})$
1: Compute a minimum-weight parity base $B \in \mathcal{B}(A, L)$ and the matrix C^*
2: Construct the matrix $\Phi^*(\theta)$ and the graph $G = G(\Phi^*)$
3: Compute the maximum weight $\hat{\delta} := \hat{\delta}(\Phi^*)$ of a perfect matching of G
4: **if** $\delta := w(B) > \hat{\delta}$ **then**
5:     **return** 0
6: **else**
7:     Compute the coefficient a of $\theta^{\hat{\delta}}$ in $\operatorname{Pf} \Phi^*(\theta)$
8:     **return** $c^{*-1}a$, where $c^* := \operatorname{Pf} C^*[U^* \setminus U, E^* \setminus E]$

Acknowledgments. The authors thank Satoru Iwata for his helpful comments, and Yusuke Kobayashi, Yutaro Yamaguchi, and Koyo Hayashi for discussions. This work was supported by JST ACT-I Grant Number JPMJPR18U9, Japan, and Grant-in-Aid for JSPS Research Fellow Grant Number JP18J22141, Japan.

References

1. Anari, N., Gharan, S.O., Vinzant, C.: Log-concave polynomials, entropy, and a deterministic approximation algorithm for counting bases of matroids. In: Proceedings of the 59th Annual IEEE Symposium on Foundations of Computer Science (FOCS 2018), pp. 35–46 (2018). https://doi.org/10.1109/FOCS.2018.00013
2. Anari, N., Liu, K., Gharan, S.O., Vinzant, C.: Log-concave polynomials II: high-dimensional walks and an FPRAS for counting bases of a matroid. In: Proceedings of the 51st Annual ACM Symposium on Theory of Computing (STOC 2019), pp. 1–12 (2019). https://doi.org/10.1145/3313276.3316385
3. Cheung, H.Y., Lau, L.C., Leung, K.M.: Algebraic algorithms for linear matroid parity problems. ACM Trans. Algorithms 10(3), 1–26 (2014). https://doi.org/10.1145/2601066
4. Colbourn, C.J., Provan, J.S., Vertigan, D.: The complexity of computing the Tutte polynomial on transversal matroids. Combinatorica 15(1), 1–10 (1995). https://doi.org/10.1007/BF01294456
5. Edmonds, J.: Matroid partition. In: Dantzig, G.B., Veinott, Jr., A.F. (eds.) Mathematics of the Decision Sciences: Part I, Lectures in Applied Mathematics, vol. 11, pp. 335–345. AMS, Providence, RI (1968). https://doi.org/10.1007/978-3-540-68279-0_7
6. Edmonds, J.: Submodular functions, matroids, and certain polyhedra. In: Guy, R., Hanani, H., Sauer, N., Schönheim, J. (eds.) Combinatorial Structures and Their Applications, pp. 69–87. Gordon and Breach, New York, NY (1970). https://doi.org/10.1007/3-540-36478-1_2
7. Frank, A.: Connections in Combinatorial Optimization. Oxford Lecture Series in Mathematics and Its Applications, Oxford University Press, New York, NY (2011)
8. Gabow, H.N., Stallmann, M.: An augmenting path algorithm for linear matroid parity. Combinatorica 6(2), 123–150 (1986). https://doi.org/10.1007/BF02579169
9. Gabow, H.N., Xu, Y.: Efficient theoretic and practical algorithms for linear matroid intersection problems. J. Comput. Syst. Sci. 53(1), 129–147 (1996). https://doi.org/10.1006/jcss.1996.0054
10. Gallai, T.: Maximum-Minimum Sätze und verallgemeinerte Faktoren von Graphen. Acta Mathematica Academiae Scientiarum Hungaricae 12, 131–173 (1964). https://doi.org/10.1007/BF02066678
11. Gessel, I., Viennot, G.: Binomial determinants, paths, and hook length formulae. Adv. Math. 58(3), 300–321 (1985). https://doi.org/10.1016/0001-8708(85)90121-5
12. Goodall, A., De Mier, A.: Spanning trees of 3-uniform hypergraphs. Adv. Appl. Math. 47(4), 840–868 (2011). https://doi.org/10.1016/j.aam.2011.04.006
13. Harvey, N.J.A.: Algebraic algorithms for matching and matroid problems. SIAM J. Comput. 39(2), 679–702 (2009). https://doi.org/10.1137/070684008
14. Ishikawa, M., Wakayama, M.: Minor summation formula of Pfaffians. Linear Multilinear Algebra 39(3), 285–305 (1995). https://doi.org/10.1080/03081089508818403
15. Iwata, S., Kobayashi, Y.: A weighted linear matroid parity algorithm. SIAM J. Comput. (to appear). https://doi.org/10.1137/17M1141709
16. Kasteleyn, P.W.: The statistics of dimers on a lattice: I. the number of dimer arrangements on a quadratic lattice. Physica, 27(12), 1209–1225 (1961). https://doi.org/10.1016/0031-8914(61)90063-5
17. Kasteleyn, P.W.: Graph theory and crystal physics. In: Harary, F. (ed.) Graph Theory and Theoretical Physics, pp. 43–110. Academic Press, New York, NY (1967)

18. Kirchhoff, G.: Ueber die Auflösung der Gleichungen, auf welche man bei der Untersuchung der linearen Vertheilung galvanischer Ströme geführt wird. Annalen der Physik **148**(12), 497–508 (1847). https://doi.org/10.1002/andp.18471481202
19. Lawler, E.L.: Combinatorial Optimization: Networks and Matroids. Holt, Rinehart and Winston, New York, NY (1976)
20. Lindström, B.: On the vector representations of induced matroids. Bull. London Math. Soc. **5**(1), 85–90 (1973). https://doi.org/10.1112/blms/5.1.85
21. Little, C.H.C.: An extension of Kasteleyn's method of enumerating the 1-factors of planar graphs. In: Holton, D.A. (ed.) Combinatorial Mathematics. LNM, vol. 403, pp. 63–72. Springer, Heidelberg (1974). https://doi.org/10.1007/BFb0057377
22. Lovász, L.: Matroid matching and some applications. J. Comb. Theor. Ser. B **28**(2), 208–236 (1980). https://doi.org/10.1016/0095-8956(80)90066-0
23. Mader, W.: Über die Maximalzahl kreuzungsfreier H-Wege. Archiv der Mathematik **31**(1), 387–402 (1978). https://doi.org/10.1007/BF01226465
24. Maurer, S.B.: Matrix generalizations of some theorems on trees, cycles and cocycles in graphs. SIAM J. Appl. Math. **30**(1), 143–148 (1976). https://doi.org/10.1137/0130017
25. Murota, K.: Computing the degree of determinants via combinatorial relaxation. SIAM J. Comput. **24**(4), 765–796 (1995)
26. Orlin, J.B.: A fast, simpler algorithm for the matroid parity problem. In: Lodi, A., Panconesi, A., Rinaldi, G. (eds.) IPCO 2008. LNCS, vol. 5035, pp. 240–258. Springer, Heidelberg (2008). https://doi.org/10.1007/978-3-540-68891-4_17
27. Robertson, N., Seymour, P.D., Thomas, R.: Permanents, Pfaffian orientations, and even directed circuits. Ann. Math. **150**(3), 929–975 (1999). https://doi.org/10.2307/121059
28. Schrijver, A.: Combinatorial Optimization, Algorithms and Combinatorics, vol. 24. Springer, Berlin (2003)
29. Snook, M.: Counting bases of representable matroids. Electron. J. Comb. **19**(4), P41 (2012)
30. Temperley, H.N.V., Fisher, M.E.: Dimer problem in statistical mechanics-an exact result. Philos. Mag. **6**(68), 1061–1063 (1961). https://doi.org/10.1080/14786436108243366
31. Tutte, W.T.: The dissection of equilateral triangles into equilateral triangles. Math. Proc. Camb. Philos. Soc. **44**(4), 463–482 (1948). https://doi.org/10.1017/S030500410002449X
32. Valiant, L.G.: The complexity of computing the permanent. Theoret. Comput. Sci. **8**(2), 189–201 (1979). https://doi.org/10.1016/0304-3975(79)90044-6
33. Vazirani, V.V.: NC algorithms for computing the number of perfect matchings in $K_{3,3}$-free graphs and related problems. Inf. Comput. **80**(2), 152–164 (1989). https://doi.org/10.1016/0890-5401(89)90017-5
34. Webb, K.P.: Counting Bases. Ph.D. thesis, University of Waterloo, Waterloo, ON (2004)
35. Yamaguchi, Y.: Shortest disjoint S-paths via weighted linear matroid parity. In: Hong, S.H. (ed.) Proceedings of the 27th International Symposium on Algorithms and Computation (ISAAC '16). Leibniz International Proceedings in Informatics, vol. 64, pp. 63:1–63:13. Schloss Dagstuhl-Leibniz-Zentrum für Informatik (2016). https://doi.org/10.4230/LIPIcs.ISAAC.2016.63

On the Recognition of {a, b, c}-Modular Matrices

Christoph Glanzer$^{(\boxtimes)}$, Ingo Stallknecht, and Robert Weismantel

ETH Zürich, Zürich, Switzerland
`christoph.glanzer@ifor.math.ethz.ch`

Abstract. Let $A \in \mathbb{Z}^{m \times n}$ be an integral matrix and $a, b, c \in \mathbb{Z}$ satisfy $a \geq b \geq c \geq 0$. The question is to recognize whether A is $\{a, b, c\}$-*modular*, i.e., whether the set of $n \times n$ subdeterminants of A in absolute value is $\{a, b, c\}$. We will succeed in solving this problem in polynomial time unless A possesses a *duplicative relation*, that is, A has nonzero $n \times n$ subdeterminants k_1 and k_2 satisfying $2 \cdot |k_1| = |k_2|$. This is an extension of the well-known recognition algorithm for totally unimodular matrices. As a consequence of our analysis, we present a polynomial time algorithm to solve integer programs in standard form over $\{a, b, c\}$-modular constraint matrices for any constants a, b and c.

Keywords: Integer optimization · Recognition algorithm · Bounded subdeterminants

1 Introduction

A matrix is called *totally unimodular* (TU) if all of its subdeterminants are equal to 0, 1 or -1. Within the past 60 years the community has established a deep and beautiful theory about TU matrices. A landmark result in the understanding of such matrices is Seymour's decomposition theorem [14]. It shows that TU matrices arise from network matrices and two special matrices using row-, column-, transposition-, pivoting-, and so-called k-sum operations. As a consequence of this theorem it is possible to recognize in polynomial time whether a given matrix is TU [13,16]. An implementation of the algorithm in [16] by Walter and Truemper [19] returns a certificate if A is not TU: For an input matrix with entries in $\{0, \pm 1\}$, the algorithm finds a submatrix $\tilde{A}$ which is minimal in the sense that $|\det(\tilde{A})| = 2$ and every proper submatrix of $\tilde{A}$ is TU. We refer to Schrijver [13] for a textbook exposition of a recognition algorithm arising from Seymour's decomposition theorem and further material on TU matrices.

There is a well-established relationship between totally unimodular and *unimodular* matrices, i.e., matrices whose $n \times n$ subdeterminants are equal to 0, 1 or -1. In analogy to this we define for $A \in \mathbb{Z}^{m \times n}$ and $m \geq n$,

$$D(A) := \{|\det(A_{I,\cdot})| : I \subseteq [m], |I| = n\},$$

© Springer Nature Switzerland AG 2021
M. Singh and D. P. Williamson (Eds.): IPCO 2021, LNCS 12707, pp. 238–251, 2021.
https://doi.org/10.1007/978-3-030-73879-2_17

the set of all $n \times n$ subdeterminants of A in absolute value, where $A_{I,\cdot}$ is the submatrix formed by selecting all rows with indices in I. It follows straightforwardly from the recognition algorithm for TU matrices that one can efficiently decide whether $D(A) \subseteq \{1, 0\}$. A technique in [2, Section 3] allows us to recognize in polynomial time whether $D(A) \subseteq \{2, 0\}$. If all $n \times n$ subdeterminants of A are nonzero, the results in [1] can be applied to calculate $D(A)$ given that $\max\{k \colon k \in D(A)\}$ is constant. Nonetheless, with the exception of these results we are not aware of other instances for which it is known how to determine $D(A)$ in polynomial time.

The main motivation for the study of matrices with bounded subdeterminants comes from integer optimization problems (IPs). It is a well-known fact that IPs of the form $\max\{c^T x \colon Ax \leq b, \ x \in \mathbb{Z}^n\}$ for $A \in \mathbb{Z}^{m \times n}$ of full column rank, $b \in \mathbb{Z}^m$ and $c \in \mathbb{Z}^n$ can be solved efficiently if $D(A) \subseteq \{1, 0\}$, i.e., if A is unimodular. This leads to the natural question whether the problem remains efficiently solvable when the assumptions on $D(A)$ are further relaxed. Quite recently in [2] it was shown that for $D(A) \subseteq \{2, 1, 0\}$ and $\operatorname{rank}(A) = n$, a general linear integer optimization problem can be solved in strongly polynomial time. Recent results have also led to understand IPs when A is *nondegenerate*, i.e., if $0 \notin D(A)$. The foundation to study the nondegenerate case was laid by Veselov and Chirkov [17]. They describe a polynomial time algorithm to solve IPs if $D(A) \subseteq \{2, 1\}$. In [1] the authors showed that IPs over nondegenerate constraint matrices are solvable in polynomial time if the largest $n \times n$ subdeterminant of the constraint matrix is bounded by a constant.

The role of bounded subdeterminants in complexity questions and in the structure of IPs and LPs has also been studied in [7–9,12], as well as in the context of combinatorial problems in [4,5,11]. The sizes of subdeterminants also play an important role when it comes to the investigation of the *diameter of polyhedra*, see [6] and [3].

1.1   Our Results

A matrix $A \in \mathbb{Z}^{m \times n}$, $m \geq n$, is called $\{a, b, c\}$-*modular* if $D(A) = \{a, b, c\}$, where $a \geq b \geq c \geq 0$.[1] The paper presents three main results. First, we prove the following structural result for a subclass of $\{a, b, 0\}$-modular matrices.

Theorem 1 (Decomposition Property). *Let $a \geq b > 0$, $\gcd(\{a, b\}) = 1$ and assume that $\{a, b\} \neq \{2, 1\}$. Using row permutations, multiplications of rows by -1 and elementary column operations, any $\{a, b, 0\}$-modular matrix $A \in \mathbb{Z}^{m \times n}$ can be brought into a block structure of the form*

$$\begin{bmatrix} L & 0 & {}^0\!/_a \\ 0 & R & {}^0\!/_b \end{bmatrix}, \tag{1}$$

in time polynomial in n, m and $\log \|A\|_\infty$, where $L \in \mathbb{Z}^{m_1 \times n_1}$ and $R \in \mathbb{Z}^{m_2 \times n_2}$ are TU, $n_1 + n_2 = n - 1$, $m_1 + m_2 = m$. In the representation above the rightmost

---

[1] For reasons of readability, we will stick to the notation that $a \geq b \geq c$ although the order of these elements is irrelevant.

column has entries in $\{0, a\}$ and $\{0, b\}$, respectively. The matrix $\begin{bmatrix} L & 0 \\ 0 & R \end{bmatrix}$ contains the $(n-1)$-dimensional unit matrix as a submatrix.

The first $n-1$ columns of (1) are TU since they form a 1-sum of two TU matrices (see [13, Chapter 19.4]). This structural property lies at the core of the following recognition algorithm. We say that a matrix A possesses a *duplicative relation* if it has nonzero $n \times n$ subdeterminants k_1 and k_2 satisfying $2 \cdot |k_1| = |k_2|$.

Theorem 2 (Recognition Algorithm). *There exists an algorithm that solves the following recognition problem in time polynomial in n, m and $\log \|A\|_\infty$: Either, it calculates $D(A)$, or gives a certificate that $|D(A)| \geq 4$, or returns a duplicative relation.*

For instance, Theorem 2 cannot be applied to check whether a matrix is $\{4, 2, 0\}$-modular, but it can be applied to check whether a matrix is $\{3, 1, 0\}$-, or $\{6, 4, 0\}$-modular. More specifically, Theorem 2 recognizes $\{a, b, c\}$-modular matrices unless $(a, b, c) = (2 \cdot k, k, 0)$, $k \in \mathbb{Z}_{\geq 1}$. In particular, this paper does not give a contribution as to whether so-called *bimodular matrices* (the case $k = 1$) can be recognized efficiently.

The decomposition property established in Theorem 1 is a major ingredient for the following optimization algorithm to solve standard form IPs over $\{a, b, c\}$-modular constraint matrices for any constant $a \geq b \geq c \geq 0$.

Theorem 3 (Optimization Algorithm). *Consider a standard form integer program of the form*

$$\max\{c^\mathsf{T} x \colon Bx = b,\ x \in \mathbb{Z}_{\geq 0}^n\}, \tag{2}$$

for $b \in \mathbb{Z}^m$, $c \in \mathbb{Z}^n$ and $B \in \mathbb{Z}^{m \times n}$ of full row rank, where $D(B^\mathsf{T})$ is constant, i.e., $\max\{k \colon k \in D(B^\mathsf{T})\}$ is constant.[2] Then, in time polynomial in n, m and the encoding size of the input data, one can solve (2) or output that $|D(B^\mathsf{T})| \geq 4$.

Notably, in Theorem 3, the assumption that $D(B^\mathsf{T})$ is constant can be dropped if B is *degenerate*, i.e., if $0 \in D(B^\mathsf{T})$.

2 Notation and Preliminaries

We quickly review the terminology used in this abstract, where we resort to standard notation whenever possible. For $k \in \mathbb{Z}_{\geq 1}$, $[k] := \{1, \ldots, k\}$. $\mathcal{I}_n$ is the n-dimensional unit matrix, where we leave out the subscript if the dimension is clear from the context. For a matrix $A \in \mathbb{Z}^{m \times n}$, we denote by $A_{i,.}$ the i-th row of A. For a subset I of $[m]$, $A_{I,.}$ is the submatrix formed by selecting all rows with indices in I, in increasing order. An analogous notation is used for the columns of A. For $k \in \mathbb{Z}$, we write $I_k := \{i \in [m] \colon A_{i,n} = k\}$, the indices of

---

[2] Note that we use $D(B^\mathsf{T})$ instead of $D(B)$ since B has full row-rank and we refer to its $m \times m$ subdeterminants.

the rows whose n-th entry is equal to k. Set $||A||_\infty := \max_{i\in[m],j\in[n]}|A_{ij}|$. For simplicity, we assume throughout the document that the input matrix A to any recognition algorithm satisfies $m \geq n$ and $\text{rank}(A) = n$ as $\text{rank}(A) < n$ implies that $D(A) = \{0\}$. Left-out entries in figures and illustrations are equal to zero.

$D(A)$ is preserved under elementary column operations, permutations of rows and multiplications of rows by -1. By a series of elementary column operations on A, any nonsingular $n \times n$ submatrix B of A can be transformed to its Hermite Normal Form (HNF), in which B becomes a lower-triangular, nonnegative submatrix with the property that each of its rows has a unique maximum entry on the main diagonal of B [13]. This can be done in time polynomial in n, m and $\log ||A||_\infty$ [15]. At various occasions we will make use of a simple adaptation of the HNF described in [1, Section 3]: For any nonsingular $n \times n$ submatrix B of A, permute the rows of A such that $A_{[n],\cdot} = B$. Apply elementary column operations to A such that B is in HNF. After additional row and column permutations and multiplications of rows by -1,

$$
A = \begin{bmatrix}
1 & & & & & & & \\
& \ddots & & & & & & \\
& & 1 & & & & & \\
* & \cdots & * & \delta_1 & & & & \\
\vdots & \vdots & \vdots & & \ddots & \ddots & & \\
* & \cdots\cdots\cdots\cdots & * & & & \delta_l & & \\
A_{n,1} & \cdots\cdots\cdots\cdots & & & A_{n,n-1} & A_{n,n} \\
A_{n+1,1} & \cdots\cdots\cdots\cdots & & & A_{n+1,n-1} & A_{n+1,n} \\
\vdots & \vdots & \vdots & \vdots & & \vdots & & \vdots \\
A_{m,1} & \cdots\cdots\cdots\cdots & & & A_{m,n-1} & A_{m,n}
\end{bmatrix}, \tag{3}
$$

where $A_{\cdot,n} \geq 0$, $|\det(B)| = \prod_{i=1}^{l} \delta_i \cdot A_{n,n}$, all entries marked by $*$ are numbers between 0 and the diagonal entry of the same row minus one, and $\delta_i \geq 2$ for all $i \in [l]$. In particular, the rows with diagonal entries strictly larger than one are at positions $n - l, \ldots, n$.

We note that it is not difficult to efficiently recognize nondegenerate matrices given a constant upper bound on $|D(A)|$. This will allow us to exclude the nondegenerate case in all subsequent algorithms. We wish to emphasize that the results in [1] can be applied to solve this task given that $\max\{k : k \in D(A)\}$ is constant.

Lemma 1. *Given a constant $d \in \mathbb{Z}$, there exists an algorithm that solves the following recognition problem in time polynomial in n, m and $\log ||A||_\infty$: Either, it calculates $D(A)$, or gives a certificate that $|D(A)| \geq d + 1$, or that $0 \in D(A)$.*

We omit the proof from this abstract as it is merely a technical counting argument.

3 Proof of Theorem 1

Transform A to (3), thus $A_{.,n} \geq 0$. As a first step we show that we can assume $A_{n,n} > 1$ without loss of generality. For this purpose, assume that $A_{n,n} = 1$, i.e., that $l = 0$. This implies that $A_{[n],.} = \mathcal{I}_n$. Consequently, any nonsingular submatrix B of A can be extended to an $n \times n$ submatrix with same determinant by (a) appending unit vectors from the topmost n rows of A to B and (b) for each unit vector appended in step (a), by appending the column to B in which this unit vector has its nonzero entry. By Laplace expansion, the $n \times n$ submatrix we obtain admits the same determinant in absolute value. Therefore, if we identify any submatrix of A with determinant larger than one in absolute value, we can transform A once more to (3) with respect to its corresponding $n \times n$ submatrix which yields $A_{n,n} > 1$ as desired. To find a subdeterminant of A of absolute value larger than one, if present, test A for total unimodularity. If the test fails, it returns a desired subdeterminant. If the test returns that A is TU, then $a = b = 1$ and A is already of the form $\begin{bmatrix} L & {}^0\!/_1 \end{bmatrix}$ for L TU.

The n-th column is not divisible by any integer larger than one as otherwise, all $n \times n$ subdeterminants of A would be divisible by this integer, contradicting $\gcd(\{a, b\}) = 1$. In particular, since $A_{n,n} > 1$, $A_{.,n}$ is not divisible by $A_{n,n}$. This implies that there exists an entry $A_{k,n} \neq 0$ such that $A_{k,n} \neq A_{n,n}$. Thus, there exist two $n \times n$ subdeterminants, $\det(A_{[n],.})$ and $\det(A_{[n-1] \cup k,.})$, of different absolute value. This allows us to draw two conclusions: First, the precondition $A_{n,n} > 1$ which we have established in the previous paragraph implies $a > b$. We may therefore assume for the rest of the proof that $a \geq 3$ as $\{a, b\} \neq \{2, 1\}$. Secondly, it follows that $l = 0$: For the purpose of contradiction assume that $l \geq 1$. Then the aforementioned two subdeterminants $\det(A_{[n],.})$ and $\det(A_{[n-1] \cup k,.})$ are both divisible by $\prod_{i=1}^{l} \delta_i > 1$. Since one of those subdeterminants must be equal to $\pm a$ and the other must be equal to $\pm b$, this contradicts our assumption of $\gcd(\{a, b\}) = 1$.

As a consequence of $l = 0$, the topmost $n - 1$ rows of A form unit vectors. Thus, any subdeterminant of rows $n, \dots, m$ of A which includes elements of the n-th column can be extended to an $n \times n$ subdeterminant of the same absolute value by appending an appropriate subset of these unit vectors. To further analyze the structure of A we will study the 2×2 subdeterminants

$$\theta_{i,j}^h := \det \begin{bmatrix} A_{i,h} & A_{i,n} \\ A_{j,h} & A_{j,n} \end{bmatrix}, \tag{4}$$

where $i, j \geq n$, $i \neq j$ and $h \leq n - 1$. It holds that $|\theta_{i,j}^h| \in \{a, b, 0\}$.

Claim 1. *There exists a sequence of elementary column operations and row permutations which turn A into the form*

$$\begin{bmatrix} \mathcal{I}_{n-1} \\ {}^0\!/\!\pm 1 \ a \\ {}^0\!/\!\pm 1 \ b \\ {}^0\!/\!\pm 1 \ 0 \end{bmatrix}, \tag{5}$$

where $\left[^{0}/_{\pm 1}\, a \right]$, $\left[^{0}/_{\pm 1}\, b \right]$ *and* $\left[^{0}/_{\pm 1}\, 0 \right]$ *are submatrices consisting of rows whose first $n - 1$ entries lie in $\{0, \pm 1\}$, and whose n-th entry is equal to a, b or 0.*

Proof (Proof of Claim 1). We start by analyzing the n-th column of A. To this end, note that $|\det(A_{[n-1] \cup k, \cdot})| = |A_{k,n}| \in \{a, b, 0\}$ for $k \geq n$. It follows that $A_{\cdot, n} \in \{a, b, 0\}^m$ as $A_{\cdot, n} \geq 0$. In addition, by what we have observed two paragraphs earlier, at least one entry of $A_{\cdot, n}$ must be equal to a and at least one entry must be equal to b. Sort the rows of A by their respective entry in the n-th column as in (5).

In the remaining proof we show that after column operations, $A_{\cdot, [n-1]} \in \{0, \pm 1\}^{m \times (n-1)}$. To this end, let $h \in [n - 1]$ be an arbitrary column index. Recall that $I_k := \{i \in [m] \colon A_{i,n} = k\}$. We begin by noting a few properties which will be used to prove the claim. For $i \in I_a$ and $j \in I_b$, it holds that $\theta^h_{i,j} = b \cdot A_{i,h} - a \cdot A_{j,h} \in \{\pm a, \pm b, 0\}$. These are Diophantine equations which are solved by

$$
\begin{aligned}
&\text{(i)} \quad (A_{i,h}, A_{j,h}) = (k \cdot a, \mp 1 + k \cdot b), \ k \in \mathbb{Z}, \\
&\text{(ii)} \quad (A_{i,h}, A_{j,h}) = (\pm 1 + k \cdot a, k \cdot b), \ k \in \mathbb{Z}, \\
&\text{(iii)} \quad (A_{i,h}, A_{j,h}) = (k \cdot a, k \cdot b), \ k \in \mathbb{Z}.
\end{aligned}
\tag{6}
$$

Furthermore, for any $i_1, i_2 \in I_a$, it holds that $\theta^h_{i_1, i_2} = a \cdot (A_{i_1, h} - A_{i_2, h})$. Since this quantity is a multiple of a and since $\theta^h_{i_1, i_2} \in \{\pm a, \pm b, 0\}$, it follows that

$$
|A_{i_1, h} - A_{i_2, h}| \leq 1, \ \forall i_1, i_2 \in I_a.
\tag{7}
$$

We now perform a column operation on $A_{\cdot, h}$: Let us fix arbitrary indices $p \in I_a$ and $q \in I_b$. The pair $(A_{p,h}, A_{q,h})$ solves one of the three Diophantine equations for a fixed k. Add $(-k) \cdot A_{\cdot, n}$ to $A_{\cdot, h}$. Now, $(A_{p,h}, A_{q,h}) \in \{(0, \mp 1), (\pm 1, 0), (0, 0)\}$. We claim that as a consequence of this column operation, $A_{\cdot, h} \in \{0, \pm 1\}^m$.

We begin by showing that $A_{I_a, h}$ and $A_{I_b, h}$ have entries in $\{0, \pm 1\}$. First, assume for the purpose of contradiction that there exists $j \in I_b$ such that $|A_{j,h}| > 1$. This implies that the pair $(A_{p,h}, A_{j,h})$ satisfies (6) for $|k| \geq 1$. As $a \geq 3$, this contradicts that $A_{p,h} \in \{0, \pm 1\}$. Secondly, assume for the purpose of contradiction that there is $i \in I_a$ such that $|A_{i,h}| > 1$.[3] As $|A_{p,h}| \leq 1$, it follows from (7) that $|A_{i,h}| = 2$ and $|A_{p,h}| = 1$. Therefore, as either $A_{p,h}$ or $A_{q,h}$ must be equal to zero, $A_{q,h} = 0$. This implies that $|\theta^h_{i,q}| = 2 \cdot b$ which is a contradiction to $\theta^h_{i,q} \in \{\pm a, \pm b, 0\}$ as A has no duplicative relation, i.e., $2 \cdot b \neq a$.

It remains to prove that $A_{I_0, h}$ has entries in $\{0, \pm 1\}$. For the purpose of contradiction, assume that there exists $i \in I_0$, $i \geq n$, such that $|A_{i,h}| \geq 2$. Choose any $s \in I_a$. Then,

$$
\theta^h_{s,i} = \det \begin{bmatrix} A_{s,h} & a \\ A_{i,h} & 0 \end{bmatrix} = -a \cdot A_{i,h},
$$

which is larger than a in absolute value, a contradiction. $\qquad\square$

---

[3] We cannot use the same approach as for the first case as b could be equal to 1 or 2.

In the next claim, we establish the desired block structure (1). We will show afterwards that the blocks L and R are TU.

Claim 2. *There exists a sequence of row and column permutations which turn A into (1) for matrices L and R with entries in $\{0, \pm 1\}$.*

Proof (Proof of Claim 2). For reasons of simplicity, we assume that A has no rows of the form $\begin{bmatrix} 0 \cdots 0 \, a \end{bmatrix}$ or $\begin{bmatrix} 0 \cdots 0 \, b \end{bmatrix}$ as such rows can be appended to A while preserving the properties stated in this claim. We construct an auxiliary graph $G = (V, E)$, where we introduce a vertex for each nonzero entry of $A_{.,[n-1]}$ and connect two vertices if they share the same row or column index. Formally, set

$$V := \{(i,j) \in [m] \times [n-1] : A_{i,j} \neq 0\},$$
$$E := \{\{(i_1, j_1), (i_2, j_2)\} \in V \times V : i_1 = i_2 \text{ or } j_1 = j_2, \ (i_1, j_1) \neq (i_2, j_2)\}.$$

Let $K_1, \ldots, K_k \subseteq V$ be the vertex sets corresponding to the connected components in G. For each $l \in [k]$, set

$$\mathbf{I}_l := \{i \in [m] : \exists j \in [n-1] \text{ s.t. } (i,j) \in K_l\},$$
$$\mathbf{J}_l := \{j \in [n-1] : \exists i \in [m] \text{ s.t. } (i,j) \in K_l\}.$$

These index sets form a partition of $[m]$, resp. $[n-1]$: Since every row of $A_{.,[n-1]}$ is nonzero, $\bigcup_{l \in [k]} \mathbf{I}_l = [m]$ and since $\text{rank}(A) = n$, every column of A contains a nonzero entry, i.e., $\bigcup_{l \in [k]} \mathbf{J}_l = [n-1]$. Furthermore, by construction, it holds that $\mathbf{I}_p \cap \mathbf{I}_q = \emptyset$ and $\mathbf{J}_p \cap \mathbf{J}_q = \emptyset$ for all $p \neq q$. The entries $A_{i,j}$ for which $(i,j) \notin \bigcup_{l \in [k]} (\mathbf{I}_l \times \mathbf{J}_l)$ are equal to zero. Therefore, sorting the rows and columns of $A_{.,[n-1]}$ with respect to the partition formed by $\mathbf{I}_l$, resp. $\mathbf{J}_l$, $l \in [k]$, yields

$$A_{.,[n-1]} = \begin{bmatrix} A_{\mathbf{I}_1, \mathbf{J}_1} & 0 & \cdots & 0 \\ 0 & \ddots & \ddots & \vdots \\ \vdots & \ddots & \ddots & 0 \\ 0 & \cdots & 0 & A_{\mathbf{I}_k, \mathbf{J}_k} \end{bmatrix}. \tag{8}$$

In what follows we show that for all $l \in [k]$, it holds that either $\mathbf{I}_l \cap I_a = \emptyset$ or that $\mathbf{I}_l \cap I_b = \emptyset$. From this property the claim readily follows: We obtain the form (1) from (8) by permuting the rows and columns such that the blocks which come first are those which correspond to the connected components K_l with $\mathbf{I}_l \cap I_a \neq \emptyset$. For the purpose of contradiction, assume that there exists $l \in [k]$ such that $\mathbf{I}_l \cap I_a \neq \emptyset$ and $\mathbf{I}_l \cap I_b \neq \emptyset$. Denote by $K_l^a := \{(i,j) \in K_l : i \in I_a\}$ and $K_l^b := \{(i,j) \in K_l : i \in I_b\}$. Among all paths in the connected component induced by K_l which connect the sets K_l^a and K_l^b, let $P := \{(i^{(1)}, j^{(1)}), \ldots, (i^{(t)}, j^{(t)})\}$ be a shortest one. By construction, $(i^{(1)}, j^{(1)})$ is the only vertex of P which lies in K_l^a and $(i^{(t)}, j^{(t)})$ is the only vertex of P which lies in K_l^b. This implies that P starts and ends with a change in the first component, i.e., $i^{(1)} \neq i^{(2)}$ and $i^{(t-1)} \neq i^{(t)}$. Furthermore, since P has minimal length, it follows from the construction of the edges that it alternates between changing first and second components, i.e., for all $s = 1,\ldots,t-2$, $i^{(s)} \neq i^{(s+1)} \Leftrightarrow j^{(s+1)} \neq j^{(s+2)}$ and $j^{(s)} \neq j^{(s+1)} \Leftrightarrow i^{(s+1)} \neq i^{(s+2)}$.

Define $B := A_{\{i^{(1)},\ldots,i^{(t)}\},\{j^{(1)},\ldots,j^{(t)},n\}} \in \mathbb{Z}^{\frac{t+2}{2} \times \frac{t+2}{2}}$. Permute the rows and columns of B such that they are ordered with respect to the order $i^{(1)},\ldots,i^{(t)}$ and $j^{(1)},\ldots,j^{(t)},n$. We claim that B is of the form

$$
B = \begin{bmatrix}
\pm 1 & & & & & a \\
\pm 1 & \pm 1 & & & & 0 \\
 & \pm 1 & \ddots & & & \vdots \\
 & & \ddots & \pm 1 & & \vdots \\
 & & & \pm 1 & \pm 1 & 0 \\
 & & & & \pm 1 & b
\end{bmatrix}.
$$

To see this, first observe that the entries on the main diagonal and the diagonal below are equal to ± 1 and that the entries $B_{1,\frac{t+2}{2}}$ and $B_{\frac{t+2}{2},\frac{t+2}{2}}$ are equal to a or b, respectively, by construction. As we have observed above, $(i^{(1)}, j^{(1)})$ is the only vertex of P which lies in K_l^a and $(i^{(t)}, j^{(t)})$ is the only vertex of P which lies in K_l^b, implying that all other entries of the rightmost column, $B_{2,\frac{t+2}{2}},\ldots,B_{\frac{t+2}{2}-1,\frac{t+2}{2}}$, are equal to zero. For the purpose of contradiction, assume that any other entry of B, say $B_{v,w}$, is nonzero. Let us consider the case that $w > v$, i.e., that $B_{v,w}$ lies above the main diagonal of B. Among all vertices of P whose corresponding entries lie in the same row as $B_{v,w}$, let $(i^{(s)}, j^{(s)})$ be the vertex with minimal s. Among all vertices of P whose corresponding entries lie in the same column as $B_{v,w}$, let $(i^{(s')}, j^{(s')})$ be the vertex with maximal s'. Then, consider the following path: Connect $(i^{(1)}, j^{(1)})$ with $(i^{(s)}, j^{(s)})$ along P; then take the edge to the vertex corresponding to $B_{v,w}$; take the edge to $(i^{(s')}, j^{(s')})$; then connect $(i^{(s')}, j^{(s')})$ with $(i^{(t)}, j^{(t)})$ along P. This path is shorter than P which is a contradiction. An analogous argument yields a contradiction if $w < v - 1$, i.e., if $B_{v,w}$ lies in the lower-triangular part of B. Thus, B is of the form stated above. By Laplace expansion applied to the last column of B, $\det(B) = \pm a \pm b \notin \{\pm a, \pm b, 0\}$ as $2 \cdot b \neq a$. Since B can be extended to an $n \times n$ submatrix of A of the same determinant in absolute value, this is a contradiction.

Regarding the computational running time of this operation, note that G can be created and its connected components can be calculated in time polynomial in n and m. A subsequent permutation of the rows and columns as noted above yields the desired form. □

It remains to show that L and R are TU. We will prove the former, the proof for the latter property is analogous. Assume for the purpose of contradiction that L is not TU. As the entries of L are all ± 1 or 0, it contains a submatrix $\tilde{A}$ of determinant ± 2 [13, Theorem 19.3]. Extend $\tilde{A}$ by appending the corresponding entries of $A_{.,n}$ and by appending an arbitrary row of $A_{I_b,.}$. We obtain a submatrix of the form

$$
\begin{bmatrix}
\tilde{A} & 0/a \\
0 & b
\end{bmatrix},
$$

whose determinant in absolute value is $2 \cdot b$, a contradiction as A has no duplicative relations. Similarly, one obtains a submatrix of determinant $\pm 2 \cdot a$ if R is not TU. □

4   Proof of Theorem 2

In this section, the following recognition problem will be of central importance: For numbers $a \geq b \geq c \geq 0$, we say that an algorithm *tests for* $\{a, b, c\}$-*modularity* if, given an input matrix $A \in \mathbb{Z}^{m \times n}$ ($m \geq n$), it checks whether $D(A) = \{a, b, c\}$. As a first step we state the following lemma which allows us to reduce the gcd in all subsequent recognition algorithms. The proof uses a similar technique as was used in [10, Remark 5]. It is deferred to a longer version of this paper.

Lemma 2 (cf. [10, Remark 5]). *Let $a \geq b > 0$ and $\gamma := \gcd(\{a, b\})$. Testing for $\{a, b, 0\}$-modularity can be reduced to testing for $\{\frac{a}{\gamma}, \frac{b}{\gamma}, 0\}$-modularity in time polynomial in n, m and $\log \|A\|_\infty$, where A is the input matrix.*

We proceed by using the decomposition established in Theorem 1 to construct an algorithm which tests for $\{a, b, 0\}$-modularity if $a \geq b > 0$ without duplicative relations, i.e., if $2 \cdot b \neq a$. We will see that this algorithm quickly yields a proof of Theorem 2. The main difference between the following algorithm and Theorem 2 is that in the former, a and b are fixed input values while in Theorem 2, a, b (and c) also have to be determined.

Lemma 3. *Let $a \geq b > 0$ such that $2 \cdot b \neq a$. There exists an algorithm with running time polynomial in n, m and $\log \|A\|_\infty$ which tests for $\{a, b, 0\}$-modularity or returns one of the following certificates: $|D(A)| \geq 4$, $\gcd(D(A)) \neq \gcd(\{a, b\})$ or a set $D' \subsetneq \{a, b, 0\}$ such that $D(A) = D'$.*

Proof. Applying Lemmata 1 and 2 allows us to assume w.l.o.g. that $0 \in D(A)$ and that $\gcd(\{a, b\}) = 1$. Note that the case $a = b$ implies $a = b = 1$. Then, testing for $\{1, 0\}$-modularity can be accomplished by first transforming A to (3) and by subsequently testing whether the transformed matrix is TU.

Thus, consider the case $a > b$. Since $\gcd(\{a, b\}) = 1$, $2 \cdot b \neq a \Leftrightarrow \{a, b\} \neq \{2, 1\}$. Therefore, the numbers a and b fulfill the prerequisites of Theorem 1. Follow the proof of Theorem 1 to transform A to (1). If the matrix is $\{a, b, 0\}$-modular, we will arrive at a representation of the form (1). Otherwise, the proof of Theorem 1 (as it is constructive) exhibits a certificate of the following form: $|D(A)| \geq 4$, $\gcd(D(A)) \neq \gcd(\{a, b\})$ or a set $D' \subsetneq \{a, b, 0\}$ such that $D(A) = D'$. Without loss of generality, we can also assume that A has at least one $n \times n$ subdeterminant equal to $\pm a$ and $\pm b$, i.e., that $\{a, b, 0\} \subseteq D(A)$.

Next, we show that i) holds if and only if both ii) and iii) hold, where

  i) every nonsingular $n \times n$ submatrix of A has determinant $\pm a$ or $\pm b$,
 ii) every nonsingular $n \times n$ submatrix of $A_{I_a \cup I_0, \cdot}$ has determinant $\pm a$,
iii) every nonsingular $n \times n$ submatrix of $A_{I_b \cup I_0, \cdot}$ has determinant $\pm b$.

$A_{I_0,\cdot}$ contains the first $n - 1$ unit vectors and hence, $A_{I_a \cup I_0,\cdot}$ and $A_{I_b \cup I_0,\cdot}$ have full column rank. We first show that ii) and iii) follow from i). Let us start with iii). For the purpose of contradiction, assume that i) holds but not iii). By construction, the n-th column of $A_{I_b \cup I_0}$ is divisible by b. Denote by A' the matrix which we obtain by dividing the last column of $A_{I_b \cup I_0,\cdot}$ by b. The entries of A' are all equal to ± 1 or 0. As we have assumed that iii) is invalid, it follows that there exists a nonsingular $n \times n$ submatrix of A' whose determinant is not equal to ± 1. In particular, A' is not TU. By [13, Theorem 19.3], as the entries of A' are all ± 1 or 0 but it is not TU, it contains a submatrix of determinant ± 2. Since $A'_{\cdot, [n-1]}$ is TU, this submatrix must involve entries of the n-th column of A'. Thus, it corresponds to a submatrix of $A_{I_b \cup I_0,\cdot}$ of determinant $\pm 2 \cdot b$. Append a subset of the first $n - 1$ unit vectors to extend this submatrix to an $n \times n$ submatrix. This $n \times n$ submatrix is also an $n \times n$ submatrix of A. Its determinant is $\pm 2 \cdot b$, which is not contained in $\{\pm a, \pm b, 0\}$ because $2 \cdot b \neq a$, a contradiction to i). For ii), the same argument yields an $n \times n$ subdeterminant of $\pm 2 \cdot a$, which is also a contradiction to i).

Next, we prove that i) holds if both ii) and iii) hold. This follows from the 1-sum structure of A. Let B be any nonsingular $n \times n$ submatrix of A. B is of the form

$$B = \begin{bmatrix} C & {}^0\!/\!a \\ D & {}^0\!/\!b \end{bmatrix},$$

where $C \in \mathbb{Z}^{m_1 \times n_1}$, $D \in \mathbb{Z}^{m_2 \times n_2}$ for n_1, n_2, m_1 and m_2 satisfying $n_1 + n_2 + 1 = n = m_1 + m_2$. As B is nonsingular, $n_i \leq m_i$ and $m_i \leq n_i + 1$, $i \in \{1, 2\}$. Thus, $n_i \leq m_i \leq n_i + 1$, $i \in \{1, 2\}$, and we identify two possible cases which are symmetric: $m_1 = n_1 + 1$, $m_2 = n_2$ and $m_1 = n_1$, $m_2 = n_2 + 1$. We start with the analysis of the former. By Laplace expansion applied to the last column,

$$\det \begin{bmatrix} C & {}^0\!/\!a \\ D & {}^0\!/\!b \end{bmatrix} = \det \begin{bmatrix} C & {}^0\!/\!a \\ D & 0 \end{bmatrix} \pm \det \begin{bmatrix} C & 0 \\ D & {}^0\!/\!b \end{bmatrix}.$$

The latter determinant is zero as $m_1 = n_1 + 1$. As the former matrix is block-diagonal, $|\det B| = |\det[C \mid {}^0\!/\!a]| \cdot |\det D|$. B is nonsingular and D is TU, therefore $|\det D| = 1$. $[C \mid {}^0\!/\!a]$ is a submatrix of $A_{I_a \cup I_0}$ which can be extended to an $n \times n$ submatrix of the same determinant in absolute value by appending a subset of the $n - 1$ unit vectors contained in $A_{I_a \cup I_0}$. Therefore by ii), $|\det B| = |\det[C \mid {}^0\!/\!a]| = a$. In the case $m_1 = n_1$, $m_2 = n_2 + 1$ a symmetric analysis leads to $|\det B| = |\det[D \mid {}^0\!/\!b]| = b$ by iii).

To test for ii), let A' be the matrix which we obtain by dividing the n-th column of $A_{I_a \cup I_0,\cdot}$ by a. Then, ii) holds if and only if every nonsingular $n \times n$ submatrix of A' has determinant ± 1. Transform A' to (3). The topmost n rows of A' must form a unit matrix. Therefore, ii) is equivalent to A' being TU, which can be tested efficiently. Testing for iii) can be done analogously. □

We now have all the necessary ingredients to prove Theorem 2. In essence, what remains is a technique to find sample values a, b and c for which we test whether A is $\{a, b, c\}$-modular using our previously established algorithms.

Proof (Proof of Theorem 2). Apply Lemma 1 for $d = 3$. Either, the algorithm calculates $D(A)$ or gives a certificate that $|D(A)| \geq 4$, or that $0 \in D(A)$. In the former two cases we are done. Assume therefore that $0 \in D(A)$. By assumption A has full column rank. Thus, find $k_1 \in D(A)$, $k_1 \neq 0$, using Gaussian Elimination. Apply Lemma 3 to check whether $D(A) = \{k_1, 0\}$. Otherwise, $|D(A)| \geq 3$. At this point, a short technical argument is needed to determine an element $k_2 \in D(A) \setminus \{k_1, 0\}$, which we defer to a longer version of this paper.

Assume w.l.o.g. that $k_1 > k_2$. If $2 \cdot k_2 = k_1$, then A has duplicative relations. If not, test A for $\{k_1, k_2, 0\}$-modularity using Lemma 3. As $\{k_1, k_2, 0\} \subseteq D(A)$, either, this algorithm returns that $D(A) = \{k_1, k_2, 0\}$ or a certificate of the form $|D(A)| \geq 4$ or $\gcd(D(A)) \neq \gcd(\{k_1, k_2\})$. In the former two cases we are done and in the third case it also follows that $|D(A)| \geq 4$.   □

5   Proof of Theorem 3

One ingredient to the proof of Theorem 3 is the following result by Gribanov, Malyshev and Pardalos [10] which reduces the standard form IP (2) to an IP in inequality form in dimension $n - m$ such that the subdeterminants of the constraint matrices are in relation.

Lemma 4 ([10, Corollary 1.1, Remark 5, Theorem 3][4]). *In time polynomial in n, m and $\log \|B\|_\infty$, (2) can be reduced to the inequality form IP*

$$\max\{h^\mathrm{T} y \colon Cy \leq g, \ y \in \mathbb{Z}^{n-m}\}, \tag{9}$$

where $h \in \mathbb{Z}^{n-m}$, $g \in \mathbb{Z}^n$ and $C \in \mathbb{Z}^{n \times (n-m)}$, with $D(C) = \frac{1}{\gcd(D(B^\mathrm{T}))} \cdot D(B^\mathrm{T})$.

To prove this reduction, the authors apply a theorem by Shevchenko and Veselov [18] which was originally published in Russian. In a longer version of this paper, we will provide an alternative elementary proof for Lemma 4.

As a second ingredient to the proof of Theorem 3, we will make use of some results for *bimodular integer programs* (BIPs). BIPs are IPs of the form $\max\{c^\mathrm{T} x \colon Ax \leq b, \ x \in \mathbb{Z}^n\}$, where $c \in \mathbb{Z}^n$, $b \in \mathbb{Z}^m$ and $A \in \mathbb{Z}^{m \times n}$ is *bimodular*, i.e., $\mathrm{rank}(A) = n$ and $D(A) \subseteq \{2, 1, 0\}$. As mentioned earlier, [2] proved that BIPs can be solved in strongly polynomial time. Their algorithm uses the following structural result for BIPs by [17] which will also be useful to us.

Theorem 4 ([17, Theorem 2], as formulated in [2, Theorem 2.1]). *Assume that the linear relaxation $\max\{c^\mathrm{T} x \colon Ax \leq b, \ x \in \mathbb{R}^n\}$ of a BIP is feasible, bounded and has a unique optimal vertex solution v. Denote by $I \subseteq [m]$ the indices of the constraints which are tight at v, i.e., $A_{I,\cdot} v = b_I$. Then, an optimal solution x^* for $\max\{c^\mathrm{T} x \colon A_I x \leq b_I, \ x \in \mathbb{Z}^n\}$ is also optimal for the BIP.*

---

[4] Note that in [10], the one-to-one correspondence between $D(C)$ and $D(B^\mathrm{T})$ is not explicitly stated in Corollary 1.1 but follows from Theorem 3.

Proof (Proof of Theorem 3). Using Lemma 4, we reduce the standard form
IP (2) to (9). Note that $\gcd(D(C)) = 1$. Let us denote (9) with objective vector
$h \in \mathbb{Z}^{n-m}$ by $\text{IP}_\leq(h)$ and its natural linear relaxation by $\text{LP}_\leq(h)$. We apply
Theorem 2 to C and perform a case-by-case analysis depending on the output.

i) The algorithm calculates and returns $D(C)$. If $0 \notin D(C)$, C is nondegenerate and $\text{IP}_\leq(h)$ can be solved using the algorithm in [1]. Thus, assume
   $0 \in D(C)$. C has no duplicative relations. As $\gcd(D(C)) = 1$, this implies
   that C is $\{a, b, 0\}$-modular for $a \geq b > 0$, where $\gcd(\{a, b\}) = 1$ and
   $\{a, b\} \neq \{2, 1\}$. Thus, C satisfies the assumptions of Theorem 1. As a consequence of Theorem 1, there exist elementary column operations which
   transform C such that its first $n - m - 1$ columns are TU, i.e., there is
   $U \in \mathbb{Z}^{(n-m) \times (n-m)}$ unimodular such that $CU = [T \mid d]$, where T is
   TU and $d \in \mathbb{Z}^n$. Substituting $z := U^{-1}y$ yields the equivalent problem
   $\max\{h^T U z : [T \mid d]z \leq g, z \in \mathbb{Z}^{n-m}\}$, which can be solved by solving its
   mixed-integer relaxation, where $z_1, \ldots, z_{n-m-1} \in \mathbb{R}$ and $z_{n-m} \in \mathbb{Z}$. This
   can be done in polynomial time [13, Chapter 18.4].
ii) The algorithm returns that $|D(C)| \geq 4$. Then, $|D(B^T)| = |D(C)| \geq 4$.
iii) The algorithm returns a duplicative relation, i.e., $\{2 \cdot k, k\} \subseteq D(C)$, $k > 0$.
   Assume w.l.o.g. that $\text{LP}_\leq(h)$ is feasible and that $\text{LP}_\leq(h)$ is bounded. Using
   standard techniques, the unbounded case can be reduced to the bounded
   case. Calculate an optimal vertex solution v to $\text{LP}_\leq(h)$. If $v \in \mathbb{Z}^{n-m}$, then v
   is also optimal for $\text{IP}_\leq(h)$. Thus, assume that $v \notin \mathbb{Z}^{n-m}$ and let $I \subseteq [n]$ be
   the indices of tight constraints at v, i.e., $C_{I,\cdot}v = g_I$. Next we prove that we
   may assume w.l.o.g. that (a) $0 \in D(C)$, (b) $k = 1$ and (c) C_I is bimodular.
   (a) From Lemma 1 applied to C for $d = 3$ we obtain three possible results:
       $|D(C)| \geq 4$, $0 \notin D(C)$ or $0 \in D(C)$. In the first case we are done and in
       the second case, C is nondegenerate and $\text{IP}_\leq(h)$ can be solved using the
       algorithm in [1]. Therefore, w.l.o.g., $0 \in D(\bar{C})$.
   (b) If $\{2 \cdot k, k, 0\} \subseteq D(C)$ for $k > 1$, it follows from $\gcd(D(C)) = 1$ that
       $|D(C)| \geq 4$. Therefore, w.l.o.g., $k = 1$.
   (c) Since $v \notin \mathbb{Z}^{n-m}$, it holds that $1 \notin D(C_I)$ as otherwise, $v \in \mathbb{Z}^{n-m}$ due to
       Cramer's rule. Apply Theorem 2 once more, but this time to C_I. If the
       algorithm returns that $|D(C_I)| \geq 4$, then $|D(C)| \geq 4$. If the algorithm
       returns a duplicative relation, i.e., $\{2 \cdot s, s\} \subseteq D(C_I)$, then $s \neq 1$ as
       $1 \notin D(C_I)$. Since by (a) and (b), $\{2, 1, 0\} \subseteq D(C)$, it follows that $\{2 \cdot
       s, 2, 1, 0\} \subseteq D(C)$. Thus, $|D(C)| \geq 4$. If the algorithm calculates and
       returns $D(C_I)$, then it has either found that $D(C_I) \subseteq \{2, 0\}$ or it returns
       an element $t \in D(C_I) \setminus \{2, 0\}$. Then, $t \neq 1$ as $1 \notin D(C_I)$, implying that
       $\{t, 2, 1, 0\} \subseteq D(C)$ and $|D(C)| \geq 4$. Thus, w.l.o.g., C_I is bimodular.
   Let $\text{IP}_\leq^{\text{cone}}(h) := \max\{h^T y : C_I y \leq g_I, y \in \mathbb{Z}^{n-m}\}$. As C_I is bimodular, this
   is a BIP. By possibly perturbing the vector h (e.g. by adding $\frac{1}{M} \cdot \sum_{i \in I} C_{i,\cdot}$
   for a sufficiently large $M > 0$), we can assume that v is the unique optimal
   solution to $\text{LP}_\leq(h)$. Solve $\text{IP}_\leq^{\text{cone}}(h)$ using the algorithm by [2]. If $\text{IP}_\leq^{\text{cone}}(h)$ is
   infeasible, so is $\text{IP}_\leq(h)$. Let $y \in \mathbb{Z}^{n-m}$ be an optimal solution for $\text{IP}_\leq^{\text{cone}}(h)$.
   We claim that either, y is also optimal for $\text{IP}_\leq(h)$ or it follows that $|D(C)| \geq$

4: If y is feasible for $\text{IP}_\le(h)$, it is also optimal since $\text{IP}_\le^{\text{cone}}(h)$ is a relaxation of $\text{IP}_\le(h)$. If C is bimodular, Theorem 4 states that y is feasible for $\text{IP}_\le(h)$, i.e., it is optimal for $\text{IP}_\le(h)$. Thus, if y is not feasible for $\text{IP}_\le(h)$, $D(C)$ contains an element which is neither 0, 1 nor 2. As $\{2,1,0\} \subseteq D(C)$ by (a) and (b), this implies that $|D(C)| \ge 4$.

□

Acknowledgments. This work was partially supported by the Einstein Foundation Berlin. We are grateful to Miriam Schlöter for proofreading the manuscript, and to the anonymous reviewers for several helpful comments.

References

1. Artmann, S., Eisenbrand, F., Glanzer, C., Oertel, T., Vempala, S., Weismantel, R.: A note on non-degenerate integer programs with small subdeterminants. Oper. Res. Lett. **44**(5), 635–639 (2016)
2. Artmann, S., Weismantel, R., Zenklusen, R.: A strongly polynomial algorithm for bimodular integer linear programming. In: Proceedings of the 49th Annual ACM SIGACT Symposium on Theory of Computing. ACM, New York (2017)
3. Bonifas, N., Di Summa, M., Eisenbrand, F., Hähnle, N., Niemeier, M.: On subdeterminants and the diameter of polyhedra. Discrete Comput. Geometry **52**(1), 102–115 (2014)
4. Conforti, M., Fiorini, S., Huynh, T., Joret, G., Weltge, S.: The stable set problem in graphs with bounded genus and bounded odd cycle packing number. In: SODA 2020: Proceedings of the Thirty-First Annual ACM-SIAM Symposium on Discrete Algorithms, pp. 2896–2915 (2020)
5. Conforti, M., Fiorini, S., Huynh, T., Weltge, S.: Extended formulations for stable set polytopes of graphs without two disjoint odd cycles. In: Bienstock, D., Zambelli, G. (eds.) IPCO 2020. LNCS, vol. 12125, pp. 104–116. Springer, Cham (2020). https://doi.org/10.1007/978-3-030-45771-6_9
6. Dyer, M., Frieze, A.: Random walks, totally unimodular matrices, and a randomised dual simplex algorithm. Math. Program. **64**(1–3), 1–16 (1994)
7. Eisenbrand, F., Vempala, S.: Geometric random edge. Math. Programm. **164**(1-2), 325–339 (2017)
8. Glanzer, C., Weismantel, R., Zenklusen, R.: On the number of distinct rows of a matrix with bounded subdeterminants. SIAM J. Discret. Math. **32**(3), 1706–1720 (2018)
9. Gribanov, D.V., Veselov, S.I.: On integer programming with bounded determinants. Optim. Lett. **10**(6), 1169–1177 (2015). https://doi.org/10.1007/s11590-015-0943-y
10. Gribanov, D., Malyshev, D., Pardalos, P.: A note on the parametric integer programming in the average case: sparsity, proximity, and FPT-algorithms. arXiv preprint arXiv:2002.01307 (2020)
11. Nägele, M., Sudakov, B., Zenklusen, R.: Submodular minimization under congruency constraints. In: Czumaj, A. (ed.) SODA pp. 849–866. SIAM (2018)
12. Paat, J., Schlöter, M., Weismantel, R.: The Integrality Number of an Integer Program. In: Bienstock, D., Zambelli, G. (eds.) IPCO 2020. LNCS, vol. 12125, pp. 338–350. Springer, Cham (2020). https://doi.org/10.1007/978-3-030-45771-6_26

13. Schrijver, A.: Theory of Linear and Integer Programming. Wiley, New York (1986)
14. Seymour, P.: Decomposition of regular matroids. J. Comb. Theory, Series B **28**(3), 305–359 (1980)
15. Storjohann, A., Labahn, G.: Asymptotically fast computation of hermite normal forms of integer matrices. In: Proceedings of the 1996 International Symposium on Symbolic and Algebraic Computation, pp. 259–266 (1996)
16. Truemper, K.: A decomposition theory for matroids. v. testing of matrix total unimodularity. Journal of Combinatorial Theory, Series B 49(2), 241–281 (1990)
17. Veselov, S., Chirkov, A.: Integer program with bimodular matrix. Discret. Optim. **6**(2), 220–222 (2009)
18. Veselov, S., Shevchenko, V.: Bounds for the maximal distance between the points of certain integer lattices (in Russian). Combinatorial-Algebraic Methods in Applied Mathematics, pp. 26–33 (1980)
19. Walter, M., Truemper, K.: Implementation of a unimodularity test. Math. Program. Comput. **5**(1), 57–73 (2013)

On the Power of Static Assignment Policies for Robust Facility Location Problems

Omar El Housni[1]([✉]), Vineet Goyal[2], and David Shmoys[3]

[1] ORIE, Cornell Tech, New York, NY, USA
oe46@cornell.edu
[2] IEOR, Columbia University, New York, NY, USA
vg2277@columbia.edu
[3] ORIE, Cornell University, Ithaca, NY, USA
david.shmoys@cornell.edu

Abstract. We consider a two-stage robust facility location problem on a metric under an uncertain demand. The decision-maker needs to decide on the (integral) units of supply for each facility in the first stage to satisfy an uncertain second-stage demand, such that the sum of first stage supply cost and the worst-case cost of satisfying the second-stage demand over all scenarios is minimized. The second-stage decisions are only assignment decisions without the possibility of adding recourse supply capacity. This makes our model different from existing work on two-stage robust facility location and set covering problems. We consider an implicit model of uncertainty with an exponential number of demand scenarios specified by an upper bound k on the number of second-stage clients. In an optimal solution, the second-stage assignment decisions depend on the scenario; surprisingly, we show that restricting to a fixed (static) fractional assignment for each potential client irrespective of the scenario gives us an $O(\log k / \log \log k)$-approximation for the problem. Moreover, the best such static assignment can be computed efficiently giving us the desired guarantee.

Keywords: Facility location · Approximation algorithms · Robust optimization

1 Introduction

We consider two-stage robust facility location problems under demand uncertainty where we are given a set of clients and a set of facilities in a common metric space. In the first stage, the decision-maker needs to select the (integral) units supply at each facility. The uncertain demand is then selected adversarially

V. Goyal—Supported in part by NSF CMMI 1636046.
D. Shmoys—Supported by CCF-1526067, CMMI-1537394, CCF-1522054, CCF-1740 822, CCF-1526067, CNS-1952063, and DMS-1839346.

and needs to be satisfied by the existing supply with the minimum assignment cost in the second stage. The goal is to determine the first-stage supply such that the sum of first-stage supply cost and the worst-case assignment cost over all demand scenarios is minimized. Our problem is motivated by settings where the lead time to procure supply is large and obtaining additional units of supply in the second stage is not feasible. The uncertain second-stage demand must then be satisfied by supply units from the first stage, a common constraint in many applications. This is a departure from existing work on two-stage robust facility location, network design and, more generally, robust covering problems that have been studied extensively in the literature [1,6,8,10] where the second-stage decisions allow for "adding more supply" (more specifically adding more sets/facilities to satisfy the requirement).

In this paper, we consider an implicit model of uncertainty with exponentially-many demand scenarios specified by an upper bound k on the number of second-stage demand clients. Since the number of scenarios is exponentially-many, we can not efficiently solve even the LP relaxation for the problem. In contrast, if the set of second-stage scenarios are explicitly specified (see, for instance, [1,6]), we can write a polynomially-sized LP relaxation with assignment decisions for each scenario. The main challenge then is related to obtaining an integral solution, which for the case of set covering and several network design problems can be reduced to deterministic versions (see, for instance, Dhamdhere et al. [6]).

In contrast, in an implicit model of uncertainty (with possibly exponentially-many scenarios), one of the fundamental challenges is to even approximately solve the linear relaxation of the problem efficiently. The implicit model of uncertainty with an upper bound on number of uncertain second-stage clients or elements has been studied extensively in the literature. Feige et al. [8] show under a reasonable complexity assumption that it is hard to solve the LP relaxation of a two-stage set covering problem within a factor better than $\Omega(\log n / \log \log n)$ under this implicit model of uncertainty. They also give an $O(\log^2 n)$-approximation for the $0 - 1$ two-stage robust set-covering problem. Gupta et al. [10] give an improved $O(\log n)$-approximation for the set-covering problem, thereby matching the deterministic approximation guarantee. El Housni and Goyal [7] show that a static policy (that is, linear in the set of second-stage elements, also referred to as an affine policy) gives an $O(\log n / \log \log n)$-approximation for the two-stage LP, thereby matching the hardness lower bound for the fractional problem. Gupta et al. [11] present approximations for several covering network design problems under the implicit model of uncertainty. Although there is a large body of work in this direction, as we mentioned earlier, our problem is different from set covering, since there is no possibility of adding recourse capacity; prior results do not imply an approximation for our model.

Khandekar et al. [15] consider uncapacitated two-stage robust facility location problems where there is a first-stage cost for opening a facility with unlimited supply and an inflated second-stage cost to open recourse facilities in the second-stage. They give a constant approximation algorithm for this model.

Our setting is different as we consider a capacitated version of the problem with a linear supply cost in the first stage as opposed to a fixed opening cost.

Other Related Work. Many variants of facility location problems have been studied extensively in the literature, including both deterministic versions as well as variants that address demand uncertainty. We refer the reader to [18] for a review of the deterministic facility location problems. Among the models that address demand uncertainty in the facility location problems, in addition to robust [2,4], there are also stochastic [12,14,17,21] and distributionally robust [3,5,16] models that have been studied extensively in the literature. In a stochastic model, there is a distribution on the second-stage demand scenarios and the goal is to minimize the total expected cost. A distributionally robust model can be thought of as a hybrid between stochastic and robust where the second-stage distribution is selected adversarially from a pre-specified set and the goal is to minimize the worst-case expected cost. We refer the reader to the survey [20] for an extensive review of facility location problems under uncertainty.

1.1   Our Contributions

Let us begin with a formal problem definition. We are given a set of n facilities $\mathcal{F}$ and m clients $\mathcal{C}$ in a common metric d where d_{ij} denotes the distance between i and j. For each facility $i \in \mathcal{F}$, there is a cost c_i per unit of supply at i. The demand uncertainty is modeled by an implicit set of scenarios $\mathcal{C}^k$ that includes all subsets of clients $\mathcal{C}$ of size at most k. The decision-maker needs to select an (integral) number of units of supply x_i for each facility $i \in \mathcal{F}$ in the first-stage. An adversary observes the first stage decisions and selects a worst-case demand scenario $S \in \mathcal{C}^k$ that must be satisfied with the first-stage supply, where each client in the realized scenario needs one unit of supply. The goal is to minimize the sum of the first-stage supply cost and the worst-case assignment cost over all second-stage demand scenarios. We refer to this problem as *soft-capacitated robust facility location* (SCRFL). Typically in the literature, *soft-capacitated* refer to settings where violations of capacity upper bounds are allowed. The analogue here is that we can add any amount of supply in a facility without upper bounds ($x_i \in \mathbb{Z}_+$) but we pay a per unit cost of supply.

We consider a class of static assignment policies, where each of the m clients has a static fractional assignment to facilities that is independent of the scenario, leading to a feasible second-stage solution for each demand scenario, while respecting supply capacities. Note this is a restriction, since the optimal second-stage assignment decisions are scenario-dependent in general. As a warm-up, we show that static assignment policies are optimal for the uncapacitated case with unlimited supply at each open facility (i.e., there is a cost c_i to open facility i with unlimited supply). We refer to this problem as *uncapacitated robust facility location* (URFL). This is based on the intuition that each client can be assigned to the closest open facilities in an optimal solution in any scenario; this leads to optimality of a static assignment policy for the LP relaxation (Theorem 1).

Theorem 1. *A static assignment policy is optimal for the linear relaxation of* (URFL).

The optimality of static assignment is not true in general when the supply at facilities is constrained (or equivalently, there is a cost per unit of supply). The main contribution in this paper is to show that static assignment policies give an $O(\log k / \log \log k)$-approximation for the LP relaxation of (SCRFL) (Theorem 2). We show this by constructing such a solution, starting from an optimal first-stage supply. The optimal static assignment policies can be computed efficiently by solving a compact LP.

Theorem 2. *A static assignment policy gives* $O(\log k / \log \log k)$-*approximation for the linear relaxation of* (SCRFL).

Furthermore, the fractional supply in the first stage can be rounded to an integral supply using ideas similar to rounding algorithms for the deterministic facility location [19]. In particular, the static assignment solution for the uncapacitated case can be rounded to give a 4-approximation algorithm for (URFL). The static assignment solution for the soft-capacitated case can be rounded within a constant factor, which results in an $O(\log k / \log \log k)$-approximation algorithm for (SCRFL). We would like to note that while the fractional assignment is static in our approximate LP solution, our integral assignment for any client in the second-stage depends on the other demand clients in the scenario; thereby, making our static assignment policy adaptive in implementation.

2   Warm-Up: Uncapacitated Robust Facility Location

2.1   Problem Formulation

In this section, we consider the *uncapacitated robust facility location problem* (URFL) where for each $i \in \mathcal{F}$, there is a cost c_i to open facility i with unlimited supply. The problem can be stated as the following integer program, where each binary variable $z_i, i \in \mathcal{F}$ indicates if facility i is opened and each $y_{ij}^S, i \in \mathcal{F}, j \in S, S \in \mathcal{C}^k$ indicates the assignment of client j to facility i in scenario S.

$$
\begin{aligned}
\min \quad & \sum_{i \in \mathcal{F}} c_i z_i + \max_{S \in \mathcal{C}^k} \sum_{i \in \mathcal{F}} \sum_{j \in S} d_{ij} y_{ij}^S \\
\text{s.t.} \quad & \sum_{i \in \mathcal{F}} y_{ij}^S \geq 1, && \forall S \in \mathcal{C}^k, \forall j \in S, \\
& z_i \geq y_{ij}^S, && \forall i \in \mathcal{F}, \forall S \in \mathcal{C}^k, \forall j \in S, \\
& z_i \in \{0,1\}, \; y_{ij}^S \geq 0, && \forall i \in \mathcal{F}, \forall S \in \mathcal{C}^k, \forall j \in S.
\end{aligned}
\tag{URFL}
$$

Note that the second-stage assignment is a transportation problem and since the demand of a client is integral (0 or 1), the optimal solutions y_{ij}^S are integral as well. The special case of (URFL) where the uncertainty set contains a single scenario corresponds to the NP-hard classical uncapacitated facility location problem, which is hard to approximate within a constant better than 1.463 under a reasonable complexity assumption [9]. We let (LP-URFL) denote the linear relaxation of (URFL), where we replace $z_i \in \{0,1\}$ by $z_i \geq 0$ for each $i \in \mathcal{F}$. While it is challenging in general to even solve the linear relaxation of a problem under the implicit model of uncertainty, we show that (LP-URFL) can be solved in polynomial time using a *Static Assignment Policy* for the second-stage variables. Moreover, we can round the fractional solution losing only a constant factor, thereby getting a constant approximation for (URFL).

2.2 Static Assignment Policy

Consider an optimal solution of (URFL). Since each open facility can have an unlimited amount of supply, each client in the realized scenario is assigned to the closest facility among the opened ones. The same observation holds as well for (LP-URFL) where each client is assigned to the same fractionally opened facilities independent of the realized scenario. Thus, the assignment of a client is static. This can be captured by the following policy.

Static Assignment Policy. There exists $y_{ij} \geq 0$ for each $i \in \mathcal{F}, j \in C$ such that

$$\forall S \in C^k, \forall i \in \mathcal{F}, \forall j \in S, \qquad y_{ij}^S = y_{ij}. \qquad (1)$$

Proof of Theorem 1. Let $(\boldsymbol{z}^*, \boldsymbol{y}^{*S}, S \in C^k)$ be an optimal solution to (LP-URFL). Since there are no capacities on facilities, each client j is assigned to the closest fractionally opened facilities. In particular, for each $j \in C$, let π_j be a permutation of $\mathcal{F} = \{1, \ldots, n\}$ such that $d_{\pi_j(1)j} \leq d_{\pi_j(2)j} \leq \cdots \leq d_{\pi_j(n)j}$, and let $\ell = \min\{p \mid z_{\pi_j(1)}^* + z_{\pi_j(2)}^* + \cdots + z_{\pi_j(p)}^* \geq 1\}$. Denote $\hat{z}_{\pi_j(\ell)} = 1 - (z_{\pi_j(1)}^* + \cdots + z_{\pi_j(\ell-1)}^*)$. The optimal solution can be written in the form (1) as follows: for $S \in C^k$ and $j \in S$; $y_{ij}^S = z_i^*$ for $i \in \{\pi_j(1), \ldots, \pi_j(\ell-1)\}$, $y_{ij}^S = \hat{z}_i$ for $i = \pi_j(\ell)$, and $y_{ij}^S = 0$, otherwise. □

Let (Static-URFL) denote the problem after restricting the second-stage variables y_{ij}^S in (LP-URFL) to a policy (1), which can then be reformulated as follows:

$$\min \sum_{i \in \mathcal{F}} c_i z_i + \max_{S \in C^k} \sum_{i \in \mathcal{F}} \sum_{j \in C} 1(j \in S) \cdot d_{ij} y_{ij}$$

$$\text{s.t.} \sum_{i \in \mathcal{F}} y_{ij} \geq 1, \qquad\qquad \forall j \in C, \quad \text{(Static-URFL)}$$

$$z_i \geq y_{ij} \geq 0, \qquad\qquad \forall i \in \mathcal{F}, \forall j \in C.$$

From Theorem 1, (Static-URFL) is equivalent to (LP-URFL). The number of variables in (Static-URFL) is reduced to a polynomial number since the y_{ij} no longer depend on the scenario S. The inner maximization problem is still taken over an exponential number of scenarios; however we can separate efficiently over these scenarios and write an efficient compact LP formulation for (Static-URFL):

$$\max_{S \in \mathcal{C}^k} \left\{ \sum_{i \in \mathcal{F}} \sum_{j \in \mathcal{C}} \mathbf{1}(j \in S) \cdot d_{ij} y_{ij} \right\} = \max_{h \in [0,1]^{|\mathcal{C}|}} \left\{ \sum_{i \in \mathcal{F}} \sum_{j \in \mathcal{C}} d_{ij} y_{ij} h_j \mid \sum_{j \in \mathcal{C}} h_j \leq k \right\}$$

$$= \min_{\mu, \omega \geq 0} \left\{ k\mu + \sum_{j \in \mathcal{C}} \omega_j \mid \mu + \omega_j \geq \sum_{i \in \mathcal{F}} d_{ij} y_{ij}, \ \forall j \in \mathcal{C} \right\}, \tag{2}$$

where the first equality holds because the optimal solution of the right maximization problem occurs at the extreme points of the k-ones polytope, which correspond to the worst-case scenarios of $\mathcal{C}^k$ and the second equality follows from strong duality. Therefore, by dropping the min and introducing μ and all ω_j as variables, we reformulate (Static-URFL) as the following linear program:

$$\min \ \sum_{i \in \mathcal{F}} c_i z_i + k\mu + \sum_{j \in \mathcal{C}} \omega_j$$

$$\text{s.t.} \ \mu + \omega_j \geq \sum_{i \in \mathcal{F}} d_{ij} y_{ij}, \qquad\qquad \forall j \in \mathcal{C},$$

$$\sum_{i \in \mathcal{F}} y_{ij} \geq 1, \qquad\qquad \forall j \in \mathcal{C}, \tag{3}$$

$$z_i \geq y_{ij}, \qquad\qquad \forall i \in \mathcal{F}, \forall j \in \mathcal{C},$$

$$z_i \geq 0, \ y_{ij} \geq 0, \ \omega_j \geq 0, \ \mu \geq 0, \qquad \forall i \in \mathcal{F}, \forall j \in \mathcal{C}.$$

Finally, we round the solution of (Static-URFL) to an integral solution for (URFL) while losing only a constant factor. This can be done using prior work on rounding techniques from the literature of deterministic facility location problems. In fact, the LP rounding technique in Shmoys et al. [19], which gives a 4-approximation algorithm to the deterministic uncapacitated problem also gives a 4-approximation algorithm for (Static-URFL). The idea is to define a ball around each client of radius equal to the fractional assignment cost of the client (which is independent of any scenario for our static policy). Then, we open facilities in non-intersecting balls of ascending radius. We state the result in Theorem 3 and defer the details of the rounding to full version of the paper [13].

Theorem 3. *(Static-URFL) can be rounded to give a 4-approximation to* (URFL).

We would like to note that Khandekar et al. [15] consider the uncapacitated robust facility location problem where additional facilities can be open in the second-stage at an inflated cost. Their results imply a 10-approximation for URFL, by taking the inflation factor to be infinite in their model. While our constant approximation is better than this previous result, our focus in this section is not about finding the best constant approximation for (URFL), but rather we introduce it as a warm-up for motivating static assignment policies before presenting our main result.

3   Soft-Capacitated Robust Facility Location

3.1   Problem Formulation

In this section, we consider the *soft-capacitated robust facility location* (SCRFL) which is similar to (URFL) except that each facility i incurs a linear supply cost, where c_i is the cost per unit of supply. We refer to x_i as the supply (or capacity) in facility i. Each client in the realized scenario needs to be satisfied by one unit of supply. The problem can be stated as the following integer program:

$$\min \sum_{i \in \mathcal{F}} c_i x_i + \max_{S \in \mathcal{C}^k} \sum_{i \in \mathcal{F}} \sum_{j \in S} d_{ij} y_{ij}^S$$

$$\text{s.t.} \sum_{i \in \mathcal{F}} y_{ij}^S \geq 1, \qquad\qquad \forall S \in \mathcal{C}^k, \forall j \in S,$$

$$\qquad\qquad\qquad\qquad\qquad\qquad\qquad\qquad\qquad\qquad\qquad\qquad \text{(SCRFL)}$$

$$x_i \geq \sum_{j \in S} y_{ij}^S, \qquad\qquad \forall i \in \mathcal{F}, \forall S \in \mathcal{C}^k, \forall j \in S,$$

$$x_i \in \mathbb{N},\ y_{ij}^S \geq 0, \qquad\qquad \forall i \in \mathcal{F}, \forall S \in \mathcal{C}^k, \forall j \in S.$$

We let (LP-SCRFL) denote the linear relaxation of (SCRFL), where we replace $x_i \in \mathbb{N}$ by $x_i \geq 0$, for each $i \in \mathcal{F}$. We would like to note that even the linear relaxation (LP-SCRFL) is challenging to solve since it has exponentially-many variables (scenarios). Unlike the uncapacitated case, the static assignment policy (1) is not optimal for (LP-SCRFL) and the optimal assignment for each client is scenario dependent. In particular, the same client could be assigned to different facilities in different scenarios. In contrast, we show the surprising result that a static assignment policy gives $O(\log k / \log \log k)$-approximation to (LP-SCRFL). Moreover, we can round the solution of the static assignment policy to an integral solution for (SCRFL) and only lose an additional constant factor. We let (Static-SCRFL) denote the problem when we restrict the second-stage variables y_{ij}^S in (LP-SCRFL) to static assignment policies (1). The problem can then be reformulated as follows:

$$\min \sum_{i \in \mathcal{F}} c_i x_i + \max_{S \in \mathcal{C}^k} \sum_{i \in \mathcal{F}} \sum_{j \in \mathcal{C}} \mathbf{1}(j \in S) \cdot d_{ij} y_{ij}$$

$$\text{s.t.} \sum_{i \in \mathcal{F}} y_{ij} \geq 1, \qquad\qquad\qquad \forall j \in \mathcal{C},$$

$$\qquad\qquad\qquad\qquad\qquad\qquad\qquad\qquad\qquad\qquad \text{(Static-SCRFL)}$$

$$x_i \geq \max_{S \in \mathcal{C}^k} \sum_{j \in \mathcal{C}} \mathbf{1}(j \in S) \cdot y_{ij}, \qquad \forall i \in \mathcal{F},$$

$$x_i \geq 0,\ y_{ij} \geq 0, \qquad\qquad\qquad \forall i \in \mathcal{F}, \forall j \in \mathcal{C}.$$

3.2 An $O(\frac{\log k}{\log \log k})$-Approximation Algorithm

Our main contribution in this section is to show that a static assignment policy (1) gives $O(\log k/\log \log k)$-approximation for (LP-SCRFL) (Theorem 2). To prove this theorem, we consider an optimal solution of (LP-SCRFL) and massage it to construct a solution of the form (1) while losing $O(\log k/\log \log k)$ factor. We first present our construction and several structural lemmas and then give the proof of Theorem 2.

Our Construction. Let $x^* : (x_i^*)_{i \in \mathcal{F}}$ be an optimal first-stage solution of (LP-SCRFL), let OPT_1 be the corresponding optimal first-stage cost and let OPT_2 be the corresponding optimal second-stage cost. We will classify the clients $\mathcal{C}$ into three subsets C_1, C_2, C_3 using Procedure 1 (below) and then specify a static assignment policy for each subset. We use the following notation in the procedure. Let $\alpha > 1$ and $r = 5 \cdot OPT_2/k$. For $\ell \geq 1$ and $j \in C$, we let B_j^ℓ denote the ball centered at client j of radius ℓr. We initialize the sets $F \leftarrow \mathcal{F}$ and $C \leftarrow \mathcal{C}$ and update them at each iteration, as explained in the procedure, until C becomes empty. We let $Cl(B)$ denote the set of clients in C that are inside the ball B and let $Sp(B)$ denote the total optimal supply of facilities F that are inside the ball B, i.e.,

$$Sp(B) = \sum_{i \in F} \mathbb{1}(i \in B) \cdot x_i^* \qquad \text{and} \qquad Cl(B) = \{j \in C \mid j \in B\}.$$

Note that both $Sp(B)$ and $Cl(B)$ depend on the current sets of facilities F and clients C, which we update at each iteration of the while loop in the procedure. But for the ease of notation, we do not refer to them with the indices F and C.

In the procedure, while the set C is not empty, we pick a client $j \in C$ and grow three balls around it: $B_j^{2\ell-1}$ (*internal* ball), $B_j^{2\ell}$ (*medium* ball) and $B_j^{2\ell+1}$ (*external* ball) starting with $\ell = 1$. For each ℓ, we check if the number of clients in the internal ball $B_j^{2\ell-1}$ is greater than k (line 4); if this is the case, we remove them from C, put them in C_1 and restart in line 2. If not, we check if the supply in the medium ball $B_j^{2\ell}$ is sufficient to satisfy half of the clients in the internal ball $B_j^{2\ell-1}$ (line 7); if that is not the case, we remove those clients from C, put them in C_2, and restart in line 2. Otherwise, we finally check if the supply in the medium ball $B_j^{2\ell}$ is sufficient to satisfy a fraction $1/2\alpha$ of the clients in the external ball $B_j^{2\ell+1}$ (line 10); if that is the case we remove all the clients in $B_j^{2\ell+1}$ and put them in C_3, we also remove all the facilities in $B_j^{2\ell}$ and restart in line 2. If none of these three conditions holds, we increase ℓ to $\ell + 1$. First, we show that after at most $\log_\alpha k$ increments (i.e., $\ell \leq \log_\alpha k$), one of three conditions must hold and therefore we will remove some clients from C and restart in line 2. Which implies that after a finite number of iterations, the set C becomes empty. In particular, We have the following lemma.

Lemma 1. *In Procedure 1, after a finite number of iterations, the set C becomes empty and $C_1 \cup C_2 \cup C_3$ is equal to $\mathcal{C}$. Moreover, ℓ is always less than $\log_\alpha k$.*

Proof. Fix a client j and let $\ell \geq 1$. If none of the three conditions holds then

$$\alpha \cdot |Cl(B_j^{2\ell-1})| \leq 2\alpha \cdot Sp(B_j^{2\ell}) < |Cl(B_j^{2\ell+1})|.$$

Therefore, the number of clients grows geometrically when we increase the radius of the balls and by induction, we have that

$$\alpha^\ell \leq \alpha^\ell \cdot |Cl(B_j^1)| < |Cl(B_j^{2\ell+1})|,$$

where $|Cl(B_j^1)| \geq 1$, since $Cl(B_j^1)$ contains at least the client j. Hence, after at most $\log_\alpha k$ increments, we will reach k clients, and must stop by the first condition, and return to line 2. Hence, we always have $\ell \leq \log_\alpha k$. Finally since, we remove at least one client at each iteration of the while loop, the set C becomes empty after at most $|\mathcal{C}|$ iterations, and finally $C_1 \cup C_2 \cup C_3 = \mathcal{C}$. $\quad\square$

Procedure 1

1: Initialize $C \leftarrow \mathcal{C}, C_1 \leftarrow \emptyset, C_2 \leftarrow \emptyset, C_3 \leftarrow \emptyset, F \leftarrow \mathcal{F}$.
2: **while** $C \neq \emptyset$ **do**
3:     Pick a client $j \in C$. Initialize $\ell = 1$
4:     **if** $|Cl(B_j^{2\ell-1})| \geq k$ **then**
5:         $C_1 \leftarrow C_1 \cup Cl(B_j^{2\ell-1}), \quad C \leftarrow C \setminus Cl(B_j^{2\ell-1})$
6:         Stop, return to line 2
7:     **if** $Sp(B_j^{2\ell}) < \frac{1}{2} \cdot |Cl(B_j^{2\ell-1})|$ **then**
8:         $C_2 \leftarrow C_2 \cup Cl(B_j^{2\ell-1}), \quad C \leftarrow C \setminus Cl(B_j^{2\ell-1})$
9:         Stop, return to line 2
10:     **if** $Sp(B_j^{2\ell}) \geq \frac{1}{2\alpha} \cdot |Cl(B_j^{2\ell+1})|$ **then**
11:         $C_3 \leftarrow C_3 \cup Cl(B_j^{2\ell+1}), \quad C \leftarrow C \setminus Cl(B_j^{2\ell+1}),$
12:         $F \leftarrow F \setminus \{i \in F \mid i \in B_j^{2\ell}\}$
13:         Stop, return to line 2
14:     **else** $\ell \leftarrow \ell + 1$, return to line 4.

Now we are ready to present our static assignment policy for (LP-SCRFL). The following three lemmas show our constructed static assignment for each client in the subsets C_1, C_2, C_3. We specify the supply used to satisfy each subset of these clients and the assignment cost. First, we know that each client in C_1 belongs to a ball with at least k clients. By the feasibility of the optimal solution, this implies that there exists k units of supply in $\boldsymbol{x}^*$ "close" to this ball. Hence, we show that we can satisfy this client with a static assignment that uses $\boldsymbol{x}^*/k$ while paying a small assignment cost (roughly a constant times the radius of the ball). We need to dedicate only one $\boldsymbol{x}^*$ for all clients in C_1 since each one is using at most $\boldsymbol{x}^*/k$ in our solution. Formally, we have the following lemma.

Lemma 2. *There exists a static assignment policy for C_1 such that each client in C_1 is using at most the supply $\boldsymbol{x}^*/k$ and has an assignment cost that is $O(\log_\alpha k/k) \cdot \mathrm{OPT}_2$, i.e., there exists $(\tilde{y}_{ij})_{i \in \mathcal{F}, j \in C_1}$ such that for each $j \in C_1$:*

$$\sum_{i \in \mathcal{F}} \tilde{y}_{ij} \geq 1, \qquad \frac{x_i^*}{k} \geq \tilde{y}_{ij} \geq 0, \ \forall i \in \mathcal{F}, \qquad and \qquad \sum_{i \in \mathcal{F}} d_{ij}\tilde{y}_{ij} = O(\log_\alpha k) \cdot \frac{\mathrm{OPT}_2}{k}.$$

Proof. Let j be a client of C_1. It is sufficient to show that the following minimization problem is feasible and its optimal cost is $O(\log_\alpha k/k) \cdot \mathrm{OPT}_2$,

$$\min \left\{ \sum_{i \in \mathcal{F}} d_{ij}y_{ij} \ \Big| \ \sum_{i \in \mathcal{F}} y_{ij} \geq 1, \ \frac{x_i^*}{k} \geq y_{ij} \geq 0, \ \forall i \in \mathcal{F} \right\}. \tag{4}$$

Problem (4) must be feasible since the total supply in $\boldsymbol{x}^*$ is greater than the total demand in any scenario, i.e., $\sum_{i \in \mathcal{F}} x_i^* \geq k$. Recall that a client j in C_1 belongs to one of the sets $Cl(B_t^{2\ell-1})$ for some $t \in \mathcal{C}$ and $\ell \leq \log_\alpha k$ (Lemma 1) such that $|Cl(B_t^{2\ell-1})| \geq k$. Consider a scenario S formed by k clients from $Cl(B_t^{2\ell-1})$. Let denote $\boldsymbol{y}^S$ the assignment of scenario S in the optimal solution of (LP-SCRFL). Consider the following candidate solution for (4):

$$y_{ij} = \frac{1}{k} \cdot \sum_{p \in S} y_{ip}^S, \ \forall i \in \mathcal{F}.$$

We have, by the feasibility of the optimal solution, $0 \leq y_{ij} \leq \frac{1}{k}x_i^*, \forall i \in \mathcal{F}$ and

$$\sum_{i \in \mathcal{F}} y_{ij} = \frac{1}{k} \cdot \sum_{i \in \mathcal{F}} \sum_{p \in S} y_{ip}^S \geq \frac{1}{k} \sum_{p \in S} 1 = 1.$$

Therefore, our solution is feasible for (4). Moreover, we have

$$\sum_{i \in \mathcal{F}} d_{ij}y_{ij} = \frac{1}{k} \cdot \sum_{i \in \mathcal{F}} \sum_{p \in S} d_{ij}y_{ip}^S$$

$$\leq \frac{1}{k} \cdot \sum_{i \in \mathcal{F}} \sum_{p \in S} d_{ip}y_{ip}^S + \frac{1}{k} \cdot \sum_{p \in S} d_{pj}$$

$$\leq \frac{1}{k} \cdot \mathrm{OPT}_2 + 2(2\ell - 1)r$$

$$\leq \frac{\mathrm{OPT}_2}{k} + 2(2\log_\alpha k - 1) \cdot 5 \cdot \frac{\mathrm{OPT}_2}{k} = O(\log_\alpha k) \cdot \frac{\mathrm{OPT}_2}{k},$$

where the first inequality follows from the triangle inequality and the fact that $\sum_{i \in \mathcal{F}} y_{ip}^S = 1$ for all $p \in S$ in the optimal solution. For the second inequality: the first term is bounded by OPT_2 by definition, the second term d_{pj} is bounded by the diameter of the ball $B_j^{2\ell-1}$ since it contains clients j and all $p \in S$. $\qquad\square$

Now, consider the set C_2. By construction, these are clients such that there is not enough supply within a distance $r = 5\text{OPT}_2/k$ to satisfy half of them. Therefore, intuitively they need to pay "large" distances in the optimal assignment cost if they show up all together in the same scenario. In the following lemma, we show that we can have no more than k of these clients. As we would show later, this would imply that we can dedicate a supply x^* to C_2 and make a static assignment of all the clients C_2 to this x^*.

Lemma 3. *The set C_2 has at most k clients.*

Proof. Suppose, for the sake of contradiction, that $|C_2| > k$. Let $G_1, G_2, \ldots, G_T$ be the disjoint subsets of clients added at each iteration in the construction of C_2 in Procedure 1. In particular, $C_2 = G_1 \cup G_2 \cup \ldots \cup G_T$ for some T where:

(i) for $t = 1, 2, \ldots, T$, $G_t = Cl(B_{j_t}^{2\ell_t - 1})$ for some client j_t and $1 \le \ell_t \le \log_\alpha k$.
(ii) the supply $Sp(B_{j_t}^{2\ell_t})$ is less than half of the clients in G_t.
(iii) each set G_t has strictly less than k clients, since the procedure has to fail the first "if" statement before adding G_t into C_2.

Recall that

$$Sp(B_{j_t}^{2\ell_t}) = \sum_{i \in F} \mathbf{1}(i \in B_{j_t}^{2\ell_t}) \cdot x_i^*,$$

where F is the current set of facilities in the procedure (and is not all $\mathcal{F}$ since some facilities have been removed previously in line 12). However, we would like to emphasize that when a facility has been removed (in line 12), all clients within distance r from this facility were removed as well (line 11). This is true, since when we remove the facilities in a medium ball, say $B_j^{2\ell}$, (line 12), we remove all clients in the corresponding external ball $B_j^{2\ell+1}$ (line 11). Hence, the remaining clients in C are at least $(2\ell + 1)r - (2\ell)r = r$ away from the removed facilities. In particular, for the clients G_t, the supply that has been removed before they were added to C_2 is at least r away from them. Therefore, all the facilities in $\mathcal{F}$ that are within a distance r from a client in G_t belong to the set F that verifies $\sum_{i \in F} \mathbf{1}(i \in B_{j_t}^{2\ell_t}) \cdot x_i^* < |G_t|/2$. This implies that the supply of all facilities within a distance r from G_t in the optimal solution, is less than half of the clients G_t. Hence, if the clients G_t show up together in a scenario, the optimal second-stage solution needs to pay an assignment cost of at least $r \cdot |G_t|/2$.

Order the sets G_t according to their cardinalities: wlog assume that $|G_1| \ge |G_2| \ge \ldots \ge |G_T|$. We construct a scenario $\hat{S}$ by taking clients from the sets $G_1, G_2, \ldots$ until we hit k. This is possible because $|C_2| > k$. Assume that

$$|G_1| + |G_2| + \ldots |G_{p-1}| + |\bar{G}_p| = k,$$

for some p, where $2 \le p \le T$. Note that $\bar{G}_p$ is a subset of G_p, since we can reach k before taking all the clients of the last set G_p. For each $t = 1, \ldots, p - 1$, the optimal second-stage decision needs to pay at least $r \cdot |G_t|/2$. Therefore,

$$\text{OPT}_2 \ge \frac{1}{2} \cdot r \cdot (|G_1| + |G_2| + \ldots |G_{p-1}|).$$

We did not include G_p, since not all these clients are necessary in the scenario $\hat{S}$, but only $|\bar{G}_p|$ of them. Since $\bar{G}_p$ has the smallest cardinality

$$|G_1| + |G_2| + \ldots |G_{p-1}| \geq \frac{1}{2}(|G_1| + |G_2| + \ldots |G_{p-1}| + |\bar{G}_p|) = \frac{k}{2}.$$

Therefore, $\mathrm{OPT}_2 \geq r \cdot k/4 = 5 \cdot \mathrm{OPT}_2/4$, which is a contradiction.   □

Finally, for clients C_3, we show that there exists $|C_3|/2\alpha$ units of supply "close" to them. In particular, we can multiply these units by 2α, dedicate them to C_3 and make a static assignment for C_3. We have the following lemma.

Lemma 4. *There exists a supply $\hat{x}$ that has a cost at most $2\alpha \cdot \mathrm{OPT}_1$ and there exists a static assignment policy such that all clients in C_3 are assigned to supply $\hat{x}$ and each client in C_3 has an assignment cost that is $O(\log_\alpha k/k) \cdot \mathrm{OPT}_2$, i.e., there exists $(\hat{y}_{ij})_{i\in\mathcal{F}, j\in C_3}$ and $(\hat{x}_i)_{i\in\mathcal{F}}$ such that for all $j \in C_3$:*

$$\sum_{i\in\mathcal{F}} \hat{y}_{ij} \geq 1, \qquad \hat{x}_i \geq \sum_{j\in C_3} \hat{y}_{ij}, \ \forall i \in \mathcal{F}, \qquad \hat{x}_i \geq 0, \ \hat{y}_{ij} \geq 0, \ \forall i \in \mathcal{F},$$

$$\sum_{i\in\mathcal{F}} c_i \hat{x}_i \leq 2\alpha \cdot \mathrm{OPT}_1 \qquad and \qquad \sum_{i\in\mathcal{F}} d_{ij} \hat{y}_{ij} = O(\log_\alpha k) \cdot \frac{\mathrm{OPT}_2}{k}.$$

Proof. Let $G_1, G_2, \ldots, G_T$ be the disjoint subsets of clients added at each iteration to construct C_3 in Procedure 1. In particular, $C_3 = G_1 \cup G_2 \cup \ldots \cup G_T$ for some T such that: for all $t = 1, 2, \ldots, T$, $G_t = Cl(B_{j_t}^{2\ell_t+1})$ for some client j_t and some integer ℓ_t with $1 \leq \ell_t \leq \log_\alpha k$. Moreover, the supply $Sp(B_{j_t}^{2\ell_t})$ is greater than a $1/2\alpha$ fraction of the clients in G_t. Hence, for each ball $B_{j_t}^{2\ell_t}$, we multiply the supply by 2α, move it to the cheapest facility in this ball and make a static assignment of all clients G_t to this cheapest facility. Since the supply in $B_{j_t}^{2\ell_t}$ is removed along with clients G_t, it will not be used by the other clients in C_3.

Formally, let i_t be the cheapest facility in the ball $B_{j_t}^{2\ell_t}$. We define, for each facility i in $B_{j_t}^{2\ell_t}$, $\hat{x}_i = 2\alpha \sum_{i'\in F} \mathbf{1}(i' \in B_{j_t}^{2\ell_t}) \cdot x_{i'}^*$ if $i = i_t$ and $\hat{x}_i = 0$ otherwise. For each client $j \in C_3$, we let $\hat{y}_{ij} = 1$ for $i = i_t$ and $j \in G_t$, and let $\hat{y}_{ij} = 0$, otherwise. Therefore, the first desired constraints in the lemma are verified. Let us check the last one. The distance between a client and its assigned facility in our solution is at most r plus the diameter of the ball $B_{j_t}^{2\ell_t}$, i.e., $r + 4\ell_t r$ which is at most $(4\log_\alpha k + 1) \cdot 5 \cdot \mathrm{OPT}_2/k = O(\log_\alpha k/k) \cdot \mathrm{OPT}_2$.   □

Proof of Theorem 2. Let $(\tilde{y}_{ij})_{i\in\mathcal{F}, j\in C_1}$ be the solution given in Lemma 2 for satisfying the clients in C_1. We dedicate a supply x^* to clients C_1. Let $(\hat{y}_{ij})_{i\in\mathcal{F}, j\in C_3}$ and $(\hat{x}_i)_{i\in\mathcal{F}}$ be the solution given in Lemma 4 for satisfying the clients in C_3. Finally, we know from Lemma 3 that C_2 has at most k clients, and therefore C_2 is a scenario. So we dedicate a supply x^* to C_2 and let the optimal assignment $y_{ij}^{C_2}$ be our static assignment solution for C_2. In particular, we give the following solution to (LP-SCRFL), where the first stage solution is $2x^* + \hat{x}$ and the static assignment policy is for all $i \in \mathcal{F}$: $y_{ij} = \tilde{y}_{ij}$ for $j \in C_1$, $y_{ij} = y_{ij}^{C_2}$ for $j \in C_2$, and $y_{ij} = \hat{y}_{ij}$ for $j \in C_3$. It is clear that $\sum_{i \in \mathcal{F}} y_{ij} \geq 1$, for each j in $C_1 \cup C_2 \cup C_3$. Moreover, for any scenario $S \in \mathcal{C}^k$ and $i \in \mathcal{F}$,

$$
\sum_{j \in S} y_{ij} = \sum_{j \in S \cap C_1} \tilde{y}_{ij} + \sum_{j \in S \cap C_2} y_{ij}^{C_2} + \sum_{j \in S \cap C_3} \hat{y}_{ij}
$$

$$
\leq \left(\sum_{j \in S \cap C_1} \frac{x_i^*}{k} \right) + x_i^* + \hat{x}_i \leq 2x_i^* + \hat{x}_i.
$$

Therefore, our solution is feasible for (LP-SCRFL). Let us evaluate its cost. The cost of the first stage is at most $2OPT_1 + 2\alpha OPT_1 = O(\alpha) \cdot OPT_1$. For the second-stage cost, consider any scenario $S \in \mathcal{C}^k$, We have

$$
\sum_{i \in \mathcal{F}} \sum_{j \in S} d_{ij} y_{ij} = \sum_{j \in S \cap C_1} \sum_{i \in \mathcal{F}} d_{ij} \tilde{y}_{ij} + \sum_{j \in S \cap C_2} \sum_{i \in \mathcal{F}} d_{ij} y_{ij}^{C_2} + \sum_{j \in S \cap C_3} \sum_{i \in \mathcal{F}} d_{ij} \hat{y}_{ij}
$$

$$
\leq \sum_{j \in S \cap C_1} O(\log_\alpha k) \cdot \frac{OPT_2}{k} + OPT_2 + \sum_{j \in S \cap C_3} O(\log_\alpha k) \cdot \frac{OPT_2}{k}
$$

$$
\leq O(\log_\alpha k) \cdot OPT_2.
$$

By balancing the terms α and $\log_\alpha k$, we choose $\alpha = \log k / \log \log k$ which gives $O(\log k / \log \log k)$-approximation to (LP-SCRFL). $\qquad \square$

Similar to the uncapacitated problem, we can solve (Static-SCRFL) efficiently using a compact linear program. In fact, we dualize the inner maximization problem in the objective function of (Static-SCRFL) in the same way as (2). In addition to that, we reformulate the second constraint in (Static-SCRFL) using the same dualization technique as follows: for each $i \in \mathcal{F}$,

$$
\max_{S \in \mathcal{C}^k} \left\{ \sum_{j \in S} y_{ij} \right\} = \max_{h \in [0,1]^{|\mathcal{C}|}} \left\{ \sum_{j \in \mathcal{C}} y_{ij} h_j \ \middle| \ \sum_{j \in \mathcal{C}} h_j \leq k \right\}
$$

$$
= \min_{\eta_i, \lambda_{ij} \geq 0} \left\{ k\eta_i + \sum_{j \in \mathcal{C}} \lambda_{ij} \ \middle| \ \eta_i + \lambda_{ij} \geq y_{ij}, \ \forall j \in \mathcal{C} \right\}.
$$

The linear program is given by

$$
\min \sum_{i \in \mathcal{F}} c_i x_i + k\mu + \sum_{j \in \mathcal{C}} \omega_j
$$

$$
\text{s.t. } \mu + \omega_j \geq \sum_{i \in \mathcal{F}} d_{ij} y_{ij}, \qquad\qquad \forall j \in \mathcal{C},
$$

$$
\sum_{i \in \mathcal{F}} y_{ij} \geq 1, \qquad\qquad \forall j \in \mathcal{C}, \qquad (5)
$$

$$
x_i \geq k\eta_i + \sum_{j \in \mathcal{C}} \lambda_{ij}, \qquad\qquad \forall i \in \mathcal{F},
$$

$$
\eta_i + \lambda_{ij} \geq y_{ij}, \qquad\qquad \forall i \in \mathcal{F}, \forall j \in \mathcal{C},
$$

$$
x_i, y_{ij}, \lambda_{ij}, \eta_i, \omega_j, \mu \geq 0, \qquad\qquad \forall i \in \mathcal{F}, \forall j \in \mathcal{C},
$$

which can be reduced, after removing the variables y_{ij}, to

$$\min \sum_{i \in \mathcal{F}} c_i x_i + k\mu + \sum_{j \in \mathcal{C}} \omega_j$$

$$\text{s.t. } \mu + \omega_j \geq \sum_{i \in \mathcal{F}} d_{ij}(\eta_i + \lambda_{ij}), \qquad \forall j \in \mathcal{C},$$

$$\sum_{i \in \mathcal{F}} \eta_i + \lambda_{ij} \geq 1, \qquad \forall j \in \mathcal{C}, \qquad (6)$$

$$x_i \geq k\eta_i + \sum_{j \in \mathcal{C}} \lambda_{ij}, \qquad \forall i \in \mathcal{F},$$

$$x_i, \lambda_{ij}, \eta_i, \omega_j, \mu \geq 0, \qquad \forall i \in \mathcal{F}, \forall j \in \mathcal{C}.$$

Finally, we round the optimal solution of (Static-SCRFL) using the filtering and rounding techniques from Shmoys et al. [19] while losing only a factor of 12. This rounding technique was designed for the deterministic problem, but the same argument works as well for (Static-SCRFL). We defer the details of the rounding to the full version of the paper [13]. Since we showed that (Static-SCRFL) gives $O(\log k / \log \log k)$-approximation to (LP-SCRFL) and we only lose a constant factor in the rounding, this results in $O(\log k / \log \log k)$-approximation algorithm for (SCRFL) (Theorem 4). Note that after rounding the supply in our static solution, the second-stage assignment for each scenario is a transportation problem and therefore its optimal solution is integral. We would like to emphasize that while our fractional assignment is static, our integral assignment is not necessarily static.

Theorem 4. *(Static-SCRFL) can be rounded to give $O(\frac{\log k}{\log \log k})$-approximation algorithm to (SCRFL).*

4 Conclusion

In this paper, we give a $O(\log k / \log \log k)$-approximation for soft-capacitated robust facility location problems with an implicit model of demand uncertainty. It is an interesting open question to study whether there exists a constant approximation algorithm for the problem, even in special cases such as the Euclidean metric. Our solution approach relies on static fractional assignment policies, which we show are optimal for the uncapacitated problem and give a strong theoretical guarantee for soft-capacitated case. Static assignment policies, while reasonable for the case of soft-capacities can be shown to be arbitrarily bad for the case of hard-capacities, where in addition to cost per unit, there is also an upper bound on supply at each facility. It is another interesting open direction to study any non-trivial approximation in this setting.

References

1. Anthony, B., Goyal, V., Gupta, A., Nagarajan, V.: A plant location guide for the unsure: approximation algorithms for min-max location problems. Math. Oper. Res. **35**(1), 79–101 (2010)
2. Atamtürk, A., Zhang, M.: Two-stage robust network flow and design under demand uncertainty. Oper. Res. **55**(4), 662–673 (2007)
3. Basciftci, B., Ahmed, S., Shen, S.: Distributionally robust facility location problem under decision-dependent stochastic demand. arXiv preprint arXiv:1912.05577 (2019)
4. Charikar, M., Khuller, S., Mount, D.M., Narasimhan, G.: Algorithms for facility location problems with outliers. In: Proceedings of the Twelfth Annual ACM-SIAM Symposium on Discrete Algorithms, pp. 642–651 (2001)
5. Delage, E., Ye, Y.: Distributionally robust optimization under moment uncertainty with application to data-driven problems. Oper. Res. **58**(3), 595–612 (2010)
6. Dhamdhere, K., Goyal, V., Ravi, R., Singh, M.: How to pay, come what may: approximation algorithms for demand-robust covering problems. In: 46th Annual IEEE Symposium on Foundations of Computer Science (FOCS 2005), pp. 367–376 (2005)
7. El Housni, O., Goyal, V.: On the optimality of affine policies for budgeted uncertainty sets. Mathematics of Operations Research (2021, to appear)
8. Feige, U., Jain, K., Mahdian, M., Mirrokni, V.: Robust combinatorial optimization with exponential scenarios. In: Fischetti, M., Williamson, D.P. (eds.) IPCO 2007. LNCS, vol. 4513, pp. 439–453. Springer, Heidelberg (2007). https://doi.org/10.1007/978-3-540-72792-7_33
9. Guha, S., Khuller, S.: Greedy strikes back: improved facility location algorithms. J. Algorithms **31**(1), 228–248 (1999)
10. Gupta, A., Nagarajan, V., Ravi, R.: Thresholded covering algorithms for robust and max-min optimization. Math. Program. **146**(1–2), 583–615 (2014)
11. Gupta, A., Nagarajan, V., Ravi, R.: Robust and maxmin optimization under matroid and knapsack uncertainty sets. ACM Trans. Algorithms (TALG) **12**(1), 10 (2016)
12. Gupta, A., Pál, M., Ravi, R., Sinha, A.: Boosted sampling: approximation algorithms for stochastic optimization. In: Proceedings of the Thirty-Sixth Annual ACM Symposium on Theory of Computing, pp. 417–426 (2004)
13. Housni, O.E., Goyal, V., Shmoys, D.: On the power of static assignment policies for robust facility location problems. arXiv preprint arXiv:2011.04925 (2020)
14. Immorlica, N., Karger, D., Minkoff, M., Mirrokni, V.S.: On the costs and benefits of procrastination: approximation algorithms for stochastic combinatorial optimization problems. In: Proceedings of the Fifteenth Annual ACM-SIAM Symposium on Discrete Algorithms, pp. 691–700 (2004)
15. Khandekar, R., Kortsarz, G., Mirrokni, V., Salavatipour, M.R.: Two-stage robust network design with exponential scenarios. Algorithmica **65**(2), 391–408 (2013)
16. Linhares, A., Swamy, C.: Approximation algorithms for distributionally-robust stochastic optimization with black-box distributions. In: Proceedings of the 51st Annual ACM Symposium on Theory of Computing, pp. 768–779 (2019)
17. Ravi, R., Sinha, A.: Hedging uncertainty: approximation algorithms for stochastic optimization problems. In: Bienstock, D., Nemhauser, G. (eds.) IPCO 2004. LNCS, vol. 3064, pp. 101–115. Springer, Heidelberg (2004). https://doi.org/10.1007/978-3-540-25960-2_8

18. Shmoys, D.B.: Approximation algorithms for facility location problems. In: Proceedings of the Third International Workshop on Approximation Algorithms for Combinatorial Optimization, pp. 27–33 (2000)
19. Shmoys, D.B., Tardos, É., Aardal, K.: Approximation algorithms for facility location problems. In: Proceedings of the Twenty-Ninth Annual ACM Symposium on Theory of Computing, pp. 265–274 (1997)
20. Snyder, L.V.: Facility location under uncertainty: a review. IIE Trans. **38**(7), 547–564 (2006)
21. Swamy, C., Shmoys, D.B.: Approximation algorithms for 2-stage stochastic optimization problems. ACM SIGACT News **37**(1), 33–46 (2006)

Robust k-Center with Two Types of Radii

Deeparnab Chakrabarty$^{(\boxtimes)}$ and Maryam Negahbani$^{(\boxtimes)}$

Dartmouth College, Hanover, NH 03755, USA
deeparnab@dartmouth.edu, maryam@cs.dartmouth.edu

Abstract. In the non-uniform k-center problem, the objective is to cover points in a metric space with specified number of balls of different radii. Chakrabarty, Goyal, and Krishnaswamy [ICALP 2016, Trans. on Algs. 2020] (CGK, henceforth) give a constant factor approximation when there are two types of radii. In this paper, we give a constant factor approximation for the two radii case in the presence of outliers. To achieve this, we need to bypass the technical barrier of bad integrality gaps in the CGK approach. We do so using "the ellipsoid method inside the ellipsoid method": use an outer layer of the ellipsoid method to reduce to stylized instances and use an inner layer of the ellipsoid method to solve these specialized instances. This idea is of independent interest and could be applicable to other problems.

Keywords: Approximation · Clustering · Outliers · Round-or-Cut

1 Introduction

In the non-uniform k-center (NUkC) problem, one is given a metric space (X, d) and balls of different radii $r_1 > \cdots > r_t$, with k_i balls of radius type r_i. The objective is to find a placement $C \subseteq X$ of centers of these $\sum_i k_i$ balls, such that they cover X with as little dilation as possible. More precisely, for every point $x \in X$ there must exist a center $c \in C$ of some radius type r_i such that $d(x, c) \leq \alpha \cdot r_i$ and the objective is to find C with α as small as possible.

Chakrabarty, Goyal, and Krishnaswamy [10] introduced this problem as a generalization to the vanilla k-center problem [16–18] which one obtains with only one type of radius. One motivation arises from source location and vehicle routing: imagine you have a fleet of t-types of vehicles of different speeds and your objective is to find depot locations so that any client point can be served as fast as possible. This can be modeled as an NUkC problem. The second motivation arises in clustering data. The k-center objective forces one towards clustering with equal sized balls, while the NUkC objective gives a more nuanced way to model the problem. Indeed, NUkC generalizes the *robust k-center problem* [13] which allows the algorithm to throw away z points as outliers. This is precisely the NUkC problem with two types of radii, $r_1 = 1$, $k_1 = k$, $r_2 = 0$, and $k_2 = z$.

Partially supported by NSF grant #1813053.

M. Singh and D. P. Williamson (Eds.): IPCO 2021, LNCS 12707, pp. 268–282, 2021.
https://doi.org/10.1007/978-3-030-73879-2_19

Chakrabarty et al. [10] give a 2-approximation for the special case of robust k-center which is the best possible [16,17]. Furthermore, they give a $(1 + \sqrt{5})$-factor approximation algorithm for the NUkC problem with two types of radii (henceforth, the 2-NUkC problem). [10] also prove that when t, the number of types of radii, is part of the input, there is no constant factor approximation algorithms unless P = NP. They explicitly leave open the case when the number of different radii types is a constant, conjecturing that constant-factor approximations should be possible. We take the first step towards this by looking at the *robust* 2-NUkC problem. That is, the NUkC problem with two kinds of radii when we can throw away z outliers. This is the case of 3-radii with $r_3 = 0$.

Theorem 1. *There is a 10-approximation for the* Robust 2-NUkC *problem.*

Although the above theorem seems a modest step towards the CGK conjecture, it is in fact a non-trivial one which bypasses multiple technical barriers in the [10] approach. To do so, our algorithm applies a two-layered round-or-cut framework, and it is foreseeable that this idea will form a key ingredient for the constantly many radii case as well. In the rest of this section, we briefly describe the [10] approach, the technical bottlenecks one faces to move beyond 2 types of radii, and our approach to bypass them. A more detailed description appears in Sect. 2.

One key observation of [10] connects NUkC with the *firefighter problem on trees* [1,12,15]. In the latter problem, one is given a tree where there is a fire at the root. The objective is to figure out if a specified number of firefighters can be placed in each layer of the tree, so that the leaves can be saved. To be precise, the objective is to select k_i nodes from layer i of the tree so that every leaf-to-root path contains at least one of these selected nodes.

Chakrabarty et al. [10] use the *integrality* of a natural LP relaxation for the firefighter problem on height-2 trees to obtain their constant factor approximation for 2-NUkC. In particular, they show how to convert a fractional solution of the standard LP relaxation of the 2-NUkC problem to a feasible fractional solution for the firefighter LP. Since the latter LP is integral for height-2 trees, they obtain an integral firefighting solution from which they construct an $O(1)$-approximate solution for the 2-NUkC problem. Unfortunately, this idea breaks down in the presence of outliers as the firefighter LP on height-2 trees *when certain leaves can be burnt* (outlier leaves, so to speak) is not integral anymore. In fact, the standard LP-relaxation for Robust 2-NUkC has unbounded integrality gap. This is the first bottleneck in the CGK approach.

Although the LP relaxation for the firefighter problem on height-2 trees is not integral when some leaves can be burnt, the problem itself (in fact for any constant height) is solvable in polynomial time using dynamic programming (DP). Using the DP, one can then obtain (see, for instance, [20]) a *polynomial sized integral* LP formulation for the firefighting problem. This suggests the following enhancement of the CGK approach using the ellipsoid method. Given a fractional solution $\mathbf{x}$ to Robust 2-NUkC, use the CGK approach to obtain a fractional solution $\mathbf{y}$ to the firefighting problem. If $\mathbf{y}$ is feasible for the integral LP formulation, then we get an integral solution to the firefighting problem which in turn gives an $O(1)$-approximation for the Robust 2-NUkC instance via the CGK approach. Otherwise, we would get a *separating hyperplane* for $\mathbf{y}$ and the poly-sized integral formulation for firefighting. If we could only use this to separate the fractional solution $\mathbf{x}$ from the integer hull of the Robust 2-NUkC problem, then we could use the ellipsoid method to approximate Robust 2-NUkC. This is the so-called "round-or-cut" technique in approximation algorithms.

Unfortunately, this method also fails and indicates a much more serious bottleneck in the CGK approach. Specifically, there is an instance of Robust 2-NUkC and an $\mathbf{x}$ in the integer hull of its solutions, such that the firefighting instance output by the CGK has *no* integral solution! Thus, one needs to enhance the CGK approach in order to obtain $O(1)$-approximations even for the Robust 2-NUkC problem. The main contribution of this paper is to provide such an approach. We show that if the firefighting instance does not have an integral solution, then we can tease out many stylized Robust 2-NUkC instances on which the round-or-cut method provably succeeds, and an $O(1)$-approximation to any one of them gives an $O(1)$-approximation to the original Robust 2-NUkC instance.

Our Approach. Any solution $\mathbf{x}$ in the integer hull of NUkC solutions gives an indication of where different radii centers are opened. As it turns out, the key factor towards obtaining algorithms for the Robust 2-NUkC problem is observing where the large radii (that is, radius r_1) balls are opened. Our first step is showing that if the fractional solution $\mathbf{x}$ tends to open the r_1-centers only on "well-separated" locations then in fact, the round-or-cut approach described above works. More precisely, if the Robust 2-NUkC instance is for some reason forced to open its r_1 centers on points which are at least cr_1 apart from each other for some constant $c > 4$, then the CGK approach plus round-or-cut leads to an $O(1)$-approximation for the Robust 2-NUkC problem. We stress that this is far from trivial and the natural LP relaxations have bad gaps even in this case. We use our approach from a previous paper [11] to handle these well-separated instances.

But how and why would such well-separated instances arise? This is where we use ideas from recent papers on fair colorful clustering [3,7,19]. If $\mathbf{x}$ suggested that the r_1-radii centers are not well-separated, then one does not need that many balls if one allows dilation. In particular, if p and q are two r_1-centers of a feasible integral solution, and $d(p,q) \leq cr_1$, then just opening one ball at either p or q with radius $(c + 1)r_1$ would cover every point that they each cover with radius r_1-balls. Thus, in this case, the approximation algorithm gets a "saving" in the budget of how many balls it can open. We exploit this savings in the budget by utilizing yet another observation from Adjiashvili, Baggio, and Zenklusen [1] on the natural LP relaxation for the firefighter problem on trees. This asserts that although the natural LP relaxation for constant height trees is not integral, one can get integral solutions by violating the constraints *additively* by a constant. The aforementioned savings allow us to get a solution without violating the budget constraints.

In summary, given an instance of the Robust 2-NUkC problem, we run an outer round-or-cut framework and use it to check whether an instance is well-separated or not. If not, we straightaway get an approximate solution via the CGK approach and the ABZ observation. Otherwise, we use enumeration (similar to [3]) to obtain $O(n)$ many different *well-separated* instances and for each, run an inner round-or-cut framework. If any of these well-separated instances are feasible, we get an approximate solution for the initial Robust 2-NUkC instance. Otherwise, we can assert a separating hyperplane for the outer round-or-cut framework.

Related Work. NUkC was introduced in [10] as a generalization to the k-center problem [16–18] and the robust k-center problem [13]. In particular CGK reduce NUkC to the firefighter problem on trees which has constant approximations [1,12,15] and recently, a quasi-PTAS [23]. NUkC has also been studied in the *perturbation resilient* [4,5,14] settings. An instance is ρ-perturbation resilient if the optimal clustering does not change even when the metric is perturbed up to factor ρ. Bandapadhyay [6] gives an exact polynomial time algorithm for 2-perturbation resilient instances with constant number of radii.

As mentioned above, part of our approach is inspired by ideas from fair colorful k-center clustering [3,7,19] problems studied recently. In this problem, the points are divided into t color classes and we are asked to cover m_i, $i \in \{1, \ldots, t\}$ many points from each color by opening k-centers. The idea of moving to well-separated instances are present in these papers. We should mention, however, that the problems are different, and their results do not imply ours.

The round-or-cut framework is a powerful approximation algorithm technique first used in a paper by Carr *et al.* [8] for the minimum knapsack problem, and since then has found use in other areas such as network design [9] and clustering [2,3,11,21,22]. Our multi-layered round-or-cut approach may find uses in other optimization problems as well.

2   Detailed Description of Our Approach

In this section, we provide the necessary technical preliminaries required for proving Theorem 1 and give a more detailed description of the CGK bottleneck and our approach. We start with notations. Let (X, d) be a metric space on a set of points X with distance function $d : X \times X \longrightarrow \mathbb{R}_{\geq 0}$ satisfying the triangle inequality. For any $u \in X$ we let $B(u, r)$ denote the set of points in a ball of radius r around u, that is, $B(u, r) = \{v \in X : d(u, v) \leq r\}$. For any set $U \subseteq X$ and function $f : U \to \mathbb{R}$, we use the shorthand notation $f(U) := \sum_{u \in U} f(u)$. For a set $U \subseteq X$ and any $v \in X$ we use $d(v, U)$ to denote $\min_{u \in U} d(v, u)$.

The 2-radii NUkC problem and the robust version are formally defined as follows.

Definition 1 (2-NUkC and Robust 2-NUkC). *The input to 2-NUkC is a metric space (X, d) along with two radii $r_1 > r_2 \geq 0$ with respective budgets $k_1, k_2 \in \mathbb{N}$. The objective of 2-NUkC is to find the minimum $\rho \geq 1$ for which* there exists subsets $S_1, S_2 \subseteq X$ such that (a) $|S_i| \leq k_i$ for $i \in \{1, 2\}$, and (b) $\bigcup_i \bigcup_{u \in S_i} B(u, \rho r_i) = X$. The input to Robust 2-NUkC contains an extra parameter $m \in \mathbb{N}$, and the objective is the same, except that condition (b) is changed to $|\bigcup_i \bigcup_{u \in S_i} B(u, \rho r_i)| \geq m$.

An instance $\mathcal{I}$ of Robust 2-NUkC is denoted as $((X, d), (r_1, r_2), (k_1, k_2), m)$. As is standard, we will focus on the *approximate feasibility* version of the problem. An algorithm for this problem takes input an instance $\mathcal{I}$ of Robust 2-NUkC, and either asserts that $\mathcal{I}$ is *infeasible*, that is, there is no solution with $\rho = 1$, or provides a solution with $\rho \leq \alpha$. Using binary search, such an algorithm implies an α-approximation for Robust 2-NUkC.

Linear Programming Relaxations. The following is the natural LP relaxation for the feasibility version of Robust 2-NUkC. For every point $v \in X$, $\mathsf{cov}_i(v)$ denotes its *coverage* by balls of radius r_i. Variable $x_{i,u}$ denotes the extent to which a ball of radius r_i is open at point u. If instance $\mathcal{I}$ is feasible, then the following polynomial sized system of inequalities has a feasible solution.

$$\{(\mathsf{cov}_i(v) : v \in X, i \in \{1, 2\}) : \quad \sum_{v \in X} \mathsf{cov}(v) \geq m \qquad \text{(Robust 2-NU}k\text{C LP)}$$

$$\sum_{u \in X} x_{i,u} \leq k_i \qquad \forall i \in \{1, 2\}$$

$$\mathsf{cov}_1(v) = \sum_{u \in B(v, r_1)} x_{1,u}, \quad \mathsf{cov}_2(v) = \sum_{u \in B(v, r_2)} x_{2,u} \qquad \forall v \in X$$

$$\mathsf{cov}(v) = \mathsf{cov}_1(v) + \mathsf{cov}_2(v) \leq 1 \qquad \forall v \in X$$

$$x_{i,u} \geq 0 \qquad \forall i \in \{1, 2\}, \forall u \in X\}$$

For our algorithm, we will work with the following integer hull of all possible fractional coverages. Fix a Robust 2-NUkC instance $\mathcal{I} = ((X, d), (r_1, r_2), (k_1, k_2), m)$ and let $\mathscr{F}$ be the set of all tuples of subsets (S_1, S_2) with $|S_i| \leq k_i$. For $v \in X$ and $i \in \{1, 2\}$, we say $\mathscr{F}$ covers v with radius r_i if $d(v, S_i) \leq r_i$. Let $\mathscr{F}_i(v) \subseteq \mathscr{F}$ be the subset of solutions that cover v with radius r_i. Moreover, we would like $\mathscr{F}_1(v)$ and $\mathscr{F}_2(v)$ to be disjoint, so if $S \in \mathscr{F}_1(v)$, we do not include it in $\mathscr{F}_2(v)$. The following is the integer hull of the coverages. If $\mathcal{I}$ is feasible, there must exist a solution in $\mathscr{P}^{\mathcal{I}}_{\mathsf{cov}}$.

$$\{(\mathsf{cov}_i(v) : v \in X, i \in \{1, 2\}) : \sum_{v \in X} (\mathsf{cov}_1(v) + \mathsf{cov}_2(v)) \geq m \qquad (\mathscr{P}^{\mathcal{I}}_{\mathsf{cov}})$$

$$\forall v \in X, i \in \{1, 2\} \quad \mathsf{cov}_i(v) - \sum_{S \in \mathscr{F}_i(v)} z_S = 0 \qquad (\mathscr{P}^{\mathcal{I}}_{\mathsf{cov}}.1)$$

$$\sum_{S \in \mathscr{F}} z_S = 1 \qquad (\mathscr{P}^{\mathcal{I}}_{\mathsf{cov}}.2)$$

$$\forall S \in \mathscr{F} \qquad z_S \geq 0\} \qquad (\mathscr{P}^{\mathcal{I}}_{\mathsf{cov}}.3)$$

Fact 1. $\mathscr{P}_{cov}^{\mathcal{I}}$ *lies inside* Robust 2-NUkC *LP.*

Firefighting on Trees. As described in Sect. 1, the CGK approach [10] is via the firefighter problem on trees. Since we only focus on Robust 2-NUkC, the relevant problem is the *weighted* 2-level fire fighter problem. The input includes a set of height-2 trees (stars) with root nodes L_1 and leaf nodes L_2. Each leaf $v \in L_2$ has a parent $p(v) \in L_1$ and an integer weight $\mathsf{w}(v) \in \mathbb{N}$. We use $\mathsf{Leaf}(u)$ to denote the leaves connected to a $u \in L_1$ (that is, $\{v \in L_2 : p(v) = u\}$). Observe that $\{\mathsf{Leaf}(u) : u \in L_1\}$ partitions L_2. So we could represent the edges of the trees by this Leaf partition. Hence the structure is identified as $(L_1, L_2, \mathsf{Leaf}, \mathsf{w})$.

Definition 2. (2-Level Fire Fighter (2-FF) Problem). *Given height-2 trees* $(L_1, L_2, \mathsf{Leaf}, \mathsf{w})$ *along with budgets* $k_1, k_2 \in \mathbb{N}$, *a feasible solution is a pair* $T = (T_1, T_2)$, $T_i \subseteq L_i$, *such that* $|T_i| \leq k_i$ *for* $i \in \{1, 2\}$. *Let* $\mathcal{C}(T) := \{v \in L_2 : v \in T_2 \vee p(v) \in T_1\}$ *be the set of leaves covered by* T. *The objective is to maximize* $\mathsf{w}(\mathcal{C}(T))$. *Hence a 2-FF instance is represented by* $((L_1, L_2, \mathsf{Leaf}, \mathsf{w}), k_1, k_2)$.

The standard LP relaxation for this problem is quite similar to the Robust 2-NUkC LP. For each vertex $u \in L_1 \cup L_2$ there is a variable $0 \leq y_u \leq 1$ that shows the extent to which u is included in the solution. For a leaf v, $Y(v)$ is the fractional amount by which v is covered through both itself and its parent.

$$\max \sum_{v \in L_2} \mathsf{w}(v)Y(v) : \quad \sum_{u \in L_i} y_u \leq k_i, \quad \forall i \in \{1, 2\}; \tag{2-FF LP}$$

$$Y(v) := y_{p(v)} + y_v \leq 1, \quad \forall v \in L_2; \quad y_u \geq 0, \quad \forall u \in L_1 \cup L_2$$

Remark 1. The following figure shows an example where the above LP relaxation has an integrality gap. However, 2-FF can be solved via dynamic programming in $O(n^3)$ time and has similar sized integral LP relaxations.

Fig. 1. A 2-FF instance with budgets $k_1 = k_2 = 1$. Multiplicity w is 1 for the circle leaves and 3 for the triangles. The highlighted nodes have $y = 1/2$ and the rest of the nodes have $y = 0$. The objective value for this y is $4 \times 1/2 + 6 = 8$ but no integral solution can get an objective value of more than 7.

2.1 CGK's Approach and Its Shortcomings

Given fractional coverages $(\mathsf{cov}_1(v), \mathsf{cov}_2(v) : v \in X)$, the CGK algorithm [10] runs the classic clustering subroutine by Hochbaum and Shmoys [17] in a greedy fashion. In English, the Hochbaum-Shmoys (HS) routine partitions a metric space such that the representatives of each part are well-separated with respect to an input parameter. The CGK algorithm obtains a 2-FF instance by applying the HS routine twice. Once on the whole metric space in decreasing order of $\mathsf{cov}(v) = \mathsf{cov}_1(v) + \mathsf{cov}_2(v)$, and the set of representatives forms the leaf layer L_2 with weights being the size of the parts. The next time on L_2 itself in decreasing order of cov_1 and the representatives form the parent layer L_1. These subroutines and the subsequent facts form a part of our algorithm and analysis.

Algorithm 1. HS

Input: Metric (U, d), parameter $r \geq 0$, and assignment $\{\mathsf{cov}(v) \in \mathbb{R}_{\geq 0} : v \in U\}$
1: $R \leftarrow \emptyset$ ▷ The set of representatives
2: **while** $U \neq \emptyset$ **do**
3:    $u \leftarrow \arg\max_{v \in U} \mathsf{cov}(v)$ ▷ The first client in U in non-increasing cov order
4:    $R \leftarrow R \cup u$
5:    $\mathsf{Child}(u) \leftarrow \{v \in U : d(u, v) \leq r\}$ ▷ Points in U at distance r from u (including u itself)
6:    $U \leftarrow U \backslash \mathsf{Child}(u)$
7: **end while**
Output: $R, \{\mathsf{Child}(u) : u \in R\}$

Algorithm 2. CGK

Input: Robust 2-NUkC instance $((X, d), (r_1, r_2), (k_1, k_2), \mathsf{m})$, dilation factors $\alpha_1, \alpha_2 > 0$, and assignments $\mathsf{cov}_1(v), \mathsf{cov}_2(v) \in \mathbb{R}_{\geq 0}$ for all $v \in X$
1: $(L_2, \{\mathsf{Child}_2(v), v \in L_2\}) \leftarrow \mathsf{HS}((X, d), \alpha_2 r_2, \mathsf{cov} = \mathsf{cov}_1 + \mathsf{cov}_2)$
2: $(L_1, \{\mathsf{Child}_1(v), v \in L_1\}) \leftarrow \mathsf{HS}((L_2, d), \alpha_1 r_1, \mathsf{cov}_1)$
3: $\mathsf{w}(v) \leftarrow |\mathsf{Child}_2(v)|$ for all $v \in L_2$
4: $\mathsf{Leaf}(u) \leftarrow \mathsf{Child}_1(u)$ for all $u \in L_1$
Output: 2-FF instance $((L_1, L_2, \mathsf{Leaf}, \mathsf{w}), (k_1, k_2))$

Definition 3. (Valuable 2-FF instances). *We call an instance $\mathcal{T}$ returned by the CGK algorithm valuable if it has an integral solution of total weight at least m. Using dynamic programming, there is a polynomial time algorithm to check whether $\mathcal{T}$ is valuable.*

Fact 2. *The following are true regarding the output of HS: (a) $\forall u \in R, \forall v \in \mathsf{Child}(u) : d(u, v) \leq r$, (b) $\forall u, v \in R : d(u, v) > r$, (c) The set $\{\mathsf{Child}(u) : u \in R\}$ partitions U, and (d) $\forall u \in R, \forall v \in \mathsf{Child}(u) : \mathsf{cov}(u) \geq \mathsf{cov}(v)$.*

Lemma 1 (rewording of Lemma 3.4. in [10]). *Let $\mathcal{I}$ be a Robust 2-NUkC instance. If for any fractional coverages $(\mathrm{cov}_1(v), \mathrm{cov}_2(v))$ the instance 2-FF created by Algorithm 2 is valuable, then one obtains an $(\alpha_1 + \alpha_2)$-approximation for $\mathcal{I}$.*

Lemma 1 suggests that if we can find fractional coverages so that the corresponding 2-FF instance $\mathcal{T}$ is valuable, then we are done. Unfortunately, the example illustrated in Fig. 2 shows that for any (α_1, α_2) there exists Robust 2-NUkC instances and fractional coverages $(\mathrm{cov}_1(v), \mathrm{cov}_2(v)) \in \mathscr{P}_{\mathrm{cov}}^{\mathcal{I}}$ in the integer hull, for which the CGK algorithm returns 2-FF instances that are not valuable.

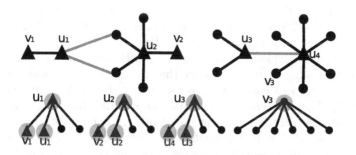

Fig. 2. At the top, there is a feasible Robust 2-NUkC instance with $k_1 = 2$, $k_2 = 3$, and m = 24. There are 6 triangles representing 3 collocated points each, along with 12 circles, each representing one point. The black edges are distance $r_1 > \alpha_2 r_2$ and the grey edges are distance $\alpha_1 r_1$. There are two integral solutions S and S' each covering exactly 24 points. $S_1 = \{u_1, u_4\}$, $S_2 = \{u_2, v_2, u_3\}$, $S_1' = \{u_2, u_3\}$, and $S_2' = \{u_1, v_1, u_4\}$. Having $z_S = z_{S'} = 1/2$ in $\mathscr{P}_{\mathrm{cov}}^{\mathcal{I}}$, gives cov_1 of $1/2$ for all the points and cov_2 of $1/2$ for the triangles. The output of Algorithm 2 is the 2-FF instance at the bottom. According to Proposition 1 the highlighted nodes have $y = 1/2$ and the rest of the nodes have $y = 0$ with objective value $12 \times 1/2 + 18 = 24$ but no integral solution can get an objective value of more than 23. (Color figure online)

2.2   Our Idea

Although the 2-FF instance obtained by Algorithm 2 from fractional coverages $(\mathrm{cov}_1(v), \mathrm{cov}_2(v) : v \in X)$ may not be valuable, [10] proved that if these coverages come from (Robust 2-NUkC LP), then there is always a *fractional* solution to (2-FF LP) for this instance which has value at least m.

Proposition 1 (rewording of Lemma 3.1. in [10]). *Let $(\mathrm{cov}_1(v), \mathrm{cov}_2(v) : v \in X)$ be any feasible solution to Robust 2-NUkC LP. As long as $\alpha_1, \alpha_2 \geq 2$, the following is a fractional solution of 2-FF LP with value at least m for the 2-FF instance output by Algorithm 2.*

$$y_v = \begin{cases} \mathrm{cov}_1(v) & v \in L_1 \\ \min\{\mathrm{cov}_2(v), 1 - \mathrm{cov}_1(p(v))\} & v \in L_2. \end{cases}$$

Therefore, the problematic instances are precisely 2-FF instances that are integrality gap examples for (2-FF LP). Our first observation stems from what Adjiashvili, Baggio, and Zenklusen [1] call "the narrow integrality gap of the firefighter LP".

Lemma 2 (From Lemma 6 of [1]). *Any basic feasible solution* $\{y_v : i \in \{1,2\}, v \in L_i\}$ *of the 2-FF LP polytope has at most 2 loose variables. A variable* y_v *is loose if* $0 < y_v < 1$ *and* $y_{p(v)} = 0$ *in case* $v \in L_2$.

In particular, if $y(L_1) \leq k_1 - 2$, then the above lemma along with Proposition 1 implies there exists an *integral* solution with value $\geq m$. That is, the 2-FF instance is valuable. Conversely, the fact that the instance is *not* valuable asserts that $y(L_1) > k_1 - 2$ which in turn implies $\mathsf{cov}_1(L_1) > k_1 - 2$. In English, the fractional coverage puts a lot of weight on the points in L_1.

This is where we exploit the ideas in [3,7,19]. By choosing $\alpha_1 > 2$ to be large enough in Proposition 1, we can ensure that points in L_1 are "well-separated". More precisely, we can ensure for any two $u, v \in L_1$ we have $d(u,v) > \alpha_1 r_1$ (from Fact 2). The well-separated condition implies that the same center cannot be fractionally covering two different points in L_1. Therefore, $\mathsf{cov}_1(L_1) > k_1 - 2$ if $(\mathsf{cov}_1, \mathsf{cov}_2) \in \mathscr{P}_{\mathsf{cov}}^{\mathcal{I}}$ is in the integer hull, then there must exist an integer solution which opens *at most* 1 center that does *not* cover points in L_1. For the time being assume in fact no such center exists and $\mathsf{cov}_1(L_1) = k_1$. Indeed, the integrality gap example in Fig. 2 satisfies this equality.

Our last piece of the puzzle is that if the cov_1's are concentrated on separated points, then indeed we can apply the round-or-cut framework to obtain an approximation algorithm. To this end, we make the following definition, and assert the following theorem.

Definition 4 (Well-Separated Robust 2-NUkC). *The input is the same as Robust 2-NUkC, along with* $Y \subseteq X$ *where* $d(u,v) > 4r_1$ *for all pairs* $u, v \in Y$, *and the algorithm is allowed to open the radius* r_1-*centers only on points in* Y.

Theorem 2. *Given a Well-Separated Robust 2-NUkC instance there is a polynomial time algorithm using the ellipsoid method that either gives a 4-approximate solution, or proves that the instance is infeasible.*

We remark the natural (Robust 2-NUkC LP) relaxation still has a bad integrality gap, and we need the round-or-cut approach. Formally, given fractional coverages $(\mathsf{cov}_1, \mathsf{cov}_2)$ we run Algorithm 2 (with $\alpha_1 = \alpha_2 = 2$) to get a 2-FF instance. If the instance is valuable, we are done by Lemma 1. Otherwise, we prove that $(\mathsf{cov}_1, \mathsf{cov}_2) \notin \mathscr{P}_{\mathsf{cov}}^{\mathcal{I}}$ by exhibiting a separating hyperplane. This crucially uses the well-separated-ness of the instance and indeed, the bad example shown in Fig. 2 is *not* well-separated. This implies Theorem 2 using the ellipsoid method.

In summary, to prove Theorem 1, we start with $(\mathsf{cov}_1, \mathsf{cov}_2)$ purported to be in $\mathscr{P}_{\mathsf{cov}}^{\mathcal{I}}$. Our goal is to either get a constant approximation, or separate $(\mathsf{cov}_1, \mathsf{cov}_2)$ from $\mathscr{P}_{\mathsf{cov}}^{\mathcal{I}}$. We first run the CGK Algorithm 2 with $\alpha_1 = 8$ and $\alpha_2 = 2$. If $\mathsf{cov}_1(L_1) \leq k_1 - 2$, we can assert that the 2-FF instance is valuable and get a 10-approximation. Otherwise, $\text{cov}_1(L_1) > k_1 - 2$, and we guess the $O(n)$ many possible centers "far away" from L_1, and obtain that many well-separated instances. We run the algorithm promised by Theorem 2 on each of them. If any one of them gives a 4-approximate solution, then we immediately get an 8-approximate[1] solution to the original instance. If *all* of them fail, then we can assert $\text{cov}_1(L_1) \leq k_1 - 2$ must be a *valid* inequality for $\mathscr{P}_{\text{cov}}^{\mathcal{I}}$, and thus obtain a hyperplane separating $(\text{cov}_1, \text{cov}_2)$ from $\mathscr{P}_{\text{cov}}^{\mathcal{I}}$. The polynomial running time is implied by the ellipsoid algorithm. Note that there are two nested runs of the ellipsoid method in the algorithm. Figure 3 below shows an illustration of the ideas.

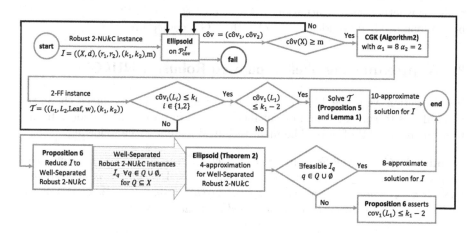

Fig. 3. Our framework for approximating Robust 2-NUkC. The three black arrows each represent separating hyperplanes we feed to the outer ellipsoid. The box in the bottom row stating "4-approximation for well-separated Robust 2-NUkC" runs the inner ellipsoid method.

2.3 Discussion

Before we move to describing algorithms proving Theorem 2 and Theorem 1, let us point out why the above set of ideas does *not* suffice to prove the full CGK conjecture, that is, give an $O(1)$-approximation for NUkC with constant many type of radii. Given fractional coverages, the CGK algorithm now returns a t-layered firefighter instance and again if such an instance is valuable (which can be checked in $n^{O(t)}$ time), we get an $O(1)$-approximation. As above, the main challenge is when the firefighter instance is not valuable. Theorem 2, in fact, does generalize if *all* layers are separated. Formally, if there are t types of radii, and there are t sets $Y_1, \ldots, Y_t$ such that (a) any two points $p, q \in Y_i$ are well-separated, that is, $d(p, q) > 4r_i$, and (b) the r_i-radii centers are only

---

[1] The factor doubles as we need to double the radius, but that is a technicality.

allowed to be opened in Y_i, then in fact there is an $O(1)$-approximation for such instances. Furthermore, if we had fractional coverages $(\mathsf{cov}_1, \mathsf{cov}_2, \ldots, \mathsf{cov}_t)$ such that in the t-layered firefighter instance returned, *all* layers have "slack", that is $\mathsf{cov}_i(L_i) \leq k_i - t$, then one can repeatedly use Lemma 2 to show that the tree instance is indeed valuable.

The issue we do not know how to circumvent is when some layers have slack and some layers do not. In particular, even with 3 kinds of radii, we do not know how to handle the case when the first layer L_1 is well-separated and $\mathsf{cov}_1(L_1) = k_1$, but the second layer has slack $\mathsf{cov}_2(L_2) \leq k_2 - 3$. Lemma 2 does not help since all the loose vertices may be in L_1, but they cannot all be picked without violating the budget. At the same time, we do not know how to separate such cov's, or whether such a situation arises when cov's are in the integer hull. We believe one needs more ideas to resolve the CGK conjecture.

3   Approximating **Well-Separated Robust 2-NUkC**

In this section we prove Theorem 2 stated in Sect. 2.2. As mentioned there, the main idea is to run the round-or-cut method, and in particular use ideas from a previous paper [11] of ours. The main technical lemma is the following.

Lemma 3. *Given* **Well-Separated Robust** *2-NUkC instance $\mathcal{I}$ and fractional coverages* $(\hat{\mathsf{cov}}_1(v), \hat{\mathsf{cov}}_2(v))$, *if the output of the CGK Algorithm 2 is not valuable, there is a hyperplane separating* $(\hat{\mathsf{cov}}_1(v), \hat{\mathsf{cov}}_2(v))$ *from* $\mathscr{P}^{\mathcal{I}}_{\mathsf{cov}}$. *Furthermore, the coefficients of this hyperplane are bounded in value by* $|X|$.

Remark 2. We need to be careful in one place. Recall that HS is used in the CGK Algorithm 2. We need to assert in HS, that points u with $d(u, Y) \leq r_1$ are prioritized over points v with $d(v, Y) > r_1$ to be taken in L_1. This is w.l.o.g. since $\mathsf{cov}_1(v) = 0$ if $d(v, Y) > r_1$ by definition of **Well-Separated Robust 2-NUkC**. Using the ellipsoid method, the above lemma implies Theorem 2 (See the full version of the paper for a detailed proof). The rest of this section is dedicated to proving Lemma 3. Fix a well-separated **Robust 2-NUkC** instance $\mathcal{I}$. Recall that $Y \subseteq X$ is a subset of points, and the radius r_1 centers are only allowed to be opened at Y. Let $\mathcal{T}$ be the 2-FF instance output by Algorithm 2 on $\mathcal{I}$ and cov with $\alpha_1 = \alpha_2 = 2$. Recall, $\mathcal{T} = ((L_1, L_2, \mathsf{Leaf}, \mathsf{w}), k_1, k_2)$. The key part of the proof is the following valid inequality in case $\mathcal{T}$ is not valuable.

Lemma 4. *If $\mathcal{T}$ is not valuable* $\sum_{v \in L_2} \mathsf{w}(v)\mathsf{cov}(v) \leq \mathsf{m} - 1$ *for any* $\mathsf{cov}(v) \in \mathscr{P}^{\mathcal{I}}_{\mathsf{cov}}$.

Before we prove Lemma 4, let us show how it proves Lemma 3. Given $(\hat{\mathsf{cov}}_1, \hat{\mathsf{cov}}_2)$ we first check[2] that $\sum_{u \in X} \hat{\mathsf{cov}}(u) \geq \mathsf{m}$, or otherwise that would be the hyperplane separating it from $\mathscr{P}^{\mathcal{I}}_{\mathsf{cov}}$. Now recall that in Algorithm 2, for $v \in L_2$,

---

[2] recall, $\hat{\mathsf{cov}}(v) = \hat{\mathsf{cov}}_1(v) + \hat{\mathsf{cov}}_2(v)$.

$w(v) = |\mathsf{Child}_2(v)|$ which is the number of points assigned to v by HS (see Line 1 of Algorithm 2). By definition of w and then parts d) and c) of Fact 2,

$$\sum_{v \in L_2} \mathsf{w}(v) \mathrm{c\hat{o}v}(v) = \sum_{v \in L_2} \sum_{u \in \mathsf{Child}(v)} \mathrm{c\hat{o}v}(v) \geq \sum_{v \in L_2} \sum_{u \in \mathsf{Child}(v)} \mathrm{c\hat{o}v}(u) = \sum_{u \in X} \mathrm{c\hat{o}v}(u) \geq \mathsf{m}.$$

That is, $(\mathrm{c\hat{o}v}_1, \mathrm{c\hat{o}v}_2)$ violates the valid inequality asserted in Lemma 4, and this would complete the proof of Lemma 3. All that remains is to prove the valid inequality lemma above.

Proof (of Lemma 4). Fix a solution $\mathsf{cov} \in \mathscr{P}^{\mathcal{I}}_{\mathsf{cov}}$ and note that this is a convex combination of coverages induced by integral feasible solutions in $\mathscr{F}$. The main idea of the proof is to use the solutions in $\mathscr{F}$ to construct *solutions* to the tree instance $\mathcal{T}$. Since $\mathcal{T}$ is *not* valuable, each of these solutions will have "small" value, and then we use this to prove the lemma. To this end, fix $S = (S_1, S_2) \in \mathscr{F}$ where $|S_i| \leq k_i$ for $i \in \{1, 2\}$. The corresponding solution $T = (T_1, T_2)$ for $\mathcal{T}$ is defined as follows: For $i \in \{1, 2\}$ and any $u \in L_i$, u is in T_i iff $S_i \in \mathscr{F}_i(u)$. That is, $d(u, S_i) \leq r_i$.

Proposition 2. *T satisfies the budget constraints $|T_i| \leq k_i$ for $i \in \{1, 2\}$.*

The next claim is the only place where we need the well-separated-ness of $\mathcal{I}$. Basically, we will argue that the leaves covered by T_1 capture all the points covered by S_1.

Proposition 3. *If $u \in L_1$ but $u \notin T_1$ then no $v \in \mathsf{Leaf}(u)$ can be covered by a ball of radius r_1 in S_1.*

Next, we can prove that overall, the leaves covered by T capture the whole set of points covered by S. Recall that $\mathcal{C}(T) = \{v \in L_2 : v \in T_2 \vee p(v) \in T_1\}$ is the set of leaves covered by T. For $v \in X$ let $\mathscr{F}(v) := \mathscr{F}_1(v) \cup \mathscr{F}_2(v)$ be the set of solutions that cover v.

Proposition 4. *Take 2-FF solution T corresponding to Well-Separated Robust 2-NUkC solution S as described earlier. We have:*

$$\sum_{v \in L_2 : S \in \mathscr{F}(v)} \mathsf{w}(v) \leq \mathsf{w}(\mathcal{C}(T)).$$

That is, the total w of the points covered by S is at most $\mathsf{w}(\mathcal{C}(T))$.

The proof of Lemma 4 now follows from the fact that $\mathcal{T}$ is not valuable thus $\mathsf{w}(\mathcal{C}(T)) \leq \mathsf{m} - 1$ and therefore, for any $S \in \mathscr{F}$ we have $\sum_{v \in L_2 : S \in \mathscr{F}(v)} \mathsf{w}(v) \leq \mathsf{m} - 1$. So we have:

$$\sum_{v \in L_2} \mathsf{w}(v)\mathsf{cov}(v) \;=_{(\mathscr{P}^{\mathcal{I}}_{\mathsf{cov}}.1)} \; \sum_{v \in L_2} \mathsf{w}(v) \sum_{S \in \mathscr{F}(v)} z_S \;=\; \sum_{S \in \mathscr{F}} z_S \sum_{\substack{v \in L_2: \\ S \in \mathscr{F}(v)}} \mathsf{w}(v)$$

$$\leq \; (\mathsf{m} - 1) \sum_{S \in \mathscr{F}} z_S \;=_{(\mathscr{P}^{\mathcal{I}}_{\mathsf{cov}}.2)} \; \mathsf{m} - 1.$$

$\square$

Proofs of Propositions 2 to 4 can be found in the full version of the paper.

4   The Main Algorithm: Proof of Theorem 1

As mentioned in Sect. 2, we focus on the feasibility version of the problem: given an instance $\mathcal{I}$ of Robust 2-NUkC we either want to prove it is infeasible, that is, there are no subsets $S_1, S_2 \subseteq X$ with (a) $|S_i| \leq k_i$ and (b) $|\bigcup_i \bigcup_{u \in S_i} B(u, r_i)| \geq$ m, or give a 10-approximation that is, open subsets S_1, S_2 that satisfy (a) and $|\bigcup_i \bigcup_{u \in S_i} B(u, 10 r_i)| \geq$ m. To this end, we apply the round-or-cut methodology on $\mathcal{P}^{\mathcal{I}}_{\mathsf{cov}}$. Given a purported côv $:= (\mathrm{c\hat{o}v}_1(v), \mathrm{c\hat{o}v}_2(v) : v \in X)$ we want to either use it to get a 10-approximate solution, or find a hyperplane separating it from $\mathcal{P}^{\mathcal{I}}_{\mathsf{cov}}$. Furthermore, we want the coefficients in the hyperplane to be poly-bounded. Using the ellipsoid method we indeed get a polynomial time algorithm thereby proving Theorem 1.

Upon receiving côv, we first check whether $\mathrm{c\hat{o}v}(X) \geq$ m or not, and if not that will be the separating hyperplane. Henceforth, we assume this holds. Then, we run CGK Algorithm 2 with $\alpha_1 = 8$ and $\alpha_2 = 2$ to get 2-FF instance $\mathcal{T} = ((L_1, L_2, \mathsf{Leaf}, \mathsf{w}), (k_1, k_2))$. Let $\{y_v : v \in L_1 \cup L_2\}$ be the solution described in Proposition 1. Next, we check if $\mathrm{c\hat{o}v}_i(L_i) = y(L_i) \leq k_i$ for both $i \in \{1, 2\}$; if not, by Proposition 1 that hyperplane would separate côv from $\mathcal{P}^{\mathcal{I}}_{\mathsf{cov}}$ (and even Robust 2-NUkC LP in fact). The algorithm then branches into two cases.

Case I: $y(L_1) \leq k_1 - 2$. In this case, we assert that $\mathcal{T}$ is valuable, and therefore by Lemma 1 we get an $\alpha_1 + \alpha_2 = 10$-approximate solution for $\mathcal{I}$ via Lemma 1, and we are done.

Proposition 5. *If* $y(L_1) \leq k_1 - 2$, *then there is an integral solution T for $\mathcal{T}$ with* $\mathsf{w}(\mathcal{C}(T)) \geq$ m.

Proof. Since $y(L_1) \leq k_1 - 2$, we see that there is a feasible solution to the slightly revised LP below.

$$\max \sum_{v \in L_2} \mathsf{w}(v) Y(v) : \quad \sum_{u \in L_1} y_u \leq k_1 - 2, \quad \sum_{u \in L_2} y_u \leq k_2,$$

$$Y(v) := y_{p(v)} + y_v \leq 1, \ \forall v \in L_2$$

Consider a basic feasible solution $\{y'_v : v \in L_1 \cup L_2\}$ for this LP, and let $T_1 := \{v \in L_1 : y'_v > 0\}$. By definition $y'(T_1) = y'(L_1) \leq k_1 - 2$. According to Lemma 2, there are at most 2 loose variables in y'. So there are at most 2 fractional vertices in T_1. This implies $|T_1| \leq k_1$. Let U be the set of leaves that are not covered by T_1, that is, $U := \{v \in L_2 : p(v) \notin T_1\}$. Let T_2 be the top k_2 members of U according to decreasing w order. We return $T = (T_1, T_2)$.

We claim T has value at least m, that is, $\mathsf{w}(\mathcal{C}(T)) \geq$ m. Note that $\mathsf{w}(\mathcal{C}(T)) = \mathsf{w}(T_2) + \sum_{u \in T_1} \mathsf{w}(\mathsf{Leaf}(u))$. By the greedy choice of T_2, $\mathsf{w}(T_2) \geq \sum_{v \in U} \mathsf{w}(v) y'_v$. Since $y'_{p(v)} = 0$ for any $v \in U$, we have $\mathsf{w}(T_2) \geq \sum_{v \in U} \mathsf{w}(v) y'_v = \sum_{v \in U} \mathsf{w}(v) Y'(v)$. Furthermore, by definition, $\sum_{u \in T_1} \mathsf{w}(\mathsf{Leaf}(u)) = \sum_{v \in L_2 \setminus U} \mathsf{w}(v)$ which in turn is at least $\sum_{v \in L_2 \setminus U} \mathsf{w}(v) Y'(v)$. Adding up proves the claim as the objective value is at least m.

$$\mathsf{w}(\mathcal{C}(T)) \geq \sum_{v \in U} \mathsf{w}(v) Y'(v) + \sum_{v \in L_2 \setminus U} \mathsf{w}(v) Y'_v = \sum_{v \in L_2} \mathsf{w}(v) Y'(v) \geq \mathsf{m}.$$

Case II, $y(L_1) > k_1 - 2$. In this case, we either get an 8-approximation or prove that the following is a valid inequality which will serve as the separating hyperplane (recall $\hat{cov}_1(L_1) = y(L_1)$).

$$cov_1(L_1) \leq k_1 - 2. \tag{1}$$

To do so, we need the following proposition which formalizes the idea stated in Sect. 2.2 that in case II, we can enumerate over $O(|X|)$ many *well-separated instances*.

Proposition 6. *Let* $(cov_1, cov_2) \in \mathscr{P}_{cov}^{\mathcal{I}}$ *be fractional coverages and suppose there is a subset* $Y \subseteq X$ *with* $d(u,v) > 8r_1$ *for all* $u,v \in Y$. *Then either* $cov_1(Y) \leq k_1 - 2$, *or at least one of the following Well-Separated Robust 2-NUkC instances are feasible*

$$\mathcal{I}_\emptyset := ((X,d),(2r_1,r_2),(k_1,k_2),Y,\mathsf{m})$$
$$\mathcal{I}_q := ((X\backslash B(q,r_1),d),(2r_1,r_2),(k_1-1,k_2),Y,\mathsf{m}-|B(q,r_1)|) \qquad \forall q \in X :$$
$$d(q,Y) > r_1.$$

The proof is left to the full version of the paper. Now we use the above proposition to complete the proof of Theorem 1. We let $Y := L_1$, and obtain the instances $\mathcal{I}_\emptyset$ and $\mathcal{I}_q$'s as mentioned in the proposition. We apply the algorithm in Theorem 2 on each of them. If any of them returns a solution, then we have an 8-approximation. More precisely, if $\mathcal{I}_\emptyset$ is feasible, Theorem 2 gives a 4-approximation for it which is indeed an 8-approximation for $\mathcal{I}$ (the extra factor 2 is because $\mathcal{I}_\emptyset$ uses $2r_1$ as its largest radius). If $\mathcal{I}_q$ is feasible for some $q \in X$ and Theorem 2 gives us a 4-approximate solution $S' = (S_1', S_2')$ for it and $S = (S_1' \cup \{q\}, S_2')$ is an 8-approximation for $\mathcal{I}$. If none of them feasible, then we see that $cov_1(L_1) \leq k_1 - 2$ indeed serves as a separating hyperplane between $\hat{cov}$ and $\mathscr{P}_{cov}^{\mathcal{I}}$. This ends the proof of Theorem 1.

References

1. Adjiashvili, D., Baggio, A., Zenklusen, R.: Firefighting on trees beyond integrality gaps. ACM Trans. Algorithms (TALG) **15**(2), 20 (2018). Also appeared in Proc., SODA 2017
2. An, H., Singh, M., Svensson, O.: LP-based algorithms for capacitated facility location. SIAM J. Comput. (SICOMP) **46**(1), 272–306 (2017). Also appeared in Proc., FOCS 2014
3. Anegg, G., Angelidakis, H., Kurpisz, A., Zenklusen, R.: A technique for obtaining true approximations for k-center with covering constraints. In: Proceedings, MPS Conference on Integer Programming and Combinatorial Optimization (IPCO), pp. 52–65 (2020)
4. Angelidakis, H., Makarychev, K., Makarychev, Y.: Algorithms for stable and perturbation-resilient problems. In: Proceedings, ACM Symposium on Theory of Computing (STOC), pp. 438–451 (2017)
5. Awasthi, P., Blum, A., Sheffet, O.: Center-based clustering under perturbation stability. Inf. Process. Lett. **112**(1–2), 49–54 (2012)

6. Bandyapadhyay, S.: On Perturbation Resilience of Non-Uniform k-Center. In: Proceedings, International Workshop on Approximation Algorithms for Combinatorial Optimization Problems (APPROX) (2020)

7. Bandyapadhyay, S., Inamdar, T., Pai, S., Varadarajan, K.R.: A constant approximation for colorful k-center. In: Proceedings, European Symposium on Algorithms (ESA), pp. 12:1–12:14 (2019)

8. Carr, R.D., Fleischer, L.K., Leung, V.J., Phillips, C.A.: Strengthening integrality gaps for capacitated network design and covering problems. In: Proceedings, ACM-SIAM Symposium on Discrete Algorithms (SODA), pp. 106–115 (2000)

9. Chakrabarty, D., Chekuri, C., Khanna, S., Korula, N.: Approximability of capacitated network design. Algorithmica **72**(2), 493–514 (2015). Also appeared in Proc., IPCO 2011

10. Chakrabarty, D., Goyal, P., Krishnaswamy, R.: The non-uniform k-center problem. ACM Trans. Algorithms (TALG) **16**(4), 1–19 (2020). Also appeared in Proc. ICALP, 2016

11. Chakrabarty, D., Negahbani, M.: Generalized center problems with outliers. ACM Trans. Algorithms (TALG) **15**(3), 1–14 (2019). Also appeared in ICALP 2018

12. Chalermsook, P., Chuzhoy, J.: Resource minimization for fire containment. In: Proceedings, ACM-SIAM Symposium on Discrete Algorithms (SODA), pp. 1334–1349 (2010)

13. Charikar, M., Khuller, S., Mount, D.M., Narasimhan, G.: Algorithms for facility location problems with outliers. In: Proceedings, ACM-SIAM Symposium on Discrete Algorithms (SODA), pp. 642–651 (2001)

14. Chekuri, C.S., Gupta, S.: Perturbation resilient clustering for k-center and related problems via LP relaxations. In: Proceedings, International Workshop on Approximation Algorithms for Combinatorial Optimization Problems (APPROX), p. 9 (2018)

15. Finbow, S., King, A., MacGillivray, G., Rizzi, R.: The firefighter problem for graphs of maximum degree three. Discret. Math. **307**(16), 2094–2105 (2007)

16. Gonzalez, T.F.: Clustering to minimize the maximum intercluster distance. Theoret. Comput. Sci. **38**, 293–306 (1985)

17. Hochbaum, D.S., Shmoys, D.B.: A best possible heuristic for the k-center problem. Math. Oper. Res. 10(2), 180–184 (1985)

18. Hochbaum, D.S., Shmoys, D.B.: A unified approach to approximation algorithms for Bottleneck problems. J. ACM **33**(3), 533–550 (1986)

19. Jia, X., Sheth, K., Svensson, O.: Fair colorful k-center clustering. In: Proceedings, MPS Conference on Integer Programming and Combinatorial Optimization (IPCO), pp. 209–222 (2020)

20. Kaibel, V.: Extended formulations in combinatorial optimization. Optima **85**, 2–7 (2011)

21. Li, S.: On uniform capacitated k-median beyond the natural LP relaxation. In: Proceedings, ACM-SIAM Symposium on Discrete Algorithms (SODA), pp. 696–707 (2015)

22. Li, S.: Approximating capacitated k-median with $(1 + \varepsilon)k$ open facilities. In: Proceedings, ACM-SIAM Symposium on Discrete Algorithms (SODA), pp. 786–796 (2016)

23. Rahgoshay, M., Salavatipour, M.R.: Asymptotic quasi-polynomial time approximation scheme for resource minimization for fire containment. In: Proceedings, International Symposium on Theoretical Aspects of Computer Science (STACS) (2020)

Speed-Robust Scheduling
Sand, Bricks, and Rocks

Franziska Eberle[1(✉)], Ruben Hoeksma[2], Nicole Megow[1], Lukas Nölke[1], Kevin Schewior[3], and Bertrand Simon[4]

[1] Faculty of Mathematics and Computer Science, University of Bremen, Bremen, Germany
{feberle,nmegow,noelke}@uni-bremen.de
[2] Department of Applied Mathematics, University of Twente, Enschede, The Netherlands
r.p.hoeksma@utwente.nl
[3] Department of Mathematics and Computer Science, Universität zu Köln, Cologne, Germany
schewior@cs.uni-koeln.de
[4] IN2P3 Computing Center, CNRS, Villeurbanne, France
bertrand.simon@cc.in2p3.fr

Abstract. The speed-robust scheduling problem is a two-stage problem where, given m machines, jobs must be grouped into at most m bags while the processing speeds of the machines are unknown. After the speeds are revealed, the grouped jobs must be assigned to the machines without being separated. To evaluate the performance of algorithms, we determine upper bounds on the worst-case ratio of the algorithm's makespan and the optimal makespan given full information. We refer to this ratio as the robustness factor. We give an algorithm with a robustness factor $2 - \frac{1}{m}$ for the most general setting and improve this to 1.8 for equal-size jobs. For the special case of infinitesimal jobs, we give an algorithm with an optimal robustness factor equal to $\frac{e}{e-1} \approx 1.58$. The particular machine environment in which all machines have either speed 0 or 1 was studied before by Stein and Zhong (SODA 2019). For this setting, we provide an algorithm for scheduling infinitesimal jobs with an optimal robustness factor of $\frac{1+\sqrt{2}}{2} \approx 1.207$. It lays the foundation for an algorithm matching the lower bound of $\frac{4}{3}$ for equal-size jobs.

1 Introduction

Scheduling problems with incomplete knowledge of the input data have been studied extensively. There are different ways to model such uncertainty, the major frameworks being *online optimization*, where parts of the input are revealed incrementally, *stochastic optimization*, where parts of the input are modeled as random variables, and *robust optimization*, where uncertainty in the

Partially supported by the German Science Foundation (DFG) under contract ME 3825/1.

M. Singh and D. P. Williamson (Eds.): IPCO 2021, LNCS 12707, pp. 283–296, 2021.
https://doi.org/10.1007/978-3-030-73879-2_20

data is bounded. Most scheduling research in this context assumes uncertainty about the job characteristics. Examples include online scheduling, where the job set is a priori unknown [1,18], stochastic scheduling, where the processing times are modeled as random variables [17], robust scheduling, where the unknown processing times are within a given interval [14], two/multi-stage stochastic and robust scheduling [5,19], and scheduling with explorable execution times [8,15].

A lot less research addresses uncertainty about the machine environment, particularly, where the processing speeds of machines change in an unforeseeable manner. A majority of such research focuses on the special case of scheduling with unknown non-availability periods, that is, machines break down temporarily [2,7] or permanently [20]. Arbitrarily changing machine speeds have been considered for scheduling on a single machine [10,16].

We consider a two-stage robust scheduling problem with multiple machines of unknown speeds. Given n jobs and m machines, we ask for a partition of the jobs into m groups, we say *bags*, that have to be scheduled on the machines after their speeds are revealed without being split up. That is, in the second stage, when the machine speeds are known, a feasible schedule assigns jobs in the same bag to the same machine. The goal is to minimize the second-stage makespan.

More formally, we define the *speed-robust scheduling* problem as follows. We are given n jobs with processing times $p_j \geq 0$, for $j \in \{1, \dots, n\}$, and the number of machines, $m \in \mathbb{N}$. Machines run in parallel but their speeds are a priori unknown. In the first stage, the task is to group jobs into at most m bags. In the second stage, the machine speeds $s_i \geq 0$, for $i \in \{1, \dots, m\}$, are revealed. The time needed to execute job j on machine i is $\frac{p_j}{s_i}$, if $s_i > 0$. If a machine has speed $s_i = 0$, then it cannot process any job; we say the machine *fails*. Given the machine speeds, the second-stage task is to assign the bags to the machines such that the makespan $C_{\max}$ is minimized, where the makespan is the maximum sum of execution times of jobs assigned to the same machine.

Given a set of bags and machine speeds, the second-stage problem emerges as classical makespan minimization on related parallel machines. It is well-known that this problem can be solved arbitrarily close to optimality by polynomial-time approximation schemes [3,12,13]. As we are interested in the information-theoretic tractability and allow superpolynomial running times, ignoring any computational concern, we assume that the second-stage problem is solved optimally. Thus, an *algorithm* for speed-robust scheduling defines a job-to-bag allocation, i.e., it gives a solution to the first-stage problem. We may use non-optimal bag-to-machine allocations to simplify the analysis.

We evaluate the performance of algorithms by a worst-case analysis, comparing an algorithm's makespan with the optimal makespan achievable when machine speeds are known in advance. We say that an algorithm is *ρ-robust* if, for any instance, its makespan is within a factor $\rho \geq 1$ of the optimal solution. The *robustness factor* of the algorithm is defined as the infimum over all such ρ.

The special case of speed-robust scheduling with machine speeds in $\{0, 1\}$ has been studied by Stein and Zhong [20]. They introduced the problem with identical machines and an unknown number of machines that fail (speed 0) in the second stage. They present a simple lower bound of $\frac{4}{3}$ on the robustness factor with equal jobs and design a general $\frac{5}{3}$-robust algorithm. For infinitesimal jobs, they give a 1.2333-robust algorithm complemented by a lower bound for each number of machines which tends to $\frac{1+\sqrt{2}}{2} \approx 1.207$ for large m. Stein and Zhong also consider the objective of minimizing the maximum difference between the most and least loaded machine, motivated by questions on fair allocation.

Our Contribution

We introduce the speed-robust scheduling problem and present robust algorithms. The algorithmic difficulty of this problem is to construct bags in the first stage that are robust under any choice of machine speeds in the second stage. The straight-forward approach of using any makespan-optimal solution on m identical machines is not sufficient. Lemma 6 shows that such an algorithm might have an arbitrarily large robustness factor. Using *Longest Processing Time* first (LPT) to create bags does the trick and is $\left(2 - \frac{1}{m}\right)$-robust for arbitrary job sizes (Theorem 4). While this was known for speeds in $\{0, 1\}$ [20], our most general result is much less obvious.

Note that LPT aims at "balancing" the bag sizes which cannot lead to a better robustness factor than $2 - \frac{1}{m}$ as we show in Lemma 7. Hence, to improve upon this factor, we need to carefully construct bags with imbalanced bag sizes. There are two major challenges with this approach: (i) finding the ideal imbalance in the bag sizes independent from the actual job processing times that would be robust for all adversarial speed settings simultaneously and (ii) to adapt bag sizes to accommodate discrete jobs.

A major contribution of this paper is an optimal solution to the first challenge by considering infinitesimal jobs (Theorem 1). One can think of this as filling bags with *sand* to the desired level. Thus, the robust scheduling problem boils down to identifying the best bag sizes as placing the jobs into bags becomes trivial. We give, for any number of machines, optimally imbalanced bag sizes and prove a robustness factor of

$$\bar{\rho}(m) = \frac{m^m}{m^m - (m-1)^m} \leq \frac{e}{e-1} \approx 1.58 \, .$$

For infinitesimal jobs in the particular machine environment in which all machines have either speed 0 or 1, we obtain an algorithm with robustness factor

$$\bar{\rho}_{01}(m) = \max_{t \leq \frac{m}{2}, \, t \in \mathbb{N}} \frac{1}{\frac{t}{m-t} + \frac{m-2t}{m}} \leq \frac{1+\sqrt{2}}{2} \approx 1.207 = \bar{\rho}_{01} \, .$$

This improves the previous upper bound of 1.233 by Stein and Zhong [20] and matches exactly their lower bound for each m. Furthermore, we show that the lower bound in [20] holds even for randomized algorithms and, thus, our algorithm is optimal for both, deterministic and randomized scheduling (Theorem 2).

The above tight results for infinitesimal jobs are crucial for our further results for discrete jobs. Following the figurative notion of sand for infinitesimal jobs,

Table 1. Summary of results on speed-robust scheduling.

	General speeds		Speeds from $\{0,1\}$	
	Lower bound	Upper bound	Lower bound	Upper bound
Discrete jobs	$\bar{\rho}(m)$	$2 - \frac{1}{m}$	$\frac{4}{3}$	$\frac{5}{3}$
(Rocks)	(Lemma 1)	(Theorem 4)	[20]	[20]
Equal-size jobs	$\bar{\rho}(m)$	1.8	$\frac{4}{3}$	
(Bricks)	(Lemma 1)	(Theorem 5)	([20], Theorem 6)	
Infinitesimal jobs	$\bar{\rho}(m) \leq \frac{e}{e-1} \approx 1.58$		$\bar{\rho}_{01}(m) \leq \frac{1+\sqrt{2}}{2} \approx 1.207$	
(Sand)	(Lemma 1, Theorem 1)		([20], Theorem 2)	

we think of equal-size jobs as *bricks* and arbitrary jobs as *rocks*. Building on those ideal bag sizes, our approaches differ substantially from the methods in [20]. When all jobs have equal processing time, we obtain a 1.8-robust solution through a careful analysis of the trade-off between using slightly imbalanced bags and a scaled version of the ideal bag sizes computed for the infinitesimal setting (Theorem 5).

When machines have only speeds in $\{0,1\}$ and jobs have arbitrary equal sizes, i.e., unit size, we give an optimal $\frac{4}{3}$-robust algorithm (Theorem 6). This is an interesting class of instances as the best known lower bound of $\frac{4}{3}$ for discrete jobs uses only unit-size jobs [20]. To achieve this result, we exploit the ideal bag sizes computed for infinitesimal jobs by using a scaled variant of these sizes. Some cases, depending on m and the optimal makespan on m machines, have to be handled individually. Here, we use a direct way of constructing bags with at most four different bag sizes and some cases can be solved by an integer linear program. We summarize our results in Table 1.

Inspired by traditional one-stage scheduling problems where jobs have *machine-dependent execution times* (unrelated machine scheduling), one might ask for such a generalization of our problem. However, it is easy to rule out any robustness factor for such a setting: Consider four machines and five jobs, where each job may be executed on a unique pair of machines. Any algorithm must build at least one bag with at least two jobs. For this bag there is at most one machine to which it can be assigned with finite makespan. If this machine fails, the algorithm cannot complete the jobs whereas an optimal solution can split this bag on multiple machines to get a finite makespan.

Due to space constraints, we omit proof details. They can be found in a full version of this paper [9].

2   Speed-Robust Scheduling with Infinitesimal Jobs

In this section, we consider speed-robust scheduling with infinitely many jobs that have infinitesimal processing times. We give optimal algorithms in both the general case and the special case with speeds in $\{0,1\}$.

2.1   General Speeds

Theorem 1. *There is an algorithm for speed-robust scheduling with infinitesimal jobs that is $\bar{\rho}(m)$-robust for all $m \geq 1$, where*

$$\bar{\rho}(m) = \frac{m^m}{m^m - (m-1)^m} \leq \frac{e}{e-1} \approx 1.58 \,.$$

This is the best possible robustness factor that can be achieved by any algorithm.

To prove Theorem 1, we first show that, even when we restrict the adversary to a particular set of m speed configurations, no algorithm can achieve a robustness factor better than $\bar{\rho}(m)$. Note that since we can scale all speeds equally by an arbitrary factor without influencing the robustness factor, we can assume that the sum of the speeds equals 1. Similarly, we assume that the total processing time of the jobs equals 1 such that the optimal makespan of the adversary is 1 and the worst-case makespan of an algorithm is equal to its robustness factor.

Intuitively, the idea behind the set of speed configurations is that the adversary can set $m - 1$ machines to equal low speeds and one machine to a high speed. The low speeds are set such that one particular bag size just fits on that machine when aiming for the given robustness factor. This immediately implies that all larger bags have to be put on the fast machine together. This way, the speed configuration can *target* a certain bag size. We provide specific bag sizes that achieve a robustness of $\bar{\rho}(m)$ and show that for the speeds targeting these bag sizes, other bag sizes would result in even larger robustness factors.

We define $U = m^m$, $L = m^m - (m-1)^m$, as well as $t_k = (m-1)^{m-k}m^{k-1}$ for $k \in \{1, \ldots, m\}$. Intuitively, these values are chosen such that the bag sizes $\frac{t_i}{L}$ are optimal and $\frac{t_i}{U}$ corresponds to the low speed of the i-th speed configuration. It is easy to verify that $\bar{\rho}(m) = \frac{U}{L}$ and for all k we have

$$\sum_{i<k} t_i = (m-1)t_k - U + L \,. \tag{1}$$

In particular, this implies that $\sum_{i \leq m} t_i = mt_m - U + L = L$, and therefore, the sum of the bag sizes is 1. Let $a_1 \leq \cdots \leq a_m$ denote the bag sizes chosen by an algorithm and $s_1 \leq \cdots \leq s_m$ the speeds chosen by the adversary.

Lemma 1. *For any $m \geq 1$, no algorithm for speed-robust scheduling with infinitesimal jobs can have a robustness factor less than $\bar{\rho}(m)$.*

Proof. We restrict the adversary to the following m speed configurations indexed by $k \in \{1, \ldots, m\}$:

$$\mathcal{S}_k := \left\{ s_1 = \frac{t_k}{U}, \ s_2 = \frac{t_k}{U}, \ \ldots, \ s_{m-1} = \frac{t_k}{U}, \ s_m = 1 - (m-1)\frac{t_k}{U} \right\}.$$

Note that for all $k \in \{1, \ldots, m\}$, we have $mt_k \leq U$ and, thus, $s_m \geq s_{m-1}$.

We show that for any bag sizes $a_1, \ldots, a_m$, the adversary can force the algorithm to have a makespan of at least $\frac{U}{L}$ with some $\mathcal{S}_k$. Since the optimal makespan is fixed to be equal to 1 by assumption, this implies a robustness factor of at least $\frac{U}{L}$.

Let $k^\star$ be the smallest index such that $a_k \geq \frac{t_k}{L}$. Such an index exists because the sum of the t_i's is equal to L (Eq. (1)) and the sum of the a_i's is equal to 1. Now, consider the speed configuration $\mathcal{S}_{k^\star}$. If one of the bags a_i for $i \geq k^\star$ is not scheduled on the m-th machine, then the makespan is at least $\frac{a_i}{s_1} \geq a_{k^\star} \frac{U}{t_{k^\star}} \geq \frac{U}{L}$. Otherwise, all a_i for $i \geq k^\star$ are scheduled on machine m. Then, using Eq. (1), the load on that machine is at least

$$\sum_{i \geq k^\star} a_i = 1 - \sum_{i < k^\star} a_i \geq 1 - \frac{1}{L} \sum_{i < k^\star} t_i = \frac{1}{L} \left(L - (m-1)t_{k^\star} + U - L\right) = \frac{U}{L} s_m .$$

Thus, either a machine $i < m$ with a bag $i' \geq k^\star$ or machine $i = m$ has a load of at least $s_i \cdot \frac{U}{L}$ and determines the makespan.   □

For given bag sizes, we call a speed configuration that maximizes the minimum makespan a *worst-case speed configuration*. Before we provide the strategy that obtains a matching robustness factor, we state a property of such best strategies for the adversary.

Lemma 2. *Given bag sizes and a worst-case speed configuration, for each machine i, there exists an optimal assignment of the bags to the machines such that only machine i determines the makespan.*

Note that, by Lemma 2, for a worst-case speed configuration, there are many different bag-to-machine assignments that obtain the same optimal makespan. Lemma 2 also implies that for such speed configurations all speeds are non-zero.

Let SAND denote the algorithm that creates m bags of the following sizes

$$a_1 = \frac{t_1}{L}, \ a_2 = \frac{t_2}{L}, \ \ldots, \ a_m = \frac{t_m}{L} .$$

Note that this is a valid algorithm since the sum of these bag sizes is equal to 1. Moreover, these bag sizes are exactly such that if we take the speed configurations from Lemma 1, placing bag j on a slow machine in configuration j results in a makespan that is equal to $\bar{\rho}(m)$. We proceed to show that SAND ensures a robustness factor of $\bar{\rho}(m)$.

Lemma 3. *For any $m \geq 1$, SAND is $\bar{\rho}(m)$-robust for speed-robust scheduling with infinitely many infinitesimal jobs.*

Proof. Let $a_1, \ldots, a_m$ be the bag sizes as specified by SAND and let $s_1, \ldots, s_m$ be a worst-case speed configuration given these bag sizes. Consider an optimal assignment of bags to machines and let $C^*_{\max}$ denote its makespan. We use one particular (optimal) assignment to obtain an upper bound on $C^*_{\max}$. By Lemma 2, an optimal assignment exists where only machine 1 determines the makespan, i.e., machine 1 has load $C^*_{\max} \cdot s_1$ and any other machine i has load strictly less than $C^*_{\max} \cdot s_i$. Consider such an assignment. If there are two bags assigned to machine 1, then there is an empty machine with speed at least s_1. Therefore, we can put one of the two bags on that machine and decrease the makespan. This contradicts that $C^*_{\max}$ is the optimal makespan, so there is exactly one bag assigned to machine 1. Let k be the index of the unique bag placed on machine 1, i.e., $C^*_{\max} = \frac{a_k}{s_1}$, and let ℓ be the number of machines of speed s_1.

If $a_k > a_\ell$, then machine $i \in \{1, \dots, \ell\}$, with speed s_1, can be assigned bag i with a load that is strictly less than $C^*_{\max} \cdot s_1$. Thus, given the current assignment, we can remove bag a_k from machine 1 and place the ℓ smallest bags on the ℓ slowest machines, one per machine, e.g., bag a_i on machine i for $i \in \{1, \dots, \ell\}$. This empties at least one machine of speed strictly larger than s_1. Then, we can place bag a_k on this (now empty) machine, which yields a makespan that is strictly smaller than $C^*_{\max}$. This contradicts the assumption that $C^*_{\max}$ is the optimal makespan, and thus, $a_k \leq a_\ell$, which implies $k \leq \ell$.

Let P_i denote the total processing time of bags assigned to machine i, and C the total *remaining capacity* of the assignment, that is, $C := \sum_{i=1}^{m}(s_i C^*_{\max} - P_i)$. We bound C, which allows us to bound $C^*_{\max}$.

Machines in the set $\{2, \dots, \ell\}$ cannot be assigned a bag of size larger than a_k since their load would be greater than $C^*_{\max} \cdot s_1$, causing a makespan greater than $C^*_{\max}$. Therefore, we assume without loss of generality that all bags $a_j < a_k$ are assigned to a machine with speed s_1. The total remaining capacity on the first k machines is therefore equal to $(k-1)a_k - \sum_{i<k} a_i$.

Consider a machine $i > k$. If the remaining capacity of this machine is greater than a_k, then we can decrease the makespan of the assignment by moving bag k to machine i. Therefore, the remaining capacity on machine i is at most a_k.

Combining the above and using (1), we obtain:

$$C \leq (m-1)a_k - \sum_{i<k} a_i = \frac{1}{L}\left((m-1)t_k - \sum_{i<k} t_i\right) = \frac{1}{L}(U - L).$$

The total processing time is $\sum_{i=1}^{m} a_i = 1$, and the maximum total processing time that the machines could process with makespan $C^*_{\max}$ is equal to $\sum_{i=1}^{m} s_i C^*_{\max} = C^*_{\max}$. Since the latter is equal to the total processing time plus the remaining capacity, we have $C^*_{\max} = 1 + C \leq \frac{U}{L}$, which proves the lemma. $\square$

The robustness factor $\bar{\rho}(m)$ is not best possible for every m when we allow algorithms that make randomized decisions and compare to an oblivious adversary. For $m = 2$, uniformly randomizing between bag sizes $a_1 = a_2 = \frac{1}{2}$ and $a_1 = \frac{1}{4}$, $a_2 = \frac{3}{4}$ yields a robustness factor that is slightly better than $\bar{\rho}(2) = \frac{4}{3}$. Interestingly, with speeds in $\{0, 1\}$ the optimal robustness factor is equal for deterministic and randomized algorithms.

2.2   Speeds in $\{0, 1\}$

Theorem 2. *For all $m \geq 1$, there is a deterministic $\bar{\rho}_{01}(m)$-robust algorithm for speed-robust scheduling with speeds in $\{0, 1\}$ and infinitesimal jobs, where*

$$\bar{\rho}_{01}(m) = \max_{t \in \mathbb{N},\, t \leq \frac{m}{2}} \frac{1}{\frac{t}{m-t} + \frac{m-2t}{m}} \leq \frac{1 + \sqrt{2}}{2} = \bar{\rho}_{01} \approx 1.207.$$

This is the best possible robustness factor that can be achieved by any algorithm, even by a randomized algorithm against an oblivious adversary.

The deterministic version of the lower bound and some useful insights were already presented in [20]. We recall some of these insights here because they are used in the proof. To do so, we introduce some necessary notation used in the remainder of this paper. The number of failing machines (i.e., machines with speed equal to 0) is referred to as t, with $t \in \{0, \ldots, m-1\}$, and we assume w.l.o.g. that these are machines $1, \ldots, t$. Furthermore, we assume for this subsection, again w.l.o.g., that the total volume of infinitesimal jobs is m, and we define bags $1, \ldots, m$ with respective sizes $a_1 \leq \cdots \leq a_m$ summing to at least m (the potential excess being unused).

Lemma 4 (Statement (3) in [20]). *For all $m \geq 1$ and $t \leq \frac{m}{2}$, there exists a makespan-minimizing allocation of bags to machines for speed-robust scheduling with speeds in $\{0, 1\}$ and infinitely many infinitesimal jobs that assigns the smallest $2t$ bags to machines $t+1, \ldots, 2t$.*

Since Lemma 4 only works for $t \leq \frac{m}{2}$, one may worry that, for larger t, there is a more difficult structure to understand. The following insight shows that this worry is unjustified. Indeed, if $m' < \frac{m}{2}$ is the number of machines that do *not* fail, one can simply take the solution for $2m'$ machines, and assign the bags from any two machines to one machine. The optimal makespan is doubled and that of the algorithm is at most doubled, so the robustness is conserved.

Lemma 5 (Proof of Theorem 2.2 in [20]). *Let $\rho > 1$. For all $m \geq 1$, if an algorithm is ρ-robust for speed-robust scheduling with speeds in $\{0, 1\}$ and infinitely many infinitesimal jobs for $t \leq \frac{m}{2}$, it is ρ-robust for $t \leq m - 1$.*

Therefore, we focus on computing bag sizes such that the makespan of a best allocation according to Lemma 4 is within a $\bar{\rho}_{01}(m)$ factor of the optimal makespan when $t \leq \frac{m}{2}$. The approach in [20] to obtain the (as we show tight) lower bound $\bar{\rho}_{01}(m)$ is as follows. Given some $t \leq \frac{m}{2}$ and a set of bags allocated according to Lemma 4,

(i) The makespan on machines $t+1, \ldots, 2t$ is at most $\bar{\rho}_{01}(m)$ times the optimal makespan $\frac{m}{m-t}$, and

(ii) The makespan on machines $2t+1, \ldots, m$ is a most $\bar{\rho}_{01}(m)$ because those machines only hold a *single* bag after a simple "folding" strategy for assigning bags to machines, which we define below.

In particular, since $t = 0$ is possible (ii) implies that all bag sizes are at most $\bar{\rho}_{01}(m)$. The fact that the total processing volume of m has to be accommodated and maximizing over t results in the lower bound given in Theorem 2.

To define bag sizes leading to a matching upper bound, we further restrict our choices when $t \leq \frac{m}{2}$ machines fail. Of course, as we match the lower bound, the restriction is no limitation but rather a simplification. When $t \leq \frac{m}{2}$ machines fail, we additionally assume that machines $t+1, \ldots, 2t$ receive exactly two bags each:

Assuming $t \leq \frac{m}{2}$, the *simple folding* of these bags onto machines assigns bags $i \geq t+1$ to machine i, and bags $i \leq t$ (recall machine i fails) to machine $2t-i+1$. Hence, bags $1, \ldots, t$ are "folded" onto machines $2t, \ldots, t+1$ (sic).

For given m, let $t^\star$ be an optimal adversarial choice for t in Theorem 2. Assuming there are bag sizes $a_1, \ldots, a_m$ that match the bound $\bar{\rho}_{01}(m)$ through simple folding, by (i) and (ii), we precisely know the makespan on all machines after folding when $t = t^\star$. That fixes $a_i + a_{2t^\star+1-i} = \bar{\rho}_{01}(m) \cdot \frac{m}{m-t^\star}$ for all $i \in \{1, \ldots, t^\star\}$ and $a_{2t^\star+1} = \cdots = a_m = \bar{\rho}_{01}(m)$. In contrast to [20], we show that defining a_i for $i \in \{1, \ldots, t^\star\}$ to be essentially a linear function of i, and thereby fixing all bag sizes, suffices to match $\bar{\rho}_{01}(m)$. The word "essentially" can be dropped when replacing $\bar{\rho}_{01}(m)$ by $\bar{\rho}_{01}$.

A clean way of thinking about the bag sizes is through *profile functions* which reflect the distribution of load over bags in the limit case $m \to \infty$. Specifically, we identify the set $\{1, \ldots, m\}$ with the interval $[0, 1]$ and define a continuous non-decreasing profile function $\bar{f} : [0, 1] \to \mathbb{R}_+$ integrating to 1. A simple way of getting back from the profile function to actual bag sizes of total size approximately m is equidistantly "sampling" $\bar{f}$, i.e., by letting $a_i = \bar{f}(\frac{i-1/2}{m})$ for all i.

Our profile function $\bar{f}$ implements the above observations and ideas in the continuous setting. Indeed, our choice

$$\bar{f}(x) = \min\left\{\frac{1}{2} + \bar{\rho}_{01} \cdot x, \bar{\rho}_{01}\right\} = \min\left\{\frac{1}{2} + \frac{(1+\sqrt{2}) \cdot x}{2}, \frac{1+\sqrt{2}}{2}\right\}$$

is linear up to $\beta = 2 - \sqrt{2} = \lim_{m\to\infty} \frac{2t^\star}{m}$ and from then on it is constantly $\bar{\rho}_{01} = \lim_{m\to\infty} \bar{\rho}_{01}(m)$. We give some intuition for why this function works using the continuous counterpart of folding: When $t \leq t^\star$ machines fail, i.e., a continuum of machines with measure $x \leq \frac{\beta}{2}$, we fold the corresponding part of $\bar{f}$ onto the interval $[x, 2x]$, yielding a rectangle of width x and height $\bar{f}(0) + \bar{f}(2x) = 2\bar{f}(x)$. We have to prove that the height does not exceed the optimal makespan $\frac{1}{1-x}$ by more than a factor of $\bar{\rho}_{01}$. Equivalently, we maximize $2\bar{f}(x)(1-x)$ (even over $x \in \mathbb{R}$) and observe the maximum of $\bar{\rho}_{01} = \frac{1+\sqrt{2}}{2}$ at $x = \frac{\beta}{2}$. When $x \in \left(\frac{\beta}{2}, \frac{1}{2}\right]$, note that by folding we *still* obtain a rectangle of height $2\bar{f}(x)$ (but width $\beta - x$), dominating the load on the other machines. Hence, the makespan is at most $\frac{\bar{\rho}_{01}}{1-x}$ for every $x \in \left[0, \frac{1}{2}\right]$.

Directly "sampling" $\bar{f}$, we obtain a weaker bound (stated below) than that in Theorem 2. The proof (and the algorithm) is substantially easier than that of the main theorem: Firstly, we translate the above continuous discussion into a discrete proof. Secondly, we exploit that $\bar{f}$ is concave to show that the total volume of the "sampled" bags is larger than m for every $m \in \mathbb{N}$. Later, we make use of the corresponding simpler algorithm. Let SAND_{01} denote the algorithm that creates m bags of size $a_i := \bar{f}(\frac{i-1/2}{m})$, for $i \in \{1, \ldots, m\}$.

Theorem 3. SAND_{01} *is $\bar{\rho}_{01}$-robust for speed-robust scheduling with speeds in $\{0, 1\}$ and infinitely many infinitesimal jobs for all $m \geq 1$.*

As the profile function disregards specific machines, obtaining bag sizes through this function seems too crude to match $\bar{\rho}_{01}(m)$ for every m. Indeed, our proof of Theorem 2 is based on a much more careful choice of the bag sizes.

3   Speed-Robust Scheduling with Discrete Jobs

In this section, we consider the most general version of Speed Robust Scheduling. While in Sects. 2 and 4 we crucially use in our algorithm design the assumptions that all jobs are infinitesimally small (sand) or are of the same size (bricks), respectively, here, their sizes can vary arbitrarily (rocks).

By a scaling argument, we may assume w.l.o.g. that the machine speeds satisfy $\sum_{i=1}^{m} s_i = \sum_{j=1}^{n} p_j$. Observe that minimizing the size of a largest bag may not yield a robust algorithm.

Lemma 6. *Algorithms for speed-robust scheduling that minimize the size of a largest bag may not have a constant robustness factor.*

Proof. Consider any integer $k \geq 1$, a number of machines $m = k^2 + 1$, one job with processing time k, and k^2 unit-size jobs. The maximum bag size is at least k, so an algorithm that builds $k + 1$ bags of size k respects the conditions of the lemma. Consider the speed configuration where k^2 machines have speed 1 and one machine has speed k. It is possible to schedule all jobs on these machines with makespan 1. However, the algorithm must either place a bag on a machine of speed 1 or all bags on the machine of speed k, which gives a makespan of k. □

For machine speeds in $\{0, 1\}$, such algorithms are $\left(2 - \frac{2}{m}\right)$-robust. Once the number m' of speed-1 machines is revealed, simply combine the two smallest bags repeatedly if $m' < m$. The makespan is then at most twice the average load on $m' + 1$ machines, i.e., $\frac{2m'}{m'+1}$ times the average load on m' machines.

The lower bound in Lemma 6 exploits the fact that bags sizes of such algorithms might be very unbalanced. An algorithm is called *balanced* if, for an instance of unit-size jobs, the bag sizes created by the algorithm differ by at most one unit. In particular, a balanced algorithm creates m bags of size k when confronted with mk unit-size jobs and m bags. For balanced algorithms, we give a lower bound in Lemma 7 and a matching upper bound in Theorem 4.

Lemma 7. *No balanced algorithm for speed-robust scheduling can obtain a better robustness factor than $2 - \frac{1}{m}$ for any $m \geq 1$.*

We now show that this lower bound is attained by a simple algorithm, commonly named as *Longest Processing Time first* (LPT) which considers jobs in non-increasing order of processing times and assigns each job to the bag that currently has the smallest size, i.e., the minimum allocated processing time.

Theorem 4. *LPT is $\left(2 - \frac{1}{m}\right)$-robust for speed-robust scheduling for all $m \geq 1$.*

Proof. While we may assume that the bags are allocated optimally to the machines once the speeds are given, we use a different allocation for the analysis. This cannot improve the robustness.

Consider the m bags and let b denote the size of a largest bag, B, that consists of at least two jobs. Consider all bags of size strictly larger than b, each containing only a single job, and place them on the same machine as OPT places the corresponding jobs. We define for each machine i with given speed s_i a capacity bound of $\left(2 - \frac{1}{m}\right) \cdot s_i$. Then, we consider the remaining bags in non-increasing order of bag sizes and iteratively assign them to the – at the time of assignment – least loaded machine with sufficient remaining capacity.

By the assumption $\sum_{i=1}^{m} s_i = \sum_{j=1}^{n} p_j$ and the capacity constraint $\left(2 - \frac{1}{m}\right) \cdot s_i$, it is sufficient to show that LPT can successfully place all bags.

The bags larger than b fit by definition as they contain a single job. Assume by contradiction that there is a bag which cannot be assigned. Consider the first such bag and let T be its size. Let $k < m$ be the number of bags that have been assigned already. Further, denote by w the size of a smallest bag. Since we used LPT to create the bags, we have $w \geq \frac{1}{2}b$. To see that, consider bag B and notice that the smallest job in it has a size at most $\frac{1}{2}b$. When this job was assigned to its bag, B was the bag with smallest size, and this size was at least $\frac{1}{2}b$ since we allocate jobs in LPT-order. Hence, the size of a smallest bag is $w \geq \frac{1}{2}b \geq \frac{1}{2}T$, where the second inequality is true as all bags larger than b can be placed.

We use this inequality to give a lower bound on the total remaining capacity on the m machines when the second-stage algorithm fails to place the $(k+1)$-st bag. The $(m-k)$ bags that were not placed have a combined volume of at least $V_\ell = (m - k - 1)w + T \geq (m - k + 1)\frac{T}{2}$. The bags that were placed have a combined volume of at least $V_p = kT$. The remaining capacity is then at least $C = (2 - \frac{1}{m})V_\ell + (1 - \frac{1}{m})V_p$, and we have

$$C = \left(2 - \frac{1}{m}\right) V_\ell + \left(1 - \frac{1}{m}\right) V_p \geq \left(2 - \frac{1}{m}\right)(m - k + 1)\frac{T}{2} + \left(1 - \frac{1}{m}\right) kT$$

$$\geq (m - k + 1)T - (m - k + 1)\frac{T}{2m} + kT - \frac{1}{m}kT \geq mT + T - \frac{m + k + 1}{2m}T$$

$$\geq mT.$$

Thus, there is a machine with remaining capacity T which contradicts the assumption that the bag of size T does not fit.   □

4   Speed-Robust Scheduling with Equal-Size Jobs

In this section we consider instances where all jobs are of equal size, i.e., bricks, as this case seems to capture the complexity of the general problem. This intuition stems from the fact that all known lower bounds already hold for this type of instances (see [20] and Lemma 10).

By a scaling argument, we may assume that all jobs have unit processing time. Before focusing on a specific speed setting, we show that in both settings we can use any algorithm for infinitesimal jobs with proper scaling to obtain a robustness which is degraded by a factor decreasing with $\frac{n}{m} =: \lambda$. Assume $\lambda > 1$, as otherwise the problem is trivial. We define the algorithm SANDFORBRICKS that builds on the optimal algorithm for infinitesimal jobs, SAND*, which is SAND for general speeds (Sect. 2.1) or SAND$_{01}$ for speeds in $\{0,1\}$ (Sect. 2.2). Let $a_1, \ldots, a_m$ be the bag sizes constructed by SAND* scaled such that a total processing volume of n can be assigned, that is, $\sum_{i=1}^{m} a_i = n$. For unit-size jobs, we define bag sizes as $a_i' = \left(1 + \frac{1}{\lambda}\right) \cdot a_i$ and assign the jobs greedily to the bags.

Lemma 8. *For n jobs with unit processing times and m machines, SANDFOR-BRICKS for speed-robust scheduling is $\left(1 + \frac{1}{\lambda}\right) \cdot \rho(m)$-robust, where $\lambda = \frac{n}{m}$ and $\rho(m)$ is the robustness factor for SAND* for m machines.*

4.1   General Speeds

For bricks, i.e., unit-size jobs, we beat the factor $2 - \frac{1}{m}$ (Theorem 4) for speed-robust scheduling and give a 1.8-robust algorithm. For $m = 2$ and $m = 3$, we give algorithms with best possible robustness factors $\frac{4}{3}$ and $\frac{3}{2}$, respectively.

Theorem 4 shows that LPT has a robustness factor of $2 - \frac{1}{m}$. We show that a slightly different algorithm, BUILDODD, has a robustness that increases with the ratio between the number of jobs and the number of machines. BUILDODD builds bags of three possible sizes: for $q \in \mathbb{N}$ such that $\lambda = \frac{n}{m} \in [2q-1, 2q+1]$, the bags sizes are $2q - 1$, $2q$ and $2q + 1$. In a manner similar to the proof of Theorem 4, we can prove BUILDODD is $(2 - \frac{1}{q+1})$-robust. The worst case happens when a bag of size $2q + 1$ is scheduled on a machine of speed $q + 1$.

Lemma 9. *For n unit-size jobs, m machines and $q \in \mathbb{N}$ with $\lambda \in [2q-1, 2q+1]$, BUILDODD is $(2 - \frac{1}{q+1})$-robust for speed-robust scheduling.*

The robustness guarantees in Lemmas 8 and 9 are decreasing and increasing, respectively, in λ. By carefully choosing between BUILDODD and SANDFOR-BRICKS, depending on the input, we obtain an improved algorithm for bricks. For $\lambda < 8$, we execute BUILDODD, which yields a robustness factor of at most 1.8 by Lemma 9, as $q \leq 4$ for $\lambda < 8$. Otherwise, when $\lambda \geq 8$, we run SANDFOR-BRICKS with a guarantee of $\frac{9}{8} \cdot \frac{e}{e-1} \approx 1.78$ by Lemma 8.

Theorem 5. *There is an algorithm for speed-robust scheduling with unit-size jobs that has a robustness factor of at most 1.8 for any $m \geq 1$.*

We give a general lower bound on the best achievable robustness factor.

Lemma 10. *For every $m \geq 3$, no algorithm for speed-robust scheduling can have a robustness factor smaller than $\frac{3}{2}$, even restricted to unit-size jobs.*

For special cases with few machines, we give best possible algorithms.

Lemma 11. *An optimal algorithm for speed-robust scheduling for unit-size jobs has robustness factor $\frac{4}{3}$ on $m = 2$ machines and $\frac{3}{2}$ on $m = 3$ machines, and larger than $\bar{\rho}(6) > \frac{3}{2}$ for $m = 6$.*

4.2   Speeds in $\{0,1\}$

When considering speeds in $\{0,1\}$, bricks (unit-size jobs) are of particular interest as the currently best known lower bound for rocks (arbitrary jobs) is $\frac{4}{3}$ and uses only bricks [20]. We present an algorithm with a matching upper bound.

Theorem 6. *There exists a $\frac{4}{3}$-robust algorithm for speed-robust scheduling with $\{0,1\}$-speeds and unit-size jobs.*

In the proof, we handle different cases depending on m and $\lceil \lambda \rceil$ by carefully tailored methods. Note that $\lceil \lambda \rceil$ is equal to the optimal makespan on m machines.

When $\lceil \lambda \rceil \geq 11$, we use SANDFORBRICKS and obtain a robustness factor of at most $\frac{4}{3}$ by Lemma 9. The proof uses a volume argument to show that jobs fit into the scaled optimal bag sizes for infinitesimal jobs, even after rounding bag sizes down to the nearest integer. When $\lceil \lambda \rceil \in \{9, 10\}$, this method is too crude. We refine it to show that for $m \geq 40$ it is still possible to scale bag sizes from SANDID and round them to integral sizes such that all jobs can be placed. The analysis exploits an amortized bound on the loss due to rounding over consecutive bags. For the case that $\lceil \lambda \rceil \leq 8$ and $m \geq 50$, we use a constructive approach and give a strategy that utilizes at most four different bag sizes. The remaining cases, $\lceil \lambda \rceil \leq 10$ and $m \leq 50$, can be verified by enumerating over all possible instances and using an integer linear program to verify that there is a solution of bag sizes that is $\frac{4}{3}$-robust.

References

1. Albers, S., Hellwig, M.: Online makespan minimization with parallel schedules. Algorithmica **78**(2), 492–520 (2017). https://doi.org/10.1007/s00453-016-0172-5
2. Albers, S., Schmidt, G.: Scheduling with unexpected machine breakdowns. Discret. Appl. Math. **110**(2–3), 85–99 (2001). https://doi.org/10.1016/s0166-218x(00)00266-3
3. Alon, N., Azar, Y., Woeginger, G.J., Yadid, T.: Approximation schemes for scheduling on parallel machines. J. Sched. **1**(1), 55–66 (1998). https://doi.org/10.1002/(sici)1099-1425(199806)1:1⟨55::aid-jos2⟩3.0.co;2-j
4. Baruah, S.K., et al.: Scheduling real-time mixed-criticality jobs. IEEE Trans. Comput. **61**(8), 1140–1152 (2012). https://doi.org/10.1109/tc.2011.142
5. Chen, L., Megow, N., Rischke, R., Stougie, L.: Stochastic and robust scheduling in the cloud. In: APPROX-RANDOM. LIPIcs, vol. 40, pp. 175–186. Schloss Dagstuhl - Leibniz-Zentrum für Informatik (2015). https://doi.org/10.4230/LIPIcs.APPROX-RANDOM.2015.175
6. Dean, J., Ghemawat, S.: MapReduce: simplified data processing on large clusters. Commun. ACM **51**(1), 107–113 (2008). https://doi.org/10.1145/1327452.1327492
7. Diedrich, F., Jansen, K., Schwarz, U.M., Trystram, D.: A survey on approximation algorithms for scheduling with machine unavailability. In: Lerner, J., Wagner, D., Zweig, K.A. (eds.) Algorithmics of Large and Complex Networks. LNCS, vol. 5515, pp. 50–64. Springer, Heidelberg (2009). https://doi.org/10.1007/978-3-642-02094-0_3

8. Dürr, C., Erlebach, T., Megow, N., Meißner, J.: An adversarial model for scheduling with testing. Algorithmica **82**(12), 3630–3675 (2020). https://doi.org/10.1007/s00453-020-00742-2

9. Eberle, F., Hoeksma, R., Megow, N., Nölke, L., Schewior, K., Simon, B.: Speed-robust scheduling. CoRR (2020). https://arxiv.org/abs/2011.05181

10. Epstein, L., et al.: Universal sequencing on an unreliable machine. SIAM J. Comput. **41**(3), 565–586 (2012). https://doi.org/10.1137/110844210

11. Graham, R.L.: Bounds for certain multiprocessing anomalies. Bell Syst. Tech. J. **45**(9), 1563–1581 (1966). https://doi.org/10.1002/j.1538-7305.1966.tb01709.x

12. Hochbaum, D.S., Shmoys, D.B.: Using dual approximation algorithms for scheduling problems theoretical and practical results. J. ACM **34**(1), 144–162 (1987). https://doi.org/10.1145/7531.7535

13. Jansen, K.: An EPTAS for scheduling jobs on uniform processors: using an MILP relaxation with a constant number of integral variables. SIAM J. Discrete Math. **24**(2), 457–485 (2010). https://doi.org/10.1137/090749451

14. Kouvelis, P., Yu, G.: Robust Discrete Optimization and Its Applications. Springer, Berlin (1997). https://doi.org/10.1007/978-1-4757-2620-6

15. Levi, R., Magnanti, T.L., Shaposhnik, Y.: Scheduling with testing. Manag. Sci. **65**(2), 776–793 (2019). https://doi.org/10.1287/mnsc.2017.2973

16. Megow, N., Verschae, J.: Dual techniques for scheduling on a machine with varying speed. SIAM J. Discret. Math. **32**(3), 1541–1571 (2018). https://doi.org/10.1137/16m105589x

17. Niño-Mora, J.: Stochastic scheduling. In: Encyclopedia of Optimization, pp. 3818–3824. Springer (2009). https://doi.org/10.1007/978-0-387-74759-0_665

18. Pruhs, K., Sgall, J., Torng, E.: Online scheduling. In: Handbook of Scheduling. Chapman and Hall/CRC (2004). https://doi.org/10.1007/978-3-319-99849-7

19. Shmoys, D.B., Sozio, M.: Approximation algorithms for 2-stage stochastic scheduling problems. In: Fischetti, M., Williamson, D.P. (eds.) IPCO 2007. LNCS, vol. 4513, pp. 145–157. Springer, Heidelberg (2007). https://doi.org/10.1007/978-3-540-72792-7_12

20. Stein, C., Zhong, M.: Scheduling when you do not know the number of machines. ACM Trans. Algorithms **16**(1), 9:1–9:20 (2020). https://doi.org/10.1145/3340320

The Double Exponential Runtime is Tight for 2-Stage Stochastic ILPs

Klaus Jansen, Kim-Manuel Klein, and Alexandra Lassota[✉]

Department of Computer Science, Kiel University, Kiel, Germany
{kj,kmk,ala}@informatik.uni-kiel.de

Abstract. We consider fundamental algorithmic number theoretic problems and their relation to a class of block structured Integer Linear Programs (ILPs) called 2-stage stochastic. A 2-stage stochastic ILP is an integer program of the form $\min\{c^T x \mid \mathcal{A}x = b, \ell \leq x \leq u, x \in \mathbb{Z}^{r+ns}\}$ where the constraint matrix $\mathcal{A} \in \mathbb{Z}^{nt \times r+ns}$ consists of n matrices $A_i \in \mathbb{Z}^{t \times r}$ on the vertical line and n matrices $B_i \in \mathbb{Z}^{t \times s}$ on the diagonal line aside. First, we show a stronger hardness result for a number theoretic problem called QUADRATIC CONGRUENCES where the objective is to compute a number $z \leq \gamma$ satisfying $z^2 \equiv \alpha \bmod \beta$ for given $\alpha, \beta, \gamma \in \mathbb{Z}$. This problem was proven to be NP-hard already in 1978 by Manders and Adleman. However, this hardness only applies for instances where the prime factorization of β admits large multiplicities of each prime number. We circumvent this necessity by proving that the problem remains NP-hard, even if each primenumber only occurs constantly often.

Then, using this new hardness result for the QUADRATIC CONGRUENCES problem, we prove a lower bound of $2^{2^{\delta(s+t)}}|I|^{O(1)}$ for some $\delta > 0$ for the running time of any algorithm solving 2-stage stochastic ILPs assuming the Exponential Time Hypothesis (ETH). Here, $|I|$ is the encoding length of the instance. This result even holds if r, $||b||_\infty$, $||c||_\infty$, $||\ell||_\infty$ and the largest absolute value Δ in the constraint matrix $\mathcal{A}$ are constant. This shows that the state-of-the-art algorithms are nearly tight. Further, it proves the suspicion that these ILPs are indeed harder to solve than the closely related n-fold ILPs where the constraint matrix is the transpose of $\mathcal{A}$.

Keywords: 2-stage stochastic ILPs · Quadratic Congruences · Lower bound · Exponential Time Hypothesis

1 Introduction

One of the most fundamental problems in algorithm theory and optimization is the INTEGER LINEAR PROGRAMMING problem. Many theoretical and practical problems can be modeled as integer linear programs (ILPs) and thus they serve as

This work was supported by DFG project JA 612/20-1.

M. Singh and D. P. Williamson (Eds.): IPCO 2021, LNCS 12707, pp. 297–310, 2021.
https://doi.org/10.1007/978-3-030-73879-2_21

a very general but powerful framework for tackling various questions. Formally, the INTEGER LINEAR PROGRAMMING problem is defined as

$$\min\{c^\top x \mid A x = b, \ell \le x \le u, x \in \mathbb{Z}^{d_2}\}$$

for some matrix $A \in \mathbb{Z}^{d_1 \times d_2}$, a right-hand side $b \in \mathbb{Z}^{d_1}$, an objective function $c \in \mathbb{Z}^{d_2}$ and some lower and upper bounds $\ell, u \in \mathbb{Z}^{d_2}$. The goal is to find a solution x such that the value of the objective function $c^\top x$ is minimized. In general, this problem is NP-hard. Thus, it is of great interest to find structures to these ILPs which make them solvable more efficiently. In this work, we consider 2-stage stochastic integer linear programs where the constraint matrix admits a specific block structure. Namely, the constraint matrix A only contains non-zero entries in the first few columns and block-wise along the diagonal aside. This yields the following form:

$$A = \begin{pmatrix} A_1 & B_1 & 0 & \cdots & 0 \\ A_2 & 0 & B_2 & \ddots & \vdots \\ \vdots & \vdots & \ddots & \ddots & 0 \\ A_n & 0 & \cdots & 0 & B_n \end{pmatrix}.$$

Thereby $A_1, \ldots, A_n \in \mathbb{Z}^{t \times r}$ and $B_1, \ldots, B_n \in \mathbb{Z}^{t \times s}$ are integer matrices themselves. The complete constraint matrix A has size $nt \times r + ns$. Let Δ denote the largest absolute entry in A.

Such 2-stage stochastic ILPs are a common tool in stochastic programming and they are often used in practice to model uncertainty of decision making over time [1,7,17,21]. Due to the applicability a lot of research has been done in order to solve these (mixed) ILPs efficiently in practice. Since we focus on the theoretical aspects of 2-stage stochastic ILPs, we only refer the reader to the surveys [9,20,24] and the references therein regarding the practical methods.

The current state-of-the-art algorithms to solve 2-stage stochastic ILPs admits a running time of $2^{(2\Delta)^{r^2 s + r s^2}} n \log^3(n) \cdot |I|$ where $|I|$ is the binary encoding length of the input [8] or respectively of $n \log^{O(rs)}(n) 2^{(2\Delta)^{O(r^2 + rs)}}$ [5] by a recent result. The first result improves upon the result in [18] due to Klein where the dependence on n was quadratic. The dependencies on the block dimensions and $|I|$ were similar. The first result in that respect was by Hemmecke and Schulz [11] who provided an algorithm with a running time of $f(r, s, t, \Delta) \cdot \text{poly}(n)$ for some computable function f. However, due to the use of an existential result from commutative algebra, no explicit bound could be stated for f.

Let us turn our attention to the n-fold ILPs for a moment, which where first introduced in [22]. These ILPs admit a constraint matrix which is the transpose of the 2-stage stochastic constraint matrix. Despite being so closely related, n-fold ILPs can be solved in time near linear in the number of blocks and only single exponentially in the block-dimensions of A_i^T, B_i^T [4,16].

Thus, it is an intrinsic questions whether we can solve 2-stage stochastic ILPs more efficient or – as the latest algorithms suggest – whether 2-stage stochastic ILPs are indeed harder to solve than the closely related n-fold ILPs. We answer this question by showing a double-exponential lower bound in the running time for any algorithm solving the 2-STAGE STOCHASTIC INTEGER LINEAR PROGRAMMING (2-STAGE ILP) problem. Here, the 2-STAGE ILP problem is the corresponding decision variant which asks whether the ILP admits a feasible solution.

To prove this hardness, we reduce from the QUADRATIC CONGRUENCES problem. This problem asks whether there exists a $z \leq \gamma$ such that $z^2 \equiv \alpha \bmod \beta$ for some $\gamma, \alpha, \beta \in \mathbb{N}$. This problem was proven to be NP-hard by Manders and Adleman [23] already in 1978 by showing a reduction from 3-SAT. This hardness even persists if the prime factorization of β is given [23]. By this result, Manders and Adleman prove that it is NP-complete to compute the solutions of diophantine equations of degree 2. However, their reduction yields large parameters. In detail, the occurrences of each prime factor in the prime factorization of β is too large to obtain the desired lower bound for the 2-STAGE ILP problem. The occurrence of each prime factor is at least linear in the number of variables and clauses of the underlying 3-SAT problem.

We give a new reduction yielding a stronger statement: The QUADRATIC CONGRUENCES problem is NP-hard even if the prime factorization of β is given and each prime factor occurs at most once (except 2 which occurs four times). Beside being useful to prove the lower bounds for solving the 2-stage stochastic ILPs, we think this results is of independent interest. We obtain a neat structure which may be helpful in various related problems or may yield stronger statements of past results which use the QUADRATIC CONGRUENCES problem.

In order to achieve the desired lower bounds on the running time, we make use of the Exponential Time Hypothesis (ETH) – a widely believed conjecture stating that the 3-SAT problem cannot be solved in subexponentially time with respect to the number of variables:

Conjecture 1 (ETH [12]). *The* 3-SAT *problem cannot be solved in time less than* $O(2^{\delta_3 n_3})$ *for some constant* $\delta_3 > 0$ *where* n_3 *is the number of variables in the instance.*

Note that we use the index 3 for all variables of the 3-SAT problem. Using the ETH, plenty lower bounds for various problems are shown, for an overview on the techniques and results see e.g. [6]. So far, the best algorithm runs in time $O(2^{0.387 n_3})$, i.e., it follows that $\delta_3 \leq 0.387$ [6].

In the following, we also need the Chinese Remainder Theorem (CRT) for some of the proofs, which states the following:

Proposition 1 (CRT [14]). *Let* $n_1, \ldots, n_k$ *be pairwise co-prime. Further, let* $i_1, \ldots, i_k$ *be some integers. Then there exists integers* x *satisfying* $x \equiv i_j \bmod n_j$ *for all* j. *Further, any two solutions* x_1, x_2 *are congruent modulo* $\prod_{j=1}^{k} n_j$.

Summary of Results

– We give a new reduction from the 3-SAT problem to the QUADRATIC CONGRUENCES problem which proves a stronger NP-hardness result: The

QUADRATIC CONGRUENCES problem remains NP-hard, even if the prime factorization of β is given and each prime number greater than 2 occurs at most once and the prime number 2 occurs four times. This does not follow from the original proof. In contrast, the original proof generates each prime factor at least $O(n_3 + m_3)$ times, where m_3 is the number of clauses in the formula. Our reduction circumvents this necessity, yet neither introduces noteworthily more nor larger prime factors. The proof is based on the original one. We believe this result is of independent interest.

- Based on this new reduction, we show strong NP-hardness for the so-called NON-UNIQUE REMAINDER problem. In this algorithmic number theoretic problem, we are given $x_1, \ldots, x_{n_{NR}}, y_1, \ldots, y_{n_{NR}}, \zeta \in \mathbb{N}$ and pairwise coprime numbers $q_1, \ldots, q_{n_{NR}}$. The question is to decide whether there exists a number $z \in \mathbb{Z}_{>0}$ with $z \leq \zeta$ satisfying $z \bmod q_i \in \{x_i, y_i\}$ for all $i \in \{1, 2, \ldots, n_{NR}\}$ simultaneously. In other words, either the residue x_i or y_i should be met for each equation. This problem is a natural generalization of the Chinese Remainder problem where $x_i = y_i$ for all i. In that case, however, the problem can be solved using the Extended Euclidean algorithm. To the best of our knowledge the NON-UNIQUE REMAINDER problem has not been considered in the literature so far.

- Finally, we show that the NON-UNIQUE REMAINDER problem can be modeled by a 2-stage stochastic ILP. Assuming the ETH, we can then conclude a doubly exponential lower bound of $2^{2^{\delta(s+t)}} |I|^{O(1)}$ on the running time for any algorithm solving 2-stage stochastic ILPs. The double exponential lower bound even holds if $r = 1$ and $\Delta, ||b||_\infty, ||c||_\infty \in O(1)$. This proves the suspicion that 2-stage stochastic ILPs are significantly harder to solve than n-fold ILPs with respect to the dimensions of the block matrices and Δ. Furthermore, it implies that the current state-of-the-art algorithms for solving 2-stage stochastic ILPs is indeed (nearly) optimal.

Further Related Work. In recent years, there was significant progress in the development of algorithms for n-fold ILPs and lower bounds on the other hand. Assume the parameters as of the transpose of the 2-stage stochastic constraint matrix, i.e., the blocks A_i^T in the first few rows have dimension $r \times t$ and the blocks B_i^T along the diagonal beneath admit a dimension of $s \times t$. The best known algorithms to solve these ILPs have a running time of $2^{O(rs^2)}(rs\Delta)^{O(r^2s+s^2)}(nt)^{1+o(1)}$ [4] or respectively a running time of $(rs\Delta)^{r^2s+s^2} L^2 (nt)^{1+o(1)}$ [16] where L denotes the encoding length of the largest number in the input. The best known lower bound is $\Delta^{\delta_{n\text{-fold}}(r+s)^2}$ for some $\delta_{n\text{-fold}} > 0$ [8].

Despite their similarity, it seems that 2-stage stochastic ILPs are significantly harder to solve than n-fold ILPs. Yet, no superexponential lower bound for the running time of any algorithm solving the 2-STAGE ILP problem was shown. There is a lower bound for a more general class of ILPs in [8] that contain 2-stage stochastic ILPs showing that the running time is double-exponential parameterized by the topological height of the treedepth decomposition of the primal or dual graph. However, the topological height of 2-stage stochastic ILPs is constant and thus no strong lower bound can be derived for this case.

If we relax the necessity of an integral solution, the 2-stage stochastic LP problem becomes solvable in time $2^{2^{\Delta^{O(t^3)}}} n \log^3(n) \log(\|u-\ell\|_\infty) \log(\|c\|_\infty)$ [2]. For the case of mixed integer linear programs there exists an algorithm solving 2-stage stochastic MILPs in time $2^{2^{\Delta^{\Delta^{t^{O(t^2)}}}}} n \log^3(n) \log(\|u-\ell\|_\infty) \log(\|c\|_\infty)$ [2]. Both results rely on the fractionality of a solution, whose size is only dependent on the parameters. This allows us to scale the problem such that it becomes an ILP (as the solution has to be integral) and thus state-of-the-art algorithms for 2-stage stochastic ILPs can be applied.

There are also studies for a more general case called 4-Block ILPs where the constraint matrix consists of non-zero entries in the first few columns, the first few rows and block-wise along the diagonal. This may be seen as the combination of n-fold and 2-stage stochastic ILPs. Only little is known about them: They are in XP [10]. Further, a lower and upper bound on the Graver Basis elements (inclusion-wise minimal kernel elements) of $O(n^r f(k, \Delta))$ was shown recently [3] where r is the number of rows in the submatrix appearing repeatedly in the first few rows and k denotes the sum of the remaining block dimensions.

Structure of this Chapter. Section 2 presents the stronger hardness result for the QUADRATIC CONGRUENCES problem we derive by giving a new reduction from the 3-SAT problem. Then, we show that the QUADRATIC CONGRUENCES problem can be modeled as a 2-stage stochastic ILP in Sect. 3. To do so, we introduce a new problem called the NON-UNIQUE REMAINDER problem as an intermediate step during the reduction. Finally, in Sect. 4, we bring the reductions together to prove the desired lower bound. This involves a construction which lowers the absolute value of Δ at the cost of slightly larger block dimensions.

Due to space restrictions, the correctness proofs including the running time analysis of the reductions are omitted as well as the proofs of Lemma 1 and Theorem 2. Instead, we give an idea of the correctness before the corresponding theorems. For all details, we refer to the full version of this paper [15], available at http://arxiv.org/abs/2008.12928.

2   Advanced Hardness for Quadratic Congruences

This section proves that every instance of the 3-SAT problem can be transformed into an equivalent instance of the QUADRATIC CONGRUENCES problem in polynomial time. Recall that the QUADRATIC CONGRUENCES problem asks whether there exists a number $z \le \gamma$ such that $z^2 \equiv \alpha \bmod \beta$ holds. This problem was proven to be NP-hard by Manders and Adleman [23] showing a reduction from 3-SAT. This hardness even persists when the prime factorization of β is given [23]. However, we aim for an even stronger statement: The QUADRATIC CONGRUENCES problem remains NP-hard even if the prime factorization of β is given and each prime number greater than 2 occurs at most once and the prime number 2 occurs four times. This does not follow from the original hardness proof. In contrast, if n_3 is the number of variables and m_3 the number of clauses in the 3-SAT formula then β admits a prime factorization with $O(n_3 + m_3)$ different prime numbers each with a multiplicity of at least $O(n_3 + m_3)$. Even though our new reduction lowers the occurrence of each prime factor greatly, we neither introduces noteworthily more nor larger prime factors.

While the structure of our proof resembles that of the original one from [23], adapting it to our needs requires various new observations concerning the behaviour of the newly generated prime factors and the functions we introduce. The original proof heavily depends on the numbers being high powers of the prime factors whereas we employ careful combinations of (new) prime factors. This requires us to introduce other number theoretical results into the arguments.

In the following, before presenting the reduction and showing its correctness formally, we want to give an idea of the hardness proof. The reduction may seem non-intuitive at first as it only shows the final result of equivalent transformations between various problems until we reach the QUADRATIC CONGRUENCES one. In the following, we list all these problems in order of their appearance whose strong NP-hardness is shown implicitly along the way. Afterwards, we give short ideas of their respective equivalence. Note that not all variables are declared at this point, but also not necessary to understand the proof sketch.

- (3-SAT) Is there a truth assignment η that satisfies all clauses σ_k of the 3-SAT formula Φ simultaneously?
- (P2) Are there values $y_k \in \{0,1,2,3\}$ and a truth assignment η such that $0 = y_k - \sum_{x_i \in \sigma_k} \eta(x_i) - \sum_{\bar{x}_i \in \sigma_k}(1 - \eta(x_i)) + 1$ for all k?
- (P3) Are there values $\alpha_j \in \{-1, +1\}$ such that $\sum_{j=0}^{\nu} \theta_j \alpha_j \equiv \tau \bmod 2^3 \cdot p^* \prod_{i=1}^{m'} p_i$ for some θ_j and τ specified in dependence on the formula later on and some prime numbers p_i and p^*?
- (P5) Is there an $x \in \mathbb{Z}$ satisfying: $0 \le |x| \le H, x \equiv \tau \bmod 2^3 \cdot p^* \prod_{i=1}^{m'} p_i, (H + x)(H - x) \equiv 0 \bmod K$ for some H dependent on the θ_j and K being a product of prime numbers?
- (P6) Is there an $x \in \mathbb{Z}$ satisfying: $0 \le |x| \le H, (\tau - x)(\tau + x) \equiv 0 \bmod 2^4 \cdot p^* \prod_{i=1}^{m'} p_i, (H + x)(H - x) \equiv 0 \bmod K$?
- (QUADRATIC CONGRUENCES) Is there a number $x \le H$ such that $(2^4 \cdot p^* \cdot \prod_{i=1}^{m'} p_i + K)x^2 \equiv K\tau^2 + 2^4 \cdot p^* \cdot \prod_{i=1}^{m'} p_i H^2 \bmod 2^4 \cdot p^* \cdot \prod_{i=1}^{m'} p_i \cdot K$?

The 3-SAT problem is transformed to Problem (P2) by using the straightforward interpretation of truth values as numbers 0 and 1 and the satisfiability of a clause as the sum of its literals being larger zero. Introducing slack variables y_k yields the above form.

Multiplying each equation of (P2) with exponentially growing factors and then forming their sum preserves the equivalence of these systems. Introducing some modulo consisting of unique prime factors larger than the outcome of the largest possible sum obviously does not influence the system. Replacing the variables $\eta(x_i)$ and y_k by variables α_j with domain $\{-1, +1\}$, re-arranging the term and defining parts of the formula as the variables θ_j and τ yields Problem (P3).

We then introduce some Problem (P4) to integrate the condition $x \leq H$. The problem asks whether there exists some $x \in \mathbb{Z}$ such that $0 \leq |x| \leq H$ and $(H + x)(H - x) \equiv 0 \mod K$ holds. By showing that each solution to the system (P4) is of form $\sum_{j=0}^{\nu} \theta_j \alpha_j$ we can combine (P3) and (P4) yielding (P5).

Using some observations about the form of solutions for the second constraint of Problem (P5) we can re-formulate it as Problem (P6).

Next, we use the fact that $p^* \prod_{i=1}^{m'} p_i$ and K are co-prime per definition and thus we can combine the second and third equation to one equivalent one. To do so, we take each left-hand side of the second and third equation and multiply the modulo of the respective other equation and form their overall sum yielding $2^4 \cdot p^* \cdot \prod_{i=1}^{m'} p_i (H^2 - x^2) + K(\tau^2 - x^2) \equiv 0 \mod 2^4 \cdot p^* \cdot \prod_{i=1}^{m'} p_i \cdot K$. Using a little re-arranging this finally yields the desired QUADRATIC CONGRUENCES problem.

Before we finally present the reduction, we first present a lemma about the size of the product of prime numbers, which comes in handy in the respective theorem. Due to space restrictions, the correctness proofs of the following lemma and theorem are omitted. They can be found in the full version [15].

Lemma 1. *Denote by q_i the ith prime number. The product of the first k prime numbers $\prod_{i=1}^{k} q_i$ is bounded by $2^{2k \log(k)}$ for all $k \geq 2$.*

Theorem 1. *The* QUADRATIC CONGRUENCES *problem is NP-hard even if the prime factorization of β is given and each prime factor greater than 2 occurs at most once and the prime factor 2 occurs 4 times.*

Proof. Transformation: We show a reduction from the well-known NP-hard problem 3-SAT where we are given a 3-SAT formula Φ with n_3 variables and m_3 clauses. First, eliminate duplicate clauses from Φ and those where some variable x_i and its negation $\bar{x}_i$ appear together. Call the resulting formula Φ', the number of occurring variables n' and denote by m' the number of appearing clauses respectively. Let $\Sigma = (\sigma_1, \ldots, \sigma_{m'})$ be some enumeration of the clauses. Denote by $p_0, \ldots, p_{2m'}$ the first $2m' + 1$ prime numbers. Compute

$$\tau_{\Phi'} = -\sum_{i=1}^{m'} \prod_{j=1}^{i} p_j.$$

Further, compute for each $i \in 1, 2, \ldots, n'$:

$$f_i^+ = \sum_{x_i \in \sigma_j} \prod_{k=1}^{j} p_k \quad \text{and} \quad f_i^- = \sum_{\bar{x}_i \in \sigma_j} \prod_{k=1}^{j} p_k.$$

Set $\nu = 2m' + n'$. Compute the coefficients c_j for all $j = 0, 1, \ldots, \nu$ as follows: Set $c_0 = 0$. For $j = 1, \ldots, 2m'$ set

$$c_j = -\frac{1}{2} \prod_{i=1}^{j} p_i \quad \text{for } j = 2k - 1 \quad \text{and} \quad c_j = -\prod_{i=1}^{j} p_i \quad \text{for } j = 2k.$$

Compute the remaining ones for $j = 1, \ldots n'$ as $c_{2m'+j} = 1/2 \cdot (f_j^+ - f_j^-)$.

Further, set $\tau = \tau_{\Phi'} + \sum_{j=0}^{\nu} c_j + \sum_{i=1}^{n'} f_i^-$.

Denote by $q_1, \ldots, q_{\nu^2 + 2\nu + 1}$ the first $\nu^2 + 2\nu + 1$ prime numbers. Let $p_{0,0}, p_{0,1}, \ldots, p_{0,\nu}, p_{1,0}, \ldots, p_{\nu,\nu}$ be the first $(\nu + 1)^2 = \nu^2 + 2\nu + 1$ prime numbers greater than $(4(\nu + 1)2^3 \prod_{i=1}^{\nu^2 + 2\nu + 1} q_i)^{1/((\nu^2 + 2\nu + 1) \log(\nu^2 + 2\nu + 1))}$ and greater than $p_{2m'}$. Define p^* as the $(\nu^2 + 2\nu + 2m' + 13)$th prime number.

Determine the parameters θ_j for $j = 0, 1, \ldots, \nu$ as the least θ_j satisfying:

$$\theta_j \equiv c_j \bmod 2^3 \cdot p^* \prod_{i=1}^{m'} p_i, \theta_j \equiv 0 \bmod \prod_{i=0, i \neq j}^{\nu} \prod_{k=0}^{\nu} p_{i,k}, \theta_j \not\equiv 0 \bmod p_{j,1}.$$

Set the following parameters:

$$H = \sum_{j=0}^{\nu} \theta_j \quad \text{and} \quad K = \prod_{i=0}^{\nu} \prod_{k=0}^{\nu} p_{i,j}.$$

Finally, set

$$\alpha = (2^4 \cdot p^* \prod_{i=1}^{m'} p_i + K)^{-1} (K\tau^2 + 2^4 \cdot p^* \prod_{i=1}^{m'} p_i \cdot H^2), \beta = 2^4 p^* \prod_{i=1}^{m'} p_i \cdot K, \gamma = H.$$

where $(2^4 \cdot p^* \prod_{i=1}^{m'} p_i + K)^{-1}$ is the inverse of $(2^4 \cdot p^* \prod_{i=1}^{m'} p_i + K) \bmod 2^4 \cdot p^* \prod_{i=1}^{m'} p_i \cdot K$.

□

Now we proved that the QUADRATIC CONGRUENCES problem is NP-hard even in the restricted case where all prime factors in β only appear at most once (except 2). Denote by $B = b_1^{\beta_1}, \ldots, b_{n_{QC}}^{\beta_{n_{QC}}}$ the prime factorization of β where $b_1, \ldots, b_{n_{QC}}$ denotes the different prime factors of β and β_i the occurrence of b_i. To apply the ETH, we also have to estimate the dimensions of the generated instance. We only use the first $O((n_3 + m_3)^2)$ prime numbers, thus their size can be bounded by $O((n_3 + m_3)^2 \log(n_3 + m_3))$. The numbers α, β, γ are products of these prime numbers. As the product of the first k prime numbers is bounded by $2^{2k \log(k)}$, see Lemma 1, we can thus bound these numbers by $2^{O((n_3 + m_3)^2 \log(n_3 + m_3))}$. This yields the following theorem, for the full proof see [15]:

Theorem 2. *An instance of the* 3-SAT *problem with n_3 variables and m_3 clauses is reducible to an instance of the* QUADRATIC CONGRUENCES *problem in polynomial time with the properties that* $\alpha, \beta, \gamma \in 2^{O((n_3+m_3)^2 \log(n_3+m_3))}$, $n_{QC} \in O((n_3+m_3)^2)$, $max_i\{b_i\} \in O((n_3+m_3)^2 \log(n_3+m_3))$, *and each prime factor in β occurs at most once except the prime factor 2 which occurs four times.*

3 Reduction from the Quadratic Congruences Problem

This sections presents the reduction from the QUADRATIC CONGRUENCES problem to the 2-STAGE ILP problem. First, we present a transformation of an instance of the QUADRATIC CONGRUENCES problem to an instance of the NON-UNIQUE REMAINDER problem. This problem was not considered so far and serves as an intermediate step in this chapter. However, it might be of independent interest as it generalizes the prominent Chinese Remainder theorem. Secondly, we show how an instance of the NON-UNIQUE REMAINDER problem can be modeled as a 2-stage stochastic ILP.

Recall that in the NON-UNIQUE REMAINDER problem, we are given numbers $x_1, \ldots, x_{n_{NR}}, y_1, \ldots, y_{n_{NR}}, q_1, \ldots, q_{n_{NR}}, \zeta \in \mathbb{N}$ where the q_is are pairwise co-prime. The question is to decide whether there exists a natural number z satisfying $z \bmod q_i \in \{x_i, y_i\}$ simultaniously for all $i \in \{1, 2, \ldots, n_{NR}\}$ and which is smaller or equal to ζ. In other words, we either should met the residue x_i or y_i. Thus, we can re-write the equation as $z \equiv x_i \bmod q_i$ or $z \equiv y_i \bmod q_i$ for all i. Indeed, this problem becomes easy if $x_i = y_i$ for all i, i.e., we know the remainder we want to satisfy for each equation [25]: First, compute s_i and r_i with $r_i \cdot q_i + s_i \cdot \prod_{j=1, j \neq i}^{n_{NR}} q_j = 1$ for all i using the Extended Euclidean algorithm. Now it holds that $s_i \cdot \prod_{j=1, j \neq i}^{n_{NR}} q_j \equiv 1 \bmod q_i$ as q_i and $\prod_{j=1, j \neq i}^{n_{NR}} q_j$ are coprime, and $s_i \cdot \prod_{j=1, j \neq i}^{n_{NR}} q_j \equiv 0 \bmod q_j$ for $j \neq i$. Thus, the smallest solution corresponds to $z = \sum_{i=1}^{n_{NR}} x_i \cdot s_i \cdot \prod_{j=1, j \neq i}^{n_{NR}} q_j$ due to the Chinese Remainder theorem [25]. Comparing z to the bound ζ finally yields the answer. Note that if n_{NR} is constant, we can solve the problem by testing all possible vectors $(v_1, \ldots, v_{n_{NR}})$ with $v_i \in \{x_i, y_i\}$ and then use the procedure explained above.

The idea of the following reduction is that we first split up the equation of the QUADRATIC CONGRUENCES problem for each prime factor of β yielding n_{NR} many equations. The equivalence is preserved by that. Then we eliminate the square by defining the remainders x_i and y_i in a way that they can only be met if the number z is a square root itself and satisfies $z^2 \equiv \alpha \bmod \beta$. Due to space restrictions, the correctness proof is omitted and can be found in [15].

Theorem 3. *The* QUADRATIC CONGRUENCES *problem is reducible to the* NON-UNIQUE REMAINDER *problem in polynomial time with the properties that* $n_{NR} \in O(n_{QC})$, $max_{i \in \{1, \ldots, n_{NR}\}}\{q_i, x_i, y_i\} = O(max_{j \in \{1, \ldots, n_{QC}\}}\{b_j^{\beta_j}\})$, *and* $\zeta \in O(\gamma)$.

Proof. Transformation: Set $q_1 = b_1^{\beta_1}, \ldots, q_{n_{NR}} = b_{n_{QC}}^{\beta_{QC}}$ and $\zeta = \gamma$ where β_i denotes the occurrence of the prime factor b_i in the prime factorization of β.

Compute $\alpha_i \equiv \alpha \bmod q_i$. Set $x_i^2 = \alpha_i$ if there exists such an $x_i \in \mathbb{Z}_{q_i}$. Further, compute $y_i = -x_i + q_i$. If there is no such number x_i and thus y_i, produce a trivial no-instance.

Instance Size: The numbers we generate in the reduction equal the prime numbers of the QUADRATIC CONGRUENCES problem including their occurrence. Hence, it holds that $\max_{i \in \{1,\dots,n_{\mathrm{NR}}\}}\{q_i\} = O(\max_{j \in \{1,\dots,n_{\mathrm{QC}}\}}\{b_j^{\beta_j}\})$. Due to the modulo, this value also bounds x_i and y_i. The upper bound on a solution equals the ones from the instance of the QUADRATIC CONGRUENCES problem, i. e., $\zeta \in O(\gamma)$, and $n_{\mathrm{NR}} = n_{\mathrm{QC}}$ holds. $\qquad\square$

Finally, we reduce the NON-UNIQUE REMAINDER problem to the 2-STAGE ILP problem. Note that the considered 2-STAGE ILP problem is a decision problem. Thus, we only seek to determine whether there exists a feasible solution. We neither optimize a solution nor are we interested in the vector itself.

The main idea is that we can re-write each equation $z \bmod q_i \in \{x_i, y_i\}$ as the system $-z + \lambda_i^1 q_i + \lambda_i^2 x_i + \lambda_i^3 y_i = 0$, $\lambda_i^2 + \lambda_i^3 = 1$ and $\lambda_i^1, \lambda_i^2, \lambda_i^3 \in \mathbb{N}$. In other words, z contains arbitrary many often q_i and exactly once x_i or y_i. Finding a solution thus corresponds to finding the number z and the corresponding values for the λ_i^js. It is easy to see that we have two equations (and lower bounds) for each $z \bmod q_i \in \{x_i, y_i\}$, only one variable z occurring in all equations and the remaining ones are exclusive for each i. This directly translates to an 2-stage stochastic ILP, which we present in the following reduction. Due to space restrictions, the correctness proof is omitted and can be found in [15].

Theorem 4. *The* NON-UNIQUE REMAINDER *problem is reducible to the 2-STAGE ILP problem in polynomial time with the properties that $n \in O(n_{NR})$, $r, s, t, \|c\|_\infty, \|b\|_\infty, \|\ell\|_\infty \in O(1)$, $\|u\|_\infty \in O(\zeta)$, and $\Delta \in O(\max_i\{q_i\})$.*

Proof. Transformation: Having the instance for the NON-UNIQUE REMAINDER problem at hand we construct our ILP as follows with $n = n_{\mathrm{NR}}$:

$$A \cdot x = \begin{pmatrix} -1 & q_1 & x_1 & y_1 & 0 & \dots & 0 & 0 & \dots & 0 \\ 0 & 0 & 1 & 1 & 0 & \dots & 0 & 0 & \dots & 0 \\ \vdots & \vdots & \ddots & \ddots & \ddots & \ddots & \ddots & \ddots & \ddots & \vdots \\ -1 & 0 & \dots & 0 & 0 & \dots & 0 & q_n & x_n & y_n \\ 0 & 0 & \dots & 0 & 0 & \dots & 0 & 0 & 1 & 1 \end{pmatrix} \cdot x = b = \begin{pmatrix} 0 \\ 1 \\ \vdots \\ 0 \\ 1 \end{pmatrix}.$$

All variables get a lower bound of 0 and an upper bound of ζ. We can set the objective function arbitrarily as we are just searching for a feasible solution, hence we set it to $c = (0, 0, \dots, 0)^\top$.

Instance Size: Due to our construction, it holds that $t = 2, r = 1, s = 3$. The number n of repeated blocks equals the number n_{NR} of equations in the instance of the NON-UNIQUE REMAINDER problem. The largest entry Δ can be bounded by $\max_i\{q_i\}$. The lower and upper bounds are at most $\|u\|_\infty = O(\zeta)$, $\|\ell\|_\infty = O(1)$. The objective function c is set to zero and is thus of constant size. The largest value in the right-hand side is $\|b\|_\infty = 1$. $\qquad\square$

4 Runtime Bounds for 2-Stage Stochastic ILPs Under ETH

This sections presents the proof that the double exponential running time in the current state-of-the-art algorithms is nearly tight assuming the Exponential Time Hypothesis (ETH). To do so, we make use of the reductions above showing that we can transform an instance of the 3-SAT problem to an instance of the 2-STAGE ILP problem.

Corollary 1. *The* 2-STAGE ILP *problem cannot be solved in time less than* $2^{\delta\sqrt{n}}$ *for some* $\delta > 0$ *assuming ETH.*

Proof. Suppose the opposite. That is, there is an algorithm solving the 2-STAGE ILP problem in time less than $2^{\delta\sqrt{n}}$. Let an instance of the 3-SAT problem with n_3 variables and m_3 clauses be given. Due to the Sparsification lemma, we may assume that $m_3 \in O(n_3)$ [13]. The Sparsification lemma states that any 3-SAT formula can be replaced by subexponentially many 3-SAT formulas, each with a linear number of clauses with respect to the number of variables. The original formula is satisfiable if at least one of the new formulas is. This yields that if we cannot decide a 3-SAT problem in subexponential time, we can also not do so for a 3-SAT problem where $m_3 \in O(n_3)$.

We can reduce such an instance to an instance of the QUADRATIC CONGRUENCES problem in polynomial time regarding n_3 such that $n_{QC} \in O(n_3^2)$, $max_i\{b_i\} \in O(n_3^2 \log(n_3))$, $\alpha, \beta, \gamma = 2^{O(n_3^2 \log(n_3))}$, see Theorems 1 and 2.

Next, we reduce this instance to an instance of the NON-UNIQUE REMAINDER problem. Using Theorem 3, this yields the parameter sizes $n_{NR} \in O(n_3^2)$, $max_{i \in \{1,\dots,n_{NR}\}}\{q_i, x_i, y_i\} = O(n_3^2 \log(n_3))$, and finally $\zeta \in 2^{O(n_3^2 \log(n_3))}$. Note that all prime numbers greater than 2 appear at most once in the prime factorization of β and 2 appears 4 times. Thus, the largest q_i, which corresponds to $max_i\{b_i^{\beta_i}\}$ equals the largest prime number in the QUADRATIC CONGRUENCES problem: The largest prime number is at least the $(\nu^2 + 2\nu + 2m' + 13) \geq 13$th prime number by a rough estimation. The 13th prime number is 41 and thus larger than $2^4 = 16$.

Finally, we reduce that instance to an instance of the 2-STAGE ILP problem with parameters $r, s, t, ||c||_\infty, ||b||_\infty, ||\ell||_\infty \in O(1)$, $||u||_\infty \in 2^{O(n_3^2 \log(n_3))}$, $n \in O(n_3^2)$, and $\Delta \in O(n_3^2 \log(n_3))$, see Theorem 4.

Hence, if there is an algorithm solving the 2-STAGE ILP problem in time less than $2^{\delta\sqrt{n}}$ this would result in the 3-SAT problem to be solved in time less than $2^{\delta\sqrt{n}} = 2^{\delta\sqrt{C_1 n_3^2}} = 2^{\delta(C_2 n_3)}$ for some constants C_1, C_2. Setting $\delta_3 \leq \delta/C_2$, this would violate the ETH. $\square$

To prove our main result, we have to reduce the size of the coefficients in the constraint matrix. To do so, we encode large coefficients into submatrices only extending the matrix dimensions slightly. A similar approach is used for example in [18] to prove a lower bound for the size of inclusion minimal kern-elements of 2-stage stochastic ILPs or in [19] to decrease the value of Δ in the matrices.

Theorem 5. *The* 2-STAGE ILP *problem cannot be solved in time less than* $2^{2^{\delta(s+t)}}|I|^{O(1)}$ *for some constant* $\delta > 0$, *even if* $r = 1$, $\Delta, ||b||_\infty, ||c||_\infty, ||b||_\infty \in O(1)$, *assuming ETH. Here* $|I|$ *denotes the encoding length of the total input.*

Proof. First, we show that we can reduce the size of Δ to $O(1)$ by altering the ILP. We do so by encoding large coefficients with base 2, which comes at the cost of enlarged dimensions of the constraint matrix. Let $enc(x)$ be the encoding of a number x with base 2. Further, let $enc_i(x)$ be the ith number of $enc(x)$. Finally, $enc_0(x)$ denotes the last significant number of the encoding. Hence, the encoding of a number x is $enc(x) = enc_0(x)enc_1(x)\ldots enc_{\lfloor \log(\Delta)\rfloor}(x)$ and x can be reconstructed by $x = \sum_{i=0}^{\lfloor \log(\Delta)\rfloor} enc_i(x) \cdot 2^i$. Let a matrix E be defined as

$$E = \begin{pmatrix} 2 & -1 & 0 & \cdots & 0 \\ 0 & 2 & -1 & 0 \cdots & 0 \\ \vdots & \ddots & \ddots & \ddots & \\ 0 & \cdots & 0 & 2 & -1 \end{pmatrix}.$$

We re-write the constraint matrix as follows: For each coefficient $a > 1$, we insert its encoding $enc(a)$ and beneath we place the matrix E (including zero rows, such that the E matrices of a block do not collide). We have to fix the dimensions for the first row in the constraint matrix, the columns without great coefficients and the right-hand side b by filling the matrix at the corresponding positions with zeros. The independent blocks consisting of $enc(a)$ and the matrix E beneath correctly encodes the number $a > 1$, i.e., it preserves the solution space: Let x_a be the number in the solution corresponding to the column with entry a of the original instance. The solution for the altered column (i.e., the sub-matrix) is $(x_a \cdot 2^0, x_a \cdot 2^1, \ldots, x_a \cdot 2^{\lfloor \log(\Delta)\rfloor})$. The additional factor of 2 for each subsequent entry is due to the diagonal of E. It is easy to see that $a \cdot x_a = \sum_{i=0}^{\lfloor \log(\Delta)\rfloor} enc_i(a) \cdot x_a \cdot 2^i$ as we can extract x_a on the right-hand side and solely the encoding of a remains. Thus, the solutions of the original matrix and the altered one transfer to each other.

Regarding the dimensions, each coefficient $a > 1$ is replaced by a $(O(\log(\Delta)) \times O(\log(\Delta)))$ matrix. Thus, the dimension expands to $t' = t \cdot O(\log(\Delta)) = O(\log(\Delta))$, $s' = s \cdot O(\log(\Delta)) = O(\log(\Delta))$, while r and n stay the same. Regarding the bounds, the lower bound for all new variables is also zero. For the upper bounds, we allow an additional factor of 2^i for the ith value of the encoding. Thus, $||u'||_\infty = 2^{\lfloor \log(\Delta)\rfloor}||u||_\infty$. We get that the largest coefficient is bounded by $\Delta' = O(1)$. The right-hand side b enlarges to b' with $O(n\log(\Delta))$ entries.

Theorem 1 shows a transformation of an instance of the 3-SAT problem with n_3 variables and m_3 clauses to a 2-stage stochastic ILP with parameters $r, s, t, ||c||_\infty, ||b||_\infty, ||\ell||_\infty \in O(1)$, $||u||_\infty \in 2^{O(n_3^2 \log(n_3))}$, $n \in O(n_3^2)$, and $\Delta \in O(n_3^2 \log(n_3))$. Further, as explained above, we can transform this ILP to an equivalent one where $\Delta' = O(1)$, $t' = O(\log(\Delta)) = O(\log(n_3^2 \log(n_3))) = O(\log(n_3))$, $s' = O(\log(\Delta)) = O(\log(n_3^2 \log(n_3))) = O(\log(n_3))$, $b' \in \mathbb{Z}^{O(n_3^2 \log(n_3))}$, and finally $||u'||_\infty = 2^{\lfloor \log(\Delta)\rfloor}||u||_\infty =$

$2^{\lfloor \log(n_3^2 \log(n_3)) \rfloor} 2^{O(n_3^2 \log(n_3))} = 2^{O(n_3^2 \log(n_3))}$, while r and n stay the same. The encoding length $|I|$ is then given by $|I| = (nt'(r + ns')) \log(\Delta') + (r + ns') \log(\||\ell\||_\infty) + (r + ns') \log(\||u'\||_\infty) + nt' \log(\||b'\||_\infty) + (r + ns') \log(\||c\||_\infty) = 2^{O(n_3^2)}$.

Hence, if there is an algorithm solving the 2-STAGE ILP problem in time less than $2^{2^{\delta(s+t)}} |I|^{O(1)}$ this would result in the 3-SAT problem to be solved in time less than $2^{2^{\delta(s+t)}} |I|^{O(1)} = 2^{2^{\delta(C_1 \log(n_3) + C_2 \log(n_3))}} 2 n_3^{O(1)} = 2^{2^{\delta C_3 \log(n_3)}} 2 n_3^{O(1)} = 2^{n_3^{\delta \cdot C_3}} 2 n_3^{O(1)} = 2^{n_3^{\delta \cdot C_4}}$ for some constants C_1, C_2, C_3, C_4. Setting $\delta = \delta'/C_4$, we get $2^{n_3^{\delta C_4}} = 2^{n_3^{\delta'}}$. As it holds for sufficient large x and $\epsilon < 1$ that $x^\epsilon < \epsilon x$, it follows that $2^{n_3^{\delta'}} < 2^{\delta' n_3}$. This violates the ETH. Note that this result even holds if $r = 1$, $\Delta, \||c\||_\infty, \||b\||_\infty, \||\ell\||_\infty \in O(1)$ as constructed by our reductions.   □

References

1. Albareda-Sambola, M., van der Vlerk, M.H., Fernández, E.: Exact solutions to a class of stochastic generalized assignment problems. Eur. J. Oper. Res. **173**(2), 465–487 (2006)
2. Brand, C., Koutecký, M., Ordyniak, S.: Parameterized algorithms for MILPs with small treedepth. CoRR, abs/1912.03501 (2019)
3. Chen, L., Koutecký, M., Xu, L., Shi, W.: New bounds on augmenting steps of block-structured integer programs. In ESA, vol. 173 of LIPIcs, pp. 33:1–33:19. Schloss Dagstuhl - Leibniz-Zentrum für Informatik (2020)
4. Cslovjecsek, J., Eisenbrand, F., Hunkenschröder, C., Weismantel, R., Rohwedder, L.: Block-structured integer and linear programming in strongly polynomial and near linear time. CoRR, abs/2002.07745v2 (2020)
5. Cslovjecsek, J., Eisenbrand, F., Pilipczuk, M., Venzin, M., Weismantel, R.: Efficient sequential and parallel algorithms for multistage stochastic integer programming using proximity. CoRR, abs/2012.11742 (2020)
6. Cygan, M., et al.: Parameterized Algorithms. Springer, Cham (2015). https://doi.org/10.1007/978-3-319-21275-3
7. Dempster, M.A.H., Fisher, M.L., Jansen, L., Lageweg, B.J., Lenstra, J.K., Kan, A.H.G.R.: Analysis of heuristics for stochastic programming: results for hierarchical scheduling problems. Math. Oper. Res. **8**(4), 525–537 (1983)
8. Eisenbrand, F., Hunkenschröder, C., Klein, K.-M., Koutecký, M., Levin, A., Onn, S.: An algorithmic theory of integer programming. CoRR, abs/1904.01361 (2019)
9. Gavenčak, T., Koutecký, M., Knop, D.: Integer programming in parameterized complexity: five miniatures. Discrete Optim., 100596 (2020)
10. Hemmecke, R., Köppe, M., Weismantel, R.: A polynomial-time algorithm for optimizing over N-Fold 4-block decomposable integer programs. In: Eisenbrand, F., Shepherd, F.B. (eds.) IPCO 2010. LNCS, vol. 6080, pp. 219–229. Springer, Heidelberg (2010). https://doi.org/10.1007/978-3-642-13036-6_17
11. Hemmecke, R., Schultz, R.: Decomposition of test sets in stochastic integer programming. Math. Program. **94**(2–3), 323–341 (2003)
12. Impagliazzo, R., Paturi, R.: On the complexity of k-SAT. J. Comput. Syst. Sci. **62**(2), 367–375 (2001)
13. Impagliazzo, R., Paturi, R., Zane, F.: Which problems have strongly exponential complexity? J. Comput. Syst. Sci. **63**(4), 512–530 (2001)

14. Ireland, K., Rosen, M.: A Classical Introduction to Modern Number Theory. GTM, vol. 84. Springer, New York (1990). https://doi.org/10.1007/978-1-4757-2103-4

15. Jansen, K., Klein, K.-M., Lassota, A.: The double exponential runtime is tight for 2-stage stochastic ILPs (2020). http://arxiv.org/abs/2008.12928arXiv:2008.12928

16. Jansen, K., Lassota, A., Rohwedder, L.: Near-linear time algorithm for n-fold ILPs via color coding. In: ICALP, vol. 132 of LIPIcs, pp. 75:1–75:13 (2019)

17. Kall, P., Wallace, S.W.: Stochastic Programming. Springer (1994). https://doi.org/10.1007/978-3-642-88272-2.pdf

18. Klein, K.-M.: About the complexity of two-stage stochastic IPs. In: Bienstock, D., Zambelli, G. (eds.) IPCO 2020. LNCS, vol. 12125, pp. 252–265. Springer, Cham (2020). https://doi.org/10.1007/978-3-030-45771-6_20

19. Knop, D., Pilipczuk, M., Wrochna, M.: Tight complexity lower bounds for integer linear programming with few constraints. In: STACS, vol. 126 of LIPIcs, pp. 44:1–44:15. Schloss Dagstuhl - Leibniz-Zentrum für Informatik (2019)

20. Küçükyavuz, S., Sen, S.: An introduction to two-stage stochastic mixed-integer programming. In: Leading Developments from INFORMS Communities, pp. 1–27. INFORMS (2017)

21. Laporte, G., Louveaux, F.V., Mercure, H.: A priori optimization of the probabilistic traveling salesman problem. Oper. Res. 42(3), 543–549 (1994)

22. De Loera, J.A., Hemmecke, R., Onn, S., Weismantel, R.: N-fold integer programming. Discret. Optim. 5(2), 231–241 (2008)

23. Manders, K.L., Adleman, L.M.: NP-complete decision problems for binary quadratics. J. Comput. Syst. Sci. 16(2), 168–184 (1978)

24. Schultz, R., Stougie, L., Van Der Vlerk, M.H.: Two-stage stochastic integer programming: a survey. Stat. Neerl. 50(3), 404–416 (1996)

25. Wagon, S.: Mathematica in action. Springer Science and Business Media (1999)

Fast Quantum Subroutines
for the Simplex Method

Giacomo Nannicini[(✉)]

IBM Quantum, IBM T. J. Watson Research Center, Yorktown Heights, NY, USA
nannicini@us.ibm.com

Abstract. We propose quantum subroutines for the simplex method that avoid classical computation of the basis inverse. For an $m \times n$ constraint matrix with at most d_c nonzero elements per column, at most d nonzero elements per column or row of the basis, basis condition number κ, and optimality tolerance ϵ, we show that pricing can be performed in $\tilde{O}(\frac{1}{\epsilon}\kappa d\sqrt{n}(d_c n + dm))$ time, where the $\tilde{O}$ notation hides polylogarithmic factors. If the ratio n/m is larger than a certain threshold, the running time of the quantum subroutine can be reduced to $\tilde{O}(\frac{1}{\epsilon}\kappa d^{1.5}\sqrt{d_c}n\sqrt{m})$. The steepest edge pivoting rule also admits a quantum implementation, increasing the running time by a factor κ^2. Classically, pricing requires $O(d_c^{0.7}m^{1.9} + m^{2+o(1)} + d_c n)$ time in the worst case using the fastest known algorithm for sparse matrix multiplication, and $O(d_c^{0.7}m^{1.9} + m^{2+o(1)} + m^2 n)$ with steepest edge. Furthermore, we show that the ratio test can be performed in $\tilde{O}(\frac{t}{\delta}\kappa d^2 m^{1.5})$ time, where t, δ determine a feasibility tolerance; classically, this requires $O(m^2)$ time in the worst case. For well-conditioned sparse problems the quantum subroutines scale better in m and n, and may therefore have a worst-case asymptotic advantage. An important feature of our paper is that this asymptotic speedup does not depend on the data being available in some "quantum form": the input of our quantum subroutines is the natural classical description of the problem, and the output is the index of the variables that should leave or enter the basis.

1 Introduction

The simplex method is one of the most impactful algorithms of the past century; to this day, it is widely used in a variety of applications. This paper studies some opportunities for quantum computers to accelerate the simplex method.

The use of quantum computers for optimization is a central research question that has attracted a significant amount of attention. Thanks to a quadratic speedup for unstructured search problems and [19] exponential speedups in the solution of linear systems [10,20], it seems natural to try to translate those speedups into faster optimization algorithms, since linear systems appear as a building block in many optimization procedures. However, few results in this direction are known. A possible reason for the paucity of results is the difficulty encountered when applying a quantum algorithm to a problem whose data is

© Springer Nature Switzerland AG 2021
M. Singh and D. P. Williamson (Eds.): IPCO 2021, LNCS 12707, pp. 311–325, 2021.
https://doi.org/10.1007/978-3-030-73879-2_22

classically described, and a classical description of the solution is required. We provide a simple example to illustrate this difficulty.

Suppose we want to solve the linear system $Ax = b$, where A is an $m \times m$ invertible matrix with at most d nonzero elements per column or row. Using the fastest known quantum linear systems algorithm [10], the gate complexity of this operation is $\tilde{O}(d\kappa \max\{T_A, T_b\})$, where T_A, T_b indicate the gate complexity necessary to describe A, b, and κ is the condition number of A. (We measure the running time for quantum subroutines as the number of basic gates, as is standard in the literature). Notice that m does not appear in the running time, as the dependence is polylogarithmic. We need $\tilde{O}(dm)$ gates to implement T_A for sparse A in the gate model, as will be discussed in the following; and $\tilde{O}(m)$ gates are necessary to implement T_b. This is natural since A has $O(dm)$ nonzero elements and b has $O(m)$ nonzero elements. To extract the solution $x = A^{-1}b$ with precision δ we can use the fast tomography algorithm of [21], yielding a total running time $\tilde{O}(\frac{1}{\delta^2}\kappa d^2 m^2)$. This is slower than a classical LU decomposition of A, which runs in time $O(d^{0.7}m^{1.9} + m^{2+o(1)})$ [28]. Thus, naive application of quantum linear system algorithms (QLSAs) does not give any advantage.

Despite the aforementioned difficulties, a few fast quantum optimization algorithms exist—not necessarily based on QLSAs; examples are the SDP algorithms of [3,6,21] and the LP algorithm of [2]. We provide a more detailed literature review in Sect. 2. Existing quantum optimization algorithms have one of these two assumptions: (i) that having quantum input/output is acceptable, ignoring the cost of a classical translation, or (ii) that qRAM, a form of quantum storage whose physical realizability is unclear, is available. qRAM allows data preparation subroutines that are exponentially faster than what would be required under the standard gate model. Both assumptions have the merit of leading to interesting algorithmic developments, but it is still an open question to find practical situations in which they are satisfied, particularly in the context of traditional optimization applications. We remark that the assumptions can be dropped and the algorithms can be implemented in the standard gate model, but the running time increases significantly. In this paper we propose quantum subroutines that may yield asymptotic speedup even without these two assumptions.

Our Results. For brevity, from now on we assume that the reader is familiar with standard linear optimization terminology; we refer to [4] for a comprehensive treatment of LPs. The simplex method aims to solve $\min c^\top x$, s.t.: $Ax = b, x \geq 0$, where $A \in \mathbb{R}^{m \times n}$ with at most d_c nonzero elements per column. It keeps a basis, i.e., a set of linearly independent columns of A, and repeatedly moves to a different basis that defines a solution with better objective function value. As is common in the literature, we use the term "basis" to refer to both the set of columns, and the corresponding submatrix of A, depending on context. We denote by B the set of basic columns, N the set of nonbasic columns, with corresponding submatrices A_B, A_N. The maximum number of nonzero elements in any column or row of the basis submatrix is denoted d. The basis change (called a *pivot*) is performed by determining a new column that should enter the basis, and removing one column from the current basis. Assessing which columns can enter the basis is called *pricing*, and it is asymptotically the most expensive step: it requires computing the basis inverse and looping over all the columns in the worst case, for a total of $O(d_c^{0.7} m^{1.9} + m^{2+o(1)} + d_c n)$ operations using the matrix multiplication algorithm of [28]. If the basis inverse is known, i.e., updated from a previous iteration, the worst-case running time is $O(m^2 + d_c n)$. With the steepest edge pivoting rule, that can achieve better performance on real-world problems by reducing the number of required iterations [15], the term $d_c n$ in the two running time expressions above increases to $m^2 n$.[1]

In the following, we denote by $T_{LS}(L, R, \epsilon)$ the running time of a QLSA on the linear system $Lx = r$ with precision ϵ, where r is any column of R. We use this notation since we frequently use a QLSA on a superposition of r.h.s. vectors, and the cost of the oracle that prepares the r.h.s. must account for the ability to prepare such superposition.

We show that we can apply Grover search to choose an entering column, so that the running time scales as $O(\sqrt{n})$ rather than $O(n)$. To apply Grover search we need a quantum oracle that determines if a column is eligible to enter the basis, i.e., if it has negative reduced cost. We propose a construction for this oracle using a QLSA, several gadgets to make state amplitudes interfere in a certain way, and amplitude estimation [7]. The construction avoids classical computation of the basis inverse. The overall running time of the oracle is $\tilde{O}(\frac{1}{\epsilon} T_{LS}(A_B, A_N, \frac{\epsilon}{2}))$, where ϵ is the precision for the reduced costs (i.e., the optimality tolerance). Using the QLSA of [10], in the circuit model and without taking advantage of the structure of A besides sparsity, this gives a total running time of $\tilde{O}(\frac{1}{\epsilon} \kappa d \sqrt{n}(d_c n + dm))$. If the ratio n/m is large, we can find a better tradeoff between the Grover speedup and the data preparation subroutines, and improve the running time of the quantum pricing algorithm to $\tilde{O}(\frac{1}{\epsilon} \kappa d^{1.5} \sqrt{d_c} n \sqrt{m})$. We can also apply the steepest edge pivoting rule increasing the running time by a factor κ^2. We summarize this below.

Theorem 1. *There exist quantum subroutines to identify if a basis is optimal, or determine a column with negative reduced cost, with running time $\tilde{O}(\frac{1}{\epsilon} \sqrt{n} T_{LS}(A_B, A_N, \frac{\epsilon}{2}))$. In the gate model without qRAM, this is $\tilde{O}(\frac{1}{\epsilon} \kappa d \sqrt{n}(d_c n + dm))$, which can be reduced to $\tilde{O}(\frac{1}{\epsilon} \kappa d^{1.5} \sqrt{d_c} n \sqrt{m})$ if the ratio n/m is larger than $2 \frac{d}{d_c}$. If qRAM to store A is available, the running time is $\tilde{O}(\frac{1}{\epsilon} \kappa \sqrt{mn})$. With the steepest edge pivoting rule, the running time of the subroutine to determine a column entering the basis increases by a factor κ^2.*

The regime with large n/m includes many natural LP formulations; e.g., the LP relaxations of cutting stock problems, vehicle routing problems, or any other formulation that is generally solved by column generation [24]. The running time of the quantum subroutines depends explicitly on the condition number of the basis and the precision of reduced costs ϵ is fixed, while with classical Gaussian elimination κ is not explicit, but the ϵ obtained would depend on

---

[1] In practical implementations of the simplex method, steepest edge is typically used only for dual simplex algorithms, due to its high computational cost; an approximate variant, with more efficient updates, is used in primal simplex instead.

it. If A is structured, the quantum running time can decrease significantly: for example, if A differs from the assignment problem constraint matrix only for a polylogarithmic number of elements, then its description in the sparse oracle access model used in this paper requires time $\tilde{O}(1)$, rather than $\tilde{O}(d_c n)$. Our running time analysis assumes that the matrix is sparse but the sparsity pattern is unstructured.

If pricing is performed via our quantum subroutine, we obtain the index of a column that has negative reduced cost with arbitrarily high probability. To determine which column should leave the basis, we have to perform the *ratio test*. Using techniques similar to those used for the pricing step, we can identify the column that leaves the basis in time $\tilde{O}(\frac{t}{\delta}\kappa d^2 m^{1.5})$, where δ and t are precision parameters of this step. In particular, t determines how well we approximate the minimum of the ratio test performed by the classical algorithm (i.e., the relative error is $\frac{2t+1}{2t-1} - 1$; there is also an absolute error controlled by t because attaining a purely relative error bound would require infinite precision). Classically, the ratio test requires time $O(m^2)$ in the worst case, because the basis inverse could be dense even if the basis is sparse. We summarize this result below.

Theorem 2. *There exists a quantum subroutine to identify if a nonbasic column proves unboundedness of the LP in time $\tilde{O}(\frac{\sqrt{m}}{\delta}T_{LS}(A_B, A_B, \delta))$. There also exists a quantum subroutine to perform the ratio test in time $\tilde{O}(\frac{t\sqrt{m}}{\delta}T_{LS}(A_B, A_B, \delta))$, returning an approximate minimizer of the ratio test with relative error $\frac{2t+1}{2t-1} - 1$ and absolute error proportional to $\frac{2}{2t-1}$. In the gate model without qRAM, the running times are respectively $\tilde{O}(\frac{1}{\delta}\kappa d^2 m^{1.5})$ and $\tilde{O}(\frac{t}{\delta}\kappa d^2 m^{1.5})$. If qRAM to store A and b is available, the running times are respectively $\tilde{O}(\frac{1}{\delta}\kappa m)$ and $\tilde{O}(\frac{t}{\delta}\kappa m)$.*

It is known that for most practical LPs the maximum number of nonzeroes in a column is essentially constant; for example, on the entire benchmark set MIPLIB2010, less than 1% of the columns have more than 200 nonzeroes. The number of nonzeroes per row of the basis is also small: on MIPLIB2010, looking at the optimal bases of the LP relaxations, less than 0.01% of the rows have more than 50 nonzeroes. As m, n increase, so typically does the sparsity. For the two largest problems in the benchmark set MIPLIB2017, with $m, n \approx 10^7$, 99.998% of the columns have less than 30 nonzero elements. Hence, we expect many bases to be extremely sparse, and it is interesting to look at the scaling of the running time under the assumption that the sparsity parameters are at most polylogarithmic in m and n. In this case, the running time of the oracle for the reduced costs in the gate model without qRAM is $\tilde{O}(\frac{\kappa}{\epsilon}(n+m))$, giving a total running time for pricing of $\tilde{O}(\frac{1}{\epsilon}\kappa\sqrt{n}(n+m))$, and the steepest edge pricing oracle is a factor κ^2 slower. For a well-conditioned basis and under the assumption that $d = O(\log mn)$, we obtain running time $\tilde{O}(\frac{1}{\epsilon}\sqrt{n}(n+m))$ for the quantum pricing subroutine, which can be reduced to $\tilde{O}(\frac{1}{\epsilon}n\sqrt{m})$ if the ratio n/m is large; and running time $\tilde{O}(\frac{t}{\delta}m^{1.5})$ for the quantum ratio test subroutine. With qRAM, the gate complexity decreases further and we achieve essentially linear scaling ($O(\sqrt{mn})$ for pricing, $O(m)$ for the ratio test, if we only look at m and n).

Summarizing, the quantum subroutines that we propose can be asymptotically faster than the best known classical version of them, under the assumption—generally verified in practice—that the LPs are extremely sparse. Indeed, while the gate complexity of the quantum subroutines depends on some optimality and feasibility tolerances in addition to the classical input parameters (m, n, and the sparsity parameters), these tolerances do not scale with m and n. For well-conditioned problems, the quantum subroutines have better scaling in m and n, and this could turn into an asymptotic advantage. To achieve this potential advantage, we never explicitly compute the basis inverse, and rely on the quantum computer to indicate which columns should enter and leave the basis at each iteration. Similar to other papers in the quantum optimization literature, we use a classical algorithm (the simplex method) and accelerate the subroutines executed at each iteration of the simplex. However, our asymptotic speedup (when the condition number is small) does not depend on the availability of qRAM or of the data in "quantum form". The key insight to obtain an asymptotic speedup even with classical input and output is to interpret the simplex method as a collection of subroutines that output only integers, avoiding the cost of extracting real vectors from the quantum computer (tomography). Our algorithms require a polylogarithmic number of fault-tolerant qubits.

We remark that even with sophisticated pivot rules, the number of iterations of the classical simplex method could be exponential. The quantum version proposed in this paper does not circumvent this issue, and its worst-case running time is slower than of the fastest quantum algorithm for LPs [2]. However, it is well established that in practice the simplex method performs much better than the worst case, in terms of the number of iterations [13,27] as well as the complexity of a single iteration, see e.g. [11]. The most attractive feature of the simplex method is its excellent practical performance, and we hope that a quantized version, closely mimicking the classical counterpart while accelerating the linear algebra, would inherit this trait.

2   Comparison with the Existing Literature

The simplex method has been extensively studied in the operations research and computer science literature. Its exponential running time in theory [22] contrasts with its excellent practical performance. This paper proposes a quantization of the simplex method with the steepest edge pivoting rule, Dantzig's rule, or a randomized rule. We are not aware on an upper bound on the expected number of pivots under these rules. The closest paper is probably [5], see also [13] for a recent discussion on pivoting.

On the quantum side, to the best of our knowledge all algorithms for LPs are derived from some classical algorithm. [3,6] are based on the multiplicative weights update method. [8,21] are based on the interior point method. [2] is based on a reduction of LPs to two-player zero-sum games and the classical algorithm of [17]. For these methods the number of iterations is polynomial (or better), and is taken directly from the classical algorithm; the computational complexity

Table 1. Summary of quantum algorithms for linear programming.

Algorithm	Iteration cost	# Iterations	qRAM size	Comments
Multiplicative weights update [1]	$\tilde{O}\left((\sqrt{m}+\sqrt{n}\frac{Rr}{\epsilon})d\left(\frac{Rr}{\epsilon}\right)^2\right)$	$O(\frac{R^2\log n}{\epsilon^2})$	$\tilde{O}\left(\left(\frac{Rr}{\epsilon}\right)^2\right)$	Outputs dual solution and (quantum) primal solution; R, r depend on n, m in general
Interior point [21]	$\tilde{O}(\kappa^2\frac{n^{2.5}}{\xi^2})$	$O(\sqrt{n}\log\frac{n}{\epsilon})$	$\tilde{O}(mn)$	κ comes from intermediate matrices
Game-theoretical [2]	$\tilde{O}\left(\sqrt{d}\left(\frac{Rr}{\epsilon}\right)^{1.5}\right)$	$\tilde{O}\left(\left(\frac{Rr}{\epsilon}\right)^2\right)$	$\tilde{O}\left(\left(\frac{Rr}{\epsilon}\right)^2\right)$	R, r depend on n, m in general
This paper	$\tilde{O}(\frac{1}{\epsilon}\kappa d\sqrt{n}(d_c n + dm))$ or $\tilde{O}(\frac{1}{\epsilon}\kappa d^{1.5}\sqrt{d_c}n\sqrt{m})$, plus $\tilde{O}(\frac{t}{\delta}\kappa d^2 m^{1.5})$	N/A (exp)	No qRAM	κ comes from inter-mediate matrices; outputs only basis information
This paper	$\tilde{O}(\frac{1}{\epsilon}\kappa\sqrt{mn})$, plus $\tilde{O}(\frac{t}{\delta}\kappa m)$	N/A (exp)	$\tilde{O}(d_c n + m)$	

of each iteration is reduced taking advantage of quantum subroutines. A faster version of the quantum multiplicative weights update method for LPs is given in [1]. This algorithm has been recently dequantized [9], with similar performance for low-rank problems.

We summarize key features of several papers in Table 1. The table highlights the main advantages of our method, in particular the fact that the iteration running time is a polynomial with very low degree even without qRAM: for all other methods, each iteration is significantly more expensive, and even more so if we consider the gate complexity in case qRAM is not available (i.e., the gate complexity increases by a factor equal to the size of the qRAM). For some methods, the steep dependence on ϵ or on the size of primal/dual solutions R, r could be a limiting factor (classical algorithms used in practice usually depend polylogarithmically on these quantities). The methods proposed in this paper suffer from the same weakness as the classical simplex method: giving a sub-exponential upper bound on the number of iterations is difficult, although this has not prevented the simplex method from being extremely efficient in practice.

As remarked in the Table 1, the papers using the multiplicative weights update framework as well as [2] have a running time that depends on a parameter $\frac{Rr}{\epsilon}$, which may in turn depend on n, m—see the discussion in [3], as well as the application to experimental design discussed in [1] to understand the necessary tradeoffs to obtain a quantum speedup in n and m. Note that ϵ-optimality of the reduced cost, as used in the simplex method, is not a global optimality guarantee, and therefore our ϵ parameter is not directly comparable to the ϵ used in the other algorithms discussed in Table 1. We also remark that in our paper, as well as in [21], the running time depends on κ of some intermediate matrices: more specifically, in our paper κ is the condition number of the basis at each iteration of the simplex method. The optimality condition and the condition number are discussed in more detail subsequently in this paper. Regarding input and output, the algorithm presented in this paper has fully classical input, and outputs the current basis at each iteration; to obtain the (primal) solution, it is necessary to classically solve a single linear system of size $m \times m$ (this is more efficient than obtaining a solution via a QLSA).

3   Overview of the Simplex Method

The simplex method solves the linear optimization problem: $\min c^\top x$, s.t.:$Ax = b, x \geq 0$, where $A \in \mathbb{R}^{m \times n}$, $c \in \mathbb{R}^n$, $b \in \mathbb{R}^m$. A basis is a subset of m linearly independent columns of A. Given a basis B, assume that it is an ordered set and let $B(j)$ be the j-th element of the set. The set $N := \{1, \ldots, n\} \setminus B$ is called the set of nonbasic variables. We denote by A_B the square invertible submatrix of A corresponding to columns in B, and A_N the remaining submatrix. The term "basis" may refer to B or A_B, depending on context. The simplex method can be described compactly as follows; see, e.g., [4] for a more detailed treatment.

- Start with any basic feasible solution. (This is w.l.o.g. because it is always possible to find one.) Let B be the current basis, N the nonbasic variables, $x = A_B^{-1} b$ the current solution.
- Repeat the following steps:
  1. Compute the reduced costs for the nonbasic variables $\bar{c}_N^\top = c_N^\top - c_B^\top A_B^{-1} A_N$. This step is called *pricing*. If $\bar{c}_N \geq 0$ the basis is optimal: the algorithm terminates. Otherwise, choose $k : \bar{c}_k < 0$. Column k is the *pivot column*.
  2. Compute $u = A_B^{-1} A_k$. If $u \leq 0$, the optimal cost is unbounded from below: the algorithm terminates.
  3. If some component of u is positive, compute (this step is called *ratio test*):

  $$r^* := \min_{j=1,\ldots,m:u_j>0} \frac{x_{B(j)}}{u_j}. \tag{1}$$

  4. Let ℓ be such that $r^* = \frac{x_{B(\ell)}}{u_\ell}$. Row ℓ is the *pivot row*. Form a new basis replacing $B(\ell)$ with k. This step is called a *pivot*. Update $x = A_B^{-1} b$.

To perform the pricing step, we compute an LU factorization of the basis A_B; this requires time $O(d_c^{0.7} m^{1.9} + m^{2+o(1)})$ using fast sparse matrix multiplication techniques [28]. (In practice, the traditional $O(m^3)$ Gaussian elimination is used instead, but the factorization is not computed from scratch at every iteration.) Then, we can compute the vector $c_B^\top A_B^{-1}$ and finally perform the $O(n)$ calculations $c_k^\top - c_B^\top A_B^{-1} A_k$ for all $k \in N$; this requires an additional $O(d_c n)$ time, bringing the total time to $O(d_c^{0.7} m^{1.9} + m^{2+o(1)} + d_c n)$. To perform the ratio test, we need the vector $u = A_B^{-1} A_k$, which takes time $O(m^2)$ assuming the LU factorization of A_B is available from pricing. As remarked earlier, A_B^{-1} may not be dense if A_B is sparse; furthermore, the vectors $c_B^\top A_B^{-1}$ and $A_B^{-1} b$ can be updated from previous iterations exploiting the factorization update, so all these step could take significantly less time in practice. Finally, since the calculations are performed with finite precision, we use an optimality tolerance ϵ and the optimality criterion becomes $\bar{c}_N \geq -\epsilon$.

It is well known that the performance of the simplex method in practice depends on the pivoting rule. One of the simplest rules is Dantzig's rule, which chooses $k = \arg\min_h \bar{c}_h$. Modern implementations of the simplex method typically rely on more sophisticated pivoting rules. Among these, the steepest edge pivoting rule has been shown to lead to a significant reduction in the number of iterations [15]. The steepest edge column selection rule is: $k = \arg\min_h \frac{\bar{c}_h}{\|A_B^{-1} A_h\|}$. This computation requires knowledge of the norms $\|A_B^{-1} A_h\|$. [15] shows how to update these norms in time $O(m^2 n)$ (dual simplex updates are slightly cheaper, but in this paper we focus on the primal simplex).

4   Quantum Implementation: Overview

Before giving an overview of our methodology, we introduce some notation and useful results from the literature. The state of a quantum computer with q qubits is a unit vector in $(\mathbb{C}^2)^{\otimes q} = \mathbb{C}^{2^q}$; we denote the standard basis vectors by $|j\rangle$, where $j \in \{0,1\}^q$ (e.g., when $q = 2$, $|01\rangle$ denotes the standard basis vector $(0,1,0,0)^\top$). A quantum state is therefore of the form $|\psi\rangle = \sum_{j \in \{0,1\}^q} \alpha_j |j\rangle$, $\sum_{j \in \{0,1\}^q} |\alpha_j|^2 = 1$. The final state is obtained by applying a unitary matrix $U \in \mathbb{C}^{2^q \times 2^q}$ to the initial state $|0^q\rangle$, where 0^q denotes the q-digit all-zero binary string. An introduction to quantum computing for non-specialists is given in [25], and a comprehensive reference is [26]. $\|\cdot\|$ denotes the spectral norm for matrices, ℓ_2-norm for vectors. Given matrices C, D (including vectors or scalars), (C, D) denotes the matrix obtained stacking C on top of D. Given $v \in \mathbb{R}^{2^q}$, we denote $|v\rangle := \sum_{j=0}^{2^q - 1} \frac{v_j}{\|v\|} |j\rangle$ its amplitude encoding.

This paper uses quantum algorithms for linear systems introduced in [10]. For the system $A_B x = b$ with integer entries, the input to the algorithm is encoded by two unitaries P_{A_B} and P_b which are queried as oracles; for lack of space, we refer to [10] for details. We define the following symbols: d_c, maximum number of nonzero entries in any column of A; d_r, maximum number of nonzero entries in any row of A_B; $d := \max\{d_c, d_r\}$, sparsity of A_B; κ, ratio of largest to smallest nonzero singular value of A_B (we assume that an upper bound on κ is known).

Theorem 3 (Thm. 5 in [10]). *Let A_B be symmetric and $\|A_B\| = 1$. Given P_{A_B}, P_b, and $\varepsilon > 0$, there exists a quantum algorithm that produces the state $|\tilde{x}\rangle$ with $\|\,|A_B^{-1}b\rangle - |\tilde{x}\rangle\| \le \varepsilon$ using $\tilde{O}(d\kappa)$ queries to P_{A_B} and P_b, with additional gate complexity $\tilde{O}(d\kappa)$.*

The restriction on A_B symmetric can be relaxed by symmetrizing the system, see [20]. Our running time analysis already takes all symmetrization costs into account, but we skip details for brevity. Throughout this paper, $\tilde{O}$ is used to suppress polylogarithmic factors in the input parameters, i.e., $\tilde{O}(f(x)) = O(f(x)\mathrm{poly}(\log n, \log m, \log \frac{1}{\epsilon}, \log \kappa, \log d, \log(\mathrm{max\_entry}(A))))$.

We now give an overview of our algorithms. As stated in the introduction, a naive application of a QLSA (with explicit classical input and output) in the context of the simplex method is slower than a full classical iteration of the method. To gain an edge, we observe that an iteration of the simplex method can be reduced to a sequence of subroutines that have integer output. Indeed, the simplex method does not require explicit knowledge of the full solution vector $A_B^{-1}b$ associated with a basis, or of the full simplex tableau $A_B^{-1}A_N$, provided that we are able to: (i) identify if the current basis is optimal or unbounded; (ii) identify a pivot, i.e., the index of a column with negative reduced cost that enters the basis, and the index of a column leaving the basis. While subroutines to perform these tasks require access to $A_B^{-1}b$ and/or $A_B^{-1}A_N$, we show that we can get an asymptotic speedup by never computing a classical description of $A_B^{-1}, A_B^{-1}b$ or $A_B^{-1}A_N$. This is because extracting amplitude-encoded vectors from a quantum computer (a process called tomography) is much more expensive than obtaining integer or binary outputs, as these can be encoded directly as basis states and read from a single measurement with high probability if the amplitudes are set correctly.

Our first objective is to implement a quantum oracle that determines if a column has negative reduced cost. Using the QLSA of Theorem 3, with straightforward data preparation we can construct an oracle that, given a column index k, outputs $|A_B^{-1}A_k\rangle$ in some register. We still need to get around three obstacles: (i) the output of the QLSA is a renormalization of the solution, rather than the (unscaled) vector $A_B^{-1}A_k$; (ii) we want to compute $c_k - c_B^\top A_B^{-1}A_k$, while so far we only have access to $|A_B^{-1}A_k\rangle$; (iii) we need the output to be a binary yes/no condition (i.e., not encoded in an amplitude) so that Grover search can be applied to it, and we are not allowed to perform measurements. We overcome the first two obstacles by: extending and properly scaling the linear system so that c_k is suitably encoded in the QLSA output; and using the inverse of the unitary that maps $|0^{\lceil \log m + 1 \rceil}\rangle$ to $|(-c_B, 1)\rangle$ to encode $c_k - c_B^\top A_B^{-1}A_k$ in the amplitude of one of the basis states. To determine the sign of such amplitude, we rely on interference to create a basis state with amplitude α such that $|\alpha| \geq \frac{1}{2}$ if and only if $c_k - c_B^\top A_B^{-1}A_k \geq 0$. At this point, we can apply amplitude estimation [7] to determine the magnitude of α up to accuracy ϵ. This requires $O(1/\epsilon)$ iterations of the amplitude estimation algorithm. We therefore obtain a unitary operation that overcomes the three obstacles. A similar scheme can be used to determine if the basis is optimal, i.e., no column with negative reduced cost exists.

Some further complications merit discussion. The first one concerns the optimality tolerance: classically, this is typically $\bar{c}_N \geq -\epsilon$ for some given ϵ. Note that an absolute optimality tolerance is not invariant to rescaling of c. In the quantum subroutines, checking $\bar{c}_k < -\epsilon$ for a column k may be too expensive if the norm of $\|(A_B^{-1}A_k, c_k)\|$ is too large, because the quantum state normalization would force us to increase the precision of amplitude estimation. We therefore use the inequality $\bar{c}_k \geq -\epsilon\|(A_B^{-1}A_k, c_k)\|$ as optimality criterion, showing that $O(\frac{1}{\epsilon})$ rounds of amplitude estimation suffice. Our approach can be interpreted as a relative optimality criterion for the reduced costs: small $\|A_B^{-1}A_k\|$ implies a small denominator in (1) and potentially a large change in the basic feasible solution, so it is reasonable to require $\bar{c}_k$ close to zero for optimality. (Although optimality of the reduced costs does not guarantee a global optimality gap, it is the criterion used in every practical implementation of the simplex method.) Since we never explicitly compute the basis inverse, we do not have classical access to $\|(A_B^{-1}A_k, c_k)\|$. To alleviate this issue, we show that there exists an efficient quantum subroutine to compute the root mean square of $\|A_B^{-1}A_k\|$ over all k, or of $\|A_B^{-1}A_k\|$ for the specific k entering the basis, providing a characterization of the optimality tolerance used in the pricing step. It is important to remark that while the optimality tolerance is relative to a norm, which in turn depends on the problem data, the discussion in Sect. 1 indicates that due to sparsity, in practice we do not expect these norms to grow with m and n (or at least, no more than polylogarithmically). To implement Dantzig's pivoting rule and determine the smallest reduced cost, as opposed to a random column with negative reduced cost, the running time is the same as stated above, using quantum minimum finding [14]. To implement the steepest edge pivoting rule the running time increases by a factor κ^2: this is due to the cost of estimating the norms $\|A_B^{-1}A_k\|$, which requires $\frac{\kappa^2}{\epsilon}$ rounds of amplitude amplification to attain accuracy ϵ. Note that steepest edge is more expensive also in the classical case, increasing the running time expression by an additive term $m^2 n$.

The second complication concerns the condition number of the basis: the running time of the quantum routines explicitly depends on it, but this is not the case for the classical algorithms based on an LU decomposition of A_B (although precision may be affected). We remark that the dependence of QLSAs cannot be improved to $\kappa^{1-\delta}$ for $\delta > 0$ unless BQP = PSPACE [20]. We do not take any specific steps to improve the worst-case condition number of the basis (e.g., [12]), but we note that similar issues affect the classical simplex method: even if the running time does not depend on κ, when κ grows large the algorithm may fail because the computation of primal solutions or pivots becomes too imprecise. Many approaches have been proposed to prevent this from happening in practice, e.g., modifications of the pivoting rule to select pivot elements that are not too small, such as the two-pass Harris ratio test (a description of which can be found in [16]). This ratio test can be quantized without increasing the asymptotic running time; the description of the corresponding algorithm is left for the full version of this paper.

With the above construction we have a quantum subroutine that determines the index of a column with negative reduced cost, if one exists. Such a column can enter the basis. To perform a basis update we still need to determine the column leaving the basis: this is our second objective. For this step we need knowledge of $A_B^{-1}A_k$. Classically, this is straightforward because the basis inverse is available, since it is necessary to compute reduced costs anyway. With the above quantum subroutines the basis inverse is not known, and in fact part of the benefit of the quantum subroutines comes from always working with the original, sparse basis, rather than its possibly dense inverse. Thus, we describe another quantum algorithm that uses a QLSA as a subroutine, and identifies the element of the basis that is an approximate minimizer of the ratio test (1). Special care must be taken in this step, because attaining the minimum of the ratio test is necessary to ensure that the basic solution after the pivot is feasible (i.e., satisfies the non negativity constraints $x \geq 0$). However, in the quantum setting we are working with continuous amplitudes, and determining if an amplitude is zero is impossible due to finite precision. Our approach to give rigorous guarantees for this step involves the use of two feasibility tolerances: a tolerance δ, that determines which

---

Algorithm 1. SIMPLEXITERATION$(A, B, c, \epsilon, \delta, t)$.

1: **Input:** Matrix A, basis B, cost vector c, precision parameters ϵ, δ, t.
2: **Output:** Flag "optimal", "unbounded", or a pair (k, ℓ) where k is a nonbasic variable with negative reduced cost, ℓ is the basic variable that should leave the basis if k enters.
3: Normalize c so that $\|c_B\| = 1$. Normalize A so that $\|A_B\| \leq 1$.
4: Apply ISOPTIMAL(A, B, ϵ) to determine if the current basis is optimal. If so, return "optimal".
5: Apply FINDCOLUMN(A, B, ϵ) to determine a column with negative reduced cost. Let k be the column index returned by the algorithm.
6: Apply ISUNBOUNDED(A_B, A_k, δ) to determine if the problem is unbounded. If so, return "unbounded".
7: Apply FINDROW(A_B, A_k, b, δ, t) to determine the index ℓ of the row that minimizes the ratio test (1). Update the basis $B \leftarrow (B \setminus \{B(\ell)\}) \cup \{k\}$.

---

components of $A_B^{-1} A_k$ will be involved in the ratio test (due to the condition $u_j > 0$ in (1)); and a precision multiplier t, that determines the approximation guarantee for the minimization in (1). In particular, our algorithm returns an approximate minimizer that attains a relative error of $\frac{2t+1}{2t-1} - 1$, plus an absolute error proportional to $\frac{2}{2t-1}$. We remark that since the minimum of the ratio test could be zero, giving a purely relative error bound seems impossible in finite precision. A similar quantum algorithm can be used to determine if column k proves unboundedness of the LP. Note that because of the inexactness of the ratio test, we could pivot to slightly infeasible solutions; recovery strategies are left for the full version of this paper.

5   Technical Discussion

A high-level overview of the algorithm is given in Algorithm 1. Due to space restrictions, we discuss only a few technical aspects; a detailed description of the quantum subroutines and proof of their correctness will be given in the full version of this paper (a preprint is available on arXiv). All data normalization is performed on line 3 of SIMPLEXITERATION$(A, B, c, \epsilon, \delta, t)$. Using the power method [23] to find the leading singular value up to accuracy ϵ', the entire normalization can be performed in time $O(\frac{1}{\epsilon'} md \log m)$. This is asymptotically negligible compared to the rest of the algorithm.

The oracles P_{A_B} and P_b can be implemented in time $\tilde{O}(dm)$ and $\tilde{O}(m)$, respectively; the construction of P_{A_B} is straightforward and amounts to a lookup table to identify indices of nonzero elements and their values (similar to classical data structures for sparse vectors), the construction of P_b can be done in a manner similar to [18], exploiting sparsity.

To determine, modulo the global phase factor, the sign of an amplitude α (i.e., a coefficient in the quantum state; we encode reduced costs in specific amplitudes) we use two subroutines called SIGNESTNFN and SIGNESTNFP. These subroutines rely on interference to create an amplitude $\frac{1}{2}(1 + \alpha)$, so that amplitude estimation allows us to distinguish, by comparing to the threshold value $\frac{1}{2}$, if $\alpha > 0$ or $\alpha < 0$. In reality, since amplitude estimation is not perfect, the analysis is more convoluted than the simple above explanation. The acronyms "NFN" an "NFP" stand for "no false negatives" and "no false positives", respectively, based on an interpretation of these subroutines as classifiers that need to assign a 0–1 label to the input. Since the quantum phase estimation, on which they are based, is a continuous transformation, a routine that has "no false negatives", i.e., with high probability it returns 1 if the input data's true class is 1 (in our case, this means that a given amplitude is $\geq -\epsilon$), may have false positives: it may also return 1 with too large probability for some input that belongs to class 0 (i.e., the given amplitude is $< -\epsilon$). The probability of these undesirable events decreases as we get away from the threshold $-\epsilon$.

The construction of reduced costs is relatively straightforward, using the approach described in Sect. 4. It yields the following result.

Theorem 4. *There exists a quantum subroutine (*FINDCOLUMN*) that returns a column $k \in N$ with reduced cost $\bar{c}_k < -\epsilon \|(A_B^{-1}A_k, \frac{c_k}{\|c_B\|})\|$, with expected number of iterations $O(\sqrt{n})$. The total gate complexity of the algorithm is $\tilde{O}(\frac{\kappa d \sqrt{n}}{\epsilon}(d_c n + dm))$. The steepest edge column selection can be implemented increasing the cost by a factor κ^2.*

We remark that Theorem 4 concerns the case in which at least one column eligible to enter the basis (i.e., with negative reduced cost) exists. Following Algorithm 1, SIMPLEXITERATION, the subroutine FINDCOLUMN is executed only if ISOPTIMAL returns false. The subroutine ISOPTIMAL can be constructed in almost the same way as FINDCOLUMN, using the counting version of Grover search to determine if there is any index k for which column A_k has negative reduced cost. The gate complexity of ISOPTIMAL is asymptotically the same as in Theorem 4. All our subroutines succeed with constant probability, that can be boosted to the desired level with polylogarithmic overhead; details on possible failures, and how to recover from them, are left for the full version of this paper. We note that when ISOPTIMAL returns 1, the current basis B is optimal but we do not have a classical description of the solution vector. To obtain a full description of the solution vector, the fastest approach is to classically solve the system $A_B x = b$ for the (known) optimal basis, which requires time $O(d_c^{0.7} m^{1.9} + m^{2+o(1)})$.

If we apply the quantum search algorithm over all columns, to find a column with negative reduced cost, we need to perform $O(\sqrt{n})$ iterations, but the unitary to prepare the data for the QLSA requires time that scales as $\tilde{O}(d_c n)$. In some cases it may be advantageous to split the set of columns into multiple sets, and apply the search algorithm to each set individually. Carrying out this analysis shows that if $\frac{n}{m} \geq 2\frac{d}{d_c}$, the running time expression in Theorem 4 can be reduced to $\tilde{O}(\frac{1}{\epsilon}\kappa d^{1.5}\sqrt{d_c}n\sqrt{m})$. (Steepest edge pivot selection is a factor κ^2 slower.)

After identifying a column to leave the basis, we must determine if the problem is unbounded, or perform the ratio test.

Theorem 5. *There exists a quantum subroutine (*IsUnbounded*) such that, with bounded probability, if the subroutine returns 1 then $(A_B^{-1}A_k)_i < \delta\|A_B^{-1}A_k\|$ for all $i = 1, \ldots, m$, with total gate complexity $\tilde{O}(\frac{\sqrt{m}}{\delta}(\kappa d^2 m))$.*

If IsUnbounded returns 1, we have a proof that the LP is unbounded from below, up to the given tolerance. Otherwise, we have to perform the ratio test.

Theorem 6. *There exists a quantum subroutine (*FindRow*) that, with bounded probability, returns ℓ such that:*

$$\frac{(A_B^{-1}A_k)_\ell}{(A_B^{-1}b)_\ell} \leq \frac{2}{2t-1}\frac{\|A_B^{-1}b\|}{\|A_B^{-1}A_k\|} + \frac{2t+1}{2t-1}\min_{h:(A_B^{-1}A_k)_h > \delta\|A_B^{-1}A_k\|}\frac{(A_B^{-1}b)_h}{(A_B^{-1}A_k)_h},$$

with total gate complexity $\tilde{O}(\frac{t}{\delta}\sqrt{m}(\kappa d^2 m))$.

Notice that the ratio test is performed approximately, i.e., the solution found after pivoting might be infeasible, but the total error in the ratio test (and hence the maximum infeasibility after pivoting) is controlled by the parameter t in Theorem 6. For example, for $t = 1000$ the index ℓ returned is within $\approx 0.1\%$ of the true minimum of the ratio test. This may lead to small infeasibilities, but we can recover from them by first using a subroutine to determine feasibility of the current basis, and then switching to Phase 1 of the simplex method in case the current basis is infeasible. This approach always succeeds, if the precision of the calculations is sufficient.

We conclude by detailing the acceleration that can be obtained with quantum-accessible storage. This is significant, and in this case the quantum subroutines achieve essentially linear scaling.

Proposition 1. *If the matrix A_B and the columns of A_N are stored in qRAM (of size $\tilde{O}(d_c n)$), the running time of* FindColumn *and* IsOptimal *is $\tilde{O}(\frac{1}{\epsilon}(\kappa\sqrt{mn}))$, whereas the running time of* FindRow *and* IsUnbounded *is $\tilde{O}(\frac{t}{\delta}\kappa m)$. The cost of preparing the data structures before the first iteration of Algorithm 1 is $\tilde{O}(d_c n)$; the time to update the data structures after the basis changes is $\tilde{O}(m)$.*

Acknowledgment. We are grateful to Sergey Bravyi, Sanjeeb Dash, Santanu Dey, Yuri Faenza, Krzysztof Onak, Ted Yoder, and to anonymous referees for useful discussions and/or comments on an early version of this manuscript. The author is partially supported by the IBM Research Frontiers Institute, Army Research Office grant W911NF-20-1-0014, and AFRL grant FA8750-C-18-0098.

References

1. van Apeldoorn, J., Gilyén, A.: Improvements in quantum SDP-solving with applications. In: Baier, C., Chatzigiannakis, I., Flocchini, P., Leonardi, S. (eds.) 46th International Colloquium on Automata, Languages, and Programming (ICALP 2019). Leibniz International Proceedings in Informatics (LIPIcs), vol. 132, pp. 99:1–99:15. Schloss Dagstuhl-Leibniz-Zentrum fuer Informatik, Dagstuhl, Germany (2019)

2. van Apeldoorn, J., Gilyén, A.: Quantum algorithms for zero-sum games. arXiv preprint arXiv:1904.03180 (2019)
3. van Apeldoorn, J., Gilyén, A., Gribling, S., de Wolf, R.: Quantum SDP-solvers: better upper and lower bounds. In: 2017 IEEE 58th Annual Symposium on Foundations of Computer Science (FOCS), pp. 403–414. IEEE (2017)
4. Bertsimas, D., Tsitsiklis, J.: Introduction to Linear Optimization. Athena Scientific, Belmont (1997)
5. Borgwardt, K.H.: The average number of pivot steps required by the simplex-method is polynomial. Zeitschrift für Oper. Res. **26**(1), 157–177 (1982)
6. Brandao, F.G., Svore, K.M.: Quantum speed-ups for solving semidefinite programs. In: 2017 IEEE 58th Annual Symposium on Foundations of Computer Science (FOCS), pp. 415–426. IEEE (2017)
7. Brassard, G., Hoyer, P., Mosca, M., Tapp, A.: Quantum amplitude amplification and estimation. Contemp. Math. **305**, 53–74 (2002)
8. Casares, P.A.M., Martin-Delgado, M.A.: A quantum interior-point predictor-corrector algorithm for linear programming. J. Phys. A Math. Theor. **53**(44), 445305 (2020)
9. Chia, N.H., Li, T., Lin, H.H., Wang, C.: Quantum-inspired sublinear algorithm for solving low-rank semidefinite programming. In: 45th International Symposium on Mathematical Foundations of Computer Science (MFCS 2020). Schloss Dagstuhl-Leibniz-Zentrum für Informatik (2020)
10. Childs, A.M., Kothari, R., Somma, R.D.: Quantum algorithm for systems of linear equations with exponentially improved dependence on precision. SIAM J. Comput. **46**(6), 1920–1950 (2017)
11. Chvátal, V.: Linear programming. Freeman, W. H (1983)
12. Clader, B.D., Jacobs, B.C., Sprouse, C.R.: Preconditioned quantum linear system algorithm. Phys. Rev. Lett. **110**(25), 250504 (2013)
13. Dadush, D., Huiberts, S.: A friendly smoothed analysis of the simplex method. In: Proceedings of the 50th Annual ACM SIGACT Symposium on Theory of Computing, pp. 390–403. ACM (2018)
14. Durr, C., Hoyer, P.: A quantum algorithm for finding the minimum. arXiv preprint quant-ph/9607014 (1996)
15. Forrest, J.J., Goldfarb, D.: Steepest-edge simplex algorithms for linear programming. Math. Program. **57**(1–3), 341–374 (1992)
16. Gill, P.E., Murray, W., Saunders, M.A., Wright, M.H.: A practical anti-cycling procedure for linearly constrained optimization. Math. Program. **45**(1), 437–474 (1989)
17. Grigoriadis, M.D., Khachiyan, L.G.: A sublinear-time randomized approximation algorithm for matrix games. Oper. Res. Lett. **18**(2), 53–58 (1995)
18. Grover, L., Rudolph, T.: Creating superpositions that correspond to efficiently integrable probability distributions. arXiv preprint quant-ph/0208112 (2002)
19. Grover, L.K.: A fast quantum mechanical algorithm for database search. In: Proceedings of the Twenty-Eighth Annual ACM Symposium on Theory of Computing, pp. 212–219. ACM (1996)
20. Harrow, A.W., Hassidim, A., Lloyd, S.: Quantum algorithm for linear systems of equations. Phys. Rev. Lett. **103**(15), 150502 (2009)
21. Kerenidis, I., Prakash, A.: A quantum interior point method for LPs and SDPs. arXiv preprint arXiv:1808.09266 (2018)
22. Klee, V., Minty, G.J.: How good is the simplex algorithm. Inequalities **3**(3), 159–175 (1972)

23. Kuczyński, J., Woźniakowski, H.: Estimating the largest eigenvalue by the power and Lanczos algorithms with a random start. SIAM J. Matrix Anal. Appl. **13**(4), 1094–1122 (1992)
24. Lübbecke, M.E., Desrosiers, J.: Selected topics in column generation. Oper. Res. **53**(6), 1007–1023 (2005)
25. Nannicini, G.: An introduction to quantum computing, without the physics. arXiv preprint arXiv:1708.03684 (2017)
26. Nielsen, M.A., Chuang, I.: Quantum Computation and Quantum Information. Cambridge University Press, Cambridge (2002)
27. Spielman, D.A., Teng, S.H.: Smoothed analysis of algorithms: why the simplex algorithm usually takes polynomial time. J. ACM (JACM) **51**(3), 385–463 (2004)
28. Yuster, R., Zwick, U.: Fast sparse matrix multiplication. ACM Trans. Algorithms (TALG) **1**(1), 2–13 (2005)

Maximum Weight Disjoint Paths in Outerplanar Graphs via Single-Tree Cut Approximators

Guyslain Naves[1], Bruce Shepherd[2(⊠)], and Henry Xia[2]

[1] Aix-Marseille University, LIS CNRS UMR 7020, Marseille, France
guyslain.naves@univ-amu.fr
[2] University of British Columbia, Vancouver, Canada
fbrucesh@cs.ubc.ca, h.xia@alumni.ubc.ca

Abstract. Since 1997 there has been a steady stream of advances for the maximum disjoint paths problem. Achieving tractable results has usually required focusing on relaxations such as: (i) to allow some bounded edge congestion in solutions, (ii) to only consider the unit weight (cardinality) setting, (iii) to only require fractional routability of the selected demands (the all-or-nothing flow setting). For the general form (no congestion, general weights, integral routing) of edge-disjoint paths (EDP) even the case of unit capacity trees which are stars generalizes the maximum matching problem for which Edmonds provided an exact algorithm. For general capacitated trees, Garg, Vazirani, Yannakakis showed the problem is APX-Hard and Chekuri, Mydlarz, Shepherd provided a 4-approximation. This is essentially the only setting where a constant approximation is known for the general form of EDP. We extend their result by giving a constant-factor approximation algorithm for general-form EDP in outerplanar graphs. A key component for the algorithm is to find a *single-tree* $O(1)$ cut approximator for outerplanar graphs. Previously $O(1)$ cut approximators were only known via distributions on trees and these were based implicitly on the results of Gupta, Newman, Rabinovich and Sinclair for distance tree embeddings combined with results of Anderson and Feige.

1 Introduction

The past two decades have seen numerous advances to the approximability of the maximum disjoint paths problem (EDP) since the seminal paper [17]. An instance of EDP consists of a (directed or undirected) "supply" graph $G = (V, E)$ and a collection of k *requests* (aka demands). Each request consists of a pair of nodes $s_i, t_i \in V$. These are sometimes viewed as a *demand graph* $H = (V(G), \{s_i t_i : i \in [k]\})$. A subset S of the requests is called *routable* if there exist edge-disjoint paths $\{P_i : i \in S\}$ such that P_i has endpoints s_i, t_i for each i. We may also be

Aix-Marseille University—Work partially supported by ANR project DISTANCIA (ANR-17-CE40-0015).

M. Singh and D. P. Williamson (Eds.): IPCO 2021, LNCS 12707, pp. 326–339, 2021.
https://doi.org/10.1007/978-3-030-73879-2_23

given a profit w_i associated with each request and the goal is to find a routable subset S which maximizes $w(S) = \sum_{i \in S} w_i$. The *cardinality version* is where we have unit weights $w_i \equiv 1$.

For directed graphs it is known [20] that there is no $O(n^{0.5-\epsilon})$ approximation, for any $\epsilon > 0$ under the assumption $P \neq NP$. Subsequently, research shifted to undirected graphs and two relaxed models. First, in the *all-or-nothing flow model* (ANF) the notion of routability is relaxed. A subset S is called routable if there is a feasible (fractional) multiflow which satisfies each request in S. In [6] a polylogarithmic approximation is given for ANF. Second, in the *congestion model* [24] one is allowed to increase the capacity of each edge in G by some constant factor. Two streams of results ensued. For general graphs, a polylogarithmic approximation is ultimately provided [5,10,11] with edge congestion 2. For planar graphs, a constant factor approximation is given [4,31] with edge congestion 2. There is also an $f(g)$-factor approximation for bounded genus g graphs with congestion 3.

As far as we know, the only congestion 1 results known for either maximum ANF or EDP are as follows; all of these apply only to the cardinality version. In [23], a constant factor approximation is given for ANF in planar graphs and for treewidth k graphs there is an $f(k)$-approximation for EDP [9]. More recent results include a constant-factor approximation in the *fully planar* case where $G+H$ is planar [16,22]. In the weighted regime, there is a factor 4 approximation for EDP in capacitated trees [8]. We remark that this problem for unit capacity "stars" already generalizes the maximum weight matching problem in general graphs. Moreover, inapproximability bounds for EDP in planar graphs are almost polynomial [12]. This lends interest to how far one can push beyond trees. Our main contribution to the theory of maximum throughput flows is the following result which is the first generalization of the (weighted) EDP result for trees [8], modulo a larger implicit constant of 224.

Theorem 1. *There is a polynomial-time 224 approximation algorithm for the maximum weight* ANF *and* EDP *problems for capacitated outerplanar graphs.*

It is natural to try to prove this is by reducing the problem in outerplanar graphs to trees and then use [8]. A promising approach is to use results of [18] – an $O(1)$ distance tree embedding for outerplanar graphs – and a *transfer theorem* [3,29] which proves a general equivalence between distance and capacity embeddings. Combined, these results imply that there is a probabilistic embedding into trees which approximates cut capacity in outerplanar graphs with constant congestion. One could then try to mimic the success of using low-distortion (distance) tree embeddings to approximate minimum cost network design problems. There is an issue with this approach however. Suppose we have a distribution on trees T_i which approximates cut capacity in expectation. We then apply a known EDP algorithm which outputs a subset of requests S_i which are routable in each T_i. While the tree embedding guarantees that the convex combination of S_i's satisfies the cut condition in G, it may be that no single S_i obeys the cut condition, even approximately. Moreover, this is also a problem for ANF. This problem persists even if we ensure that each T_i is dominated (or dominating) by G. For instance, if capacity in each T_i is upper bounded by capacity in G, then in expectation the T_i's will cover at least some constant fraction of G. There is no guarantee, however, that any of the S_i's would as well.

We overcome these issues by computing a **single** tree which approximates the cut capacity in G – see Theorem 3. Our algorithmic proof is heavily inspired by work of Gupta [19] which gives a method for eliminating Steiner nodes in probabilistic (distance) tree embeddings for general graphs.

It turns out that having a single-tree is not enough for us and we need additional technical properties to apply the algorithm from [8]. First, our single tree T should have integer capacities and be non-expansive, i.e., $\hat{u}(\delta_T(S)) \le u(\delta_G(S))$ (where $\hat{u}/u$ are the edge capacities in T/G and δ is used to denote the edges in the cut induced by S). To see why it is useful that T is an under-estimator of G's cut capacity, consider the classical grid example of [17]. They give an instance with a set of $\sqrt{n}$ requests which satisfy the cut condition in $2 \cdot G$, but for which one can only route a single request in the capacity of G.

If our tree is an under-estimator, then we can ultimately obtain a "large" weight subset of requests satisfying the cut condition in G itself. However, even this is not generally sufficient for (integral) routability. For a multiflow instance G/H one normally also requires that $G + H$ is Eulerian, even for easy instances such as when G is a 4-cycle. The final ingredient we use is that our single tree T is actually a **subtree** of G which allows us to invoke the following result – see Sect. 3.1.

Theorem 2. *Let G be an outerplanar graph with integer edge capacities $u(e)$. Let H denote a demand graph such that $G + H = (V(G), E(G) \cup E(H))$ is outerplanar. If G, H satisfies the cut condition, then H is routable in G, and an integral routing can be found in polynomial-time.*

The key point here is that we can avoid the usual parity condition needed, such as in [15, 26, 32]. We are not presently aware of the above result's existence in the literature.

1.1   A Single-Subtree Cut Sparsifier and Related Results

Our main cut approximation theorem is the following which may be of independent interest.

Theorem 3. *There is a polynomial-time algorithm that, for any connected outerplanar graph $G = (V, E)$ with integer edge capacities $u(e) > 0$, finds a subtree T of G with integer edge weights $\hat{u}(e) \ge 0$ such that*

$$\frac{1}{14} u(\delta_G(X)) \le \hat{u}(\delta_T(X)) \le u(\delta_G(X)) \text{ for each proper subset } X \subseteq V$$

We discuss some connections of this result to prior work on sparsifiers and metric embeddings. Celebrated work of Räcke [28] shows the existence of a single capacitated tree T (not a subtree) which behaves as a flow sparsifier for a given graph G. In particular, routability of demands on T implies fractional routability in G with edge congestion $polylog(n)$; this bound was further improved to $O(\log^2 n \log \log n)$ [21]. Such single-tree results were also instrumental in an application to maximum throughput flows: a polylogarithmic approximation for the maximum all-or-nothing flow problem in general graphs [7]. Even more directly to Theorem 3 is work on cut sparsifiers; in [30] it is shown that there is a single tree (again, not subtree) which approximates cut capacity in a general graph G within a factor of $O(\log^{1.5} \log \log n)$. As far as we know, our result is the only global-constant factor single-tree cut approximator for a family of graphs.

Räcke improved the bound for flow sparsification to an optimal congestion of $O(\log n)$ [29]. Rather than a single tree, this work requires a convex combination of (general) trees to simulate the capacity in G. His work also revealed a beautiful equivalence between the existence of good (low-congestion) distributions over trees for capacities, and the existence of good (low-distortion) distributions over trees for distances [3]. This *transfer theorem* states very roughly that for a graph G the following are equivalent for a given $\rho \geq 1$. (1) For any edge lengths $\ell(e) > 0$, there is a (distance) embedding of G into a distribution of trees which has stretch at most ρ. (2) For any edge capacities $u(e) > 0$, there is a (capacity) embedding of G into a distribution of trees which has congestion at most ρ. This work has been applied in other related contexts such as flow sparsifiers for proper subsets of terminals [14].

The transfer theorem uses a very general setting where there are a collection of valid *maps*. A map M sends an edge of G to an abstract "path" $M(e) \subseteq E(G)$. The maps may be refined for the application of interest. In the so-called *spanning tree setting*, each M is associated with a subtree T_M of G (the setting most relevant to Theorem 3). $M(e)$ is then the unique path which joins the endpoints of e in T_M. For an edge e, its *stretch* under M is $(\sum_{e' \in M(e)} \ell(e'))/\ell(e)$. In the context of distance tree embeddings this model has been studied in [1,2,13]. In capacity settings, the *congestion* of an edge under M is $(\sum_{e':e \in M(e')} c(e'))/c(e)$. One can view this as simulating the capacity of G using the tree's edges with bounded congestion. The following result shows that we cannot guarantee a single subtree with $O(1)$ congestion even for outerplanar graphs. This appears in the long version [25] and the example was found independently by Anastasios Sidiropoulos [33].

Theorem 4. *There is an infinite family $\mathcal{O}$ of outerplanar graphs such that for every $G \in \mathcal{O}$ and every spanning tree T of G:*

$$\max_X \frac{u(\delta_G(X))}{u(\delta_T(X))} = \Omega(\log |V(G)|),$$

where the max is taken over fundamental cuts of T.

This suggests that the single-subtree result Theorem 3 is a bit lucky and critically requires the use of tree capacities different from u. Of course a single tree is sometimes unnecessarily restrictive. For instance, outerplanar graphs also have an $O(1)$-congestion embedding using a distribution of subtrees by the transfer theorem (although we are not aware of one explicitly given in the literature). This follows implicitly due to existence of an $O(1)$-stretch embedding into subtrees [18].

Finally we remark that despite the connections between distance and capacity tree embeddings, Theorem 3 stands in contrast to the situation for distance embeddings. Every embedding of the n point cycle into subtrees suffers distortion $\Omega(n)$, and indeed this also holds for embedding into arbitrary (using Steiner nodes etc.) trees [27].

2   Single Spanning Tree Cut Approximator in Outerplanar Graphs

In this section we first show the existence of a single-tree which is an $O(1)$ cut approximator for an outerplanar graph G. Subsequently we show that there is such a tree with two additional properties. First, its capacity on every cut is at most the capacity in G, and second, all of its weights are integral. These additional properties (integrality and conservativeness) are needed in our application to EDP. The formal statement we prove is as follows.

Theorem 3. *There is a polynomial-time algorithm that, for any connected outerplanar graph $G = (V, E)$ with integer edge capacities $u(e) > 0$, finds a subtree T of G with integer edge weights $\hat{u}(e) \geq 0$ such that*

$$\frac{1}{14}u(\delta_G(X)) \leq \hat{u}(\delta_T(X)) \leq u(\delta_G(X)) \text{ for each proper subset } X \subseteq V$$

In Sect. 2.1, we show how to view capacity approximators in G as (constrained) distance tree approximators in the planar dual graph. From then on, we look for distance approximators in the dual which correspond to trees in G. In Sect. 2.2 we prove there exists a single-subtree cut approximator. In the long version of the paper [25] we show how to make this conservative while maintaining integrality of the capacities on the tree. This is essential for our application to disjoint paths.

2.1   Converting Flow-Sparsifiers in Outerplanar Graphs to Distance-Sparsifiers in Trees

Let $G = (V, E)$ be an outerplanar graph with capacities $u : E \to \mathbb{R}^+$. Without loss of generality, we can assume that G is 2-node connected, so the boundary of the outer face of G is a cycle that contains each node exactly once. Let G^* be the dual of G; we assign weights to the dual edges in G^* equal to the capacities on the corresponding edges in G. Let G_z be the graph obtained by adding an apex node z to G which is connected to each node of G, that is $V(G_z) = V \cup \{z\}$ and $E(G_z) = E \cup \{(z, v) : v \in V\}$. We may embed z into the outer face of G, so G_z is planar. Let G_z^* denote the planar dual of G_z.

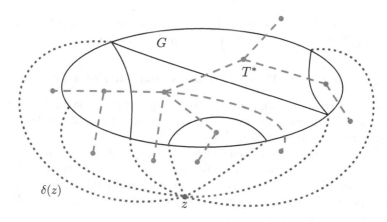

Fig. 1. The solid edges form the outerplanar graph G, and the dotted edges are the edges incident to the apex node z in G_z. The dashed edges form the dual tree T^*.

Note that $\delta(z) = \{(z,v) : v \in V\}$ are the edges of a spanning tree of G_z, so $E(G_z)^* \setminus \delta(z)^*$ are the edges of a spanning tree T^* of G_z^*. Each non-leaf node of T^* corresponds to an inner face of G, and each leaf of T^* corresponds to a face of G_z whose boundary contains the apex node z. Also note that we obtain G^* if we combine all the leaves of T^* into a single node (which would correspond to the outer face of G). We will call T^* the dual tree of the outerplanar graph G (Fig. 1).

Let a central cut of G be a cut $\delta(S)$ such that both of its shores S and $V \setminus S$ induced connected subgraphs of G. Hence, the shores of a central cut are subpaths of the outer cycle, so the dual of $\delta(S)$ is a leaf-to-leaf path in T^*. Since the edges of any cut in a connected graph is a disjoint union of central cuts, it suffices to only consider central cuts.

We want to find a strictly embedded cut-sparsifier $T = (V, F, u^*)$ of G (i.e. a spanning tree T of G with edges weights u^*) such that for any nonempty $X \subsetneq V$, we have

$$\alpha u(\delta_G(X)) \leq u^*(\delta_T(X)) \leq \beta u(\delta_G(X)). \tag{1}$$

In the above inequality, we can replace $u^*(\delta_T(X))$ with $u^*(\delta_G(X))$ if we set $u^*(e) = 0$ for each edge $e \notin E(T)$. In the dual tree (of G), $\delta_G(X)^*$ is a leaf-to-leaf path for any central cut $\delta(X)$, so inequality (1) is equivalent to

$$\alpha u(P) \leq u^*(P) \leq \beta u(P) \tag{2}$$

for any leaf-to-leaf path P in T^*.

Finally, we give a sufficient property on the weights u^* assigned to the edges such that all edges of positive weight are in the spanning tree of G. Recall that the dual of the edges not in the spanning tree of G would form a spanning tree of G^*. Since we assign weight 0 to edges not in the spanning tree of G, it is sufficient for the 0 weight edges to form a spanning subgraph of G^*. Since G^*

is obtained by combining the leaves of T^* into a single node, it suffices for each node $v \in V(T^*)$ to have a 0 weight path from v to a leaf of T^*.

2.2 An Algorithm to Build a Distance-Sparsifier of a Tree

In this section, we present an algorithm to obtain a distance-sparsifier of a tree. In particular, this allows us to obtain a cut-approximator of an outerplanar graph from a distance-sparsifier of its dual tree.

Let $T = (V, E, u)$ be a weighted tree where $u : E \to \mathbb{R}^+$ is the length function on T. Let $L \subset V$ be the leaves of T. We assign non-negative weights u^* to the edges of T. Let d be the shortest path metric induced by the original weights u, and let d^* be the shortest path metric induced by the new weights u^*. We want the following two conditions to hold:

1. there exists a 0 weight path from each $v \in V$ to a leaf of L.
2. for any two leaves $x, y \in L$, we have

$$\frac{1}{4} d(x, y) \leq d^*(x, y) \leq 2d(x, y). \tag{3}$$

We define u^* recursively as follows. If $|L| \leq 1$, we are done by setting $u^* = 0$ on every edge. If every node is in L, we are done by setting $u = u^*$.

Let $r \in T \setminus L$ be a non-leaf node, and consider T to be rooted at r. For $v \in V$, let $T(v)$ denote the subtree rooted at v, and let $h(v)$ denote the *height* of v, defined by $h(v) = \min\{d(v, x) : x \in L \cap T(v)\}$. Now, let $r_1, \dots, r_k$ be the points in T that are at distance exactly $h(r)/2$ from r. Without loss of generality, suppose that each r_i is a node (otherwise we can subdivide the edge to get a node), and order the r_i's by increasing $h(r_i)$, that is $h(r_{i-1}) \leq h(r_i)$ for each $i = 2, \dots, k$. Furthermore, suppose that we have already assigned weights to the edges in each subtree $T(r_i)$ using this algorithm recursively with $L_i = L \cap V(T(r_i))$, so it remains to assign weights to the edges not in any of these subtrees. We assign a weight of $h(r_i)$ to the first edge on the path from r_i to r for each $i = 2, \dots, k$, and weight 0 to all other edges (Fig. 2). In particular, all edges on the path from r_1 to r receive weight 0.

The algorithm terminates on components with at most one vertex in L. Let $w \in L$ be a leaf closest to r, $d(r, w) = h(r)$, and let l be the length of the edge incident to w. As the length of the longest path from the root to w is halved at each recursive step, w will be isolated after at most $\log_2 \frac{h(r)}{l}$ recursive steps.

Since we assign 0 weight to edges on the $r_1 r$ path, Condition 1 is satisfied for all nodes above the r_i's in the tree by construction. It remains to prove Condition 2. We use the following upper and lower bounds. For each leaf $x \in L$,

$$d^*(x, r) \leq 2d(x, r) - h(r), \tag{4}$$
$$d^*(x, r) \geq d(x, r) - h(r). \tag{5}$$

We prove the upper bound in (4) by induction. We are done if T only has 0 weight edges, and the cases that cause the algorithm to terminate will only have

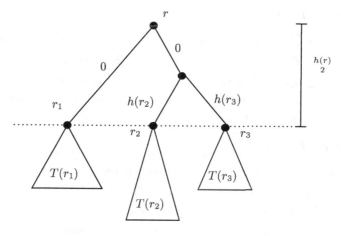

Fig. 2. The algorithm assigns weights to the edges above $r_1, \ldots, r_k$, and is run recursively on the subtrees $T(r_1), \ldots, T(r_k)$.

0 weight edges. For the induction, we consider two separate cases depending on whether $x \in T(r_1)$.

Case 1: $x \in T(r_1)$.

$$
\begin{aligned}
d^*(x, r) &= d^*(x, r_1) + d^*(r_1, r) && (r_1 \text{ is between } x \text{ and } r) \\
&= d^*(x, r_1) && (\text{by definition of } u^*) \\
&\leq 2d(x, r_1) - h(r_1) && (\text{by induction}) \\
&= 2d(x, r) - 2d(r, r_1) - h(r_1) && (\; r_1 \text{ is between } x \text{ and } r) \\
&= 2d(x, r) - \frac{3}{2} h(r) && (h(r_1) = h(r)/2 \text{ by definition of } r_1 \;) \\
&\leq 2d(x, r) - h(r)
\end{aligned}
$$

Case 2: $x \in T(r_i)$ for some $i \neq 1$.

$$
\begin{aligned}
d^*(x, r) &= d^*(x, r_i) + d^*(r_i, r) && (r_i \text{ is between } x \text{ and } r) \\
&= d^*(x, r_i) + h(r_i) && (\text{by definition of } u^* \;) \\
&\leq 2d(x, r_i) - h(r_i) + h(r_i) && (\text{by induction}) \\
&= 2d(x, r) - 2d(r_i, r) && (r_i \text{ is between } x \text{ and } r) \\
&= 2d(x, r) - h(r) && (d(r_i, r) = h(r)/2 \text{ by definition of } r_i)
\end{aligned}
$$

This proves inequality (4).

We prove the lower bound in (5) similarly.

Case 1: $x \in T(r_1)$.

$$
\begin{aligned}
d^*(x,r) &= d^*(x,r_1) + d^*(r_1,r) && (r_1 \text{ is between } x \text{ and } r)\\
&= d^*(x,r_1) && (\text{by definition of } u^*)\\
&\geq d(x,r_1) - h(r_1) && (\text{by induction})\\
&= d(x,r) - d(r,r_1) - h(r_1) && (r_1 \text{ is between } x \text{ and } r)\\
&= d(x,r) - h(r) && (\text{by definition of } r_1)
\end{aligned}
$$

Case 2: $x \in T(r_i)$ for some $i \neq 1$.

$$
\begin{aligned}
d^*(x,r) &= d^*(x,r_i) + d^*(r_i,r) && (r_i \text{ is between } x \text{ and } r)\\
&= d^*(x,r_i) + h(r_i) && (\text{by definition of } u^*)\\
&\geq d(x,r_i) - h(r_i) + h(r_i) && (\text{by induction})\\
&= d(x,r) - d(r_i,r) && (r_i \text{ is between } x \text{ and } r)\\
&= d(x,r) - h(r)/2 && (d(r_i,r) = h(r)/2 \text{ by definition of } r_i)\\
&\geq d(x,r) - h(r)
\end{aligned}
$$

This proves inequality (5).

Finally, we prove property 2, that is inequality (3), by induction. Let $x, y \in L$ be two leaves of T. Suppose that $x \in T(r_i)$ and $y \in T(r_j)$. By induction, we may assume that $i \neq j$, so without loss of generality, suppose that $i < j$.

We prove the upper bound.

$$
\begin{aligned}
d^*(x,y) &= d^*(x,r_i) + d^*(r_i,r_j) + d^*(r_j,y)\\
&\leq 2d(x,r_i) - h(r_i) + 2d(y,r_j) - h(r_j) + d^*(r_i,r_j) && (\text{by (4)})\\
&\leq 2d(x,r_i) - h(r_i) + 2d(y,r_j) - h(r_j) + h(r_i) + h(r_j) && (\text{by definition of } u^*)\\
&= 2d(x,r_i) + 2d(y,r_j)\\
&\leq 2d(x,y)
\end{aligned}
$$

We prove the lower bound.

$$
\begin{aligned}
d(x,y) &= d(x,r_i) + d(r_i,r_j) + d(r_j,y)\\
&\leq d(x,r_i) + d(r_j,y) + h(r_i) + h(r_j)\\
&\qquad (\text{because } d(r,r_i) = h(r)/2 \leq h(r_i) \text{ for all } i \in [k])\\
&\leq 2d(x,r_i) + 2d(r_j,y) && (\text{by definition of } h)\\
&\leq 2d^*(x,r_i) + 2h(r_i) + 2d^*(y,r_j) + 2h(r_j) && (\text{by (5)})\\
&= 2d^*(x,y) - 2d^*(r_i,r_j) + 2h(r_i) + 2h(r_j).
\end{aligned}
$$

Now we finish the proof of the lower bound by considering two cases.

Case 1: $i = 1$, that is x is in the first subtree.

$$
\begin{aligned}
d(x,y) &\leq 2d^*(x,y) - 2d^*(r_1,r_j) + 2h(r_1) + 2h(r_j)\\
&= 2d^*(x,y) - 2h(r_j) + 2h(r_1) + 2h(r_j) && (\text{by definition of } u^*)\\
&\leq 2d^*(x,y) + 2h(r_1)\\
&\leq 4d^*(x,y)
\end{aligned}
$$

Case 2: $i > 1$, that is neither x nor y is in the first subtree.

$$d(x,y) \leq 2d^*(x,y) - 2d^*(r_i, r_j) + 2h(r_i) + 2h(r_j)$$
$$= 2d^*(x,y) - 2h(r_i) - 2h(r_j) + 2h(r_i) + 2h(r_j) \quad \text{(by definition of } u^*\text{)}$$
$$= 2d^*(x,y)$$

This completes the proof of property 2.

3 Maximum Weight Disjoint Paths

In this section we prove our main result for EDP, Theorem 1.

3.1 Required Elements

We first prove the following result which establishes conditions for when the cut condition implies routability.

Theorem 2. *Let G be an outerplanar graph with integer edge capacities $u(e)$. Let H denote a demand graph such that $G + H = (V(G), E(G) \cup E(H))$ is outerplanar. If G, H satisfies the cut condition, then H is routable in G, and an integral routing can be found in polynomial-time.*

The novelty in this statement is that we do not require the Eulerian condition on $G + H$. This condition is needed in virtually all classical results for edge-disjoint paths. In fact, even when G is a 4-cycle and H consists of a matching of size 2, the cut condition need not be sufficient to guarantee routability. The main exception is the case when G is a tree and a trivial greedy algorithm suffices to route H. We prove the theorem by giving a simple (but not so simple) algorithm to compute a routing.

To prove this theorem, we need the following 2-node reduction lemma which is generally known.

Lemma 1. *Let G be a graph and let H be a collection of demands that satisfies the cut condition. Let $G_1, \ldots, G_k$ be the blocks of G (the 2-node connected components and the cut edges (aka bridges) of G). Let H_i be the collection of nontrivial (i.e., non-loop) demands after contracting each edge $e \in E(G) \backslash E(G_i)$. Then each G_i, H_i satisfies the cut condition. Furthermore, if G (or $G + H$) was outerplanar (or planar), then each G_i (resp. $G_i + H_i$) is outerplanar (resp. planar). Moreover, if each H_i is routable in G_i, then H is routable in G (Fig. 3).*

Proof. Consider the edge contractions to be done on $G + H$ to obtain $G_i + H_i$. Then, any cut in $G_i + H_i$ was also a cut in $G + H$. Since G, H satisfies the cut condition, then G_i, H_i must also satisfy the cut condition. Furthermore, edge contraction preserves planarity and outerplanarity.

For each $st \in H$ and each G_i, the reduction process produces a request $s_i t_i$ in G_i. If this is not a loop, then s_i, t_i lie in different components of G after deleting the edges of G_i. In this case, we say that st spawns $s_i t_i$. Let J be the set of edges spawned by a demand st. It is easy to see that the edges of J form an st path. Hence if each H_i is routable in G_i, we have that H is routable in G.

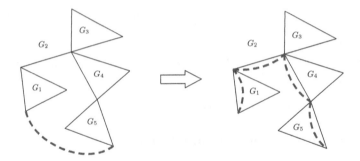

Fig. 3. The new demand edges that replace a demand edge whose terminals belong in different blocks. Solid edges represent edges of G and dashed edges represent demand edges.

Proof (Proof of Theorem 2).

Without loss of generality, we may assume that the edges of G (resp. H) have unit capacity (resp. demand). Otherwise, we may place $u(e)$ (resp. $d(e)$) parallel copies of such an edge e. In the algorithmic proof, we may also assume that G is 2-node connected. Otherwise, we may apply Lemma 1 and consider each 2-node connected component of G separately. When working with 2-node connected G, the boundary of its outer face is a simple cycle. So we label the nodes $v_1, \ldots, v_n$ by the order they appear on this cycle.

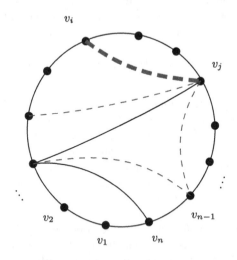

Fig. 4. The solid edges form the outerplanar graph G. The dashed edges are the demand edges. The thick dashed edge is a valid edge to route because there are no terminals v_k with $i < k < j$.

If there are no demand edges, then we are done. Otherwise, since $G + H$ is outerplanar, without loss of generality there exists $i < j$ such that $v_i v_j \in E(H)$

and no v_k is a terminal for $i < k < j$ (Fig. 4). Consider the outer face path $P = v_i, v_{i+1}, \ldots, v_j$. We show that the cut condition is still satisfied after removing both the path P and the demand $v_i v_j$. This represents routing the demand $v_i v_j$ along the path P.

Consider a central cut $\delta_G(X)$. Suppose that v_i and v_j are on opposite sides of the cut. Then, we decrease both $\delta_G(X)$ and $\delta_H(X)$ by 1, so the cut condition holds. Suppose that $v_i, v_j \notin X$, that is v_i and v_j are on the same side of the cut. Then, either $X \subset V(P) \setminus \{v_i, v_j\}$ or $X \cap V(P) = \emptyset$. We are done if $X \cap V(P) = \emptyset$ because $\delta_G(X) \cap E(P) = 0$. Otherwise, $X \subset V(P) \setminus \{v_i, v_j\}$ contains no terminals, so we cannot violate the cut condition.

We also need the following result from [8].

Theorem 5. *Let T be a tree with integer edge capacities $u(e)$. Let H denote a demand graph such that each fundamental cut of H induced by an edge $e \in T$ contains at most $ku(e)$ edges of H. We may then partition H into at most $4k$ edges sets $H_1, \ldots, H_{4k}$ such that each H_i is routable in T.*

3.2 Proof of the Main Theorem

Theorem 1. *There is a polynomial-time 224 approximation algorithm for the maximum weight* ANF *and* EDP *problems for capacitated outerplanar graphs.*

Proof. We first run the algorithms to produce an integer-capacitated tree $T, \hat{u}$ which is an 14 cut approximator for G. In addition T is a subtree and it is a conservative approximator for each cut in G. First, we prove that the maximum weight routable in T is not too much smaller than for G (in either the EDP or ANF model). To see this let S be an optimal solution in G, whose value is OPT(G). Clearly S satisfies the cut condition in G and hence by Theorem 3 it satisfies, up to a factor of 14, the cut condition in $T, \hat{u}$. Thus by Theorem 5 there are 56 sets such that $S = \cup_{i=1}^{56} S_i$ and each S_i is routable in T. Hence one of the sets S_i accrues at least $\frac{1}{56}^{th}$ the profit from OPT(G).

Now we use the factor 4 approximation [8] to solve the maximum EDP=ANF problem for $T, \hat{u}$. Let S be a subset of requests which are routable in T and have weight at least $\frac{1}{4}$ OPT(T) $\geq \frac{1}{224}$ OPT(G). Since T is a subtree of G we have that $G + T$ is outerplanar. Since $T, \hat{u}$ is an under-estimator of cuts in G, we have that the edges of T (viewed as requests) satisfies the cut condition in G. Hence by Theorem 2 we may route these single edge requests in G. Hence since S can route in T, we have that S can also route in G, completing the proof.

4 Conclusions

The technique of finding a single-tree constant-factor cut approximator (for a global constant) appears to hit a limit at outerplanar graphs. It would be interesting to find a graph parameter k which ensures a single-tree $O(f(k))$ cut approximator.

The authors thank Nick Harvey for his valuable feedback on this article. We also thank the conference reviewers for their helpful remarks. The authors Shepherd and Xia are grateful for support from the Natural Sciences and Engineering Research Council of Canada.

References

1. Abraham, I., Bartal, Y., Neiman, O.: Nearly tight low stretch spanning trees. In: FOCS, pp. 781–790 (2008)
2. Alon, N., Karp, R.M., Peleg, D., West, D.: A graph-theoretic game and its application to the k-server problem. SIAM J. Comput. **24**(1), 78–100 (1995)
3. Andersen, R., Feige, U.: Interchanging distance and capacity in probabilistic mappings. arXiv preprint arXiv:0907.3631 (2009)
4. Chekuri, C., Khanna, S., Shepherd, F.B.: Edge-disjoint paths in planar graphs with constant congestion. SIAM J. Comput. **39**, 281–301 (2009)
5. Chekuri, C., Ene, A.: Poly-logarithmic approximation for maximum node disjoint paths with constant congestion. In: Proceedings of the 24th Annual ACM-SIAM Symposium on Discrete Algorithms, pp. 326–341. SIAM (2013)
6. Chekuri, C., Khanna, S., Bruce Shepherd, F.: The all-or-nothing multicommodity flow problem. In: Proceedings of the 36th Annual ACM Symposium on Theory of Computing, STOC 2004, pp. 156–165. ACM, New York (2004)
7. Chekuri, C., Khanna, S., Shepherd, F.B.: The all-or-nothing multicommodity flow problem. SIAM J. Comput. **42**(4), 1467–1493 (2013)
8. Chekuri, C., Mydlarz, M., Shepherd, F.B.: Multicommodity demand flow in a tree and packing integer programs. ACM Trans. Algorithms (TALG) **3**(3), 27-es (2007)
9. Chekuri, C., Naves, G., Shepherd, F.B.: Maximum edge-disjoint paths in k-sums of graphs. In: Fomin, F.V., Freivalds, R., Kwiatkowska, M., Peleg, D. (eds.) ICALP 2013. LNCS, vol. 7965, pp. 328–339. Springer, Heidelberg (2013). https://doi.org/10.1007/978-3-642-39206-1_28
10. Chuzhoy, J., Li, S.: A polylogarithimic approximation algorithm for edge-disjoint paths with congestion 2. arXiv preprint arXiv:1208.1272 (2012)
11. Chuzhoy, J., Li, S.: A polylogarithimic approximation algorithm for edge-disjoint paths with congestion 2. In: Proceedings of IEEE FOCS (2012)
12. Chuzhoy, J., Kim, D.H.K., Nimavat, R.: New hardness results for routing on disjoint paths. In: Proceedings of the 49th Annual ACM SIGACT Symposium on Theory of Computing, pp. 86–99 (2017)
13. Elkin, M., Emek, Y., Spielman, D.A., Teng, S.-H.: Lower-stretch spanning trees. SIAM J. Comput. **38**(2), 608–628 (2008)
14. Englert, M., Gupta, A., Krauthgamer, R., Racke, H., Talgam-Cohen, I., Talwar, K.: Vertex sparsifiers: new results from old techniques. SIAM J. Comput. **43**(4), 1239–1262 (2014)
15. Frank, A.: Edge-disjoint paths in planar graphs. J. Comb. Theor. Ser. B **39**(2), 164–178 (1985)
16. Garg, N., Kumar, N., Sebő, A.: Integer plane multiflow maximisation: flow-cut gap and one-quarter-approximation. In: Bienstock, D., Zambelli, G. (eds.) IPCO 2020. LNCS, vol. 12125, pp. 144–157. Springer, Cham (2020). https://doi.org/10.1007/978-3-030-45771-6_12
17. Garg, N., Vazirani, V.V., Yannakakis, M.: Primal-dual approximation algorithms for integral flow and multicut in trees. Algorithmica **18**(1), 3–20 (1997)

18. Gupta, A., Newman, I., Rabinovich, Y., Sinclair, A.: Cuts, trees and ℓ_1-embeddings of graphs. Combinatorica **24**(2), 233–269 (2004)
19. Gupta, A.: Steiner points in tree metrics don't (really) help. SODA **1**, 220–227 (2001)
20. Guruswami, V., Khanna, S., Rajaraman, R., Shepherd, B., Yannakakis, M.: Near-optimal hardness results and approximation algorithms for edge-disjoint paths and related problems. J. Comput. Syst. Sci. **67**(3), 473–496 (2003)
21. Harrelson, C., Hildrum, K., Rao, S.: A polynomial-time tree decomposition to minimize congestion. In: Proceedings of the 15th Annual ACM Symposium on Parallel Algorithms and Architectures, pp. 34–43 (2003)
22. Huang, C.-C., Mari, M., Mathieu, C., Schewior, K., Vygen, J.: An approximation algorithm for fully planar edge-disjoint paths. arXiv preprint arXiv:2001.01715 (2020)
23. Kawarabayashi, K., Kobayashi, Y.: All-or-nothing multicommodity flow problem with bounded fractionality in planar graphs. SIAM J. Comput. **47**(4), 1483–1504 (2018)
24. Kleinberg, J., Tardos, E.: Approximations for the disjoint paths problem in high-diameter planar networks. J. Comput. Syst. Sci. **57**(1), 61–73 (1998)
25. Naves, G., Shepherd, B., Xia, H.: Maximum weight disjoint paths in outerplanar graphs via single-tree cut approximators. arXiv preprint arXiv:2007.10537 (2020)
26. Okamura, H., Seymour, P.D.: Multicommodity flows in planar graphs. J. Comb. Theor. Ser. B **31**(1), 75–81 (1981)
27. Rabinovich, Y., Raz, R.: Lower bounds on the distortion of embedding finite metric spaces in graphs. Discrete Comput. Geom **19**(1), 79–94 (1998)
28. Räcke, H.: Minimizing congestion in general networks. In: Proceedings of IEEE FOCS, pp. 43–52 (2002)
29. Räcke, H.: Optimal hierarchical decompositions for congestion minimization in networks. In: STOC, pp. 255–264 (2008)
30. Räcke, H., Shah, C.: Improved guarantees for tree cut sparsifiers. In: Schulz, A.S., Wagner, D. (eds.) ESA 2014. LNCS, vol. 8737, pp. 774–785. Springer, Heidelberg (2014). https://doi.org/10.1007/978-3-662-44777-2_64
31. Séguin-Charbonneau, L., Shepherd, F.B.: Maximum edge-disjoint paths in planar graphs with congestion 2. Math. Program., 1–23 (2020). https://doi.org/10.1007/s10107-020-01513-1
32. Seymour, P.D.: Matroids and multicommodity flows. Eur. J. Comb. **2**(3), 257–290 (1981)
33. Sidiropoulos, A.: Private communication (2014)

A Tight Approximation Algorithm for the Cluster Vertex Deletion Problem

Manuel Aprile[1]([⊠]), Matthew Drescher[2], Samuel Fiorini[2], and Tony Huynh[3]

[1] Dipartimento di Matematica, Università degli Studi di Padova, Padua, Italy
[2] Département de Mathématique, Université libre de Bruxelles, Brussels, Belgium
sfiorini@ulb.ac.be
[3] School of Mathematics, Monash University, Melbourne, Australia

Abstract. We give the first 2-approximation algorithm for the cluster vertex deletion problem. This is tight, since approximating the problem within any constant factor smaller than 2 is UGC-hard. Our algorithm combines the previous approaches, based on the local ratio technique and the management of true twins, with a novel construction of a "good" cost function on the vertices at distance at most 2 from any vertex of the input graph.

As an additional contribution, we also study cluster vertex deletion from the polyhedral perspective, where we prove almost matching upper and lower bounds on how well linear programming relaxations can approximate the problem.

Keywords: Approximation algorithm · Cluster vertex deletion · Linear programming relaxation · Sherali-Adams hierarchy

1 Introduction

A *cluster graph* is a graph that is a disjoint union of complete graphs.

Let G be any graph. A set $X \subseteq V(G)$ is called a *hitting set* if $G - X$ is a cluster graph. Given a graph G and (vertex) cost function $c : V(G) \to \mathbb{Q}_{\geqslant 0}$, the *cluster vertex deletion* problem (CLUSTER-VD) asks to find a hitting set X whose cost $c(X) := \sum_{v \in X} c(v)$ is minimum. We denote by $\mathrm{OPT}(G, c)$ the minimum cost of a hitting set.

If G and H are two graphs, we say that G *contains* H if some *induced* subgraph of G is isomorphic to H. Otherwise, G is said to be *H-free*. Denoting by P_k the path on k vertices, we easily see that a graph is a cluster graph if and only if it is P_3-free. Hence, $X \subseteq V(G)$ is a hitting set if and only if X contains a vertex from each induced P_3.

CLUSTER-VD has applications in graph modeled data clustering in which an unknown set of samples may be contaminated. An optimal solution for CLUSTER-VD can recover a clustered data model, retaining as much of the original data as

This project was supported by ERC Consolidator Grant 615640-ForEFront. Moreover, the paper was also supported by the Belgian FNRS through grant T008720F-35293308-BD-OCP, and by the Australian Research Council.

M. Singh and D. P. Williamson (Eds.): IPCO 2021, LNCS 12707, pp. 340–353, 2021.
https://doi.org/10.1007/978-3-030-73879-2_24

possible [15]. Vertex deletion problems such as CLUSTER-VD, where one seeks to locate vertices whose removal leaves a graph with desirable properties, often arise when measuring robustness and attack tolerance of real-life networks [1, 2, 16].

From what precedes, CLUSTER-VD is a hitting set problem in a 3-uniform hypergraph, and as such has a "textbook" 3-approximation algorithm. Moreover, the problem has an approximation-preserving reduction from VERTEX COVER, hence obtaining a $(2 - \varepsilon)$-approximation algorithm for some $\varepsilon > 0$ would contradict either the Unique Games Conjecture or P $\neq$ NP.

The first non-trivial approximation algorithm for CLUSTER-VD was a 5/2-approximation due to You, Wang and Cao [24]. Shortly afterward, Fiorini, Joret and Schaudt gave a 7/3-approximation [9], and subsequently a 9/4-approximation [10].

1.1 Our Contribution

In this paper, we close the gap between 2 and $9/4 = 2.25$ and prove the following *tight* result.

Theorem 1. CLUSTER-VD *has a 2-approximation algorithm.*

All previous approximation algorithms for CLUSTER-VD are based on the local ratio technique. See the survey of Bar-Yehuda, Bendel, Freund, and Rawitz [13] for background on this standard algorithmic technique. Our algorithm is no exception, see Algorithm 1 below. However, it significantly differs from previous algorithms in its crucial step, namely, Step 14. In fact, almost all our efforts in this paper focus on that particular step of the algorithm, see Theorem 2 below.

Let H be an induced subgraph of G, and let $c_H : V(H) \rightarrow \mathbb{Q}_{\geqslant 0}$. The weighted graph (H, c_H) is said to be α-*good in* G (for some factor $\alpha \geqslant 1$) if c_H is not identically 0 and $c_H(X \cap V(H)) \leqslant \alpha \cdot \text{OPT}(H, c_H)$ holds for every (inclusionwise) minimal hitting set X of G. We overload terminology and say that an induced subgraph H is α-*good in* G if there exists a cost function c_H such that (H, c_H) is α-good in G. We stress that the local cost function c_H is defined obliviously of the global cost function $c : V(G) \rightarrow \mathbb{Q}_{\geqslant 0}$.

We will use two methods to establish α-goodness of induced subgraphs. We say that (H, c_H) is *strongly* α-*good* if c_H is not identically 0 and $c_H(V(H)) \leqslant \alpha \cdot \text{OPT}(H, c_H)$. Clearly, if (H, c_H) is strongly α-good then (H, c_H) is α-good in G for every graph G which contains H. We say that H itself is *strongly* α-*good* if (H, c_H) is strongly α-good for some cost function c_H.

If we cannot find a strongly α-good induced subgraph of G, we will find an induced subgraph H that has a special vertex v_0 whose neighborhood $N(v_0)$ is entirely contained in H, and a cost function $c_H : V(H) \rightarrow \mathbb{Z}_{\geqslant 0}$ such that $c_H(v) \geqslant 1$ for all vertices v in the closed neighborhood $N[v_0]$ and $c_H(V(H)) \leqslant \alpha \cdot \text{OPT}(H, c_H) + 1$. Since no minimal hitting set X can contain all the vertices of $N[v_0]$, $c_H(X \cap V(H)) \leqslant c_H(V(H)) - 1 \leqslant \alpha \cdot \text{OPT}(H, c_H)$ and so (H, c_H) is α-good in G. We say that (H, c_H) (sometimes simply H) is *centrally* α-*good (in* G) *with respect to* v_0. Moreover, we call v_0 the *root vertex*.

Algorithm 1. CLUSTER-VD-APX(G, c)

Input: (G, c) a weighted graph
Output: X a minimal hitting set of G
1: **if** G is a cluster graph **then**
2:    $X \leftarrow \varnothing$
3: **else if** there exists $u \in V(G)$ with $c(u) = 0$ **then**
4:    $G' \leftarrow G - u$
5:    $c'(v) \leftarrow c(v)$ for $v \in V(G')$
6:    $X' \leftarrow$ CLUSTER-VD-APX(G', c')
7:    $X \leftarrow X'$ if X' is a hitting set of G; $X \leftarrow X' \cup \{u\}$ otherwise
8: **else if** there exist true twins $u, u' \in V(G)$ **then**
9:    $G' \leftarrow G - u'$
10:    $c'(u) \leftarrow c(u) + c(u'); c'(v) \leftarrow c(v)$ for $v \in V(G' - u)$
11:    $X' \leftarrow$ CLUSTER-VD-APX(G', c')
12:    $X \leftarrow X'$ if X' does not contain u; $X \leftarrow X' \cup \{u'\}$ otherwise
13: **else**
14:    find a weighted induced subgraph (H, c_H) that is 2-good in G
15:    $\lambda^* \leftarrow \max\{\lambda \mid \forall v \in V(H) : c(v) - \lambda c_H(v) \geqslant 0\}$
16:    $G' \leftarrow G$
17:    $c'(v) \leftarrow c(v) - \lambda^* c_H(v)$ for $v \in V(H)$; $c'(v) \leftarrow c(v)$ for $v \in V(G) \setminus V(H)$
18:    $X \leftarrow$ CLUSTER-VD-APX(G', c')
19: **end if**
20: **return** X

In order to illustrate these ideas, consider the following two examples (see Fig. 1). First, let H be a C_4 (that is, a 4-cycle) contained in G and $\mathbf{1}_H$ denote the unit cost function on $V(H)$. Then $(H, \mathbf{1}_H)$ is strongly 2-good, since $\sum_{v \in V(H)} \mathbf{1}_H(v) = 4 = 2\,\mathrm{OPT}(H, \mathbf{1}_H)$. Second, let H be a P_3 contained in G, starting at a vertex v_0 that has degree-1 in G. Then $(H, \mathbf{1}_H)$ is centrally 2-good with respect to v_0, but it is not strongly 2-good.

Each time we find a 2-good weighted induced subgraph in G, the local ratio technique allows us to recurse on an induced subgraph G' of G in which at least one vertex of H is deleted from G. For example, the 2-good induced subgraphs mentioned above allow us to reduce to input graphs G that are C_4-free and have minimum degree at least 2.

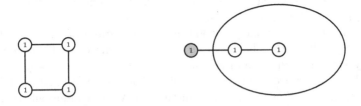

Fig. 1. $(C_4, \mathbf{1}_{C_4})$ on the left is strongly 2-good. $(P_3, \mathbf{1}_{P_3})$, on the right, is centrally 2-good in G with respect to the gray vertex, which has degree 1 in G.

In order to facilitate the search for α-good induced subgraphs, it greatly helps to assume that G is *twin-free*. That is, G has no two distinct vertices u, u' such that $uu' \in E(G)$ and for all $v \in V(G - u - u')$, $uv \in E(G)$ if and only if $u'v \in E(G)$. Equivalently, u and u' are such that $N[u] = N[u']$. Two such vertices u, u' are called *true twins*. As in [9,10], our algorithm reduces G whenever it has a pair of true twins u, u' (see Steps 8–12). The idea is simply to add the cost of u' to that of u and delete u'.

The crux of our algorithm, Step 14, relies on the following structural result. Below, we denote by $N_{\leqslant i}[v]$ (resp. $N_i(v)$) the set of vertices at distance at most (resp. equal to) i from vertex v (we omit the subscript if $i = 1$).

Theorem 2. *Let G be a twin-free graph, let v_0 be any vertex of G, and let H be the subgraph of G induced by $N_{\leqslant 2}[v_0]$. There exists a cost function $c_H : V(H) \to \mathbb{Z}_{\geqslant 0}$ such that (H, c_H) is either strongly 2-good, or centrally 2-good in G with respect to v_0. Moreover, c_H can be constructed in polynomial time.*

We also study CLUSTER-VD from the polyhedral point of view. In particular we investigate how well linear programming (LP) relaxations can approximate the optimal value of CLUSTER-VD.

Letting $\mathcal{P}_3(G)$ denote the collection of all vertex sets $\{u, v, w\}$ that induce a P_3 in G, we define $P(G) := \{x \in [0,1]^{V(G)} \mid \forall \{u, v, w\} \in \mathcal{P}_3(G) : x_u + x_v + x_w \geqslant 1\}$. We let $\mathsf{SA}_r(G)$ denote the relaxation obtained from $P(G)$ by applying r rounds of the Sherali-Adams hierarchy [20]. If a cost function $c : V(G) \to \mathbb{Q}_{\geqslant 0}$ is provided, we let $\mathsf{SA}_r(G, c) := \min\{\sum_{v \in V(G)} c(v)x_v \mid x \in \mathsf{SA}_r(G)\}$ denote the optimum value of the corresponding linear programming relaxation.

It is not hard to see that the straightforward LP relaxation $P(G)$ has worst case integrality gap equal to 3 (by *worst case*, we mean that we take the supremum over all graphs G). Indeed, for a random n-vertex graph, $\mathrm{OPT}(G, 1_G) = n - O(\log^2 n)$ with high probability, while $\mathrm{LP}(G, 1_G) \leqslant n/3$. We show that the worst case integrality gap drops from 3 to 2.5 after applying one round of Sherali-Adams.

Theorem 3. *For every graph G, the integrality gap of $\mathsf{SA}_1(G)$ is at most $5/2$. Moreover, for every $\varepsilon > 0$ there is some instance (G, c) of CLUSTER-VD such that $\mathrm{OPT}(G, c) \geqslant (5/2 - \varepsilon)\,\mathsf{SA}_1(G, c)$.*

By relying on Theorem 2, we further show that the integrality gap decreases to $2 + \varepsilon$ after applying $\mathrm{poly}(1/\varepsilon)$ rounds.

Theorem 4. *For every fixed $\varepsilon > 0$, performing $r = \mathrm{poly}(1/\varepsilon)$ rounds of the Sherali-Adams hierarchy produces an LP relaxation of CLUSTER-VD whose integrality gap is at most $2 + \varepsilon$. That is, $\mathrm{OPT}(G, c) \leqslant (2 + \varepsilon)\,\mathsf{SA}_r(G, c)$ for all weighted graphs (G, c).*

On the negative side, applying known results on VERTEX COVER [5], we show that no polynomial-size LP relaxation of CLUSTER-VD can have integrality gap at most $2 - \varepsilon$ for some $\varepsilon > 0$. This result is unconditional: it does not rely on P $\neq$ NP nor the Unique Games Conjecture. We refer the reader to the full version of our paper [4] for precise definitions and for the proofs of Theorems 3 and 4.

1.2   Comparison to Previous Works

We now revisit all previous approximation algorithms for CLUSTER-VD [9,10, 24]. The presentation given here slightly departs from [9,24], and explains in a unified manner what is the bottleneck in each of the algorithms.

Fix $k \in \{3, 4, 5\}$, and let $\alpha := (2k - 1)/(k - 1)$. Notice that $\alpha = 5/2$ if $k = 3$, $\alpha = 7/3$ if $k = 4$ and $\alpha = 9/4$ if $k = 5$. In [10, Lemma 3], it is shown that if a twin-free graph G contains a k-clique, then one can find an induced subgraph H containing the k-clique and a cost function c_H such that (H, c_H) is strongly α-good.

Therefore, in order to derive an α-approximation for CLUSTER-VD, one may assume without loss of generality that the input graph G is twin-free and has no k-clique. Let v_0 be a maximum degree vertex in G, and let H denote the subgraph of G induced by $N_{\leqslant 2}[v_0]$. In [10], it is shown by a tedious case analysis that one can construct a cost function c_H such that (H, c_H) is 2-good in G, using the fact that G has no k-clique.

In this paper we show that the latter assumption is not needed: one can *always* construct a cost function c_H that makes (H, c_H) 2-good in G, provided that G is twin-free, see Theorem 2. This result allows us to avoid laborious case-checking, and single-handedly closes the approximability status of CLUSTER-VD.

1.3   Other Related Works

CLUSTER-VD has also been widely studied from the perspective of *fixed parameter tractability*. Given a graph G and parameter k as input, the task is to decide if G has a hitting set X of size at most k. A $2^k n^{\mathcal{O}(1)}$-time algorithm for this problem was given by Hüffner, Komusiewicz, Moser, and Niedermeier [15]. This was subsequently improved to a $1.911^k n^{\mathcal{O}(1)}$-time algorithm by Boral, Cygan, Kociumaka, and Pilipczuk [7], and a $1.811^k n^{\mathcal{O}(1)}$-time algorithm by Tsur [23]. By the general framework of Fomin, Gaspers, Lokshtanov, and Saurabh [11], these parametrized algorithms can be transformed into exponential algorithms which compute the size of a minimum hitting set for G exactly, the fastest of which runs in time $O(1.488^n)$.

For polyhedral results, [14] gives some facet-defining inequalities of a polytope that is affinely equivalent to the CLUSTER-VD polytope, as well as complete linear descriptions for special classes of graphs.

Another related problem is the *feedback vertex set* problem in tournaments (FVST). Given a tournament T with costs on the vertices, the task is to find a minimum cost set of vertices X such that $T - X$ does not contain a directed cycle.

For unit costs, note that CLUSTER-VD is equivalent to the problem of deleting as few elements as possible from a *symmetric* relation to obtain a transitive relation, while FVST is equivalent to the problem of deleting as few elements as possible from an *antisymmetric* and complete relation to obtain a transitive relation.

In a tournament, hitting all directed cycles is equivalent to hitting all directed triangles, so FVST is also a hitting set problem in a 3-uniform hypergraph. Moreover, FVST is also UCG-hard to approximate to a constant factor smaller than 2. Cai, Deng, and Zang [8] gave a 5/2-approximation algorithm for FVST, which was later improved to a 7/3-approximation algorithm by Mnich, Williams, and Végh [19]. Lokshtanov, Misra, Mukherjee, Panolan, Philip, and Saurabh [18] recently gave a *randomized* 2-approximation algorithm, but no deterministic (polynomial-time) 2-approximation algorithm is known. For FVST, one round of the Sherali-Adams hierarchy actually provides a 7/3-approximation [3]. This is in contrast with Theorem 3.

Among other related covering and packing problems, Fomin, Le, Lokshtanov, Saurabh, Thomassé, and Zehavi [12] studied both CLUSTER-VD and FVST from the *kernelization* perspective. They proved that the unweighted versions of both problems admit subquadratic kernels: $\mathcal{O}(k^{\frac{5}{3}})$ for CLUSTER-VD and $\mathcal{O}(k^{\frac{3}{2}})$ for FVST.

1.4 Overview of the Proof

We give a sketch of the proof of Theorem 2. Recall that $H = G[N_{\leqslant 2}[v_0]]$. If the subgraph induced by $N(v_0)$ contains a hole (that is, an induced cycle of length at least 4), then H contains a wheel, which makes H strongly 2-good, see Lemma 1. If the subgraph induced by $N(v_0)$ contains an induced $2P_3$ (that is, two disjoint and anticomplete copies of P_3), then H is strongly 2-good, see Lemma 2. This allows us to reduce to the case where the subgraph induced by $N(v_0)$ is chordal and $2P_3$-free.

Lemma 5 then gives a direct construction of a cost function c_H which certifies that (H, c_H) is centrally 2-good, provided that the subgraph induced by $N[v_0]$ is *twin-free*. This is the crucial step of the proof. It serves as the base case of the induction. Here, we use a slick observation due to Lokshtanov [17]: since the subgraph induced by $N(v_0)$ is chordal and $2P_3$-free, it has a hitting set that is a clique. In a previous version, our proof of Theorem 2 was slightly more complicated.

We show inductively that we can reduce to the case where the subgraph induced by $N[v_0]$ is twin-free. The idea is to delete vertices from H to obtain a smaller graph H', while preserving certain properties, and then compute a suitable cost function c_H for H, given a suitable cost function $c_{H'}$ for H'. We delete vertices at distance 2 from v_0. When this creates true twins in H, we delete one vertex from each pair of true twins. At the end, we obtain a twin-free induced subgraph of $H[N[v_0]]$, which corresponds to our base case.

We devote Sect. 2 to the main technical ingredients of the proof, and in Sect. 2.4 we prove Theorem 2. Due to space constraints we omit some proofs and details, for which we refer to the full version of our paper [4]. In particular, [4] contains the proof that Theorem 2 implies Theorem 1, which can be proven similarly as in [10, Proof of Theorem 1], as well as a complexity analysis of Algorithm 1.

2   Finding 2-good Induced Subgraphs

The goal of this section is to prove Theorem 2. Our proof is by induction on the number of vertices in $H = G[N_{\leq 2}[v_0]]$. First, we quickly show that we can assume that the subgraph induced by $N(v_0)$ is chordal and $2P_3$-free. Using this, we prove the theorem in the particular case where the subgraph induced by $N[v_0]$ is twin-free. Finally, we prove the theorem in the general case by showing how to deal with true twins.

2.1   Restricting to Chordal, $2P_3$-free Neighborhoods

As pointed out earlier in the introduction, 4-cycles are strongly 2-good. This implies that wheels of order 5 are strongly 2-good (putting a zero cost on the apex). Recall that a *wheel* is a graph obtained from a cycle by adding an apex vertex (called the *center*). We now show that *all* wheels of order at least 5 are strongly 2-good. This allows our algorithm to restrict to input graphs such that the subgraph induced on each neighborhood is chordal. In a similar way, we show that we can further restrict such neighborhoods to be $2P_3$-free.

Lemma 1. *Let $H := W_k$ be a wheel on $k \geq 5$ vertices and center v_0, let $c_H(v_0) := k - 5$ and $c_H(v) := 1$ for $v \in V(H - v_0)$. Then (H, c_H) is strongly 2-good.*

Proof. Notice that $\mathrm{OPT}(H, c_H) \geq k - 3$ since a hitting set either contains v_0 and at least 2 more vertices, or does not contain v_0 but contains $k - 3$ other vertices. Hence, $\sum_{v \in V(H)} c_H(v) = k - 5 + k - 1 = 2(k - 3) \leq 2\,\mathrm{OPT}(H, c_H)$. □

Lemma 2. *Let H be the graph obtained from $2P_3$ by adding a universal vertex v_0. Let $c_H(v_0) := 2$ and $c_H(v) := 1$ for $v \in V(H - v_0)$. Then (H, c_H) is strongly 2-good.*

Proof. It is easy to check that $\mathrm{OPT}(H, c_H) \geq 4$. Thus, $\sum_{v \in V(H)} c_H(v) = 8 \leq 2\,\mathrm{OPT}(H, c_H)$. □

2.2   The Twin-Free Case

Throughout this section, we assume that H is a *twin-free* graph with a universal vertex v_0 such that $H - v_0$ is chordal and $2P_3$-free. Our goal is to construct a cost function c_H that certifies that H is centrally 2-good.

It turns out to be easier to define the cost function on $V(H - v_0) = N(v_0)$ first, and then adjust the cost of v_0. This is the purpose of the next lemma. Below, $\omega(G, c)$ denotes the maximum weight of a clique in weighted graph (G, c).

Lemma 3. *Let H be a graph with a universal vertex v_0 and $H' := H - v_0$. Let $c_{H'} : V(H') \to \mathbb{Z}_{\geqslant 1}$ be a cost function such that*

(i) $c_{H'}(V(H')) \geqslant 2\omega(H', c_{H'})$ and
(ii) $\mathrm{OPT}(H', c_{H'}) \geqslant \omega(H', c_{H'}) - 1$.

Then we can extend $c_{H'}$ to a function $c_H : V(H) \to \mathbb{Z}_{\geqslant 1}$ such that $c_H(V(H)) \leqslant 2\,\mathrm{OPT}(H, c_H) + 1$. In other words, (H, c_H) is centrally 2-good with respect to v_0.

Proof. Notice that $\mathrm{OPT}(H, c_H) = \min(c_H(v_0) + \mathrm{OPT}(H', c_{H'}), c_{H'}(V(H')) - \omega(H', c_{H'}))$, since if X is hitting set of H that does not contain v_0, then $H - X$ is a clique.

Now, choose $c_H(v_0) \in \mathbb{Z}_{\geqslant 1}$ such that

$$\max(1, c_{H'}(V(H')) - 2\,\mathrm{OPT}(H', c_{H'}) - 1) \leqslant c_H(v_0) \leqslant c_{H'}(V(H')) - 2\omega(H', c_{H'}) + 1. \quad (1)$$

It is easy to check that such $c_H(v_0)$ exists because the lower bound in (1) is at most the upper bound, thanks to conditions (i) and (ii).

This choice satisfies $c_H(V(H)) \leqslant 2\,\mathrm{OPT}(H, c_H) + 1$ since it holds both in case $\mathrm{OPT}(H, c_H) = c_H(v_0) + \mathrm{OPT}(H', c_{H'})$ by the upper bound on $c_{H'}(V(H'))$ given by (1), and in case $\mathrm{OPT}(H, c_H) = c_{H'}(V(H')) - \omega(H', c_{H'})$ by the upper bound on $c_H(v_0)$ given by (1). $\qquad \square$

We abuse notation and regard a clique X of a graph as both a set of vertices and a subgraph. We call a hitting set X of a graph G a *hitting clique* if X is also a clique.

Lemma 4. *Every chordal, $2P_3$-free graph contains a hitting clique.*

Proof. Let G be a chordal, $2P_3$-free graph. Since G is chordal, G admits a *clique tree* T [6]. In T, the vertices are the maximal cliques of G and, for every two maximal cliques K, K', the intersection $K \cap K'$ is contained in every clique of the K–K' path in T. For an edge $e := KK'$ of T, Let T_1 and T_2 be the components of $T - e$ and G_1 and G_2 be the subgraphs of G induced by the union of all the cliques in T_1 and T_2, respectively. It is easy to see that deleting $K \cap K'$ separates G_1 from G_2 in G. Now, since G is $2P_3$-free, at least one of $G'_1 := G_1 - (K \cap K')$ or $G'_2 := G_2 - (K \cap K')$ is a cluster graph. If both G'_1 and G'_2 are cluster graphs, we are done since $K \cap K'$ is the desired hitting clique. Otherwise, if G'_i is not a cluster graph, then we can orient e towards T_i. Applying this argument on each edge, we define an orientation of T, which must have a sink K_0. But then removing K_0 from G leaves a cluster graph, and we are done. Since the clique tree of a chordal graph can be constructed in polynomial time [6], the hitting clique can be found in polynomial time. $\qquad \square$

We are ready to prove the base case for Theorem 2.

Lemma 5. *Let H be a twin-free graph with universal vertex v_0 such that $H - v_0$ is chordal and $2P_3$-free. There exists a cost function c_H such that (H, c_H) is centrally 2-good with respect to v_0. Moreover, c_H can be found in time polynomial in the size of H.*

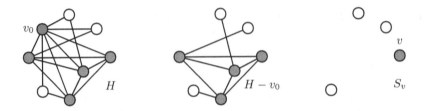

Fig. 2. Here H is twin-free and the blue vertices form a hitting clique K_0 for $H - v_0$, which is chordal and $2P_3$-free. For $v \in K_0$, the set S_v defined as in the proof of Lemma 5 consists of the unique maximal independent set containing v. We obtain $c_H = (6, 1, 1, 1, 1, 3, 3, 3)$, which is easily seen to be centrally 2-good with respect to v_0.

Proof. By Lemma 4, some maximal clique of $H - v_0$, say K_0, is a hitting set.

We claim that there is a family of stable sets $\mathcal{S} = \{S_v \mid v \in K_0\}$ of $H - v_0$ satisfying the following properties:

(P1) every vertex of $H - v_0$ is contained in some S_v;
(P2) for each $v \in K_0$, S_v contains v and at least one other vertex;
(P3) for every two distinct vertices $v, v' \in K_0$, $H[S_v \cup S_{v'}]$ contains a P_3.

Before proving the claim, we prove that it implies the lemma, making use of Lemma 3. Consider the cost function $c_{H'} := \sum_{v \in K_0} \chi^{S_v}$ on the vertices of $H' := H - v_0$ defined by giving to each vertex u a cost equal to the number of stable sets S_v that contain u (see Fig. 2). Let us show that $c_{H'}$ satisfies the conditions of Lemma 3 and can therefore be extended to a cost function c_H on $V(H)$ such that (H, c_H) is centrally 2-good with respect to v_0.

First, by (P1), we have $c_{H'}(u) \in \mathbb{Z}_{\geqslant 1}$ for all $u \in V(H')$. Second, condition (i) of Lemma 3 follows from (P2) since each stable set S_v contributes at least two units to $c_{H'}(V(H'))$ and at most one unit to $\omega(H', c_{H'})$. Third, (P3) implies that every hitting set of H' meets every stable set S_v, except possibly one. Hence, $\mathrm{OPT}(H', c_{H'}) \geqslant |K_0| - 1$. Also, every clique of H' meets every stable set S_v in at most one vertex, implying that $\omega(H', c_{H'}) \leqslant |K_0|$, and equality holds since $c_{H'}(K_0) = |K_0|$. Putting the last two observations together, we see that $\mathrm{OPT}(H', c_{H'}) \geqslant |K_0| - 1 = \omega(H', c_{H'}) - 1$ and hence condition (ii) of Lemma 3 holds.

Now, we prove that our claim holds. Let $K_1, \ldots, K_t$ denote the clusters (maximal cliques) of cluster graph $H - v_0 - K_0$. For $i \in [t]$, consider the submatrix A_i of the adjacency matrix $A(H)$ with rows indexed by the vertices of K_0 and columns indexed by the vertices of K_i.

Notice that A_i contains neither $\begin{pmatrix} 1 & 0 \\ 0 & 1 \end{pmatrix}$ nor $\begin{pmatrix} 0 & 1 \\ 1 & 0 \end{pmatrix}$ as a submatrix, as this would give a C_4 contained in $H - v_0$, contradicting the chordality of $H - v_0$. Hence, after permuting its rows and columns if necessary, A_i can be assumed to be staircase-shaped. That is, every row of A_i is nonincreasing and every column nondecreasing. Notice also that A_i does not have two equal columns, since these would correspond to two vertices of K_i that are true twins.

For each K_i that is not complete to $v \in K_0$, define $\varphi_i(v)$ as the vertex $u \in K_i$ whose corresponding column in A_i is the *first* containing a 0 in row v. Now, for each v, let S_v be the set including v, and $\varphi_i(v)$, for each K_i that is not complete to v.

Because K is maximal, no vertex $u \in K_i$ is complete to K_0. Since no two columns of A_i are identical, we must have $u = \varphi_i(v)$ for some $v \in K_0$. This proves (P1).

Notice that $v \in S_v$ by construction and that $|S_v| \geqslant 2$ since otherwise, v would be universal in H and thus a true twin of v_0. Hence, (P2) holds.

Finally, consider any two distinct vertices $v, v' \in K_0$. Since v, v' are not true twins, the edge vv' must be in a P_3 contained in $H - v_0$. Assume, without loss of generality, that there is a vertex $u \in K_i$ adjacent to v and not to v' for some $i \in [t]$. Then $\{v, v', \varphi_i(v')\}$ induces a P_3 contained in $H[S_v \cup S_{v'}]$, proving (P3). This concludes the proof of the claim.

We remark that the cost function c_H can be computed in polynomial time. We first obtain efficiently the collection $\mathcal{S}$, hence the restriction of c_H to H', and then just let $c_H(v_0) := c_{H'}(V(H')) - 2\omega(H', c_{H'}) + 1 = c_{H'}(V(H')) - 2|K_0| + 1$. This sets $c_H(v_0)$ to its upper bound in (1), see the proof of Lemma 3.   □

2.3   Handling True Twins in $G[N[v_0]]$

We now deal with the general case where $G[N[v_0]]$ contains true twins. We start with an extra bit of terminology relative to true twins. Let G be a twin-free graph, and $v_0 \in V(G)$. Suppose that u, u' are true twins in $G[N[v_0]]$. Since G is twin-free, there exists a vertex v that is adjacent to exactly one of u, u'. We say that v is a *distinguisher* for the edge uu' (or for the pair $\{u, u'\}$). Notice that either $uu'v$ or $u'uv$ is an induced P_3. Notice also that v is at distance 2 from v_0.

Now, consider a graph H with a special vertex $v_0 \in V(H)$ (the root vertex) such that

(H1) every vertex is at distance at most 2 from v_0, and
(H2) every pair of vertices that are true twins in $H[N[v_0]]$ has a distinguisher.

Let v be any vertex that is at distance 2 from v_0. Consider the equivalence relation $\equiv$ on $N[v_0]$ with $u \equiv u'$ whenever $u = u'$ or u, u' are true twins in $H - v$. Observe that the equivalence classes of $\equiv$ are of size at most 2 since, if u, u', u'' are distinct vertices with $u \equiv u' \equiv u''$, then v cannot distinguish every edge of the triangle on u, u' and u''. Hence, two of these vertices are true twins in H, which contradicts (H2).

From what precedes, the edges contained in $N[v_0]$ that do not have a distinguisher in $H - v$ form a matching $M := \{u_1 u_1', \ldots, u_k u_k'\}$ (possibly, $k = 0$). Let H' denote the graph obtained from H by deleting v and exactly one endpoint from each edge of M. Notice that the resulting subgraph is the same, up to isomorphism, no matter which endpoint is chosen.

The lemma below states how we can obtain a cost function c_H that certifies that H is centrally 2-good from a cost function $c_{H'}$ that certifies that H' is centrally 2-good. It is inspired by [10, Lemma 3]. We defer the proof to [4]. See Fig. 3 for an example.

Lemma 6. *Let H be any graph satisfying (H1) and (H2) for some $v_0 \in V(H)$. Let $v \in N_2(v_0)$. Let $M := \{u_1 u_1', \ldots, u_k u_k'\}$ be the matching formed by the edges in $N[v_0]$ whose unique distinguisher is v, where $u_i' \neq v_0$ for all i (we allow the case $k = 0$). Let $H' := H - u_1' - \cdots - u_k' - v$. Given a cost function $c_{H'}$ on $V(H')$, define a cost function c_H on $V(H)$ by letting $c_H(u_i') := c_{H'}(u_i)$ for $i \in [k]$, $c_H(v) := \sum_{i=1}^{k} c_{H'}(u_i) = \sum_{i=1}^{k} c_H(u_i')$, and $c_H(u) := c_{H'}(u)$ otherwise. First, H' satisfies (H1) and (H2). Second, if $(H', c_{H'})$ is centrally 2-good, then (H, c_H) is centrally 2-good.*

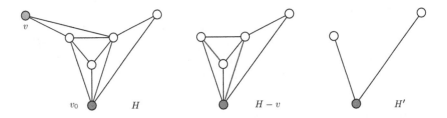

Fig. 3. $H - v$ violates (H2), and contains two pairs of true twins, indicated by the red edges. Lemma 6 applies. We see that H' is a P_3, for which Lemma 5 gives $c_{H'} = 1_{H'}$. In (H, c_H), all vertices get a unit cost except v, which gets a cost of 2, since there are 2 pairs of true twins in $H - v$. Thus we obtain $c_H = (1, 2, 1, 1, 1, 1)$. (Color figure online)

2.4 Putting Things Together

We are ready to prove Theorem 2.

Proof (of Theorem 2). We can decide in polynomial time (see for instance [22]) if $H[N(v_0)]$ is chordal, and if not, output a hole of $H[N(v_0)]$. If the latter holds, we are done by Lemma 1. If the former holds, we can decide in polynomial time (see [21], and the proof of Lemma 4) whether H contains a $2P_3$. If it does, we are done by Lemma 2.

From now on, assume that the subgraph induced by $N(v_0)$ is chordal and $2P_3$-free. This is done without loss of generality. Notice that hypotheses (H1) and (H2) from Sect. 2.3 hold for H. This is obvious for (H1). To see why (H2) holds, remember that G is twin-free. Hence, every edge uu' contained in $N[v_0]$ must have a distinguisher in G, which is in $N_{\leqslant 2}[v_0]$. (In fact, notice that if u and u' are true twins in $H[N[v_0]]$ then the distinguisher is necessarily in $N_2(v_0)$.)

We repeatedly apply Lemma 6 in order to delete each vertex of $N_2(v_0)$ one after the other and reduce to the case where H is a twin-free graph for which v_0 is universal. We can then apply Lemma 5. The whole process takes polynomial time. □

3 Conclusion

In this paper we provide a tight approximation algorithm for the cluster vertex deletion problem (CLUSTER-VD). Our main contribution is the efficient construction of a local cost function on the vertices at distance at most 2 from any vertex v_0 such that every minimal hitting set of the input graph has local cost at most *twice* the local optimum. If the subgraph induced by $N(v_0)$ (the first neighborhood of v_0) contains a hole, or a $2P_3$, then this turns out to be straightforward. The most interesting case arises when the local subgraph H is twin-free, has radius 1, and moreover $H[N(v_0)] = H - v_0$ is chordal and $2P_3$-free. Such graphs are very structured, which we crucially exploit.

Lemma 3 allows us to define the local cost function on the vertices distinct from v_0 and then later adjust the cost of v_0. We point out that condition (ii) basically says that the local cost function should define a hyperplane that "almost" separates the hitting set polytope and the clique polytope of the chordal, $2P_3$-free graph $H - v_0$. This was a key intuition which led us to the proof of Theorem 2. If these polytopes were disjoint, this would be easy. But actually this is not the case since they have a common vertex (as we show, $H - v_0$ has a hitting clique).

One natural question arising from our approach of CLUSTER-VD in general graphs is the following: is the problem polynomial-time solvable on *chordal* graphs? What about chordal, $2P_3$-*free* graphs? We propose this last question as our first open question.

Our second contribution is to study the CLUSTER-VD problem from the polyhedral point of view, in particular with respect to the tightness of the Sherali-Adams hierarchy. Our results on Sherali-Adams fail to match the 2-approximation factor of our algorithm (by epsilon), and we suspect this is not by chance. We believe that, already for certain classes of triangle-free graphs, the LP relaxation given by a bounded number of rounds of the Sherali-Adams hierarchy has an integrality gap strictly larger than 2. Settling this is our second open question.

Acknowledgements. We are grateful to Daniel Lokshtanov for suggesting Lemma 4, which allowed us to simplify our algorithm and its proof.

References

1. Albert, R., Jeong, H., Barabási, A.-L.: Error and attack tolerance of complex networks. Nature **406**(6794), 378–382 (2000)
2. Aprile, M., Castro, N., Ferreira, G., Piccini, J., Robledo, F., Romero, P.: Graph fragmentation problem: analysis and synthesis. Int. Trans. Oper. Res. **26**(1), 41–53 (2019)
3. Aprile, M., Drescher, M., Fiorini, S., Huynh, T.: A simple 7/3-approximation algorithm for feedback vertex set in tournaments. arXiv preprint arXiv:2008.08779 (2020)
4. Aprile, M., Drescher, M., Fiorini, S., Huynh, T.: A tight approximation algorithm for the cluster vertex deletion problem. arXiv preprint arXiv:2007.08057 (2020)

5. Bazzi, A., Fiorini, S., Pokutta, S., Svensson, O.: No small linear program approximates vertex cover within a factor $2 - \epsilon$. Math. Oper. Res. **44**(1), 147–172 (2019)
6. Blair, J.R.S., Peyton, B.: An introduction to chordal graphs and clique trees. In: George, A., Gilbert, J.R., Liu, J.W.H. (eds.) Graph Theory and Sparse Matrix Computation. The IMA Volumes in Mathematics and its Applications, vol. 56. Springer, New York (1993). https://doi.org/10.1007/978-1-4613-8369-7_1
7. Boral, A., Cygan, M., Kociumaka, T., Pilipczuk, M.: A fast branching algorithm for cluster vertex deletion. Theor. Comput. Syst. **58**(2), 357–376 (2016)
8. Cai, M.-C., Deng, X., Zang, W.: An approximation algorithm for feedback vertex sets in tournaments. SIAM J. Comput. **30**(6), 1993–2007 (2001)
9. Fiorini, S., Joret, G., Schaudt, O.: Improved approximation algorithms for hitting 3-vertex paths. In: Louveaux, Q., Skutella, M. (eds.) IPCO 2016. LNCS, vol. 9682, pp. 238–249. Springer, Cham (2016). https://doi.org/10.1007/978-3-319-33461-5_20
10. Fiorini, S., Joret, G., Schaudt, O.: Improved approximation algorithms for hitting 3-vertex paths. Math. Program. **182**(1–2, Ser. A), 355–367 (2020)
11. Fomin, F.V., Gaspers, S., Lokshtanov, D., Saurabh, S.: Exact algorithms via monotone local search. J. ACM **66**(2), 23 (2019)
12. Fomin, F.V., Le, T., Lokshtanov, D., Saurabh, S., Thomassé, S., Zehavi, M.: Subquadratic kernels for implicit 3-hitting set and 3-set packing problems. ACM Trans. Algorithms **15**(1), 13:1–13:44 (2019)
13. Freund, A., Bar-Yehuda, R., Bendel, K.: Local ratio: a unified framework for approximation algorithms. ACM Comput. Surv. **36**, 422–463 (2005)
14. Hosseinian, S., Butenko, S.: Polyhedral properties of the induced cluster subgraphs. arXiv preprint arXiv:1904.12025 (2019)
15. Hüffner, F., Komusiewicz, C., Moser, H., Niedermeier, R.: Fixed-parameter algorithms for cluster vertex deletion. Theor. Comput. Syst. **47**(1), 196–217 (2010)
16. Jahanpour, E., Chen, X.: Analysis of complex network performance and heuristic node removal strategies. Commun. Nonlinear Sci. Numer. Simul. **18**(12), 3458–3468 (2013)
17. Lokshtanov, D.: Personal communication
18. Lokshtanov, D., Misra, P., Mukherjee, J., Panolan, F., Philip, G., Saurabh, S.: 2-approximating feedback vertex set in tournaments. In: Proceedings of the 14th Annual ACM-SIAM Symposium on Discrete Algorithms, pp. 1010–1018. SIAM (2020)
19. Mnich, M., Williams, V.V., Végh, L.A.: A 7/3-approximation for feedback vertex sets in tournaments. In: Sankowski, P., Zaroliagis, C.D. (eds.) 24th Annual European Symposium on Algorithms, ESA 2016. LIPICS, Aarhus, Denmark, 22–24 August 2016, vol. 57, pp. 67:1–67:14. Schloss Dagstuhl - Leibniz-Zentrum für Informatik (2016)
20. Sherali, H.D., Adams, W.P.: A hierarchy of relaxations between the continuous and convex hull representations for zero-one programming problems. SIAM J. Discrete Math. **3**(3), 411–430 (1990)
21. Tarjan, R.E.: Decomposition by clique separators. Discret. Math. **55**(2), 221–232 (1985)

22. Tarjan, R.E., Yannakakis, M.: Simple linear-time algorithms to test chordality of graphs, test acyclicity of hypergraphs, and selectively reduce acyclic hypergraphs. SIAM J. Comput. **13**(3), 566–579 (1984)
23. Tsur, D.: Faster parameterized algorithm for cluster vertex deletion. CoRR, abs/1901.07609 (2019)
24. You, J., Wang, J., Cao, Y.: Approximate association via dissociation. Discret. Appl. Math. **219**, 202–209 (2017)

Fixed Parameter Approximation Scheme
for Min-Max k-Cut

Karthekeyan Chandrasekaran and Weihang Wang$^{(\boxtimes)}$

University of Illinois Urbana-Champaign, Urbana, IL 61801, USA
{karthe,weihang3}@illinois.edu

Abstract. We consider the graph k-partitioning problem under the min-max objective, termed as MINMAX k-CUT. The input here is a graph $G = (V, E)$ with non-negative edge weights $w : E \to \mathbb{R}_+$ and an integer $k \geq 2$ and the goal is to partition the vertices into k non-empty parts $V_1, \ldots, V_k$ so as to minimize $\max_{i=1}^{k} w(\delta(V_i))$. Although minimizing the sum objective $\sum_{i=1}^{k} w(\delta(V_i))$, termed as MINSUM k-CUT, has been studied extensively in the literature, very little is known about minimizing the max objective. We initiate the study of MINMAX k-CUT by showing that it is NP-hard and W[1]-hard when parameterized by k, and design a parameterized approximation scheme when parameterized by k. The main ingredient of our parameterized approximation scheme is an exact algorithm for MINMAX k-CUT that runs in time $(\lambda k)^{O(k^2)} n^{O(1)}$, where λ is the value of the optimum and n is the number of vertices. Our algorithmic technique builds on the technique of Lokshtanov, Saurabh, and Suria-narayanan (FOCS, 2020) who showed a similar result for MINSUM k-CUT. Our algorithmic techniques are more general and can be used to obtain parameterized approximation schemes for minimizing ℓ_p-norm measures of k-partitioning for every $p \geq 1$.

Keywords: k-cut · Min-max objective · Parameterized approximation scheme

1 Introduction

Graph partitioning problems are fundamental for their intrinsic theoretical value as well as applications in clustering. In this work, we consider graph partitioning under the *minmax* objective. The input here is a graph $G = (V, E)$ with non-negative edge weights $w : E \to \mathbb{R}_+$ along with an integer $k \geq 2$ and the goal is to partition the vertices of G into k non-empty parts $V_1, \ldots, V_k$ so as to minimize $\max_{i=1}^{k} w(\delta(V_i))$; here, $\delta(V_i)$ is the set of edges which have exactly one end-vertex in V_i and $w(\delta(V_i)) := \sum_{e \in \delta(V_i)} w(e)$ is the total weight of edges in $\delta(V_i)$. We refer to this problem as MINMAX k-CUT.

Motivations. Minmax objective for optimization problems has an extensive literature in approximation algorithms. It is relevant in scenarios where the goal is

Supported in part by NSF grants CCF-1814613 and CCF-1907937.

M. Singh and D. P. Williamson (Eds.): IPCO 2021, LNCS 12707, pp. 354–367, 2021.
https://doi.org/10.1007/978-3-030-73879-2_25

to achieve fairness/balance—e.g., load balancing in multiprocessor scheduling, discrepancy minimization, min-degree spanning tree, etc. In the context of graph cuts and partitioning, recent works (e.g., see [1,7,19]) have proposed and studied alternative minmax objectives that are different from MINMAX k-CUT.

The complexity of MINMAX k-CUT was also raised as an open problem by Lawler [22]. Given a partition $V_1, \ldots, V_k$ of the vertex set of an input graph, one can measure the quality of the partition in various natural ways. Two natural measures are (i) the max objective given by $\max_{i=1}^{k} w(\delta(V_i))$ and (ii) the sum objective given by $\sum_{i=1}^{k} w(\delta(V_i))$. We will discuss other ℓ_p-norm measures later. Once a measure is defined, a corresponding optimization problem involves finding a partition that minimizes the measure. We will denote the optimization problem where the goal is to minimize the sum objective as MINSUM k-CUT.

MINSUM k-CUT *and Prior Works.* The objectives in MINMAX k-CUT and MINSUM k-CUT for $k = 2$ coincide owing to the symmetric nature of the graph cut function (i.e., $w(\delta(S)) = w(\delta(V \setminus S))$ for all $S \subseteq V$) but the objectives differ for $k \geq 3$. MINSUM k-CUT has been studied extensively in the algorithms community leading to fundamental graph structural results. We briefly recall the literature on MINSUM k-CUT.

Goldschmidt and Hochbaum [13,14] showed that MINSUM k-CUT is NP-hard when k is part of input by a reduction from CLIQUE and designed the first polynomial time algorithm for fixed k. Their algorithm runs in time $n^{O(k^2)}$, where n is the number of vertices in the input graph. Subsequently, Karger and Stein [20] gave a random contraction based algorithm that runs in time $\tilde{O}(n^{2k-2})$. Thorup [30] gave a tree-packing based deterministic algorithm that runs in time $\tilde{O}(n^{2k})$. The last couple of years has seen renewed interests in MINSUM k-CUT with exciting progress [8,10,15–18,23,25]. Very recently, Gupta, Harris, Lee, and Li [15,18] have shown that the Karger-Stein algorithm in fact runs in $\tilde{O}(n^k)$ time; $n^{(1-o(1))k}$ seems to be a lower bound on the run-time of any algorithm [23]. The hardness result of Goldschmidt and Hochbaum as well as their algorithm inspired Saran and Vazirani [27] to consider MINSUM k-CUT when k is part of input from the perspective of approximation. They showed the first polynomial-time 2-approximation for MINSUM k-CUT. Alternative 2-approximations have also been designed subsequently [26,31]. For k being a part of the input, Manurangsi [25] showed that there does not exist a polynomial-time $(2-\epsilon)$-approximation for any constant $\epsilon > 0$ under the Small Set Expansion Hypothesis.

MINSUM k-CUT has also been investigated from the perspective of fixed-parameter algorithms. It is known that MINSUM k-CUT when parameterized by k is W[1]-hard and does not admit a $f(k)n^{o(1)}$-time algorithm for any function $f(k)$ [9,12]. Motivated by this hardness result and Manurangsi's $(2 - \epsilon)$-inapproximability result, Gupta, Lee, and Li [16] raised the question of whether there exists a *parameterized approximation algorithm* for MINSUM k-CUT when parameterized by k, i.e., can one obtain a $(2 - \epsilon)$-approximation in time $f(k)n^{O(1)}$ for some constant $\epsilon > 0$? As a proof of concept, they designed a 1.9997-approximation algorithm that runs in time $2^{O(k^6)}n^{O(1)}$ [16] and a

$(1 + \epsilon)$-approximation algorithm that runs in time $(k/\epsilon)^{O(k)} n^{k+O(1)}$ [17]. Subsequently, Kawarabayashi and Lin [21] designed a $(5/3 + \epsilon)$-approximation algorithm that runs in time $2^{O(k^2 \log k)} n^{O(1)}$. This line of work culminated in a *parameterized approximation scheme* when parameterized by k—Lokshtanov, Saurabh, and Surianarayanan [24] designed a $(1 + \epsilon)$-approximation algorithm that runs in time $(k/\epsilon)^{O(k)} n^{O(1)}$. We emphasize that, from the perspective of algorithm design, a parameterized approximation scheme is more powerful than a parameterized approximation algorithm.

Fixed-terminal Variants. A natural approach to solve both MINMAX k-CUT and MINSUM k-CUT is to solve their fixed-terminal variants: The input here is a graph $G = (V, E)$ with non-negative edge weights $w : E \rightarrow \mathbb{R}_+$ along with k terminals $v_1, \ldots, v_k \in V$ and the goal is to partition the vertices into k parts $V_1, \ldots, V_k$ such that $v_i \in V_i$ for every $i \in [k]$ so as to minimize the measure of interest for the partition. The fixed-terminal variant of MINSUM k-CUT, popularly known as MULTIWAY CUT, is NP-hard for $k \geq 3$ [11] and has a rich literature. It admits a 1.2965 approximation [28] and does not admit a $(1.20016 - \epsilon)$-approximation for any constant $\epsilon > 0$ under the unique games conjecture [4]. The fixed-terminal variant of MINMAX k-CUT, known as MINMAX MULTIWAY CUT, is NP-hard for $k \geq 4$ [29] and admits an $O(\sqrt{\log n \log k})$-approximation [2]. Although fixed-terminal variants are natural approaches to solve global cut problems (similar to using min $\{s, t\}$-cut to solve global min-cut), they have two limitations: (1) they are not helpful when k is part of input and (2) even for fixed k, they do not give the best algorithms (e.g., even for $k = 3$, MULTIWAY CUT is NP-hard while MINSUM k-CUT is solvable in polynomial time as discussed above).

MINMAX k-CUT *vs* MINSUM k-CUT. There is a fundamental structural difference between MINMAX k-CUT and MINSUM k-CUT. Optimal solutions to MINSUM k-CUT satisfy a nice property: assuming that the input graph is connected, every part in an optimal partition for MINSUM k-CUT induces a connected subgraph. Hence, MINSUM k-CUT is also phrased as the problem of finding a minimum weight subset of edges to remove so that the resulting graph contains at least k connected components. However, this nice property does not hold for MINMAX k-CUT as illustrated by the example in Fig. 1.

MINMAX k-CUT *for Fixed* k. For fixed k, there is an easy approach to solve MINMAX k-CUT based on the following observation: For a given instance, an optimum solution to MINMAX k-CUT is a k-approximate optimum to MINSUM k-CUT. The randomized algorithm of Karger and Stein implies that the number of k-approximate solutions to MINSUM k-CUT is $n^{O(k^2)}$ and they can all be enumerated in polynomial time [15,18,20] (also see [8]). These two facts immediately imply that MINMAX k-CUT can be solved in $n^{O(k^2)}$ time. We recall that the graph cut function is symmetric and submodular.[1] In an upcoming work, Chandrasekaran and Chekuri [5] show that the more general problem of min-max

---

[1] A function $f : 2^V \rightarrow \mathbb{R}$ is symmetric if $f(S) = f(V \setminus S)$ for all $S \subseteq V$ and is submodular if $f(A) + f(B) \geq f(A \cap B) + f(A \cup B)$.

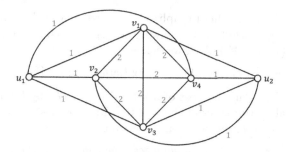

Fig. 1. An example where the unique optimum partition for MINMAX k-CUT for $k = 5$ induces a disconnected part. The edge weights are as shown. The unique optimum partition for MINMAX 5-CUT is $(\{u_1, u_2\}, \{v_1\}, \{v_2\}, \{v_3\}, \{v_4\})$.

symmetric submodular k-partition[2] is also solvable in time $n^{O(k^2)}T$, where n is the size of the ground set and T is the time to evaluate the input submodular function on a given set.

1.1 Results

In this work, we focus on MINMAX k-CUT when k is part of input. We first show that MINMAX k-CUT is strongly NP-hard. Our reduction also implies that it is W[1]-hard when parameterized by k, i.e., there does not exist a $f(k)n^{O(1)}$-time algorithm for any function $f(k)$.

Theorem 1. MINMAX k-CUT *is strongly NP-hard and W[1]-hard when parameterized by* k.

Our hardness reduction also implies that MINMAX k-CUT does not admit an algorithm that runs in time $n^{o(k)}$ assuming the exponential time hypothesis. Given the hardness result, it is natural to consider approximations and fixed-parameter tractability. Using the known 2-approximation for MINSUM k-CUT and the observation that the optimum value of MINSUM k-CUT is at most k times the optimum value of MINMAX k-CUT, it is easy to get a $(2k)$-approximation for MINMAX k-CUT. An interesting open question is whether we can improve the approximability.

The hardness results also raise the question of whether MINMAX k-CUT admits a parameterized approximation algorithm when parameterized by k or, going a step further, does it admit a parameterized approximation scheme when parameterized by k? We resolve this question affirmatively by designing a parameterized approximation scheme. Let $G = (V, E)$ be a graph with non-negative edge weights $w : E \rightarrow \mathbb{R}_+$. We write G to denote the unit-weight version of

---

[2] In the min-max symmetric submodular k-partition problem, the input is a symmetric submodular function $f : 2^V \rightarrow \mathbb{R}$ given by an evaluation oracle, and the goal is to partition the ground set V into k non-empty parts $V_1, \ldots, V_k$ so as to minimize $\max_{i=1}^{k} f(V_i)$.

the graph (i.e., the unweighted graph) and G_w to denote the graph with edge weights w. We emphasize that the unweighted graph could have parallel edges. For a partition $(V_1, \ldots, V_k)$ of V, we define

$$\text{cost}_{G_w}(V_1, \ldots, V_k) := \max\{w(\delta(V_i)) : i \in [k]\}.$$

We will denote the minimum cost of a k-partition in G_w by $\text{OPT}(G_w, k)$. The following is our algorithmic result showing that MINMAX k-CUT admits a parameterized approximation scheme when parameterized by k.

Theorem 2. *There exists a randomized algorithm that takes as input an instance of* MINMAX k-CUT, *namely an n-vertex graph $G = (V, E)$ with edge weights $w : E \to \mathbb{R}_{\geq 0}$ and an integer $k \geq 2$, along with an $\epsilon \in (0, 1)$, and runs in time $(k/\epsilon)^{O(k^2)} n^{O(1)} \log(\max_{e \in E} w(e))$ to return a partition $\mathcal{P}$ of the vertices of G with k non-empty parts such that $\text{cost}_{G_w}(\mathcal{P}) \leq (1 + \epsilon)\text{OPT}(G_w, k)$ with high probability.*

We note that $\log(\max_{e \in E} w(e))$ is polynomial in the size of the input. Theorem 2 can be viewed as the counterpart of the parameterized approximation scheme for MINSUM k-CUT due to Lokshtanov, Saurabh, and Surianarayanan [24] but for MINMAX k-CUT. The central component of our parameterized-approximation scheme given in Theorem 2 is the following result which shows a fixed-parameter algorithm for MINMAX k-CUT in unweighted graphs when parameterized by k and the solution size.

Theorem 3. *There exists a deterministic algorithm that takes as input an unweighted instance of* MINMAX k-CUT, *namely an n-vertex graph $G = (V, E)$ and an integer $k \geq 2$, along with an integer λ, and runs in time $(k\lambda)^{O(k^2)} n^{O(1)}$ to determine if there exists a k-partition $(V_1, \ldots, V_k)$ of V such that $\text{cost}_G(V_1, \ldots, V_k) \leq \lambda$ and if so, then finds an optimum.*

We emphasize that the algorithm in Theorem 3 is deterministic while the algorithm in Theorem 2 is randomized.

1.2 Outline of Techniques

Our NP-hardness and W[1]-hardness results for MINMAX k-CUT are based on a reduction from the clique problem. Our reduction is an adaptation of Downey et al.'s reduction [12] from the clique problem to MINSUM k-CUT.

Our randomized algorithm for Theorem 2 essentially reduces the input weighted instance of MINMAX k-CUT to an instance where Theorem 3 can be applied: we reduce the instance to an unweighted instance with optimum value $O((k/\epsilon^3)) \log n$, i.e., the optimum value is logarithmic in the number of vertices. Moreover, the reduction runs in time $2^{O(k)}(n/\epsilon)^{O(1)} \log \text{OPT}(G_w, k)$. Applying Theorem 3 to the reduced instance yields a run-time of

$$\left(\left(\frac{k^2}{\epsilon^3}\right) \log n\right)^{O(k^2)} n^{O(1)} = \left(\frac{k}{\epsilon}\right)^{O(k^2)} (\log n)^{O(k^2)} n^{O(1)}$$

$$= \left(\frac{k}{\epsilon}\right)^{O(k^2)} (k^{O(k^2)} + n) n^{O(1)} = \left(\frac{k}{\epsilon}\right)^{O(k^2)} n^{O(1)}.$$

Hence, the total run-time is $(k/\epsilon)^{O(k^2)}n^{O(1)}\log \text{OPT}(G_w, k)$ (including the reduction time), thereby proving Theorem 2.

We now briefly describe the reduction to an unweighted instance with logarithmic optimum: (i) Firstly, we do a standard knapsack PTAS-style rounding procedure to convert the instance to an unweighted instance with a $(1+\epsilon)$-factor loss. (ii) Secondly, we delete cuts with small value to ensure that all connected components in the graph have large min-cut value, i.e., have min-cut value at least $\epsilon\text{OPT}/k$—this deletion procedure can remove at most ϵOPT edges and hence, a $(1+\epsilon)$-approximate solution in the resulting graph gives a $(1+O(\epsilon))$-approximate solution in the original graph. (iii) Finally, we do a random sampling of edges with probability $q := \Theta(k\log n/(\epsilon^3\text{OPT}))$. This gives a subgraph that preserves all cut values within a $(1 \pm \epsilon)$-factor when scaled by q with high probability [3]. The preservation of all cut values also implies that the optimum value to MINMAX k-CUT is also preserved within a $(1 \pm \epsilon)$-factor. The scaling factor of q allows us to conclude that the optimum in the subsampled graph is $O((k/\epsilon^3))\log n$. We note that this three step reduction recipe follows the ideas of [24] who designed a parameterized approximation scheme for MINSUM k-CUT. Our contribution to the reduction is simply showing that their reduction ideas also apply to MINMAX k-CUT.

The main contribution of our work is in proving Theorem 3, i.e., giving a fixed-parameter algorithm for MINMAX k-CUT when parameterized by k and the solution size. We discuss this now. At a high-level, we exploit the tools developed by [24] who designed a dynamic program based fixed-parameter algorithm for MINSUM k-CUT when parameterized by k and the solution size. Our algorithm for MINMAX k-CUT is also based on a dynamic program. However, since we are interested in the minmax objective, the subproblems in our dynamic program are completely different from that of [24]. We begin with the observation that an optimum solution to MINMAX k-CUT is a k-approximate optimum to MINSUM k-CUT. This observation and the tree packing approach for MINSUM k-CUT due to [8] allows us to obtain, in polynomial time, a spanning tree T of the input graph such that the number of edges of the tree crossing a MINMAX k-CUT optimum partition is $O(k^2)$. We will call a partition Π with $O(k^2)$ edges of the tree T crossing Π to be a T-feasible partition. Next, we use the tools of [24] to generate, in polynomial time, a suitable tree decomposition of the input graph—let us call this a *good* tree decomposition. The central intuition underlying our algorithm is to use the spanning tree T to guide a dynamic program on the good tree decomposition.

As mentioned before, our dynamic program is different from that of [24]. We now sketch the sub-problems in our dynamic program. For simplicity, we assume that we have a value $\lambda \geq \text{OPT}(G, k)$. We call a partition $\mathcal{P}$ of a set S to be a k-*subpartition* if $\mathcal{P}$ has at most k non-empty parts. The *adhesion* of a tree node t in the tree decomposition, denoted A_t, is the intersection of the bag corresponding to t with that of its parent (the adhesion of the root node of the tree decomposition is the empty set). The good tree decomposition that we generate has low adhesion, i.e., the adhesion size is $O(\lambda k)$ for every tree node. In order to define our sub-problems for a tree node t, we consider the set $\mathcal{F}^{A_t}$ of all possible k-subpartitions of the adhesion A_t and which can be extended to a T-feasible partition of the entire vertex set. A simple counting argument shows that $|\mathcal{F}^{A_t}| = (\lambda k)^{O(k^2)}$. Now consider a Boolean function f_t : $\mathcal{F}^{A_t} \times \{0, 1, \ldots, \lambda\}^k \times \{0, 1, \ldots, 2k^2\} \to \{0, 1\}$. We note that the domain of the function is small, i.e., the domain has size $(\lambda k)^{O(k^2)}$. Let $(\mathcal{P}_{A_t}, \bar{x}, d)$ denote an argument/input to the function. The function aims to determine if there exists a k-subpartition $\mathcal{P}$ of the union of the bags descending from t in the tree decomposition (call this set of vertices to be V_t) so that (i) the restriction of the partition $\mathcal{P}$ to A_t is exactly $\mathcal{P}_{A_t}$ (ii) the number of edges crossing the i'th part of $\mathcal{P}$ in the subgraph $G[V_t]$ is exactly x_i for all $i \in [k]$, and (iii) the number of tree edges crossing the partition $\mathcal{P}$ whose both endpoints are in V_t is at most d. It is easy to see that if we can compute such a function f_r for the root node r of the tree decomposition, then it can be used to find the optimum value of MINMAX k-CUT, namely OPT(G, k).

However, we are unable to solve the sub-problem (i.e., compute such a function f_t) based on the sub-problem values of the children of t. On the one hand, given an arbitrary optimal partition Ω to MINMAX k-CUT, restricting Ω to V_t yields a partition $\mathcal{P}$ that satisfies (i) (ii), and (iii) above for some choice of $\mathcal{P}_{A_t}$, $\bar{x}$, and d. On the other hand, a partition $\mathcal{P}$ of V_t satisfying (i) (ii), and (iii) for some choice of $\mathcal{P}_{A_t}$, $\bar{x}$, and d does not necessarily extend to an optimal partition of V. Therefore, identifying all inputs $(\mathcal{P}_{A_t}, \bar{x}, d)$ for which there exists a partition satisfying (i) (ii), and (iii) is not necessary for our purpose. Instead, for a fixed optimal partition Ω, it suffices if our f_t function evaluates to 1 on inputs $(\mathcal{P}_{A_t}, \bar{x}, d)$ for which (i) (ii), and (iii) are satisfied by the partition obtained by restricting Ω to V_t.

To identify partitions of V_t that potentially extend to Ω, for tree node t with bag $\chi(t)$, we construct a family $\mathcal{D}$ of *nice decompositions* such that $\mathcal{D}$ is small-sized (i.e., with $|\mathcal{D}| = (\lambda k)^{O(k^2)} n^{O(1)}$). A nice decomposition is a triple of the form $(O, \mathcal{P}_{\chi(t)}, \mathcal{Q}_{\chi(t)})$ satisfying certain properties (which are slightly different from the properties used in [24]). Here, O is a subset of $\chi(t)$ and $\mathcal{P}_{\chi(t)}, \mathcal{Q}_{\chi(t)}$ are partitions of $\chi(t)$, where $\mathcal{Q}_{\chi(t)}$ refines $\mathcal{P}_{\chi(t)}$. The constructed family $\mathcal{D}$ is such that at least one of the $\mathcal{Q}_{\chi(t)}$ partitions in $\mathcal{D}$ is a refinement of a restriction of Ω to $\chi(t)$. By induction hypothesis, we know that $f_{t'}(\mathcal{P}_{A_{t'}}, \bar{x}', d') = 1$ for all children t' of t and all inputs $(\mathcal{P}_{A_{t'}}, \bar{x}', d')$ for which the restriction of Ω to $V_{t'}$ satisfies (i) (ii), and (iii). In order to set $f_t(\mathcal{P}_{A_t}, \bar{x}, d) = 1$ for all inputs $(\mathcal{P}_{A_t}, \bar{x}, d)$ for which the restriction of Ω to V_t satisfies (i) (ii), and (iii), it suffices to consider all inputs $(\mathcal{P}_{A_t}, \bar{x}, d)$ for which there exists a partition $\mathcal{P}$ of V_t satisfying (i) (ii), and (iii) such that (a) the restriction of $\mathcal{P}$ to $\chi(t)$ coarsens $\mathcal{Q}_{\chi(t)}$ for some $\mathcal{Q}_{\chi(t)}$ in $\mathcal{D}$ and (b) for all children t' of t, the restriction of $\mathcal{P}$ to $\chi(t')$ gives a partition satisfying (i) (ii), and (iii) for some input $(\mathcal{P}_{A_{t'}}, \bar{x}', d')$. Therefore, the family $\mathcal{D}$ of nice decompositions and the $f_{t'}$ values for all children t' of t together suffice to identify all inputs $(\mathcal{P}_{A_t}, \bar{x}, d)$ that correspond to a restriction of Ω to V_t satisfying (i) (ii), and (iii). To expedite this process, for each nice decomposition $(O, \mathcal{P}_{\chi(t)}, \mathcal{Q}_{\chi(t)})$, we use the set O and the partition $\mathcal{P}_{\chi(t)}$ to limit the number of coarsenings of $\mathcal{Q}_{\chi(t)}$ to be considered. During this process, every time $f_t(\mathcal{P}_{A_t}, \bar{x}, d)$ is set to 1 for some input $(\mathcal{P}_{A_t}, \bar{x}, d)$, we can only guarantee that there is an actual partition of V_t satisfying (i) (ii), and (iii) (but that partition is not necessarily the restriction of Ω to V_t). However, if the restriction of Ω to V_t indeed satisfies (i) (ii), and (iii) for some input $(\mathcal{P}_{A_t}, \bar{x}, d)$, then the process will indeed set $f_t(\mathcal{P}_{A_t}, \bar{x}, d)$ to 1.

To formalize our tolerance of this one-sided error in the dynamic program, we define the notion of f-correctness and f-soundness for the function f_t (see Definition 6 and Proposition 1). We show that this weaker goal of computing an f-correct and f-sound function f_t based on f-correct and f-sound functions $f_{t'}$ for all children t' of t can be achieved in time $(\lambda k)^{O(k^2)} n^{O(1)}$ (see Lemma 4). Since the domain of the function is of size $(\lambda k)^{O(k^2)}$ and the tree decomposition is polynomial in the size of the input, the total number of sub-problems that we solve in the dynamic program is $(\lambda k)^{O(k^2)} n^{O(1)}$, thus proving Theorem 3.

One of our main contributions beyond the techniques of [24] is the introduction of the clean notions of f-correctness and f-soundness in the sub-problems of the dynamic program and combining them to address the minmax objective. An advantage of our dynamic program (in contrast to that of [24]) is that it is also applicable for alternative norm-based measures of k-partitions: here, the goal is to find a k-partition of the vertex set of the given edge-weighted graph so as to minimize $(\sum_{i=1}^{k} w(\delta(V_i))^p)^{1/p}$—we call this as MIN ℓ_p-NORM k-CUT. We note that MINMAX k-CUT is exactly MIN ℓ_∞-NORM k-CUT while MINSUM k-CUT is exactly MIN ℓ_1-NORM k-CUT. Our dynamic program can also be used to obtain the counterpart of Theorem 3 for MIN ℓ_p-NORM k-CUT for every $p \geq 1$. This result in conjunction with the reduction to unweighted instances (which can be shown to hold for MIN ℓ_p-NORM k-CUT) also leads to a parameterized approximation scheme for MIN ℓ_p-NORM k-CUT for every $p \geq 1$.

Organization. We set up the tools to prove Theorem 3 in Sect. 2. We sketch a proof of Theorem 3 in Sect. 3. We refer the reader to the full version of this work [6] for all missing proofs (including Theorems 1 and 2). We conclude with a few open questions in Sect. 4.

2   Tools for the Fixed-Parameter Algorithm

Let $G = (V, E)$ be a graph. Throughout this work, we consider a partition to be an ordered tuple of non-empty subsets. An ordered tuple of subsets $(S_1, \ldots, S_k)$, where $S_i \subseteq V$ for all $i \in [k]$, is a k-subpartition of V if $S_1 \cup \ldots \cup S_k = V$ and $S_i \cap S_j = \emptyset$ for every pair of distinct $i, j \in [k]$. We emphasize the distinction between partitions and k-subpartitions—in a partition, all parts are required to be non-empty but the number of parts can be fewer than k while a k-subpartition allows for empty parts but the number of parts is exactly k.

For a subgraph $H \subseteq G$, a subset $X \subseteq V$, and a partition/k-subpartition $\mathcal{P}$ of X, we use $\delta_H(\mathcal{P})$ to denote the set of edges in $E(H)$ whose end-vertices are in different parts of $\mathcal{P}$. For a subgraph H of G and a subset $S \subseteq V(H)$, we use

$\delta_H(S)$ to denote the set of edges in H with exactly one end-vertex in S. We will denote the set of (exclusive) neighbors of a subset S of vertices in the graph G by $N_G(S)$. We need the notion of a tree decomposition.

Definition 1. *Let $G = (V, E)$ be a graph. A pair (τ, χ), where τ is a tree and $\chi : V(\tau) \to 2^V$ is a mapping of the nodes of the tree to subsets of vertices of the graph, is a* tree decomposition *of G if the following conditions hold:*

(i) $\cup_{t \in V(\tau)} \chi(t) = V$,
(ii) for every edge $e = uv \in E$, there exists some $t \in V(\tau)$ such that $u, v \in \chi(t)$, and
(iii) for every $v \in V$, the set of nodes $\{t \in V(\tau) : v \in \chi(t)\}$ induces a connected subtree of τ.

For each $t \in V(\tau)$, we call $\chi(t)$ to be a bag *of the tree decomposition.*

We now describe certain notations that will be helpful while working with the tree decomposition. Let (τ, χ) be the tree decomposition of the graph $G = (V, E)$. We root τ at an arbitrary node $r \in V(\tau)$. For a tree node $t \in V(\tau) \backslash \{r\}$, there is a unique edge between t and its parent. Removing this edge disconnects τ into two subtrees τ_1 and τ_2, and we say that the set $A_t := \chi(\tau_1) \cap \chi(\tau_2)$ is the *adhesion associated with t*. For the root node r, we define $A_r := \emptyset$. For a tree node $t \in V(\tau)$, we denote the subgraph induced by all vertices in bags descending from t as G_t (here, the node t is considered to be a descendant of itself). We need the notions of compactness and edge-unbreakability.

Definition 2. *A tree decomposition (τ, χ) of a graph G is* compact *if for every tree node $t \in V(\tau)$, the set of vertices $V(G_t) \backslash A_t$ induces a connected subgraph in G and $N_G(V(G_t) \backslash A_t) = A_t$.*

Definition 3. *Let $G = (V, E)$ be a graph and let $S \subseteq V$. The subset S is (a, b)-edge-unbreakable if for every nonempty proper subset S' of V satisfying $|E[S', V \backslash S']| \le b$, we have that either $|S \cap S'| \le a$ or $|S \backslash S'| \le a$.*

Informally, a subset S is (a, b)-edge-unbreakable if every non-trivial 2-partition of $G[S]$ either has large cut value or one side of the partition is small in size. With these definitions, we have the following result from [24].

Lemma 1. *[24] There exists a polynomial time algorithm that takes a graph $G = (V, E)$, an integer $k \ge 2$, and an integer λ as input and returns a compact tree decomposition (τ, χ) of G such that*

(i) each adhesion has size at most λk, and
(ii) for every tree node $t \in V(\tau)$, the bag $\chi(t)$ is $((\lambda k + 1)^5, \lambda k)$-edge-unbreakable.

Next, we need the notion of α-respecting partitions.

Definition 4. *Let $G = (V, E)$ be a graph and G' be a subgraph of G. A partition $\mathcal{P}$ of V α-respects G' if $|\delta_{G'}(\mathcal{P})| \le \alpha$.*

The following lemma shows that we can efficiently find a spanning tree T of a given graph such that there exists an optimum k-partition that $2k^2$-respects T. It follows from Lemma 7 of [8] and the observation that an optimum solution to MINMAX k-CUT is a k-approximate optimum to MINSUM k-CUT.

Lemma 2. *[8] There exists a polynomial time algorithm that takes a graph G as input and returns a spanning tree T of G such that there exists an optimum min-max k-partition Π that $(2k^2)$-respects T.*

The next definition allows us to handle partitions of subsets that are crossed by a spanning tree at most $2k^2$ times. For $S \subseteq U$ and a partition/k-subpartition $\mathcal{P}$ of U, a partition/k-subpartition $\mathcal{P}'$ of S is a *restriction of $\mathcal{P}$ to S* if for every $u, v \in S$, u and v are in the same part of $\mathcal{P}'$ if and only if they are in the same part of $\mathcal{P}$.

Definition 5. *Let $G = (V, E)$ be a graph, T be a spanning tree of G, and $X \subseteq V$. A partition $\mathcal{P}$ of X is T-feasible if there exists a partition $\mathcal{P}'$ of V such that*

(i) The restriction of $\mathcal{P}'$ to X is $\mathcal{P}$, and
(ii) $\mathcal{P}'$ $(2k^2)$-respects T.

Moreover, a k-subpartition $\mathcal{P}'$ of X is T-feasible if the partition obtained from $\mathcal{P}'$ by discarding the empty parts of $\mathcal{P}'$ is T-feasible.

3   Fixed-Parameter Algorithm Parameterized by k and Solution Size

In this section we prove Theorem 3. Let $(G = (V, E), k)$ be the input instance of MINMAX k-CUT with n vertices. The input graph G could possibly have parallel edges. We assume that G is connected. Let $\mathrm{OPT} = \mathrm{OPT}(G, k)$ (i.e., OPT is the optimum objective value of MINMAX k-CUT on input G) and let λ be the input such that $\lambda \geq \mathrm{OPT}$. We will design a dynamic programming algorithm that runs in time $(\lambda k)^{O(k^2)} n^{O(1)}$ to compute OPT.

Given the input, we first use Lemma 1 to obtain a tree decomposition (τ, χ) of G satisfying the conditions of the lemma. Since the algorithm in the lemma runs in polynomial time, the size of the tree decomposition (τ, χ) is polynomial in the input size. Next, we use Lemma 2 to obtain a spanning tree T such that there exists an optimum min-max k-partition $\Omega = (\Omega_1, \ldots, \Omega_k)$ of V that $(2k^2)$-respects T, and moreover T is a subgraph of G. We fix the tree decomposition (τ, χ), the spanning tree T, and the optimum solution Ω with these choices in the rest of this section. We note that $\Omega_i \neq \emptyset$ for all $i \in [k]$ and $\max_{i \in [k]} |\delta_G(\Omega_i)| = \mathrm{OPT}$. We emphasize that the choice of Ω is fixed only for the purposes of the correctness of the algorithm and is not known to the algorithm explicitly.

Our algorithm is based on dynamic program (DP). We now state the subproblems in our dynamic program (DP), bound the number of subproblems in the DP, and prove Theorem 3. For a tree node $t \in V(\tau)$, let $\mathcal{F}^{A_t}$ be the collection of partitions of the adhesion A_t that are (i) T-feasible and (ii) have at most

k parts. We emphasize that elements of $\mathcal{F}_{A_t}$ are of the form $\mathcal{P}_{A_t} = (\tilde{P}_1, \ldots, \tilde{P}_{k'})$ for some $k' \in \{0, 1, \ldots, k\}$, where $\tilde{P}_i \neq \emptyset$ for all $i \in [k']$. The following lemma bounds the size of $\mathcal{F}^{A_t}$, which in turn, will be helpful in bounding the number of subproblems to be solved in our dynamic program.

Lemma 3. *For every tree node* $t \in V(\tau)$, *we have* $|\mathcal{F}^{A_t}| = (\lambda k)^{O(k^2)}$. *Moreover, the collection* $\mathcal{F}^{A_t}$ *can be enumerated in* $(\lambda k)^{O(k^2)}$ *time.*

The proof of the lemma appears in the full version [6]. The following definition will be useful in identifying the subproblems of the DP.

Definition 6. *Let* $t \in V(\tau)$ *be a tree node, and* $f_t : \mathcal{F}^{A_t} \times \{0, 1, \ldots, \lambda\}^k \times \{0, 1, \ldots, 2k^2\} \to \{0, 1\}$ *be a Boolean function.*

1. *(**Correctness**) The function* f_t *is f-correct if we have* $f_t(\mathcal{P}_{A_t}, \bar{x}, d) = 1$ *for all* $\mathcal{P}_{A_t} = (\tilde{P}_1, \ldots, \tilde{P}_{k'}) \in \mathcal{F}^{A_t}$, $\bar{x} \in \{0, 1, \ldots, \lambda\}^k$, *and* $d \in \{0, 1, \ldots, 2k^2\}$ *for which there exists a k-subpartition* $\mathcal{P} = (P_1', \ldots, P_k')$ *of* $V(G_t)$ *satisfying the following conditions:*
   (i) $P_i' \cap A_t = \tilde{P}_i$ *for all* $i \in [k']$,
   (ii) $|\delta_{G_t}(P_i')| = x_i$ *for all* $i \in [k]$,
   (iii) $|\delta_T(\mathcal{P})| \leq d$, *and*
   (iv) $\mathcal{P}$ *is a restriction of* Ω *to* $V(G_t)$.
   A k-subpartition of $V(G_t)$ *satisfying the above four conditions is said to witness f-correctness of* $f_t(\mathcal{P}_{A_t}, \bar{x}, d)$.
2. *(**Soundness**) The function* f_t *is f-sound if for all* $\mathcal{P}_{A_t} = (\tilde{P}_1, \ldots, \tilde{P}_{k'}) \in \mathcal{F}^{A_t}$, $\bar{x} \in \{0, 1, \ldots, \lambda\}^k$ *and* $d \in \{0, 1, \ldots, 2k^2\}$, *we have* $f_t(\mathcal{P}_{A_t}, \bar{x}, d) = 1$ *only if there exists a k-subpartition* $\mathcal{P} = (P_1', \ldots, P_k')$ *of* $V(G_t)$ *satisfying conditions (i) (ii) and (iii) above. A k-subpartition of* $V(G_t)$ *satisfying (i) (ii) and (iii) is said to witness f-soundness of* $f_t(\mathcal{P}_{A_t}, \bar{x}, d)$.

We emphasize the distinction between correctness and soundness: correctness relies on all four conditions while soundness relies only on three conditions. We crucially need distinct correctness and soundness definitions in our sub-problems in order for Bellman's principle of optimality to hold.

The next proposition shows that an f-correct and f-sound function for the root node of the tree decomposition can be used to recover the optimum value. Its proof appears in the full version [6].

Proposition 1. *If we have a function* $f_r : \mathcal{F}^{A_r} \times \{0, 1, \ldots, \lambda\}^k \times \{0, 1, \ldots, 2k^2\} \to \{0, 1\}$ *that is both f-correct and f-sound, where r is the root of the tree decomposition* τ, *then*

$$OPT = \min\left\{ \max_{i \in [k]}\{x_i\} : f_r(\mathcal{P}_\emptyset, \bar{x}, 2k^2) = 1, \bar{x} \in [\lambda]^k \right\},$$

where $\mathcal{P}_\emptyset$ *is the 0-tuple that denotes the trivial partition of* $A_r = \emptyset$.

By Proposition 1, it suffices to compute an f-correct and f-sound function f_r, where r is the root of the tree decomposition τ. The next technical lemma allows us to compute this in a bottom-up fashion on the tree decomposition.

Lemma 4. *There exists an algorithm that takes as input (τ, χ), a tree node $t \in V(\tau)$, Boolean functions $f_{t'} : \mathcal{F}^{A_{t'}} \times \{0, 1, \dots, \lambda\}^k \times \{0, 1, \dots, 2k^2\} \to \{0, 1\}$ for every child t' of t that are f-correct and f-sound, and runs in time $(\lambda k)^{O(k^2)} n^{O(1)}$ to return a function $f_t : \mathcal{F}^{A_t} \times \{0, 1, \dots, \lambda\}^k \times \{0, 1, \dots, 2k^2\} \to \{0, 1\}$ that is f-correct and f-sound.*

The proof of Lemma 4 is the most technical part of our contribution. Its proof appears in the full version [6]. We now complete the proof of Theorem 3 using Lemma 4 and Proposition 1.

Proof (Proof of Theorem 3). In order to compute a function $f_r : \mathcal{F}^{A_r} \times \{0, 1, \dots, \lambda\}^k \times \{0, 1, \dots, 2k^2\} \to \{0, 1\}$ that is both f-correct and f-sound, we can apply Lemma 4 on each tree node $t \in V(\tau)$ in a bottom up fashion starting from leaf nodes of the tree decomposition. Therefore, using Lemmas 3 and 4, the total run time to compute f_r is

$$(\lambda k)^{O(k^2)} n^{O(1)} \cdot |V(\tau)| = (\lambda k)^{O(k^2)} n^{O(1)} \cdot \text{poly}(n, \lambda, k) = (\lambda k)^{O(k^2)} n^{O(1)}.$$

Using Proposition 1, we can compute OPT from the function f_r. Consequently, the total time to compute OPT is $(\lambda k)^{O(k^2)} n^{O(1)}$. □

4   Conclusion

Our work adds to the exciting recent collection of works aimed at improving the algorithmic understanding of alternative objectives in graph partitioning [1,7,19]. We addressed the graph k-partitioning problem under the minmax objective. Our algorithmic ideas generalize in a natural manner to also lead to a parameterized approximation scheme for MIN ℓ_p-NORM k-CUT for every $p \geq 1$.

Based on prior works in approximation literature for minmax and minsum objectives, it is a commonly held belief that the minmax objective is harder to approximate than the minsum objective. Our results suggest that for the graph k-partitioning problem, the complexity/approximability of the two objectives are perhaps the same. A relevant question towards understanding if the two objectives exhibit a complexity/approximability gap is the following: When k is part of input, is MINMAX k-CUT constant-approximable? We recall that when k is part of input, MINSUM k-CUT does not admit a $(2 - \epsilon)$-approximation for any constant $\epsilon > 0$ under the Small Set Expansion Hypothesis [25] and admits a 2-approximation [27]. The 2-approximation for MINSUM k-CUT is based on solving the same problem in the *Gomory-Hu* tree of the given graph. We are aware of examples where this approach for MINMAX k-CUT only leads to a $\Theta(n)$-approximation, where n is the number of vertices in the input graph (see full version [6]). The best approximation factor that we know currently for MINMAX k-CUT is $2k$ (see Sect. 1.1). A reasonable stepping stone would be to show that MINMAX k-CUT is APX-hard.

References

1. Ahmadi, S., Khuller, S., Saha, B.: Min-max correlation clustering via multicut. In: Integer Programming and Combinatorial Optimization, pp. 13–26. IPCO (2019)
2. Bansal, N., et al.: Min-max graph partitioning and small set expansion. SIAM J. Comput. **43**(2), 872–904 (2014)
3. Benczúr, A., Karger, D.: Randomized approximation schemes for cuts and flows in capacitated graphs. SIAM J. Comput. **44**(2), 290–319 (2015)
4. Bérczi, K., Chandrasekaran, K., Király, T., Madan, V.: Improving the Integrality Gap for Multiway Cut. Mathematical Programming (2020)
5. Chandrasekaran, K., Chekuri, C.: Min-max partitioning of hypergraphs and symmetric submodular functions. In: Proceedings of the 32nd ACM-SIAM Symposium on Discrete Algorithms (to appear). SODA (2021)
6. Chandrasekaran, K., Wang, W.: Fixed Parameter Approximation Scheme for Min-max k-cut. arXiv: https://arxiv.org/abs/2011.03454 (2020)
7. Charikar, M., Gupta, N., Schwartz, R.: Local guarantees in graph cuts and clustering. In: Integer Programming and Combinatorial Optimization, pp. 136–147. IPCO (2017)
8. Chekuri, C., Quanrud, K., Xu, C.: LP relaxation and tree packing for minimum k-cuts. In: 2nd Symposium on Simplicity in Algorithms, pp. 7:1–7:18. SOSA (2019)
9. Cygan, M., et al.: Parameterized Algorithms. Springer, Cham (2015). https://doi.org/10.1007/978-3-319-21275-3
10. Cygan, M., et al.: Randomized contractions meet lean decompositions. arXiv: https://arxiv.org/abs/1810.06864 (2018)
11. Dahlhaus, E., Johnson, D., Papadimitriou, C., Seymour, P., Yannakakis, M.: The complexity of multiterminal cuts. SIAM J. Comput. **23**(4), 864–894 (1994)
12. Downey, R., Estivill-Castro, V., Fellows, M., Prieto, E., Rosamund, F.: Cutting up is hard to do: the parameterised complexity of k-cut and related problems. Electron. Notes Theor. Comput. Sci. **78**, 209–222 (2003)
13. Goldschmidt, O., Hochbaum, D.: Polynomial algorithm for the k-cut problem. In: Proceedings of the 29th Annual Symposium on Foundations of Computer Science, pp. 444–451. FOCS (1988)
14. Goldschmidt, O., Hochbaum, D.: A polynomial algorithm for the k-cut problem for fixed k. Math. Oper. Res. **19**(1), 24–37 (1994)
15. Gupta, A., Harris, D., Lee, E., Li, J.: Optimal Bounds for the k-cut Problem. arXiv: https://arxiv.org/abs/2005.08301 (2020)
16. Gupta, A., Lee, E., Li, J.: An FPT algorithm beating 2-approximation for k-Cut. In: Proceedings of the 29th Annual ACM-SIAM Symposium on Discrete Algorithms, pp. 2821–2837. SODA (2018)
17. Gupta, A., Lee, E., Li, J.: Faster exact and approximate algorithms for k-cut. In: Proceedings of the 59th IEEE Annual Symposium on Foundations of Computer Science, pp. 113–123. FOCS (2018)
18. Gupta, A., Lee, E., Li, J.: The karger-stein algorithm is optimal for k-cut. In: Proceedings of the 52nd Annual ACM SIGACT Symposium on Theory of Computing, pp. 473–484. STOC (2020)
19. Kalhan, S., Makarychev, K., Zhou, T.: Correlation clustering with local objectives. Adv. Neural Inform. Process. Syst. **32**, 9346–9355 (2019)
20. Karger, D., Stein, C.: A new approach to the minimum cut problem. J. ACM **43**(4), 601–640 (1996)

21. Kawarabayashi, K.i., Lin, B.: A nearly 5/3-approximation FPT Algorithm for Min-k-Cut. In: Proceedings of the 31st ACM-SIAM Symposium on Discrete Algorithms, pp. 990–999. SODA (2020)
22. Lawler, E.: Cutsets and partitions of hypergraphs. Networks **3**, 275–285 (1973)
23. Li, J.: Faster minimum k-cut of a simple graph. In: Proceedings of the 60th Annual Symposium on Foundations of Computer Science, pp. 1056–1077. FOCS (2019)
24. Lokshtanov, D., Saurabh, S., Surianarayanan, V.: A parameterized approximation scheme for min k-Cut. In: Proceedings of the 61st IEEE Annual Symposium on Foundations of Computer Science (to appear). FOCS (2020)
25. Manurangsi, P.: Inapproximability of maximum biclique problems, minimum k-Cut and densest at-least-k-Subgraph from the small set expansion hypothesis. Algorithms **11**(1), 10 (2018)
26. Ravi, R., Sinha, A.: Approximating k-cuts using network strength as a lagrangean relaxation. Eur. J. Oper. Res. **186**(1), 77–90 (2008)
27. Saran, H., Vazirani, V.: Finding k cuts within twice the optimal. SIAM J. Comput. **24**(1), 101–108 (1995)
28. Sharma, A., Vondrák, J.: Multiway cut, pairwise realizable distributions, and descending thresholds. In: Proceedings of the Forty-Sixth Annual ACM Symposium on Theory of Computing, pp. 724–733. STOC (2014)
29. Svitkina, Z., Tardos, É.: Min-max multiway cut. In: Approximation, Randomization, and Combinatorial Optimization. Algorithms and Techniques, pp. 207–218. APPROX (2004)
30. Thorup, M.: Minimum k-way cuts via deterministic greedy tree packing. In: Proceedings of the 40th Annual ACM Symposium on Theory of Computing, pp. 159–166. STOC (2008)
31. Zhao, L., Nagamochi, H., Ibaraki, T.: Greedy splitting algorithms for approximating multiway partition problems. Math. Program. **102**(1), 167–183 (2005)

Computational Aspects of Relaxation Complexity

Gennadiy Averkov[1] , Christopher Hojny[2] , and Matthias Schymura[1]([✉])

[1] BTU Cottbus-Senftenberg, Platz der Deutschen Einheit 1, 03046 Cottbus, Germany
{averkov,schymura}@b-tu.de
[2] TU Eindhoven, PO Box 513, 5600 MB Eindhoven, The Netherlands
c.hojny@tue.nl

Abstract. The relaxation complexity $\mathrm{rc}(X)$ of the set of integer points X contained in a polyhedron is the smallest number of facets of any polyhedron P such that the integer points in P coincide with X. It is an important tool to investigate the existence of compact linear descriptions of X. In this article, we derive tight and computable upper bounds on $\mathrm{rc}_{\mathbb{Q}}(X)$, a variant of $\mathrm{rc}(X)$ in which the polyhedra P are required to be rational, and we show that $\mathrm{rc}(X)$ can be computed in polynomial time if X is 2-dimensional. We also present an explicit formula for $\mathrm{rc}(X)$ of a specific class of sets X and present numerical experiments on the distribution of $\mathrm{rc}(X)$ in dimension 2.

Keywords: Integer programming formulation · Relaxation complexity

1 Introduction

A successful approach for solving discrete optimization problems is based on integer programming techniques. To this end (i) a suitable encoding $X \subseteq \mathbb{Z}^d$ of the discrete problem's solutions together with an objective $c \in \mathbb{R}^d$ has to be selected, and (ii) a linear system $Ax \leq b$, $Cx = f$ defining a polyhedron $P \subseteq \mathbb{R}^d$ has to be found such that $P \cap \mathbb{Z}^d = X$. In the following, we refer to such a polyhedron P as a *relaxation* of X. Then, the discrete problem can be tackled by solving the integer program $\max\{c^\mathsf{T} x : Ax \leq b,\ Cx = f,\ x \in \mathbb{Z}^d\}$, which can be solved, e.g., by branch-and-bound or branch-and-cut techniques (see Schrijver [16]).

To solve these integer programs efficiently, the focus in Step (ii) was mostly on identifying facet defining inequalities of the integer hull of P. Such inequality systems, however, are typically exponentially large and one may wonder about the minimum number of facets of any relaxation of X, which allows to compare different encodings X of the discrete problem. Kaibel and Weltge [11] called this quantity the *relaxation complexity* of X, denoted $\mathrm{rc}(X)$, and showed that certain encodings of, e.g., the traveling salesman problem or connected subgraphs, have exponentially large relaxation complexity. They also introduced the quantity $\mathrm{rc}_{\mathbb{Q}}(X)$, which is the smallest number of facets of a *rational* relaxation of X, and posed the question whether $\mathrm{rc}(X) = \mathrm{rc}_{\mathbb{Q}}(X)$ holds in general. Recently, this

© Springer Nature Switzerland AG 2021
M. Singh and D. P. Williamson (Eds.): IPCO 2021, LNCS 12707, pp. 368–382, 2021.
https://doi.org/10.1007/978-3-030-73879-2_26

was answered affirmatively for $d \leq 4$, see [1]. The same authors also showed that $rc(X)$ is computable if $d \leq 3$ using an algorithm with potentially superexponential worst case running time. Computability for $d \geq 4$ and explicit formulas for $rc(X)$ for specific sets X, however, are still open problems.

In this article, we follow the line of research started in [1] and derive a tight upper bound on $rc_{\mathbb{Q}}(X)$ for arbitrary dimensions, which is based on a robustification of $rc_{\mathbb{Q}}(X)$ against numerical errors (Sect. 2). We also point out when this upper bound can be used to compute $rc_{\mathbb{Q}}(X)$ exactly. In Sect. 3 we focus on dimension $d = 2$ and show that there is a polynomial time algorithm to compute $rc(X)$ in this case. Sect. 4 derives an explicit formula for the relaxation complexity for integer points in rectangular boxes. We conclude the article in Sect. 5 with a discussion of numerical experiments that are based on the first practically applicable implementation to compute $rc(X)$ in cases where this quantity is known to have a finite certificate.

Notation and Terminology. For a set $X \subseteq \mathbb{R}^d$, we denote by $conv(X)$ its convex hull and by $aff(X)$ its affine hull. The boundary and interior of X are denoted by $bd(X)$ and $int(X)$, respectively. A set $X \subseteq \mathbb{Z}^d$ is called *lattice-convex* if $conv(X) \cap \mathbb{Z}^d = X$. The dimension of a set X is the dimension of its affine hull, and we say that a full-dimensional set $X \subseteq \mathbb{Z}^d$ is in *convex position* if $X \subseteq bd(conv(X))$. If $X \subseteq \mathbb{Z}^2$ is 2-dimensional and in convex position, we can define a cyclic order of X by starting a closed walk on $bd(conv(X))$ and labeling the points of X in the order visited during the walk. For two sets $X, Y \subseteq \mathbb{R}^d$, $X + Y = \{x + y : x \in X, y \in Y\}$ is their Minkowski sum. We define $[n] := \{1, \ldots, n\}$, $[n]_0 = \{0, 1, \ldots, n\}$, and e_i to be the i-th standard unit vector in $\mathbb{R}^d$.

A crucial concept in [1] is that of observers. For $X \subseteq \mathbb{Z}^d$ with $conv(X) \cap \mathbb{Z}^d = X$, an *observer* is a point $z \in \mathbb{Z}^d \setminus X$ such that $conv(X \cup \{z\}) \cap \mathbb{Z}^d = X \cup \{z\}$. The set of all observers of X is denoted by $Obs(X)$. The relevance of the observers for finding $rc(X)$ comes from the fact that any linear system $Ax \leq b$ separating X and $Obs(X)$ also separates X and $\mathbb{Z}^d \setminus X$, see [1]. This motivates, for $X, Y \subseteq \mathbb{Z}^d$, to introduce $rc(X, Y)$ (resp. $rc_{\mathbb{Q}}(X, Y)$) as the smallest number of inequalities in a (rational) system $Ax \leq b$ separating X and $Y \setminus X$.

Related Literature. Kaibel and Weltge [11] introduced the notion of relaxation complexity and derived a concept for deriving lower bounds on $rc(X)$. Computability of $rc(X)$, for $d = 2$, has been shown by Weltge [19] who also derived a lower bound on $rc(X)$ only depending on the dimension of X. In [1], this lower bound has been improved and computability also for $d = 3$ has been established. The interplay between the number of inequalities in a relaxation and the size of their coefficients has been investigated in [8]; see also [9] for a lower bound on the relative size of coefficients in a relaxation. For $X \subseteq \{0, 1\}^d$, Jeroslow [10] derived an upper bound on $rc(X, \{0, 1\}^d)$, which is an important subject in the area of social choice, see, e.g., Hammer et al. [7] and Taylor and Zwicker [17].

2   Computable Bounds on the Relaxation Complexity

Let $X \subseteq \mathbb{Z}^d$ be finite and lattice-convex. In contrast to the lower bound on $\mathrm{rc}(X)$ provided by Kaibel and Weltge [11], to the best of our knowledge no systematic way for deriving algorithmically computable lower and upper bounds on $\mathrm{rc}(X)$ and $\mathrm{rc}_\mathbb{Q}(X)$ has been discussed so far. With the aim of making progress on this problem, we introduce a robustification of $\mathrm{rc}(X)$ by enforcing that a relaxation is not allowed to support $\mathrm{conv}(X)$. To make this precise, for $\varepsilon > 0$, let $X_\varepsilon := X + \mathcal{B}_\varepsilon^1$, where $\mathcal{B}_\varepsilon^1 = \{0, \pm\varepsilon e_1, \dots, \pm\varepsilon e_d\}$ is the discrete ℓ_1-ball with radius ε. By construction, we clearly have $X \subseteq \mathrm{int}(\mathrm{conv}(X_\varepsilon))$.

Definition 1. Let $X \subseteq \mathbb{Z}^d$ be finite, full-dimensional and lattice-convex, and let $\varepsilon > 0$. A polyhedron $Q \subseteq \mathbb{R}^d$ is called an ε-*relaxation* of X if $X_\varepsilon \subseteq Q$ and $\mathbb{Z}^d \setminus \mathrm{int}(Q) = \mathbb{Z}^d \setminus X$. The ε-*relaxation complexity* $\mathrm{rc}_\varepsilon(X)$ is the smallest number of facets of an ε-relaxation of X.

Alternatively speaking, an ε-relaxation of X is a polyhedron Q with $\mathrm{int}(Q) \cap \mathbb{Z}^d = X$, and for which the minimum of $\|x - z\|_1$, over any $x \notin \mathrm{int}(Q)$ and $z \in X$, is at least ε. Further, in complete analogy to $\mathrm{rc}(X, Y)$, we define $\mathrm{rc}_\varepsilon(X, Y)$ to be the smallest number of inequalities necessary to separate X_ε and $Y \setminus X_\varepsilon$. Note also that we defined $\mathrm{rc}_\varepsilon(X)$ only for full-dimensional sets X. All results below generalize to lower-dimensional cases, however, for the ease and brevity of presentation, we discuss only the full-dimensional case in this article.

Our first observation on the robustification $\mathrm{rc}_\varepsilon(X)$ of $\mathrm{rc}(X)$ is that it always admits a *finite* set $Y^\varepsilon \subseteq \mathbb{Z}^d$ with $\mathrm{rc}_\varepsilon(X) = \mathrm{rc}_\varepsilon(X, Y^\varepsilon)$. The existence of such a finite certificate Y^ε is noteworthy, because neither for $\mathrm{rc}(X)$ nor $\mathrm{rc}_\mathbb{Q}(X)$ is it known whether a finite certificate exists. In general, $\mathrm{rc}(X) = \mathrm{rc}(X, \mathrm{Obs}(X))$ and $\mathrm{rc}_\mathbb{Q}(X) = \mathrm{rc}_\mathbb{Q}(X, \mathrm{Obs}(X))$, but $\mathrm{Obs}(X)$ can be infinite for $d \geq 3$ (see [19, Sect. 7.5]).

Lemma 1. *Let $X \subseteq \mathbb{Z}^d$ be finite, full-dimensional, and lattice-convex and let $\varepsilon > 0$ be such that X admits an ε-relaxation. Then, there is an explicitly computable finite set $Y^\varepsilon \subseteq \mathbb{Z}^d$ such that $\mathrm{rc}_\varepsilon(X) = \mathrm{rc}_\varepsilon(X, Y^\varepsilon)$.*

Proof. We show that there exists an explicit constant $c_{d,\varepsilon,X} > 0$ such that every ε-relaxation Q of X is contained in $P_{X,\varepsilon} := \mathrm{conv}(X) + c_{d,\varepsilon,X} \cdot \mathcal{B}_d^2$, where $\mathcal{B}_d^2 \subseteq \mathbb{R}^d$ is the Euclidean unit ball. Once this is established, $Y^\varepsilon := P_{X,\varepsilon} \cap \mathbb{Z}^d$ serves as the explicitly computable finite set such that $\mathrm{rc}_\varepsilon(X) = \mathrm{rc}_\varepsilon(X, Y^\varepsilon)$.

In order to find $P_{X,\varepsilon}$, we first note that the inradius of $\mathrm{conv}(\mathcal{B}_\varepsilon^1)$ equals $\delta := \varepsilon/\sqrt{d}$, and so $X + \delta \cdot \mathcal{B}_d^2 \subseteq \mathrm{conv}(X_\varepsilon)$. Now, let Q be an arbitrary ε-relaxation of X and let $q \in Q$ be some point therein. As Q weakly separates X from $\mathbb{Z}^d \setminus X$, the set $\mathrm{conv}(X \cup \{q\})$ contains at most $|X| + 1$ integer points. For $v \in X$, let

$$K_{v,q} := \mathrm{conv}\left((v + \delta \cdot \mathcal{B}_d^2) \cup \{q, 2v - q\}\right) \text{ and } K'_{v,q} := \mathrm{conv}\left((v + \delta \cdot \mathcal{B}_d^2) \cup \{q\}\right).$$

The set $K_{v,q}$ is convex and centrally symmetric around the integer point v. Moreover, $K'_{v,q} \subseteq Q$, so that we would get a contradiction if $|K'_{v,q} \cap \mathbb{Z}^d| > |X| + 1$. A classical extension of Minkowski's first fundamental theorem in the geometry of numbers is due to van der Corput [3] who proved that, for every convex body $K \subseteq \mathbb{R}^d$ that is centrally symmetric around an integer point, the inequality $\mathrm{vol}(K) \leq 2^d \cdot |K \cap \mathbb{Z}^d|$ holds. Using this result and noting that $K_{v,q}$ contains a double pyramid over a $(d-1)$-dimensional Euclidean ball, we have

$$|K'_{v,d} \cap \mathbb{Z}^d| \geq \tfrac{1}{2} |K_{v,d} \cap \mathbb{Z}^d| \geq \tfrac{1}{2^{d+1}} \mathrm{vol}(K_{v,q})$$

$$\geq \tfrac{1}{2^{d+1}} \cdot \tfrac{2}{d} \|q - v\|_2 \, \mathrm{vol}_{d-1}(\delta \cdot \mathcal{B}^2_{d-1}) = \frac{\delta^{d-1} \kappa_{d-1}}{d \, 2^d} \cdot \|q - v\|_2,$$

where $\kappa_d = \mathrm{vol}(\mathcal{B}^2_d)$ denotes the volume of the Euclidean unit ball. To avoid the discussed contradiction above, we know that the last term is upper bounded by $|X| + 1$, which translates to the inequality

$$\|q - v\|_2 \leq \frac{d \, 2^d \cdot (|X| + 1)}{\delta^{d-1} \kappa_{d-1}} \leq \frac{d^{(d+1)/2} \, 2^{d+1}}{\varepsilon^{d-1} \kappa_{d-1}} |X| =: c_{d,\varepsilon,X}.$$

Since $q \in Q$ and $v \in X$ were arbitrary, this indeed shows that the ε-relaxation Q of X is contained in $\mathrm{conv}(X) + c_{d,\varepsilon,X} \cdot \mathcal{B}^2_d = P_{X,\varepsilon}$, as desired. $\qquad \square$

We will see shortly that $\mathrm{rc}_\varepsilon(X)$ is an upper approximation of the relaxation complexity of X. In order to also give a lower approximation, for $t \in \mathbb{N}$, we let $B_t = [-t, t]^d \cap \mathbb{Z}^d$ be the set of integer points with coordinates bounded by t in absolute value. Now, it is clear that $\mathrm{rc}_\varepsilon(X) \geq \mathrm{rc}_{\varepsilon'}(X)$, for every $\varepsilon \geq \varepsilon' > 0$, and likewise that $\mathrm{rc}(X, B_t) \leq \mathrm{rc}(X, B_{t'})$, for every $t \leq t'$. Thus, the parameters

$$\mathrm{rc}_\square(X) = \max_{t \in \mathbb{N}} \mathrm{rc}(X, B_t) \quad \text{and} \quad \mathrm{rc}_0(X) = \min_{\varepsilon > 0} \mathrm{rc}_\varepsilon(X)$$

are well-defined. The main advantage of these numbers is that they sandwich $\mathrm{rc}(X)$, and that they are limits of explicitly computable bounds.

Theorem 1. *Let $X \subseteq \mathbb{Z}^d$ be finite, full-dimensional and lattice-convex.*

1. If $\varepsilon > 0$ is rational, then $\mathrm{rc}_\varepsilon(X)$ can be computed in finite time.
2. It holds $\mathrm{rc}_\square(X) \leq \mathrm{rc}(X) \leq \mathrm{rc}_\mathbb{Q}(X) = \mathrm{rc}_0(X)$.

Proof. Lemma 1 provides us with a finite and explicitly computable set $Y^\varepsilon \subseteq \mathbb{Z}^d$ such that $\mathrm{rc}_\varepsilon(X) = \mathrm{rc}_\varepsilon(X, Y^\varepsilon)$. If $\varepsilon \in \mathbb{Q}$, then $X_\varepsilon \subseteq \mathbb{Q}^d$, which means that we ask to minimally separate the finite rational sets X_ε and $Y^\varepsilon \setminus X_\varepsilon$ from one another. The arguments in [1, Prop. 4.9] show how this can be done via a bounded mixed-integer program and Claim 1 follows.

For Claim 2, we first observe that for every $t \in \mathbb{N}$ we clearly have $\mathrm{rc}(X, B_t) \leq \mathrm{rc}(X)$, and thus $\mathrm{rc}_\square(X) \leq \mathrm{rc}(X) \leq \mathrm{rc}_\mathbb{Q}(X)$. We thus need to show that $\mathrm{rc}_\mathbb{Q}(X) = \mathrm{rc}_0(X)$. For one inequality, let $Q \subseteq \mathbb{R}^d$ be a rational relaxation of X having $\mathrm{rc}_\mathbb{Q}(X)$ facets and facet description $Ax \leq b$. Since Q is a rational polyhedron, Q is necessarily bounded (otherwise it would contain infinitely many integral points). Thus, there exists $\delta > 0$ such that for each $y \in \mathbb{Z}^d \setminus X$ there exists an inequality $a^\mathsf{T} x \leq \beta$ in $Ax \leq b$ with $a^\mathsf{T} y \geq \beta + \delta$. Consequently, we can increase β slightly to get another relaxation Q' with $\mathrm{rc}_\mathbb{Q}(X)$ facets. This shows that there exists $\varepsilon' > 0$ such that $X_{\varepsilon'} \subseteq Q'$ and $\mathbb{Z}^d \setminus \text{int}(Q') = \mathbb{Z}^d \setminus X$, i.e., Q' is an ε'-relaxation of X, and thus $\text{rc}_0(X) \le \text{rc}_{\varepsilon'}(X) \le \text{rc}_{\mathbb{Q}}(X)$.

For the reverse inequality, we fix an $\varepsilon > 0$ and note that every ε-relaxation of X is bounded (see the proof of Lemma 1). We may thus perturb any such ε-relaxation slightly into a rational relaxation of X with equally many facets. As the result, we get $\text{rc}_{\mathbb{Q}}(X) \le \text{rc}_{\varepsilon}(X)$, for every $\varepsilon > 0$. $\square$

The main message of Theorem 1 is that, for every rational $\varepsilon > 0$, the number $\text{rc}_\varepsilon(X)$ is an explicitly computable upper bound on $\text{rc}(X)$ and $\text{rc}_{\mathbb{Q}}(X)$, and that these upper bounds converge to $\text{rc}_{\mathbb{Q}}(X)$ with $\varepsilon \to 0$. However, without further information we do not know whether the computed value $\text{rc}_\varepsilon(X)$ agrees with $\text{rc}_{\mathbb{Q}}(X)$. A situation in which we are sure when to stop computing $\text{rc}_\varepsilon(X)$ for decreasing values of $\varepsilon > 0$ is, when we can also compute an eventually matching lower bound:

Theorem 2. *Let $X \subseteq \mathbb{Z}^d$ be finite, full-dimensional and lattice-convex. If we have $\text{rc}_{\square}(X) = \text{rc}_{\mathbb{Q}}(X)$, then there is a finite algorithm that computes $\text{rc}(X)$.*

Proof. Let $(\varepsilon_t)_{t \in \mathbb{N}} \subseteq \mathbb{Q}$ be a rational strictly decreasing null sequence. Then, by Theorem 1, Part 2, there exists $t' \in \mathbb{N}$ such that $\text{rc}_{\mathbb{Q}}(X) = \text{rc}_{\varepsilon_t}(X)$ for every $t \ge t'$. If $\text{rc}_{\square}(X) = \text{rc}_{\mathbb{Q}}(X)$, there exists $t^\star \in \mathbb{N}$ such that $\text{rc}_{\mathbb{Q}}(X) = \text{rc}(X, B_t)$ for every $t \ge t^\star$. Thus, for $t = \max\{t', t^\star\}$, $\text{rc}(X, B_t) = \text{rc}_{\varepsilon_t}(X)$. Since both quantities can be computed in finite time due to Theorem 1, Part 1 and [1, Cor. 4.9], the assertion follows. $\square$

Whether $\text{rc}(X)$ and $\text{rc}_{\mathbb{Q}}(X)$ are computable and whether the equality $\text{rc}(X) = \text{rc}_{\mathbb{Q}}(X)$ is true has already been asked by Kaibel and Weltge [11]. In light of the above results, there are two other basic questions intimately related to these two: Is $\text{rc}_{\square}(X)$ computable and does $\text{rc}_{\square}(X) = \text{rc}(X)$ hold in general? We emphasize that the confirmation of $\text{rc}_{\square}(X) = \text{rc}(X)$ and $\text{rc}(X) = \text{rc}_{\mathbb{Q}}(X)$ would resolve all of the questions above, because these two identities would imply $\text{rc}_{\square}(X) = \text{rc}_{\mathbb{Q}}(X)$ and also the computability of $\text{rc}(X)$ in view of Theorem 2. Our constructions suggest infinite iterative procedures that produce sequences of integer values converging to $\text{rc}_{\square}(X)$ and $\text{rc}_{\mathbb{Q}}(X)$ after finitely many steps. However, the existence of such procedures per se does not resolve the computability of $\text{rc}_{\square}(X)$ or $\text{rc}_{\mathbb{Q}}(X)$. For computability, one would additionally need to be able to decide, when the integer sequence achieves the value it finitely converges to.

3   Computational Complexity in Dimension 2

Let $X \subseteq \mathbb{Z}^d$ be finite and lattice-convex such that $\text{Obs}(X)$ is finite. By the separation theorems for convex sets, $\mathcal{I}(X) := \{I \subseteq \text{Obs}(X) : \text{conv}(X) \cap \text{conv}(I) = \emptyset\}$ contains all subsets of observers that can be separated from X by a single hyperplane. Let $\mathcal{I}_{\max}(X)$ be the family of all inclusionwise maximal sets in $\mathcal{I}(X)$.

Observation 3. *For a finite lattice-convex set $X \subseteq \mathbb{Z}^d$ with finitely many observers, $\text{rc}(X)$ is the smallest number k of sets $I_1, \ldots, I_k \in \mathcal{I}_{\max}(X)$ such that $\text{Obs}(X) = \bigcup_{i=1}^k I_i$.*

In the following, we use this observation to show that the relaxation complexity of a finite lattice-convex set in $\mathbb{Z}^2$ can be computed in polynomial time. Note that Weltge [19] has already shown that $\mathrm{rc}(X)$ is computable if $X \subseteq \mathbb{Z}^2$. His algorithm is based on finite so-called guard sets (which are supersets of $\mathrm{Obs}(X)$), and he showed how a brute-force algorithm can compute $\mathrm{rc}(X)$. We complement this result by providing an algorithm with very low polynomial complexity.

Since Observation 3 is based on observers, a crucial step is to find the set $\mathrm{Obs}(X)$ of observers in the case that $X \subseteq \mathbb{Z}^2$ is two-dimensional. The main reason why this can be done efficiently is that $\mathrm{Obs}(X)$ is the set of integer points in the boundary of enlarging the lattice polygon $P = \mathrm{conv}(X)$ by lattice-distance one over each of its edges (see Weltge [19, Prop. 7.5.6]). To make this precise, let $P = \{x \in \mathbb{R}^2 : a_i x_1 + b_i x_2 \leq c_i, i \in [k]\}$ be a polyhedron described by k inequalities with the assumption that $a_i, b_i, c_i \in \mathbb{Z}$ with a_i and b_i coprime, for all $i \in [k]$. Then, following Castryck [2] we write $P^{(-1)} := \{x \in \mathbb{R}^2 : a_i x_1 + b_i x_2 \leq c_i + 1, i \in [k]\}$ and say that $P^{(-1)}$ is obtained from P by *moving out the edges*. Note that neither does $P^{(-1)}$ need to be a lattice polygon again, nor may it have as many edges as P. With this notation the previous discussion can be formulated as

$$\mathrm{Obs}(X) = \mathrm{bd}(P^{(-1)}) \cap \mathbb{Z}^2.$$

In particular, this means that $\mathrm{Obs}(X)$ is in convex position and that we can efficiently list its elements in counterclockwise order.

Lemma 2. *Let $V \subseteq \mathbb{Z}^2$ be a finite 2-dimensional set and let $X = \mathrm{conv}(V) \cap \mathbb{Z}^2$. There is an algorithm that determines $\mathrm{Obs}(X) = \{y_0, \ldots, y_\ell\}$, with the labeling in counterclockwise order, and which runs in $\mathcal{O}(\ell + k \log k + \gamma k)$ time, where $k = |V|$ and γ is an upper bound on the binary encoding size of any point in V.*

Proof. Let $P = \mathrm{conv}(V)$. The algorithm consists of the following four steps:

1. Compute an irredundant inequality description $P = \{x \in \mathbb{R}^2 : a_i x_1 + b_i x_2 \leq c_i, i \in [k]_0\}$, where $a_i, b_i, c_i \in \mathbb{Z}$ with a_i and b_i coprime, and the outer normal vectors $(a_i, b_i)^\intercal$ labeled in counterclockwise order.
2. Move out the edges of P and let $P^{(-1)} = \{x \in \mathbb{R}^2 : a_{i_j} x_1 + b_{i_j} x_2 \leq c_{i_j} + 1, j \in [m]_0\}$ be such that all redundancies are removed.
3. Compute the set of vertices $\{w_0, w_1, \ldots, w_m\}$ of $P^{(-1)}$ in counterclockwise order.
4. For each $i \in [m]_0$, compute the integer points on the segment $[w_i, w_{i+1}]$, with the index taken modulo $m + 1$.

Some detailed comments are in order:

In Step 1, we first use a standard convex hull algorithm in the plane, e.g., Graham's scan (cf. [4, Ch. 8] for details) to compute the vertices $\{v_0, \ldots, v_r\}$ of P in counterclockwise order. Since P is a lattice polygon, each of its edges, say $[v_j, v_{j+1}]$, corresponds to an integer vector $(\eta_1, \eta_2)^\intercal = v_{j+1} - v_j$. Because of the counterclockwise ordering, $(\eta_2, -\eta_1)^\intercal$ is an outer normal vector to the edge at hand, and dividing out by the greatest common divisor of η_1 and η_2 leads to the desired inequality description. When we use Euclid's Algorithm in this last step, we obtain a running time of $\mathcal{O}(k \log k + \gamma r) \subseteq \mathcal{O}(k \log k + \gamma k)$.

The only thing to do in Step 2, besides increasing all the right hand sides by one unit, is to remove the redundancies. One way to do this is to use duality between convex hulls and intersections of hyperplanes, and again invoke, e.g., Graham's scan. This can be done in time $\mathcal{O}(k \log k)$.

Step 3 just amounts to an iterative computation of the intersection point w_j of the pair of equations $a_{i_j}x_1 + b_{i_j}x_2 = c_{i_j} + 1$ and $a_{i_{j+1}}x_1 + b_{i_{j+1}}x_2 = c_{i_{j+1}} + 1$, $j \in [m]_0$. In this computation we also record the normal vector $(a_{i_{j+1}}, b_{i_{j+1}})^\mathsf{T}$ that corresponds to the edge with endpoints w_j and w_{j+1}. This needs $\mathcal{O}(k)$ steps.

For Step 4 we may use Euclid's Algorithm on the defining data of the edge of $P^{(-1)}$ that contains w_i and w_{i+1}, and determine an affine unimodular transformation $A_i : \mathbb{R}^2 \to \mathbb{R}^2$ such that $A_i[w_i, w_{i+1}] = [\omega_i e_1, \omega_{i+1} e_1]$, for some $\omega_i < \omega_{i+1}$. The integer points on the latter segment in increasing order of the first coordinate are given by

$$z_j = \frac{\omega_{i+1} - \lceil \omega_i \rceil - j}{\omega_{i+1} - \omega_i} \omega_i e_1 + \frac{\lceil \omega_i \rceil + j - \omega_i}{\omega_{i+1} - \omega_i} \omega_{i+1} e_1, \text{ for } j \in \{0, \ldots, \lfloor \omega_{i+1} \rfloor - \lceil \omega_i \rceil\}.$$

Using the inverse transformation A_i^{-1} then leads to the correctly ordered list of integer points on the segment $[w_i, w_{i+1}]$. For a given edge of $P^{(-1)}$ these steps can be performed in time proportional to the number of integer points that it contains, leading to a total running time of $\mathcal{O}(\ell + \gamma k)$ for this step.

Conclusively, we saw that the outlined algorithm terminates with the correctly computed list of observers after $\mathcal{O}(\ell + k \log k + \gamma k)$ iterations.   □

Based on this result, we can show that $\mathrm{rc}(X)$ can be computed efficiently.

Theorem 4. *Let $V \subseteq \mathbb{Z}^2$ be a finite 2-dimensional set, let $X = \mathrm{conv}(V) \cap \mathbb{Z}^2$, and $Y = \mathrm{Obs}(X)$. Then, the relaxation complexity $\mathrm{rc}(X)$ can be computed in time $\mathcal{O}(|V| \cdot \log|V| + |V| \cdot |Y| \cdot \log|Y| + \gamma \cdot |V|)$, where γ is an upper bound on the binary encoding size of any point in V.*

Proof. Assume that the set of observers $\mathrm{Obs}(X)$ is given in counterclockwise order $y_0, \ldots, y_\ell$. Then, for each $I \in \mathcal{I}_{\max}(X)$, there exist $r, s \in [\ell]_0$ such that $I = \{y_r, y_{r+1}, \ldots, y_{r+s}\}$, where indices are modulo $\ell + 1$. That is, the sets in $\mathcal{I}_{\max}(X)$ form "discrete intervals" of observers, see Fig. 1. In particular, $\mathcal{I}_{\max}(X)$ contains at most $\ell + 1$ intervals.

Because of Observation 3, we can determine $\mathrm{rc}(X)$ by finding the smallest number of intervals in $\mathcal{I}_{\max}(X)$ that is sufficient to cover $\mathrm{Obs}(X)$. This problem can be solved in $\mathcal{O}(\ell \log \ell)$ time using the minimum circle-covering algorithm by Lee and Lee [12]. Thus, the assertion follows if the observers and the intervals in $\mathcal{I}_{\max}(X)$ can be computed in $\mathcal{O}(|V| \cdot \log|V| + |V| \cdot |Y| \cdot \log|Y| + \gamma \cdot |V|)$ time.

The list of observers $\mathrm{Obs}(X)$ in counterclockwise order can be found in $\mathcal{O}(|Y| + |V| \cdot \log|V| + \gamma \cdot |V|)$ time by Lemma 2. To find the sets $I \in \mathcal{I}_{\max}(X)$, note that we can use binary search on y_{r+s} to find a maximum interval $\{y_r, \ldots, y_{r+s}\}$

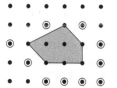 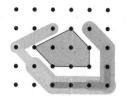 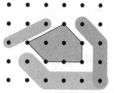

Fig. 1. A lattice polygon conv(X) and its observers (left), $\mathcal{I}_{max}(X)$ (center), and a minimum interval covering (right).

such that conv($\{y_r, y_{r+s}\}$) $\cap$ conv(V) = $\emptyset$. In each of the $\mathcal{O}(\log|Y|)$ steps of the binary search, we have to check whether the line segment conv($\{y_r, y_{r+s}\}$) intersects one of the $\mathcal{O}(|V|)$ edges of conv(V). Thus, a single set I can be computed in $\mathcal{O}(|V| \cdot \log|Y|)$ time. Combining these running times concludes the proof. $\square$

Remark 1. One can show that the number $|Y|$ of observers of X in which means that the presented algorithm Theorem 4 can be of order $\Theta(2^\gamma)$, which means that the presented algorithm is not polynomial in the input size. However, if the points in V are encoded in unary, then the algorithm is indeed polynomial.

A question related to computing the relaxation complexity in the plane has been studied by Edelsbrunner and Preparata [5]: Given two finite sets $X, Y \subseteq \mathbb{R}^2$, they describe an algorithm to find a convex polygon $Q \subseteq \mathbb{R}^2$ with the minimal possible number of edges such that $X \subseteq Q$ and int(Q)$\cap Y = \emptyset$, or to decide that no such polygon exists. That is, if we apply their algorithm to a lattice-convex set X and its observers Y, we can find a polygon that *weakly* separates X and Y. Their algorithm runs in $\mathcal{O}(|X \cup Y| \cdot \log|X \cup Y|)$ time.

The computational problem of *strictly* separating two finite point sets in any dimension by a given number of hyperplanes has been the focus of Megiddo's work [14]. He introduces the *k-separation problem* as the decision problem on whether finite sets $X, Y \subseteq \mathbb{Z}^d$ can be separated by k hyperplanes. The case $k = 1$ reduces to the linear separation problem and can be solved by linear programming in polynomial time. Megiddo [14] proves the following for $k \geq 2$:

(a) If d is arbitrary, but k is fixed, then the k-separation problem is NP-complete. This even holds for $k = 2$.
(b) If $d = 2$, but k is arbitrary, then the k-separation problem is NP-complete.
(c) If both d and k are fixed, then the k-separation problem is solvable in polynomial time.

With regard to computing the relaxation complexity, comparing Theorem 4 with Part (b) shows that (at least in the plane) deciding rc(X) $\leq k$ is a computationally easier problem than the general k-separation problem. Part (c) is applicable to computing rc(X) in polynomial time via a binary search whenever $X \subseteq \mathbb{Z}^d$ has finitely many observers whose cardinality is polynomially bounded as a function of $|X|$. Relevant families of lattice-convex sets X with this property have been identified in [1, Thm. 4.4 & Thm. 4.5]: First, if $X \subseteq \mathbb{Z}^d$ contains a representative of every residue class in $(\mathbb{Z}/2\mathbb{Z})^d$, then $\mathrm{Obs}(X) \subseteq 2X - X$, and thus $|\mathrm{Obs}(X)| \in \mathcal{O}(|X|^2)$. Second, if $\mathrm{conv}(X)$ contains an interior integer point, then $|\mathrm{Obs}(X)| \leq c_d \cdot |X|$, with c_d a constant only depending on the dimension d.

However, since in dimensions $d \geq 3$ not every lattice-convex set has finitely many observers, the following question might have an affirmative answer:

Question 1. For $d \geq 3$ fixed, is it NP-hard to compute $\mathrm{rc}_\mathbb{Q}(X)$?

4   Discrete Rectangular Boxes

The exact value of $\mathrm{rc}(X)$, or any of its variants, is known only for very few classes of lattice-convex sets $X \subseteq \mathbb{Z}^d$: For the discrete 0/1-cube it was shown by Weltge [19, Thm. 8.1.3] that it always admits a simplex relaxation, that is, $\mathrm{rc}(\{0,1\}^d) = d+1$. The conjectured value $\mathrm{rc}(\Delta_d) = d+1$ for the discrete standard simplex $\Delta_d = \{0, e_1, \ldots, e_d\}$ could only be affirmed for dimensions $d \leq 4$ so far (see [1, Cor. 3.8]). Jeroslow [10, Thm. 7] showed that, for every $1 \leq k \leq 2^{d-1}$ there is a k-element subset X_k of $\{0,1\}^d$ such that $\mathrm{rc}(X_k, \{0,1\}^d) = k$. Further, for $k = 2^{d-1}$ one can choose $X_{2^{d-1}} = X_{\mathrm{even}} = \left\{ x \in \{0,1\}^d : \sum_{i=1}^d x_i \text{ is even} \right\}$, however the value $\mathrm{rc}(X_{\mathrm{even}})$ is not known. Besides some very specific examples that were needed to establish computability of $\mathrm{rc}(X)$ for 3-dimensional lattice-convex sets $X \subseteq \mathbb{Z}^3$ (see [1, Sect. 6]), we are not aware of any further classes of examples for which the relaxation complexity is known exactly.

In this section, we provide a full characterization of the relaxation complexity of an additional parameterized class of lattice-convex sets and thus add to the short list of exact results given above. To this end, we call a set $\{a, \ldots, b\}$ with $a, b \in \mathbb{Z}$ and $a \leq b$ a *discrete segment* of length $b - a$, and we call the Cartesian product of finitely many discrete segments a *discrete rectangular box*. These examples contain a representative of every residue class in $(\mathbb{Z}/2\mathbb{Z})^d$, so that the relaxation complexity agrees with the rational relaxation complexity in view of [1, Thm. 1.4].

The most basic example of a discrete rectangular box is the discrete unit cube $\{0,1\}^d$, with $\mathrm{rc}(\{0,1\}^d) = d + 1$ as shown by Weltge [19, Thm. 8.1.3]. It turns out that Weltge's (rational) relaxation with $d+1$ facets can be generalized to the case that one segment of the cube is allowed to have arbitrary length.

Lemma 3. *For $b \in \mathbb{N}$, let $X_b := \{0,1\}^\ell \times \{0, 1, \ldots, b\} \subseteq \mathbb{Z}^{\ell+1}$. Then, the simplex*

$$P := \left\{ x \in \mathbb{R}^{\ell+1} : x_k \leq 1 + \sum_{i=k+1}^{\ell+1} (b+1)^{-i} x_i, \text{ for } k \in [\ell], \right.$$

$$\left. x_{\ell+1} \leq b \quad \text{and} \quad x_1 + \sum_{i=2}^{\ell+1} (b+1)^{-i} x_i \geq 0 \right\}$$

is a relaxation of X_b. In particular, $\mathrm{rc}(X_b) \leq \ell + 2$.

Proof. The proof is just an adjustment of Weltge's arguments in [19, Lem. 7.2.1]. We need to show that $X_b = P \cap \mathbb{Z}^{\ell+1}$. The inclusion $X_b \subseteq P \cap \mathbb{Z}^{\ell+1}$ is quickly checked, so we give the details for the reverse inclusion.

Let $x \in P \cap \mathbb{Z}^{\ell+1}$. First, we show that $x_i \leq 1$, for all $i \in [\ell]$. To this end, let $k \in [\ell]$ be the largest index for which $x_k > 1$. Then, the first defining inequalities of P together with basic facts about geometric series give

$$1 < x_k \leq 1 + \sum_{i=k+1}^{\ell+1} (b+1)^{-i} x_i \leq \sum_{i=k+1}^{\ell} (b+1)^{-i} + \frac{b}{(b+1)^{\ell+1}}$$

$$< 1 + \frac{1}{b} + \frac{b-1}{(b+1)^{\ell+1}} < 1 + \frac{1}{b} + \frac{b-1}{b} = 2,$$

in contradiction to $x_k \in \mathbb{Z}$.

Second, we show that x_1 is non-negative. Indeed, by the last defining inequality of P, the just established fact that $x_i \leq 1$, for $i \in [\ell]$, and $x_{\ell+1} \leq b$, we have

$$x_1 \geq -\sum_{i=2}^{\ell+1} (b+1)^{-i} x_i \geq -\sum_{i=2}^{\ell} (b+1)^{-i} - \frac{b}{(b+1)^{\ell+1}} > -1.$$

It remains to show that $x_i \geq 0$, for every $i \in \{2, \ldots, \ell+1\}$. To this end, let $j \in \{2, \ldots, \ell+1\}$ be the smallest index such that $x_j \leq -1$. We first show that $x_i = 0$, for every $i < j$: Let $k < j$ be the largest index with $x_k > 0$, which means that $x_k = 1$. Then, by the first defining inequalities of P, we have

$$1 = x_k \leq 1 + \sum_{i=k+1}^{\ell+1} (b+1)^{-i} x_i = 1 + (b+1)^{-j} x_j + \sum_{i=j+1}^{\ell+1} (b+1)^{-i} x_i$$

$$\leq 1 - (b+1)^{-j} + \sum_{i=j+1}^{\ell} (b+1)^{-i} + \frac{b}{(b+1)^{\ell+1}} < 1.$$

The last inequality follows from the closed form expression for geometric sums and basic algebraic manipulations.

Now, knowing that $x_i = 0$, for every $i < j$, we use the last defining inequality of P and get

$$0 \leq x_1 + \sum_{i=2}^{\ell+1} (b+1)^{-i} x_i = (b+1)^{-j} x_j + \sum_{i=j+1}^{\ell+1} (b+1)^{-i} x_i$$

$$\leq -(b+1)^{-j} + \sum_{i=j+1}^{\ell} (b+1)^{-i} + \frac{b}{(b+1)^{\ell+1}} < 0.$$

This contradiction finishes the proof. $\qquad \square$

This simplicial relaxation of X_b quickly leads to an upper bound on the relaxation complexity of general discrete rectangular boxes. To see that these upper bounds are tight, we employ a concept that was introduced by Kaibel and Weltge [11] to establish their exponential lower bound for the traveling salesman polytope. Let $X \subseteq \mathbb{Z}^d$ be lattice-convex. A set $H \subseteq (\mathrm{aff}(X) \cap \mathbb{Z}^d) \setminus X$ is called a *hiding set* for X if, for all distinct $x, y \in H$, we have $\mathrm{conv}(\{x, y\}) \cap \mathrm{conv}(X) \neq \emptyset$. Then, no valid inequality for X can separate x and y simultaneously, showing $\mathrm{rc}(X) \geq |H|$, for every hiding set H.

Theorem 5 (Box theorem). *For integers $k > 0$ and $\ell \geq 0$, let $S_1, \ldots, S_k$ be discrete segments, each having length at least 2, and let $T_1, \ldots, T_\ell$ be discrete segments of length 1. Consider the discrete box $X = S_1 \times \ldots \times S_k \times T_1 \times \ldots \times T_\ell$ in $\mathbb{Z}^{k+\ell}$. Then, $\mathrm{rc}(X) = 2k + \ell$.*

Proof. Without loss of generality let $S_i = \{a_i, \ldots, b_i\}$ be such that $a_i < 0$ and $b_i > 0$, and let $T_j = \{0, 1\}$, for all $1 \leq j \leq \ell$. We form a $2k + \ell$ element set H by attaching to each S_i two points $(a_i - 1)e_i$ and $(b_i + 1)e_i$ in $\mathbb{Z}^{k+\ell}$, and to each T_j the point $2e_{k+j}$ in $\mathbb{Z}^{k+\ell}$. It is straightforward to check that H is a hiding set of X. Indeed, by construction the midpoint of any two $p, q \in H$ with $p \neq q$ belongs to $\mathrm{conv}(X)$. This shows $\mathrm{rc}(X) \geq 2k + \ell$.

To prove the upper bound $\mathrm{rc}(X) \leq 2k + \ell$ it suffices to verify the case $k = 1$. Indeed, if $k > 0$, a relaxation Q' of $S_k \times T_1 \times \ldots \times T_\ell$ with $2 + \ell$ facets gives rise to the relaxation $\mathrm{conv}(S_1 \times \ldots \times S_{k-1}) \times Q'$ of X with $2k + \ell$ facets. The case $k = 1$ is however exactly the content of Lemma 3. $\square$

5   Numerical Experiments

In the previous sections, we discussed several ways to find upper bounds on or the exact value of the relaxation complexity. Moreover, we have seen that a lower bound is given by a maximum cardinality hiding set. Although the general question on the computability of $\mathrm{rc}(X)$ is not settled, we know that $\mathrm{rc}(X) = \mathrm{rc}(X, \mathrm{Obs}(X))$ can be computed if the set $\mathrm{Obs}(X)$ of observers is finite (see [1, Thm. 5.1]).

In the sequel, we describe the first practically applicable implementation of this computation and discuss several experiments that we conducted. As we mentioned in the beginning of Sect. 4, the exact value of the relaxation complexity is known only in very few particular cases. With this in mind we compiled a database of exact values for lattice polygons with a bounded number of interior integer points, and also investigated the quality of the hiding set bound in dimension 2. Besides these 2-dimensional instances, we also looked at a few more specific computations in higher dimensions.

Implementation Details. Let X be a finite lattice-convex set. Our implementation essentially consists of one method for computing $\mathrm{rc}(X)$ and another method to compute a maximum hiding set for X. We did not use the algorithm discussed in Sect. 3 to compute $\mathrm{rc}(X)$ in dimension 2. Instead we implemented a variation of [1, Alg. 1] to find $\mathrm{Obs}(X)$ and used the mixed-integer programming model suggested in [1] to compute $\mathrm{rc}(X, \mathrm{Obs}(X)) = \mathrm{rc}(X)$. The reason is that we aimed for a method that works in arbitrary dimension.

We tested our implementation not only for the 2-dimensional instances, which we discuss in more detail below, but also for discrete rectangular boxes and for the set of integer points in the standard crosspolytope. Using the basic mixed-integer program from [1], it is already in dimension 4 computationally challenging to compute the relaxation complexity of such examples. For this reason, we enhanced the mixed-integer program by additional cutting planes. The idea of the basic model is to introduce $k \geq \mathrm{rc}(X)$ inequalities and, for each $y \in \mathrm{Obs}(X)$, binary variables s_{yi}, $i \in [k]$, that indicate whether inequality i separates the point y from X. The task is then to minimize the number of indices $i \in [k]$ for which $s_{yi} = 1$ holds for at least one $y \in \mathrm{Obs}(X)$.

To enhance this formulation, we can add additional cutting planes to the model that are based on the idea of hiding sets. If $y, z \in \mathrm{Obs}(X)$ are distinct endpoints of a segment that contains a point of $\mathrm{conv}(X)$, then y and z cannot be separated by the same inequality, i.e., $s_{yi} + s_{zi} \leq 1$ is a valid cut for every $i \in [k]$. Adding these cuts and handling symmetries in the definition of the k inequalities, drastically reduced the running time.

To be able to report on the quality of the hiding set bound $\mathrm{rc}(X) \geq |H|$, we need to be able to compute a *maximum size* hiding set of X. To this end, we first observe that we can restrict the search to within the set of observers of X.

Lemma 4. *Let $X \subseteq \mathbb{Z}^d$ be finite and lattice-convex, let H be a hiding set for X, and assume that there exists an $h \in H \setminus \mathrm{Obs}(X)$. Then, for every $y \in \mathrm{Obs}(X) \cap \mathrm{conv}(\{h\} \cup X)$, the set $H \setminus \{h\} \cup \{y\}$ is a hiding set for X as well.*

In particular, there is a maximum size hiding set contained in $\mathrm{Obs}(X)$.

Proof. For a point $z \notin \mathrm{conv}(X)$, we let $C_z := z + \{\sum_{x \in X} \lambda_x (x - z) : \lambda_x \geq 0\}$ be the smallest convex cone with apex z that contains X. Further, the set $U_z := C_z \setminus \mathrm{conv}(\{z\} \cup X)$ contains all points $w \notin \mathrm{conv}(X)$ such that $\mathrm{conv}(\{z, w\}) \cap \mathrm{conv}(X) \neq \emptyset$. By the choice of y, we have $\mathrm{conv}(\{y\} \cup X) \subseteq \mathrm{conv}(\{h\} \cup X)$, and in particular $U_h \subseteq U_y$. This shows that indeed $H' := H \setminus \{h\} \cup \{y\}$ is a hiding set for X, and since $\mathrm{conv}(\{y, h\}) \cap \mathrm{conv}(X) = \emptyset$, we also have $|H'| = |H|$. $\square$

With this in mind, we define an auxiliary graph $G(X) = (V(X), E(X))$ with node set $V(X) = \mathrm{Obs}(X)$, and where two distinct nodes x and y are adjacent if and only if $\mathrm{conv}(\{x, y\}) \cap \mathrm{conv}(X) = \emptyset$. Then, there is a one-to-one correspondence between hiding sets consisting of observers and stable sets in $G(X)$. To find a maximum stable set in $G(X)$, we solved the binary program $\max\{\sum_{x \in V(X)} z_x : z_x + z_y \leq 1, \forall \{x, y\} \in E(X), z \in \{0, 1\}^{V(X)}\}$.

We have implemented the aforementioned methods in Python 3.7.8, using SageMath 9.1 [18] to implement [1, Alg. 1] and to construct the graph $G(X)$. All mixed-integer programs were solved using SCIP 7.0.0 [6], which has been called via its Python interface [13]. Note that SCIP is not an exact solver and thus the results reported below are only correct up to numerical tolerances.

Numerical Results. For our experiments in dimension 2, we used the representatives of all lattice polygons P with at least one and at most 12 interior integer points provided by Castryck [2]. In our first experiment, we compared

Table 1. Distribution of deviation of maximum hiding set sizes from relaxation complexity in percent.

Size max. Hiding set	Number of facets							
	3	4	5	6	7	8	9	10
rc	100.00	45.19	36.70	45.79	55.09	52.71	45.90	100.00
rc - 1	–	54.81	63.30	54.21	44.91	47.29	54.10	–

Table 2. Distribution of relaxation complexities compared to number of facets in percent.

Relaxation complexity	Number of facets							
	3	4	5	6	7	8	9	10
3	100.00	5.20	3.89	3.40	3.71	2.61	1.64	–
4	–	94.80	66.92	54.79	51.32	42.17	40.98	100.00
5	–	–	29.19	40.39	43.49	54.16	55.74	–
6	–	–	–	1.42	1.48	1.06	1.64	–

the exact value of the relaxation complexity with the hiding set lower bound. Table 1 shows the deviation of these two values parameterized by the number of edges of P. We can see that the hiding set bound deviates by at most one. For $k \in \{4, 6, 7, 8, 9\}$, the hiding set bound is distributed relatively equally between $rc(X)$ and $rc(X) - 1$. Thus, if $H(X)$ is the hiding set bound for X, one may wonder whether the inequalities $rc(X) \geq H(X) \geq rc(X) - 1$ hold for every $X \subseteq \mathbb{Z}^2$.

The second experiment compares the relaxation complexity with the number of edges of P, see Table 2. Although P can be rather complex with up to 10 edges (where 10 is realized by a single instance), the relaxation complexity is at most six, where for the majority of all tested instances the relaxation complexity is either four or five. Moreover, there exist polygons with up to nine edges that admit simplicial relaxations. Thus, already in dimension 2, the difference between the number of facets and the relaxation complexity can be very large. In high dimensions this phenomenon does not come as a surprise, since there are knapsack polytopes whose integer hull has super-polynomially many facets (cf. Pokutta and Van Vyve [15, Cor. 3.8]).

Our third set of experiments concerned discrete rectangular boxes and the discrete crosspolytope $\lozenge_d = \{0, \pm e_1, \ldots, \pm e_d\}$. We managed to computationally determine the relaxation complexity of at most 4-dimensional discrete rectangular boxes with various length of the involved discrete segments. It was these experiments that actually led us to guess and eventually prove the exact values provided in Theorem 5. Weltge [19, Prop. 7.2.2] proved that $rc(\lozenge_d) \leq 2d$, for all $d \geq 4$ and wondered whether this bound is best possible. The $2d$ bound is already quite surprising, as the crosspolytope has 2^d facets. Using our efficient implementation for rc(X) described above, we obtained *simplex* relaxations of $\Diamond_d$, for the dimensions $d = 3, 4, 5$. This of course raises the question whether rc($\Diamond_d$) = $d + 1$ holds for every dimension $d \geq 3$.

References

1. Averkov, G., Schymura, M.: Complexity of linear relaxations in integer programming. Mathematical Programming (2021). https://doi.org/10.1007/s10107-021-01623-4, (online first)
2. Castryck, W.: Moving out the edges of a lattice polygon. Discrete Comput. Geom. **47**(3), 496–518 (2012)
3. van der Corput, J.G.: Verallgemeinerung einer Mordellschen Beweismethode in der Geometrie der Zahlen II. Acta Arith. **2**, 145–146 (1936)
4. Edelsbrunner, H.: Algorithms in Combinatorial Geometry. Monographs in Theoretical Computer Science, vol. 10. Springer-Verlag, Berlin Heidelberg (1987). https://doi.org/10.1007/978-3-642-61568-9
5. Edelsbrunner, H., Preparata, F.P.: Minimum polygonal separation. Inform. Comput. **77**, 218–232 (1988)
6. Gamrath, G., et al.: The SCIP Optimization Suite 7.0. Technical report, Optimization Online, March 2020. http://www.optimization-online.org/DB_HTML/2020/03/7705.html
7. Hammer, P.L., Ibaraki, T., Peled, U.N.: Threshold numbers and threshold completions. In: Studies on Graphs and Discrete Programming (Brussels, 1979), Annals of Discrete Mathematics, vol. 11, pp. 125–145. North-Holland, Amsterdam-New York (1981)
8. Hojny, C.: Polynomial size IP formulations of knapsack may require exponentially large coefficients. Oper. Res. Lett. **48**(5), 612–618 (2020)
9. Hojny, C.: Strong IP formulations need large coefficients. Discrete Optim. **39**, 100624 (2021). https://doi.org/10.1016/j.disopt.2021.100624
10. Jeroslow, R.G.: On defining sets of vertices of the hypercube by linear inequalities. Discrete Math. **11**, 119–124 (1975)
11. Kaibel, V., Weltge, S.: Lower bounds on the sizes of integer programs without additional variables. Math. Program. **154**(1–2, Ser. B), 407–425 (2015)
12. Lee, C., Lee, D.: On a circle-cover minimization problem. Inf. Process. Lett. **18**(2), 109–115 (1984)
13. Maher, S., Miltenberger, M., Pedroso, J.P., Rehfeldt, D., Schwarz, R., Serrano, F.: PySCIPOpt: mathematical programming in Python with the SCIP optimization suite. In: Greuel, G.-M., Koch, T., Paule, P., Sommese, A. (eds.) ICMS 2016. LNCS, vol. 9725, pp. 301–307. Springer, Cham (2016). https://doi.org/10.1007/978-3-319-42432-3_37
14. Megiddo, N.: On the complexity of polyhedral separability. Discrete Comput. Geom. **3**, 325–337 (1988)
15. Pokutta, S., Van Vyve, M.: A note on the extension complexity of the knapsack polytope. Oper. Res. Lett. **41**(4), 347–350 (2013)
16. Schrijver, A.: Theory of Linear and Integer Programming. Wiley-Interscience Series in Discrete Mathematics, A Wiley-Interscience Publication, John Wiley & Sons Ltd, Chichester (1986)
17. Taylor, A.D., Zwicker, W.S.: Simple Games: Desirability Relations, Trading, Pseudoweightings. Princeton University Press, Princeton (1999)

18. The Sage Developers: SageMath, the Sage Mathematics Software System (Version 9.1) (2020). https://www.sagemath.org
19. Weltge, S.: Sizes of Linear Descriptions in Combinatorial Optimization. Ph.D. thesis, Otto-von-Guericke-Universität Magdeburg (2015). http://dx.doi.org/10.25673/4350

Complexity of Branch-and-Bound and Cutting Planes in Mixed-Integer Optimization - II

Amitabh Basu[1]([✉])[ID], Michele Conforti[2][ID], Marco Di Summa[2][ID], and Hongyi Jiang[1][ID]

[1] Johns Hopkins University, Baltimore, USA
{basu.amitabh,hjiang32}@jhu.edu
[2] Università degli Studi Padova, Padua, Italy
{conforti,disumma}@math.unipd.it

Abstract. We study the complexity of cutting planes and branching schemes from a theoretical point of view. We give some rigorous underpinnings to the empirically observed phenomenon that combining cutting planes and branching into a branch-and-cut framework can be orders of magnitude more efficient than employing these tools on their own. In particular, we give general conditions under which a cutting plane strategy and a branching scheme give a provably exponential advantage in efficiency when combined into branch-and-cut. The efficiency of these algorithms is evaluated using two concrete measures: number of iterations and sparsity of constraints used in the intermediate linear/convex programs. To the best of our knowledge, our results are the first mathematically rigorous demonstration of the superiority of branch-and-cut over pure cutting planes and pure branch-and-bound.

Keywords: Integer programming · Cutting planes · Branching schemes · Proof complexity

1 Introduction

In this paper, we consider the following mixed-integer optimization problem:

$$\sup \langle c, x \rangle$$
$$\text{s.t.} \quad x \in C \cap (\mathbb{Z}^n \times \mathbb{R}^d) \tag{1}$$

where C is a closed, convex set in $\mathbb{R}^{n+d}$.

State of the art algorithms for integer optimization are based on two ideas that are at the origin of mixed-integer programming and have been constantly refined: *cutting planes* and *branch-and-bound*. Decades of theoretical and experimental research into both these techniques is at the heart of the outstanding

Supported by ONR Grant N000141812096, NSF Grant CCF2006587, AFOSR Grant FA95502010341, and SID 2019 from University of Padova.

M. Singh and D. P. Williamson (Eds.): IPCO 2021, LNCS 12707, pp. 383–398, 2021.
https://doi.org/10.1007/978-3-030-73879-2_27

success of integer programming solvers. Nevertheless, we feel that there is lot of scope for widening and deepening our understanding of these tools. We have recently started building foundations for a rigorous, quantitative theory for analyzing the strengths and weaknesses of cutting planes and branching [3]. We continue this project in the current manuscript.

In particular, we provide a theoretical framework to explain an empirically observed phenomenon: algorithms that make a combined use of both cutting planes and branching techniques are more efficient (sometimes by orders of magnitude), compared to their stand alone use in algorithms. Not only is a theoretical understanding of this phenomenon lacking, a deeper understanding of the interaction of these methods is considered to be important by both practitioners and theoreticians in the mixed-integer optimization community. To quote an influential computational survey [35] "... it seems that a tighter coordination of the two most fundamental ingredients of the solvers, branching and cutting, can lead to strong improvements."

The main computational burden in any cutting plane or branch-and-bound or branch-and-cut algorithm is the solution of the intermediate convex relaxations. Thus, there are two important aspects to deciding how efficient such an algorithm is: 1) How many linear programs (LPs) or convex optimization problems are solved? 2) How computationally challenging are these convex problems? The first aspect has been widely studied using the concepts of proof size and rank; see [6,10–12,16,19–21,25,45] for a small sample of previous work. Formalizing the second aspect is somewhat tricky and we will focus on a very specific aspect: the *sparsity* of the constraints describing the linear program. The collective wisdom of the optimization community says that sparsity of constraints is a highly important aspect in the efficiency of linear programming [5,26,44,48]. Additionally, most mixed-integer optimization solvers use sparsity as a criterion for cutting plane selection; see [22–24] for an innovative line of research. Compared to cutting planes, sparsity considerations have not been as prominent in the choice of branching schemes. This is primarily because for variable disjunctions sparsity is not an issue, and there is relatively less work on more general branching schemes; see [1,17,18,33,36–40]. In our analysis, we are careful about the sparsity of the disjunctions as well – see Definition 3 below.

1.1   Framework for Mathematical Analysis

We now present the formal details of our approach. A *cutting plane* for the feasible region of (1) is a halfspace $H = \{x \in \mathbb{R}^{n+d} : \langle a, x \rangle \leq \delta\}$ such that $C \cap (\mathbb{Z}^n \times \mathbb{R}^d) \subseteq H$. The most useful cutting planes are those that are not valid for C, i.e., $C \nsubseteq H$. There are several procedures used in practice for generating cutting planes, all of which can be formalized by the general notion of a *cutting plane paradigm*. A cutting plane paradigm is a function CP that takes as input any closed, convex set C and outputs a (possibly infinite) family $CP(C)$ of cutting planes valid for $C \cap (\mathbb{Z}^n \times \mathbb{R}^d)$. Two well-studied examples of cutting plane paradigms are the *Chvátal-Gomory cutting plane paradigm* [46, Chapter 23] and the *split cut paradigm* [14, Chapter 5]. We will assume that all cutting planes are rational in this paper.

State-of-the-art solvers embed cutting planes into a systematic enumeration scheme called *branch-and-bound*. The central notion is that of a *disjunction*, which is a union of polyhedra $D = Q_1 \cup \ldots \cup Q_k$ such that $\mathbb{Z}^n \times \mathbb{R}^d \subseteq D$, i.e., the polyhedra together cover all of $\mathbb{Z}^n \times \mathbb{R}^d$. One typically uses a (possibly infinite) family of disjunctions for potential deployment in algorithms. A well-known example is the family of *split disjunctions* that are of the form $D_{\pi,\pi_0} := \{x \in \mathbb{R}^{n+d} : \langle \pi, x \rangle \leq \pi_0\} \cup \{x \in \mathbb{R}^{n+d} : \langle \pi, x \rangle \geq \pi_0 + 1\}$, where $\pi \in \mathbb{Z}^n \times \{0\}^d$ and $\pi_0 \in \mathbb{Z}$. When the first n coordinates of π correspond to a standard unit vector, we get *variable disjunctions*, i.e., disjunctions of the form $\{x : x_i \leq \pi_0\} \cup \{x : x_i \geq \pi_0 + 1\}$, for $i = 1, \ldots, n$.

A family of disjunctions $\mathcal{D}$ can also form the basis of a cutting plane paradigm. Given any disjunction D, any halfspace H such that $C \cap D \subseteq H$ is a cutting plane, since $C \cap (\mathbb{Z}^n \times \mathbb{R}^d) \subseteq C \cap D$ by definition of a disjunction. The corresponding cutting plane paradigm $\mathcal{CP}(C)$, called *disjunctive cuts based on* $\mathcal{D}$, is the family of all such cutting planes derived from disjunctions in $\mathcal{D}$. A well-known example is the family of *lift-and-project cuts* derived from variable disjunctions.

In the following we assume that all convex optimization problems that need to be solved have an optimal solution or are infeasible.

Definition 1. *A branch-and-cut algorithm based on a family $\mathcal{D}$ of disjunctions and a cutting plane paradigm $\mathcal{CP}$ maintains a list $\mathcal{L}$ of convex subsets of the initial set C which are guaranteed to contain the optimal point, and a lower bound LB that stores the objective value of the best feasible solution found so far (with $LB = -\infty$ if no feasible solution has been found). At every iteration, the algorithm selects one of these subsets $N \in \mathcal{L}$ and solves the convex optimization problem $\sup\{\langle c, x \rangle : x \in N\}$ to obtain x^N. If the objective value is less than or equal to LB, then this set N is discarded from the list $\mathcal{L}$. Else, if x^N satisfies the integrality constraints, LB is updated with the value of x^N and N is discarded from the list. Otherwise, the algorithm makes a decision whether to branch or to cut. In the former case, a disjunction $D = (Q_1 \cup \ldots \cup Q_k) \in \mathcal{D}$ is chosen such that $x^N \notin D$ and the list is updated $\mathcal{L} := \mathcal{L} \setminus \{N\} \cup \{Q_1 \cap N, \ldots, Q_k \cap N\}$. If the decision is to cut, then the algorithm selects a cutting plane $H \in \mathcal{CP}(P)$ such that $x^N \notin H$, and updates the relaxation N by adding the cut H, i.e., updates $\mathcal{L} := \mathcal{L} \setminus \{N\} \cup \{N \cap H\}$.*

Motivated by the above, we will refer to a family $\mathcal{D}$ of disjunctions also as a *branching scheme*. In a branch-and-cut algorithm, if one always chooses to add a cutting plane and never uses a disjunction to branch, then it is said to be a *(pure) cutting plane algorithm* and if one does not use any cutting planes ever, then it is called a *(pure) branch-and-bound algorithm*. We note here that in practice, when a decision to cut is made, several cutting planes are usually added as opposed to just one single cutting plane like in Definition 1. In our mathematical framework, allowing only a single cut makes for a seamless generalization from pure cutting plane algorithms, and also makes quantitative analysis easier.

Definition 2. *The execution of any branch-and-cut algorithm on a mixed-integer optimization instance can be represented by a tree. Every convex relaxation N processed by the algorithm is denoted by a node in the tree. If the optimal value for N is not better than the current lower bound, or is integral, N is a leaf. Otherwise, in the case of a branching, its children are $Q_1 \cap N, \ldots, Q_k \cap N$, and in the case of a cutting plane, there is a single child representing $N \cap H$ (we use the same notation as in Definition 1). This tree is called the* branch-and-cut tree *(branch-and-bound tree, if no cutting planes are used). If no branching is done, this tree (which is really a path) is called a* cutting plane proof. *The* size *of the tree or proof is the total number of nodes.*

Proof Versus Algorithm. Although we use the word "algorithm" in Definition 1, it is technically a *non-deterministic* algorithm, or equivalently, a proof schema or proof system for optimality [2] (leaving aside the question of finite termination for now). This is because no indication is given on how the important decisions are made: Which set N to process from $\mathcal{L}$? Branch or cut? Which disjunction or cutting plane to use? Nevertheless, the proof system is very useful for obtaining information theoretic lower bounds on the efficiency of any deterministic branch-and-cut algorithm. Moreover, one can prove the validity of any upper bound on the objective, i.e., the validity of $\langle c, x \rangle \leq \gamma$ by exhibiting a branch-and-cut tree where this inequality is valid for all the leaves. If γ is the optimal value, this is a proof of optimality, but one may often be interested in the branch-and-cut/branch-and-bound/cutting plane proof complexity of other valid inequalities as well. The connections between integer programming and proof complexity has a long history; see [7,8,13,15,27–29,31,34,41–43], to cite a few. Our results can be interpreted in the language of proof complexity as well.

Recall that we quantify the complexity of any branch-and-bound/cutting plane/branch-and-cut algorithm using two aspects: the number of LP relaxations processed and the sparsity of the constraints defining the LPs. The number of LP relaxations processed is given precisely by the number of nodes in the corresponding tree (Definition 2). Sparsity is formalized in the following definitions.

Definition 3. *Let $1 \leq s \leq n + d$ be a natural number that we call the* sparsity parameter. *Then the pair $(\mathcal{CP}, s)$ will denote the restriction of the paradigm $\mathcal{CP}$ that only reports the sub-family of cutting planes that can be represented by inequalities with at most s non-zero coefficients; the notation $(\mathcal{CP}, s)(C)$ will be used to denote this sub-family for any particular convex set C. Similarly, $(\mathcal{D}, s)$ will denote the sub-family of the family of disjunctions $\mathcal{D}$ such that each polyhedron in the disjunction has an inequality description where every inequality has at most s non-zero coefficients.*

1.2 Our Results

Sparsity Versus Size. Our first set of results considers the trade-off between the sparsity parameter s and the number of LPs processed, i.e., the size of the tree. There are several avenues to explore in this direction. For example, one could compare pure branch-and-bound algorithms based on $(\mathcal{D}, s_1)$ and $(\mathcal{D}, s_2)$, i.e., fix a particular disjunction family $\mathcal{D}$ and consider the effect of sparsity on the branch-and-bound tree sizes. One could also look at two different families of disjunctions $\mathcal{D}_1$ and $\mathcal{D}_2$ and look at their relative tree sizes as one turns the knob on the sparsity parameter. Similar questions could be asked about cutting plane paradigms $(\mathcal{CP}_1, s_1)$ and $(\mathcal{CP}_2, s_2)$ for interesting paradigms $\mathcal{CP}_1, \mathcal{CP}_2$. Even more interestingly, one could compare pure branch-and-bound and pure cutting plane algorithms against each other.

We first focus on pure branch-and-bound algorithms based on the family $\mathcal{S}$ of split disjunctions. A very well-known example of pure integer instances (i.e., $d = 0$) due to Jeroslow [32] shows that if the sparsity of the splits used is restricted to be 1, i.e., one uses only variable disjunctions, then the branch-and-bound algorithm will generate an exponential (in the dimension n) sized tree. On the other hand, if one allows fully dense splits, i.e., sparsity is n, then there is a tree with just 3 nodes (one root, and two leaves) that solves the problem. We ask what happens in Jeroslow's example if one uses split disjunctions with sparsity $s > 1$. Our first result shows that unless the sparsity parameter $s = \Omega(n)$, one cannot get constant size trees, and if the sparsity parameter $s = O(1)$, then the tree is of exponential size.

Theorem 1. *Let H be the halfspace defined by inequality $2\sum_{i=1}^{n} x_i \leq n$, where n is an odd number. Consider the instances of (1) with $d = 0$, the objective $\sum_{i=1}^{n} x_i$ and $C = H \cap [0,1]^n$. The optimum is $\lfloor \frac{n}{2} \rfloor$, and any branch-and-bound proof with sparsity $s \leq \lfloor \frac{n}{2} \rfloor$ that certifies $\sum_{i=1}^{n} x_i \leq \lfloor \frac{n}{2} \rfloor$ has size at least $\Omega(2^{\frac{n}{2s}})$.*

The bounds in Theorem 1 give a constant lower bound when $s = \Omega(n)$. We establish another lower bound which does better in this regime.

Theorem 2. *Let H be the halfspace defined by inequality $2\sum_{i=1}^{n} x_i \leq n$, where n is an odd number. Consider the instances of (1) with $d = 0$, the objective $\sum_{i=1}^{n} x_i$ and $C = H \cap [0,1]^n$. The optimum is $\lfloor \frac{n}{2} \rfloor$, and any branch-and-bound proof with sparsity $s \leq \lfloor \frac{n}{2} \rfloor$ that certifies $\sum_{i=1}^{n} x_i \leq \lfloor \frac{n}{2} \rfloor$ has size at least $\Omega\left(\sqrt{\frac{n(n-s)}{s}}\right)$.*

Next we consider the relative strength of cutting planes and branch-and-bound. Our previous work has studied conditions under which one method can dominate the other, depending on which cutting plane paradigm and branching scheme one chooses [3]. For this paper, the following result from [3] is relevant: for every convex 0/1 pure integer instance, any branch-and-bound proof based on variable disjunctions can be "simulated" by a lift-and-project cutting plane proof without increasing the size of the proof (versions of this result for *linear* 0/1 programming were known earlier; see [19,20]). Moreover, in [3] we constructed a family of stable set instances where lift-and-project cuts give exponentially shorter proofs than branch-and-bound. This is interesting because lift-and-project cuts are disjunctive cuts based on the same family of variable disjunctions, so it is not a priori clear that they have an advantage. These results were obtained with no regard for sparsity. We now show that once we also track the sparsity parameter, this advantage can disappear.

Theorem 3. *Let H be the halfspace defined by inequality $2\sum_{i=1}^{n} x_i \leq n$, where n is an odd number. Consider the intances of (1) with $d = 0$, the objective $\sum_{i=1}^{\lceil \frac{n}{2} \rceil} x_i$ and $C = H \cap [0,1]^n$. The optimum is $\lfloor \frac{n}{2} \rfloor$, and there is a branch-and-bound algorithm based on variable disjunctions, i.e., the family of split disjunctions with sparsity 1, that certifies $\sum_{i=1}^{\lceil \frac{n}{2} \rceil} x_i \leq \lfloor \frac{n}{2} \rfloor$ in $O(n)$ steps. However, any cutting plane for C with sparsity $s \leq \lfloor \frac{n}{2} \rfloor$ is trivial, i.e., valid for $[0,1]^n$, no matter what cutting plane paradigm is used to derive it.*

Superiority of Branch-and-Cut. We next consider the question of when combining branching and cutting planes is *provably* advantageous. For this question, we leave aside the complications arising due to sparsity considerations and focus only on the size of proofs. The following discussion and results can be extended to handle the issue of sparsity as well, but we leave it out of this extended abstract.

Given a cutting plane paradigm $\mathcal{CP}$, and a branching scheme $\mathcal{D}$, are there families of instances where branch-and-cut based on $\mathcal{CP}$ and $\mathcal{D}$ does provably better than pure cutting planes based on $\mathcal{CP}$ alone and pure branch-and-bound based on $\mathcal{D}$ alone? If a cutting plane paradigm $\mathcal{CP}$ and a branching scheme $\mathcal{D}$ are such that either for every instance, $\mathcal{CP}$ gives cutting plane proofs of size at most a polynomial factor larger than the shortest branch-and-bound proofs with $\mathcal{D}$, or vice versa, for every instance $\mathcal{D}$ gives proofs of size at most polynomially larger than the shortest cutting plane proofs based on $\mathcal{CP}$, then combining them into branch-and-cut is likely to give no substantial improvement since one method can always do the job of the other, up to polynomial factors. As mentioned above, prior work [3] had shown that disjunctive cuts based on variable disjunctions (with no restriction on sparsity) dominate branch-and-bound based on variable disjunctions for 0/1 instances, and as a consequence branch-and-cut based on these paradigms is dominated by pure cutting planes. In this paper, we show that the situation completely reverses if one considers a broader family of disjunctions; in particular, the family of disjunctions must be rich enough to contain all split disjunctions.

Theorem 4. *Let $C \subseteq \mathbb{R}^n$ be a closed, convex set. Let $k \in \mathbb{N}$ be a fixed natural number and let $\mathcal{D}$ be any family of disjunctions that contains all split disjunctions, such that all disjunctions in $\mathcal{D}$ have at most k terms in the disjunction. If a valid inequality $\langle c, x \rangle \leq \delta$ for $C \cap \mathbb{Z}^n$ has a cutting plane proof of size L using disjunctive cuts based on $\mathcal{D}$, then there exists a branch-and-bound proof of size at most $(k+1)L$ based on $\mathcal{D}$. Moreover, there is a family of instances where branch-and-bound based on split disjunctions solves the problem in $O(1)$ time whereas there is a polynomial lower bound on split cut proofs.*

The above discussion and theorem motivate the following definition which formalizes the situation where no method dominates the other. To make things precise, we assume that there is a well-defined way to assign a concrete *size* to any instance of (1); see [30] for a discussion on how to make this formal. Additionally, when we speak of an instance, we allow the possibility of proving the validity of any inequality valid for $C \cap (\mathbb{Z}^n \times \mathbb{R}^d)$, not necessarily related to an upper bound on the objective value. Thus, an instance is a tuple (C, c, γ) such that $\langle c, x \rangle \leq \gamma$ for all $x \in C \cap (\mathbb{Z}^n \times \mathbb{R}^d)$.

Definition 4. *A cutting plane paradigm $\mathcal{CP}$ and a branching scheme $\mathcal{D}$ are complementary if there is a family of instances where $\mathcal{CP}$ gives polynomial (in the size of the instances) size proofs and the shortest branch-and-bound proof based on $\mathcal{D}$ is exponential (in the size of the instances), and there is another family of instances where $\mathcal{D}$ gives polynomial size proofs while $\mathcal{CP}$ gives exponential size proofs.*

We wish to formalize the intuition that branch-and-cut is expected to be exponentially better than branch-and-bound or cutting planes alone for complementary pairs of branching schemes and cutting plane paradigms. But we need to make some mild assumptions about the branching schemes and cutting plane paradigms. *All known branching schemes and cutting plane methods from the literature satisfy these conditions.*

Definition 5. *A branching scheme is said to be* regular *if no disjunction involves a continuous variable, i.e., each polyhedron in the disjunction is described using inequalities that involve only the integer constrained variables.*

A branching scheme $\mathcal{D}$ is said to be embedding closed *if disjunctions from higher dimensions can be applied to lower dimensions. More formally, let n_1, n_2, d_1, $d_2 \in \mathbb{N}$. If $D \in \mathcal{D}$ is a disjunction in $\mathbb{R}^{n_1} \times \mathbb{R}^{d_1} \times \mathbb{R}^{n_2} \times \mathbb{R}^{d_2}$ with respect to $\mathbb{Z}^{n_1} \times \mathbb{R}^{d_1} \times \mathbb{Z}^{n_2} \times \mathbb{R}^{d_2}$, then the disjunction $D \cap (\mathbb{R}^{n_1} \times \mathbb{R}^{d_1} \times \{0\}^{n_2} \times \{0\}^{d_2})$, interpreted as a set in $\mathbb{R}^{n_1} \times \mathbb{R}^{d_1}$, is also in $\mathcal{D}$ for the space $\mathbb{R}^{n_1} \times \mathbb{R}^{d_1}$ with respect to $\mathbb{Z}^{n_1} \times \mathbb{R}^{d_1}$ (note that $D \cap (\mathbb{R}^{n_1} \times \mathbb{R}^{d_1} \times \{0\}^{n_2} \times \{0\}^{d_2})$, interpreted as a set in $\mathbb{R}^{n_1} \times \mathbb{R}^{d_1}$, is certainly a disjunction with respect to $\mathbb{Z}^{n_1} \times \mathbb{R}^{d_1}$; we want $\mathcal{D}$ to be closed with respect to such restrictions).*

A cutting plane paradigm $\mathcal{CP}$ is said to be regular *if it has the following property, which says that adding "dummy variables" to the formulation of the instance should not change the power of the paradigm. Formally, let $C \subseteq \mathbb{R}^n \times \mathbb{R}^d$ be any closed, convex set and let $C' = \{(x, t) \in \mathbb{R}^n \times \mathbb{R}^d \times \mathbb{R} : x \in C, \ t = \langle f, x \rangle\}$ for some $f \in \mathbb{R}^n$. Then if a cutting plane $\langle a, x \rangle \leq b$ is derived by $\mathcal{CP}$ applied to C, i.e., this inequality is in $\mathcal{CP}(C)$, then it should also be in $\mathcal{CP}(C')$, and conversely, if $\langle a, x \rangle + \mu t \leq b$ is in $\mathcal{CP}(C')$, then the equivalent inequality $\langle a + \mu f, x \rangle \leq b$ should be in $\mathcal{CP}(C)$.*

A cutting plane paradigm $\mathcal{CP}$ is said to be embedding closed *if cutting planes from higher dimensions can be applied to lower dimensions. More formally, let $n_1, n_2, d_1, d_2 \in \mathbb{N}$. Let $C \subseteq \mathbb{R}^{n_1} \times \mathbb{R}^{d_1}$ be any closed, convex set. If the inequality $\langle c_1, x_1 \rangle + \langle a_1, y_1 \rangle + \langle c_2, x_2 \rangle + \langle a_2, y_2 \rangle \leq \gamma$ is a cutting plane for $C \times \{0\}^{n_2} \times \{0\}^{d_2}$ with respect to $\mathbb{Z}^{n_1} \times \mathbb{R}^{d_1} \times \mathbb{Z}^{n_2} \times \mathbb{R}^{d_2}$ that can be derived by applying $\mathcal{CP}$ to $C \times \{0\}^{n_2} \times \{0\}^{d_2}$, then the cutting plane $\langle c_1, x_1 \rangle + \langle a_1, y_1 \rangle \leq \gamma$ that is valid for $C \cap (\mathbb{Z}^{n_1} \times \mathbb{R}^{d_1})$ should also belong to $\mathcal{CP}(C)$.*

A cutting plane paradigm $\mathcal{CP}$ is said to be inclusion closed, *if for any two closed convex sets $C \subseteq C'$, we have $\mathcal{CP}(C') \subseteq \mathcal{CP}(C)$. In other words, any cutting plane derived for C' can also be derived for a subset C.*

Theorem 5. *Let $\mathcal{D}$ be a regular, embedding closed branching scheme and let $\mathcal{CP}$ be a regular, embedding closed, and inclusion closed cutting plane paradigm such that $\mathcal{D}$ includes all variable disjunctions and $\mathcal{CP}$ and $\mathcal{D}$ form a complementary pair. Then there exists a family of instances of (1) which have polynomial size branch-and-cut proofs, whereas any branch-and-bound proof based on $\mathcal{D}$ and any cutting plane proof based on $\mathcal{CP}$ is of exponential size.*

Example 1. As a concrete example of a complementary pair that satisfies the other conditions of Theorem 5, consider $\mathcal{CP}$ to be the Chvátal-Gomory paradigm and $\mathcal{D}$ to be the family of variable disjunctions. From their definitions, they are both regular and $\mathcal{D}$ is embedding closed. The Chvátal-Gomory paradigm is also embedding closed and inclusion closed. For the Jeroslow instances from Theorem 1, the single Chvátal-Gomory cut $\sum_{i=1}^{n} x_i \leq \lfloor \frac{n}{2} \rfloor$ proves optimality, whereas variable disjunctions produce a tree of size $2^{\lfloor \frac{n}{2} \rfloor}$. On the other hand, consider the set T, where $T = \mathrm{conv}\{(0,0),(1,0),(\frac{1}{2},h)\}$ and the valid inequality $x_2 \leq 0$ for $T \cap \mathbb{Z}^2$. Any Chvátal-Gomory paradigm based proof has size exponential in the size of the input, i.e., every proof has length at least $\Omega(h)$ [46]. On the other hand, a single disjunction on the variable x_1 solves the problem.

In [3], we also studied examples of disjunction families $\mathcal{D}$ such that disjunctive cuts based on $\mathcal{D}$ are complementary to branching schemes based on $\mathcal{D}$.

Example 1 shows that the classical Chvátal-Gomory cuts and variable branching are complementary and thus give rise to a superior branch-and-cut routine when combined by Theorem 5. As discussed above, for 0/1 problems, lift-and-project cuts and variable branching do *not* form a complementary pair, and neither do split cuts and split disjunctions by Theorem 4. It would be nice to establish the converse of Theorem 5: if there is a family where branch-and-cut is exponentially superior, then the cutting plane paradigm and branching scheme are complementary. In Theorem 6 below, we prove a partial converse along these lines in the pure integer case. This partial converse requires the disjunction family to include all split disjunctions. It would be more satisfactory to establish similar results without this assumption. More generally, it remains an open question if our definition of complementarity is an exact characterization of when branch-and-cut is superior.

Theorem 6. *Let $\mathcal{D}$ be a branching scheme that includes all split disjunctions and let $\mathcal{CP}$ be any cutting plane paradigm. Suppose that for every pure integer instance and any cutting plane proof based on $\mathcal{CP}$ for this instance, there is a branch-and-bound proof based on $\mathcal{D}$ of size at most a polynomial factor (in the size of the instance) larger. Then for any branch-and-cut proof based on $\mathcal{D}$ and $\mathcal{CP}$ for a pure integer instance, there exists a pure branch-and-bound proof based on $\mathcal{D}$ that has size at most polynomially larger than the branch-and-cut proof.*

The high level message that we extract from our results is the formalization of the following simple intuition. For branch-and-cut to be superior to pure cutting planes or pure branch-and-bound, one needs the cutting planes and branching scheme to do "sufficiently different" things. For example, if they are both based on the same family of disjunctions (such as lift-and-project cuts and variable branching, or the setting of Theorem 4), then we may not get any improvements with branch-and-cut. The definition of a complementary pair attempts to make the notion of "sufficiently different" formal and Theorem 5 derives the concrete superior performance of branch-and-cut from this formalization.

2   Proofs

We present the proofs of Theorems 1, 2, 5 and 6 in the subsections below. The proofs of Theorems 3 and 4 are excluded from this extended abstract. All missing proofs from this extended abstract can be found in the full version of the paper [4].

2.1   Proof of Theorem 1

Definition 6. *Consider the instances in Theorem 1, and the branch-and-bound tree T produced by split disjunctions to solve it. Assume node N of T contains at least one integer point in $\{0,1\}^n$, and $D_1, D_2, \ldots, D_r$ are the split disjunctions used to derive N from the root of T. For $1 \leq j \leq r$, D_j is a true split disjunction of N if both of the two halfspaces of D_j have a nonempty intersection with the integer hull of the corresponding parent node, i.e. the parent node's integer hull is split into two nonempty parts by D_j. Otherwise, it is called a false split disjunction of N. We define the generation variable set of N as the index set $I \subseteq \{1, 2, \ldots, n\}$ such that it consists of all the indices of the variables involved in the true split disjunctions of N. The generation set of the root node is empty.*

The proof of the following lemma is excluded from this extended abstract.

Lemma 1. *Consider the instances in Theorem 1, and the branch-and-bound tree T produced by split disjunctions with sparsity parameter $s < \lfloor \frac{n}{2} \rfloor$ to solve it. For any node N of T with at least one feasible integer point $v = (v_1, v_2, \ldots, v_n) \in \{0,1\}^n$, let P, P_I and I denote the relaxation, the integer hull and the generation variable set corresponding to N. Define $V := \{(x_1, x_2, \ldots, x_n) \in \{0,1\}^n : x_i = v_i \text{ for } i \in I, \sum_{j=1}^n x_j = \lfloor \frac{n}{2} \rfloor\}$.*
    If $|I| \leq \lfloor \frac{n}{2} \rfloor - s$, then we have:

(i) $V \neq \emptyset$ and $V \subseteq P_I \cap \{0,1\}^n$;
(ii) the objective LP value of N is $\frac{n}{2}$.

Proof (Proof of Theorem 1). For a node N of the branch-and-bound tree containing at least one integer point, if it is derived by exactly m true split disjunctions, then we say it is a node of generation m. By Lemma 1, if $m \leq \frac{1}{s} \lfloor \frac{n}{2} \rfloor - 1$, then a node N of generation m has LP objective value $\frac{n}{2}$, and in the subtree rooted at N there must exist at least two descendants from generation $m + 1$, since the leaf nodes must have LP values less than or equal to $\lfloor \frac{n}{2} \rfloor$. Therefore, there are at least 2^m nodes of generation m when $m \leq \frac{1}{s} \lfloor \frac{n}{2} \rfloor - 1$. This finishes the proof. □

2.2   Proof of Theorem 2

The following lemma follows from an application of Sperner's theorem [47].

Lemma 2. *Let $w_1, \ldots, w_k \in \mathbb{Z} \setminus \{0\}$ and $W \in \mathbb{Z}$. Then the number of 0/1 solutions to $\sum_{j=1}^{k} w_j x_j = W$ is at most $\binom{k}{\lfloor k/2 \rfloor}$.*

Proof (Proof of Theorem 2). We consider the instance from Theorem 2. For any split disjunction $D := \{x : \langle a, x \rangle \leq b\} \cup \{x : \langle a, x \rangle \geq b+1\}$, we define $V(D)$ to be the set of all the optimal LP vertices (of the original polytope) that lie strictly in the corresponding split set $\{x : b \leq \langle a, x \rangle \leq b+1\}$. Let the support of a be given by $T \subseteq \{1, \ldots, n\}$ with $t := |T| \leq s \leq \lfloor n/2 \rfloor$. Since $a \in \mathbb{Z}^n$ and $b \in \mathbb{Z}$, $V(D)$ is precisely the subset of the optimal LP vertices $\hat{x}$ such that $\langle a, \hat{x} \rangle = b + \frac{1}{2}$. Fix some $\ell \in T$ and consider those optimal LP vertices $\hat{x} \in V(D)$ where $\hat{x}_\ell = \frac{1}{2}$. This means that $\sum_{j \in T \setminus \{\ell\}} a_j \hat{x}_j = b + \frac{1}{2} - \frac{a_\ell}{2}$. Let r_i be the number of 0/1 solutions to $\sum_{j \in T \setminus \{\ell\}} a_j \hat{x}_j = b + \frac{1}{2} - \frac{a_\ell}{2}$ with exactly i coordinates set to 1. Then the number of vertices from $V(D)$ with the ℓ-th coordinate equal to $\frac{1}{2}$ is

$$\sum_{i=0}^{t-1} r_i \binom{n-t}{\lfloor n/2 \rfloor - i} \leq \left(\sum_{i=0}^{t-1} r_i \right) \binom{n-t}{\lfloor n/2 \rfloor - \lfloor t/2 \rfloor}.$$

since $\binom{n-t}{\lfloor n/2 \rfloor - i} \leq \binom{n-t}{\lfloor n/2 \rfloor - \lfloor t/2 \rfloor}$ for all $i \in \{0, \ldots, t-1\}$. Using Lemma 2, $\sum_{i=0}^{t-1} r_i \leq \binom{t-1}{\lfloor t/2 \rfloor}$ and we obtain the upper bound $\binom{t-1}{\lfloor t/2 \rfloor} \binom{n-t}{\lfloor n/2 \rfloor - \lfloor t/2 \rfloor}$ on the number of vertices from $V(D)$ with the ℓ-th coordinate equal to $\frac{1}{2}$. Therefore, $|V(D)| \leq t \binom{t-1}{\lfloor t/2 \rfloor} \binom{n-t}{\lfloor n/2 \rfloor - \lfloor t/2 \rfloor} =: p(t)$. Since n is odd, we have

$$p(t) = \begin{cases} \dfrac{t!(n-t)!}{(t/2)!(t/2-1)!((n-t-1)/2)!((n-t+1)/2)!} & \text{if } t \text{ is even,} \\[2em] \dfrac{t!(n-t)!}{((t-1)/2)!((t-1)/2)!((n-t)/2)!((n-t)/2)!} & \text{if } t \text{ is odd.} \end{cases}$$

A direct calculation then shows that

$$\frac{p(t+1)}{p(t)} = \begin{cases} \dfrac{(t+1)(n-t+1)}{t(n-t)} & \text{if } t \text{ is even,} \\[1.5em] 1 & \text{if } t \text{ is odd.} \end{cases}$$

Let h be the largest even number not exceeding s. Since $p(1) = \binom{n-1}{\lfloor n/2 \rfloor}$, we obtain, for every $t \in \{1, \ldots, s\}$,

$$p(t) \leq p(s) = \binom{n-1}{\lfloor n/2 \rfloor} \prod_{\substack{1 \leq q \leq s \\ q \text{ even}}} \frac{q+1}{q} \cdot \frac{n-q+1}{n-q} = \binom{n-1}{\lfloor n/2 \rfloor} \cdot \frac{(h+1)!!}{h!!} \cdot \frac{(n-1)!!}{(n-2)!!} \cdot \frac{(n-h-2)!!}{(n-h-1)!!},$$

where $m!!$ denotes the product of all integers from 1 up to m of the same parity as m. Using the fact that, for every even positive integer ℓ,

$$\sqrt{\frac{\pi \ell}{2}} < \frac{\ell!!}{(\ell-1)!!} < \sqrt{\frac{\pi(\ell+1)}{2}}$$

(see, e.g., [9, 49]), we have (for $h \geq 1$, i.e., $s \geq 2$)

$$
\begin{aligned}
p(t) &\leq \binom{n-1}{\lfloor n/2 \rfloor} \cdot \frac{(h+1)(h-1)!!}{h!!} \cdot \frac{(n-1)!!}{(n-2)!!} \cdot \frac{(n-h-2)!!}{(n-h-1)!!} \\
&\leq \binom{n-1}{\lfloor n/2 \rfloor}(h+1) \sqrt{\frac{2}{\pi h}} \cdot \sqrt{\frac{\pi n}{2}} \cdot \sqrt{\frac{2}{\pi(n-h-1)}} \\
&= \binom{n-1}{\lfloor n/2 \rfloor} \sqrt{\frac{2n(h+1)^2}{\pi h(n-h-1)}} \\
&= \binom{n-1}{\lfloor n/2 \rfloor} O\left(\sqrt{\frac{ns}{n-s}}\right).
\end{aligned}
$$

Thus, this is an upper bound on $|V(D)|$. Since the total number of optimal LP vertices of the instance is $n\binom{n-1}{\lfloor n/2 \rfloor}$, we obtain the following lower bound of on the size of a branch-and-bound proof: $\frac{n\binom{n-1}{\lfloor n/2 \rfloor}}{|V(D)|} = \Omega\left(\sqrt{\frac{n(n-s)}{s}}\right).$ ☐

2.3 Proofs of Theorems 5 And 6

Lemmas 3–5 below are straightforward consequences of the definitions, and the proofs are omitted from this extended abstract.

Lemma 3. *Let $C \subseteq C'$ be two closed, convex sets. Let D be any branching scheme and let CP be an inclusion closed cutting plane paradigm. If there is a branch-and-bound proof with respect to C' based on D for the validity of an inequality $\langle c, x \rangle \leq \gamma$, then there is a branch-and-bound proof with respect to C based on D for the validity of $\langle c, x \rangle \leq \gamma$ of the same size. The same holds for cutting plane proofs based on CP.*

Lemma 4. *Let D and CP be both embedding closed and let $C \subseteq \mathbb{R}^{n_1} \times \mathbb{R}^{d_1}$ be a closed, convex set. Let $\langle c, x \rangle \leq \gamma$ be a valid inequality for $C \cap (\mathbb{Z}^{n_1} \times \mathbb{R}^{d_1})$. If there is a branch-and-bound proof with respect to $C \times \{0\}^{n_2} \times \{0\}^{d_2}$ based on D for the validity of $\langle c, x \rangle \leq \gamma$ interpreted as a valid inequality in $\mathbb{R}^{n_1} \times \mathbb{R}^{d_1} \times \mathbb{R}^{n_2} \times \mathbb{R}^{d_2}$ for $(C \times \{0\}^{n_2} \times \{0\}^{d_2}) \cap (\mathbb{Z}^{n_1} \times \mathbb{R}^{d_1} \times \mathbb{Z}^{n_2} \times \mathbb{R}^{d_2})$, then there is a branch-and-bound proof with respect to C based on D for the validity of $\langle c, x \rangle \leq \gamma$ of the same size. The same holds for cutting plane proofs based on CP.*

Lemma 5. *Let $C \subseteq \mathbb{R}^{n+d}$ be a polytope and let $\langle c, x \rangle \leq \gamma$ be a valid inequality for $C \cap (\mathbb{Z}^n \times \mathbb{R}^d)$. Let $X := \{(x, t) \in \mathbb{R}^{n+d} \times \mathbb{R} : x \in C, \ t = \langle c, x \rangle\}$. Then, for any regular branching scheme D or a regular cutting plane paradigm CP, any proof of validity of $\langle c, x \rangle \leq \gamma$ with respect to $C \cap (\mathbb{Z}^n \times \mathbb{R}^d)$ can be changed into a proof of validity of $t \leq \gamma$ with respect to $X \cap (\mathbb{Z}^n \times \mathbb{R}^d \times \mathbb{R})$ with no change in length, and vice versa.*

Proof (Proof of Theorem 5). Let $\{P_k \subseteq \mathbb{R}^{n_k} \times \mathbb{R}^{d_k} : k \in \mathbb{N}\}$ be a family of closed, convex sets, and $\{(c_k, \gamma_k) \in \mathbb{R}^{n_k} \times \mathbb{R}^{d_k} \times \mathbb{R} : k \in \mathbb{N}\}$ be a family of tuples such that $\langle c_k, x \rangle \le \gamma_k$ is valid for $P_k \cap (\mathbb{Z}^{n_k} \times \mathbb{R}^{d_k})$, and $\mathcal{CP}$ has polynomial size proofs for this family of instances, whereas $\mathcal{D}$ has exponential size proofs. Similarly, let $\{P'_k \subseteq \mathbb{R}^{n'_k} \times \mathbb{R}^{d'_k} : k \in \mathbb{N}\}$ be a family of closed, convex sets, and $\{(c'_k, \gamma'_k) \in \mathbb{R}^{n'_k} \times \mathbb{R}^{d'_k} \times \mathbb{R} : k \in \mathbb{N}\}$ be a family of tuples such that $\langle c'_k, x \rangle \le \gamma'_k$ is valid for $P'_k \cap (\mathbb{Z}^{n'_k} \times \mathbb{R}^{d'_k})$, and $\mathcal{D}$ has polynomial size proofs for this family of instances, whereas $\mathcal{CP}$ has exponential size proofs.

Below, we are going to combine these two families into a single family of instances on which branch-and-cut gives polynomial size proofs, and pure cutting plane or pure branch-and-bound proofs are of exponential size. For this, the growth rates of sizes of instances in the two families need to be polynomially comparable. Otherwise, this can create issues, e.g., polynomial size proofs as measured by the sizes of the first family may become exponential size when compared with the sizes in the second family. It is not hard to show that given any two families of instances whose sizes are not bounded, there exist infinite subfamilies of the two families such that the growth rates of the two subfamilies are polynomially comparable. Since the polynomial or exponential behaviour of the proof sizes were defined with respect to the sizes of the instances, passing to infinite subfamilies maintains this behaviour. For details, please see the full version of this paper [4]. From now on, we will assume the two families (P_k, c_k, γ_k) and (P'_k, c'_k, γ'_k) are polynomially comparable in size (by passing to subfamilies if necessary).

We first embed P_k and P'_k into a common ambient space for each $k \in \mathbb{N}$. This is done by defining $\bar{n}_k = \max\{n_k, n'_k\}$, $\bar{d}_k = \max\{d_k, d'_k\}$, and embedding both P_k and P'_k into the space $\mathbb{R}^{\bar{n}_k} \times \mathbb{R}^{\bar{d}_k}$ by defining $Q_k := P_k \times \{0\}^{\bar{n}_k - n_k} \times \{0\}^{\bar{d}_k - d_k}$ and $Q'_k := P'_k \times \{0\}^{\bar{n}_k - n'_k} \times \{0\}^{\bar{d}_k - d'_k}$. By Lemma 4, $\mathcal{D}$ has an exponential lower bound on sizes of proofs for the inequality $\langle c_k, x \rangle \le \gamma_k$, interpreted as an inequality in $\mathbb{R}^{\bar{n}_k} \times \mathbb{R}^{\bar{d}_k}$, valid for $Q_k \cap (\mathbb{Z}^{\bar{n}_k} \times \mathbb{R}^{\bar{d}_k})$. By Lemma 4, $\mathcal{CP}$ has an exponential lower bound on sizes of proofs for the inequality $\langle c'_k, x \rangle \le \gamma'_k$, interpreted as an inequality in $\mathbb{R}^{\bar{n}_k} \times \mathbb{R}^{\bar{d}_k}$, valid for $Q'_k \cap (\mathbb{Z}^{\bar{n}_k} \times \mathbb{R}^{\bar{d}_k})$.

We now make the objective vector common for both families of instances. Define $X_k := \{(x, t) \in \mathbb{R}^{\bar{n}_k} \times \mathbb{R}^{\bar{d}_k} \times \mathbb{R} : x \in Q_k, \ t = \langle c_k, x \rangle\}$ and $X'_k := \{(x, t) \in \mathbb{R}^{\bar{n}_k} \times \mathbb{R}^{\bar{d}_k} \times \mathbb{R} : x \in Q'_k, \ t = \langle c'_k, x \rangle\}$. By Lemma 5, the inequality $t \le \gamma_k$ has an exponential lower bound on sizes of proofs based on $\mathcal{D}$ for X_k and the inequality $t \le \gamma'_k$ has an exponential lower bound on sizes of proofs based on $\mathcal{CP}$ for X'_k.

We next embed these families as faces of the same closed convex set. Define $Z_k \subseteq \mathbb{R}^{\bar{n}_k} \times \mathbb{R}^{\bar{d}_k} \times \mathbb{R} \times \mathbb{R}$, for every $k \in \mathbb{N}$, as the convex hull of $X_k \times \{0\}$ and $X'_k \times \{0\}$.

The key point to note is that these constructions combine two families whose sizes are polynomially comparable and therefore the new family that is created has sizes that are polynomially comparable to the original two families.

We let (x, t, y) denote points in the new space $\mathbb{R}^{\bar{n}_k} \times \mathbb{R}^{\bar{d}_k} \times \mathbb{R} \times \mathbb{R}$, i.e., y denotes the last coordinate. Consider the family of inequalities $t - \gamma_k(1 - y) - \gamma'_k y \le 0$ for every $k \in \mathbb{N}$. Note that this inequality reduces to $t \le \gamma_k$ when $y = 0$ and it reduces to $t \le \gamma'_k$ when $y = 1$. Thus, the inequality is valid for $Z_k \cap (\mathbb{Z}^{\bar{n}_k} \times \mathbb{R}^{\bar{d}_k} \times \mathbb{R} \times \mathbb{Z})$, i.e., when we constrain y to be an integer variable.

Since $X_k \times \{0\} \subseteq Z_k$, by Lemma 3, proofs of $t - \gamma_k(1-y) - \gamma'_k y \leq 0$ based on $\mathcal{D}$ have an exponential lower bound on their size. Similarly, since $X'_k \times \{0\} \subseteq Z_k$, by Lemma 3, proofs of $t - \gamma_k(1-y) - \gamma'_k y \leq 0$ based on $\mathcal{CP}$ have an exponential lower bound on their size.

However, for branch-and-cut based on $\mathcal{CP}$ and $\mathcal{D}$, we can first branch on the variable y (recall from the hypothesis that $\mathcal{D}$ allows branching on any integer variable). Since $\mathcal{CP}$ has a polynomial proof for P_k and (c_k, γ_k) and therefore for the valid inequality $t \leq \gamma_k$ for $X_k \times \{0\}$, we can process the $y = 0$ branch in polynomial time with cutting planes. Similarly, $\mathcal{D}$ has a polynomial proof for P'_k and (c'_k, γ'_k) and therefore for the valid inequality $t \leq \gamma'_k$ for $X'_k \times \{0\}$, we can process the $y = 1$ branch also in polynomial time. Thus, branch-and-cut runs in polynomial time overall for this family of instances.   □

Proof (Proof of Theorem 6). Recall that we restrict ourselves to the pure integer case, i.e., $d = 0$. Consider any branch-and-cut proof for some instance. If no cutting planes are used in the proof, this is a pure branch-and-bound proof and we are done. Otherwise, let N be a node of the proof tree where a cutting plane $\langle a, x \rangle \leq \gamma$ is used. Since we assume all cutting planes are rational, we may assume $a \in \mathbb{Z}^n$ and $\gamma \in \mathbb{Z}$. Thus, $N' = N \cap \{x : \langle a, x \rangle \geq \gamma + 1\}$ is integer infeasible. Since $\langle a, x \rangle \leq \gamma$ is in $\mathcal{CP}(N)$, by our assumption, there must be a branch-and-bound proof of polynomial size based on $\mathcal{D}$ for the validity of $\langle a, x \rangle \leq \gamma$ with respect to N. Since $N' \subseteq N$, by Lemma 3, there must be a branch-and-bound proof for the validity of $\langle a, x \rangle \leq \gamma$ with respect to N', thus proving the infeasibility of N'. In the branch-and-cut proof, one can replace the child of N by first applying the disjunction $\{x : \langle a, x \rangle \leq \gamma\} \cup \{x : \langle a, x \rangle \geq \gamma + 1\}$ on N, and then on N', applying the above branch-and-bound proof of infeasibility. We now have a branch-and-cut proof for the original instance with one less cutting plane node. We can repeat this for all nodes where a cutting plane is added and convert the entire branch-and-cut tree into a pure branch-and-bound tree with at most a polynomial blow up in size.   □

References

1. Aardal, K., Bixby, R.E., Hurkens, C.A., Lenstra, A.K., Smeltink, J.W.: Market split and basis reduction: towards a solution of the cornuéjols-dawande instances. INFORMS J. Comput. **12**(3), 192–202 (2000)
2. Arora, S., Barak, B.: Computational Complexity: a Modern Approach. Cambridge University Press, Cambridge (2009)
3. Basu, A., Conforti, M., Di Summa, M., Jiang, H.: Complexity of cutting plane and branch-and-bound algorithms for mixed-integer optimization (2020). https://arxiv.org/abs/2003.05023
4. Basu, A., Conforti, M., Di Summa, M., Jiang, H.: Complexity of cutting plane and branch-and-bound algorithms for mixed-integer optimization – II (2020). https://arxiv.org/abs/2011.05474
5. Bixby, R.E.: Solving real-world linear programs: a decade and more of progress. Oper. Res. **50**(1), 3–15 (2002)

6. Bockmayr, A., Eisenbrand, F., Hartmann, M., Schulz, A.S.: On the Chvátal rank of polytopes in the 0/1 cube. Discrete Appl. Math. **98**(1–2), 21–27 (1999)
7. Bonet, M., Pitassi, T., Raz, R.: Lower bounds for cutting planes proofs with small coefficients. J. Symbolic Logic **62**(3), 708–728 (1997)
8. Buss, S.R., Clote, P.: Cutting planes, connectivity, and threshold logic. Arch. Math. Logic **35**(1), 33–62 (1996)
9. Chen, C.P., Qi, F.: Completely monotonic function associated with the gamma functions and proof of Wallis' inequality. Tamkang J. Math. **36**(4), 303–307 (2005)
10. Chvátal, V.: Hard knapsack problems. Oper. Res. **28**(6), 1402–1411 (1980)
11. Chvátal, V.: Cutting-plane proofs and the stability number of a graph, Report Number 84326-OR. Universität Bonn, Bonn, Institut für Ökonometrie und Operations Research (1984)
12. Chvátal, V., Cook, W.J., Hartmann, M.: On cutting-plane proofs in combinatorial optimization. Linear Algebra Appl. **114**, 455–499 (1989)
13. Clote, P.: Cutting planes and constant depth Frege proofs. In: Proceedings of the Seventh Annual IEEE Symposium on Logic in Computer Science, pp. 296–307 (1992)
14. Conforti, M., Cornuéjols, G., Zambelli, G.: Integer programming models. Integer Programming. GTM, vol. 271, pp. 45–84. Springer, Cham (2014). https://doi.org/10.1007/978-3-319-11008-0_2
15. Cook, W.J., Coullard, C.R., Turán, G.: On the complexity of cutting-plane proofs. Discrete Appl. Math. **18**(1), 25–38 (1987)
16. Cook, W.J., Dash, S.: On the matrix-cut rank of polyhedra. Math. Oper. Res. **26**(1), 19–30 (2001)
17. Cornuéjols, G., Liberti, L., Nannicini, G.: Improved strategies for branching on general disjunctions. Math. Program. **130**(2), 225–247 (2011)
18. Dadush, D., Tiwari, S.: On the complexity of branching proofs. arXiv preprint arXiv:2006.04124 (2020)
19. Dash, S.: An exponential lower bound on the length of some classes of branch-and-cut proofs. In: Cook, W.J., Schulz, A.S. (eds.) IPCO 2002. LNCS, vol. 2337, pp. 145–160. Springer, Heidelberg (2002). https://doi.org/10.1007/3-540-47867-1_11
20. Dash, S.: Exponential lower bounds on the lengths of some classes of branch-and-cut proofs. Math. Oper. Res. **30**(3), 678–700 (2005)
21. Dash, S.: On the complexity of cutting-plane proofs using split cuts. Oper. Res. Lett. **38**(2), 109–114 (2010)
22. Dey, S.S., Iroume, A., Molinaro, M.: Some lower bounds on sparse outer approximations of polytopes. Oper. Res. Lett. **43**(3), 323–328 (2015)
23. Dey, S.S., Molinaro, M., Wang, Q.: Approximating polyhedra with sparse inequalities. Math. Program. **154**(1–2), 329–352 (2015)
24. Dey, S.S., Molinaro, M., Wang, Q.: Analysis of sparse cutting planes for sparse MILPs with applications to stochastic MILPs. Math. Oper. Res. **43**(1), 304–332 (2018)
25. Eisenbrand, F., Schulz, A.S.: Bounds on the Chvátal rank of polytopes in the 0/1-cube. Combinatorica **23**(2), 245–261 (2003)
26. Eldersveld, S.K., Saunders, M.A.: A block-LU update for large-scale linear programming. SIAM J. Matrix Anal. Appl. **13**(1), 191–201 (1992)
27. Goerdt, A.: Cutting plane versus Frege proof systems. In: Börger, E., Kleine Büning, H., Richter, M.M., Schönfeld, W. (eds.) CSL 1990. LNCS, vol. 533, pp. 174–194. Springer, Heidelberg (1991). https://doi.org/10.1007/3-540-54487-9_59

28. Goerdt, A.: The cutting plane proof system with bounded degree of falsity. In: Börger, E., Jäger, G., Kleine Büning, H., Richter, M.M. (eds.) CSL 1991. LNCS, vol. 626, pp. 119–133. Springer, Heidelberg (1992). https://doi.org/10.1007/BFb0023762

29. Grigoriev, D., Hirsch, E.A., Pasechnik, D.V.: Complexity of semi-algebraic proofs. In: Alt, H., Ferreira, A. (eds.) STACS 2002. LNCS, vol. 2285, pp. 419–430. Springer, Heidelberg (2002). https://doi.org/10.1007/3-540-45841-7_34

30. Grötschel, M., Lovász, L., Schrijver, A.: Geometric Algorithms and Combinatorial Optimization, Algorithms and Combinatorics: Study and Research Texts, vol. 2. Springer-Verlag, Berlin (1988). https://doi.org/10.1007/978-3-642-78240-4

31. Impagliazzo, R., Pitassi, T., Urquhart, A.: Upper and lower bounds for tree-like cutting planes proofs. In: Proceedings Ninth Annual IEEE Symposium on Logic in Computer Science, pp. 220–228. IEEE (1994)

32. Jeroslow, R.G.: Trivial integer programs unsolvable by branch-and-bound. Math. Program. **6**(1), 105–109 (1974). https://doi.org/10.1007/BF01580225, https://doi.org/10.1007/BF01580225

33. Karamanov, M., Cornuéjols, G.: Branching on general disjunctions. Math. Program. **128**(1–2), 403–436 (2011)

34. Krajíček, J.: Discretely ordered modules as a first-order extension of the cutting planes proof system. J. Symbolic Logic **63**(4), 1582–1596 (1998)

35. Lodi, A.: Mixed integer programming computation. In: Jünger, M., Liebling, T.M., Naddef, D., Nemhauser, G.L., Pulleyblank, W.R., Reinelt, G., Rinaldi, G., Wolsey, L.A. (eds.) 50 Years of Integer Programming 1958-2008, pp. 619–645. Springer, Heidelberg (2010). https://doi.org/10.1007/978-3-540-68279-0_16

36. Mahajan, A., Ralphs, T.K.: Experiments with branching using general disjunctions. In: Operations Research and Cyber-Infrastructure, pp. 101–118. Springer (2009). https://doi.org/10.1007/978-0-387-88843-9_6

37. Mahmoud, H., Chinneck, J.W.: Achieving MILP feasibility quickly using general disjunctions. Comput. Oper. Res. **40**(8), 2094–2102 (2013)

38. Ostrowski, J., Linderoth, J., Rossi, F., Smriglio, S.: Constraint orbital branching. In: Lodi, A., Panconesi, A., Rinaldi, G. (eds.) IPCO 2008. LNCS, vol. 5035, pp. 225–239. Springer, Heidelberg (2008). https://doi.org/10.1007/978-3-540-68891-4_16

39. Owen, J.H., Mehrotra, S.: Experimental results on using general disjunctions in branch-and-bound for general-integer linear programs. Comput. Optim. Appl. **20**(2), 159–170 (2001)

40. Pataki, G., Tural, M., Wong, E.B.: Basis reduction and the complexity of branch-and-bound. In: Proceedings of the Twenty-First Annual ACM-SIAM Symposium on Discrete Algorithms, pp. 1254–1261. SIAM (2010)

41. Pudlák, P.: Lower bounds for resolution and cutting plane proofs and monotone computations. J. Symbolic Logic **62**(3), 981–998 (1997)

42. Pudlák, P.: On the complexity of the propositional calculus. London Mathematical Society Lecture Note Series, pp. 197–218 (1999)

43. Razborov, A.A.: On the width of semialgebraic proofs and algorithms. Math. Oper. Res. **42**(4), 1106–1134 (2017)

44. Reid, J.K.: A sparsity-exploiting variant of the Bartels-Golub decomposition for linear programming bases. Math. Program. **24**(1), 55–69 (1982). https://doi.org/10.1007/BF01585094

45. Rothvoß, T., Sanitá, L.: 0/1 polytopes with quadratic Chvátal rank. In: Goemans, M., Correa, J. (eds.) IPCO 2013. LNCS, vol. 7801, pp. 349–361. Springer, Heidelberg (2013). https://doi.org/10.1007/978-3-642-36694-9_30

46. Schrijver, A.: Theory of Linear and Integer Programming. John Wiley and Sons, New York (1986)
47. Sperner, E.: Ein satz über untermengen einer endlichen menge. Math. Z. **27**(1), 544–548 (1928)
48. Venderbei,   R.J.:   https://vanderbei.princeton.edu/tex/talks/IDA_CCR/SparsityMatters.pdf (2017)
49. Watson, G.N.: A note on gamma functions. Edinburgh Math. Notes **42**, 7–9 (1959)

Face Dimensions of General-Purpose Cutting Planes for Mixed-Integer Linear Programs

Matthias Walter[✉]

Department of Applied Mathematics, University of Twente,
Enschede, The Netherlands
m.walter@utwente.nl

Abstract. Cutting planes are a key ingredient to successfully solve mixed-integer linear programs. For specific problems, their strength is often theoretically assessed by showing that they are facet-defining for the corresponding mixed-integer hull. In this paper we experimentally investigate the dimensions of faces induced by general-purpose cutting planes generated by a state-of-the-art solver. Therefore, we relate the dimension of each cutting plane to its impact in a branch-and-bound algorithm.

1 Introduction

We consider the mixed-integer program

$$\max c^\intercal x \tag{1a}$$
$$\text{s.t.} \quad Ax \le b \tag{1b}$$
$$x_i \in \mathbb{Z} \quad \forall i \in I \tag{1c}$$

for a matrix $A \in \mathbb{R}^{m \times n}$, vectors $c \in \mathbb{R}^n$ and $b \in \mathbb{R}^m$ and a subset $I \subseteq \{1, 2, \dots, n\}$ of integer variables. Let $P := \text{conv}\{x \in \mathbb{R}^n : x \text{ satisfies (1b) and (1c)}\}$ denote the corresponding *mixed-integer hull*. A *cutting plane* for (1) is an inequality $a^\intercal x \le \beta$ that is valid for P (and possibly invalid for some point computed during branch-and-cut). Such a valid inequality *induces the face* $F := \{x \in P : a^\intercal x = \beta\}$ of P. For more background on polyhedra we refer to [12].

To tackle specific problems, one often tries to find *facet-defining* inequalities, which are inequalities whose induced face F satisfies $\dim(F) = \dim(P) - 1$. This is justified by the fact that any system $Cx \le d$ with $P = \{x : Cx \le d\}$ has to contain an inequality that induces F. Since the dimension of a face can vary between -1 (no $x \in P$ satisfies $a^\intercal x = \beta$) and $\dim(P)$ (all $x \in P$ satisfy $a^\intercal x = \beta$), the following hypothesis is a reasonable generalization of facetness as a strength indicator.

Hypothesis 1. *The practical strength of an inequality correlates with the dimension of its induced face.*

© Springer Nature Switzerland AG 2021
M. Singh and D. P. Williamson (Eds.): IPCO 2021, LNCS 12707, pp. 399–412, 2021.
https://doi.org/10.1007/978-3-030-73879-2_28

There is no unified notion of the practical strength of an inequality, but we will later define one that is related to its impact in a branch-and-bound algorithm. The main goal of this paper is to computationally test this hypothesis for general-purpose cutting planes used in MIP solvers.

Outline. In Sect. 2 we present the algorithm we used to compute dimensions. Sect. 3 is dedicated to the score we used to assess a cutting plane's impact, and in Sect. 4 we present our findings, in particular regarding Hypothesis 1.

2   Computing the Dimension of a Face

This section is concerned about how to effectively compute the dimension of P or of one of its faces F induced by some valid inequality $a^\mathsf{T} x \leq \beta$. In our experiments we will consider instances with several hundreds variables, and hence the enumeration of all vertices of P or F is typically impossible. Instead, we use an oracle-based approach. An *optimization oracle* for P is a black-box subroutine that can solve any linear program over the polyhedron P, i.e., for any given $w \in \mathbb{R}^n$ it can solve

$$\max w^\mathsf{T} x \text{ s.t. } x \in P. \tag{2}$$

In case (2) is feasible and bounded, the oracle shall return an optimal solution, and in case it is feasible and unbounded, it shall return an unbounded direction. Note that for such an oracle we neither require all vertices of P nor all valid (irredundant) inequalities. Indeed, we can use any MIP solver and apply it to problem (1) with $c := w$. We now discuss how one can compute $\dim(P)$ by only accessing an optimization oracle for P. The basic algorithm is known (see Lemma 6.5.3 in [4]), but we provide a slightly improved version that requires at most $2n$ oracle queries (in contrast to the bound of $3n$ for the cited one). In fact, one can prove that there is no oracle-based algorithm that always requires less than $2n$ [13].

To keep the presentation simple we assume that P is bounded, although our implementation can handle unbounded polyhedra as well. The algorithm maintains a set $X \subseteq P$ of affinely independent points and a system $Dx = e$ of valid equations, where D has full row-rank r. Hence, $\dim(X) \leq \dim(P) \leq n - r$ holds throughout and the algorithm works by either increasing $\dim(X)$ or r in every iteration. The details are provided in Algorithm 1.

Proposition 1. *For an optimization oracle for a non-empty polytope $P \subseteq \mathbb{R}^n$, Algorithm 1 requires $2n$ oracle queries to compute a set $X \subseteq P$ with $|X| = \dim(P) + 1$ and a system $Dx = e$ of $n - \dim(P)$ equations satisfying*

$$\mathrm{aff}(P) = \mathrm{aff}(X) = \{x \in \mathbb{R}^n : Dx = e\}.$$

Proof. Since every point $x \in X$ was computed by the optimization oracle, we have $X \subseteq P$. Moreover, $Dx = e$ is valid for P since for each equation $d^\mathsf{T} x = \gamma$, $\min\{d^\mathsf{T} x : x \in P\} = \gamma = \max\{d^\mathsf{T} x : x \in P\}$ holds. We now prove by induction on the number of iterations that in every iteration of the algorithm, the points $x \in X$ are affinely independent and that the rows of D are linearly independent.

Initially, this invariant holds because $X = \emptyset$ and D has no rows. Whenever an equation $d^\mathsf{T} x = d^\mathsf{T} x^+$ is added to $Dx = e$ in step 7, the vector d is linearly independent to the rows of D (see step 3). If in step 10, X is initialized with a single point, it is clearly affinely independent. Similarly, the two points in step 14 also form an affinely independent set since $d^\mathsf{T} x^+ > d^\mathsf{T} x^-$ implies $x^+ \neq x^-$. Suppose X is augmented by $\bar{x} := x^+$ in step 19 or by $\bar{x} := x^-$ in step 22. Note that by the choice of d in step 3, $d^\mathsf{T} x = \gamma$ holds for all $x \in X$. Due to $d^\mathsf{T} \bar{x} \neq \gamma$, $X \cup \{\bar{x}\}$ remains affinely independent.

After the first iteration, we have $|X| + |\operatorname{rows}(D)| = 2$, and in every further iteration, this quantity increases by 1. Hence, the algorithm requires n iterations, each of which performs 2 oracle queries. The result follows.   □

Algorithm 1: Affine hull of a polytope via optimization oracle

Input: Optimization oracle for a polytope $\emptyset \neq P \subseteq \mathbb{R}^n$.
Output: Affine basis X of $\operatorname{aff}(P)$, non-redundant system $Dx = e$ with
$\operatorname{aff}(P) = \{x \in \mathbb{R}^n : Dx = e\}$.

1 Initialize $X := \emptyset$ and $Dx = e$ empty.
2 **while** $|X| + 1 < n - |\operatorname{rows}(D)|$ **do**
3 $\quad$ Compute a direction vector $d := \operatorname{aff}(X)^\perp \setminus \operatorname{span}(\operatorname{rows}(D))$.
4 $\quad$ Query the oracle to maximize $d^\mathsf{T} x$ over $x \in P$ and let
$\quad\quad x^+ := \arg\max\{d^\mathsf{T} x : x \in P\}$.
5 $\quad$ Query the oracle to maximize $-d^\mathsf{T} x$ over $x \in P$ and let
$\quad\quad x^- := \arg\min\{d^\mathsf{T} x : x \in P\}$.
6 $\quad$ **if** $d^\mathsf{T} x^+ = d^\mathsf{T} x^-$ **then**
7 $\quad\quad$ Add equation $d^\mathsf{T} x = d^\mathsf{T} x^+$ to system $Dx = e$.
8 $\quad\quad$ **if** $X = \emptyset$ **then**
9 $\quad\quad\quad$ set $X := \{x^+\}$.
10 $\quad\quad$ **end**
11 $\quad$ **end**
12 $\quad$ **else if** $X = \emptyset$ **then**
13 $\quad\quad$ set $X := \{x^+, x^-\}$.
14 $\quad$ **end**
15 $\quad$ **else**
16 $\quad\quad$ Let $\gamma := d^\mathsf{T} x$ for some $x \in X$.
17 $\quad\quad$ **if** $d^\mathsf{T} x^+ \neq \gamma$ **then**
18 $\quad\quad\quad$ set $X := X \cup \{x^+\}$.
19 $\quad\quad$ **end**
20 $\quad\quad$ **else**
21 $\quad\quad\quad$ set $X := X \cup \{x^-\}$.
22 $\quad\quad$ **end**
23 $\quad$ **end**
24 **end**
25 **return** $(X, Dx = e)$.

Implementation Details. We now describe some details of the implementation within the software framework IPO [14]. In the description of Algorithm 1 we did not specify step 3 precisely. While one may enforce the requirement $d \notin \text{span}(\text{rows}(D))$ via $d \in \text{rows}(D)^{\perp}$, it turned out that this is numerically less stable than first computing a basis of $\text{aff}(X)^{\perp}$ and selecting a basis element d that is not in the span of D's rows. Moreover, we can take d's sparsity and other numerical properties into account. Sparsity can speed-up the overall computation since very sparse objective vectors are sometimes easier to optimize for a MIP solver. In theory, for all $x \in X$, the products $d^{\mathsf{T}}x$ have the same value. However, due to floating-point arithmetic the computed values may differ. It turned out that preferring directions d for which the range of these products is small helps to avoid numerical difficulties.

In our application we compute the dimension of P and of several of its faces. We can exploit this by caching all points $x \in P$ returned by an optimization oracle in a set $\bar{X}$. Then, for each face F induced by an inequality $a^{\mathsf{T}}x \leq \beta$ we can then compute the set $\bar{X}_F := \{x \in \bar{X} : a^{\mathsf{T}}x = \beta\}$. Before querying the oracle we can then check the set $\bar{X}_F$ for a point with sufficiently large objective value, which saves two calls to the MIP solvers.

Let $\bar{D}x = \bar{e}$ be the equation system returned by Algorithm 1 for P. Now, for a face F induced by inequality $a^{\mathsf{T}}x \leq \beta$ we can initialize $Dx = e$ in the algorithm by $\bar{D}x = \bar{e}$. Moreover, if $a^{\mathsf{T}}x = \beta$ is not implied by $Dx = e$ we add this equation to $Dx = e$ as well.

Since the algorithm is implemented in floating-point arithmetic, errors can occur which may lead to wrong dimension results. We checked the results using an exact arithmetic implementation of Algorithm 1 for easier instances (with less than 200 variables). For these, the computed dimension varied by at most 2 from the true dimension. We conclude that the relative dimension errors are sufficiently small.

In a first implementation, our code frequently reported dimension -1, and it turned out that often the right-hand side was only slightly larger than needed to make the inequality supporting. Thus, for each inequality $a^{\mathsf{T}}x \leq \beta$, normalized to $||a||_2 = 1$, we computed $\beta^{\text{true}} := \max\{a^{\mathsf{T}}x : x \in P\}$ by a single oracle query. Whenever we observed $\beta < \beta^{\text{true}} - 10^{-4}$, we considered the cut $a^{\mathsf{T}}x \leq \beta$ as invalid (indicated by the symbol ⚡). If $\beta > \beta^{\text{true}} + 10^{-4}$, we declare the cut to be non-supporting. Otherwise, we replace β by β^{true} before running Algorithm 1.

3   Measuring the Strength of a Single Inequality

In this section we introduce our *cut impact measure* for indicating, for a given cutting plane $a^{\mathsf{T}}x \leq \beta$, how useful its addition in the context of branch-and-cut is. The main goal of solving an LP at a branch-and-bound node is to determine a dual bound of the current subproblem. If this bound is at most the value of the best feasible solution known so far, then the subproblem can be discarded. Thus, we consider the value of such a bound (after adding a certain cut) in relation to the problem's optimum $z^{\star} := \max\{c^{\mathsf{T}}x : Ax \leq b, \ x_i \in \mathbb{Z} \ \forall i \in I\}$ and the value $z^{\text{LP}} := \max\{c^{\mathsf{T}}x : Ax \leq b\}$ of the LP relaxation.

Our first approach was to just evaluate the dual bound obtained from the LP relaxation $Ax \leq b$ augmented by $a^\mathsf{T}x \leq \beta$. However, adding a single inequality often does not cut off the optimal face of the LP relaxation, which means that the bound does not change. In a second attempt we tried to evaluate the dual bound of the LP relaxation augmented by a random selection of k cutting planes. However, the variance of the resulting cut impact measure was very large even after averaging over 10.000 such selections.

As a consequence, we discarded cut impact measures based on the combined impact of several cutting planes. Instead we carried out the following steps for given cutting planes $a_1^\mathsf{T}x \leq \beta_1$, $a_2^\mathsf{T}x \leq \beta_2$, ..., $a_k^\mathsf{T}x \leq \beta_k$:

1. Compute the optimum $z^\star$ and optimum solution $x^\star \in P$.
2. Compute z^{LP}.
3. For $i = 0, 1, 2, \ldots, k$, solve

$$\max c^\mathsf{T}x \text{ s.t. } Ax \leq b, \; x_j \in \mathbb{Z} \; \forall j \in I \text{ and } a_i^\mathsf{T}x \leq \beta_i \text{ if } i \geq 1$$

with $x^\star$ as an initial incumbent, with presolve, cutting planes and heuristics disabled and where a time limit of 60 s is enforced.
4. Let N denote the minimum number of branch-and-bound nodes used in any these $k + 1$ runs.
5. For $i = 0, 1, \ldots, k$, let z^i be the dual bound obtained in run i when stopping after N branch-and-bound nodes. For $i \geq 1$, the *closed gap of cut i* is defined as $(z^i - z^{\mathrm{LP}})/(z^\star - z^{\mathrm{LP}})$. For $i = 0$, this yields the *closed gap without cuts*.

Remarks. The closed gap is essentially the dual bound, normalized such that a value of 0 means no bound improvement over the root LP bound without cuts and a value of 1 means that the instance was solved to optimality. The effective limit of N branch-and-bound nodes was introduced such that all runs reach this limit. This circumvents the question of how to compare runs in which the problem was solved to optimality with those that could not solve it. For all runs, presolve and domain propagation were disabled due to our focus on the branch-and-bound algorithm itself. To avoid interaction with heuristics, the latter were disabled, but an optimal solution $x^\star$ was provided.

We are aware that it is debatable how meaningful this cut measure is with respect to the actual importance of a single inequality in a practical setting. However, we think that it is highly nontrivial to design a cut measure that is meaningful, robust and computable with reasonable effort. Therefore, we see our proposed measure as one contribution and hope that further research leads to the development of more measures, which in turn can lead to more reliable statements.

4   Computational Study

In order to test Hypothesis 1 we considered the 65 instances from the MIPLIB 3 [10]. We did not use the more recent MIPLIB 2017 [11] because we had to restrict ourselves to problems with a decent number of variables. For each of them we computed the dimension of the mixed-integer hull P, imposing a time limit of 10 min[1]. Moreover, we ran the state-of-the-art solver SCIP [3] and collected all cutting planes generated in the root node, including those that were discarded by SCIP's cut selection routine[2]. For each of the cuts we computed the dimension of its induced face (see Sect. 2) as well as the closed gap after processing N (as defined in Sect. 3) branch-and-bound nodes.

While for instances air03, air05, nw04, no cuts were generated by SCIP, our implementation of Algorithm 1 ran into numerical difficulties during the computation of dim(P) for set1ch. Moreover, dim(P) could not be computed within 10 min for instances air04, arki001, cap6000, dano3mip, danoint, dsbmip, fast0507, gesa2_0, gesa3_0, l152lav, mas74, misc06, mitre, mkc, mod010, mod011, pk1, pp08aCUTS, pp08a, qnet1, rentacar, rout and swath.

We evaluated the remaining 37 instances, whose characteristic data is shown in Table 1. We distinguish how many cuts SCIP found by which cut separation method, in particular to investigate whether certain separation routines tend to generate cuts with low- or high-dimensional faces.

We verified some of the invalid cutting planes manually, i.e., checked that these cuts are generated by SCIP and that there exists a feasible solution that is indeed cut off. The most likely reason for their occurrence is that SCIP performed dual reductions although we disabled them via corresponding parameters[3].

Since the results on face dimensions and cut strength turned out to be very instance-specific, we created one plot per instance. We omit the ones for p2756 (too many failures, see Table 1), blend2 (only 7 cuts analyzed), enigma (dim(P) = 3 is very small), and for markshare1, markshare2 and noswot (all cuts were ineffective). Moreover, we present several plots in the Appendix A since these are similar to those of other instances.

The plots show the dimension of the cuts (horizontal axis, rounded to 19 groups) together with their closed gap (vertical axis, 14 groups) according to Sect. 3. Each circle corresponds to a nonempty set of cuts, where the segments depict the respective cut classes (see Table 1) and their color depicts the number k of cuts, where the largest occurring number L is specified in the caption. The colors are red ($0.9L < k \leq L$), orange ($0.7L < k \leq 0.9L$), yellow ($0.5L < k \leq 0.7L$), green ($0.3L < k \leq 0.5L$), turquoise ($0.1L < k \leq 0.3L$) and blue ($1 \leq k < 0.1L$). For instance, the circle for fixnet6 containing a red and a turquoise segment subsumes cutting planes with face dimensions between 730 and 777, and closed gap of approximately 0.45. As the legend next to the plot indicates, this circle represents $k \in [109, 121]$ c-MIR cuts and $k' \in [13, 36]$ multicommodity flow cuts. The dashed horizontal line indicates the closed gap without cuts (see Sect. 3).

---

[1] All experiments were carried out on a single core of an Intel Core i3 CPU running at 2.10 GHz with 8 GB RAM.

[2] We disabled presolve, domain propagation, dual reductions, symmetry, and restarts.

[3] We set misc/allowweakdualreds and misc/allowstrongdualreds to false.

Table 1. Characteristics of the relevant 37 instances with number of successfully analyzed cuts, failures (numerical problems 0/0, timeouts ◔, invalid cuts ϟ), dimension of the mixed-integer hull, and number N of branch-and-bound nodes (see Sect. 3). ◐ – Lifted extended weight inequalities [9,15,16] ◕ – Complemented Mixed-Integer Rounding (c-MIR) inequalities [8,16] ◔ – {0,1/2}-Chvátal-Gomory inequalities [2,6] ◓ – Strengthened Chvátal-Gomory inequalities [7] ◒ – Lifted flow-cover inequalities [5] ◑ – Multi-commodity flow inequalities [1]

Instance	Cuts total	Analyzed by class						Failed			Dim. of P	B& B nodes
		◐	◕	◔	◓	◒	◑	0/0	◔	ϟ		
bell3a	25		25									121	53075
bell5	53		36				1		11		5	97	7815
blend2	8		7								1	245	597
dcmulti	172		137						14		21	467	773
egout	125		97		1				24	3		41	1
enigma	82		59	1	17	5						3	1
fiber	533	127	139	7	4	9	207		33	6	1	946	23395
fixnet6	772		612					83	42		35	779	53900
flugpl	42		42									9	1753
gen	28	17	8						3			540	21
gesa2	470	16	419	2	6				27			1176	21114
gesa3	259		194	2	6	15			38	4		1104	729
gt2	143		92	3	39	8				1		188	1
harp2	1028	659	210	9	4	112			5	28	1	1300	13748
khb05250	122		78						5		39	1229	425
lseu	125	35	85	2	3							89	3495
markshare1	107		104		3							50	318260
markshare2	84		79		3					2		60	286347
mas76	225		204							1	20	151	211381
misc03	634	3	283	33	311					4		116	13
misc07	808		286	34	440				36	12		204	22839
mod008	441	119	272							3	47	319	1158
modglob	268		186						3	1	78	327	112516
noswot	164		151		11				2			120	160344
p0033	94	14	40		10				30			27	127
p0201	263	12	152	14	78				7			139	2
p0282	904	368	441	6	15	1			4		69	282	57
p0548	577	159	213	10	16	62			117			520	921
p2756	1000	82	96	22	48	55			26	671		2716	15149
qiu	63		51						2	10		709	10406
qnet1	162		95	10	47				6	4		1233	5
rgn	278		168		65				45			160	1691
seymour	6246		656	53	5528				9	0		1255	9
stein27	886		517	9	360							27	3673
stein45	1613		1221	10	382							45	45371
vpm1	281		129		32				117		3	288	27881
vpm2	353		251	1	30				63		8	286	172271

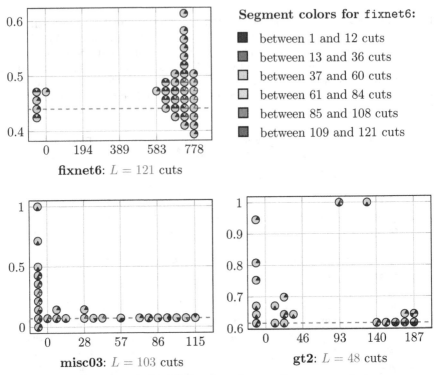

fixnet6: $L = 121$ cuts

Segment colors for `fixnet6`:

- ■ between 1 and 12 cuts
- ■ between 13 and 36 cuts
- □ between 37 and 60 cuts
- □ between 61 and 84 cuts
- ■ between 85 and 108 cuts
- ■ between 109 and 121 cuts

misc03: $L = 103$ cuts

gt2: $L = 48$ cuts

The first three plots already highlight that the results are very heterogeneous: while the faces of the strongest cuts in `fixnet6` have a high dimension, the strongest ones for `misc03` are not even supporting. Even when considering non-supporting cuts as outliers, the dimension does not indicate practical strength, as the plot for `gt2` shows. A quick look at the other plots lets us conclude that Hypothesis 1 is false—at least for the strength measure from Sect. 3.

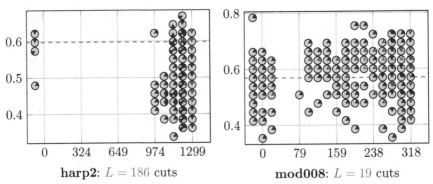

harp2: $L = 186$ cuts

mod008: $L = 19$ cuts

The dashed line for `harp2` shows that adding a single cut does not necessarily help in branch-and-bound, which may be due to side-effects such as different branching decisions. For some instances, such as `mod008`, the cuts' face dimensions are well distributed. In contrast to this, some instances exhibit only very few distinct dimension values, e.g., only non-supporting cuts for `qiu`. Interestingly, the 6246 cuts for `seymour` induce only empty faces as well as faces with dimensions between 1218 and 1254. This can partially be explained via the cut classes. On the one hand, all generated strengthened Chvátal-Gomory cuts are non-supporting. On the other hand, some of the c-MIR cuts and $\{0, 1/2\}$-cuts are non-supporting while others induce faces of very high dimension.

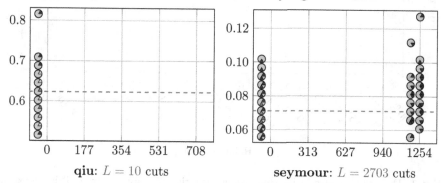

qiu: $L = 10$ cuts   seymour: $L = 2703$ cuts

In general we do not see an indication that cuts from certain classes induce higher dimensional faces than others. At first glance, such a pattern is apparent for `misc07` (at dimensions 130–160), however one has to keep in mind that these blue segments constitute a minority of the cuts. In line with that, the majority of the cuts for `fiber` is concentrated around dimension 900 with a closed gap similar to that without cuts.

Despite the heterogeneity of the results, one observation is common to many instances: the distribution of the face dimensions is biased towards -1 and high-dimensions, i.e., not many cuts inducing low dimensional faces are generated.

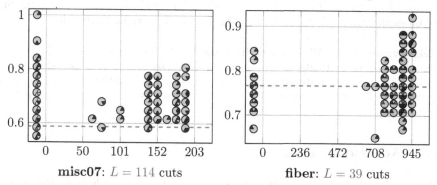

misc07: $L = 114$ cuts   fiber: $L = 39$ cuts

A corresponding histogram is depicted in Fig. 1. We conjecture that the high dimensions occur because lifting and strengthening techniques for cutting planes are quite evolved.

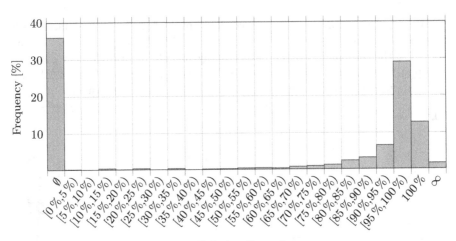

Fig. 1. Distribution of relative dimensions over the 37 instances from Table 1 except for p2756 (avoiding a bias due to many failures dimension computations). A cut of dimension k in an instance having ℓ cuts and with $\dim(P) = d$ contributes $1/(36\ell)$ to its bar. For $k = -1$ (resp. $k = d$), this is $\emptyset$ (resp. ∞), and it is $k/(d-1)$ otherwise.

The first bar in Fig. 1 indicates that for an instance chosen uniformly at random among the ones we considered and then a randomly chosen cut for this instance, this cut is non-supporting with probability greater than 35%. This is remarkably high and thus we conclude this paper by proposing to investigate means to (heuristically) test for such a situation, with the goal of strengthening a non-supporting cutting plane by a reduction of its right-hand side.

Acknowledgments. We thank R. Hoeksma and M. Uetz as well as the SCIP development team, in particular A. Gleixner, C. Hojny and M. Pfetsch for valuable suggestions on the computational experiments and their presentation. Finally, we thank three anonymous referees for raising interesting discussion points, especially regarding Sect. 3.

A   Additional Plots

Here we provide additional instance-specific plots. This underlines the conclusions drawn in Sect. 4 and allows inspection of results for instances with certain characteristics (see Table 1).

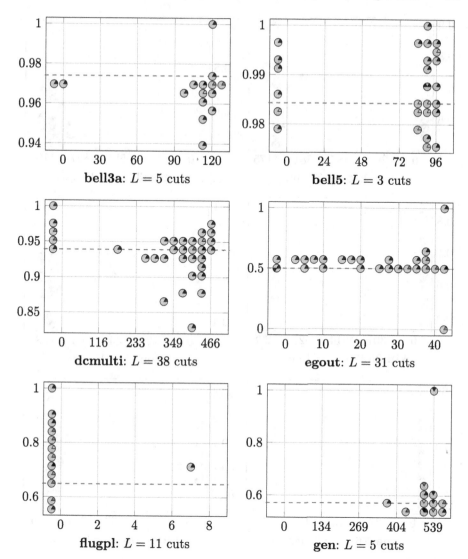

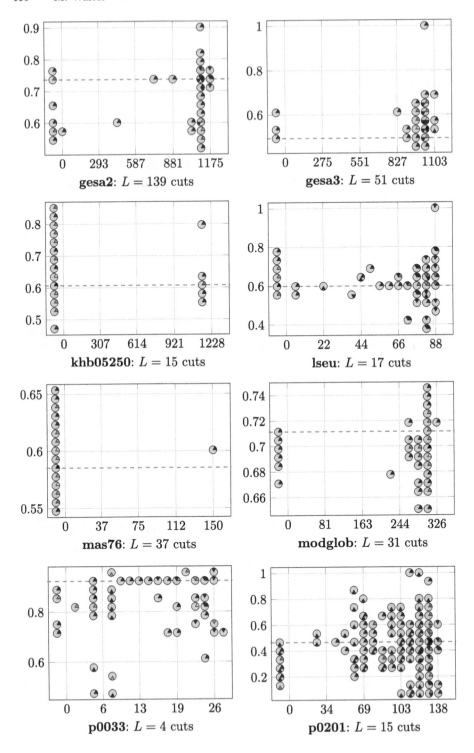

gesa2: $L = 139$ cuts

gesa3: $L = 51$ cuts

khb05250: $L = 15$ cuts

lseu: $L = 17$ cuts

mas76: $L = 37$ cuts

modglob: $L = 31$ cuts

p0033: $L = 4$ cuts

p0201: $L = 15$ cuts

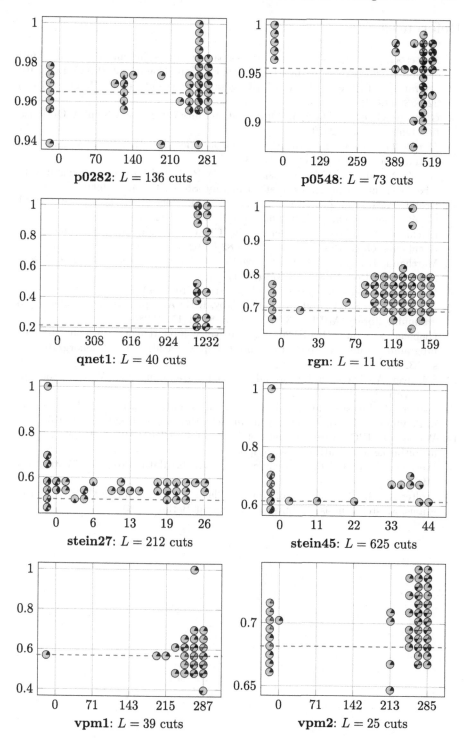

References

1. Achterberg, T., Raack, C.: The MCF-separator: detecting and exploiting multi-commodity flow structures in MIPs. Math. Program. Comput. **2**(2), 125–165 (2010)
2. Caprara, A., Fischetti, M.: {0, 1/2}-Chvátal-Gomory cuts. Math. Program. **74**(3), 221–235 (1996)
3. Gamrath, G., et al.: The SCIP Optimization Suite 7.0. Technical report, Optimization Online, March 2020
4. Martin, G., Lovász, L., Schrijver, A.: Geometric Algorithms and Combinatorial Optimizations. Springer-Verlag (1993)
5. Gu, Z., Nemhauser, G.L., Savelsbergh, M.W.P.: Lifted flow cover inequalities for mixed 0–1 integer programs. Math. Program. **85**(3), 439–467 (1999)
6. Koster, A.M.C.A., Zymolka, A., Kutschka, M.: Algorithms to separate $\{0, \frac{1}{2}\}$-Chvátal-Gomory cuts. Algorithmica **55**(2), 375–391 (2009)
7. Letchford, A.N., Lodi, A.: Strengthening Chvátal-Gomory cuts and Gomory fractional cuts. Oper. Res. Lett. **30**(2), 74–82 (2002)
8. Marchand, H., Wolsey, L.A.: Aggregation and mixed integer rounding to solve MIPs. Oper. Res. **49**(3), 363–371 (2001)
9. Martin, A.: Integer programs with block structure (1999)
10. Mczeal, C.M., Savelsbergh, M.W.P., Bixby, R.E.: An updated mixed integer programming library: MIPLIB 3.0. Optima, **58**, 12–15 (1998)
11. Gleixner, A., et al.: MIPLIB 2017: data-driven compilation of the 6th mixed-integer programming library. Math. Program. Comput. (2020). https://doi.org/10.1007/s12532-020-00194-3
12. Schrijver, A.: Theory of Linear and Integer Programming. John Wiley & Sons Inc, New York, NY, USA (1986)
13. Walter, M.: Investigating Polyhedra by Oracles and Analyzing Simple Extensions of Polytopes. PhD thesis, Otto-von-Guericke-Universität Magdeburg (2016)
14. Walter, M.: IPO - Investigating Polyhedra by Oracles (2016). Software available at: bitbucket.org/matthias-walter/ipo/
15. Weismantel, R.: On the 0/1 knapsack polytope. Math. Program. **77**(3), 49–68 (1997)
16. Wolter, K.: Implementation of Cutting Plane Separators for Mixed Integer Programs. Master's thesis, Technische Universität Berlin (2006)

Proximity Bounds for Random Integer Programs

Marcel Celaya[1]([⊠]) and Martin Henk[2]

[1] ETH Zurich Institute for Operations Research Department
of Mathematics Rämistrasse, 101 8092 Zurich, Switzerland
marcel.celaya@ifor.math.ethz.ch
[2] Technische Universität Berlin Institut für Mathematik,
Sekr. MA4-1 Straße des 17 Juni 136, 10623 Berlin, Germany
henk@math.tu-berlin.de

Abstract. We study proximity bounds within a natural model of random integer programs of the type $\max c^\top x : Ax = b, x \in \mathbb{Z}_{\geq 0}$, where $A \in \mathbb{Z}^{m \times n}$ is of rank m, $b \in \mathbb{Z}^m$ and $c \in \mathbb{Z}^n$. In particular, we seek bounds for proximity in terms of the parameter $\Delta(A)$, which is the square root of the determinant of the Gram matrix $AA^\top$ of A. We prove that, up to constants depending on n and m, the proximity is "generally" bounded by $\Delta(A)^{1/(n-m)}$, which is significantly better than the best deterministic bounds which are, again up to dimension constants, linear in $\Delta(A)$.

1 Introduction

Given an linear program of the form

$$\max \ c^\top x \ : \ Ax = b \tag{1}$$
$$x \geq 0,$$

where A is a full-row-rank $m \times n$ integral matrix, $b \in \mathbb{Z}^m$, and $c \in \mathbb{Z}^n$, the proximity problem seeks to understand how far away an optimal vertex x^* of the feasible region can be to a nearby feasible integer solution z^*. Typically it is further required that z^* is itself optimal; we do not impose this requirement in this manuscript. Assuming the feasible region has at least one such integral point, bounds for proximity are typically given in terms of the largest possible absolute value $\Delta_m(A)$ of any $m \times m$ subdeterminant of A. Note that this parameter is within a factor of $\binom{n}{m}$ of $\Delta(A) := \sqrt{\det(AA^\top)}$. This is a well-studied problem which goes back to the classic Cook et al. result [4] bounding the proximity of the dual of (1). See, for instance, the recent works of Eisenbrand and Weismantel [5] and of Aliev, Henk, and Oertel [1] and the references therein.

The first author was funded by the Deutsche Forschungsgemeinschaft (DFG, German Research Foundation) under Germany's Excellence Strategy—The Berlin Mathematics Research Center MATH+(EXC-2046/1, project ID: 390685689).

M. Singh and D. P. Williamson (Eds.): IPCO 2021, LNCS 12707, pp. 413–426, 2021.
https://doi.org/10.1007/978-3-030-73879-2_29

In this manuscript, we would like to understand the worst-possible proximity, which we denote by $\mathrm{dist}(A)$, over all choices of b and c, when the matrix A is chosen *randomly*. The model of randomness we consider is the following: we choose the matrix A up to left-multiplication by unimodular matrices, and we choose A uniformly at random subject to the condition that the greatest common divisor of the maximal minors of A is 1, and that $\Delta(A)$ is at most some sufficiently large (with respect to m and n) integer T. This is a natural model to study from a geometric point of view, since $\Delta(A)$ is the determinant of the lattice of integer points in the kernel of A. This is also the model considered by Aliev and Henk in [2], in their investigation of diagonal Frobenius numbers.

Our main result concerns not $\mathrm{dist}(A)$ but rather a related random variable we denote by $\mathrm{dist}^*(A)$. This is an asymptotic version of $\mathrm{dist}(A)$ that further imposes some mild restrictions on b. Our main result is that it satisfies the following Markov-type inequality:

$$\mathbf{P}\left(\mathrm{dist}^*(A) > t\Delta(A)^{1/(n-m)}\right) \ll t^{-2/3}. \tag{2}$$

Here $\ll$ means less than, up to constants which only depend on n and m. In particular, this shows that proximity generally depends only on $\Delta^{1/(n-m)}$ in our random setting, for "almost all" choices of b in a certain precise sense. This is significantly better than the linear dependency on Δ_m in the deterministic case, that is known to be tight [1, Theorem 1]. A similar result, with a slightly different random model, was obtained in [1] the so-called knapsack scenario, where $m = 1$. We also mention recent work of Oertel, Paat, and Weismantel in [8], which considers a random model that allows b to vary but keeps A fixed.

The proof of this result combines ideas of [2] and [1] using facts from the geometry of numbers, some results of Schmidt from [9] on random sublattices of $\mathbb{Z}^n$ of fixed dimension, and computations of the measure of certain distinguished regions of the real Grassmannian $\mathrm{Gr}(d, n)$ of d-dimensional subspaces of $\mathbb{R}^n$, where $d = n - m$. The idea is two-fold. First, we use the results of Schmidt to relate the discrete measure in our model to the continuous $O(n)$-invariant probability measure ν of $\mathrm{Gr}(d, n)$, where $O(n)$ here denotes the group of orthogonal $n \times n$ real matrices. We show that there are essentially two distinct "bad" regions of $\mathrm{Gr}(d, n)$, both parameterized by t, in which $\mathrm{dist}^*(A)$ could be large, but whose measure with respect to ν gets smaller as t gets larger.

We remark that the exponent of $-2/3$ is mainly an artifact of the proof, and we expect that it can be further improved. The problem of finding an inequality analogous to (2) for $\mathrm{dist}(A)$ is more challenging and remains open, as the polyhedral combinatorics of (1) may interfere with our analysis.

2   Main Result and Notation

2.1   Notation

Throughout this manuscript we assume fixed positive integers d, m, n such that $n = m+d$. For a subset $\sigma \subseteq [n]$ and $x \in \mathbb{R}^n$, we let x_σ denote the vector obtained by orthogonally projecting x onto the coordinates indexed by σ. Similarly, if A is a matrix, then we denote by A_σ the submatrix of A whose columns are those indexed by σ. In particular, if $k \in [n]$ then A_k denotes the corresponding column of A. If A_σ is an invertible square matrix we say σ is a *basis* of A. We denote the complement of σ by $\bar{\sigma} := [n] \setminus \sigma$. Given a d-dimensional subspace $L \subseteq \mathbb{R}^n$, the m-dimensional orthogonal complement of L is denoted by $L^\perp$. If $\Lambda \subset \mathbb{R}^n$, let $\Lambda_\mathbb{R}$ denote the linear subspace of $\mathbb{R}^n$ spanned by Λ. We say $\sigma \subseteq [n]$ is a *coordinate basis* of Λ or $\Lambda_\mathbb{R}$ if the coordinate projection map

$$\Lambda_\mathbb{R} \to \mathbb{R}^\sigma$$
$$x \mapsto x_\sigma$$

is an isomorphism. This is equivalent to saying that σ is a basis of A for any full-row-rank matrix A such that $\ker(A) = \Lambda_\mathbb{R}$,

2.2 Definition of dist(A)

Let $A \in \mathbb{Z}^{m \times n}$ be a full-row-rank matrix. For a basis σ of A, we define the semigroup

$$S_\sigma(A) := \left\{ x \geq 0 : x_\sigma = A_\sigma^{-1} A g,\ x_{\bar{\sigma}} = 0,\ \mathbf{g} \in \mathbb{Z}^n \right\}. \tag{3}$$

For a vector $b \in \mathbb{Z}^m$, we define the polyhedron

$$\mathcal{P}(A, b) := \{ x \in \mathbb{R}^n : Ax = b,\ x \geq 0 \}.$$

The idea behind these definitions is that if $x^* \in S_\sigma(A)$, then $b := Ax^*$ is an integral vector, and $\mathcal{P}(A, b)$ is a nonempty polyhedron containing x^* as a vertex. Now given a basis σ of A and $x^* \in S_\sigma(A)$, we define the distance

$$\text{dist}(A, \sigma, x^*) := \inf_{z^* \in \mathbb{Z}^n \cap \mathcal{P}(A,b)} \|x^* - z^*\|_2.$$

where $b := Ax^*$. We then define the worst-case distance over all choices of bases σ of A and elements $x^* \in S_\sigma(A)$ as

$$\text{dist}(A) := \sup_\sigma \sup_{x^*} \text{dist}(A, \sigma, x^*). \tag{4}$$

This definition has the disadvantage that it is stated in terms of the matrix A. Since we may replace $Ax = b$ with $UAx = Ub$ for any $m \times m$ integral matrix U, it is not so clear from this formulation how to define our random model. This motivates an alternative, more geometric definition of dist(A) which we now state.

2.3 Definition of dist(Λ)

Suppose instead we start with a d-dimensional sublattice Λ of $\mathbb{Z}^n$. Suppose σ is a coordinate basis of Λ. Then we may define the semigroup

$$S_\sigma(\Lambda) := \{ x \geq 0 : x \in \Lambda_\mathbb{R} + \mathbf{g},\ x_{\bar{\sigma}} = 0,\ \mathbf{g} \in \mathbb{Z}^n \}. \tag{5}$$

For $\boldsymbol{x}^* \in \mathcal{S}_\sigma(\Lambda)$, define the distance

$$\operatorname{dist}(\Lambda, \sigma, \boldsymbol{x}^*) := \sup_{\boldsymbol{g} \in (\Lambda_{\mathbb{R}} + \boldsymbol{x}^*) \cap \mathbb{Z}^n} \inf_{\boldsymbol{z}^* \in (\Lambda + \boldsymbol{g}) \cap \mathbb{R}^n_{\geqslant 0}} \|\boldsymbol{x}^* - \boldsymbol{z}^*\|_2. \tag{6}$$

The extra sup accounts for the fact that, if Λ is not primitive, then there are multiple ways to embed Λ into $\Lambda_{\mathbb{R}} + \boldsymbol{x}^*$ as an integral translate of Λ. Finally, define the worst case distance

$$\operatorname{dist}(\Lambda) := \sup_\sigma \sup_{\boldsymbol{x}^*} \operatorname{dist}(\Lambda, \sigma, \boldsymbol{x}^*), \tag{7}$$

where the supremum is taken over all coordinate bases of Λ and elements $\boldsymbol{x}^* \in \mathcal{S}_\sigma(\Lambda)$.

We now explain the relationship between definitions (4) and (7). First note that if $\boldsymbol{A}$ is any integral matrix such that $\Lambda_{\mathbb{R}} = \ker(\boldsymbol{A})$, then the two definitions (3) and (5) of $\mathcal{S}_\sigma(\boldsymbol{A})$ and $\mathcal{S}_\sigma(\Lambda)$ coincide. Moreover, if Λ is a *primitive* lattice, that is, if $\Lambda = \Lambda_{\mathbb{R}} \cap \mathbb{Z}^n$, then we have

$$\operatorname{dist}(\Lambda, \sigma, \boldsymbol{x}^*) = \operatorname{dist}(\boldsymbol{A}, \sigma, \boldsymbol{x}^*)$$

and therefore

$$\operatorname{dist}(\Lambda) = \operatorname{dist}(\boldsymbol{A}).$$

Definition (7) also makes sense when Λ is non-primitive, however, and it is immediate from the definitions that in general,

$$\operatorname{dist}(\Lambda) \geq \operatorname{dist}(\Lambda_{\mathbb{R}} \cap \mathbb{Z}^n).$$

The key advantage of definition (7) is that there are only finitely many d-dimensional sublattices of $\mathbb{Z}^n$ whose determinant is at most some fixed positive integer T. Thus, we may consider the uniform distribution over these bounded-determinant lattices.

2.4   An Asymptotic Version of dist(Λ)

We next consider a modification of dist (Λ). Choose any full-row-rank matrix $\boldsymbol{A}$ such that $\ker(\boldsymbol{A}) = \Lambda_{\mathbb{R}}$, the particular choice of $\boldsymbol{A}$ is not important. Let $B_2^n \subset \mathbb{R}^n$ denote the n-dimensional Euclidean ball of radius 1. Define the vector $\boldsymbol{w} \in \mathbb{R}^n$ as follows: for each $i \in [n]$, set

$$\boldsymbol{w}_i := \sqrt{1 - \boldsymbol{A}_i^\top (\boldsymbol{A}\boldsymbol{A}^\top)^{-1} \boldsymbol{A}_i}. \tag{8}$$

This vector $\boldsymbol{w}$ measures, for each $i \in [n]$, the largest possible value of $\boldsymbol{x}_i$ for any $\boldsymbol{x} \in B_2^n \cap \Lambda_{\mathbb{R}}$. Denote by $\mu := \mu(\Lambda, B_2^n)$ the covering radius of B_2^n with respect to Λ. That is,

$$\mu := \inf\{t > 0 : \Lambda + t B_2^n \text{ contains } \Lambda_{\mathbb{R}}\}.$$

If σ is a basis of $\boldsymbol{A}$ then define the following subsemigroup of $\mathcal{S}_\sigma(\Lambda)$:

$$\mathcal{S}_\sigma^*(\Lambda) := \left\{\boldsymbol{x} \in \mathcal{S}_\sigma(\Lambda) : \boldsymbol{x}_\sigma \geq \mu\boldsymbol{w}_\sigma + \boldsymbol{A}_\sigma^{-1}\boldsymbol{A}_{\bar\sigma}\boldsymbol{w}_{\bar\sigma}\right\}.$$

The next proposition shows that if we further restrict $\boldsymbol{x}^*$ so that it can only lie in $\mathcal{S}_\sigma^*(\Lambda)$, then we can guarantee that $\mathcal{P}(\boldsymbol{A}, \boldsymbol{b})$ contains an integral point reasonably close to $\boldsymbol{x}^*$. We prove it in Sect. 5.

Proposition 1. *For a basis σ of $\boldsymbol{A}$ and $\boldsymbol{x}^* \in \mathcal{S}_\sigma^*(\Lambda)$, let $\boldsymbol{b} = \boldsymbol{A}\boldsymbol{x}^*$. Then $\mathcal{P}(\boldsymbol{A}, \boldsymbol{b})$ contains a translate of the scaled ball $\mu \cdot (B_2^n \cap \Lambda_\mathbb{R})$, which in turn contains an integral vector.*

Now set

$$\text{dist}^*(\Lambda) := \sup_\sigma \sup_{\boldsymbol{x}^*} \text{dist}(\Lambda, \sigma, \boldsymbol{x}^*), \tag{9}$$

where the supremum is taken over all bases σ of $\boldsymbol{A}$ and elements $\boldsymbol{x}^*$ of the semigroup $\mathcal{S}_\sigma^*(\Lambda)$.

2.5 Main Result

We are now ready to state the main theorem.

Theorem 1. *For $T \gg 1$, let Λ be a sublattice of $\mathbb{Z}^n$ of dimension d and determinant at most T, chosen uniformly at random. Then for all $t > 1$,*

$$\mathbf{P}\left(\text{dist}^*(\Lambda) > t\left(\Delta(\Lambda)\right)^{1/d}\right) \ll t^{-2/3}.$$

What we would like to do is translate this statement into a statement about integer programs, and in particular derive inequality (2). For this we use a known result on the ratio between primitive sublattices and all sublattices with a fixed determinant upper bound, a consequence of Theorems 1 and 2 in [9]:

Lemma 1. *Suppose there are exactly $N(d, n, T)$ d-dimensional sublattices of $\mathbb{Z}^n$ with determinant at most T, of which exactly $P(d, n, T)$ are primitive. Then*

$$\lim_{T \to \infty} \frac{P(d, n, T)}{N(d, n, T)} = \frac{1}{\zeta(d+1) \cdots \zeta(n)},$$

where $\zeta(\cdot)$ denotes the Riemann zeta function.

Recall from the introduction our probability model. We start with a sufficiently large integer T relative to m and n, and consider the set of all $m \times n$ integral matrices $\boldsymbol{A}$ such that the greatest common divisor of all maximal minors of $\boldsymbol{A}$ equals 1, and that $\Delta(\boldsymbol{A}) \leq T$. The group of $m \times m$ unimodular matrices acts on this set of matrices by mulitplication on the left, and there are finitely many orbits of this action. We consider the uniform distribution on these orbits. We define

$$\text{dist}^*(\boldsymbol{A}) := \text{dist}^*(\ker(\boldsymbol{A}) \cap \mathbb{Z}^n).$$

Note that this definition depends not on $\boldsymbol{A}$ but only on the orbit of $\boldsymbol{A}$. The greatest common divisor condition ensures that $\Delta(\boldsymbol{A})$ equals the determinant of the lattice $\ker(\boldsymbol{A}) \cap \mathbb{Z}^n$. We derive the next corollary by combining Theorem 1, Lemma 1, and the simple conditional probability inequality $\mathbf{P}(E \mid F) \leq \mathbf{P}(E)/\mathbf{P}(F)$, where E is the event that $\text{dist}^*(\Lambda) > t\left(\Delta(\Lambda)\right)^{1/d}$ and F is the event that Λ is primitive.

Corollary 1. *For $T \gg 1$, choose $\boldsymbol{A}$ randomly as above, with determinant at most T. Then for all $t > 1$,*

$$\mathbf{P}\left(\mathrm{dist}^*\left(\boldsymbol{A}\right) > t\left(\Delta\left(\boldsymbol{A}\right)\right)^{1/d}\right) \ll t^{-2/3}.$$

We remark that the question of deriving the constants in this bound remains unexplored.

3   A Theorem of Schmidt

In this section we state a result that is fundamental to the proof, which follows from the results of Schmidt in [9]. We continue with our assumption that $d = n - m$. Let $\mathrm{Gr}\,(d, n)$ denote the set of d-dimensional subspaces of $\mathbb{R}^n$. Let ν denote the unique $O(n)$-invariant probability measure on the real Grassmannian $\mathrm{Gr}\,(d, n)$.

Definition 1 ([9, p. 40]). *A subset $\xi \subset \mathrm{Gr}\,(d, n)$ is Jordan measurable if for all $\varepsilon > 0$ there exists continuous functions $f_1 \leq \mathbf{1}_\xi \leq f_2$ such that*

$$\int (f_2 - f_1)\, d\nu < \varepsilon.$$

Here $\mathbf{1}_\xi$ denotes the indicator function of ξ.

In the next definition, $\lambda_i\,(\Lambda)$ denotes the ith successive minimum of the d-dimensional Euclidean ball of radius 1 with respect to the lattice Λ.

Definition 2. *Let $\boldsymbol{a} = (a_1, \ldots, a_d) \in \mathbb{R}^d$, with each $a_i \geq 1$. Let T be a positive integer, and let $\xi \subset \mathrm{Gr}\,(d, n)$. Then we define $G\,(\boldsymbol{a}, \xi, T)$ to be the set of sublattices Λ of $\mathbb{Z}^n$ of dimension d with determinant at most T, such that*

$$\frac{\lambda_{i+1}\,(\Lambda)}{\lambda_i\,(\Lambda)} \geq a_i \text{for all} i = 1, 2, \ldots, d - 1,$$

and $\Lambda_{\mathbb{R}} \in \xi$.

The result of Schmidt that we intend to use is a combination of Theorems 3 and 5 in [9]:

Theorem 2. *Assuming $\xi \subset \mathrm{Gr}\,(d, n)$ is Jordan measurable, we have*

$$|G\,(\boldsymbol{a}, \xi, T)| \asymp \left(\prod_{i=1}^{d-1} a_i^{-i(d-i)}\right) \nu\,(\xi)\, T^n,$$

where $f \asymp g$ means $f \ll g$ and $g \ll f$.

Let $G(d, n, T)$ denote the set of all sublattices of $\mathbb{Z}^n$ of dimension d with determinant at most T. Let $\mathbf{P} = \mathbf{P}_{d,n,T}$ denote the uniform probability distribution over $G(d, n, T)$.

Corollary 2. *For $t > 1$, we have*

$$\mathbf{P}\left(\max_{i \in [d-1]} \left\{\frac{\lambda_{i+1}(\Lambda)}{\lambda_i(\Lambda)}\right\} \geq t\right) \ll (d-1)\, t^{-(d-1)}.$$

Proof. Following Aliev and Henk in [2], let

$$\boldsymbol{\delta}_i(t) := \left(1, \ldots, 1, \underset{i}{t}, 1, \ldots, 1\right)^{\top} \in \mathbb{R}^d.$$

Applying the union bound to Theorem 2, this probability is at most

$$\sum_{i=1}^{d-1} \frac{|G\left(\boldsymbol{\delta}_i(t), \operatorname{Gr}(d,n), T\right)|}{|G\left(\boldsymbol{\delta}_i(1), \operatorname{Gr}(d,n), T\right)|} \ll \sum_{i=1}^{d-1} t^{-i(d-i)} \leq (d-1)\, t^{-(d-1)}.$$

4 Typical Cramer's Rule Ratios

We see in the next section that the proximity can be bounded from above by an expression involving the largest absolute value of the entries of $A_\sigma^{-1} A_{\bar{\sigma}}$, as σ ranges over all bases of A, and A is chosen randomly. Hence, we would like to show that that the largest absolute value of any entry of the matrix $A_\sigma^{-1} A_{\bar{\sigma}}$ is typically not too large, where for our purposes the subspace $L := \ker A$ is chosen uniformly at random from $\operatorname{Gr}(d,n)$. Note that the matrix $A_\sigma^{-1} A_{\bar{\sigma}}$ depends only on L and σ. We remark that the entries of the matrix $A_\sigma^{-1} A_{\bar{\sigma}}$ are explicitly computed using Cramer's rule: for $i \in \sigma$ and $j \notin \sigma$, we have

$$\left(A_\sigma^{-1} A_j\right)_i = \frac{\det\left(A_{\sigma-i+j}\right)}{\det\left(A_\sigma\right)}.$$

As before, we let $\nu : \mathscr{G} \to [0,1]$ denote the $O(n)$-invariant probability measure on $\operatorname{Gr}(d,n)$. The precise statement we show is the following: Fix $\sigma \subseteq [n]$, $i \in \sigma$, $j \in [n] \setminus \sigma$. Then, as a function of a parameter $s > 1$, we have

$$\nu\left(\ker(A) : A_\sigma \text{ is nonsingular}, \left|\left(A_\sigma^{-1} A_j\right)_i\right| > s\right) = \frac{2}{\pi s} + \mathcal{O}\left(s^{-3}\right). \tag{10}$$

The proof proceeds in the three subsections below. First, we get a handle on ν by relating it to another probability distribution, namely the Gaussian distribution γ on the matrix space $\mathbb{R}^{m \times n}$, where the entries are i.i.d. normally distributed with mean 0 and variance 1. This is done via the kernel map, which is introduced in Subsect. 4.1 and related to γ in Subsect. 4.2. Equation 10 is then derived in Subsect. 4.3.

4.1 The Real Grassmannian

For a general introduction to matrix groups and Grassmannians, we refer the reader to [3]. There is a right action of the orthogonal group $O(n)$ on $\operatorname{Gr}(d,n)$ defined as follows: if $\ker(A) \in \operatorname{Gr}(d,n)$, where $A \in \mathbb{R}^{m \times n}$, then

$$(\ker(A)) \cdot U = \ker(AU). \tag{11}$$

This is well-defined, since if $\ker(\boldsymbol{A}) = \ker(\boldsymbol{A}')$ for some $\boldsymbol{A}' \in \mathbb{R}^{m \times n}$, then $\boldsymbol{A} = \boldsymbol{D}\boldsymbol{A}'$ for some invertible $m \times m$ matrix $\boldsymbol{D}$, and hence

$$\ker(\boldsymbol{A}\boldsymbol{U}) = \ker(\boldsymbol{D}\boldsymbol{A}'\boldsymbol{U}) = \ker(\boldsymbol{A}'\boldsymbol{U}).$$

Let $\mathrm{St}^{m \times n} := \{\boldsymbol{A} \in \mathbb{R}^{m \times n} : \mathrm{rank}(\boldsymbol{A}) = m\}$. Call this the *Stiefel manifold*. Again, there is a right action of $O(n)$ on $\mathrm{St}^{m \times n}$ which in this case is simply right multiplication:

$$\boldsymbol{A} \cdot \boldsymbol{U} = \boldsymbol{A}\boldsymbol{U}.$$

The only thing to check here is that $\boldsymbol{A}\boldsymbol{U}$ indeed lies in $\mathrm{St}^{m \times n}$, but this is indeed the case since

$$\boldsymbol{A}\boldsymbol{U}(\boldsymbol{A}\boldsymbol{U})^\top = \boldsymbol{A}\boldsymbol{U}\boldsymbol{U}^\top \boldsymbol{A}^\top = \boldsymbol{A}\boldsymbol{A}^\top,$$

thus $\boldsymbol{A}$ and $\boldsymbol{A}\boldsymbol{U}$ have the same Gram matrix $\boldsymbol{A}\boldsymbol{A}^\top$, and an $m \times n$ matrix has full-row-rank if and only if its Gram matrix does.

The kernel map gives rise to a surjective map

$$\ker : \mathrm{St}^{m \times n} \to \mathrm{Gr}(d, n)$$
$$\boldsymbol{A} \mapsto \ker(\boldsymbol{A})$$

Thus, we see from (11) that the following statement holds:

Proposition 2. *The map* $\ker : \mathrm{St}^{m \times n} \to \mathrm{Gr}(d, n)$ *is equivariant with respect to the right actions of* $O(n)$ *on* $\mathrm{St}^{m \times n}$ *and* $\mathrm{Gr}(d, n)$; *that is,* $(\ker(\boldsymbol{A})) \cdot \boldsymbol{U} = \ker(\boldsymbol{A} \cdot \boldsymbol{U})$.

4.2 Probability Spaces

Consider the probability space $(\mathbb{R}^{m \times n}, \mathscr{B}(\mathbb{R}^{m \times n}), \gamma)$ where $\mathscr{B}(\mathbb{R}^{m \times n})$ is the Borel σ-algebra, and the measure γ is defined so that each $\boldsymbol{A} \in \mathbb{R}^{m \times n}$ has iid $N(0, 1)$ entries. In other words, γ is the standard Gaussian probability measure on the mn-dimensional real vector space $\mathbb{R}^{m \times n}$ with mean zero and identity covariance matrix. By restricting to $\mathrm{St}^{m \times n}$, we get the probability space $(\mathrm{St}^{m \times n}, \mathscr{B}(\mathrm{St}^{m \times n}), \gamma)$. We can do this because $\mathbb{R}^{m \times n} \backslash \mathrm{St}^{m \times n}$ is an algebraic hypersurface in $\mathbb{R}^{m \times n}$, and therefore has measure zero with respect to γ. Let $\mathscr{B} := \mathscr{B}(\mathrm{St}^{m \times n})$.

The Grassmannian $\mathrm{Gr}(d, n)$ is endowed with the topology where $E \subseteq \mathrm{Gr}(d, n)$ is open iff $\ker^{-1}(E)$ is open in $\mathrm{St}^{m \times n}$. Let $\mathscr{G}$ denote the associated Borel σ-algebra. The measure ν on $\mathrm{Gr}(d, n)$ is characterized as follows:

Proposition 3 ([7, Corollary 3.1.3]). *The measure ν is the unique measure on* $\mathrm{Gr}(d, n)$ *satisfying*

$$\nu(E \cdot \boldsymbol{U}) = \nu(E) \ \text{ for all } E \in \mathscr{G} \text{ and } \boldsymbol{U} \in O(n) \tag{12}$$
$$\nu(\mathrm{Gr}(d, n)) = 1.$$

The map $\ker : \mathrm{St}^{m \times n} \to \mathrm{Gr}\,(d, n)$ thus defines a map of probability spaces:

$$\ker : \left(\mathrm{St}^{m \times n}, \mathscr{B}, \gamma\right) \to \left(\mathrm{Gr}\,(d, n), \mathscr{G}, \nu\right).$$

Proposition 4. *The measure ν is the pushforward measure of γ under this map. That is, $\nu(E) = \gamma(\ker^{-1}(E))$ for each $E \in \mathscr{G}$.*

Proof. We establish the conditions of (12). By surjectivity, and the fact that γ is a probability measure, we have

$$\gamma(\ker^{-1}(\mathrm{Gr}\,(d, n))) = \gamma\left(\mathrm{St}^{m \times n}\right) = 1.$$

It therefore remains to show $\gamma(\ker^{-1}(E \cdot U)) = \gamma(\ker^{-1}(E))$ for each $E \in \mathscr{G}$ and $U \in O(n)$. By Proposition 2, we have

$$\ker^{-1}(E \cdot U) = \ker^{-1}(E) \cdot U. \tag{13}$$

Now, $\mathbb{R}^{m \times n}$ has the inner product $\langle A, B \rangle = \mathrm{trace}\left(A B^{\top}\right)$. With respect to this inner product we may consider the subgroup $O\,(m \times n)$ of $\mathrm{GL}\,(\mathbb{R}^{m \times n})$ which is given by

$$O\,(m \times n) := \left\{\varphi \in \mathrm{GL}\,\left(\mathbb{R}^{m \times n}\right) : \langle \varphi\,(A), \varphi\,(B)\rangle = \langle A, B\rangle\right\}.$$

Observe that, for a fixed $U \in O\,(n)$, the linear map $\varphi_U \in \mathrm{GL}\,(\mathbb{R}^{m \times n})$ given by

$$\varphi_U\,(A) = A U \tag{14}$$

lies in $O\,(m \times n)$, since

$$\langle \varphi\,(A), \varphi\,(B)\rangle = \mathrm{trace}\left(A U\,(B U)^{\top}\right) = \mathrm{trace}\left(A B^{\top}\right) = \langle A, B\rangle.$$

Now the probability measure γ on $\mathbb{R}^{m \times n}$ is defined so that the coordinates $A_{i,j}$ of a randomly chosen $A \in \mathbb{R}^{m \times n}$ are iid $N(0, 1)$ normally distributed. In particular this measure is invariant under isometry, in that for all $\mathcal{K} \in \mathscr{B}\,(\mathbb{R}^{m \times n})$ and $\varphi \in O\,(m \times n)$, we have

$$\gamma\,(\varphi\,(\mathcal{K})) = \gamma\,(\mathcal{K}). \tag{15}$$

The same is therefore true for the restricted probability measure γ on $\mathrm{St}^{m \times n}$. It follows that if $U \in O(n)$ and $E \in \mathscr{G}$, then, using (13), (14), and (15), we have

$$\gamma\left(\ker^{-1}(E \cdot U)\right) = \gamma\left(\ker^{-1}(E) \cdot U\right) = \gamma\left(\varphi_U\left(\ker^{-1}(E)\right)\right) = \gamma\left(\ker^{-1}(E)\right).$$

4.3 Cramer's Rule Ratios

Let $\sigma \subset [n]$ of size m, and define

$$\mathrm{St}^{m \times n}_{\sigma} := \left\{A \in \mathrm{St}^{m \times n} : A_{\sigma} \text{ is nonsingular}\right\}.$$
$$\mathrm{Gr}\,(d, n)_{\sigma} := \left\{\ker\,(A) \in \mathrm{Gr}\,(d, n) : A_{\sigma} \text{ is nonsingular}\right\}.$$

Note that $\gamma\left(\mathrm{St}^{m \times n}_{\sigma}\right) = \nu\left(\mathrm{Gr}\,(d, n)_{\sigma}\right) = 1$. Also define, for $s > 1$, $i \in \sigma$, and $j \notin \sigma$,

$$\xi_{\sigma, i, j}\,(s) := \left\{\ker\,(A) \in \mathrm{Gr}\,(d, n)_{\sigma} : \left|\left(A_{\sigma}^{-1} A_j\right)_i\right| > s\right\}.$$

Proposition 5. *For $s > 1$ and σ, i, j as above, we have*

$$\nu\left(\xi_{\sigma,i,j}\left(s\right)\right) = \frac{2}{\pi s} + \mathcal{O}\left(s^{-3}\right).$$

Proof. Let $\boldsymbol{A}$ be a random element of $\mathrm{St}_{\sigma}^{m \times n}$, and let H denote the (random) hyperplane spanned by the columns of $\boldsymbol{A}_{\sigma \setminus \{i\}}$, and let ℓ denote the line perpendicular to H. Let $\boldsymbol{u}_{\ell}$ denote the unit normal vector to H whose first nonzero coordinate is positive. Thus,

$$\ell = \mathbb{R}\boldsymbol{u}_{\ell} = \{\lambda \boldsymbol{u}_{\ell} : \lambda \in \mathbb{R}\}.$$

Let $\alpha \in \{-1, +1\}$ denote the sign of the first nonzero entry of $\boldsymbol{e}_i^{\top} \boldsymbol{A}_{\sigma}^{-1}$. Then we can write

$$\boldsymbol{u}_{\ell}^{\top} = \frac{\alpha \boldsymbol{e}_i^{\top} \boldsymbol{A}_{\sigma}^{-1}}{\left\|\boldsymbol{e}_i^{\top} \boldsymbol{A}_{\sigma}^{-1}\right\|_2},$$

since for all $k \in \sigma \setminus \{i\}$ we have

$$\alpha \boldsymbol{e}_i^{\top} \boldsymbol{A}_{\sigma}^{-1} \boldsymbol{A}_k = \alpha \boldsymbol{e}_i^{\top} \boldsymbol{A}_{\sigma}^{-1} \boldsymbol{A}_{\sigma} \boldsymbol{e}_k = 0,$$

and $\alpha \boldsymbol{e}_i^{\top} \boldsymbol{A}_{\sigma}^{-1}$ has first nonzero component positive by definition of α.

Now let k be any element of $[n]$ outside of $\sigma \setminus \{i\}$. Since $\boldsymbol{u}_{\ell}$ depends only on $\boldsymbol{A}_{\sigma \setminus \{i\}}$, and the entries of $\boldsymbol{A}$ are mutually independent, we have that $\boldsymbol{u}_{\ell}$ and $\boldsymbol{A}_k$ are independent random vectors. Now, for any fixed unit vector $\boldsymbol{v} \in \mathbb{S}^{n-1}$, as $\boldsymbol{A}_k$ has $N(0,1)$ iid entries, then the dot product $\boldsymbol{v}^{\top} \boldsymbol{A}_k$ also has distribution $N(0,1)$. Thus, for any fixed $t \in \mathbb{R}$, the random variable

$$\gamma\left(\boldsymbol{u}_{\ell}^{\top} \boldsymbol{A}_k \le t \mid \ell\right)$$

(i.e. the conditional probability in terms of the σ-algebra generated by ℓ) is in fact constant. Evaluating at the line $\ell = \mathbb{R}\boldsymbol{e}_1$, for example, this constant is given by

$$\gamma\left(\boldsymbol{A}_{1,k} \le t\right).$$

This shows that the random quantity $\boldsymbol{u}_{\ell}^{\top} \boldsymbol{A}_k$ has distribution $N(0,1)$. We have

$$\left(\boldsymbol{A}_{\sigma}^{-1} \boldsymbol{A}_j\right)_i = \frac{\boldsymbol{e}_i^{\top} \boldsymbol{A}_{\sigma}^{-1} \boldsymbol{A}_j}{\boldsymbol{e}_i^{\top} \boldsymbol{A}_{\sigma}^{-1} \boldsymbol{A}_i} = \frac{\boldsymbol{u}_{\ell}^{\top} \boldsymbol{A}_j}{\boldsymbol{u}_{\ell}^{\top} \boldsymbol{A}_i}.$$

The independence of $\boldsymbol{u}_{\ell}^{\top} \boldsymbol{A}_i$ and $\boldsymbol{u}_{\ell}^{\top} \boldsymbol{A}_j$ imply that $\left(\boldsymbol{A}_{\sigma}^{-1} \boldsymbol{A}_j\right)_i$ has the Cauchy distribution, that is, the ratio of two iid $N(0,1)$ random variables. In particular, the cdf of $\left(\boldsymbol{A}_{\sigma}^{-1} \boldsymbol{A}_j\right)_i$ is given by

$$\gamma\left(\left(\boldsymbol{A}_{\sigma}^{-1} \boldsymbol{A}_j\right)_i \le t\right) = \frac{1}{\pi} \arctan(t) + \frac{1}{2}.$$

See [6, p. 50] for more on the Cauchy distribution. Using the series expansion

$$\arctan\left(t\right) = \frac{\pi}{2} - \frac{1}{t} + \frac{1}{3t^3} - \frac{1}{5t^5} + \cdots,$$

we get

$$\gamma\left(\left(A_\sigma^{-1}A_j\right)_i \le t\right) = 1 - \left(\frac{1}{\pi t} - \frac{1}{3\pi t^3} + \frac{1}{5\pi t^5} - \cdots\right).$$

Hence, using Proposition 4 and the fact $s > 1$, we conclude

$$\begin{aligned}
\nu\left(\xi_{\sigma,i,j}\left(s\right)\right) &= \gamma\left(\left|\left(A_\sigma^{-1}A_j\right)_i\right| > s\right) \\
&= 2 \cdot \gamma\left(\left(A_\sigma^{-1}A_j\right)_i > s\right) \\
&= 2\left(1 - \gamma\left(\left(A_\sigma^{-1}A_j\right)_i \le s\right)\right) \\
&= 2\left(\frac{1}{\pi s} - \frac{1}{3\pi s^3} + \frac{1}{5\pi s^5} - \cdots\right) \\
&= \frac{2}{\pi s} + \mathcal{O}\left(s^{-3}\right).
\end{aligned}$$

5   Proof of Main Result

In this final section we prove the main result of this paper, Theorem 1.

Definition 3. *Define the constant*

$$\tilde{\omega}_d := \frac{\omega_d^{1/d}}{d},$$

where ω_d denotes the volume of the d-dimensional Euclidean ball of radius 1. This constant $\tilde{\omega}_d$ is of the order $d^{-3/2}$.

Definition 4. *Assume $\Lambda_{\mathbb{R}} = \ker(A)$. Given positive real numbers s and u, we say Λ is (σ, s, u)-controlled if σ is a basis of A and:*

1. *The largest entry of $A_\sigma^{-1}A_{\bar{\sigma}}$ is at most s, and*
2. *The successive minima ratios of Λ are not too large: we have*

$$\frac{\lambda_{i+1}\left(\Lambda\right)}{\lambda_i\left(\Lambda\right)} < \left(\tilde{\omega}_d u\right)^{2/(d-1)}$$

for all $i = 1, 2, \ldots, d-1$.

Lemma 2 ([2, **Proof of Lemma 5.2**]). *Suppose that*

$$\frac{\lambda_{i+1}\left(\Lambda\right)}{\lambda_i\left(\Lambda\right)} < \left(\tilde{\omega}_d u\right)^{2/(d-1)}$$

for all $i = 1, 2, \ldots, d-1$. Then

$$\mu < u\left(\Delta\left(\Lambda\right)\right)^{1/d}.$$

Lemma 3. *If σ is a basis of A and Λ is (σ, s, u)-controlled, then for all $x^* \in S_\sigma\left(\Lambda\right)$ we have*

$$\mathrm{dist}\left(\Lambda, \sigma, x^*\right) \le 2n^{3/2}su\left(\Delta\left(\Lambda\right)\right)^{1/d}.$$

Proof. Let $\boldsymbol{b} = \boldsymbol{A}\boldsymbol{x}^*$, let $B = B_2^n \cap \Lambda_{\mathbb{R}}$, and let μ denote the covering radius of B with respect to Λ. Define the vector $\boldsymbol{v} \in \mathbb{R}^n$ so that:

$$\boldsymbol{v}_j = \mu \boldsymbol{w}_j \text{ for all } j \in \bar{\sigma}$$
$$\boldsymbol{A}\boldsymbol{v} = \boldsymbol{b}.$$

We show that the scaled, translated ball $\mu B + \boldsymbol{v}$ is contained in $\mathcal{P}(\boldsymbol{A}, \boldsymbol{b})$. Since $B \subseteq \Lambda_{\mathbb{R}}$, we have that each $\boldsymbol{x} \in \mu B + \boldsymbol{v}$ satisfies $\boldsymbol{A}\boldsymbol{x} = \boldsymbol{b}$. For each $j \in [n]$, let $\boldsymbol{x}^{(j)}$ be the unique point in $\mu B + \boldsymbol{v}$ such that $\boldsymbol{x}_j^{(j)}$ is minimized. If $j \in \bar{\sigma}$, then

$$\boldsymbol{x}_j^{(j)} = \mu(-\boldsymbol{w}_j) + \boldsymbol{v}_j = \mu(-\boldsymbol{w}_j) + \mu \boldsymbol{w}_j = 0.$$

If $j \in \sigma$, then since $\boldsymbol{x}^* \in \mathcal{S}_\sigma(\Lambda)$ we have

$$\begin{aligned}
\boldsymbol{x}_j^{(j)} &= \mu(-\boldsymbol{w}_j) + \boldsymbol{v}_j \\
&= \mu(-\boldsymbol{w}_j) + \left(\boldsymbol{A}_\sigma^{-1}\boldsymbol{b} - \boldsymbol{A}_\sigma^{-1}\boldsymbol{A}_{\bar{\sigma}}\boldsymbol{w}_{\bar{\sigma}}\right)_j \\
&\geq \mu(-\boldsymbol{w}_j) + \mu \boldsymbol{w}_j \\
&= 0.
\end{aligned}$$

This concludes the proof that $\mu B + \boldsymbol{v} \subseteq \mathcal{P}(\boldsymbol{A}, \boldsymbol{b})$.

Let $\boldsymbol{g} \in (\Lambda_{\mathbb{R}} + \boldsymbol{x}^*) \cap \mathbb{Z}^n$. Since μ is the covering radius of B with respect to Λ, there exists $\boldsymbol{z}^* \in (\Lambda + \boldsymbol{g}) \cap (\mu B + \boldsymbol{v})$ such that

$$\|\boldsymbol{x}^* - \boldsymbol{z}^*\|_2 \leq \|\boldsymbol{x}^* - \boldsymbol{v}\|_2 + \|\boldsymbol{v} - \boldsymbol{z}^*\|_2 \leq \mu \|\tilde{\boldsymbol{w}}\|_2 + \mu. \tag{16}$$

where we define $\tilde{\boldsymbol{w}} := (\boldsymbol{v} - \boldsymbol{x}^*)/\mu$. That is, $\tilde{\boldsymbol{w}}$ satisfies

$$\boldsymbol{A}\tilde{\boldsymbol{w}} = 0$$
$$\tilde{\boldsymbol{w}} = \boldsymbol{w}_j \text{ for all } j \in \bar{\sigma}.$$

Observe that

$$\tilde{\boldsymbol{w}}_\sigma = -\boldsymbol{A}_\sigma^{-1}\boldsymbol{A}_{\bar{\sigma}}\tilde{\boldsymbol{w}}_{\bar{\sigma}}.$$

Using the fact $\boldsymbol{w} \in [0, 1]^n$, we therefore have

$$\begin{aligned}
\|\tilde{\boldsymbol{w}}\|_2^2 &= \|\tilde{\boldsymbol{w}}_\sigma\|_2^2 + \|\tilde{\boldsymbol{w}}_{\bar{\sigma}}\|_2^2 \\
&= \left\|\boldsymbol{A}_\sigma^{-1}\boldsymbol{A}_{\bar{\sigma}}\tilde{\boldsymbol{w}}_{\bar{\sigma}}\right\|_2^2 + \|\tilde{\boldsymbol{w}}_{\bar{\sigma}}\|_2^2 \\
&\leq m \left\|\boldsymbol{A}_\sigma^{-1}\boldsymbol{A}_{\bar{\sigma}}\right\|_\infty^2 \|\tilde{\boldsymbol{w}}_{\bar{\sigma}}\|_1^2 + \|\tilde{\boldsymbol{w}}_{\bar{\sigma}}\|_2^2 \\
&\leq \left(ms^2 + 1\right) d^2.
\end{aligned}$$

Thus we conclude

$$\begin{aligned}
\|\boldsymbol{x}^* - \boldsymbol{z}^*\|_2 &\leq \mu \left(\|\tilde{\boldsymbol{w}}\|_2 + 1\right) \\
&\leq u\Delta^{1/d}\left(\sqrt{(ms^2 + 1)\,d^2} + 1\right) \\
&\leq 2n^{3/2} su\Delta^{1/d}.
\end{aligned}$$

Proof (Proof of Theorem 1). Let Λ be a uniformly chosen lattice from $G\left(d, n, T\right)$. Let $t > 1$, and let $s := t^{2/3}/(2n^{3/2})$ and $u := t^{1/3}$, so that $t = 2n^{3/2}su$ as in Lemma 3. We have

$$\mathbf{P}\left(\text{dist}\left(\Lambda\right) > t\left(\Delta\left(\Lambda\right)\right)^{1/d}\right)$$

$$\leq \sum_{\sigma} \mathbf{P}\left(\sigma \text{ basis of } \boldsymbol{A}, \text{dist}\left(\Lambda, \sigma, \boldsymbol{x}^*\right) > t\left(\Delta\left(\Lambda\right)\right)^{1/d} \text{ for some } \boldsymbol{x}^* \in \mathcal{S}_\sigma\left(\Lambda\right)\right)$$

$$\leq \sum_{\sigma} \mathbf{P}\left(\sigma \text{ basis of } \boldsymbol{A}, \Lambda \text{ is not}(\sigma, s, u)\text{-controlled}\right)$$

where the sums are over all subsets $\sigma \subseteq [n]$ of size m. It therefore suffices to show, for each such σ,

$$\mathbf{P}\left(\sigma \text{ basis of } \boldsymbol{A}, \Lambda \text{ is not } (\sigma, s, u)\text{-controlled}\right) \ll t^{-2/3}.$$

By definition, this probability is at most

$$\mathbf{P}\left(\max_{i \in [d-1]}\left\{\frac{\lambda_{i+1}\left(\Lambda\right)}{\lambda_i\left(\Lambda\right)}\right\} \geq \left(\tilde{\omega}_d u\right)^{2/(d-1)}\right) + \sum_{\substack{i \in \sigma \\ j \notin \sigma}} \mathbf{P}\left(\sigma \text{ basis of } \boldsymbol{A}, \left(\boldsymbol{A}_\sigma^{-1}\boldsymbol{A}_j\right)_i \geq s\right).$$

$$(17)$$

By Theorem 2, we have

$$\mathbf{P}\left(\sigma \text{ basis of } \boldsymbol{A}, \left(\boldsymbol{A}_\sigma^{-1}\boldsymbol{A}_j\right)_i \geq s\right) = \frac{\left|G\left(1, \xi_{\sigma,i,j}\left(s\right), T\right)\right|}{\left|G\left(1, \text{Gr}\left(d, n\right), T\right)\right|} \asymp \nu\left(\xi_{\sigma,i,j}\left(s\right)\right).$$

Hence, applying Corollary 2 and Proposition 5, for T sufficiently large, we may estimate up to constants the quantity (17) by

$$u^{-2} + s^{-1} \ll t^{-2/3}.$$

Acknowledgements. The authors wish to thank the anonymous referees for their helpful comments and suggestions.

References

1. Aliev, I., Henk, M., Oertel, T.: Distances to lattice points in knapsack polyhedra. Math. Program, **182**(1–2, Ser. A), 175–198 (2020)
2. Aliev, I., Henk, M.: Feasibility of integer knapsacks. SIAM J. Optim. **20**(6), 2978–2993 (2010)
3. Baker, A.: Matrix Groups: An Introduction to Lie Group Theory. Springer Undergraduate Mathematics Series. Springer, London (2003)
4. Cook, W., Gerards, A.M.H., Schrijver, A., Tardos, É.: Sensitivity theorems in integer linear programming. Math. Programm. **34**(3), 251–264 (1986)
5. Eisenbrand, F., Weismantel, R.: Proximity results and faster algorithms for integer programming using the Steinitz lemma. ACM Trans. Algorithms **16**(1), 1–14 (2019)
6. Feller, V., Feller, W.: An Introduction to Probability Theory and Its Applications, vol. 1. A Wiley publication in mathematical statistics, Wiley (1968)

7. Krantz, S.G., Parks, H.R.: Geometric Integration Theory. Cornerstones, Birkhäuser Boston (2008)
8. Oertel, T., Paat, J., Weismantel, R.: The distributions of functions related to parametric integer optimization. SIAM J. Appl. Algebra Geometry **4**(3), 422–440 (2020)
9. Schmidt, W.M.: The distribution of sublattices of $\mathbf{Z}^m$. Monatshefte für Mathematik, **125**(1), 37–81 (1998)

On the Integrality Gap of Binary Integer Programs with Gaussian Data

Sander Borst, Daniel Dadush, Sophie Huiberts$^{(\boxtimes)}$, and Samarth Tiwari

Centrum Wiskunde & Informatica (CWI), Amsterdam, The Netherlands
{sander.borst,dadush,s.huiberts,samarth.tiwari}@cwi.nl

Abstract. For a binary integer program (IP) $\max c^{\mathsf{T}}x, Ax \leq b, x \in \{0,1\}^n$, where $A \in \mathbb{R}^{m \times n}$ and $c \in \mathbb{R}^n$ have independent Gaussian entries and the right-hand side $b \in \mathbb{R}^m$ satisfies that its negative coordinates have ℓ_2 norm at most $n/10$, we prove that the gap between the value of the linear programming relaxation and the IP is upper bounded by $\mathrm{poly}(m)(\log n)^2/n$ with probability at least $1 - 1/n^7 - 2^{-\mathrm{poly}(m)}$. Our results give a Gaussian analogue of the classical integrality gap result of Dyer and Frieze (Math. of O.R., 1989) in the case of random packing IPs. In contrast to the packing case, our integrality gap depends only polynomially on m instead of exponentially. By recent breakthrough work of Dey, Dubey and Molinaro (SODA, 2021), the bound on the integrality gap immediately implies that branch and bound requires $n^{\mathrm{poly}(m)}$ time on random Gaussian IPs with good probability, which is polynomial when the number of constraints m is fixed.

Keywords: Integer programming · Integrality gap · Branch and bound

1 Introduction

Consider the following linear program with n variables and m constraints

$$\mathsf{val_{LP}} = \max_{x} \mathsf{val}(x) = c^{\mathsf{T}}x$$

$$\text{s.t. } Ax \leq b \qquad\qquad \text{(Primal LP)}$$

$$x \in [0,1]^n$$

Let $\mathsf{val_{IP}}$ be the value of the same optimization problem with the additional restriction that x is integral, i.e., $x \in \{0,1\}^n$. We denote the integrality gap as $\mathsf{IPGAP} := \mathsf{val_{LP}} - \mathsf{val_{IP}}$. The integrality gap of integer linear programs forms an

S. Borst, D. Dadush and S. Tiwari—This project has received funding from the European Research Council (ERC) under the European Union's Horizon 2020 research and innovation programme (grant agreement QIP–805241).

D. Dadush and S. Huiberts—This work was done while the author was participating in a program at the Simons Institute for the Theory of Computing.

© Springer Nature Switzerland AG 2021
M. Singh and D. P. Williamson (Eds.): IPCO 2021, LNCS 12707, pp. 427–442, 2021.
https://doi.org/10.1007/978-3-030-73879-2_30

important measure for the complexity of solving said problem in a number of works on the average-case complexity of integer programming [1,3–5,7,11].

So far, probabilistic analyses of the integrality gap have focussed on 0–1 packing IPs and the generalized assignment problem. In particular, the entries of $A \in \mathbb{R}^{m \times n}, b \in \mathbb{R}^m, c \in \mathbb{R}^n$ in these problems are all non-negative, and the entries of b were assumed to scale linearly with n.

In this paper, we analyze the integrality gap of (Primal LP) under the assumption that the entries of A and c are all independent Gaussian $\mathcal{N}(0,1)$ distributed, and that the negative part of b is small: $\|b^-\|_2 \le n/10$.

We prove that, with high probability, the integrality gap IPGAP is small, i.e., (Primal LP) admits a solution $x \in \{0,1\}^n$ with value close to the optimum.

Theorem 1. *There exists an absolute constant $C \ge 1$, such that, for $m \ge 1$, $n \ge Cm$, $b \in \mathbb{R}^m$ with $\|b^-\|_2 \le n/10$, if (Primal LP) is sampled with independent $\mathcal{N}(0,1)$ entries in c and A, we have that*

$$\Pr\left(\mathsf{IPGAP} \ge 10^{15} \cdot t \cdot \frac{m^{2.5}(m + \log n)^2}{n} \right) \le 4 \cdot \left(1 - \frac{1}{25} \right)^t + n^{-7},$$

for all $1 \le t \le \frac{n}{Cm^{2.5}(m + \log n)^2}$.

In the previous probabilistic analyses by [4,5,11], it is assumed that $b_i = \beta_i n$ for fixed $\beta_1, \ldots, \beta_m \in (0, 1/2)$ and the entries of (A, c) are independently distributed uniformly in the interval $[0, 1]$. Those works prove a similar bound as above, except that in their results the dependence on m is exponential instead of polynomial. Namely, they require $n \ge 2^{O(m)}$ and the integrality gap scales like $2^{O(m)} \log^2 n/n$. Furthermore, the integrality gap in [5] also had a $O(1/\beta^m)$ dependence, where $\beta := \min_{i \in [m]} \beta_i$, whereas the integrality gap in Theorem 1 does not depend on the "shape" of b (other than requiring $\|b^-\|_2 \le n/10$). Due to a recent breakthrough [3], the integrality gap above also implies that branch and bound applied to the above IP produces a branching tree of size at most $n^{\mathrm{poly}(m)}$ with good probability.

In the rest of the introduction, we begin with an overview of the main techniques we use to prove Theorem 1, describing the similarities and differences with the analysis of Dyer and Frieze [5], and highlight several open problems. We continue by explaining the relation between the integrality gap and the complexity of branch and bound (Subsect. 1.2 below), and conclude with a discussion of related work.

1.1 Techniques

Our proof strategy follows along similar lines to that of Dyer and Frieze [5], which we now describe. In their strategy, one first solves an auxiliary LP $\max c^\mathsf{T} x, Ax \le b - \epsilon 1_m$, for $\epsilon > 0$ small, to get its optimal solution x^*, which is both feasible and nearly optimal for the starting LP (proved by a simple scaling argument), together with its optimal dual solution $u^* \ge 0$ (see Subsect. 2.2 for

the formulation of the dual). From here, they round down the fractional components of x^* to get a feasible IP solution $x' := \lfloor x^* \rfloor$. We note that the feasibility
of x' depends crucially on the packing structure of the LPs they work with, i.e.,
that A has non-negative entries (which does not hold in the Gaussian setting).
Lastly, they construct a nearly optimal integer solution x'', by carefully choosing
a subset of coordinates $T \subset \{i \in [n] : x'_i = 0\}$ of size $O(\text{poly}(m) \log n)$, where
they flip the coordinates of x' in T from 0 to 1 to get x''. The coordinates of
T are chosen accordingly the following criteria. Firstly, the coordinates should
be *very cheap* to flip, which is measured by the absolute value of their *reduced
costs*. Namely, they enforce that $|c_i - A_{.,i}^\mathsf{T} u^*| = O(\log n/n)$, $\forall i \in T$. Secondly, T
is chosen to make the *excess slack* $\|A(x^* - x'')\|_\infty = 1/\text{poly}(n)$, i.e., negligible.
We note that guaranteeing the existence of T is highly non-trivial. Crucial to the
analysis is that after conditioning on the exact value of x^* and u^*, the columns
of $W := \begin{bmatrix} c^\mathsf{T} \\ A \end{bmatrix} \in \mathbb{R}^{(m+1)\times n}$ (the objective extended constraint matrix), indexed
by $N_0 := \{i \in [n] : x_i^* = 0\}$, are independently distributed subject to having
negative reduced cost, i.e., subject to $c_i - A_{.,i}^\mathsf{T} u^* < 0$ for $i \in N_0$ (see Lemma 5). It
is the large amount of left-over randomness in these columns that allowed Dyer
and Frieze to show the existence of the subset T via a discrepancy argument
(more on this below). Finally, given a suitable T, a simple sensitivity analysis is
used to show the bound on the gap between $c^\mathsf{T} x''$ and the (Primal LP) value.
This analysis uses the basic formula for the optimality gap between primal and
dual solutions (see (Gap Formula) in Subsect. 2.2), and relies upon bounds on
the size of the reduced costs of the flipped variables, the total excess slack and
the norm of the dual optimal solution u^*.

As a first difference with the above strategy, we are able to work directly
with the optimal solution x^* of the original LP without having to replace b by
$b' := b - \epsilon 1_m$. The necessity of working with this more conservative feasible
region in [5] is that flipping 0 coordinates of x' to 1 can only *decrease* $b - Ax'$.
In particular, if the coordinates of $b - Ax' \geq 0$ are too small, it becomes difficult
to find a set T that doesn't force x'' to be infeasible. By working with b' instead
of b, they can insure that $b - Ax' \geq \epsilon 1_m$, which avoids this problem. In the
Gaussian setting, it turns out that we have equal power to both increase and
decrease the slack of $b - Ax'$, due to the fact that the Gaussian distribution is
symmetric about 0. We are in fact able to simultaneously fix both the feasibility
and optimality error of x', which gives us more flexibility. In particular, we will
be able to use randomized rounding when we move from x^* to x', which will
allow us to start with a smaller initial slack error than is achievable by simply
rounding x^* down.

Our main quantitative improvement – the reduction from an exponential to
a polynomial dependence in m – arises from two main sources. The first source
of improvement is a substantially improved version of a discrepancy lemma of
Dyer and Frieze [5, Lemma 3.4]. This lemma posits that for any large enough set
of "suitably random" columns in $\mathbb{R}^m$ and any not too big target vector $t \in \mathbb{R}^m$,
then with non-negligible probability there exists a set containing half the columns

whose sum is very close to t. This is the main lemma used to show the existence of the subset T, chosen from a suitably filtered subset of the columns of A in N_0, used to reduce the excess slack. The non-negligible probability in their lemma was of order $2^{-O(m)}$, which implied that one had to try $2^{O(m)}$ disjoint subsets of the filtered columns before having a constant probability of success of finding a suitable T. In our improved variant of the discrepancy lemma, given in Lemma 8, we show that by sub-selecting a $1/(2\sqrt{m})$-fraction of the columns instead of $1/2$-fraction, we can increase the success probability to constant, with the caveat of requiring a slightly larger set of initial columns.

The second source of improvement is the use of a much milder filtering step mentioned above. In both the uniform and Gaussian case, the subset T is chosen from a subset of N_0 associated with columns of A having reduced costs of absolute value at most some parameter $\Delta > 0$. The probability of finding a suitable T increases as Δ grows larger, since we have more columns to choose from, and the target integrality gap scales linearly with Δ, as the columns we choose from become more expensive as Δ grows. Depending on the distribution of c and A, the reduced cost filtering induces non-trivial correlations between the entries of the corresponding columns of A, which makes it difficult to use them within the context of the discrepancy lemma. To deal with this problem in the uniform setting, Dyer and Frieze filtered much more aggressively, by additionally restricting to columns of A lying in a sub-cube $[\alpha, \beta]^m$, where $\alpha = \Omega(\log^3 n/n)$ and $\beta := \min_{i \in [m]} \beta_i$ as above. By doing the reduced cost filtering more carefully, this allowed them to ensure that the distribution of the filtered columns in A is in fact uniform in $[\alpha, \beta]^m$, thereby removing the unwanted correlations. With this aggressive filtering, the columns in N_0 only pass the filtering step with probability $O(\beta^m \Delta)$, which is the source of the $(1/\beta)^m$ dependence in their integrality gap. In the Gaussian context, we show how to work directly with the columns of A with only reduced cost filtering, which increases the success probability of the filtering test to $\Theta(\Delta)$. While the entries of the filtered columns of A do indeed correlate, using the rotational symmetry of the Gaussian distribution, we show that after applying a suitable rotation R, the coordinates of filtered columns of RA are all independent. This allows us to apply the discrepancy lemma in a "rotated space", thereby completely avoiding the correlation issues in the uniform setting.

As already mentioned, we are also able to substantially relax the rigid requirements on the right hand side b and to remove any stringent "shape-dependence" of the integrality gap on b. Specifically, for $b_i = \beta_i n$, $\beta_i \in (0, 1/2)$, $\forall i \in [m]$, the shape parameter $\beta := \min_{i \in m} \beta_i$, is used to both lower bound $|N_0|$ by roughly $\Omega((1 - 2\beta)n)$, the number of zeros in x^*, as well as upper bound the ℓ_1 norm of the optimal dual solution u^* by $O(1/\beta)$ (this a main reason for the choice of the $[\alpha, \beta]^m$ sub-cube above). These bounds are both crucial for determining the existence of T. In the Gaussian setting, we are able to establish $|N_0| = \Omega(n)$ and $\|u^*\|_2 = O(1)$, using only that $\|b^-\|_2 \leq n/10$. Due to the different nature of the distributions we work with, our arguments to establish these bounds are completely different from those used by Dyer and Frieze. Firstly, the lower bound on

$|N_0|$, which is strongly based on the packing structure of the IP in [5], is replaced by a sub-optimality argument. Namely, we show that the objective value of any LP basic solution with too few zero coordinates must be sub-optimal, using the concentration properties of the Gaussian distribution. The upper bound on the ℓ_1 norm of u^* in [5] is deterministic and based on packing structure; namely, that the objective value of a (Primal LP) of packing-type is at most $\sum_{i=1}^{n} c_i \leq n$ (since $c_i \in [0,1], \forall i \in [m]$). In the Gaussian setting, we prove our bound on the norm of u^* by first establishing simple upper and lower bounds on the dual objective function, which hold with overwhelming probability, and optimizing over these simple approximations (see Lemma 4).

We note that we expect the techniques we develop here to yield improvements to the analysis of random packing IPs as well. The main technical difficulty at present is showing that a milder filtering step (e.g., just based on the reduced costs) is also sufficient in the packing case. This in essence reduces to understanding whether Lemma 8 can be generalized to handle random columns whose entries are allowed to have non-trivial correlations. Another possible approach to obtain improvements, in particular with respect to further relaxing the restrictions on b, is to try and flip both 0s to 1 and 1s to 0 in the rounding of x' to x''. The columns of W associated with the one coordinates of x^* are no longer independent however. A final open question is whether these techniques can be extended to handle discrete distributions on A and c.

1.2   Relation to Branch and Bound

In a recent breakthrough work, Dey, Dubey and Molinaro [3] proved that, if the entries of A, c are independently distributed in $[0,1]$ and for all $0 < \alpha \min\{30m, \frac{\log n}{a_2}$ one has

$$\Pr\left(\mathsf{IPGAP} \geq \alpha a_1 \frac{\log^2 n}{n}\right) \leq 4 \cdot 2^{-\alpha a_2} + \frac{1}{n}$$

for some $a_1, a_2 > 0$, then with probability at least $1 - \frac{2}{n} - 4 \cdot 2^{-\alpha a_2}$, a branch-and-bound algorithm that branches on variables, always selecting the node with the largest LP value, will produce a tree of size at most

$$n^{O(ma_1 \log a_1 + \alpha a_1 \log m)}$$

for all $\alpha \leq \min\{30m, \frac{\log n}{a_2}\}$. The values of a_1 and a_2, they get from [5].

In fact, their analysis goes through for the entries of A and c independently Gaussian $\mathcal{N}(0,1)$ distributed as well with minor modifications. Specifically, one needs to condition on the columns of $\begin{bmatrix} c^\mathsf{T} \\ A \end{bmatrix}$ having bounded norm, after which the final net argument needs to be slightly adapted. The details of this adaptation are give in the full version.

Taking $a_1 = 2 \cdot 10^{15} m^{4.5}$, $a_2 = 1/30$ and $\alpha = 30 \cdot \min\{m, \log n\}$, this result together with Theorem 1, proves that branch and bound can find the best integer solution to (Primal LP) with a branching tree of size $n^{\text{poly}(m)}$ with probability $1 - \frac{6}{n} - 4 \cdot 2^{-m}$. Here, the fact that Theorem 1 depends only polynomially on m results in a much better upper bound than the exponential dependence on m from [5].

1.3 Related Work

The worst-case complexity of solving $\max\{c^{\mathsf{T}}x : Ax = b, x \geq 0, x \in \mathbb{Z}^n\}$ scales as $n^{O(n)}$ times a polynomial factor in the bit complexity of the problem. This is a classical result due to Lenstra [10] and Kannan [9].

If we restrict to IPs with integer data, a dynamic programming algorithm can solve $\max\{c^{\mathsf{T}}x : Ax = b, x \geq 0, x \in \mathbb{Z}^n\}$ in time $O(\sqrt{m}\Delta)^{2m} \log(\|b\|_\infty) + O(nm)$, where Δ is the largest absolute value of entries in the input matrix A [6,8,12]. Integer programs of the form $\max\{c^{\mathsf{T}}x : Ax = b, 0 \leq x \leq u, x \in \mathbb{Z}^n\}$ can similarly be solved in time

$$n \cdot O(m)^{(m+1)^2} \cdot O(\Delta)^{m \cdot (m+1)} \log^2(m \cdot \Delta),$$

which was proved in [6]. Note that integer programs of the form $\max\{c^{\mathsf{T}}x : Ax \leq b, x \in \{0,1\}^n\}$ can be rewritten in this latter form by adding m slack variables.

In terms of random inputs, we mention the work of Röglin and Vöcking [13]. They prove that a class of IPs satisfying some minor conditions has polynomial smoothed complexity if and only if that class admits a pseudopolynomial time algorithm. An algorithm has polynomial smoothed complexity if its running time is polynomial with high probability when its input has been perturbed by adding random noise, where the polynomial may depend on the magnitude φ^{-1} of the noise as well as the dimensions n, m of the problem. An algorithm runs in pseudopolynomial time if the running time is polynomial when the numbers are written in unary, i.e., when the input data consists of integers of absolute value at most Δ and the running time is bounded by a polynomial $p(n, m, \Delta)$. In particular, they prove that solving the randomly perturbed problem requires only polynomially many calls to the pseudopolynomial time algorithm with numbers of size $(nm\varphi)^{O(1)}$ and considering only the first $O(\log(nm\varphi))$ bits of each of the perturbed entries.

If, for the sake of comparison, we choose $b \in \mathbb{R}^m$ in (Primal LP) from a $\mathcal{N}(0,1)$ Gaussian distribution independently of c and A, then the result of [13] proves that, with high probability, it is sufficient to solve polynomially many problems with integer entries of size $(nm)^{O(1)}$. Since $\Delta = (nm)^{O(m)}$ in this setting (by Hadamard's inequality), the result of [6] tells us that this problem can be solved in time $(nm)^{O(m^3)}$.

1.4 Organization

In Sect. 2, we give preliminaries on probability theory, linear programming and integer rounding. In Sect. 3, we prove properties of the LP optimal solution x^*, and in Sect. 4 we look at the distribution of the columns of $\begin{bmatrix} c^\mathsf{T} \\ A \end{bmatrix}$ corresponding to indices $i \in [n]$ with $x_i^* = 0$. Then in Sect. 5, we prove Theorem 1, using a discrepancy result that we prove in the full version of this paper [2]. All other proofs can be found in the full version as well.

2   Preliminaries

2.1   Basic Notation

We denote the reals and non-negative reals by $\mathbb{R}, \mathbb{R}^+$ respectively, and the integers and positive integers by $\mathbb{Z}, \mathbb{N}$ respectively. For $k \geq 1$ an integer, we let $[k] := \{1, \ldots, k\}$. For $s \in \mathbb{R}$, we let $s^+ := \max\{s, 0\}$ and $s^- := \min\{s, 0\}$ denote the positive and negative part of s. We extend this to a vector $x \in \mathbb{R}^n$ by letting $x^{+(-)}$ correspond to applying the positive (negative) part operator coordinate-wise. We let $\|x\|_2 = \sqrt{\sum_{i=1}^n x_i^2}$ and $\|x\|_1 = \sum_{i=1}^n |x_i|$ denote the ℓ_2 and ℓ_1 norm respectively. We use $\log x$ to denote the base e natural logarithm. We use $0_m, 1_m \in \mathbb{R}^m$ to denote the all zeros and all ones vector respectively, and $e_1, \ldots, e_m \in \mathbb{R}^m$ denote the standard coordinate basis. We write $\mathbb{R}_+^m := [0, \infty)^m$. For a random variable $X \in \mathbb{R}$, we let $\mathbb{E}[X]$ denote its expectation and $\mathrm{Var}[X] := \mathbb{E}[X^2] - \mathbb{E}[X]^2$ denote its variance.

2.2   The Dual Program, Gap Formula and the Optimal Solutions

A convenient formulation of the dual of (Primal LP) is given by

$$\min \mathrm{val}^*(u) := b^\mathsf{T} u + \sum_{i=1}^n (c - A^\mathsf{T} u)_i^+ \qquad \text{(Dual LP)}$$

$$\text{s.t. } u \geq 0.$$

To keep the notation concise, we will often use the identity $\|(c - A^t u)^+\|_1 = \sum_{i=1}^n (c - A^\mathsf{T} u)_i^+$.

For any primal solution x and dual solution u to the above pair of programs, we have the following standard formula for the primal-dual gap:

$$\mathrm{val}^*(u) - \mathrm{val}(x) := b^\mathsf{T} u + \sum_{i=1}^n (c - A^\mathsf{T} u)_i^+ - c^\mathsf{T} x \qquad \text{(Gap Formula)}$$

$$= (b - Ax)^\mathsf{T} u + \left(\sum_{i=1}^n x_i (A^\mathsf{T} u - c)_i^+ + (1 - x_i)(c - A^\mathsf{T} u)_i^+ \right).$$

Throughout the rest of the paper, we let x^* and u^* denote primal and dual optimal basic feasible solutions for (Primal LP) and (Dual LP) respectively, which we note are unique with probability 1. We use the notation

$$W := \begin{bmatrix} c^\mathsf{T} \\ A \end{bmatrix} \in \mathbb{R}^{(m+1) \times n}, \qquad (1)$$

to denote the objective extended constraint matrix. We will frequently make use of the sets $N_b := \{i \in [n] : x_i^* = b\}$, $b \in \{0,1\}$, the 0 and 1 coordinates of x^*, and $S := \{i \in [n] : x_i^* \in (0,1)\}$, the fractional coordinates of x^*. We will also use the fact that $|S| \le m$, which follows since x^* is a basic solution to (Primal LP) and A has m rows.

2.3 Gaussian and Sub-Gaussian Random Variables

The standard, mean zero and variance 1, Gaussian $\mathcal{N}(0,1)$ has density function $\varphi(x) := \frac{1}{\sqrt{2\pi}} e^{-x^2/2}$. A standard Gaussian vector in $\mathbb{R}^d$, denoted $\mathcal{N}(0, I_d)$, has probability density $\prod_{i=1}^d \varphi(x_i) = \frac{1}{\sqrt{2\pi}^d} e^{-\|x\|^2/2}$ for $x \in \mathbb{R}^d$. A random variable $Y \in \mathbb{R}$ is σ-sub-Gaussian if for all $\lambda \in \mathbb{R}$, we have

$$\mathbb{E}[e^{\lambda Y}] \le e^{\sigma^2 \lambda^2/2}. \tag{2}$$

A standard normal random variable $X \sim \mathcal{N}(0,1)$ is 1-sub-Gaussian. If variables $Y_1, \ldots, Y_k \in \mathbb{R}$ are independent and respectively σ_i-sub-Gaussian, $i \in [k]$, then $\sum_{i=1}^k Y_i$ is $\sqrt{\sum_{i=1}^k \sigma_i^2}$-sub-Gaussian.

For a σ-sub-Gaussian random variable $Y \in \mathbb{R}$ we have the following standard tailbound:

$$\max\{\Pr[Y \le -\sigma s], \Pr[Y \ge \sigma s]\} \le e^{-\frac{s^2}{2}}, s \ge 0. \tag{3}$$

For $X \sim \mathcal{N}(0, I_d)$, we will use the following higher dimensional analogue:

$$\Pr[\|X\|_2 \ge s\sqrt{d}] \le e^{-\frac{d}{2}(s^2 - 2\log s - 1)} \le e^{-\frac{d}{2}(s-1)^2}, s \ge 1. \tag{4}$$

We will use this bound to show that the columns of A corresponding to the fractional coordinates in the (almost surely unique) optimal solution x^* are bounded.

Lemma 1. *Letting $S := \{i \in [n] : x_i^* \in (0,1)\}$, we have that*

$$\Pr[\exists i \in S : \|A_{\cdot,i}\|_2 \ge (4\sqrt{\log(n)} + \sqrt{m})] \le n^{-7}.$$

2.4 A Local Limit Theorem

Before we can state the discrepancy lemma, we introduce the concept of Gaussian convergence.

Definition 1. *Suppose $X_1, X_2, \ldots$ is a sequence of i.i.d copies of a random variable X with density f. X is said to be (γ, k)-Gaussian convergent if the density f_n of $\sum_{i=1}^n X_i / \sqrt{n}$ satisfies:*

$$|f_n(x) - \varphi(x)| \le \frac{\gamma}{n} \quad \forall x \in \mathbb{R}, n \ge k,$$

where $\varphi := \frac{1}{\sqrt{2\pi}} e^{-x^2/2}$ is the Gaussian probability density function.

The above definition quantifies the speed of convergence in the context of the central limit theorem. The rounding strategy used to obtain the main result utilizes random variables that are the weighted sum of a uniform and an independent normal variable. Crucially, the given convergence estimate will hold for these random variables:

Lemma 2. *Let U be uniform on $[-\sqrt{3}, \sqrt{3}]$ and let $Z \sim \mathcal{N}(0, 1)$. Then there exists a universal constant $k_0 \geq 1$ such that $\forall \epsilon \in [0, 1]$, the random variable $\sqrt{\epsilon}U + \sqrt{1 - \epsilon}Z$ is $(1/10, k_0)$-Gaussian convergent and has maximum density at most 1.*

2.5   Rounding to Binary Solutions

In the proof of Theorem 1, we will take our optimal solution x^* and round it to an integer solution x', by changing the fractional coordinates. Note that as x^* is a basic solution, it has at most m fractional coordinates. One could round to a integral solution by setting all of them to 0, i.e., $x' = \lfloor x^* \rfloor$. If we assume that the Euclidean norm of every column of A is bounded by C, then we have $\|A(x^* - x')\|_2 \leq mC$, since x^* has at most m fractional variables. However, by using randomized rounding we can make this bound smaller, as stated in the next lemma. We use this to obtain smaller polynomial dependence in Theorem 1.

Lemma 3. *Consider an $m \times n$ matrix A with $\|A_{.,i}\|_2 \leq C$ for all $i \in [m]$ and $y \in [0, 1]^n$. Let $S = \{i \in [n] : y_i \in (0, 1)\}$. There exists a vector $y' \in \{0, 1\}^n$ with $\|A(y - y')\|_2 \leq C\sqrt{|S|}/2$ and $y_i' = y_i$ for all $i \notin S$.*

3   Properties of the Optimal Solutions

The following lemma is the main result of this section, which gives principal properties we will need of the optimal primal and dual LP solutions. Namely, we prove an upper bound on the norm of the optimal dual solution u^* and a lower bound on the number of zero coordinates of the optimal primal solution x^*.

Lemma 4. *Given $\delta := \frac{\sqrt{2\pi}}{n}\|b^-\|_2 \in [0, 1/2)$, $\epsilon \in (0, 1/5)$, let x^*, u^* denote the optimal primal and dual LP solutions, and let $\alpha := \frac{1}{\sqrt{2\pi}}\sqrt{\left(\frac{1-3\epsilon}{1-\epsilon}\right)^2 - \delta^2}$ and choose $\beta \in [1/2, 1]$ with $H(\beta) = \frac{\alpha^2}{4}$. Then, with probability at least $1 - 2\left(1 + \frac{2}{\epsilon}\right)^{m+1}e^{-\frac{\epsilon^2 n}{8\pi}} - e^{-\frac{\alpha^2 n}{4}}$, the following holds:*

1. *$c^\mathsf{T}x^* \geq \alpha n$.*
2. *$\|u^*\|_2 \leq \frac{1+\epsilon}{1-3\epsilon-(1-\epsilon)\delta}$.*
3. *$|\{i \in [n] : x_i^* = 0\}| \geq (1 - \beta)n - m$.*

4   Properties of the 0 Columns

For $Y := (c, a_1, \ldots, a_m) \sim \mathcal{N}(0, I_{m+1})$ and $u \in \mathbb{R}^m$, let Y^u denote the random variable Y conditioned the event $c - \sum_{i=1}^m u_i a_i \leq 0$. We will crucially use the following lemma directly adapted from Dyer and Frieze [5, Lemma 2.1], which shows that the columns of W associated with the 0 coordinates of x^* are independent subject to having negative reduced cost.

Recall that $N_b = \{i \in [n] : x_i^* = b\}$, that $S = \{i \in [n] : x_i^* \in (0,1)\}$ and that

$$W := \begin{bmatrix} c^\top \\ A \end{bmatrix} \in \mathbb{R}^{(m+1) \times n} \text{ is the objective extended constraint matrix.}$$

Lemma 5. *Let $N_0' \subseteq [n]$. Conditioning on $N_0 = N_0'$, the submatrix $W_{\cdot, [n] \setminus N_0'}$ uniquely determines x^* and u^* almost surely. If we further condition on the exact value of $W_{\cdot, [n] \setminus N_0'}$, assuming x^* and u^* are uniquely defined, then any column $W_{\cdot, i}$ with $i \in N_0'$ is distributed according to Y^{u^*} and independent of $W_{\cdot, [n] \setminus \{i\}}$.*

To make the distribution of the columns $A_{\cdot, i}$ easier to analyze we rotate them.

Lemma 6. *Let R be a rotation that sends the vector u to the vector $\|u\|_2 e_m$. Suppose $(c, a) \sim Y^u$. Define $a' := Ra$. Then $(c, Ra) \sim (c', a')$, where (c', a') is the value of $(\bar{c}', \bar{a}') \sim \mathcal{N}(0, I_{m+1})$ conditioned on $\|u\|_2 \bar{a}_m' - \bar{c}' \geq 0$.*

We will slightly change the distribution of the (c', a_m') above using rejection sampling, as stated in the next lemma. This will make it easier to apply the discrepancy result of Lemma 8, which is used to round x^* to an integer solution of nearby value. In what follows, we denote the probability density function of a random variable X by f_X. In the following lemma, we use $\mathrm{unif}(0, \nu)$ to denote the uniform distribution on the interval $[0, \nu]$, for $\nu \geq 0$.

Lemma 7. *For any $\omega \geq 0$, $\nu > 0$, let $X, Y \sim \mathcal{N}(0,1)$ be independent random variables and let $Z = \omega Y - X$. Let X', Y', Z' be these variables conditioned on $Z \geq 0$. We apply rejection sampling on (X', Y', Z') with acceptance probability*

$$\Pr[\mathit{accept} | Z' = z] = \frac{2\varphi(\nu/\sqrt{1 + \omega^2}) \mathbf{1}_{z \in [0, \nu]}}{2\varphi(z/\sqrt{1 + \omega^2})}.$$

Let X'', Y'', Z'' be the variables X', Y', Z' conditioned on acceptance. Then:

1. $\Pr[\mathit{accept}] = 2\nu \varphi(\nu/\sqrt{1 + \omega^2})/\sqrt{1 + \omega^2}$.
2. $Y'' \sim W + V$ where $W \sim \mathcal{N}(0, \frac{1}{1+\omega^2})$, $V \sim \mathrm{unif}(0, \frac{\nu\omega}{1+\omega^2})$ and W, V are independent.

5   Proof of Theorem 1

Recall that $S = \{i \in [n] : x_i^* \in (0,1)\}$ and $N_0 = \{i \in [n] : x_i^* = 0\}$. To prove Theorem 1, we will assume that the following three conditions hold:

1. $\|A_{.,i}\|_2 \le 4\sqrt{\log(n)} + \sqrt{m}, \ \forall i \in S$.
2. $\|u^*\| \le 3$.
3. $|N_0| \ge n/500$.

Using Lemmas 1 and 4 we can show that these events hold with probability $1 - n^{-\Omega(1)}$. Now we take our optimal basic solution x^* and round it to an integral vector x' using Lemma 3. Then we can generate a new solution x'' from x' by flipping the values at indices $T \subseteq N_0$ to one. In Lemma 9 we show that with high probability there is such a set T, such that x'' is a feasible solution to our primal problem and that $\text{val}(x^*) - \text{val}(x'')$ is small.

We do this by looking at t disjoint subsets of N_0 with small reduced costs. Then we show for each of these sets that with constant probability it contains a subset T such that for x'' obtained from T, x'' is feasible and all constraints that are tight for x^* are close to being tight for x''. This argument relies on the following improved discrepancy lemma.

Lemma 8. *For $k, m \in \mathbb{N}$, let $a = \lceil 2\sqrt{m} \rceil$ and $\theta > 0$ satisfy $\left(\frac{2\theta}{\sqrt{2\pi k}}\right)^m \binom{ak}{k} = 1$. Let $Y_1, \dots, Y_{ak} \in \mathbb{R}^m$ be i.i.d. random vectors whose coordinates are independent random variables. For $k_0 \in \mathbb{N}, \gamma \ge 0, M > 0$, assume that $Y_{1,i}, i \in [m]$, is (γ, k_0)-Gaussian convergent and admits a probability density $g_i : \mathbb{R} \to \mathbb{R}_+$ satisfying $\max_{x \in \mathbb{R}} g_i(x) \le M$. Then, if*

$$k \ge \max\{(4\sqrt{m}+2)k_0, 144m^{\frac{3}{2}}(\log M + 3), 150\,000(\gamma+1)m^{\frac{7}{4}}\},$$

then for any vector $A \in \mathbb{R}^m$ with $\|A\|_2 \le \sqrt{k}$ the following holds:

$$\Pr\left[\exists K \subset [ak] : |K| = k, \|(\sum_{j \in K} Y_j) - A\|_\infty \le \theta\right] \ge \frac{1}{25}. \tag{5}$$

If a suitable T exists, then using the gap formula we show that $\text{val}(x^*) - \text{val}(x'')$ is small. Because the t sets independent the probability of failure decreases exponentially with t. Hence, we can make the probability of failure arbitrarily small by increasing t. We know $\text{val}(x^*) = \text{val}_{\text{LP}}$ and because $x'' \in \{0,1\}^n$ we have $\text{val}_{\text{IP}} \ge \text{val}(x'')$, so $\text{IPGAP} = \text{val}_{\text{LP}} - \text{val}_{\text{IP}} \le \text{val}(x^*) - \text{val}(x'')$, which is small with high probability.

Lemma 9. *If $n \ge \exp(k_0)$ for k_0 from Lemma 2 and conditions 1, 2 and 3 above hold, then*

$$\Pr\left[\text{IPGAP} > 10^{15}t \cdot \frac{m^{2.5}(\log n + m)^2}{n}\right] \le 2 \cdot \left(1 - \frac{1}{25}\right)^t \tag{6}$$

for $1 \le t \le \frac{n}{20\,000\sqrt{m}k^2}$, where $k := \lceil 165\,000m(\log(n) + m)\rceil$.

Proof. It suffices to condition on N_0 and $W_{.,[n]\setminus N_0}$, subject to the conditions 1–3. Now let R be a rotation that sends the vector u^* to the vector $\|u^*\|_2 e_m$. Define:

$$\Delta := 10\,000\sqrt{mk}/n,$$

$$B_i := RA_{.,i}, \qquad\qquad\qquad\qquad \text{for } i \in N_0,$$

$$Z_t := \{i \in N_0 : \|u^*\|_2(B_i)_m - c_i \in [0, t\Delta]\}, \qquad \text{for } 1 \le t \le \frac{1}{2\Delta k}.$$

We consider a (possibly infeasible) integral solution x' to the LP, generated by rounding the fractional coordinates from x^*. By Lemma 3 we can find such a solution with $\|A(x^* - x')\|_2 \le (4\sqrt{\log n} + \sqrt{m})\sqrt{|S|}/2 \le (4\sqrt{\log n} + \sqrt{m})\sqrt{m}/2$. We will select a subset $T \subseteq Z_t$ of size $|T| = k$ of coordinates to flip from 0 to 1 to obtain $x'' \in \{0,1\}^n$ from x', so $x'' := x' + \sum_{i\in T} e_i$. By complementary slackness, we know for $i \in [n]$ that $x_i^*(A^\mathsf{T} u^* - c)_i^+ = (1 - x_i^*)(c - A^\mathsf{T} u^*)^+ = 0$ and that $x_i^* \notin \{0,1\}$ implies $(c - A^\mathsf{T} u^*)_i = 0$, and for $j \in [m]$ that $u_j^* > 0$ implies $b_j = (Ax^*)_j$. This observation allows us to prove the following key bound for the integrality gap of (Primal LP)

$$\begin{aligned}
\mathrm{val}(x^*) - \mathrm{val}(x'') &= \mathrm{val}^\star(u^*) - \mathrm{val}(x'') \\
&= (b - Ax'')^\mathsf{T} u^* \\
&\quad + \left(\sum_{i=1}^n x_i''(A^\mathsf{T} u^* - c)_i^+ + (1 - x_i'')(c - A^\mathsf{T} u^*)_i^+ \right) \quad \text{(by Gap Formula)} \\
&= (x^* - x'')^\mathsf{T} A^\mathsf{T} u^* + \sum_{i\in T}(A^\mathsf{T} u^* - c)_i \quad \text{(by complementary slackness)} \\
&\le \sqrt{m}\|u^*\|_2 \|A(x'' - x^*)\|_\infty + t\Delta k \quad \text{(since } T \subseteq Z_t\text{)}.
\end{aligned}$$

Condition 2 tells us that $\|u^*\|_2 \le 3$, and by definition we have

$$t\Delta k \le 27226 \cdot 10^{10} t \cdot \frac{m^{2.5}(\log(n) + m)^2}{n},$$

so the rest of this proof is dedicated to showing the existence of a set $T \subseteq Z_t$ such that $\|A(x'' - x^*)\|_\infty \le O(1/n)$ and $Ax'' \le b$.

By applying Lemma 5, we see that $\{(c_i, A_{.,i})\}_{i\in N_0}$ are independent vectors, distributed as $\mathcal{N}(0, I_{m+1})$ conditioned on $c_i - A_{.,i}^\mathsf{T} u^* \le 0$. This implies that the vectors $\{(c_i, B_i)\}_{i\in N_0}$ are also independent. By Lemma 6, it follows that $(c_i, B_i) \sim \mathcal{N}(0, I_{m+1}) \mid \|u^*\|_2(B_i)_m - c_i \ge 0$. Note that the coordinates of B_i are therefore independent and $(B_i)_j \sim \mathcal{N}(0,1)$ for $j \in [m-1]$.

To simplify the upcoming calculations, we apply rejection sampling as specified in Lemma 7 with $\nu = \Delta t$ on $(c_i, (B_i)_m)$, for each $i \in N_0$. Let $Z_t' \subseteq N_0$ denote the indices which are accepted by the rejection sampling procedure. By the guarantees of Lemma 7, we have that $Z_t' \subseteq Z_t$ and

$$\Pr[i \in Z_t' \mid i \in N_0] = \frac{2\Delta t\varphi(\Delta t/\sqrt{1 + \|u^*\|_2^2})}{\sqrt{1 + \|u^*\|_2^2}} \ge \frac{2\Delta t\varphi(1/2)}{\sqrt{10}} \ge \Delta t/5.$$

Furthermore, for $i \in Z'_t$ we know that $(B_i)_m$ is distributed as a sum of independent $N(0, \frac{1}{1+\|u^*\|_2^2})$ and $\text{unif}(0, t\Delta)$ random variables, and thus $(B_i)_m$ has mean and variance

$$\mu_t := \mathbb{E}[(B_i)_m | i \in Z'_t] = \Delta t/2,$$

$$\sigma_t^2 := \text{Var}[(B_i)_m | i \in Z'_t] = \frac{1}{1 + \|u^*\|_2^2} + \frac{1}{12}\left(\frac{\|u^*\|_2 \Delta t}{1 + \|u^*\|_2^2}\right)^2 \in [1/10, 2].$$

Now define $\Sigma^{(t)}$ to be the diagonal matrix with $\Sigma_{j,j}^{(t)} = 1$, $j \in [m-1]$, and $\Sigma_{m,m}^{(t)} = \sigma_t$. Conditional on $i \in Z'_t$, define $B_i^{(t)}$ as the random variable

$$B_i^{(t)} := (\Sigma^{(t)})^{-1}(B_i - \mu_t e_m) \mid i \in Z'_t.$$

This ensures that all coordinates of $B^{(t)}$ are independent, mean zero and have variance one.

We have assumed that $|N_0| \geq n/500$ and we know $\Pr[i \in Z'_t | i \in N_0] \geq \Delta t/5$. Now, using the Chernoff bound we find that:

$$\Pr[|Z'_t| < 2t\sqrt{m}k] \leq \Pr\left[|Z'_t| < \frac{1}{5}t\Delta|N_0|/2\right]$$

$$\leq \exp\left(-\frac{1}{8} \cdot \frac{1}{5}t\Delta|N_0|\right)$$

$$\leq \left(1 - \frac{1}{25}\right)^t. \tag{7}$$

Now we define:

$$\theta := \frac{\sqrt{2\pi k}}{2}\left(\frac{\lceil 2\sqrt{m}k \rceil}{k}\right)^{-1/m}, \qquad\qquad d := A(x^* - x').$$

$$\theta' := 2\sqrt{m}\theta. \qquad\qquad\qquad\qquad\qquad d' := d - 1_m\theta'.$$

Observe that

$$\theta = \frac{\sqrt{2\pi k}}{2}\left(\frac{\lceil 2\sqrt{m}k \rceil}{k}\right)^{-1/m} \leq \frac{\sqrt{2\pi k}}{2}(2\sqrt{m})^{-k/m} \leq \frac{1}{32m^2 n}.$$

So $\theta' \leq 1/8$.

If $|Z'_t| \geq \lceil 2\sqrt{m}\rceil kt$, then we can take t disjoint subsets $Z_t^{(1)}, \ldots Z_t^{(k)}$ of Z'_t of size $\lceil 2\sqrt{m}\rceil k$. Conditioning on this event, we wish to apply Lemma 8 to each set $\{B_i^{(t)}\}_{i \in Z_t^{(l)}}$, $l \in [t]$, to help us find a candidate rounding of x' to a "good" integer solution x''.

Now we check that all conditions of Lemma 8 are satisfied. By definition we have $\left(\frac{2\theta}{\sqrt{2\pi k}}\right)^m \binom{ak}{k} = 1$, and we can bound

$$
\begin{aligned}
\left\|(\Sigma^{(t)})^{-1}(Rd' - e_m k\mu_t)\right\|_2 &\leq \max(1, 1/\sigma_t)(\|Rd\|_2 + \theta' + k\mu_t) \\
&\leq \sqrt{10}(\|RA(x^* - x')\|_2 + \theta' + k\Delta t/2) \\
&\leq \sqrt{10}\left(\sqrt{m}(4\sqrt{\log n} + \sqrt{m})/2 + \frac{1}{8} + \frac{1}{4}\right) \\
&\leq 4\sqrt{10m(\log n + m)} \leq \sqrt{k}.
\end{aligned}
$$

We now show that the conditions of Lemma 8 for $M = 1, \gamma = 1/10$, and k_0 specified below, are satisfied by $\{B_i^{(t)}\}_{i \in Z_t^{(l)}}, \forall l \in [t]$.

First, we observe that the $B_i^{(t)}$ are distributed as $(B_i^{(t)})_m \sim \sqrt{\epsilon}V + \sqrt{1 - \epsilon}U$ for $\epsilon = \frac{1}{(1+\|u^*\|_2^2)\sigma_t^2}$, where U is uniform on $[-\sqrt{3}, \sqrt{3}]$ and $V \sim \mathcal{N}(0,1)$. By Lemma 2, $(B_i^{(t)})_m$ is $(1/10, k_0)$-Gaussian convergent for some k_0 and has maximum density at most 1. Recalling that the coordinates of $B_i^{(t)}$, $i \in Z_t'$, are independent and $(B_i^{(t)})_j \sim \mathcal{N}(0,1), \forall j \in [m-1]$, we see that $B_i^{(t)}$ has independent $(1/10, k_0)$-Gaussian convergent entries of maximum density at most 1. Lastly, we note that

$$
\begin{aligned}
k = 165\,000m(\log(n) + m) &\geq 165\,000(m^2 + k_0 m) \\
&\geq \max\{(4\sqrt{m} + 2)k_0, 144m^{\frac{3}{2}}(\log 1 + 3), 150\,000(\gamma + 1)m^{\frac{7}{4}}\}
\end{aligned}
$$

as needed to apply Lemma 8, using that $n \geq \exp(k_0)$.

Therefore, applying Lemma 8, for each $l \in [t]$, with probability at least $1 - 1/25$, there exists a set $T_l \subseteq Z_t^{(l)}$ of size k such that:

$$
\left\|\sum_{i \in T_l} B_i^{(t)} - (\Sigma^{(t)})^{-1}(Rd' - e_m k\mu_t)\right\|_\infty \leq \theta. \tag{8}
$$

Call the event that (8) is valid for any of the t sets E_t. Because the success probabilities for each of the t sets are independent, we get:

$$
\Pr[\neg E_t \mid |Z_t'| \geq \lceil 2\sqrt{m}\rceil tk] \leq \left(1 - \frac{1}{25}\right)^t.
$$

Combining this with Eq. (7), we see that $\Pr[\neg E_t] \le 2 \cdot (1 - \frac{1}{25})^t$. If E_t occurs, we choose $T \subseteq Z'_t$, $|T| = k$, satisfying (8). Then,

$$\left\| \sum_{i \in T} A_{.,i} - d' \right\|_\infty \le \left\| \sum_{i \in T} A_{.,i} - d' \right\|_2 = \left\| \sum_{i \in T} B_{.,i} - Rd' \right\|_2$$

$$= \left\| \sum_{i \in T} (\Sigma^{(t)}) B_{.,i}^{(t)} + k\mu_t e_m - Rd' \right\|_2$$

$$\le \max(1, \sigma_t)\sqrt{m} \left\| \sum_{i \in T} B_{.,i}^{(t)} - (\Sigma^{(t)})^{-1}(Rd' - e_m k \mu_t) \right\|_\infty$$

$$\le 2\sqrt{m}\theta = \theta'.$$

Now we will show that when E_t occurs, x'' is feasible and $\|A(x'' - x^*)\|_\infty = O(1/n)$. First we check feasibility:

$$\sum_{i=1}^m x_i'' a_{ji} = (Ax')_j + \sum_{i \in T} a_{ji} \le (Ax')_j + d_j' + \theta'$$

$$= (Ax')_j + (A(x^* - x'))_j = (Ax^*)_j \le b_j.$$

Hence the solution is feasible for our LP. We also have

$$\|A(x'' - x^*)\|_\infty = \|Ax'' - Ax' - d\|_\infty$$

$$= \| \sum_{i \in T} A_{.,i} - d'\|_\infty \le \| \sum_{i \in T} A_{.,i} - d\|_\infty + \theta' \le 2\theta'.$$

Now we can finalize our initial computation:

$$\mathrm{val}(x^*) - \mathrm{val}(x'') \le \sqrt{m}\|u^*\|_2 \|A(x'' - x^*)\|_\infty + t\Delta k$$

$$\le 6\sqrt{m}\theta' + 10\,000 \cdot \frac{\sqrt{m} \cdot t \cdot k^2}{n}$$

$$\le \frac{12}{32mn} + 27226 \cdot 10^{10} t \cdot \frac{m^{2.5}(\log n + m)^2}{n}$$

$$\le 10^{15} t \cdot \frac{m^{2.5}(\log n + m)^2}{n}.$$

$\square$

References

1. Beier, R., Vöcking, B.: Probabilistic analysis of knapsack core algorithms. In: Munro, J.I. (ed.) Proceedings of the Fifteenth Annual ACM-SIAM Symposium on Discrete Algorithms, SODA 2004, New Orleans, Louisiana, USA, 11–14 Jan 2004, pp. 468–477. SIAM (2004)
2. Borst, S., Dadush, D., Huiberts, S., Tiwari, S.: On the Integrality Gap of Binary Integer Programs with Gaussian Data. arXiv:2012.08346 [cs, math] (Dec 2020)

3. Dey, S.S., Dubey, Y., Molinaro, M.: Branch-and-bound solves random binary IPs in polytime. In: Proceedings of the 2021 ACM-SIAM Symposium on Discrete Algorithms (SODA), pp. 579–591. Society for Industrial and Applied Mathematics (Jan 2021). https://doi.org/10.1137/1.9781611976465.35
4. Dyer, M., Frieze, A.: Probabilistic analysis of the generalised assignment problem. Math. Program. **55**(1–3), 169–181 (1992). https://doi.org/10.1007/bf01581197
5. Dyer, M., Frieze, A.: Probabilistic analysis of the multidimensional knapsack problem. Math. OR **14**(1), 162–176 (1989). https://doi.org/10.1287/moor.14.1.162
6. Eisenbrand, F., Weismantel, R.: Proximity results and faster algorithms for integer programming using the Steinitz lemma. ACM Trans. Algorithms **16**(1), 1–14 (2020). https://doi.org/10.1145/3340322
7. Goldberg, A., Marchetti-Spaccamela, A.: On finding the exact solution of a zero-one knapsack problem. In: Proceedings of the Sixteenth Annual ACM Symposium on Theory of Computing - STOC 1984. ACM Press (1984). https://doi.org/10.1145/800057.808701
8. Jansen, K., Rohwedder, L.: Integer programming (2019). https://doi.org/10.1002/9781119454816.ch10
9. Kannan, R.: Minkowski's convex body theorem and integer programming. Math. OR **12**(3), 415–440 (1987). https://doi.org/10.1287/moor.12.3.415
10. Lenstra, H.: Integer programming with a fixed number of variables. Math. OR **8**(4), 538–548 (1983). https://doi.org/10.1287/moor.8.4.538
11. Lueker, G.S.: On the average difference between the solutions to linear and integer knapsack problems. In: Applied Probability-Computer Science: The Interface, vol. 1, pp. 489–504. Birkhäuser, Boston (1982). https://doi.org/10.1007/978-1-4612-5791-2
12. Papadimitriou, C.H.: On the complexity of integer programming. J. ACM **28**(4), 765–768 (1981). https://doi.org/10.1145/322276.322287
13. Röglin, H., Vöcking, B.: Smoothed analysis of integer programming. Math. Program. **110**(1), 21–56 (2007). https://doi.org/10.1007/s10107-006-0055-7

Linear Regression with Mismatched Data: A Provably Optimal Local Search Algorithm

Rahul Mazumder$^{(\boxtimes)}$ (iD) and Haoyue Wang (iD)

Massachusetts Institute of Technology, Cambridge, MA 02139, USA
{rahulmaz,haoyuew}@mit.edu

Abstract. Linear regression is a fundamental modeling tool in statistics and related fields. In this paper, we study an important variant of linear regression in which the predictor-response pairs are partially mismatched. We use an optimization formulation to simultaneously learn the underlying regression coefficients and the permutation corresponding to the mismatches. The combinatorial structure of the problem leads to computational challenges, and we are unaware of any algorithm for this problem with both theoretical guarantees and appealing computational performance. To this end, in this paper, we propose and study a simple greedy local search algorithm. We prove that under a suitable scaling of the number of mismatched pairs compared to the number of samples and features, and certain assumptions on the covariates; our local search algorithm converges to the global optimal solution with a linear convergence rate under the noiseless setting.

Keywords: Linear regression · Mismatched data · Local search method · Learning permutations

1 Introduction

Linear regression and its extensions are among the most useful models in statistics and related fields. In the classical and most common setting, we are given n samples with features $x_i \in \mathbb{R}^d$ and response $y_i \in \mathbb{R}$, where i denotes the sample indices. We assume that the features and responses are perfectly matched i.e., x_i and y_i correspond to the same record or sample. An interesting twist to this problem—also the focus of this paper—is when the feature-response pairs are partially mismatched due to errors in the data merging process [6,7,10]. Here, we consider a mismatched linear model with responses $y = [y_1, ..., y_n] \in \mathbb{R}^n$ and covariates $X = [x_1, ..., x_n]^\top \in \mathbb{R}^{n \times d}$ satisfying

$$P^* y = X \beta^* + \epsilon \qquad (1.1)$$

Supported by grants from the Office of Naval Research: ONR-N000141812298 (YIP) and National Science Foundation: NSF-IIS-1718258.

M. Singh and D. P. Williamson (Eds.): IPCO 2021, LNCS 12707, pp. 443–457, 2021.
https://doi.org/10.1007/978-3-030-73879-2_31

where $\beta^* \in \mathbb{R}^d$ are the regression coefficients, $\epsilon = [\epsilon_1, ..., \epsilon_n]^\top \in \mathbb{R}^n$ is the noise term, and $P^* \in \mathbb{R}^{n \times n}$ is an unknown permutation matrix. We consider the setting where $n > d$ and X has full rank; and seek to estimate both β^* and P^* based on the n observations $\{(y_i, x_i)\}_1^n$. The main computational difficulty arises in learning the unknown permutation. Linear regression with mismatched/permuted data (model (1.1)) has a long history in statistics dating back to 1960s [6].

Recently, this problem has garnered significant attention from the statistics and machine learning communities. A series of recent works [1–5,7–12,14] have studied the statistical and computational aspects of this model. To learn the coefficients β^* and the matrix P^*, one can consider the following natural optimization problem:

$$\min_{\beta, P} \ \|Py - X\beta\|^2 \quad \text{s.t. } P \in \Pi_n \tag{1.2}$$

where Π_n is the set of permutation matrices in $\mathbb{R}^{n \times n}$. Solving problem (1.2) is difficult as there are combinatorially many choices for $P \in \Pi_n$. Given P, it is easy to estimate β via least squares. [12] shows that in the noiseless setting ($\epsilon = 0$), a solution $(\hat{P}, \hat{\beta})$ of problem (1.2) equals (P^*, β^*) with probability one if $n \geq 2d$ and the entries of X are i.i.d. from a distribution that is absolutely continuous with respect to the Lebesgue measure. [5,8] studies the recovery of (P^*, β^*) under the noisy setting.

It is shown in [8] that Problem (1.2) is NP-hard for $d \geq 2$. A polynomial-time approximation algorithm appears in [5] for a fixed d, though this does not appear to result in a practical algorithm. Several heuristics have been proposed for (1.2): Examples include, alternating minimization [4,14], Expectation Maximization [1] but they lack theoretical guarantees. [10] uses robust regression methods to approximate solutions to (1.2) and discuss statistical properties of the corresponding estimator when the number of mismatched pairs is small.

Problem (1.2) can be formulated as a mixed integer program (MIP) with $O(n^2)$ binary variables. Solving this MIP with off-the-shelf MIP solvers (e.g., Gurobi) becomes computationally expensive for even a small value of n (e.g. $n \approx 50$). To our knowledge, there is no computationally practical algorithm that *provably* solves the original problem (1.2) under suitable statistical assumptions. Addressing this gap is the main focus of this paper: We propose and study a novel greedy local search method for Problem (1.2). Loosely speaking, our algorithm at every step swaps a pair of indices in the permutation in an attempt to improve the cost function. This algorithm is typically efficient in practice based on our preliminary numerical experiments. Suppose r denotes the number of mismatched pairs i.e., the Hamming distance between P^* and the identity matrix I_n. We establish theoretical guarantees on the convergence of the proposed method, under the assumption that r is small compared to n (we make this notion precise later), an assumption appearing in [10] (see also references therein). We consider the noiseless setting (i.e., $\epsilon = 0$) and establish that under some assumptions on the problem data, our local search method converges to an optimal solution of Problem (1.2).

Notation and Preliminaries: For a vector a, we let $\|a\|$ denote the Euclidean norm, $\|a\|_\infty$ the ℓ_∞-norm and $\|a\|_0$ the ℓ_0-pseudo-norm (i.e., number of nonzeros) of a. We let $\||\cdot\||_2$ denote the operator norm for matrices. Let $\{e_1, ..., e_n\}$ be the natural orthogonal basis of $\mathbb{R}^n$. For an interval $[a, b] \subseteq \mathbb{R}$, we let $|[a, b]| = b - a$. For a finite set S, we let $\#S$ denote its cardinality. For any permutation matrix P, let π_P be the corresponding permutation of $\{1, 2, ..., n\}$, that is, $\pi_P(i) = j$ if and only if $e_i^\top P = e_j^\top$ if and only if $P_{ij} = 1$. For two different permutation matrices P and Q, we define the distance between them as

$$\mathsf{dist}(P, Q) = \# \left\{ i \in [n] : \pi_P(i) \neq \pi_Q(i) \right\}. \tag{1.3}$$

For any permutation matrix $P \in \Pi_n$, we define its support as:

$$\mathsf{supp}(P) := \left\{ i \in [n] : \pi_P(i) \neq i \right\}. \tag{1.4}$$

For a real symmetric matrix A, let $\lambda_{\max}(A)$ and $\lambda_{\min}(A)$ denote the largest and smallest eigenvalues of A, respectively.

For two positive scalar sequences $\{a_n\}, \{b_n\}$, we write $a_n = \widetilde{O}(b_n)$ or equivalently, $a_n/b_n = \widetilde{O}(1)$, if a_n/b_n is bounded by a polynomial (of finite degree) in $\log(n)$. In particular, we view any value that can be bounded by a polynomial of $\log(n)$ as a constant.

2 A Local Search Method

Here we present our local search method for (1.2). For any fixed $P \in \Pi_n$, by minimizing the objective function in (1.2) with respect to β, we have an equivalent formulation

$$\min_P \quad \|Py - HPy\|^2 \quad \text{s.t.} \quad P \in \Pi_n \tag{2.1}$$

where $H = X(X^\top X)^{-1} X^\top$. To simplify the notation, denote $\widetilde{H} := I_n - H$, then Problem (2.1) is equivalent to

$$\min_P \quad \|\widetilde{H} Py\|^2 \quad \text{s.t.} \quad P \in \Pi_n . \tag{2.2}$$

For a given permutation matrix P, define the R-neighbourhood of P as

$$\mathcal{N}_R(P) := \left\{ Q \in \Pi_n : \mathsf{dist}(P, Q) \leq R \right\}. \tag{2.3}$$

It is easy to check that $\mathcal{N}_1(P) = \{P\}$, and for any $R \geq 2$, $\mathcal{N}_R(P)$ has more than one element. Algorithm 1 introduces our proposed local search method with search width R, which is an upper bound on the number of mismatched pairs.

Algorithm 1. Local search method with search width R for Problem (2.2).

Input: Initial permutation $P^{(0)} = I_n$. Search width R.
For $k = 0, 1, 2, \ldots$.

$$P^{(k+1)} \in \operatorname{argmin} \left\{ \|\widetilde{H}Py\|^2 : P \in \mathcal{N}_2(P^{(k)}) \cap \mathcal{N}_R(I_n) \right\}. \tag{2.4}$$

If $\|\widetilde{H}P^{(k+1)}y\|^2 = \|\widetilde{H}P^{(k)}y\|^2$, output $P^{(k)}$.

Algorithm 1 uses an explicit constraint on the search width at every step. When $R \geq n$, we perform local search without any constraint on the search width or neighborhood size. In this paper, we focus on the case where the underlying P^* is close to I_n, i.e., $r \ll n$. Under this assumption, it is reasonable to set $R = cr \ll n$ for some constant $c > 1$. See Sect. 3 for more details.

Let us examine the per-iteration cost of (2.4). The cardinality of $\mathcal{N}_2(P^{(k)})$ is upper bounded by $O(n^2)$. Furthermore, we note that

$$
\begin{aligned}
\|\widetilde{H}Py\|^2 &= \|\widetilde{H}(P - P^{(k)})y + \widetilde{H}P^{(k)}y\|^2 \\
&= \|\widetilde{H}(P - P^{(k)})y\|^2 + 2\langle (P - P^{(k)})y, \widetilde{H}P^{(k)}y \rangle + \|\widetilde{H}P^{(k)}y\|^2 .
\end{aligned} \tag{2.5}
$$

For each $P \in \mathcal{N}_2(P^{(k)})$, the vector $(P - P^{(k)})y$ has at most two nonzero entries. So the computation of the first term in (2.5) costs $O(1)$ operations. As we retain a copy of $\widetilde{H}P^{(k)}y$ in memory, computing the second term in (2.5) also costs $O(1)$ operations. Therefore, computing (2.4) using the procedure outlined above requires $O(n^2)$ operations.

3   Theoretical Guarantees for Local Search

In this section, we present theoretical guarantees for Algorithm 1. Our theory is based on the assumption that the data X is "well-behaved" (See Assumption 1). In particular, we assume that the projection matrix $\widetilde{H}$ satisfies a "restricted eigenvalue condition" (RE). (We caution the reader that despite nomenclature similarities, our notion of RE is different than what appears in the high-dimensional statistics literature [13]). To give an example, our RE condition is satisfied with high probability, when the rows of X are independent draws from a well-behaved multivariate distribution and when the sample size n is sufficiently large—see Sect. 3.1 for details. Under this RE condition, our analysis is completely deterministic in nature. The RE assumption on $\widetilde{H}$ allows us to relate the objective function $\|\widetilde{H}P^{(k)}y\|^2$ to a simple function $\|(P^{(k)} - P^*)y\|^2$. Then our analysis reduces to an analysis of the local structure of Π_n in terms of minimizing $\|(P^{(k)} - P^*)y\|^2$.

3.1   A Restricted Eigenvalue (RE) Condition

A main building block of our analysis is a RE property of $\widetilde{H}$. Define

$$\mathcal{B}_m := \{ w \in \mathbb{R}^n : \|w\|_0 \leq m \}. \tag{3.1}$$

We say that $\widetilde{H}$ satisfies a RE condition with parameter (δ, m) (denoted by the shorthand $\mathrm{RE}(\delta, m)$) if the following holds true

$$\mathrm{RE}(\delta, m): \qquad \|\widetilde{H}u\|^2 \geq (1 - \delta)\|u\|^2 \ \forall \ u \in \mathcal{B}_m. \tag{3.2}$$

To provide some intuition on the RE condition, we show (cf Lemma 1) that this condition is satisfied with high probability when the rows of X are drawn independently from a mean-zero distribution with finite support and a well-conditioned covariance matrix.

Lemma 1. *(Restricted eigenvalue property) Suppose $x_1, \ldots, x_n$ are i.i.d. zero-mean random vectors in $\mathbb{R}^d$ with covariance matrix $\Sigma \in \mathbb{R}^{d \times d}$. Suppose there exist constants $\gamma, b, V > 0$ such that $\lambda_{\min}(\Sigma) \geq \gamma$, $\|x_i\| \leq b$ and $\|x_i\|_\infty \leq V$ almost surely. Given any $\tau > 0$, define*

$$\delta_n := 16V^2\left(\frac{d}{n\gamma}\log(2d/\tau) + \frac{dm}{n\gamma}\log(3n^2)\right).$$

Suppose n is large enough such that $\sqrt{\delta_n} \geq 2/n$ and $\sqrt{3b^2 \log(2d/\tau)/(n \|\|\Sigma\|\|_2)} \leq 1/2$. Then with probability at least $1 - 2\tau$, condition $\mathrm{RE}(\delta_n, m)$ holds true.

The proof of this lemma is presented in Appendix 5.1. For simplicity, we state Lemma 1 for bounded x_i's; though this can be generalized to sub-Gaussian x_i's.

Lemma 1 implies that: given a pre-specified probability level (e.g., $1 - 2\tau = 0.99$), RE parameters δ, m, and other data parameters d, b, γ, Σ, we can choose $n = \widetilde{O}(dm/\delta)$ such that $\mathrm{RE}(\delta, m)$ holds with high probability. In the following, while presenting the scaling of (n, d, r) in the guarantees for Algorithm 1, when we say that data is generated from the setting of Lemma 1, we make the default assumption that there exist universal constants $\bar{c} > 0$ and $\bar{C} > 0$ such that the parameters $(\gamma, V, b, \|\|\Sigma\|\|_2, \tau)$ in Lemma 1 satisfy $\bar{c} \leq \gamma, V, b, \|\|\Sigma\|\|_2, \tau \leq \bar{C}$.

In Algorithm 1, we use a constraint on the search width, i.e., $P^{(k)} \in \mathcal{N}_R(I_n)$. Suppose $r = \mathrm{dist}(P^*, I_n) \ll n$ and we set $R = cr$ for some constant $c > 1$, then it holds that $\mathrm{dist}(P^{(k)}, P^*) \leq (c + 1)r$. This implies $P^{(k)}y - P^*y \in \mathcal{B}_{(c+1)r}$. In the noiseless setting with $\epsilon = 0$, we have $\widetilde{H}P^*y = 0$, and hence $\|\widetilde{H}P^{(k)}y\|^2 = \|\widetilde{H}(P^{(k)} - P^*)y\|^2$. Suppose the $\mathrm{RE}(\delta_n, (c+1)r)$ condition in (3.2) holds, because $P^{(k)}y - P^*y \in \mathcal{B}_{(c+1)r}$, we have

$$(1 - \delta_n)\|(P^{(k)} - P^*)y\|^2 \leq \|\widetilde{H}P^{(k)}y\|^2 \leq \|(P^{(k)} - P^*)y\|^2 \tag{3.3}$$

where, the second inequality is because $\|\|\widetilde{H}\|\|_2 \leq 1$. In light of (3.3), when δ_n is small, the objective function $\|\widetilde{H}P^{(k)}y\|^2$ can be approximately replaced by a simpler function $\|(P^{(k)} - P^*)y\|^2$. In what follows, we analyze the local search method on this simple approximation.

3.2 One-Step Decrease

We prove elementary lemmas on the one-step decrease property. Recall that for a given permutation matrix P, $\mathrm{supp}(P) = \{i \in [n] : \pi_P(i) \neq i\}$.

Lemma 2. *Given* $y \in \mathbb{R}^n$ *and a permutation matrix* $P \in \Pi_n$, *there exists a permutation matrix* $\widetilde{P} \in \Pi_n$ *such that* $\mathsf{dist}(P, \widetilde{P}) = 2$, $\mathsf{supp}(\widetilde{P}) \subseteq \mathsf{supp}(P)$ *and*

$$\|Py - y\|^2 - \|\widetilde{P}y - y\|^2 \geq (1/2)\|Py - y\|_\infty^2 .$$

The proof of Lemma 2 is presented in Sect. 5.2. Applying Lemma 2 with y replaced by P^*y and P replaced by $P(P^*)^{-1}$, we have the following corollary.

Corollary 1. *Given* $y \in \mathbb{R}^n$ *and* $P, P^* \in \Pi_n$, *there exists a permutation matrix* $\widetilde{P} \in \Pi_n$ *such that* $\mathsf{dist}(\widetilde{P}, P) = 2$, $\mathsf{supp}(\widetilde{P}(P^*)^{-1}) \subseteq \mathsf{supp}(P(P^*)^{-1})$ *and*

$$\|Py - P^*y\|^2 - \|\widetilde{P}y - P^*y\|^2 \geq (1/2)\|Py - P^*y\|_\infty^2 .$$

Corollary 1 provides a lower bound on the change of the (approximate) objective value as one moves from permutation P to $\widetilde{P}$, and will be used in the analysis of the local search algorithm. When $Py - P^*y$ is sparse, Corollary 1 translates to a contraction in the ℓ_2-norm of $Py - P^*y$, as shown below.

Corollary 2. *Let* $y \in \mathbb{R}^n$ *and* $P, P^* \in \Pi_n$; *and suppose* $\|Py - P^*y\|_0 \leq m$. *Let* $\widetilde{P} \in \Pi_n$ *be the permutation matrix appearing in Corollary 1. Then*

$$\|\widetilde{P}y - P^*y\|^2 \leq (1 - 1/(2m)) \|Py - P^*y\|^2 . \tag{3.4}$$

Proof. Since $\|Py - P^*y\|_0 \leq m$, it holds $\|Py - P^*y\|^2 \leq m\|Py - P^*y\|_\infty^2$. Using Corollary 1, we have:

$$\|Py - P^*y\|^2 - \|\widetilde{P}y - P^*y\|^2 \geq (1/2)\|Py - P^*y\|_\infty^2 \geq (1/(2m))\|Py - P^*y\|^2 ,$$

which results in the conclusion (3.4). □

3.3 Main Results

Here we state and prove the main theorem on the convergence of Algorithm 1. We first state the assumptions used in our proof. Recall that $r = \mathsf{dist}(P^*, I_n)$.

Assumption 1. *(1.) We consider a linear model (1.1) with noise term* $\epsilon = 0$.
(2). There exist constants $U > L > 0$ *such that* $U \geq |(P^*y)_i - y_i| \geq L$ *for all* $i \in \mathsf{supp}(P^*)$.
(3). In Algorithm 1, we set $R = 10C_1 r U^2/L^2 + 4$ *for some constant* $C_1 > 1$.
(4). For some $\delta_n < 1/(4(r + R))$ *the condition* $RE(\delta_n, R + r)$ *holds.*

Note that the lower bound in Assumption 1 (2) ensures that any two mismatched responses are not too close. Assumption 1 (3) requires that R be set to a constant multiple of r. This constant can be large ($\geq 10U^2/L^2$), and is an artifact of our proof techniques. Our numerical experience however, suggests that this constant can be much smaller in practice. Assumption 1 (4) is a restricted eigenvalue condition. This property holds true under the settings stated in Lemma 1 when $n \geq Cdr^2$ for some constant $C > 0$.

We first present a technical result used in the proof of Theorem 1.

Lemma 3. *Suppose Assumption 1 holds. Let* $\{P^{(k)}\}_k$ *be the permutation matrices generated by Algorithm 1. Suppose* $\|P^{(k)}y - P^*y\|_\infty \geq L$ *for some* $k \geq 1$. *If for all* $t \leq k - 1$, *at least one of the two conditions holds: (i)* $t \leq R/2 - 1$; *or (ii)* $\mathsf{supp}(P^*) \subseteq \mathsf{supp}(P^{(t)})$, *then for all* $t \leq k - 1$, *we have*

$$\|P^{(t+1)}y - P^*y\|^2 - \|P^{(t)}y - P^*y\|^2 \leq -L^2/5 . \tag{3.5}$$

We omit the proof of Lemma 3 due to space constraints. Lemma 3 is used for technical reasons. In our analysis, we make heavy use of the one-step decrease condition in Corollary 2. Note that if the permutation matrix at the current iteration, denoted by $P^{(k)}$, is on the boundary i.e. $\mathsf{dist}(P^{(k)}, I_n) = R$, it is not clear whether the permutation found by Corollary 2 is within the search region $\mathcal{N}_R(I_n)$. Lemma 3 helps address this issue (See the proof of Theorem 1 below for details).

We now state and prove the linear convergence of Algorithm 1.

Theorem 1. *Suppose Assumption 1 holds with* R *being an even number. Let* $\{P^{(k)}\}$ *be the permutation matrices generated by Algorithm 1. Then*

1) For all $k \geq R/2$, *we have that* $\mathsf{supp}(P^*) \subseteq \mathsf{supp}(P^{(k)})$.
2) For any $k \geq 0$,

$$\|\widetilde{H}P^{(k)}y\|^2 \leq \left(1 - \frac{1}{4(R+r)}\right)^k \|\widetilde{H}P^{(0)}y\|^2 .$$

Proof. Part 1) We show this result by contradiction. Suppose that there exists a $k \geq R/2$ such that $\mathsf{supp}(P^*) \not\subseteq \mathsf{supp}(P^{(k)})$. Let $T \geq R/2$ be the first iteration such that $\mathsf{supp}(P^*) \not\subseteq \mathsf{supp}(P^{(T)})$, i.e.,

$$\mathsf{supp}(P^*) \not\subseteq \mathsf{supp}(P^{(T)}) \quad \text{and} \quad \mathsf{supp}(P^*) \subseteq \mathsf{supp}(P^{(k)}) \ \forall \ R/2 \leq k \leq T - 1 .$$

Let $i \in \mathsf{supp}(P^*)$ but $i \notin \mathsf{supp}(P^{(T)})$, then by Assumption 1 (2),

$$\|P^{(T)}y - P^*y\|_\infty \geq |e_i^\top(P^{(T)}y - P^*y)| = |e_i^\top(y - P^*y)| \geq L.$$

By Lemma 3, we have $\|P^{(k+1)}y - P^*y\|^2 - \|P^{(k)}y - P^*y\|^2 \leq -L^2/5$ for all $k \leq T - 1$. Summing up these inequalities, we have

$$\|P^{(T)}y - P^*y\|^2 - \|P^{(0)}y - P^*y\|^2 \leq -TL^2/5 \leq -RL^2/10 \tag{3.6}$$

where the last inequality follows from our assumption that $T \geq R/2$. From Assumption 1 (2) and noting that $P^{(0)} = I_n$, we have

$$\|P^{(0)}y - P^*y\|^2 = \|y - P^*y\|^2 \leq rU^2, \tag{3.7}$$

where we use that $(y - P^*y)$ is r-sparse. Using (3.6) and (3.7), we have:

$$\|P^{(T)}y - P^*y\|^2 \leq rU^2 - RL^2/10 \overset{(a)}{\leq} rU^2 - \frac{L^2}{10}\frac{10C_1rU^2}{L^2} = (1-C_1)U^2 \overset{(b)}{<} 0, \tag{3.8}$$

where above, the inequality (a) uses $R = 10C_1rU^2/L^2 + 4$; and (b) uses $C_1 > 1$. Note that (3.8) leads to a contradiction, so such an iteration counter T does not exist; and for all $k \geq R/2$, we have $\mathsf{supp}(P^*) \subseteq \mathsf{supp}(P^{(k)})$.

Part 2) By Corollary 2, there exists a permutation matrix $\widetilde{P}^{(k)} \in \Pi_n$ such that $\mathsf{dist}(\widetilde{P}^{(k)}, P^{(k)}) \leq 2$, $\mathsf{supp}(\widetilde{P}^{(k)}(P^*)^{-1}) \subseteq \mathsf{supp}(P^{(k)}(P^*)^{-1})$ and

$$\|\widetilde{P}^{(k)}y - P^*y\|^2 \leq \left(1 - \frac{1}{2\|P^{(k)}y - P^*y\|_0}\right)\|P^{(k)}y - P^*y\|^2 .$$

Since $\|P^{(k)}y - P^*y\|_0 \leq \mathsf{dist}(P^{(k)}, I_n) + \mathsf{dist}(P^*, I_n) \leq r + R$, we have

$$\|\widetilde{P}^{(k)}y - P^*y\|^2 \leq \left(1 - \frac{1}{2(R+r)}\right)\|P^{(k)}y - P^*y\|^2. \tag{3.9}$$

Note that $\widetilde{H}P^*y = \widetilde{H}X\beta^* = 0$ and $\|\|\widetilde{H}\|\|_2 \leq 1$, so we have

$$\|\widetilde{H}\widetilde{P}^{(k)}y\|^2 = \|\widetilde{H}(\widetilde{P}^{(k)}y - P^*y)\|^2 \leq \|\widetilde{P}^{(k)}y - P^*y\|^2 . \tag{3.10}$$

In the following, we use the shorthand notation $\widetilde{R} = R + r$. Combining (3.9) and (3.10) we have

$$\|\widetilde{H}\widetilde{P}^{(k)}y\|^2 \leq \|\widetilde{P}^{(k)}y - P^*y\|^2 \leq (1 - (2\widetilde{R})^{-1})\|P^{(k)}y - P^*y\|^2. \tag{3.11}$$

By Assumption 1 (4), we have $\|\widetilde{H}(P^{(k)} - P^*)y\|^2 \geq (1 - \delta_n)\|(P^{(k)} - P^*)y\|^2$. Combining this with (3.11) we have

$$\begin{aligned} \|\widetilde{H}\widetilde{P}^{(k)}y\|^2 &\leq (1 - \delta_n)^{-1}(1 - (2\widetilde{R})^{-1})\|\widetilde{H}(P^{(k)} - P^*)y\|^2 \\ &= (1 - \delta_n)^{-1}(1 - (2\widetilde{R})^{-1})\|\widetilde{H}P^{(k)}y\|^2, \end{aligned} \tag{3.12}$$

where the last line uses $\widetilde{H}P^*y = 0$. Since $\delta_n \leq 1/(4\widetilde{R})$, we have

$$(1 - \delta_n)^{-1}(1 - (2\widetilde{R})^{-1}) \leq 1 - (4\widetilde{R})^{-1}$$

which when used in (3.12) leads to:

$$\|\widetilde{H}\widetilde{P}^{(k)}y\|^2 \leq (1 - (4\widetilde{R})^{-1})\|\widetilde{H}P^{(k)}y\|^2. \tag{3.13}$$

To complete the proof, we will make use of the following claim, the proof of this claim is presented in Appendix 5.3.

Claim. For any $k \geq 0$ it holds $\widetilde{P}^{(k)} \in \mathcal{N}_R(I_n) \cap \mathcal{N}_2(P^{(k)})$. $\tag{3.14}$

Starting with the definition of $P^{(k+1)}$, we have the following inequalities:

$$\|\widetilde{H}P^{(k+1)}y\|^2 = \min_{P \in \mathcal{N}_2(P^{(k)}) \cap \mathcal{N}_R(I_n)} \|\widetilde{H}Py\|^2 \overset{(a)}{\leq} \|\widetilde{H}\widetilde{P}^{(k)}y\|^2$$

$$\overset{(b)}{\leq} (1 - (4\widetilde{R})^{-1})\|\widetilde{H}P^{(k)}y\|^2,$$

where, (a) makes use of the above claim $\widetilde{P}^{(k)} \in \mathcal{N}_R(I_n) \cap \mathcal{N}_2(P^{(k)})$; and (b) uses inequality (3.13). Therefore, we have:

$$\|\widetilde{H}P^{(k+1)}y\|^2 \leq (1 - (4\widetilde{R})^{-1})\|\widetilde{H}P^{(k)}y\|^2,$$

which leads to the conclusion in part 2.   □

Theorem 1 shows that the sequence of objective values generated by Algorithm 1 converges to zero (the optimal objective value of (2.2)) at a linear rate. The parameter for the linear rate of convergence depends upon $r = \text{dist}(P^*, I_n)$ and the search width R. The proof is based on the assumption that the RE condition holds (Assumption 1 (4)) with some $\delta_n \leq 1/(4(R+r))$. This RE condition holds under the setting of Lemma 1 when $n \geq Cdr^2$ for some constant $C > 0$ (See Sect. 3.1). The sample-size requirement is more stringent than that needed in order for the model to be identifiable ($n \geq 2d$) [12]. In particular, when $n/d = \widetilde{O}(1)$, the number of mismatched pairs r needs to be bounded by a constant. While our theory appears to suggest that n needs to be quite large to learn P^*, numerical evidence presented in Sect. 4 suggests that one can recover P^* with a smaller sample size.

4   Experiments

We numerically study the convergence performance of Algorithm 1. We consider the noiseless setup $P^*y = X\beta^*$ where entries of $X \in \mathbb{R}^{n \times d}$ are iid $N(0,1)$; all coordinates of $\beta^* \in \mathbb{R}^d$ are iid $N(0,1)$ (β^* is independent of X). To generate P^*, we fix $r \geq 1$ and select r coordinates uniformly from $\{1, \ldots, n\}$, then generate a uniformly distributed random permutation on these r coordinates[1].

We test the performance of Algorithm 1 with different combinations of (d, r, n). We simply set $R = n$ in Algorithm 1. Even though this setting is not covered by our theory, in practice when r is small, the algorithm converges to optimality with the number of iterations being bounded by a small constant multiple of r (e.g., for $r = 50$, the algorithm converges to optimality within around 60 iterations). We set the maximum number of iterations as 1000. For the results presented below, we consider 50 independent trials and present the averaged results.

Figure 1 presents the results on examples with $n = 500$, $d \in \{20, 50, 100, 200\}$, and 40 roughly equispaced values of $r \in [10, 400]$. In Fig. 1 [left panel], we plot the Hamming distance of the solution $\hat{P}$ computed by Algorithm 1 and the underlying permutation P^* (i.e. $\text{dist}(\hat{P}, P^*)$) versus r. In Fig. 1 [right panel], we present error in estimating β versus r. More precisely, let $\hat{\beta}$ be the solution of computed by Algorithm 1 (i.e. $\hat{\beta} = (X^\top X)^{-1} X^\top \hat{P} y$), then the *"beta error"* is defined as $\|\hat{\beta} - \beta^*\|/\|\beta^*\|$. For each choice of (r, d), the point on the line is the average of 50 independent replications, and the vertical error bar shows the

---

[1] This permutation P^* may not satisfy $\text{dist}(P^*, I_n) = r$, but $\text{dist}(P^*, I_n)$ will be close to r.

standard deviation of the mean (the error bars are small and hardly visible in the figures). As shown in Fig. 1, when r is small, the underlying permutation P^* can be exactly recovered, and thus the corresponding beta error is also 0. As r becomes larger, Algorithm 1 fails to recover P^* exactly; and $\text{dist}(P^*, \hat{P})$ is close to the maximal possible value 500. In contrast, the estimation error of β^* behaves in a continuous way: As the value of r increases, the value of $\|\hat{\beta} - \beta^*\|/\|\beta^*\|$ increases continuously. We also observe that the recovery of P^* depends upon the number of covariates d. This is consistent with our analysis that the performance of our algorithm depends upon both r and d.

Figure 2 presents similar results where we exchange the roles of r and d. It shows examples with $n = 500$, $r \in \{20, 50, 100, 200\}$, and 40 different values of d ranging from 10 to 400. When d is small, Algorithm 1 is able to recover P^* exactly. But when d exceeds a certain threshold, $\text{dist}(\hat{P}, P^*)$ increases quickly. The threshold for larger r is smaller. From Fig. 2 [left panel], it is interesting to note a non-monotone behavior of the Hamming distance as d increases. In contrast, the beta error increases continuously as d increases (see Fig. 2 [right panel]).

In terms of the speed of Algorithm 1, we note that for an instance with $n = 500$, $d = 100$ and $r = 50$, Algorithm 1 outputs the solution within around 60 iterations and 0.25 s on the Julia 1.2.0 platform. The total computational time scales approximately as $O(n^2 r)$ when exact recovery is achieved.

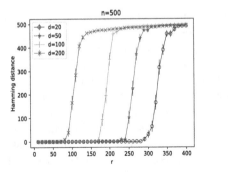 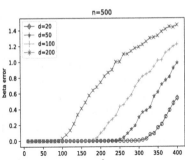

Fig. 1. Left: values hamming distance $\text{dist}(\hat{P}, P^*)$ versus r. Right: values of beta error $\|\hat{\beta} - \beta^*\|/\|\beta^*\|$ versus r.

5   Appendix: Proofs and Technical Results

Lemma 4. *Suppose rows $x_1, ..., x_n$ of the matrix of covariates X are i.i.d. zero-mean random vectors in $\mathbb{R}^d$ with covariance matrix $\Sigma \in \mathbb{R}^{d \times d}$. Suppose $\|x_i\| \leq b$ almost surely. Then for any $t > 0$, it holds*

$$\mathbb{P}\left(\| \tfrac{1}{n} X^\top X - \Sigma \|_2 \geq t \| \Sigma \|_2 \right) \leq 2d \exp\left(- \frac{nt^2 \| \Sigma \|_2}{2b^2(1+t)} \right).$$

See e.g. Corollary 6.20 of [13] for a proof.

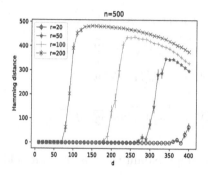
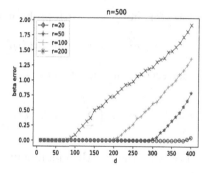

Fig. 2. Left: values of hamming distance $\mathrm{dist}(\hat{P}, P^*)$ vs r. Right: values of beta error $\|\hat{\beta} - \beta^*\|/\|\beta^*\|$ vs r.

5.1 Proof of Lemma 1

Proof. It suffices to prove that for any $u \in \mathcal{B}_m$ (cf definition (3.1)),

$$\|Hu\|^2 = \|X(X^{\top}X)^{-1}X^{\top}u\|^2 \le \delta_n\|u\|^2 . \tag{5.1}$$

Take $t_n := \sqrt{3b^2\log(2d/\tau)/(n \||\Sigma\||_2)}$. When n is large enough, we have $t_n \le 1/2$, then from Lemma 4 and some simple algebra we have

$$\|| \frac{1}{n}X^{\top}X - \Sigma \||_2 \le t_n \||\Sigma\||_2 \tag{5.2}$$

with probability at least $1 - \tau$. When (5.2) holds, we have

$$\lambda_{\min}(X^{\top}X)/n \ge (1 - t_n)\lambda_{\min}(\Sigma) \ge (1 - t_n)\gamma \ge \gamma/2$$

where, we use $t_n \le 1/2$. Hence we have $\lambda_{\max}((X^{\top}X)^{-1}) \le 2/(n\gamma)$ and

$$\|| X(X^{\top}X)^{-1}\||_2 = \sqrt{\lambda_{\max}((X^{\top}X)^{-1})} \le \sqrt{2/(n\gamma)} . \tag{5.3}$$

Let $\mathcal{B}_m(1) := \{u \in \mathcal{B}_m : \|u\| \le 1\}$, and let $u^1, ..., u^M$ be an $(\sqrt{\delta_n}/2)$-net of $\mathcal{B}_m(1)$, that is, for any $u \in \mathcal{B}_m(1)$, there exists some u^j such that $\|u^j - u\| \le \sqrt{\delta_n}/2$. Since the $(\sqrt{\delta_n}/2)$-covering number of $\mathcal{B}_m(1)$ is bounded by $(6/\sqrt{\delta_n})^m\binom{n}{m}$, we can take

$$M \le (6/\sqrt{\delta_n})^m\binom{n}{m} \le (3n)^m n^m = (3n^2)^m$$

where the second inequality is from our assumption that $\sqrt{\delta_n} \ge 2/n$. By Hoeffding inequality, for each fixed $j \in [M]$, and for all $k \in [d]$, we have

$$\mathbb{P}\left(\frac{1}{\sqrt{n}} |e_k^{\top}X^{\top}u^j| > t \right) \le 2\exp\left(-\frac{nt^2}{2\|u^j\|^2 U^2} \right) .$$

Therefore, for any $\rho > 0$, with probability at least $1 - \rho$, we have

$$|e_k^\top X^\top u^j|/\sqrt{n} \le \sqrt{2\log(2d/\rho)/n}\, V\|u^j\| \le V\sqrt{2\log(2d/\rho)/n}\ ,$$

where the second inequality is because each $u^j \in \mathcal{B}_m(1)$. As a result,

$$\frac{1}{\sqrt{n}}\|X^\top u^j\| = \Big(\sum_{k=1}^{d} (|e_k^\top X^\top u^j|/\sqrt{n})^2 \Big)^{1/2} \le V\sqrt{2d\log(2d/\rho)/n}\ .$$

Take $\rho = \tau/M$, then by the union bound, with probability at least $1-\tau$, it holds

$$\|X^\top u^j\|/\sqrt{n} \le V\sqrt{2d\log(2dM/\tau)/n} \quad \forall\, j \in [M]\ . \tag{5.4}$$

Combining (5.4) with (5.3), we have that for all $j \in [M]$,

$$\|X(X^\top X)^{-1}X^\top u^j\| \le \||\, X(X^\top X)^{-1}\,\||_2 \cdot \|X^\top u^j\| \tag{5.5}$$
$$\le 2V\sqrt{(d/n\gamma)\log(2dM/\tau)}\ .$$

Recall that $M \le (3n^2)^m$, so we have

$$2V\sqrt{(d/n\gamma)\log(2dM/\tau)} \le 2V\Big(\frac{d}{n\gamma}\log(2d/\tau) + \frac{dm}{n\gamma}\log(3n^2)\Big)^{1/2} \le \frac{\sqrt{\delta_n}}{2}.$$

where the last inequality follows the definition of δ_n. Using the above bound in (5.5), we have

$$\|X(X^\top X)^{-1}X^\top u^j\| \le \sqrt{\delta_n}/2.$$

For any $u \in \mathcal{B}_m(1)$, there exists some $j \in [M]$ such that $\|u - u^j\| \le \sqrt{\delta_n}/2$, hence

$$\|X(X^\top X)^{-1}X^\top u\| \le \|X(X^\top X)^{-1}X^\top u^j\| + \|X(X^\top X)^{-1}X^\top (u - u^j)\|$$
$$\le \sqrt{\delta_n}/2 + \|u - u^j\|_2 \le \sqrt{\delta_n}\ . \tag{5.6}$$

Since both (5.2) and (5.4) have failure probability of at most τ, we know that (5.6) holds with probability at least $1 - 2\tau$. This proves the conclusion for all $u \in \mathcal{B}_m(1)$. For a general $u \in \mathcal{B}_m$, $u/\|u\| \in \mathcal{B}_m(1)$, hence we have

$$\|Hu\| = \|X(X^\top X)^{-1}X^\top u\| \le \sqrt{\delta_n}\|u\|$$

which is equivalent to what we had set out to prove (5.1).   □

5.2   Proof of Lemma 2

Proof. For any $k \in [n]$, let $k_+ := \pi_P(k)$. Let i be an index such that

$$\big(y_{i_+} - y_i\big)^2 = \|Py - y\|_\infty^2\ .$$

Without loss of generality, we can assume $y_{i_+} > y_i$. Denote $i_0 = i$ and $i_1 = i_+$. By the structure of a permutation, there exists a cycle that

$$i_0 \xrightarrow{P} i_1 \xrightarrow{P} \cdots \xrightarrow{P} i_t \xrightarrow{P} \cdots \xrightarrow{P} i_s = i_0 \qquad (5.7)$$

where $q_1 \xrightarrow{P} q_2$ means $q_2 = \pi_P(q_1)$. By moving from y_i to y_{i_+}, the first step in the cycle (5.7) "upcrosses" the value $(y_i + y_{i_+})/2$. Since the cycle (5.7) returns to i_0 finally, there must exist one step that "downcrosses" the value $(y_i + y_{i_+})/2$. In other words, there exists $j \in [n]$ with $(j, j_+) \neq (i, i_+)$ such that $y_{j_+} < y_j$ and $(y_i + y_{i_+})/2 \in [y_{j_+}, y_j]$. Define $\widetilde{P}$ as follows:

$$\pi_{\widetilde{P}}(i) = j_+, \quad \pi_{\widetilde{P}}(j) = i_+, \quad \pi_{\widetilde{P}}(k) = \pi_P(k) \ \forall k \neq i, j .$$

We immediately know $\mathsf{dist}(P, \widetilde{P}) = 2$ and $\mathsf{supp}(\widetilde{P}) \subseteq \mathsf{supp}(P)$. Since

$$y_{i_+} - y_i = \|Py - y\|_\infty \geq y_j - y_{j_+},$$

there are 3 cases depending upon the ordering of $y_i, y_{i_+}, y_j, y_{j_+}$. We consider these cases to arrive at the final inequality in Lemma 2.

Case 1: $(y_j \geq y_{i_+} \geq y_{j_+} \geq y_i)$ In this case, let $a = y_j - y_{i_+}$, $b = y_{i_+} - y_{j_+}$ and $c = y_{j_+} - y_i$. Then $a, b, c \geq 0$, and

$$
\begin{aligned}
\|Py - y\|^2 - \|\widetilde{P}y - y\|^2 &= (y_i - y_{i_+})^2 + (y_j - y_{j_+})^2 - (y_i - y_{j_+})^2 - (y_j - y_{i_+})^2 \\
&= (b+c)^2 + (a+b)^2 - a^2 - c^2 \\
&= 2b^2 + 2ab + 2bc .
\end{aligned}
$$

Since $(y_i + y_{i_+})/2 \in [y_{j_+}, y_j]$, we have

$$b = y_{i_+} - y_{j_+} \geq y_{i_+} - \frac{y_i + y_{i_+}}{2} = \frac{y_{i_+} - y_i}{2} ,$$

and hence

$$\|Py - y\|^2 - \|\widetilde{P}y - y\|^2 \geq 2b^2 \geq \frac{(y_{i_+} - y_i)^2}{2} = \frac{1}{2}\|Py - y\|_\infty^2 .$$

Case 2: $(y_{i_+} \geq y_j \geq y_i \geq y_{j_+})$. In this case, let $a = y_{i_+} - y_j$, $b = y_j - y_i$ and $c = y_i - y_{j_+}$. Then $a, b, c \geq 0$, and

$$
\begin{aligned}
\|Py - y\|^2 - \|\widetilde{P}y - y\|^2 &= (y_i - y_{i_+})^2 + (y_j - y_{j_+})^2 - (y_i - y_{j_+})^2 - (y_j - y_{i_+})^2 \\
&= (a+b)^2 + (b+c)^2 - a^2 - c^2 \\
&= 2b^2 + 2ab + 2bc .
\end{aligned}
$$

Since $(y_i + y_{i_+})/2 \in [y_{j_+}, y_j]$, we have

$$b = y_j - y_i \geq \frac{y_i + y_{i_+}}{2} - y_i = \frac{y_{i_+} - y_i}{2} ,$$

and hence

$$\|Py - y\|^2 - \|\widetilde{P}y - y\|^2 \geq 2b^2 \geq \frac{(y_{i_+} - y_i)^2}{2} = \frac{1}{2}\|Py - y\|_\infty^2 .$$

Case 3: $(y_{i_+} \geq y_j \geq y_{j_+} \geq y_i)$. In this case, let $a = y_{i_+} - y_j$, $b = y_j - y_{j_+}$ and $c = y_{j_+} - y_i$. Then $a, b, c \geq 0$, and

$$\begin{aligned}\|Py - y\|^2 - \|\widetilde{P}y - y\|^2 &= (y_i - y_{i_+})^2 + (y_j - y_{j_+})^2 - (y_i - y_{j_+})^2 - (y_j - y_{i_+})^2 \\ &= (a + b + c)^2 + b^2 - a^2 - c^2 \\ &= 2b^2 + 2ab + 2bc + 2ac .\end{aligned}$$

Note that $\|Py - y\|_\infty^2 = (y_i - y_{i_+})^2 = (a+b+c)^2$. Because $(y_i + y_{i_+})/2 \in [y_{j_+}, y_j]$, we know that $a \leq (a + b + c)/2$ and $c \leq (a + b + c)/2$. So we have

$$\|Py - y\|^2 - \|\widetilde{P}y - y\|^2 \geq w\|Py - y\|_\infty^2 ,$$

where

$$w := \min\left\{\frac{2b^2 + 2ab + 2bc + 2ac}{(a + b + c)^2} : a, b, c \geq 0; \ a, c \leq (a + b + c)/2\right\}.$$

This is equivalent to

$$\begin{aligned}w &= \min\left\{2b^2 + 2ab + 2bc + 2ac : a, b, c \geq 0; \ a, c \leq 1/2; \ a + b + c = 1\right\} \\ &= \min\left\{2b + 2ac : a, b, c \geq 0; \ a, c \leq 1/2; \ a + b + c = 1\right\} \\ &= \min\left\{2(1 - a - c) + 2ac : a, c \geq 0; \ a, c \leq 1/2\right\} \\ &= \min\left\{2(1 - a)(1 - c) : a, c \geq 0; \ a, c \leq 1/2\right\} \\ &= 1/2\end{aligned}$$

$\square$

5.3   Proof of Claim (3.14) in Theorem 1

Proof. To prove this claim, we just need to prove that $\widetilde{P}^{(k)} \in \mathcal{N}_R(I_n)$, i.e. $\mathsf{dist}(\widetilde{P}^{(k)}, I_n) \leq R$. If $k \leq R/2 - 1$, because $\mathsf{dist}(P^{(t+1)}, P^{(t)}) \leq 2$ for all $t \geq 0$ and $P^{(0)} = I_n$, we have $\mathsf{dist}(P^{(k)}, I_n) \leq 2k \leq R - 2$. Hence

$$\mathsf{dist}(\widetilde{P}^{(k)}, I_n) \leq \mathsf{dist}(\widetilde{P}^{(k)}, P^{(k)}) + \mathsf{dist}(P^{(k)}, I_n) \leq R .$$

We consider the case when $k \geq R/2$. By Part (1) of Theorem 1, it holds $\mathsf{supp}(P^*) \subseteq \mathsf{supp}(P^{(k)})$. We will show that $\mathsf{supp}(\widetilde{P}^{(k)}) \subseteq \mathsf{supp}(P^{(k)})$. Equivalently, we just need to show that for any $i \notin \mathsf{supp}(P^{(k)})$, we have $i \notin \mathsf{supp}(\widetilde{P}^{(k)})$. Let $i \notin \mathsf{supp}(P^{(k)})$, then $e_i^\top P^{(k)} = e_i^\top$. Since $\mathsf{supp}(P^*) \subseteq \mathsf{supp}(P^{(k)})$, we also have $e_i^\top P^* = e_i^\top$. So it holds $e_i^\top P^{(k)}(P^*)^{-1} = e_i^\top$ or equivalently $i \notin \mathsf{supp}(P^{(k)}(P^*)^{-1})$. Because $\mathsf{supp}(\widetilde{P}^{(k)}(P^*)^{-1}) \subseteq \mathsf{supp}(P^{(k)}(P^*)^{-1})$, we have $i \notin \mathsf{supp}(\widetilde{P}^{(k)}(P^*)^{-1})$, or equivalently $e_i^\top \widetilde{P}^{(k)}(P^*)^{-1} = e_i^\top$. This implies $e_i^\top \widetilde{P}^{(k)} = e_i^\top P^* = e_i^\top$, or equivalently, $i \notin \mathsf{supp}(\widetilde{P}^{(k)})$.

$\square$

References

1. Abid, A., Zou, J.: Stochastic EM for shuffled linear regression. arXiv preprint arXiv:1804.00681 (2018)
2. Dokmanić, I.: Permutations unlabeled beyond sampling unknown. IEEE Signal Process. Lett. **26**(6), 823–827 (2019)
3. Emiya, V., Bonnefoy, A., Daudet, L., Gribonval, R.: Compressed sensing with unknown sensor permutation. In: 2014 IEEE International Conference on Acoustics, Speech and Signal Processing (ICASSP), pp. 1040–1044. IEEE (2014)
4. Haghighatshoar, S., Caire, G.: Signal recovery from unlabeled samples. IEEE Trans. Signal Process. **66**(5), 1242–1257 (2017)
5. Hsu, D.J., Shi, K., Sun, X.: Linear regression without correspondence. In: Advances in Neural Information Processing Systems, pp. 1531–1540 (2017)
6. Neter, J., Maynes, E.S., Ramanathan, R.: The effect of mismatching on the measurement of response errors. J. Am. Stat. Assoc. **60**(312), 1005–1027 (1965)
7. Pananjady, A., Wainwright, M.J., Courtade, T.A.: Denoising linear models with permuted data. In: 2017 IEEE International Symposium on Information Theory (ISIT), pp. 446–450. IEEE (2017)
8. Pananjady, A., Wainwright, M.J., Courtade, T.A.: Linear regression with shuffled data: statistical and computational limits of permutation recovery. IEEE Trans. Inf. Theory **64**(5), 3286–3300 (2017)
9. Shi, X., Li, X., Cai, T.: Spherical regression under mismatch corruption with application to automated knowledge translation. J. Am. Stat. Assoc., 1–12 (2020)
10. Slawski, M., Ben-David, E., Li, P.: Two-stage approach to multivariate linear regression with sparsely mismatched data. J. Mach. Learn. Res. **21**(204), 1–42 (2020)
11. Tsakiris, M.C., Peng, L., Conca, A., Kneip, L., Shi, Y., Choi, H., et al.: An algebraic-geometric approach to shuffled linear regression. arXiv preprint arXiv:1810.05440 (2018)
12. Unnikrishnan, J., Haghighatshoar, S., Vetterli, M.: Unlabeled sensing with random linear measurements. IEEE Trans. Inf. Theory **64**(5), 3237–3253 (2018)
13. Wainwright, M.J.: High-Dimensional Statistics: A Non-asymptotic Viewpoint, vol. 48. Cambridge University Press, Cambridge (2019)
14. Wang, G., et al.: Signal amplitude estimation and detection from unlabeled binary quantized samples. IEEE Trans. Signal Process. **66**(16), 4291–4303 (2018)

A New Integer Programming Formulation of the Graphical Traveling Salesman Problem

Robert D. Carr[1] and Neil Simonetti[2(✉)]

[1] Computer Science Department, University of New Mexico,
Albuquerque, NM 87131, USA
bobcarr@unm.edu

[2] Business, Computer Science, and Mathematics Department, Bryn Athyn College,
Bryn Athyn, PA 19009-0717, USA
neil.simonetti@brynathyn.edu

Abstract. In the Traveling Salesman Problem (TSP), a salesman wants to visit a set of cities and return home. There is a cost c_{ij} of traveling from city i to city j, which is the same in either direction for the Symmetric TSP. The objective is to visit each city exactly once, minimizing total travel costs. In the Graphical TSP, a city may be visited more than once, which may be necessary on a sparse graph. We present a new integer programming formulation for the Graphical TSP requiring only two classes of constraints that are either polynomial in number or polynomially separable, while addressing an open question proposed by Denis Naddef.

Keywords: Linear program · Relaxation · TSP · Traveling Salesman Problem · GTSP · Graphical Traveling Salesman Problem

1 Introduction

The Traveling Salesman Problem (TSP), is one of the most studied problems in combinatorial optimization [9,10,13]. In its classic form, a salesman wants to visit each of a set of cities exactly once and return home while minimizing travel costs. Costs of traveling between cities are stored in a matrix where entry c_{ij} indicates the cost of traveling from city i to city j. Units may be distance, time, money, etc.

If the underlying graph for the TSP is sparse, a complete cost matrix can still be constructed by setting c_{ij} equal to the shortest path between city i and city j for each pair of cities. However, this has the disadvantage of turning a sparse graph $G = (V, E)$ where the edge set E could be of size $O(|V|)$ into a complete graph $G' = (V, E')$, where the edge set E' is $\Omega(|V|^2)$.

R. D. Carr—This material is based upon research supported in part by the U. S. Office of Naval Research under award number N00014-18-1-2099.

© Springer Nature Switzerland AG 2021
M. Singh and D. P. Williamson (Eds.): IPCO 2021, LNCS 12707, pp. 458–472, 2021.
https://doi.org/10.1007/978-3-030-73879-2_32

Ratliff and Rosenthal [18] were the first to consider a case where the edge set is not expanded to a complete graph, but left sparse, while soon after, Fleischmann [7] and Cornuéjols, Fonlupt, and Naddef [4] examined this in a more general case, the latter giving this its name: the Graphical Traveling Salesman Problem (GTSP). As a consequence, a city may be visited more than once, since there is no guarantee the underlying graph will be Hamiltonian. While the works of Fleischmann and Cornuéjols et al. focused on cutting planes and facet-defining inequalities, this paper will look at a new compact formulation. While the theoretical bound for the integrality gap is not improved, our initial computational results show that in some practial cases, the integrality gap is reduced when solving a linear programming relaxation of the problem.

2   Basic Formulations

This paper will investigate the symmetric GTSP, where the cost of traveling between two cities is the same, regardless of direction, which allows the following notation to be used:

$G = (V, E)$: The graph G with vertex set V and edge set E.

c_e : The cost of using edge $e \in E$, replaces c_{ij}.

x_e : The variable indicating the use of edge $e \in E$, replaces x_{ij}
which is used in most general TSP formulations.

$\delta(v)$: The set of edges incident to vertex v.

$\delta(S)$: The set of edges with exactly one endpoint in vertex set S.

$x(F)$: The sum of variables x_e for all $e \in F \subset E$.

If given a formulation on a complete graph K_n, a formulation for a sparse graph G can be created by simply setting $x_e = 0$ for any edge e in the graph K_n but not in the graph G.

2.1   Symmetric TSP

The standard formulation for the TSP, attributed to Dantzig, Fulkerson, and Johnson [5], contains constraints that guarantee the degree of each node in a solution is exactly two (degree constraints) and constraints that prevent a solution from being a collection of disconnected subtours (subtour elimination constraints).

$$
\begin{aligned}
\text{minimize} \quad & \sum_{e \in E} c_e x_e \\
\text{subject to} \quad & \sum_{e \in \delta(v)} x_e = 2 \ \forall v \in V \\
& \sum_{e \in \delta(S)} x_e \geq 2 \ \forall S \subset V, \ S \neq \emptyset \\
& x_e \in \{0, 1\} \quad \forall e \in E.
\end{aligned}
$$

When this integer program is relaxed, the integer constraints $x_e \in \{0, 1\}$ are replaced by the boundary constraints $0 \leq x_e \leq 1$.

It should also be noted that while the subtour elimination constraints are only needed for the cases where $3 \leq |S| \leq \frac{|V|}{2}$, there are still exponentially many of these constraints. Using a similar technique to Martin [15], and Carr and Lancia [1], these constraints can be replaced by a polynomial number of flow constraints which ensure the solution is a 2 edge-connected graph.

2.2 Symmetric GTSP

This formulation for the Graphical TSP comes from Cornuéjols, Fonlupt, and Naddef [4], and differs from the formulation above by allowing the degree of a node to be any even integer, and by removing any upper bound on the variables.

$$
\begin{array}{lll}
\text{minimize} & \sum_{e \in E} c_e x_e & \\
\text{subject to} & \sum_{e \in \delta(v)} x_e \text{ is positive and even } \forall v \in V & \\
& \sum_{e \in \delta(S)} x_e \geq 2 & \forall S \subset V,\ S \neq \emptyset \\
& x_e \geq 0 & \forall e \in E. \\
& x_e \text{ is integer} & \forall e \in E.
\end{array}
$$

When this program is relaxed, the integer constraints at the end are removed, and the disjunctive constraints that require the degree of each node to be any positive and even integer are effectively replaced by a lower bound of two on the degree of each node.

The disjunctive constraints for the formulation above are unusual for two reasons. Firstly, in most mixed-integer programs, only individual variables are constrained to be integers, not sums of variables. Secondly, the sum is required not to be just integer, but an even integer. In terms of a mixed-integer formulation, the second peculiarity could be addressed with:

$$
\sum_{e \in \delta(v)} \frac{x_e}{2} \in \mathbb{Z} \ \forall v \in V
$$

To our knowledge, no other integer programming formulation for a graph theory application uses constraints of this kind on non-binary integer variables. The T-join problem uses binary variables, and its relationship to our problem will be discussed in the next section.

Addressing the first peculiarity, that integer and mixed-integer programs only allow integrality of variables, we could set these sums to new variables indexed on the vertices of the graph, d_v.

$$
\sum_{e \in \delta(v)} \frac{x_e}{2} = d_v \ \forall v \in V
$$
$$
d_v \in \mathbb{Z} \ \forall v \in V
$$

While this approach works, it feels unsatisfying. The addition of these d_v variables is purely cosmetic. When solving the relaxation, there is nothing preventing us from waiting until a solution is generated before defining the values of d_v using the sums above. Thus the new variables do not facilitate the addition of any new constraints, and do nothing to strengthen the LP relaxation in any way.

When solving the integer program, we can bypass the d_v variables by branching with constraints based on the degree sums. For example, if the solution from a relaxation creates a graph where node i has odd degree q, we branch with constraints of the form:

$$\sum_{e \in \delta(i)} x_e \leq q - 1 \quad \text{and} \quad \sum_{e \in \delta(i)} x_e \geq q + 1$$

At a conference, Denis Naddef proposed a challenge of finding a set of constraints for a mixed-integer formulation of GTSP, where integrality constraints are limited to only $x_e \in \mathbb{Z}$ [16]. We address the state of this challenge in Sect. 4.

3   New Constraints

Cornuéjols et al. proved that an upper bound of two on each x_e is implied if all the edge costs are positive [4] (and also note that without this additional bound, graphs with negative weight edges would not have finite optimal solutions). This fact allows us to dissect the variables x_e into two components y_e and z_e such that, for each edge $e \in E$:

$$y_e = 1 \text{ if edge } e \text{ is used exactly once,} \quad y_e = 0 \text{ otherwise}$$
$$z_e = 1 \text{ if edge } e \text{ is used exactly twice,} \quad z_e = 0 \text{ otherwise}$$

Note that
$$x_e = y_e + 2z_e \tag{1}$$

Additionally, we can add the constraint $y_e + z_e \leq 1$ for each edge $e \in E$, since using both would imply an edge being used three times in a solution. But more importantly, we now have a way to enforce even degree without using disjunctions, since only the y_e variables matter in determining if the degree of a node is odd or even with respect to the x_e variables.

3.1   Enforcing Even Degree Without Disjunctions

Since the upper bound on the y_e variables is one, the following constraints will enforce even degree:

$$\sum_{e \in F}(1 - y_e) + \sum_{e \in \delta(v) \setminus F} y_e \geq 1 \quad \forall v \in V \text{ and } F \subset \delta(v) \text{ with } |F| \text{ odd.} \tag{2}$$

This type of constraint was used by Yannakakis et al. [21] and Lancia et al. [12] when working with the parity polytope, and has appeared in exact formulations of parity constrained problems such as T-joins [19], mentioned below.

Note that for integer values of y_e, if the y-degree of a node v is odd, then when F is the set of edges incident to v indicated by y, the expression in the left-hand side of the constraint above will be zero. If the y-degree of a node v is even, then for any set F with $|F|$ odd, the left-hand side must be at least one.

For sparse graphs, this adds $O(|V|2^{\Delta-1})$ constraints, where Δ is the maximum degree in G. Typical graphs from roadmaps usually have $\Delta \leq 6$, while graphs from highway maps might have $\Delta \leq 8$. Also note that Euler's formula for planar graphs guarantees that $|E| \leq 3|V| - 6$, and so the average degree of a node in a planar graph cannot be more than six.

One can generalize these constraints on the values of y_e by using sets of nodes instead of individual nodes.

$$\sum_{e \in F}(1 - y_e) + \sum_{e \in \delta(S) \setminus F} y_e \geq 1 \quad \forall S \subset V \text{ and } F \subset \delta(S) \text{ with } |F| \text{ odd.} \quad (3)$$

However, we lose the bound on the number of constraints from sparse graphs, since there may be several nodes in the set S. These constraints are T-join constraints on the y_e variables, where the set T is empty, since we require every node to have even degree [6]. Therefore, every set of nodes, S, will contain an even number, namely zero, of nodes from the set of odd degree nodes, T, which is the empty set.

Unfortunately, in the relaxation of the linear program with these constraints, odd degree nodes (or sets) can still happen by allowing a path of nodes where, for each edge e in the path, $y_e = 0$ and $z_e = \frac{1}{2}$. See Fig. 1.

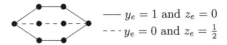

$$\begin{array}{ll} \text{——} & y_e = 1 \text{ and } z_e = 0 \\ \text{- - -} & y_e = 0 \text{ and } z_e = \frac{1}{2} \end{array}$$

Fig. 1. A path where $y_e = 0$ and $z_e = \frac{1}{2}$

3.2   Spanning Tree Constraints

One method to discourage this half-z path is to require the edges indicated by $\boldsymbol{y}$ and $\boldsymbol{z}$ contain a spanning tree. This is different than demanding that $\boldsymbol{x}$ contains a spanning tree since each unit of z_e contributes two units to x_e. For our spanning tree constraint, each z_e contributes only one unit toward a spanning tree. This means that for any node whose y-degree is zero, the z-degree must be at least one, and in the case where two nodes with y-degree zero are connected using an edge in the spanning tree, the z-degree of one of those nodes must be at least two, which effectively prevents this half-z path.

Place constraints on binary variables $\boldsymbol{t}$ such that the edges where $t_e = 1$ indicate a tree that spans all nodes of G (the graph must be connected and contain no cycles). This is done by the well-known partition inequalities that will be discussed in Sect. 4.

As with the subtour elimination constraints, Martin describes a compact set of constraints that ensure t indicates a tree (or a convex combination of trees) [15]. Then, add the constraint:

$$t_e \leq y_e + z_e, \quad \forall e \in E \tag{4}$$

Only the connectedness of the graph indicated by t_e is important, since the constraint only requires that $y + z$ dominate a spanning tree, so the constraints that would prevent cycles are unnecessary.

This constraint is valid since any tour that visits every node has within it a spanning tree that touches each node.

In the case of Fig. 1, we use a partition of three sets to find a violated constraint: the six nodes in the outer cycle form one set, and the individual interior nodes form the other two sets. Define P to be the set of all edges that have endpoints in different sets of the partition. Then the partition constraint demands that $\sum_{e \in P} t_e \geq 2$. But along the three non-zero edges that connect this partition, $y_e = 0$ and $z_e = \frac{1}{2}$, so $y_e + z_e$ on these edges only adds to $\frac{3}{2}$.

4   A New Mixed IP Formulation

4.1   Proving the Formulation

The tree constraints are sufficient, when combined with constraint (2) and integrality constraints $y_e \in \{0, 1\}$, to find an optimal integer solution value, making the subtour elimination constraints unnecessary. The following new mixed integer programming formulation therefore does not include these optional constraints. Note that all GTSP tours will satisfy the constraints in this formulation. The notation $\delta(V_1, ..., V_k)$ refers to the set of edges with endpoints in different vertex sets.

$$
\begin{aligned}
\text{minimize} \quad & \sum_{e \in E} c_e x_e \\
\text{subject to} \quad & x_e = y_e + 2z_e & \forall e \in E & \tag{4.1} \\
& \sum_{e \in F}(1 - y_e) + \sum_{e \in \delta(v) \setminus F} y_e \geq 1 \ \forall v \in V \text{ and } F \subset \delta(v) \text{ with } |F| \text{ odd} & \tag{4.2} \\
& \sum_{e \in \delta(V_1, ..., V_k)} t_e \geq k - 1 & \forall \text{ partitions } V_1, ..., V_k \text{ of } V & \tag{4.3} \\
& t_e \leq y_e + z_e \leq 1 & \forall e \in E & \tag{4.4} \\
& 0 \leq t_e \leq 1 & \forall e \in E & \tag{4.5} \\
& 0 \leq z_e \leq 1 & \forall e \in E & \tag{4.6} \\
& y_e \in \{0, 1\} & \forall e \in E & \tag{4.7}
\end{aligned}
$$

Theorem 1. *Given a MIP solution (y^*, z^*) to the GTSP formulation above, then $x^* = y^* + 2z^*$ will indicate an edge set that is an Euler tour, or a convex combination of Euler tours.*

Proof. It should be noted that it is sufficient for the MIP solution to dominate an Euler tour (or a convex combination of them), since if there is some edge e where x_e^* is larger than necessary by ϵ units (for some $0 < \epsilon \leq x_e^* \leq 2$), one can add edge e twice to a collection of Euler tours with total weight $\frac{\epsilon}{2}$.

Let $(\boldsymbol{y}^*, \boldsymbol{z}^*)$ be a feasible solution to the GTSP formulation specified. By constraints (4.3) and (4.4), we know that $\boldsymbol{y}^* + \boldsymbol{z}^*$ dominates a convex combination of spanning trees, thus we have $\boldsymbol{y}^* + \boldsymbol{z}^* \geq \sum_i \lambda_i \boldsymbol{T}^i$, where each $\boldsymbol{T}^i$ is an edge incidence vector of a spanning tree. Define $\boldsymbol{R}^i$ by $R_e^i = 1$ if both $T_e^i = 1$ and $y_e^* = 0$, and $R_e^i = 0$ otherwise. So $\boldsymbol{R}^i$ becomes the remnant of tree $\boldsymbol{T}^i$ when edges indicated by $\boldsymbol{y}^*$ are removed. Since $\boldsymbol{y}^*$ only contains integer values, the constraint $y_e + z_e \leq 1$ guarantees that $z_e = 0$ whenever $y_e = 1$, and guarantees $y_e = 0$ whenever $z_e > 0$. This means that for all edges where $y_e = 0$, we have $\boldsymbol{z}^* \geq \sum_i \lambda_i \boldsymbol{T}^i = \sum_i \lambda_i \boldsymbol{R}^i$. Since $\boldsymbol{R}^i$ is the result of removing edges from $\boldsymbol{T}^i$ where $y_e = 1$, we are guaranteed that $R_e^i = 0$ for every i where $y_e = 1$, and thus $\boldsymbol{z}^* \geq \sum_i \lambda_i \boldsymbol{R}^i$ over all edges.

Hence, $\boldsymbol{x}^* = \boldsymbol{y}^* + 2\boldsymbol{z}^* \geq \boldsymbol{y}^* + 2\sum_i \lambda_i \boldsymbol{R}^i = \sum_i \lambda_i (\boldsymbol{y}^* + 2\boldsymbol{R}^i)$, where for each i, $\boldsymbol{y}^* + 2\boldsymbol{R}^i$ is an Euler tour, because constraint (4.2) ensures the graph indicated by $\boldsymbol{y}^* + 2\boldsymbol{R}^i$ will have even degree at every node, and constraints (4.3) and (4.4) ensure the graph indicated by $\boldsymbol{y}^* + \boldsymbol{R}^i$, and thus $\boldsymbol{y}^* + 2\boldsymbol{R}^i$, is connected. $\qquad \square$

Constraints (4.3), commonly referred to as partition inequalities [17] [20], are exponential in number, but these can also be reduced to a compact set of constraints using the techniques from Martin [15]. The constraints we used are below, and require a model using directed edges to regulate flow variables ϕ. Assume $V = \{1, 2, ..., n\}$ and designate city n as the home city.

In our formulation (inspired by Martin), we use directed flow variables ϕ^k that carry one unit of flow from any node with index higher than k to node k and supported by the values $\overrightarrow{t}_{ij}$ as edge capacities. From any feasible integral solution (directed spanning tree), it is not hard to derive such a set of unit flows by directing the tree from the home node n. For this flow, we can now set flow going into any node j with $j > k$ to zero in flow problem ϕ^k, and flow balancing constraints among the nodes numbered $k + 1$ or higher are also unnecessary. Finally, $t_e = \overrightarrow{t}_{ij} + \overrightarrow{t}_{ji}$ to create variables for the undirected spanning tree.

$$\phi_{i,j}^k = 0 \qquad \forall j \in V, k \in V \setminus \{n\} \text{ with } j > k, \{i,j\} \in E$$

$$\phi_{k,i}^k = 0 \qquad \forall k \in V \setminus \{n\}, \{i,k\} \in E$$

$$\sum_{i \in \delta(k)} \phi_{i,k}^k = 1 \qquad \forall k \in V \setminus \{n\}$$

$$\sum_{i \in \delta(j)} \phi_{i,j}^k - \sum_{i \in \delta(j)} \phi_{j,i}^k = 0 \qquad \forall j \in V, k \in V \setminus \{n\} \text{ with } j < k$$

$$0 \leq \phi_{i,j}^k \leq \overrightarrow{t}_{ij} \qquad \forall k \in V \setminus \{n\}, \{i,j\} \in E$$

$$t_e = \overrightarrow{t}_{ij} + \overrightarrow{t}_{ji} \, \forall e = \{i,j\} \in E$$

$$\sum_{e \in E} t_e \leq n - 1$$

$$t_e \leq y_e + z_e \qquad \forall e \in E$$

Constraints (4.2) are exponential in Δ, the maximum degree of the graph. This is not a concern if the graph is sparse, leading to a compact formulation. If the graph is not sparse, identifying when a constraint from the set (4.2) is violated, a process called separation, can be done efficiently, even if the solution is from a relaxation and thus contains fractional values for some y_e variables.

Theorem 2. *Given a solution to the relaxation of the GTSP formulation above without constraints (4.2), if a constraint from (4.2) is violated, it can be found in $O(|V|^2)$ time.*

Proof. For each node $v \in V$, minimize the left-hand side of constraint (4.2) over all possible sets $F \subset \delta(v)$ ($|F|$ even or odd), by placing edges with $y_e > \frac{1}{2}$ in F and leaving edges with $y_e < \frac{1}{2}$ for $\delta(v) \setminus F$. Edges with $y_e = \frac{1}{2}$ could go in either set.

- If this minimum is not less than 1, no constraint from (4.2) will be violated for this node.
- If the minimum is less than 1, and $|F|$ is odd, this is a violated constraint from (4.2).
- If the minimum is less than 1, and $|F|$ is even, find the edge e where $|y_e - \frac{1}{2}|$ is smallest. Then flip the status of the membership of edge e in F. This will create the minimum left-hand side over all sets F with $|F|$ odd.

For each node, this requires summing or searching items indexed by $\delta(v)$ a constant number of times, and since $|\delta(v)| < |V|$ this requires $O(|V|^2)$ time. □

4.2 Addressing the Naddef Challenge

We would have preferred to simply require the values in x to be integer and allow y and z to hold fractional values, which addresses the challenge that Denis Naddef proposed [16]. He wished to know if one could find a simple formulation for the GTSP that finds optimal solutions by only requiring integrality of the decision variables x^*, and nothing else. But this cannot be done (for polynomially-sized or polynomially-separable classes of inequalities unless $P = NP$), which can be seen by the following theorem.

Theorem 3. *Let G be a 3-regular graph, and let G' be the result of adding one vertex to the middle of each edge in G. Consider a solution x^*, where $x_e^* = 1$ for each edge $e \in G'$. Then x^* is in the GTSP polytope iff G is Hamiltonian.*

Proof. If G is Hamiltonian, let P be the set of edges in a Hamilton cycle of G, and let P' be the set of corresponding edges in the graph G'. Note that in G' every degree-two node is adjacent to two degree-three nodes, and that the cycle P' reaches every degree-three node in G'. One GTSP tour in G' can be created by adding an edge of weight two on exactly one of the two edges adjacent to each degree-two node in G' but not used in P'. The other GTSP tour can be created by adding an edge of weight two on the edges not chosen by the first tour. The convex combination of each of these tours with weight $\frac{1}{2}$ will create a solution where $x_e^* = 1$ for each edge $e \in G'$

Now suppose we have a solution $\boldsymbol{x}^*$, where $x_e^* = 1$ for each edge $e \in G'$ and $\boldsymbol{x}^*$ is in the GTSP polytope. Express $\boldsymbol{x}^* = \sum_k \lambda_k \boldsymbol{\chi}^k$ as a convex combination of GTSP tours. Consider any degree-two vertex v in G'. Since v has degree two, $\boldsymbol{x}^*(\delta(v)) = 2$. Also $\boldsymbol{\chi}^k(\delta(v)) \geq 2$ must be true for any GTSP tour $\boldsymbol{\chi}^k$, and so, by the convex combination, it must be that $\boldsymbol{\chi}^k(\delta(v)) = 2$ for each $\boldsymbol{\chi}^k$. If the neighbors of v are nodes i and j, then $\boldsymbol{\chi}^k(\delta(v)) = 2$ implies either $\chi_{i,v}^k = 1$ and $\chi_{j,v}^k = 1$, or $\chi_{i,v}^k = 2$ and $\chi_{j,v}^k = 0$, or $\chi_{i,v}^k = 0$ and $\chi_{j,v}^k = 2$. The edges of weight one in $\boldsymbol{\chi}^k$ form disjoint cycles, so pick one such cycle C, and let S_1 be the set of vertices in C. (If there are no edges of weight one in $\boldsymbol{\chi}^k$, let S_1 be a set containing any single degree three vertex.) Let S_2 be the set of degree two vertices v such that $\chi_{i,v}^k = 2$ for some $i \in S_1$. Notice that $\boldsymbol{\chi}^k(\delta(S_1 \cup S_2)) = 0$, since the degree of any node, $v \in S_2$ is exactly two in $\boldsymbol{\chi}^k$, and the edge connecting v to S_1 has weight two. $\boldsymbol{\chi}^k$ is a tour and thus must be connected, which is only possible if $S_1 \cup S_2$ is the entire vertex set of G', and therefore the cycle C must visit every degree three node in G'. The corresponding cycle in the graph G would therefore be a Hamilton cycle.   □

If one knew when the integer solution $\boldsymbol{x}^*$ were in the GTSP polytope, then this theorem would imply a polynomial time algorithm to determine if a 3-regular graph is Hamiltonian, which is an NP-complete problem [8].

The challenge that Naddef proposed never specifically defined what makes a formulation simple. Certainly having all constraint sets be polynomially-sized or polynomially-separable (in terms of n, the number of nodes) would qualify as simple, but there may be other normal sets of constraints that could satisfy the spirit of Naddef's challenge. One such example is Naddef's conjecture that simply using the three classes of inequalities from his 1985 paper (path, wheelbarrow, and bicycle inequalities) [4] with integrality constraints only on the variables $\boldsymbol{x}$, would be sufficient to formulate the problem. Since it is not known if these three classes of inequalities can be separated in polynomial time, the theorem above does not directly address this conjecture.

However, if Naddef's proposed solution above were sufficient to describe the GTSP polytope, then the following reasoning will hold: If we are given an arbitrary constraint of the form $\boldsymbol{ax} \geq b$, it can be recognized in polynomial time whether or not this constraint belongs to a particular class of inequality (such as path, wheelbarrow, or bicycle). And it can be recognized whether or not a potential solution $\boldsymbol{x}^*$, corresponding to the graph G from the theorem above, violates this constraint. If that potential solution $\boldsymbol{x}^*$ were not in the GTSP polytope, then the violated inequality (path, wheelbarrow, or bicycle) would serve as a polynomially sized verification that the graph G is not Hamiltonian, which would imply $NP = $ co-NP.

This would apply to any integer programming formulation with a finite number of inequality classes that contain inequalities that are normal. In this case, we define normal to mean that the membership of any individual constraint in a class can be verified in polynomial time.

This implies Naddef's challenge cannot be completed successfully using normal inequalities, unless $NP = \text{co-}NP$. However, our formulation follows its spirit, as the integer constrained variables in our formulation $\boldsymbol{y}$ have a one-to-one correspondence to the integer constrained variables $\boldsymbol{x}$ in the challenge.

5   Relaxations and Steiner Nodes

5.1   Symmetric GTSP with Steiner Nodes

Cornuéjols et al. also proposed a variant of the GTSP where only a subset of nodes are visited [4]. As in most road networks, one may travel through many intersections that are not also destinations when traveling from one place to another. Cornuéjols et al. referred to these intersection nodes as Steiner nodes. This creates a formulation on a graph $G = (V_d \cup V_s, E)$ with $V_d \cap V_s = \emptyset$ where V_d represents the set of destination nodes and V_s the set of Steiner nodes.

$$
\begin{aligned}
\text{minimize} \quad & \sum_{e \in E} c_e x_e \\
\text{subject to} \quad & \sum_{e \in \delta(v)} x_e \text{ is positive and even } \forall v \in V_d \\
& \sum_{e \in \delta(v)} x_e \text{ is even} && \forall v \in V_s \\
& \sum_{e \in \delta(S)} x_e \geq 2 && \forall S \subset V \text{ where } S \cap V_d \neq \emptyset, \neq V_d \\
& x_e \geq 0 && \forall e \in E. \\
& x_e \text{ is integer} && \forall e \in E.
\end{aligned}
$$

Note that only sets that include destination nodes need to have corresponding cut constraints, and these can be limited to sets where the intersection is $\frac{|V_d|}{2}$ or smaller. Again, these can be replaced by the flow constraints in the style proposed by Martin [15]. The constraints used in our computational results are similar to those in the multi-commodity flow formulation found by Letchford, Nasiri, and Theis [14]. We were able to reduce the number of variables used by Letchford, et al. by a factor of 2, by setting many variables to 0, as we did with the flow variables in the formulation of Sect. 4.1. Assume $V_d = \{1, 2, ..., d\}$ and $V_s = \{d+1, d+2, ...n\}$ and designate city d as the home city. We include $d-1$ flow problems, where each problem requires that 2 units of flow pass from nodes $S = \{k+1, k+2, ..., d\}$ to node k using the values of $\boldsymbol{x}$ as edge capacities. (We use 2 units of flow to prevent any cut separating destination nodes having a value less than 2 in $\boldsymbol{x}$.) Since nodes in S are all sources, we can set flow into these nodes to zero, as well as setting the flow coming out of node k to zero.

$$
\begin{aligned}
f_{i,j}^{k} &= 0 \quad \forall j \in V_d, k \in V_d \setminus \{d\} \text{ with } j > k, \{i, j\} \in E \\
f_{k,i}^{k} &= 0 \quad \forall k \in V_d \setminus \{d\}, \{i, k\} \in E \\
\sum_{i \in \delta(k)} f_{i,k}^{k} &= 2 \quad \forall k \in V_d \setminus \{d\}
\end{aligned}
$$

$$\sum_{i \in \delta(j)} f_{i,j}^k - \sum_{i \in \delta(j)} f_{j,i}^k = 0 \quad \forall j \in V, k \in V_d \setminus \{d\} \text{ with either } j < k \text{ or } j \in V_s$$

$$f_{i,j}^k + f_{j,i}^k \leq x_e \quad \forall k \in V_d \setminus \{d\}, \{i,j\} = e \in E$$

5.2 Preventing the Half-z Path Without Spanning Trees

While the spanning tree constraints (4) of Sect. 3.2 can prevent half-z paths (see Fig. 1) when integrality of $\boldsymbol{y}$ is enforced, for the computational results in the next section, the integrality gaps for many of our instances were better when using the subtour elimination constraints plus the following, which prevents half-z paths (with three or more edges) without requiring the integrality of $\boldsymbol{y}$.

$$\sum_{e' \in \delta(i)} x_{e'} + \sum_{e' \in \delta(j)} x_{e'} - 2z_e \geq 4 \quad \forall e \in E \qquad (5)$$

where i and j are endpoints of edge e.

If $z_e = 1$, this constraint is the subtour elimination constraint for the set $\{i, j\}$. If $z_e = 0$, this is the sum of the degree constraints (lower bound) for nodes i and j. But in the middle of a path of length three or longer with edges that have $y_e = 0$ and $z_e = \frac{1}{2}$, the left side of this constraint will only add to three.

It should be noted that this constraint can only be used when both endpoints of e are destination nodes, since Steiner nodes do not have a lower bound of degree 2, but could be degree zero.

It should also be noted that if the GTSP instance is composed only of three paths of length three between two specific nodes (see Fig. 1 from Sect. 3.1) constraints (2) from Sect. 3.1 (those that enforce even degree) and (5) (defined above) will be enough to close the entire integrality gap using an LP relaxation. If the paths are all four or more edges long, this constraint will not eliminate the integrality gap, but will help. (See Fig. 2)

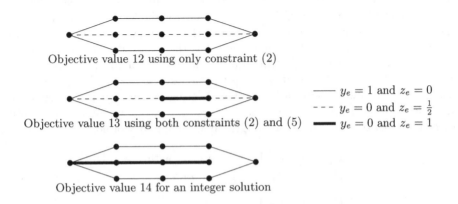

Objective value 12 using only constraint (2)

Objective value 13 using both constraints (2) and (5)

—— $y_e = 1$ and $z_e = 0$
- - - $y_e = 0$ and $z_e = \frac{1}{2}$
—— $y_e = 0$ and $z_e = 1$

Objective value 14 for an integer solution

Fig. 2. Solutions from a 3-path configuration of four edges each

As the paths get longer, the integrality gap slowly grows. The spanning tree constraints will be useful once the paths reach a length of at least seven. In our computational experiments, these spanning tree constraints never were useful in reducing the integrality gap.

5.3   Removing Steiner Nodes

Removing Steiner nodes increases the effectiveness of constraints in (5). A graph with Steiner nodes $G = (V_d \cup V_s, E)$ can be transformed into a graph without Steiner nodes $G' = (V_d, E')$ by doing the following:

For each pair of nodes $i, j \in V_d$, if the shortest path from i to j in G contains no other nodes in V_d, then add an edge connecting i to j to E' with a cost equal to the cost of this shortest path; otherwise, add no edge from i to j to E'.

In all but one of our test problems (see Sect. 6), removing Steiner nodes resulted in fewer, not more, edges in the original instance. That removing Steiner nodes often reduces the total edges in a graph was also observed by Corberán, Letchford, and Sanchis [3].

6   Computational Results

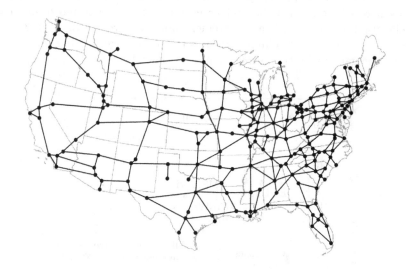

Fig. 3. Map of basic United States highway system

Our search for a reasonable sized data set based on the interstate highways of the United States led us to a text file uploaded by Sergiy Kolodyazhnyy on GitHub [11]. After a few errors were corrected and additions made, we had a highway network with 216 nodes and 358 edges, with a maximum degree node of seven

(Indianapolis). See Fig. 3. Data for this graph and the instances in this section, may be found at https://ns.faculty.brynathyn.edu/interstate/.

Instances were created from this map by choosing a subset of cities as destination nodes, and adding any cities along a shortest path between destinations as Steiner nodes. Alternate versions of these instances were constructed by removing the Steiner nodes as indicated in Sect. 5.3. Table 1 gives the basic information for several instances we used. Table 2 shows the results from running the relaxation of the formulation from Cornuéjols et al. [4]. It should be noted that the solutions found by this relaxation were the same whether Steiner nodes were removed or not.

Running times on a 2.1 GHz Xeon processor for all of the relaxations were under 10 s, while the running times to generate the integer solutions never exceeded five minutes. We wish to point out that the value of the new formulation is not a faster running time, but the reduced integrality gap.

In this section, the integrality gap refers to the difference in objective values between the program where integer constraints are enforced and the program where the integer constraints are relaxed. The percentage is the gap size expressed as a percentage of the integer solution value. This is different than the ratio definitions of integrality gap used in most other contexts. (e.g. [2])

When our constraints were added, the spanning tree constraints (4) were not useful when (2) and (5) were present. In most cases, removing Steiner nodes did not change the optimal values found by our relaxation. In one case, the relaxation was better when the Steiner nodes were removed, and in one case, the relaxation was worse when the Steiner nodes were removed. Table 3 shows our results, where the last column indicates the percentage that our formulation closed of the gap left by the formulation of Cornuéjols et al. [4].

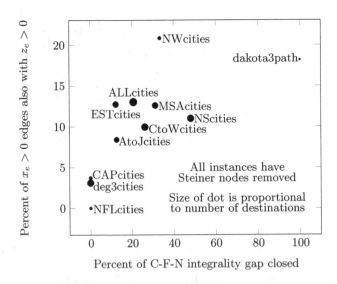

Fig. 4. Scatter plot of integrality gap closure and percent of variables with $z_e > 0$

We noticed that in instances where the z_e variables were rarely positive, our relaxation fared no better than that of Cornuéjols et al. But when the number of edges with values of $z_e > 0$ reached about 10% of the total of edges where $x_e > 0$, we were able to shave anywhere from 10% to almost 50% of the gap left behind by Cornuéjols et al. (See Fig. 4)

Table 1. GTSP instances

Name	Description	Number of Destinations	Solution (miles)
dakota3path	3-path configuarion in northern plains	11	2682
NFLcities	Cities with National Football League teams	29	11050
NWcities	Cities in the Northwest region	43	8119
CAPcities	48 state capitals plus Washington D.C	49	14878
AtoJcities	Cities beginning with letters from A to J	101	17931
ESTcities	Cities east of the Mississippi River	139	13251
MSAcities	Centers of 145 metropolitan statistical areas	145	22720
deg3cities	Cities in original graph with degree ≥ 3	171	18737
NScities	Cities that Neil Simonetti has visited	174	22127
CtoWcities	Cities beginning with letters from C to W	182	24389
ALLcities	Entire graph	216	26410

Table 2. Relaxations from Cornuéjols et al. formulation [4]

Name	Destinations (Steiner Nodes)	Edges (w/o Steiner)	Integrality Gap (%)
dakota3path	11 (0)	12 (12)	139 (5.18%)
NFLcities	29 (152)	304 (135)	35 (0.32%)
NWcities	43 (4)	63 (59)	12 (0.15%)
CAPcities	49 (132)	301 (199)	34 (0.23%)
AtoJcities	101 (95)	326 (289)	261.5 (1.45%)
ESTcities	139 (2)	243 (240)	61.4 (0.46%)
MSAcities	145 (63)	348 (317)	143 (0.63%)
deg3cities	171 (16)	321 (305)	70 (0.37%)
NScities	174 (29)	341 (324)	93.5 (0.42%)
CtoWcities	182 (30)	353 (358)	151 (0.62%)
ALLcities	216 (0)	358 (358)	274.8 (1.04%)

Table 3. Relaxations from our additional constraints

Name	Integrality Gap with Steiner Nodes (%)	Integrality Gap w/o Steiner Nodes (%)	Best % of Gap Closed from Formulation in [4]
dakota3path	–	0 (0%)	100%
NFLcities	35 (0.32%)	Same as Steiner	0%
NWcities	8 (0.10%)	Same as Steiner	33.3%
CAPcities	34 (0.23%)	Same as Steiner	0%
AtoJcities	228.5 (1.45%)	Same as Steiner	12.6%
ESTcities	53.9 (0.46%)	Same as Steiner	12.2%
MSAcities	114.5 (0.50%)	98.5 (0.43%)	31.1%
deg3cities	70 (0.37%)	Same as Steiner	0%
NScities	48.5 (0.22%)	Same as Steiner	48.1%
CtoWcities	103 (0.42%)	111.5 (0.46%)	31.8%
ALLcities	–	217.8 (0.82%)	20.7%

R. D. Carr and N. Simonetti

References

1. Carr, R.D., Lancia, G.: Compact vs. Exponential-size LP relaxations. Oper. Res. Lett. **30**, 57–66 (2002)
2. Carr, R.D., Vempala, S.: On the held-Karp relaxation for the asymmetric and symmetric traveling salesman problems. Math. Program. **100**, 569–587 (2004)
3. Corberán, A., Letchford, A.N., Sanchis, J.M.: A cutting plane algorithm for the general routing problem. Math. Program. **90**, 291–316 (2001)
4. Cornuéjols, G., Fonlupt, J., Naddef, D.: The traveling salesman problem on a graph and some related integer Polyhedra. Math. Program. **33**, 1–27 (1985)
5. Dantzig, G., Fulkerson, R., Johnson, S.: Solution of a large-scale traveling salesman problem. Oper. Res. **2**, 393–410 (1954)
6. Edmonds, J., Johnson, E.L.: Matching, Euler tours and the Chinese postman. Math. Program. **5**, 88–124 (1973)
7. Fleischmann, B.: A cutting plane procedure for the traveling salesman problem on road networks. Eur. J. Oper. Res. **21**(3), 307–317 (1985)
8. Garey, M.R., Johnson, D.S., Tarjan, E.: The planar hamiltonian circuit problem is NP-complete. SIAM J. Comput. **5**, 704–714 (1976)
9. Guten, G., Punnen, A.P. (eds.): The Traveling Salesman Problem and its Variations. Springer, New York (2007) https://doi.org/10.1007/b101971
10. Junger, M., Reinelt, G., Rinaldi, G.: The traveling salesman problem. In: Ball, M.O., Magnanti, T.L., Monma, C.L., Nemhauser, G.L. (eds.) Handbooks in Operations Research and Management Science, vol. 7, Network Models Elsevier, Amsterdam (1995)
11. Kolodyazhnyy, S.: Dijkstra Algorithm for Shortest Path. https://github.com/SergKolo/MSUD-CS2050-SPRING-2016/blob/master/input_for_dijkstra.txt (web) Accessed June 2018
12. Lancia, G., Serafini, P.: The parity polytope. Compact Extended Linear Programming Models. EATOR, pp. 113–121. Springer, Cham (2018). https://doi.org/10.1007/978-3-319-63976-5_8
13. Lawler, E.L., Lenstra, J.K., Rinnooy Kan, A.H.G., Shmoys, D.B.: Sequencing and scheduling: algorithms and complexity. Handbooks Oper. Res. Manage. Sci. **4(C)**, 445–522 (1993)
14. Letchford, A.N., Nasiri, S.D., Theis, D.O.: Compact formulations of the Steiner TSP and related problems. Eur. J. Oper. Res. **228**, 83–92 (2013)
15. Martin, R.K.: Using separation algorithms to generate mixed integer model reformulations. Oper. Res. Lett. **10**, 119–128 (1991)
16. Naddef, D.: Personal Communication (2019)
17. Nash-Williams, C.S.J.A.: Edge-disjoint spanning trees of finite graphs. J. London Math. Soc. **36**, 445–450 (1961)
18. Ratliff, H.D., Rosenthal, A.: Order-picking in a rectangular warehouse: a solvable case of the traveling salesman problem. Oper. Res. **31**(3), 507–521 (1983)
19. Schrijver, A.: Combinatorial Optimization - Polyhedra and Efficiency. Springer, Berlin (2003)
20. Tutte, W.T.: On the problem of decomposing a graph into n connected factors. J. London Math. Soc. **36**, 221–230 (1961)
21. Yannakakis, M.: Expressing combinatorial optimization problems by Linear Programs. J. Comput. Syst. Sci. **43**, 441–466 (1991)

Implications, Conflicts, and Reductions for Steiner Trees

Daniel Rehfeldt[1,2]([envelope]) [iD] and Thorsten Koch[1,2] [iD]

[1] Chair of Software and Algorithms for Discrete Optimization, TU Berlin,
Straße des 17. Juni 135, 10623 Berlin, Germany
[2] Applied Algorithmic Intelligence Methods Department, Zuse Institute Berlin,
Takustraße 7, 14195 Berlin, Germany
{rehfeldt,koch}@zib.de

Abstract. The Steiner tree problem in graphs (SPG) is one of the most studied problems in combinatorial optimization. In the last decade, there have been significant advances concerning approximation and complexity of the SPG. However, the state of the art in (practical) exact solution of the SPG has remained largely unchallenged for almost 20 years.

The following article seeks to once again advance exact SPG solution. The article is based on a combination of three concepts: Implications, conflicts, and reductions. As a result, various new SPG techniques are conceived. Notably, several of the resulting techniques are provably stronger than well-known methods from the literature, used in exact SPG algorithms. Finally, by integrating the new methods into a branch-and-cut algorithm we obtain an exact SPG solver that outperforms the current state of the art on a large collection of benchmark sets. Furthermore, we can solve several instances for the first time to optimality.

Keywords: Steiner tree problem · Exact solution · Reduction techniques · Branch-and-cut

1 Introduction

Given an undirected connected graph $G = (V, E)$, edge costs $c : E \to \mathbb{Q}_{>0}$ and a set $T \subseteq V$ of *terminals*, the *Steiner tree problem in graphs* (SPG) is to find a tree $S \subseteq G$ with $T \subseteq V(S)$ that minimizes $c(E(S))$. The SPG is a classic $\mathcal{NP}$-hard problem [15], and one of the most studied problems in combinatorial optimization. Part of its theoretical appeal might be attributed to the fact that the SPG generalizes two other classic combinatorial optimization problems: Shortest paths, and minimum spanning trees. On the practical side, many applications can be modeled as SPG or closely related problems, see e.g. [4,19].

The SPG has seen numerous theoretical advances in the last 10 years, bringing forth significant improvements in complexity and approximability. See e.g. [3,10] for approximation, and [16,34] for complexity results. However, when

Supported by Research Campus Modal, and DFG Cluster of Excellence MATH+.

M. Singh and D. P. Williamson (Eds.): IPCO 2021, LNCS 12707, pp. 473–487, 2021.
https://doi.org/10.1007/978-3-030-73879-2_33

it comes to (practical) exact algorithms, the picture is bleak. After flourishing in the 1990s and early 2000s, algorithmic advances came to a staggering halt with the joint PhD theses of Polzin and Vahdati Daneshmand almost 20 years ago [21,33]. The authors introduced a wealth of new results and algorithms for SPG, and combined them in an exact solver that drastically outperformed all previous results from the literature. Their work is also published in a series of articles [22–26]. However, their solver is not publicly available.

The 11th DIMACS Challenge in 2014, dedicated to Steiner tree problems, brought renewed interest to the field of exact algorithms. In the wake of the challenge, several new exact SPG solvers were introduced in the literature, e.g. [8, 9,20]. Overall, the 11th DIMACS Challenge brought notable progress on the solution of notoriously hard SPG instances that had been designed to defy known solution techniques, see [18,30]. However, on the vast majority of instances from the literature, [21,33] stayed out of reach: For many benchmark instances, their solver is even two orders of magnitude or more faster, and it can furthermore solve substantially more instances to optimality—including those introduced at the DIMACS Challenge [27]. In 2018, the 3rd PACE Challenge [2] took place, dedicated to fixed-parameter tractable algorithms for SPG. Thus, the PACE Challenge considered mostly instances with a small number of terminals, or with small tree-width. Still, even for these special problem types, the solver by [21,33] remained largely unchallenged, see e.g. [13].

The following article aims to once again advance the state of the art in exact SPG solution.

1.1 Contribution

This article is based on a combination of three concepts: Implications, conflicts, and reductions. As a result, various new SPG techniques are conceived. The main contributions are as follows.

- By using a new implication concept, a distance function is conceived that provably dominates the well-known bottleneck Steiner distance. As a result, several reduction techniques that are stronger than results from the literature can be designed.
- We show how to derive conflict information between edges from the above methods. Further, we introduce a new reduction operation whose main purpose is to introduce additional conflicts. Such conflicts can for example be used to generate cuts for the IP formulation.
- We introduce a more general version of the powerful so-called extended reduction techniques. We furthermore enhance this framework by using both the previously introduced new distance concept, and the conflict information.
- Finally, we integrate the components into a branch-and-cut algorithm. Besides preprocessing, domain propagation, and cuts, also primal heuristics can be improved (by using the new implication concept). The practical implementation is realized as an extension of the branch-and-cut based Steiner tree solver SCIP-JACK [9].

The resulting exact SPG solver outperforms the current state-of-the-art solver from [21,33] on a wide range of well-established benchmark sets from the literature. Furthermore, it can solve several instances for the first time to optimality. Proofs of the new results are given in the extended version of this article [28].

1.2 Preliminaries and Notation

We write $G := (V, E)$ for an undirected graph, with vertices V and edges E. We set $n := |V|$ and $m := |E|$. For $S \subseteq G$ write $V(S)$ and $E(S)$ for its vertices and edges. For a walk W we write $V(W)$ and $E(W)$ for the set of included vertices and edges. For $U \subseteq V$ we define $\delta(U) := \{\{u, v\} \in E \mid u \in U, v \in V \setminus U\}$. We define the *neighborhood* of $v \in V$ as $N(v) := \{w \in W \mid \{v, w\} \in \delta(v)\}$. Note that $v \notin N(v)$.

Given edge costs $c : E \mapsto \mathbb{Q}_{\geq 0}$, the triplet (V, E, c) is referred to as *network*. By $d(v, w)$ we denote the cost of a shortest path (with respect to c) between vertices $v, w \in V$. For any (distance) function $\tilde{d} : \binom{V}{2} \mapsto \mathbb{Q}_{\geq 0}$, and any $U \subseteq V$ we define the $\tilde{d}$-*distance graph* on U as the network

$$D_G(U, \tilde{d}) := (U, \binom{U}{2}, \tilde{c}), \tag{1}$$

with $\tilde{c}(\{v, w\}) := \tilde{d}(v, w)$ for all $v, w \in U$. If $\tilde{d} = d$, we write $D_G(U)$ instead of $D_G(U, d)$. If a given $\tilde{d} : \binom{V}{2} \mapsto \mathbb{Q}_{\geq 0}$ is symmetric, we occasionally write $\tilde{d}(e)$ instead of $\tilde{d}(v, w)$ for an edge $e = \{v, w\}$.

2 From Implications to Reductions

Reduction techniques have been a key ingredient in exact SPG solvers, see e.g. [5, 17,23,32]. Among these techniques, the *bottleneck Steiner distance* introduced in [7] is arguably the most important one, being the backbone of several powerful reduction methods.

2.1 The Bottleneck Steiner Distance

Let P be a simple path with at least one edge. The *bottleneck length* [7] of P is

$$bl(P) := \max_{e \in E(P)} c(e). \tag{2}$$

Let $v, w \in V$. Let $\mathcal{P}(v, w)$ be the set of all simple paths between v and w. The *bottleneck distance* [7] between v and w is defined as

$$b(v, w) := \inf\{bl(P) \mid P \in \mathcal{P}(v, w)\}, \tag{3}$$

with the common convention that $\inf \emptyset = \infty$. Consider the distance graph $D :=$ $D_G(T \cup \{v, w\})$. Let b_D be the bottleneck distance in D. Define the *bottleneck Steiner distance* [7] between v and w as

$$s(v, w) := b_D(v, w). \tag{4}$$

The arguably best known bottleneck Steiner distance reduction method is based on the following criterion, which allows for edge deletion [7].

Theorem 1. *Let* $e = \{v, w\} \in E$. *If* $s(v, w) < c(e)$, *then no minimum Steiner tree contains* e.

2.2   A Stronger Bottleneck Concept

Initially, for an edge $e = \{v, w\}$ define the *restricted bottleneck distance* $\bar{b}(e)$ [23] as the bottleneck distance between v and w on $(V, E \setminus \{e\}, c)$.

The basis of the new bottleneck Steiner distance concept is formed by a node-weight function that we introduce below. For any $v \in V \setminus T$ and $F \subseteq \delta(v)$ define

$$p^+(v, F) = \max \left\{ 0, \sup \left\{ \bar{b}(e) - c(e) \mid e \in \delta(v) \cap F, e \cap T \neq \emptyset \right\} \right\}. \tag{5}$$

We call $p^+(v, F)$ the *F-implied profit* of v. The following observation motivates the subsequent usage of the implied profit. Assume that $p^+(v, \{e\}) > 0$ for an edge $e \in \delta(v)$. If a Steiner tree S contains v, but not e, then there is a Steiner tree S' with $e \in E(S')$ such that $c(E(S')) + p^+(v, \{e\}) \leq c(E(S))$.

Let $v, w \in V$. Consider a finite walk $W = (v_1, e_1, v_2, e_2, ..., e_r, v_r)$ with $v_1 = v$ and $v_r = w$. We say that W is a (v, w)-walk. For any $k, l \in \mathbb{N}$ with $1 \leq k \leq l \leq r$ define the subwalk $W(k, l) := (v_k, e_k, v_{k+1}, e_{k+1}, ..., e_l, v_l)$. W will be called *Steiner walk* if $V(W) \cap T \subseteq \{v, w\}$ and v, w are contained exactly once in W. The set of all Steiner walks from v to w will be denoted by $\mathcal{W}_T(v, w)$. With a slight abuse of notation we define $\delta_W(u) := \delta(u) \cap E(W)$ for any walk W and any $u \in V$. Define the *implied Steiner cost* of a Steiner walk $W \in \mathcal{W}_T(v, w)$ as

$$c_p^+(W) := \sum_{e \in E(W)} c(e) - \sum_{u \in V(W) \setminus \{v, w\}} p^+ \left(u, \delta(u) \setminus \delta_W(u) \right). \tag{6}$$

Further, set

$$P_W^+ := \{ u \in V(W) \mid p^+ \left(u, \delta(u) \setminus \delta_W(u) \right) > 0 \} \cup \{v, w\}. \tag{7}$$

Define the *implied Steiner length* of W as

$$l_p^+(W) := \max \{ c_p^+(W(v_k, v_l)) \mid 1 \leq k \leq l \leq r, \ v_k, v_l \in P_W^+ \}. \tag{8}$$

Define the *implied Steiner distance* between v and w as

$$d_p^+(v, w) := \min \{ l_p^+(W) \mid W \in \mathcal{W}_T(v, w) \}. \tag{9}$$

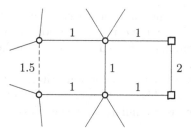

Fig. 1. Segment of a Steiner tree instance. Terminals are drawn as squares. The dashed edge can be deleted by employing Theorem 2.

Note that $d_p^+(v, w) = d_p^+(w, v)$. At last, consider the distance graph $D^+ := D_G(T \cup \{v, w\}, d_p^+)$. Let b_{D^+} be the bottleneck distance in D^+. Define the *implied bottleneck Steiner distance* between v and w as

$$s_p(v, w) := b_{D^+}(v, w). \tag{10}$$

Note that $s_p(v, w) \le s(v, w)$ and that $\frac{s(v,w)}{s_p(v,w)}$ can become arbitrarily large. Thus, the next result provides a stronger reduction criterion than Theorem 1.

Theorem 2. *Let $e = \{v, w\} \in E$. If $s_p(v, w) < c(e)$, then no minimum Steiner tree contains e.*

Figure 1 shows a segment of an SPG instance for which Theorem 2 allows for the deletion of an edge, but Theorem 1 does not. The implied bottleneck Steiner distance between the endpoints of the dashed edge is 1—corresponding to a walk along the four non-terminal vertices. The edge can thus be deleted. In contrast, the (standard) bottleneck Steiner distance between the endpoints is 1.5 (corresponding to the edge itself). Unfortunately, already computing the implied Steiner distance is hard:

Proposition 1. *Computing the implied Steiner distance is $\mathcal{NP}$-hard.*

Despite this $\mathcal{NP}$-hardness, one can devise heuristics that provide useful upper bounds on s_p, as discussed in the extended version of this article.

2.3 Bottleneck Steiner Reductions Beyond Edge Deletion

This section discusses applications of the implied bottleneck Steiner distance that allow for additional reduction operations: Edge contraction and node replacement. We start with the former. For an edge e and vertices v, w define $b_e(v, w)$ as the bottleneck distance between v and w on $(V, E \setminus \{e\}, c \restriction_{E \setminus \{e\}})$. With this definition, we define a generalization of the classic *NSV* reduction test from [6].

Proposition 2. *Let $\{v, w\} \in E$, and $t_1, t_2 \in T$ with $t_1 \neq t_2$. If*

$$s_p(v, t_1) + c(\{v, w\}) + s_p(w, t_2) \le b_{\{v,w\}}(t_1, t_2), \tag{11}$$

then there is a minimum Steiner tree S with $\{v, w\} \in E(S)$.

If criterion (11) is satisfied, one can contract edge $\{v, w\}$ and make the resulting vertex a terminal. Note that in the original criterion by [6] the standard distance is used, and no strengthening is achieved when the (standard) bottleneck Steiner distance is used instead.

This section closes with a new reduction criterion based on the standard bottleneck Steiner distance. This result also serves to highlight the complications that arise if one attempts to formulate similar conditions based on the implied bottleneck Steiner distance.

Proposition 3. *Let* $D := D_G(T, d)$. *Let* Y *be a minimum spanning tree in* D. *Write its edges* $\{e_1^Y, e_2^Y, ..., e_{|T|-1}^Y\} := E(Y)$ *in non-ascending order with respect to their weight in* D. *Let* $v \in V \setminus T$. *If for all* $\Delta \subseteq \delta(v)$ *with* $|\Delta| \geq 3$ *it holds that:*

$$\sum_{i=1}^{|\Delta|-1} d(e_i^Y) \leq \sum_{e \in \Delta} c(e), \tag{12}$$

then there is at least one minimum Steiner tree S *such that* $|\delta_S(v)| \leq 2$.

If the conditions (12) are satisfied for a vertex $v \in V \setminus T$, one can *pseudo-eliminate* [6] or *replace* [21] vertex v, i.e., delete v and connect any two vertices $u, w \in N(v)$ by a new edge $\{u, w\}$ of weight $c(\{v, u\}) + c(\{v, w\})$.

The SPG depicted in Fig. 2 exemplifies why Proposition 3 cannot be formulated by using the implied bottleneck Steiner distance. The weight of the minimum spanning tree Y for $D_G(T, d)$ is 4, but the weight of a minimum spanning tree with respect to s_p is 2. Similarly also the BD_m reduction technique from [6] cannot be directly formulated by using the implied bottleneck Steiner distance.

Fig. 2. SPG instance. Terminals are drawn as squares

3   From Reductions to Conflicts

In this section we use the concept of *conflicts* between edges. We say that a set $E' \subset E$ with $|E'| \geq 2$ is in conflict if no minimum Steiner tree contains more than one edge of E'. Recall that we have seen three types of reductions so far: Edge deletion, edge contraction, and node replacement. For simplicity, we assume in the following that a reduction is only performed if it retains *all* optimal solutions. We say that such a reduction is *valid*. We start with an SPG instance $I = (G, T, c)$, and consider a series of subsequent, valid reductions (of one of the three above types) that are applied to I. In each reduction step $i \geq 0$, the current instance $I^{(i)} = (G^{(i)}, T^{(i)}, c^{(i)})$ is transformed to instance $I^{(i+1)} = (G^{(i+1)}, T^{(i+1)}, c^{(i+1)})$. We set $I^{(0)} := I$.

3.1   Node Replacement

In [24] the authors observe that two edges that originate from a common edge by a series of replacements cannot both be contained in a minimum Steiner tree. We introduce an edge conflict criterion that is strictly stronger. For a series of $i = 0, 1, ..., k$ valid reductions, we define sets of *replacement ancestors* $\Lambda^{(i)}$: $E^{(i)} \to \mathcal{P}(\{1, ..., k\})$, and $\Lambda^{(i)}_{FIX} \in \mathcal{P}(\{1, ..., k\})$. We set $\Lambda^{(0)}(e) := \emptyset$ for all $e \in E$, and $\Lambda^{(0)}_{FIX} := \emptyset$. Further, we define $\lambda^{(0)} := 0$. Consider a reduced instance $I^{(i)}$. If we contract an edge $e \in E^{(i)}$, we set $\Lambda^{(i+1)}_{FIX} := \Lambda^{(i)}_{FIX} \cup \Lambda^{(i)}(e)$. If we replace a vertex $v \in V^{(i)}$, we set $\lambda^{(i+1)} := \lambda^{(i)} + 1$. Further, we define for each newly inserted edge $\{u, w\}$, with $u, w \in N(v)$:

$$\Lambda^{(i+1)}(\{u, w\}) := \Lambda^{(i)}(\{v, u\}) \cup \Lambda^{(i)}(\{v, w\}) \cup \{\lambda^{(i)}\}.$$

If no node replacement is performed, we set $\lambda^{(i+1)} := \lambda^{(i)}$.

Proposition 4. *Let I be an SPG and let $I^{(k)}$ be the SPG obtained from performing a series of k valid reductions on I. Further, let $e_1, e_2 \in E^{(k)}$. If $\Lambda^{(k)}(e_1) \cap \Lambda^{(k)}(e_2) \neq \emptyset$, then no minimum Steiner tree $S^{(k)}$ for $I^{(k)}$ contains both e_1 and e_2.*

Corollary 1. *Let I, $I^{(k)}$ as in Proposition 4, and let $e \in E^{(k)}$. If $\Lambda^{(k)}(e) \cap \Lambda^{(k)}_{FIX} \neq \emptyset$, then no minimum Steiner tree $S^{(k)}$ for $I^{(k)}$ contains e.*

Note that any edge e as in Corollary 1 can be deleted.

3.2   Edge Replacement

This subsection introduces a new replacement operation, whose primary benefit lies in the conflicts it creates.

Proposition 5. *Let $e = \{v, w\} \in E$ with $e \cap T = \emptyset$. Define*

$$\mathcal{D} := \{\Delta \subseteq (\delta(v) \cup \delta(w)) \setminus \{e\} \mid \Delta \cap \delta(v) \neq \emptyset, \Delta \cap \delta(w) \neq \emptyset\}.$$

For any $\Delta \in \mathcal{D}$ let

$$U_\Delta := \{u \in V \mid \{u, v\} \in \Delta \vee \{u, w\} \in \Delta\}.$$

If for all $\Delta \in \mathcal{D}$ with $|\Delta| \geq 3$ the weight of a minimum spanning tree on $D_G(U_\Delta, s)$ is smaller than $c(\Delta)$, then each minimum Steiner tree S satisfies $|\delta_S(v)| \leq 2$ and $|\delta_S(w)| \leq 2$.

If the condition of Proposition 5 is successful, we can perform what we will call a *path replacement* of e: We delete e and add for each pair $p, q \in V$ with $p \in N(v) \setminus \{w\}$, $q \in N(w)) \setminus \{v\}$, $p \neq q$ an edge $\{p, q\}$ with weight $c(\{p, v\}) + c(\{v, w\}) + c(\{q, w\})$. The apparent increase in the number of edges by this operations seems highly disadvantageous. However, due to the increased weight, the new edges can often be deleted by using the criterion from Theorem 2. We only perform a path replacement if at most one of the new edges needs to be inserted. If exactly one new edge remains, we create new replacement ancestors as follows: Let $\hat{e} = \{p, q\}$ be the newly inserted edge. Initially, set $\lambda^{(i+1)} := \lambda^{(i)}$ and $\Lambda^{(i+1)}(\hat{e}) := \Lambda^{(i)}(\{p, v\}) \cup \Lambda^{(i)}(\{v, w\}) \cup \Lambda^{(i)}(\{v, q\})$. Next, for each $e' \in (\delta(v) \cup \delta(w)) \setminus \{e\}$ increment $\lambda^{(i+1)}$, and add $\lambda^{(i+1)}$ to $\Lambda^{(i+1)}(\hat{e})$ and $\Lambda^{(i+1)}(e')$. One can show that Proposition 4 remains valid if path replacement is added to the list of valid reduction operations. As we will see in the remainder of this article, conflicts cannot only be used for further reductions, but also for generating cuts in an IP model.

4   From Steiner Distances and Conflicts to Extended Reduction Techniques

At the end of the last section we have seen a reduction method that inspects a number of trees (of depth 3) that extend an edge considered for replacement. This section continues along this path, based on the reduction concepts introduced so far. In the following, we introduce new so-called *extended reduction algorithms* that (provably) dominate the strongest ones from the literature, due to [24].

4.1   The Framework

For a tree Y in G, let $L(Y) \subseteq V(Y)$ be the set of its leafs. We start with several definitions from [24]. Let Y' be a tree with $Y' \subseteq Y$. The *linking set* between Y and Y' is the set of all vertices $v \in V(Y')$ such that there is a path $Q \subseteq Y$ from v to a leaf of Y with $V(Q) \cap V(Y') = \{v\}$. Note that Q can consist of a single vertex. Y' is *peripherally contained* in Y if the linking set between Y and Y' is $L(Y')$. For any $P \subseteq V(Y)$ with $|P| > 1$ let Y_P be the union of the (unique) paths between any $v, w \in P \cup V(Y)$ in Y. Note that Y_P is a tree, and that $Y_P \subseteq Y$ holds. P is called *pruning set* if it contains the linking set between Y_P and Y. Additionally, we will use the following new definition: P is called *strict pruning set* if it is equal to the linking set between Y_P and Y.

Additionally, we define a stronger, and new, inclusion concept. Consider a tree $Y \subseteq G$, and a subtree Y'. Let P be a pruning set for Y'. We say that Y' is *P-peripherally contained* in Y if P is a pruning set for Y. Now let P be a strict pruning set for Y'. We say that Y' is *strictly P-peripherally contained* in Y if P is a strict pruning set for Y. One obtains the following important property.

Observation 1. *Let $Y \subseteq G$ be a tree, let $Y' \subseteq Y$ be a subtree, and let P be a pruning set for Y'. If Y' is peripherally contained in Y, then Y' is also P-peripherally contained in Y.*

Note that an equivalent property holds for strict pruning sets. Given a tree Y and a set $E' \subseteq E$, we write with a slight abuse of notation $Y + E'$ for the subgraph with the edge set $E(Y) \cup E'$. Algorithm 1 shows a high level description of the

extended reduction framework used in this article. The framework is similar to
the one introduced in [24], but more general. A possible input for Algorithm 1 is
an SPG instance together with a single edge. If the algorithm returns *true*, the
edge can be deleted. Besides EXTENSIONSETS, which is described in Algorithm 2,
the extended reduction framework contains the following subroutines, with SPG
$I = (G, T, c)$, tree $Y \subseteq G$, and a pruning set P for Y.

- RULEDOUT(I, Y, P) returns *true* if Y is shown to not be P-peripherally contained in any minimum Steiner tree. Otherwise, returns *false*.
- RULEDOUTSTRICT(I, Y, P) same as above, but with strict pruning set P.
- STRICTPRUNINGSETS(I, Y) returns a subset of all strict pruning sets for Y.
- TRUNCATE(I, Y) returns *true* if no further extensions of Y should be performed; otherwise returns *false*.
- PROMISING(I, Y, v) is given additionally a vertex $v \in L(Y)$. Returns *true* if further extensions of Y from v should be performed; otherwise returns *false*.

Algorithm 1: EXTENDED-RULEDOUT

Data: SPG instance $I = (G, T, c)$, tree Y with $Y \cap T \subseteq L(Y)$
Result: *true* if Y is shown to not be peripherally contained in any
minimum Steiner tree; *false* otherwise

1 **foreach** $P \in$ STRICTPRUNINGSETS(I, Y) **do**
2 $\quad$ **if** RULEDOUTSTRICT(I, Y, P) **then return** *true*
3 **if** TRUNCATE(I, Y) **then return** *false*
4 **foreach** $v \in L(Y)$ **do**
5 $\quad$ **if** $v \in T$ **or not** PROMISING(I, Y, v) **then continue**
6 $\quad$ *success* := *true*
7 $\quad$ **foreach** $E' \in$ EXTENSIONSETS(I, Y, v) **do**
8 $\quad\quad$ **if not** EXTENDED-RULEDOUT$(I, Y + E')$ **then**
9 $\quad\quad\quad$ *success* := *false*
10 $\quad\quad\quad$ **break**
11 $\quad$ **if** *success* **then return** *true*
12 **return** *false*

In Lines 1–2 of Algorithm 1, we try to peripherally rule-out tree Y. If that is
not possible, we try to recursively extend Y in Lines 4–11. Since (given positive
edge weights) no minimum Steiner tree has a non-terminal leaf, we can extend
from any of the non-terminal leaves of Y. Note that ruling-out all extensions
along one single leaf is sufficient to rule-out Y.

4.2   Reduction Criteria

In this section we introduce several elimination criteria used within RULEDOUT
and RULEDOUTSTRICT. Note that any criterion that is valid for RULEDOUT is
also valid for RULEDOUTSTRICT. We also note that several of the criteria in this
section are similar to results from [21,24], but are all stronger. Throughout this
section we consider a graph $G = (V, E)$ and an SPG instance $I = (G, T, c)$.

Algorithm 2: EXTENSIONSETS

Data: SPG instance $I = (G, T, c)$, tree Y, vertex $v \in V(Y)$
Result: Set $\Gamma \subseteq \mathcal{P}(\delta(v))$ such that for all $\gamma \in \mathcal{P}(\delta(v)) \setminus \Gamma$, tree $Y + \gamma$ is not
       peripherally contained in any minimum Steiner tree.

1   $Q := \emptyset,\ R := \emptyset$
2   **foreach** $e := \{v, w\} \in \delta(v) \setminus E(Y)$ **do**
3   **if** RULEDOUT$(I, Y + \{e\}, L(Y) \cup \{w\})$ **then continue**
4   **if** RULEDOUTSTRICT$(I, Y + \{e\}, L(Y) \cup \{w\})$ **then**
5         $R := R \cup \{e\}$
6   **continue**
7      $Q := Q \cup \{e\}$
8 **return** $\mathcal{P}(Q) \cup R$

Consider a tree $Y \subseteq G$, and a pruning set P for Y such that $V(Y_P) \cap T \subseteq L(Y_P)$. For each $p \in P$ let $\overline{Y}_p \subset Y$ such that $V(\overline{Y}_p)$ is exactly the set of vertices $v \in V(Y_P)$ that satisfy the following: For any $q \in P \setminus \{p\}$ the (unique) path in Y from v to q contains p. Note that when removing $E(Y_P)$ from Y, each non-trivial connected component equals one $\overline{Y}_p$. Further, note that $p \in V(\overline{Y}_p)$ for all $p \in P$. Let $G_{Y,P} = (V_{Y,P}, E_{Y,P})$ be the graph obtained from $G = (V, E)$ by contracting for each $p \in P$ the subtree $\overline{Y}_p$ into p. For any parallel edges, we keep only one of minimum weight. We identify the contracted vertices $V(\overline{Y}_p)$ with the original vertex p. Overall, we thus have $V_{Y,P} \subseteq V$. Let $c_{Y,P}$ be the edge weights on $G_{Y,P}$ derived from c. Let

$$T_{Y,P} := \big(T \cap V_{Y,P}\big) \cup \{p \in P \mid T \cap V(\overline{Y}_p) \neq \emptyset\}. \tag{13}$$

Finally, let $s_{Y,P}$ be the bottleneck Steiner distance on $(G_{Y,P}, T_{Y,P}, c_{Y,P})$. The next theorem generalizes a number of results from the literature. See [11, 21] for similar, but weaker, conditions.

Theorem 3. *Let $Y \subseteq G$ be a tree, and let P be a pruning set for Y such that $V(Y_P) \cap T \subseteq L(Y_P)$. Let $I_{Y,P}$ be the SPG on the distance network $D_{G_{Y,P}}(V_{Y,P}, s_{Y,P})$ with terminal set P. If the weight of a minimum Steiner tree for $I_{Y,P}$ is smaller than $c(E(Y_P))$, then Y is not P-peripherally contained in any minimum Steiner tree for I.*

If computing (or even approximating) a minimum Steiner tree on $D_{G_{Y,P}}(V_{Y,P}, s_{Y,P})$ is deemed to expensive, one can use the weight of a minimum spanning tree on $D_{G_{Y,P}}(P, s_{Y,P})$ instead.

Next, let $Y \subseteq G$ be a tree with pruning set P, and let $v, w \in V(Y)$ and let Q be the path between v, w in Y. We define a *pruned tree bottleneck* between v and w as a subpath $Q(a, b)$ of Q that satisfies $|\delta_Y(u)| = 2|$ and $u \notin P$ for all $u \in V(Q(a,b)) \setminus \{a, b\}$, $V(Q(a,b)) \cap T \subseteq \{a, b\}$, and maximizes $c(V(Q(a, b)))$. The weight $c(V(Q(a, b)))$ of such a pruned tree bottleneck is denoted by $b_{Y,P}(v, w)$. Using this definition and the implied bottleneck Steiner distance, we obtain the following result.

Proposition 6. *Let Y be a tree, let P be a pruning set for Y, and let $v, w \in V(Y)$. If $s_p(v, w) < b_{Y,P}(v, w)$, then Y is not P-peripherally contained in any minimum Steiner tree.*

In the extended version of this article we also give an reduction criteria based on reduced costs of an IP formulation, which can only be used for the RULED-OUTSTRICT routine. Finally, another important reduction criteria is constituted by edge conflicts between edges of an enumerated tree.

5   Exact Solution

This section describes the usage of the previous techniques for exact SPG solution. They have been implemented as an extension of the solver SCIP-JACK [9].

5.1   Branch-and-cut

We enhance several vital components of branch-and-cut algorithms. The most natural application of reduction methods is within presolving. However, we also use (limited versions of) them for domain propagation during branch-and-bound, translating the deletion of edges into variable fixings in the integer programming model. The edge conflicts described in this article can be used for generating clique cuts [1]. Note that SCIP-JACK also separates the Steiner cuts from the bidirected cut formulations, as well as the flow-balance constraints from [17]. Finally, also primal heuristics are improved. First, the stronger reduction methods enhance primal heuristics that involve the solution of auxiliary SPG instances, such as from the combination of several Steiner trees. Second, the implication concept introduced in this article can be used to directly improve a classic SPG heuristic by [31], as shown in the extended version of this article.

5.2   Computational Results

This section provides computational results for the new solver. In particular, we compare its performance with the updated results of the solver by [21,33] published in [27]. The computational experiments were performed on Intel Xeon CPUs E3-1245 with 3.40 GHz and 32 GB RAM. This machine obtains a score of 488.993589 with the benchmark software of the 11th DIMACS Challenge (with the same compiler as [27]). Thus, this computer is roughly 1.59 times faster than the machine used in [27]. We have scaled the run-times reported in the following accordingly. We use the same LP solver as [27]: CPLEX 12.6 [12]. For the comparison we use a diverse range of well-established benchmark sets from the STEINLIB [18] and the 11th DIMACS Challenge.

Table 1 provides results of the two solvers for a time limit of two hours. The second column shows the number of instances in the test-set. Column three shows the mean time taken by the solver of [21,33], column four shows the mean time of the new solver. The next column gives the relative speedup of the new

Table 1. Comparison of the new solver and the solver by [21,33].

Test set	#	Mean time (sh. geo. mean)			Maximum time			# Solved	
		P. & V. [s]	New [s]	Speedup	P. & V. [s]	New [s]	Speedup	P. & V.	New
E	20	0.3	0.3	1.0	**3.4**	4.8	0.7	20	20
2R	27	5.0	**3.0**	1.7	43.9	**12.2**	3.6	27	27
ALUE	15	1.4	3.2	0.4	**14.8**	35.3	0.4	15	15
vienna-s	85	7.8	**3.9**	2.0	623.5	**74.0**	8.4	85	85
ES10000FST	1	138.0	**122.4**	1.1	138.0	**122.4**	1.1	1	1
SP	8	81.0	**19.7**	4.1	>7200	**598.4**	>12.0	6	**8**
GEO-adv	23	158.7	**62.7**	2.5	6476.5	**926.6**	7.0	23	23
Cophag14	21	27.7	**12.6**	2.2	>7200	**3300.2**	>2.2	20	**21**
WRP3	63	22.8	**18.8**	1.2	6073.2	**3646.2**	1.7	63	63
PUC	50	2458.1	**1910.2**	1.3	>7200	>7200	1.0	12	**14**

solver. The next three columns provide the same information for the maximum run-time, the last two columns give the number of solved instances. For the mean time we use the well-established shifted geometric mean [1] with a shift of 10.

Overall, the new solver performs better on eight of the ten test-sets. Often by a significant margin, such as for *GEO-adv* or *SP*. Only on *ALUE* the new solver performs significantly worse. A possible reason is the existence of small node-separators on these instances. These separators are heavily exploited by partitioning methods in [21,33]. However, our solver includes no algorithms yet to exploit such separators. It should be noted that the same behavior can be observed for the related test-set *ALUT* which is not included in Table 1. On the other hand, there are also several other benchmark sets for which the new solver is faster, but which are not contained in Table 1, since they are similar to already included ones (e.g. *GEO-org*, *1R*, *WRP3*, *LIN*). We also note that in [27] specialized settings are used for individual test-sets. In contrast, we run all test-sets with default settings, although a considerable speed-up (usually more than 50 percent) could be achieved when using specialized settings.

For most of the test-sets in Table 1, the new solver is around one order of magnitude or more faster than the previous version of SCIP-JACK described in [29], both with respect to the mean and maximum time. Even if one merely considers the enhancements described in this article (as compared to methods already known in the literature), the speed-up is still huge, as shown in the extended version of this article. In particular, the s_p-based methods have a significant impact, as we demonstrate in the following. To this end, we use three benchmark sets from the DIMACS Challenge, and three from the STEINLIB. Table 2 shows in the first column the name of the test-set, followed by its number of instances. The next columns show the percentual average number of nodes and edges of the instances after the preprocessing without (column three and four), and with (columns five and six) the s_p based methods. The last two columns reports the percentual relative change between the previous results. It can be seen that the s_p methods allow for a significant additional reduction of the problem size. This behavior is rather remarkable, given the variety of powerful reduction methods already included in SCIP-Jack. Furthermore, the instances from the *GEO-adv* set come already in preprocessed form; by means of the reduction package from [32]. We note that the overall run-time of the preprocessing notably decreases when the s_p based methods are used.

Table 2. Average remaining nodes and edges after preprocessing.

Test set	#	Base preprocessing		$+s_p$ techniques		Relative change	
		Nodes [%]	Edges [%]	Nodes [%]	Edges [%]	Nodes [%]	Edges [%]
ALUE	15	1.4	1.4	0.8	0.9	**−75.0**	**−55.6**
vienna-s	85	8.5	8.1	6.1	5.8	**−39.3**	**−39.6**
WRP3	63	49.5	50.6	45.6	45.8	**−8.6**	**−10.4**
Copenhag14	21	39.4	37.1	35.4	32.1	**−11.3**	**−15.6**
GEO-adv	23	29.1	30.7	26.9	28.2	**−8.2**	**−8.9**
ES10000FST	1	47.2	52.1	38.1	42.2	**−23.9**	**−21.1**

Finally, we provide results for several large-scale Euclidean Steiner tree problems. For solving such problems, the bottleneck is usually the full Steiner tree concatanation [14]. This concatanation can also be solved as an SPG, however [25]. We report results for Euclidean instances from [14] with 25 thousand (*EST-25k*) and 50 thousand (*EST-50k*) points in the plane. Both test-sets contain 15 instances. For *EST-25k* the mean and maximum times of the new solver are 66.2 and 92.4 s—between one and two orders of magnitude faster than those of the well-known geometric Steiner tree solver GeoSteiner 5.1 [14]. The *EST-50k* instances can also be solved quickly, with a mean of 286.1 s. Moreover, 7 of the 15 instances are solved for the first time to optimality—in at most 390 s. On the other hand, GeoSteiner cannot solve these instances even after seven days of computation. Unfortunately, [27] does not report results for these instances. However, the solver by [20], which won the heuristic SPG category at the 11th DIMACS Challenge, does not reach the upper bounds from GeoSteiner on any of the *EST-25k* and *EST-50k* instances.

6   Outlook

There are several promising routes for further improvement. First, one could enhance the newly introduced methods. For example, by using full-backtracking in the extended reduction methods, by improving the approximation of the implied bottleneck Steiner distance, or by adapting the latter for replacement techniques. Second, several powerful methods described in [21,33] could be added to the new solver, e.g. a stronger IP formulation realized via price-and-cut, or additional reduction techniques via partitioning.

Unlike the solver by [21,33], the new SCIP-Jack will be made freely available for academic use—as part of the next SCIP release.

References

1. Achterberg, T.: Constraint Integer Programming. Ph.D. thesis, Technische Universität Berlin (2007)
2. Bonnet, É., Sikora, F.: The PACE 2018 parameterized algorithms and computational experiments challenge: the third iteration. In: Paul, C., Pilipczuk, M. (eds.) 13th International Symposium on Parameterized and Exact Computation (IPEC 2018), Leibniz International Proceedings in Informatics (LIPIcs), vol. 115, pp. 26:1–26:15. Schloss Dagstuhl-Leibniz-Zentrum fuer Informatik, Dagstuhl, Germany (2019). https://doi.org/10.4230/LIPIcs.IPEC.2018.26
3. Byrka, J., Grandoni, F., Rothvoß, T., Sanità, L.: Steiner tree approximation via iterative randomized rounding. J. ACM **60**(1), 6 (2013). https://doi.org/10.1145/2432622.2432628
4. Cheng, X., Du, D.Z.: Steiner trees in industry. In: Du, D.Z., Pardalos, P.M. (eds.) Handbook of Combinatorial Optimization, vol. 11, pp. 193–216. Springer, Boston https://doi.org/10.1007/0-387-23830-1_4
5. Duin, C.: Steiner Problems in Graphs. Ph.D. thesis, University of Amsterdam (1993)
6. Duin, C.W., Volgenant, A.: Reduction tests for the Steiner problem in graphs. Networks **19**(5), 549–567 (1989). https://doi.org/10.1002/net.3230190506
7. Duin, C., Volgenant, A.: An edge elimination test for the Steiner problem in graphs. Oper. Re. Lett. **8**(2), 79–83 (1989). https://doi.org/10.1016/0167-6377(89)90005-9
8. Fischetti, M., et al.: Thinning out Steiner trees: a node-based model for uniform edge costs. Math. Program. Comput. **9**(2), 203–229 (2016). https://doi.org/10.1007/s12532-016-0111-0
9. Gamrath, G., Koch, T., Maher, S.J., Rehfeldt, D., Shinano, Y.: SCIP-Jack—a solver for STP and variants with parallelization extensions. Math. Program. Comput. **9**(2), 231–296 (2016). https://doi.org/10.1007/s12532-016-0114-x
10. Goemans, M.X., Olver, N., Rothvoß, T., Zenklusen, R.: Matroids and integrality gaps for Hypergraphic Steiner tree relaxations. In: Proceedings of the Forty-Fourth Annual ACM Symposium on Theory of Computing, STOC 2012, pp. 1161–1176. Association for Computing Machinery, New York, NY, USA (2012). https://doi.org/10.1145/2213977.2214081
11. Hwang, F., Richards, D., Winter, P.: The Steiner Tree Problem. Elsevier Science, Annals of Discrete Mathematics (1992)
12. IBM: Cplex (2020). https://www.ibm.com/analytics/cplex-optimizer
13. Iwata, Y., Shigemura, T.: Separator-based pruned dynamic programming for steiner tree. In: Proceedings of the AAAI Conference on Artificial Intelligence, vol. 33, pp. 1520–1527 (2019). https://doi.org/10.1609/aaai.v33i01.33011520
14. Juhl, D., Warme, D.M., Winter, P., Zachariasen, M.: The GeoSteiner software package for computing Steiner trees in the plane: an updated computational study. Math. Program. Comput. **10**(4), 487–532 (2018). https://doi.org/10.1007/s12532-018-0135-8
15. Karp, R.: Reducibility among combinatorial problems. In: Miller, R., Thatcher, J. (eds.) Complexity of Computer Computations, pp. 85–103. Plenum Press (1972)
16. Kisfaludi-Bak, S., Nederlof, J., Leeuwen, E.J.V.: Nearly ETH-tight algorithms for planar Steiner tree with terminals on few faces. ACM Trans. Algorithms (TALG) **16**(3), 1–30 (2020). https://doi.org/10.1145/3371389

17. Koch, T., Martin, A.: Solving Steiner tree problems in graphs to optimality. Networks **32**, 207–232 (1998). https://doi.org/10.1002/(SICI)1097-0037(199810)32:3 %3C207::AID-NET5%3E3.0.CO;2-O
18. Koch, T., Martin, A., Voß, S.: SteinLib: An updated library on Steiner tree problems in graphs. In: Du, D.Z., Cheng, X. (eds.) Steiner Trees in Industries, pp. 285–325. Kluwer (2001)
19. Leitner, M., Ljubic, I., Luipersbeck, M., Prossegger, M., Resch, M.: New Real-world Instances for the Steiner Tree Problem in Graphs. Technical Report, ISOR, Uni Wien (2014)
20. Pajor, T., Uchoa, E., Werneck, R.F.: A robust and scalable algorithm for the Steiner problem in graphs. Math. Program. Comput. **10**(1), 69–118 (2017). https://doi.org/10.1007/s12532-017-0123-4
21. Polzin, T.: Algorithms for the Steiner problem in networks. Ph.D. thesis, Saarland University (2003)
22. Polzin, T., Daneshmand, S.V.: A comparison of Steiner tree relaxations. Discrete Appl. Math. **112**(1–3), 241–261 (2001)
23. Polzin, T., Daneshmand, S.V.: Improved Algorithms for the Steiner Problem in Networks. Discrete Appl. Math. **112**(1–3), 263–300 (2001). https://doi.org/10.1016/S0166-218X(00)00319-X
24. Polzin, T., Daneshmand, S.V.: Extending reduction techniques for the Steiner tree problem. In: Möhring, R., Raman, R. (eds.) ESA 2002. LNCS, vol. 2461, pp. 795–807. Springer, Heidelberg (2002). https://doi.org/10.1007/3-540-45749-6_69
25. Polzin, T., Daneshmand, S.V.: On Steiner trees and minimum spanning trees in hypergraphs. Oper. Res. Lett. **31**(1), 12–20 (2003). https://doi.org/10.1016/S0167-6377(02)00185-2
26. Polzin, T., Daneshmand, S.V.: Practical partitioning-based methods for the Steiner problem. In: Àlvarez, C., Serna, M. (eds.) WEA 2006. LNCS, vol. 4007, pp. 241–252. Springer, Heidelberg (2006). https://doi.org/10.1007/11764298_22
27. Polzin, T., Vahdati-Daneshmand, S.: The Steiner Tree Challenge: An updated Study (2014), unpublished manuscript at http://dimacs11.cs.princeton.edu/downloads.html
28. Rehfeldt, D., Koch, T.: Implications, conflicts, and reductions for Steiner trees. Technical Report 20–28, ZIB, Takustr. 7, 14195 Berlin (2020)
29. Rehfeldt, D., Shinano, Y., Koch, T.: SCIP-jack: an exact high performance solver for Steiner tree problems in graphs and related problems. In: Bock, H.G., Jäger, W., Kostina, E., Phu, H.X. (eds.) Modeling, Simulation and Optimization of Complex Processes HPSC 2018. LNCS, pp. 201–223. Springer, Cham (2021). https://doi.org/10.1007/978-3-030-55240-4_10
30. Rosseti, I., de Aragão, M., Ribeiro, C., Uchoa, E., Werneck, R.: New benchmark instances for the Steiner problem in graphs. In: Extended Abstracts of the 4th Metaheuristics International Conference (MIC 2001), pp. 557–561. Porto (2001)
31. Takahashi, H., Matsuyama, A.: An approximate solution for the Steiner problem in graphs. Math. Japonicae **24**, 573–577 (1980)
32. Uchoa, E., Poggi de Aragão, M., Ribeiro, C.C.: Preprocessing Steiner problems from VLSI layout. Networks **40**(1), 38–50 (2002). https://doi.org/10.1002/net.10035
33. Vahdati Daneshmand, S.: Algorithmic approaches to the Steiner problem in networks. Ph.D. thesis, Universität Mannheim (2004)
34. Vygen, J.: Faster algorithm for optimum Steiner trees. Inf. Process. Lett. **111**(21), 1075–1079 (2011). https://doi.org/10.1016/j.ipl.2011.08.005

Author Index

Printed in the United States
by Baker & Taylor Publisher Services